Stochastic Averaging

Wei-Qiu Zhu · Mao-Lin Deng · Guo-Qiang Cai

Stochastic Averaging

Methods and Applications, Volume 2

Science Press
Beijing

Springer

Wei-Qiu Zhu
Department of Mechanics
Zhejiang University
Hangzhou, Zhejiang, China

Mao-Lin Deng
Department of Mechanics
Zhejiang University
Hangzhou, Zhejiang, China

Guo-Qiang Cai
Florida Atlantic University
Boca Raton, FL, USA

ISBN 978-981-96-3832-1 ISBN 978-981-96-3833-8 (eBook)
https://doi.org/10.1007/978-981-96-3833-8

Jointly published with Science Press
The print edition is not for sale in China (Mainland). Customers from China (Mainland) please order the print book from: Science Press.

Translation from the Chinese Simplified language edition: "Stochastic Averaging Method and Its Applications, Volume 2" by Wei-Qiu Zhu et al., © Science Press 2023. Published by Science Press. All Rights Reserved.

This Springer imprint is published by the registered company Springer Nature Singapore Pte Ltd.
The registered company address is: 152 Beach Road, #21-01/04 Gateway East, Singapore 189721, Singapore

Preface

Nonlinear stochastic dynamic systems are prevalent the natural, technical sciences, and social sciences. The study on nonlinear stochastic dynamics began in the early 1960s. So far, it has been developed to a well-established discipline. Several exact, approximate analytical and numerical solution methods have been proposed and developed. The only exact solution method is based on that the excitation is Gaussian white noise and the response is Markov diffusion process. The exact probability density and statistics of system response are obtained from the formulation and solution of Fokker-Planck-Kolmogorov (FPK) equation. However, the exact solution method has limited application since real nonlinear stochastic dynamical system has no exact solution and the FPK equation is very difficult to solve. The approximate analytical solution methods include equivalent (statistical) linearization, equivalent nonlinear system method, moment equation and its truncation scheme, stochastic averaging method, etc. The numerical methods include Monte Carlo simulation, cell mapping and path integral methods, etc. Each method has its advantages and drawbacks. Relatively, the stochastic averaging methods are the most powerful and widespread applied approximate analytical methods. These methods have stochastic averaging principle as their sound mathematical foundation. They extend the application range of exact solution method. They simplify the system and reduce the dimension of system while retain the essential nonlinearity of original system. Physically, the stochastic averaging methods transform the study of nonlinear stochastic dynamical system into the study of amplitude or energy of system or its subsystems, and the obtained results for amplitude or energy can be translated back into the probability and statistics for the original system. Stochastic averaging methods can be used to predict the response, to study the stability, reliability and stochastic optimal control.

Starting from the early 1990s, the first author and his collaborators and students developed the stochastic averaging methods of quasi-Hamiltonian systems under Gaussian white noise excitation and then extend them to non-Gaussian, non-white noise excitation, to quasi-generalized Hamiltonian systems and some cases involving non-Gaussian and non-white noise excitations, etc. Meanwhile, Prof. Y. K. Lin and the second author at the Center for Applied Stochastics Research of Florida Atlantic

University, USA, derived the non-smooth and smooth averaged equations for general nonlinear stochastic dynamical systems and applied them to numerous single-degree-of-freedom and two-dimensional nonlinear stochastic systems, particularly ecological systems. Although most research results have been published as technical papers, to write a book is still necessary for the readers to understand easily and systematically. Chapters 2–4 in Volume 1 and Chap. 4 in Volume 2 were drafted by the third author, while the remaining chapters (except Chap. 1 in Volume 1) were drafted by the second author. The first author revised and finalized the manuscript.

We are grateful to the National Natural Science Foundation of China for funding our research. We also extend gratitude to our colleagues Dr. Y. Q. Yang, Prof. Z. L. Huang, Prof. Z. G. Ying, Prof. Z. H. Liu, Prof. R. H. Huan, Prof. L. C. Chen, Associate Profs. Y. Wang, Y. Zeng, W. T. Jia, Y. J. Wu, Q. F. Lü, J. J. Sun for their invaluable contribution in developing the stochastic averaging methods. The support and help of Science Press and the financial support for publication of the book from China Classics International Publishing Company and Institute of Applied Mechanics of Zhejiang University are greatly appreciated.

Hangzhou, China — Wei-Qiu Zhu
Hangzhou, China — Mao-Lin Deng
Boca Raton, USA — Guo-Qiang Cai

Contents

Chapter 1
Stochastic Averaging Methods of Quasi-integrable Hamiltonian Systems Excited by Colored Noises

The white noise does not exist in reality due to its infinite energy, and it is merely an idealized mathematical model. Real noises are always colored noises. One class of models for colored noises is generated from Gaussian white noise using linear or nonlinear filters (see Sects. 2.6.1 and 2.6.2, vol. 1). A noise generated from a linear filter is often referred to as a rational noise because its power spectral density is a rational function of frequency. Fractional Gaussian noises are another type of colored noises generated by Gaussian white noise through a special filter. Colored noise can be wideband or narrowband. A colored noise can also be wideband in one frequency band and narrowband in the others, for instance, fractional Gaussian noises (see Sect. 2.5.3, vol. 1). Narrowband noises can also be synthesized by combining harmonic and white noise or wideband noise, or by randomizing harmonic functions (see Sect. 2.6.3, vol. 1). This chapter presents the stochastic averaging methods of quasi-integrable Hamiltonian systems excited by four classes of colored noises: stationary wideband noise, fractional Gaussian noise in the wideband frequency domain, combination of harmonic and wideband noise, and narrowband randomized harmonic noise.

1.1 Stationary Wideband Noise Excitation

Consider a quasi-integrable Hamiltonian system excited by stationary wideband noises (Deng and Zhu 2007). The motion equations are

W. Zhu et al., *Stochastic Averaging*, https://doi.org/10.1007/978-981-96-3833-8_1

$$
\begin{aligned}
\dot{Q}_i &= \frac{\partial H}{\partial P_i},\\
\dot{P}_i &= -\frac{\partial H}{\partial Q_i} - \varepsilon \sum_{j=1}^{n} c_{ij}(\mathbf{Q},\mathbf{P})\frac{\partial H}{\partial P_j} + \varepsilon^{1/2}\sum_{k=1}^{m} f_{ik}(\mathbf{Q},\mathbf{P})\xi_k(t),\\
i &= 1, 2, \ldots, n,
\end{aligned}
\tag{1.1}
$$

where $\mathbf{Q} = [Q_1, Q_2, \ldots, Q_n]^{\mathrm{T}}$ and $\mathbf{P} = [P_1, P_2, \ldots, P_n]^{\mathrm{T}}$ are vectors of generalized displacements and momenta; $\xi_k(t)$ are stationary wideband noises with correlation functions $R_{kl}(\tau)$ and power spectral density functions $S_{kl}(\omega)$; $c_{ij}(\mathbf{Q}, \mathbf{P})$ represent linear and/or nonlinear damping coefficients; $f_{ij}(\mathbf{Q}, \mathbf{P})$ represent random excitation amplitudes; ε is a small parameter. $H = H(\mathbf{q}, \mathbf{p})$ is a separable Hamiltonian, i.e.,

$$
H = \sum_{i=1}^{n} H_i(q_i, p_i). \tag{1.2}
$$

For most dynamical systems, it can be further assumed that

$$
H_i(q_i, p_i) = p_i^2 \big/ 2 + U_i(q_i), \quad i = 1, 2, \ldots, n. \tag{1.3}
$$

Denote $\partial H_i \big/ \partial q_i = g_i(q_i)$, Eq. (1.1) can be rewritten in the form

$$
\begin{aligned}
\dot{Q}_i &= P_i,\\
\dot{P}_i &= -g_i(Q_i) - \varepsilon \sum_{j=1}^{n} c_{ij}(\mathbf{Q},\mathbf{P})P_j + \varepsilon^{1/2}\sum_{k=1}^{m} f_{ik}(\mathbf{Q},\mathbf{P})\xi_k(t),\\
i &= 1, 2, \ldots, n.
\end{aligned}
\tag{1.4}
$$

Equation (1.4) describes a quasi-integrable Hamiltonian system excited by stationary wideband noises. In the following, the stochastic averaging method will be developed for this system in two cases, single degree-of-freedom (SDOF) and multiple degree-of-freedom (MDOF) systems.

1.1.1 SDOF System

The amplitude-envelope stochastic averaging method and the energy-envelope stochastic averaging method for SDOF nonlinear systems excited by stationary wideband noises have been developed in Sect. 4.4, vol. 1. Here, an alternative procedure of the stochastic averaging method will be presented for SDOF quasi-Hamiltonian systems excited by stationary wideband noises (Zhu et al. 2001). The motion equations of the system are

$$\dot{Q} = P,$$
$$\dot{P} = -g(Q) - \varepsilon c(Q, P)P + \varepsilon^{1/2} \sum_{k=1}^{m} f_k(Q, P)\xi_k(t). \tag{1.5}$$

System (1.5) reduces to a SDOF Hamiltonian system when $\varepsilon = 0$. Assume that the restoring force function $g(q)$ and the potential functions $U(q) = \int_0^q g(x)dx$ satisfying the following four conditions: (i) $g(0) = 0$, (ii) there exists $q_r > b$ so that $g(q_r) \neq 0$ and $U(q_r) > 0$, (iii) there exists $q_l < b$ so that $g(q_l) \neq 0$ and $U(q_l) = U(q_r)$, and (iv) for all $q \in (q_l, q_r)$, $U(q) < U(q_r)$. Then, this Hamiltonian system has a family of periodic solutions in the vicinity V of the point $(b, 0)$ (see Fig. 1.1). The periodic solution can be described as follows

$$q(t) = a\cos\phi(t) + b, \quad p(t) = -a\nu(a, \phi)\sin\phi(t),$$
$$\phi(t) = \gamma(t) + \theta. \tag{1.6}$$

where a is the amplitude,

$$\nu(a, \phi) = \frac{\mathrm{d}\gamma}{\mathrm{d}t} = \sqrt{\frac{2[U(a+b) - U(a\cos\phi + b)]}{a^2\sin^2\phi}} \tag{1.7}$$

represents the instantaneous frequency, and

$$U(a+b) = U(-a+b) = H \tag{1.8}$$

describes the relationship between a, b and the Hamiltonian H. Note that b depends on a, or b is a function of a. Because the frequency ν depends on a, ϕ, $\sin\phi$ and $\cos\phi$ are called generalized harmonic functions (Xu and Chung 1994). Expand ν into a Fourier series

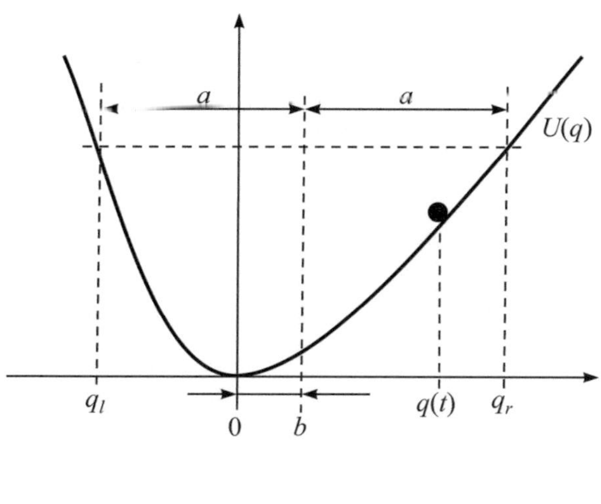

Fig. 1.1 Schematic diagram of potential function $U(q)$

$$\nu(a, \phi) = \omega(a) + \sum_{r=1}^{\infty} \omega_r(a) \cos r\phi. \tag{1.9}$$

Performing average operation on Eq. (1.9) yields

$$\frac{1}{2\pi} \int_0^{2\pi} \nu(a, \phi) \mathrm{d}\phi = \omega(a). \tag{1.10}$$

Equation (1.10) indicates that $\omega(a)$ is the mean frequency. For the hereafter averaging operation, the following approximation will be used

$$\phi(t) = \omega(a)t + \theta. \tag{1.11}$$

For quasi-linear systems, $\nu = \omega$ is a constant.

Return to quasi-Hamiltonian system (1.5). The periodic solution (1.6) become randomized periodic solution

$$Q(t) = A \cos \Phi(t) + B, \quad P(t) = -A\nu(A, \Phi) \sin \Phi(t), \quad \Phi(t) = \Gamma(t) + \Theta(t), \tag{1.12}$$

where

$$\nu(A, \Phi) = \frac{\mathrm{d}\Gamma}{\mathrm{d}t} = \sqrt{\frac{2[U(A+B) - U(A\cos\Phi + B)]}{A^2 \sin^2 \Phi}}. \tag{1.13}$$

A, Φ, Γ, Θ are all random processes. Regarding Eq. (1.12) as a generalized van der Pol transformation from Q, P to A, Φ, differentiating the first equation in Eq. (1.12) with respect to t and then subtracting the second equation lead to

$$\dot{A}(\cos\Phi + d) - \dot{\Theta} A \sin\Phi = 0, \tag{1.14}$$

where

$$d = \frac{\mathrm{d}B}{\mathrm{d}A} = \frac{g(-A+B) + g(A+B)}{g(-A+B) - g(A+B)}. \tag{1.15}$$

Differentiating the second equation in Eq. (1.12) with respect to t and then substituting the result into the second equation of system (1.5) yield

$$\begin{aligned}
&\dot{A}\{\nu(A,\Phi)\sin\Phi + A\frac{\partial}{\partial A}[\nu(A,\Phi)\sin\Phi]\} + \dot{\Theta}A\frac{\partial}{\partial \Gamma}[\nu(A,\Phi)\sin\Phi] \\
&= \varepsilon c(A\cos\Phi + B, -A\nu(A,\Phi)\sin\Phi)(-A\nu(A,\Phi)\sin\Phi) \\
&\quad - \varepsilon^{1/2} f_k(A\cos\Phi + B, -A\nu(A,\Phi)\sin\Phi)\xi_k(t).
\end{aligned} \tag{1.16}$$

Combining Eqs. (1.14) and (1.16) results in

$$\frac{\mathrm{d}A}{\mathrm{d}t} = \varepsilon F_1(A,\Phi) + \varepsilon^{1/2}\sum_{k=1}^{m} G_{1k}(A,\Phi)\xi_k(t),$$
$$\frac{\mathrm{d}\Phi}{\mathrm{d}t} = \nu(A,\Phi) + \varepsilon F_2(A,\Phi) + \varepsilon^{1/2}\sum_{k=1}^{m} G_{2k}(A,\Phi)\xi_k(t). \tag{1.17}$$

where

$$F_1(A,\Phi) = \frac{-A^2}{g(A+B)(1+d)}c(A\cos\Phi + B, -A\nu(A,\Phi)\sin\Phi) \times \nu^2(A,\Phi)\sin^2\Phi,$$
$$F_2(A,\Phi) = \frac{-A}{g(A+B)(1+d)}c(A\cos\Phi + B, -A\nu(A,\Phi)\sin\Phi) \times \nu^2(A,\Phi)(\cos\Phi + d)\sin\Phi,$$

$$G_{1k}(A,\Phi) = \frac{-A}{g(A+B)(1+d)}f_k(A\cos\Phi + B, -A\nu(A,\Gamma)\sin\Phi) \times \nu(A,\Phi)\sin\Phi,$$
$$G_{2k}(A,\Phi) = \frac{-1}{g(A+B)(1+d)}f_k(A\cos\Phi + B, -A\nu(A,\Phi)\sin\Phi) \times \nu(A,\Phi)(\cos\Phi + d). \tag{1.18}$$

It is seen from Eq. (1.17) that $\Phi(t)$ is a fast-varying process, and $A(t)$ is a slowly-varying process. According to Khasminskii's theorems (Khasminskii 1966, 1968), when $\varepsilon \to 0, A(t)$ converges weakly to a one-dimensional Markov diffusion process. This limiting process can be described by the following averaged Itô SDE,

$$\mathrm{d}A = m(A)\mathrm{d}t + \sigma(A)\mathrm{d}B(t), \tag{1.19}$$

where $B(t)$ is the unit Wiener process. According to Eqs. (4.27) and (4.28) in vol. 1, the drift and diffusion coefficients of Eq. (1.19) are

$$m(A) = \varepsilon\left\langle F_1 + \sum_{k,l=1}^{m}\int_{-\infty}^{0}\left(\left.\frac{\partial G_{1k}}{\partial A}\right|_t G_{1l}|_{t+\tau} + \left.\frac{\partial G_{1k}}{\partial \Phi}\right|_t G_{2l}|_{t+\tau}\right)R_{kl}(\tau)\mathrm{d}\tau\right\rangle_t,$$
$$\sigma^2(A) = \varepsilon\left\langle \sum_{k,l=1}^{m}\int_{-\infty}^{\infty} G_{1k}|_t G_{1l}|_{t+\tau}R_{kl}(\tau)\mathrm{d}\tau\right\rangle_t. \tag{1.20}$$

Due to the periodicity with respect to Φ, the time averaging $\langle[\cdot]\rangle_t$ can be replaced by the averaging with respect to Φ.

To obtain an explicit expressions for $m(A)$ and $\sigma^2(A)$, expand εF_i and $\varepsilon^{1/2}G_{ik}$ into Fourier series, i.e.,

$$
\begin{aligned}
\varepsilon F_1(A,\Phi) &= F_{10} + \sum_{r=1}^{\infty}\Big(F_{1r}^{(c)}\cos r\Phi + F_{1r}^{(s)}\sin r\Phi\Big),\\
\varepsilon^{1/2}G_{jk}(A,\Phi) &= G_{jk0} + \sum_{r=1}^{\infty}\Big(G_{jkr}^{(c)}\cos r\Phi + G_{jkr}^{(s)}\sin r\Phi\Big),\\
j &= 1,2;\quad k = 1,2,\ldots,m,
\end{aligned}
\tag{1.21}
$$

where each coefficient is a function of A. Substituting Eq. (1.21) into Eq. (1.20) and completing integration with respect to τ and averaging over Φ yield

$$
\begin{aligned}
m(A) = {}& F_{10}(A) + \pi\frac{\mathrm{d}G_{1k0}}{\mathrm{d}A}G_{1l0}S_{kl}(0) + \frac{\pi}{2}\sum_{r=1}^{\infty}\{[\frac{\mathrm{d}G_{1kr}^{(c)}}{\mathrm{d}A}G_{1lr}^{(c)}\\
&+ \frac{\mathrm{d}G_{1kr}^{(s)}}{\mathrm{d}A}G_{1lr}^{(s)} + r(G_{1kr}^{(s)}G_{2lr}^{(c)} - G_{1kr}^{(c)}G_{2lr}^{(s)})]S_{kl}(r\omega(A))\\
&+ [\frac{\mathrm{d}G_{1kr}^{(c)}}{\mathrm{d}A}G_{1lr}^{(s)} - \frac{\mathrm{d}G_{1kr}^{(s)}}{\mathrm{d}A}G_{1lr}^{(c)} + r(G_{1kr}^{(s)}G_{2lr}^{(s)}\\
&+ G_{1kr}^{(c)}G_{2lr}^{(c)})]I_{kl}(r\omega(A))\}.
\end{aligned}
$$

$$
\begin{aligned}
\sigma^2(A) = {}& 2\pi G_{1k0}G_{1l0}S_{kl}(0) + \pi\sum_{r=1}^{\infty}\Big[\Big(G_{1kr}^{(c)}G_{1lr}^{(c)} + G_{1kr}^{(s)}G_{1lr}^{(s)}\Big)S_{kl}(r\omega(A))\\
&+ \Big(G_{1kr}^{(c)}G_{1lr}^{(s)} - G_{1kr}^{(s)}G_{1lr}^{(c)}\Big)I_{kl}(r\omega(A))\Big],
\end{aligned}
\tag{1.22}
$$

where

$$
S_{kl}(\omega) = \frac{1}{\pi}\int_{-\infty}^{0} R_{kl}(\tau)\cos\omega\tau\,\mathrm{d}\tau,\quad I_{kl}(\omega) = \frac{1}{\pi}\int_{-\infty}^{0} R_{kl}(\tau)\sin\omega\tau\,\mathrm{d}\tau. \tag{1.23}
$$

The averaged FPK equation associated with Eq. (1.19) is

$$
\frac{\partial p}{\partial t} = -\frac{\partial}{\partial a}\big[m(a)p\big] + \frac{1}{2}\frac{\partial^2}{\partial a^2}\big[\sigma^2(a)p\big] \tag{1.24}
$$

where the first and second derivative moments are

$$
m(a) = m(A)|_{A=a},\quad \sigma^2(a) = \sigma^2(A)\big|_{A=a}. \tag{1.25}
$$

In Eq. (1.24), $p = p(a, t|a_0)$ can be the transition PDF of $A(t)$ with initial condition

$$p(a, 0|a_0) = \delta(a - a_0). \tag{1.26}$$

Or $p = p(a, t)$ can also be the PDF of $A(t)$ with initial condition

$$p(a, 0) = p(a_0). \tag{1.27}$$

The boundary conditions for Eq. (1.24) depend on domain of V. If domain V is $[0, \infty)$, the boundary conditions read

$$p = \text{finite}, \quad a = 0, \tag{1.28}$$

$$p, \partial p/\partial a \to 0, \quad a \to \infty. \tag{1.29}$$

If domain V is finite with a boundary h_c, and the Hamiltonian system with $h \geq h_c$ has no periodic solution, such as Duffing oscillator with soft spring. FPK Eq. (1.24) has no stationary solution in $0 \leq h < h_c$ since the system (1.5) will passage boundary $h = h_c$ sooner or later.

Under the boundary conditions (1.28) and (1.29), the stationary solution of Eq. (1.24) is

$$p(a) = \frac{C}{\sigma^2(a)} \exp\left[\int_0^a \frac{2m(a)}{\sigma^2(u)} du\right], \tag{1.30}$$

where C is a normalization constant. The stationary PDF of the Hamiltonian $H(t)$ is

$$p(h) = p(a)\left|\frac{da}{dh}\right| = \left.\frac{p(a)}{g(a)(1+d)}\right|_{a=U^{-1}(h)-b}. \tag{1.31}$$

Here, U^{-1} is the inverse function of U. According to Eq. (5.18) in vol. 1, the approximate stationary PDF of the generalized displacement and generalized momentum of the original system (1.5) is

$$p(q, p) = \left.\frac{p(h)}{T(h)}\right|_{h=p^2/2+U(q)}, \tag{1.32}$$

where $T(h)$ is the period of the Hamiltonian system associated with system (1.5) and can be expressed as

$$T(h) = 4\int_0^a \frac{dq}{\sqrt{2h - 2U(q)}}. \tag{1.33}$$

According to the relationship between H and a in Eq. (1.8), the following averaged Itô SDE for $H(t)$ can be derived.

$$\mathrm{d}H = \overline{m}(H)\mathrm{d}t + \overline{\sigma}(H)\mathrm{d}B(t), \tag{1.34}$$

where $B(t)$ is the unit Wiener process, and the drift and diffusion coefficients are

$$\begin{aligned} \overline{m}(H) &= \left\{ g(A+B)(1+d)m(A) + \frac{\sigma^2(A)}{2}\frac{\mathrm{d}\left[g(A+B)(1+d)\right]}{\mathrm{d}A} \right\}\Bigg|_{A=U^{-1}(H)-B}, \\ \overline{\sigma}^2(H) &= g^2(A+B)\,(1+d)^2\sigma^2(A)\big|_{A=U^{-1}(H)-B}. \end{aligned} \tag{1.35}$$

Solving the reduced FPK equation associated with Itô Eq. (1.34) yields $p(h)$ in Eq. (1.31).

Example 1.1 Consider a Duffing-van der Pol oscillator excited externally and parametrically by stationary wideband noises (Zhu et al. 2001). The motion equations are

$$\begin{aligned} \dot{Q} &= P \\ \dot{P} &= -\omega_0^2 Q - \alpha Q^3 - \left(-\beta_1 + \beta_2 Q^2\right)P + Q\xi_1(t) + \xi_2(t) \end{aligned} \tag{1.36}$$

where $\omega_0, \alpha, \beta_1, \beta_2$ are constants, $\xi_k(t)$ $(k = 1, 2)$ are independent stationary second-order rational noises with zero mean-value and the following power spectral densities (PSDs)

$$S_k(\omega) = \frac{D_k}{\pi}\frac{1}{\left(\omega^2 - \omega_k^2\right)^2 + 4\zeta_k^2\omega_k^2\omega^2}, \quad k = 1, 2, \tag{1.37}$$

where ζ_k, ω_k, D_k are constant; β_i, D_k are small quantities of ε order.

The Hamiltonian system associated with system (1.36) has a family of periodic solutions on entire phase plane (q, p) if $\alpha > 0$. Thus, the stochastic averaging method can be applied to system (1.36). For this example,

$$\begin{aligned} & g(q) = \omega_0^2 q + \alpha q^3, \quad U(q) = \omega_0^2 q^2/2 + \alpha q^4/4, \\ & b = d = 0, \\ & \nu(a,\phi) = \left[\left(\omega_0^2 + 3\alpha a^2/4\right)(1 + \eta\cos 2\phi)\right]^{1/2}, \\ & \eta = \left(\alpha a^2/4\right)/\left(\omega_0^2 + 3\alpha a^2/4\right) \le 1/3, \end{aligned} \tag{1.38}$$

Expand $\nu(a, \phi)$ with respective to ϕ as Fourier series and take the following approximation,

$$\nu(a,\phi) = b_0(a) + b_2(a)\cos 2\phi + b_4(a)\cos 4\phi + b_6(a)\cos 6\phi \tag{1.39}$$

where

$$\begin{aligned}
b_0(a) &= \left(\omega_0^2 + 3\alpha a^2/4\right)^{1/2}\left(1 - \eta^2/16\right),\\
b_2(a) &= \left(\omega_0^2 + 3\alpha a^2/4\right)^{1/2}\left(\eta/2 + 3\eta^3/64\right),\\
b_4(a) &= \left(\omega_0^2 + 3\alpha a^2/4\right)^{1/2}\left(-\eta^2/16\right),\\
b_6(a) &= \left(\omega_0^2 + 3\alpha a^2/4\right)^{1/2}\left(\eta^3/64\right).
\end{aligned} \tag{1.40}$$

The mean frequency is

$$\omega(a) = b_0(a). \tag{1.41}$$

After the transformation given in Eq. (1.12), system (1.36) becomes

$$\begin{aligned}
\frac{\mathrm{d}A}{\mathrm{d}t} &= \varepsilon F_1(A,\Phi) + \varepsilon^{1/2}G_{11}(A,\Phi)\xi_1(t) + \varepsilon^{1/2}G_{12}(A,\Phi)\xi_2(t),\\
\frac{\mathrm{d}\Phi}{\mathrm{d}t} &= \nu(A,\Phi) + \varepsilon F_2(A,\Phi) + \varepsilon^{1/2}G_{21}(A,\Phi)\xi_1(t)\\
&\quad + \varepsilon^{1/2}G_{22}(A,\Phi)\xi_2(t)
\end{aligned} \tag{1.42}$$

where

$$\begin{aligned}
\varepsilon F_1(A,\Phi) &= -\frac{A^2}{g(A)}\left(-\beta_1 + \beta_2 A^2\cos^2\Phi\right)\nu^2(A,\Phi)\sin^2\Phi\\
\varepsilon F_2(A,\Phi) &= -\frac{A}{g(A)}\left(-\beta_1 + \beta_2 A^2\cos^2\Phi\right)\nu^2(A,\Phi)\sin\Phi\cos\Phi\\
\varepsilon^{1/2}G_{11}(A,\Phi) &= -\frac{A^2}{g(A)}\nu^2(A,\Phi)\sin\Phi\cos\Phi,\quad \varepsilon^{1/2}G_{12}(A,\Phi)\\
&= -\frac{A}{g(A)}\nu(A,\Phi)\sin\Phi,
\end{aligned} \tag{1.43}$$

$$\varepsilon^{1/2}G_{21}(A,\Phi) = -\frac{A}{g(A)}\nu^2(A,\Phi)\cos^2\Phi,\quad \varepsilon^{1/2}G_{22}(A,\Phi) = -\frac{1}{g(A)}\nu(A,\Phi)\cos\Phi.$$

Assume that the PSDs $S_k(\omega)$ of the noises $\xi_k(t)$ $(k = 1, 2)$ change slowly over a wide frequency band. Then, the stochastic averaging method for SDOF quasi Hamiltonian systems under wideband random excitation can be applied. $A(t)$ weakly converges to a one-dimensional Markov diffusion process. The averaged Itô SDE for $A(t)$ is of the form of Eq. (1.19) and the associated averaged FPK equation is the form of Eq. (1.24). The first and second derivative moments can be calculated according to Eqs. (1.22) and (1.25) as follows

$$
\begin{aligned}
m(a) = & -\frac{a^2}{8g(a)}\left[-\beta_1(4\omega_0^2 + 5\alpha a^2/2) + \beta_2 a^2(\omega_0^2 + 3\alpha a^2/4)\right] \\
& + \frac{\pi a^2}{32g(a)}\{[2b_0(a) - b_4(a)]S_1(2\omega(a))\frac{\mathrm{d}}{\mathrm{d}a}[(2b_0(a) - b_4(a)) \\
& \times \frac{a^2}{g(a)}\Big] + [b_2(a) - b_4(a)]S_1(4\omega(a))\frac{\mathrm{d}}{\mathrm{d}a}\left[(b_2(a) - b_4(a))\frac{a^2}{g(a)}\right] \\
& + b_4(a)S_1(6\omega(a))\frac{\mathrm{d}}{\mathrm{d}a}\left[b_4(a)\frac{a^2}{g(a)}\right] + b_6(a)S_1(8\omega(a)) \\
& \times \frac{\mathrm{d}}{\mathrm{d}a}\left[b_6(a)\frac{a^2}{g(a)}\right]\Big\} + \frac{\pi a^3}{16g^2(a)}\{[2b_0(a) - b_4(a)][2b_0(a) + 2b_2(a) \\
& + b_4(a)]S_1(2\omega(a)) + 2[b_2(a) - b_6(a)][b_2(a) + 2b_4(a) + b_6(a)] \\
& \times S_1(4\omega(a)) + 3b_4(a)[b_4(a) + 2b_6(a)]S_1(6\omega(a)) \\
& + 4b_6^2(a)S_1(8\omega(a))\} + \frac{\pi a}{8g(a)}\{[2b_0(a) - b_2(a)]S_2(\omega(a)) \\
& \times \frac{\mathrm{d}}{\mathrm{d}a}\left[(2b_0(a) - b_2(a))\frac{a}{g(a)}\right] + [b_2(a) - b_4(a)]S_2(3\omega(a)) \\
& \times \frac{\mathrm{d}}{\mathrm{d}a}\left[(b_2(a) - b_4(a))\frac{a}{g(a)}\right] + [b_4(a) - b_6(a)]S_2(5\omega(a)) \\
& \times \frac{\mathrm{d}}{\mathrm{d}a}\left[(b_4(a) - b_6(a))\frac{a}{g(a)}\right] + b_6(a)S_2(7\omega(a)) \\
& \times \frac{\mathrm{d}}{\mathrm{d}a}\left[b_6(a)\frac{a}{g(a)}\right]\Big\} + \frac{\pi a}{8g^2(a)}\{\left[4b_0^2(a) - b_2^2(a)\right] \\
& \times S_2(\omega(a)) + 3\left[b_2^2(a) - b_4^2(a)\right]S_2(3\omega(a)) \\
& + 5\left[b_4^2(a) - b_6^2(a)\right]S_2(5\omega(a)) + 7b_6^2(a)S_2(7\omega(a))\},
\end{aligned}
\tag{1.44}
$$

$$
\begin{aligned}
\sigma^2(a) = & \frac{\pi a^4}{16g^2(a)}\{[2b_0(a) - b_4(a)]^2 S_1(2\omega(a)) \\
& + [b_2(a) - b_6(a)]^2 S_1(4\omega(a)) + b_4^2(a)S_1(6\omega(a)) \\
& + b_6^2(a)S_1(8\omega(a))\} + \frac{\pi a^2}{4g^2(a)}\{[2b_0(a) - b_2(a)]^2 S_2(\omega(a)) \\
& + [b_2(a) - b_4(a)]^2 S_2(3\omega(a)) + [b_4(a) - b_6(a)]^2 S_2(5\omega(a)) \\
& + b_6^2(a)S_2(7\omega(a))\}.
\end{aligned}
$$

The stationary PDFs $p(a)$ and $p(q, p)$ for stationary response of system (1.36) can be obtained using Eqs. (1.30) and (1.32). Some numerical results obtained by using the stochastic averaging method and Monte Carlo simulation for two sets of parameters are shown in Figs. 1.2 and 1.3. It can be seen that the two results are in good agreement.

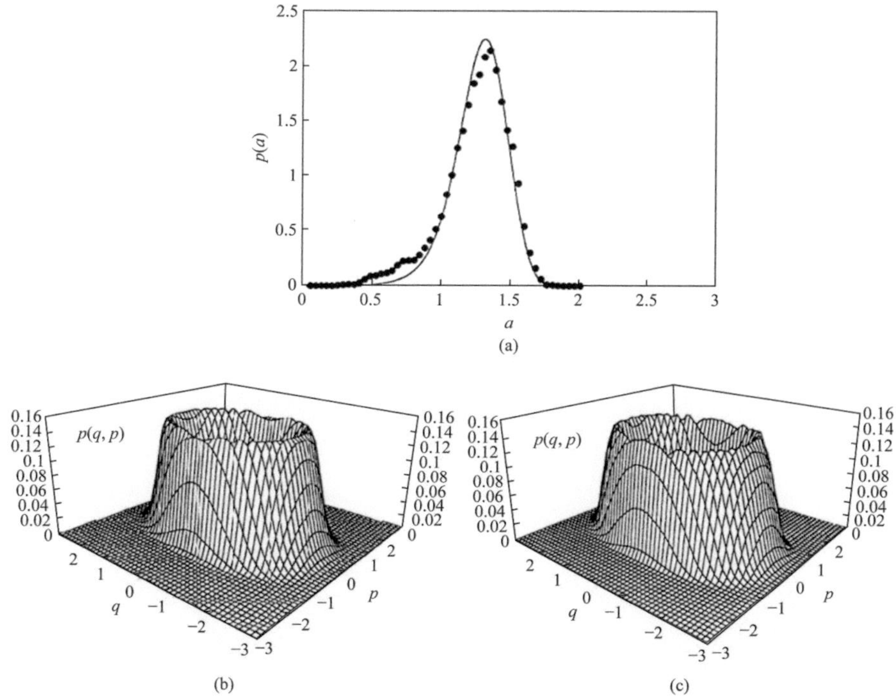

Fig. 1.2 Stationary PDFs $p(a)$ and $p(q, p)$ of system (1.36). **a** $p(a)$, ——, result from the stochastic averaging method; ●, result from Monte Carlo simulation of system (1.36); **b** $p(q, p)$, result from the stochastic averaging method; **c** $p(q, p)$, result from Monte Carlo simulation of system (1.36). System parameter are $\omega_0 = 1$, $\omega_1 = \omega_2 = 2$, $\alpha = 1$, $\beta_1 = 0.01$, $\beta_2 = 0.02$, $\xi_1 = \xi_2 = 0.3$, $D_1 = D_2 = 0.01$ (Zhu et al. 2001)

The averaged Itô SDE for $H(t)$ can be derived according to Eqs. (1.34) and (1.35). For the Duffing oscillator with linear damping excited by stationary wideband external excitations, the averaged FPK equation is of the form of Eq. (1.24) and the first and second derivative moments can be obtained from Eq. (1.44) with negative β_1. $\beta_2 = 0$ and $S_1(\omega) = 0$.

Example 1.2 Consider a SDOF vibration-impact system subjected to a stationary rational noise (Huang et al. 2003). The motion equations are given by

$$\begin{aligned} \dot{Q} &= P, \\ \dot{P} &= -g(Q) - 2\zeta\omega_0 P + \xi(t), \end{aligned} \tag{1.45}$$

where

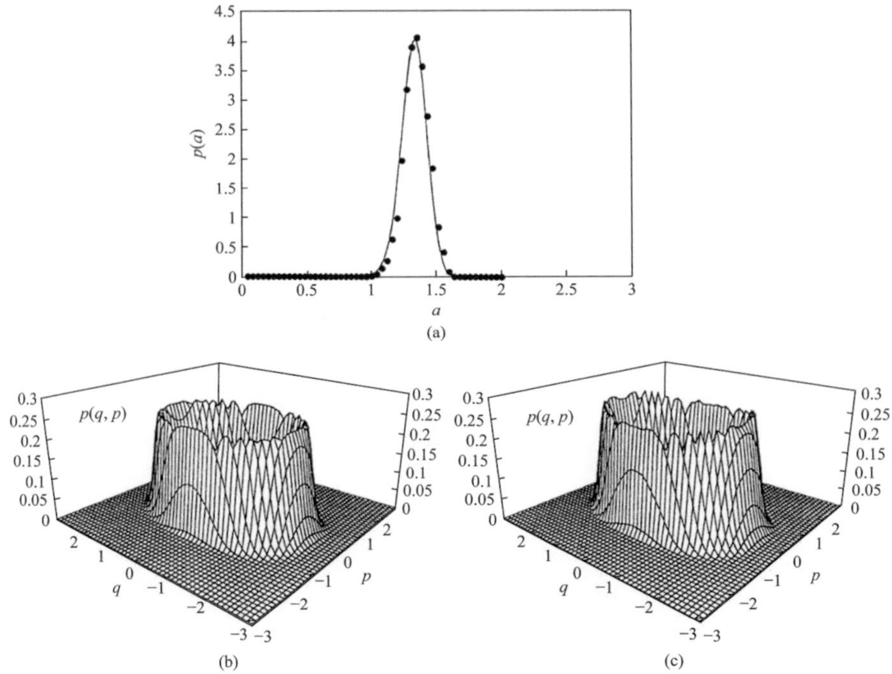

Fig. 1.3 Stationary PDFs $p(a)$ and $p(q, p)$ of system (1.36). **a** $p(a)$, ——, result from the stochastic averaging method; ●, result from Monte Carlo simulation of system (1.36); **b** $p(q, p)$, result from the stochastic averaging method; **c** $p(q, p)$, result from Monte Carlo simulation of system (1.36). System parameter are $\omega_0 = 1$, $\omega_1 = \omega_2 = 5$, $\alpha = 1$, $\beta_1 = 0.01$, $\beta_2 = 0.02$, $\xi_1 = \xi_2 = 0.5$, $D_1 = D_2 = 0.2$ (Zhu et al. 2001)

$$g(Q) = \begin{cases} \omega_0^2 Q - B_l(-Q-\delta_l)^{3/2} & Q < -\delta_l, \\ \omega_0^2 Q & -\delta_l < Q < \delta_r, \\ \omega_0^2 Q + B_r(Q-\delta_r)^{3/2} & Q > \delta_r, \end{cases} \tag{1.46}$$

in which δ_l, δ_r are the distance between mass and the left and right elastic walls, respectively; B_l, B_r are the stiffness parameters of the left and right walls, which are obtained from the Hertz contact law and depend on the geometry and material properties of the mass and walls; $\xi(t)$ is stationary rational noise with PSD of the form in Eq. (1.37).

Substituting $U(Q) = \int_0^Q g(u)\mathrm{d}u$ into Eq. (1.13) yields $\nu(A, \Theta)$. The averaged Itô SDE and its corresponding averaged FPK equation can be derived from Eq. (1.14) through Eq. (1.25). The drift and diffusion coefficients are calculated as follows

$$m(A) = \left\langle F_1 + \int_{-\infty}^{0} \left(\left.\frac{\partial G_{11}}{\partial A}\right|_t G_{11}|_{t+\tau} + \left.\frac{\partial G_{11}}{\partial \Phi}\right|_t G_{21}|_{t+\tau} \right) R(\tau)\mathrm{d}\tau \right\rangle_t,$$

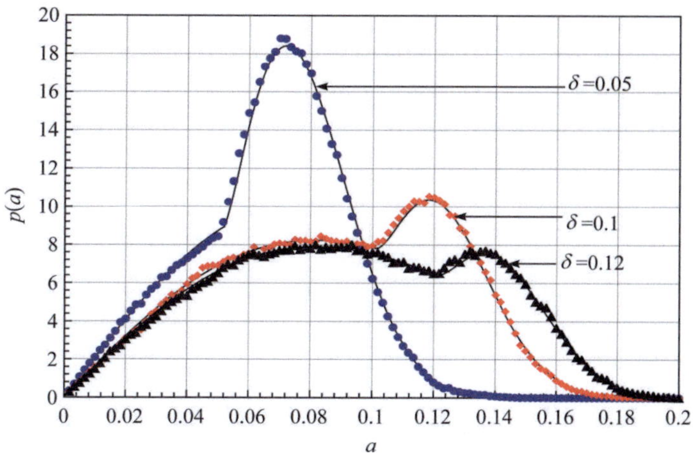

Fig. 1.4 Stationary PDF $p(a)$ of displacement amplitude of single-DOF double-wall vibration-impact system (1.45). ——, results from the stochastic averaging method; ●▲■, results from Monte Carlo simulation of system (1.45). System parameter are $\omega_0 = 1.0$, $\zeta = 0.1$, $\delta_l = \delta_r = \delta = 0.05, 0.1, 0.12$, $B_l = B_r = 50$, $D_1 = 0.8$, $\omega_1 = 5.0$, $\zeta_1 = 0.5$ (Huang et al. 2003)

$$\sigma^2(A) = \left\langle \int_{-\infty}^{\infty} G_{11}|_t G_{11}|_{t+\tau} R(\tau)\mathrm{d}\tau \right\rangle_t, \tag{1.47}$$

where

$$\begin{aligned} F_1 &= -2\zeta\omega A^2\nu^2(A,\Phi)\sin^2\Phi\Big/g(A+B)(1+d), \\ G_{11} &= -A\nu(A,\Phi)\sin\Phi\Big/g(A+B)(1+d), \\ G_{21} &= -\nu(A,\Phi)\cos\Phi\Big/g(A+B)(1+d). \end{aligned} \tag{1.48}$$

The vibration-impact system (1.45) is highly nonlinear as the excitation intensity increases, the mass-wall distance decreases, the wall stiffness increases, or a combination of these factors. The comparison between the two results from the stochastic averaging method and from Monte Carlo simulation shown in Fig. 1.4 indicates that the stochastic averaging method yields very good results.

1.1.2 MDOF System

Return to the system (1.4). Assume that when $\varepsilon = 0$, its n Hamiltonian subsystems satisfy the four conditions below Eq. (1.5) and in the domain V_i nearing equilibrium point, and each Hamiltonian subsystem has a family of periodic solutions of the form (1.6). Based on this assumption, the following generalized van der Pol transformation can be applied to system (1.4),

$$
\begin{aligned}
&Q_i(t) = A_i \cos\Phi_i(t) + B_i, \quad P_i(t) = -A_i \nu_i(A_i, \Phi_i) \sin\Phi_i(t), \\
&\Phi_i(t) = \Gamma_i(t) + \Theta_i(t),
\end{aligned}
\tag{1.49}
$$

where

$$
\nu_i(A_i, \Phi_i) = \frac{\mathrm{d}\Gamma_i}{\mathrm{d}t} = \sqrt{\frac{2[U_i(A_i + B_i) - U_i(A_i\cos\Phi_i + B_i)]}{A_i^2 \sin^2\Phi_i}}. \tag{1.50}
$$

The relationship between A_i, B_i and H_i is

$$
U_i(A_i + B_i) = U_i(-A_i + B_i) = H_i. \tag{1.51}
$$

Following the derivation from Eqs. (1.14) to (1.18) leads to the following equations

$$
\begin{aligned}
&\frac{\mathrm{d}A_i}{\mathrm{d}t} = \varepsilon F_i^{(1)}(\mathbf{A}, \mathbf{\Phi}) + \varepsilon^{1/2}\sum_{k=1}^{m} G_{ik}^{(1)}(\mathbf{A}, \mathbf{\Phi})\xi_k(t), \\
&\frac{\mathrm{d}\Phi_i}{\mathrm{d}t} = \nu_i(A_i, \Phi_i) + \varepsilon F_i^{(2)}(\mathbf{A}, \mathbf{\Phi}) + \varepsilon^{1/2}\sum_{k=1}^{m} G_{ik}^{(2)}(\mathbf{A}, \mathbf{\Phi})\xi_k(t), \\
&i = 1, 2, \ldots, n.
\end{aligned}
\tag{1.52}
$$

where $\mathbf{A} = [A_1, A_2, \ldots, A_n]^{\mathrm{T}}$, $\mathbf{\Phi} = [\Phi_1, \Phi_2, \ldots, \Phi_n]^{\mathrm{T}}$.

$$
F_i^{(1)}(\mathbf{A}, \mathbf{\Phi}) = \frac{-A_i \nu_i(A_i, \Phi_i)\sin\Phi_i}{g_i(A_i + B_i)(1 + d_i)} \sum_{j=1}^{n} c_{ij}(\mathbf{A}, \mathbf{\Phi}) A_j \nu_j(A_j, \Phi_j) \sin\Phi_j,
$$

$$
F_i^{(2)}(\mathbf{A}, \mathbf{\Phi}) = \frac{-\nu_i(A_i, \Phi_i)(\cos\Phi_i + d_i)}{g_i(A_i + B_i)(1 + d_i)} \sum_{j=1}^{n} c_{ij}(\mathbf{A}, \mathbf{\Phi}) A_j \nu_j(A_j, \Phi_j) \sin\Phi_j,
$$

$$
\begin{aligned}
G_{ik}^{(1)}(\mathbf{A}, \mathbf{\Phi}) &= \frac{-A_i \nu_i(A_i, \Phi_i)\sin\Phi_i}{g_i(A_i + B_i)(1 + d_i)} f_{ik}(\mathbf{A}, \mathbf{\Phi}), \\
G_{ik}^{(2)}(\mathbf{A}, \mathbf{\Phi}) &= \frac{-\nu_i(A_i, \Phi_i)(\cos\Phi_i + d_i)}{g_i(A_i + B_i)(1 + d_i)} f_{ik}(\mathbf{A}, \mathbf{\Phi}).
\end{aligned}
\tag{1.53}
$$

where

$$
d_i = \frac{\mathrm{d}B_i}{\mathrm{d}A_i} = \frac{g_i(-A_i + B_i) + g_i(A_i + B_i)}{g_i(-A_i + B_i) - g_i(A_i + B_i)} \tag{1.54}
$$

and $c_{ij}(\mathbf{A}, \mathbf{\Phi}), f_{ik}(\mathbf{A}, \mathbf{\Phi})$ are obtained from system (1.4) through the transformation (1.49).

It is seen from Eq. (1.52) that $\Phi_i(t)$ are fast-varying processes while $A_i(t)$ are slowly-varying processes. The dimension and form of the averaged Itô equations depend on if there is internal resonance in system (1.4).

(i) **Non-internal resonant case**

In this case, only $A_i(t)$ $(i = 1, 2, \ldots, n)$ are slowly-varying processes. According to the Khasminskii theorem (Khasminskii 1966, 1968), when $\varepsilon \to 0$, $\mathbf{A}(t)$ weakly converges to a n-dimensional vector Markov diffusion process. The averaged Itô SDEs describing this limiting process are

$$\begin{aligned} \mathrm{d}A_i &= m_i(\mathbf{A})\mathrm{d}t + \sum_{k=1}^{m} \sigma_{ik}(\mathbf{A})\mathrm{d}B_k(t), \\ i &= 1, 2, \ldots, n, \end{aligned} \tag{1.55}$$

where $B_k(t)$ $(k = 1, 2, \ldots, m)$ are independent unit Wiener processes. The drift and diffusion coefficients can be derived according to Eqs. (4.27) and (4.28) in vol. 1 as follows

$$m_i(\mathbf{A}) = \varepsilon\left\langle F_i^{(1)} + \sum_{k,l=1}^{m}\int_{-\infty}^{0}\sum_{j=1}^{n}\left(\frac{\partial G_{ik}^{(1)}}{\partial A_j}\bigg|_t G_{jl}^{1}\bigg|_{t+\tau} + \frac{\partial G_{ik}^{(1)}}{\partial \Phi_j}\bigg|_t G_{jl}^{(2)}\bigg|_{t+\tau}\right)R_{kl}(\tau)\mathrm{d}\tau\right\rangle_t,$$

$$b_{ij}(\mathbf{A}) = \sum_{k=1}^{m}\sigma_{ik}(\mathbf{A})\sigma_{jk}(\mathbf{A}) = \varepsilon\left\langle\sum_{k,l=1}^{m}\int_{-\infty}^{\infty}(G_{ik}^{(1)}\Big|_t G_{jl}^{(1)}\Big|_{t+\tau})R_{kl}(\tau)\mathrm{d}\tau\right\rangle_t. \tag{1.56}$$

where the fast-varying terms are periodic functions of $\mathbf{\Phi}$ so that the time averaging can be replaced by averaging with respect to $\mathbf{\Phi}$. Before performing averaging, it is necessary to expand the functions in Eq. (1.53) into Fourier series as follows

$$\varepsilon F(\mathbf{A}, \mathbf{\Phi}) = F_0 + \sum_{r=1}^{\infty}\sum_{|\mathbf{s}|=r}^{\infty}\left(F_{\mathbf{s}}^{(c)}\cos(\mathbf{s}, \mathbf{\Phi}) + F_{\mathbf{s}}^{(s)}\sin(\mathbf{s}, \mathbf{\Phi})\right), \tag{1.57}$$

where the coefficients F_0, $F_s^{(c)}$, $F_s^{(s)}$ are functions of $\mathbf{A}$, notation F stands for any of $F_i^{(1)}, F_i^{(2)}, G_{ik}^{(1)}, G_{ik}^{(2)}$, $s = [s_1, s_2, \ldots, s_n]^{\mathrm{T}}$ is an integer vector, $|\mathbf{s}| = \sum_{i=1}^{n}|s_i|$, and $(\mathbf{s}, \mathbf{\Phi}) = \sum_{i=1}^{n} s_i\Phi_i$. After expanding all the functions in Eq. (1.56) according to Eq. (1.57), completing averaging with respect to $\mathbf{\Phi}$ and integrating with respect to τ, the specific expressions for $m_i(\mathbf{A})$ and $b_{ij}(\mathbf{A})$ in Eq. (1.56) can be obtained. The associated averaged FPK equation is then

$$\frac{\partial p}{\partial t} = -\frac{\partial}{\partial a_i}\left[m_i(\mathbf{a})p\right] + \frac{1}{2}\sum_{i,j=1}^{n}\frac{\partial^2}{\partial a_i\partial a_j}\left[b_{ij}(\mathbf{a})p\right], \tag{1.58}$$

where the first and second derivative moments are

$$m_i(\mathbf{a}) = m_i(\mathbf{A})|_{\mathbf{A}=\mathbf{a}}, \quad b_{ij}(\mathbf{a}) = \sum_{k=1}^{m} \sigma_{ik}\sigma_{jk}(\mathbf{A})\Big|_{\mathbf{A}=\mathbf{a}} . \tag{1.59}$$

In Eq. (1.58), $p = p(\mathbf{a}, t|\mathbf{a}_0)$ with the initial condition

$$p(\mathbf{a}, 0|\mathbf{a}_0) = \delta(\mathbf{a} - \mathbf{a}_0), \tag{1.60}$$

or $p = p(\mathbf{a}, t)$ with the initial condition

$$p(\mathbf{a}, 0) = p(\mathbf{a}_0). \tag{1.61}$$

The boundary conditions for Eq. (1.58) depend on the domain V_i. If all V_i are entire phase space (q_i, p_i), then the boundary conditions are

$$\begin{aligned} &p = \text{finite}, \quad a_i = 0, \\ &p,\ \partial p/\partial a_i \to 0, \quad |\mathbf{a}| \to \infty, \\ &i = 1, 2, \ldots, n. \end{aligned} \tag{1.62}$$

The probability flow in Eq. (1.58) contains only the probability potential flow and no probability circulating flow. The system belongs to the class of stationary potential and an exact stationary $p(\mathbf{a})$ may be obtained by using the method of exact stationary solution for the stochastically excited and dissipated Hamiltonian systems (Zhu 2003).

Once the stationary solution $p(\mathbf{a})$ of the FPK Eq. (1.58) is obtained, the stationary PDF for $H(t)$ can be obtained as follow

$$p(\mathbf{h}) = p(\mathbf{a})\left|\frac{\partial \mathbf{a}}{\partial \mathbf{h}}\right| = p(\mathbf{a})\prod_{i=1}^{n}\frac{\mathrm{d}a_i}{\mathrm{d}h_i} = \left[p(\mathbf{a})\Big/\prod_{i=1}^{n} g_i(a_i)(1+d_i)\right]\Bigg|_{a_i=U_i^{-1}(h_i)-b_i} . \tag{1.63}$$

Similar to Eq. (1.32), the approximate stationary PDF for the generalized displacements and generalized momenta of system (1.4) is

$$p(\mathbf{q}, \mathbf{p}) = \frac{p(\mathbf{h})}{\prod_{i=1}^{n} T_i(h_i)}\Bigg|_{h_i=p_i^2/2+U_i(q_i)} , \tag{1.64}$$

where

$$T_i(h_i) = 4\int_0^{a_i} \frac{\mathrm{d}q_i}{\sqrt{2h_i - 2U_i(q_i)}}. \tag{1.65}$$

The following averaged Itô SDEs of H_i can be derived from the averaged Itô SDEs of A_i in Eq. (1.55) by using Eq. (1.51) and the Itô stochastic differential rule.

$$\mathrm{d}H_i = \overline{m}_i(\mathbf{H})\mathrm{d}t + \sum_{k=1}^{m} \overline{\sigma}_{ik}(\mathbf{H})\mathrm{d}B_k(t), \quad i = 1, 2, \ldots, n, \tag{1.66}$$

where $\mathbf{H} = [H_1, H_2, \ldots, H_n]^{\mathrm{T}}$, and the drift and diffusion coefficients are

$$\begin{aligned}
\overline{m}_i(\mathbf{H}) &= \left\{ g_i(A_i + B_i)(1 + d_i)m_i(\mathbf{A}) + \frac{1}{2}\frac{\mathrm{d}\left[g_i(A_i + B_i)(1 + d_i)\right]}{\mathrm{d}A} \right. \\
&\quad \left. \sum_{k=1}^{m} \sigma_{ik}\sigma_{ik}(\mathbf{A}) \right\}\Bigg|_{A_i = U_i(H_i) - B_i}, \\
\sum_{k=1}^{m} \overline{\sigma}_{ik}\overline{\sigma}_{jk}(\mathbf{H}) &= g_i(A_i + B_i)g_j(A_j + B_j)(1 + d_i)(1 + d_j) \\
&\quad \times \sum_{k=1}^{m} \sigma_{ik}\sigma_{jk}(\mathbf{A})\Bigg|_{A_i = U_i(H_i) - B_i}.
\end{aligned} \tag{1.67}$$

The stationary PDF (1.63) can also be obtained by solving the averaged FPK equation associated with Itô SDEs in Eq. (1.66).

(ii) **Internal resonant case**

Assume that there exist internal resonant relationships among n mean frequencies of n Hamiltonian subsystems, i.e., $\sum_{r=1}^{n} k_r^u \overline{\omega}_r = O_u(\varepsilon)$, $u = 1, 2, \ldots, \alpha$ $(1 \le \alpha \le n-1)$ where $\overline{\omega}_r(a_r) = \int_0^{2\pi} v_r(a_r, \phi_r)\mathrm{d}\phi_r / 2\pi$, $r = 1, 2, \ldots, n$. Introduce α combinations of angle variables $\Psi_u = \sum_{r=1}^{n} k_r^u \Phi_r$, $u = 1, 2, \ldots, \alpha$. The following governing motion equations can be derived from Eq. (1.52).

$$\begin{aligned}
\frac{\mathrm{d}A_i}{\mathrm{d}t} &= \varepsilon \overline{F}_i^{(1)}(\mathbf{A}, \boldsymbol{\Psi}, \boldsymbol{\Phi}_1) + \varepsilon^{1/2} \sum_{k=1}^{m} \overline{G}_{ik}^{(1)}(\mathbf{A}, \boldsymbol{\Psi}, \boldsymbol{\Phi}_1)\xi_k(t), \\
\frac{\mathrm{d}\Psi_u}{\mathrm{d}t} &= \sum_{r=1}^{n} k_r^u v_r(A_r, \Phi_r) + \varepsilon \sum_{r=1}^{n} k_r^u \overline{F}_r^{(2)}(\mathbf{A}, \boldsymbol{\Psi}, \boldsymbol{\Phi}_1)
\end{aligned}$$

$$
\begin{aligned}
&+\varepsilon^{1/2}\sum_{r=1}^{n}\sum_{k=1}^{m}k_r^u\overline{G}_{rk}^{(2)}(\mathbf{A},\boldsymbol{\Psi},\boldsymbol{\Phi}_1)\xi_k(t),\\
&\frac{\mathrm{d}\Phi_s}{\mathrm{d}t}=\nu_s(A_r,\Phi_r)+\varepsilon\overline{F}_s^{(2)}(\mathbf{A},\boldsymbol{\Psi},\boldsymbol{\Phi}_1)+\varepsilon^{1/2}\sum_{k=1}^{m}\overline{G}_{sk}^{(2)}(\mathbf{A},\boldsymbol{\Psi},\boldsymbol{\Phi}_1)\xi_k(t),\\
&i=1,2,\ldots,n;\quad u=1,2,\ldots,\alpha;\quad s=\alpha+1,\alpha+2,\ldots,n.
\end{aligned}
\tag{1.68}
$$

where $\boldsymbol{\Psi}=[\Psi_1,\Psi_2,\cdots,\Psi_\alpha]^{\mathrm{T}}$, $\boldsymbol{\Phi}_1=[\Phi_{\alpha+1},\Phi_{\alpha+2},\ldots,\Phi_n]^{\mathrm{T}}$; $\overline{F}_i^{(1)},\overline{F}_i^{(2)},\overline{G}_{ik}^{(1)},\overline{G}_{ik}^{(2)}$ are obtained from $F_i^{(1)},F_i^{(2)},G_{ik}^{(1)},G_{ik}^{(2)}$ in Eq. (1.53) by replacing $\Phi_1,\Phi_2,\ldots,\Phi_\alpha$ by $\boldsymbol{\Psi}$ and $\boldsymbol{\Phi}_1$. In system (1.68), $\boldsymbol{\Phi}_1$ is fast-varying vector process, while $\mathbf{A}$, Ψ are slowly-varying vector processes. According to the Khasminskii theorem (Khasminskii 1966, 1968), when $\varepsilon\to 0$, $[\mathbf{A}^{\mathrm{T}},\Psi^{\mathrm{T}}]^{\mathrm{T}}$ converges weakly to a $(n+\alpha)$-dimensional Markov diffusion process. The averaged Itô SDEs describing this limiting vector process is of the form

$$
\begin{aligned}
&\mathrm{d}A_i=m_i(\mathbf{A},\boldsymbol{\Psi})\mathrm{d}t+\sum_{k=1}^{m}\sigma_{ik}(\mathbf{A},\boldsymbol{\Psi})\mathrm{d}B_k(t),\\
&\mathrm{d}\Psi_u=[O_u(\varepsilon)+m_u^{\Psi}(\mathbf{A},\boldsymbol{\Psi})]\mathrm{d}t+\sum_{k=1}^{m}\sigma_{uk}^{\Psi}(\mathbf{A},\boldsymbol{\Psi})\mathrm{d}B_k(t),\\
&i=1,2,\ldots,n;\quad u=1,2,\ldots,\alpha.
\end{aligned}
\tag{1.69}
$$

where $B_k(t)$ $(k=1,2,\ldots,m)$ are independent unit Wiener processes. The drift and diffusion coefficients can be obtained according to Eqs. (4.27) and (4.28) in vol. 1 as follows

$$
\begin{aligned}
m_i(\mathbf{A},\boldsymbol{\Psi})=\varepsilon\Bigg\langle\overline{F}_i^{(1)}+\int_{-\infty}^{0}\sum_{k,l=1}^{m}\Bigg[\sum_{j=1}^{n}\left(\frac{\partial\overline{G}_{ik}^{(1)}}{\partial A_j}\bigg|_t\overline{G}_{jl}^{(1)}\bigg|_{t+\tau}\right)\\
+\sum_{u=1}^{\alpha}\sum_{r=1}^{n}\left(k_r^u\frac{\partial\overline{G}_{ik}^{(1)}}{\partial\Psi_u}\bigg|_t\overline{G}_{rl}^{(2)}\bigg|_{t+\tau}\right)+\sum_{s=\alpha+1}^{n}\left(\frac{\partial\overline{G}_{ik}^{(1)}}{\partial\Phi_s}\bigg|_t\overline{G}_{sl}^{(2)}\bigg|_{t+\tau}\right)\Bigg]R_{kl}(\tau)\mathrm{d}\tau\Bigg\rangle_t,\\
m_u^{\Psi}(\mathbf{A},\boldsymbol{\Psi})=\varepsilon\Bigg\langle\sum_{r=1}^{n}k_r^u\overline{F}_r^{(2)}+\int_{-\infty}^{0}\sum_{k,l=1}^{m}\Bigg[\sum_{j,r=1}^{n}\left(k_r^u\frac{\partial\overline{G}_{rk}^{(2)}}{\partial A_j}\bigg|_t\overline{G}_{jl}^{(1)}\bigg|_{t+\tau}\right)\\
+\sum_{v=1}^{\alpha}\sum_{j,r=1}^{n}\left(k_r^uk_j^v\frac{\partial\overline{G}_{rk}^{(2)}}{\partial\Psi_v}\bigg|_t\overline{G}_{jl}^{(2)}\bigg|_{t+\tau}\right)\\
+\sum_{s=\alpha+1}^{n}\sum_{r=1}^{n}\left(k_r^u\frac{\partial\overline{G}_{rk}^{(2)}}{\partial\Phi_s}\bigg|_t\overline{G}_{sl}^{(2)}\bigg|_{t+\tau}\right)\Bigg]R_{kl}(\tau)\mathrm{d}\tau\Bigg\rangle_t,
\end{aligned}
$$

$$
\begin{aligned}
b_{ij}(\mathbf{A},\boldsymbol{\Psi}) &= \sum_{k=1}^{m} \sigma_{ik}\sigma_{jk} = \varepsilon\left\langle \int_{-\infty}^{\infty} \sum_{k,l=1}^{m} \left(\overline{G}_{ik}^{(1)}\Big|_t \overline{G}_{jl}^{(1)}\Big|_{t+\tau}\right) R_{kl}(\tau)\mathrm{d}\tau \right\rangle_t, \\
b_{uv}^{\Psi}(\mathbf{A},\boldsymbol{\Psi}) &= \sum_{k=1}^{m} \sigma_{uk}^{\Psi}\sigma_{vk}^{\Psi} = \varepsilon\left\langle \int_{-\infty}^{\infty} \sum_{k,l=1}^{m}\sum_{j,r=1}^{n} \left(k_r^u k_j^v \overline{G}_{jk}^{(2)}\Big|_t \overline{G}_{rl}^{(2)}\Big|_{t+\tau}\right) R_{kl}(\tau)\mathrm{d}\tau \right\rangle_t, \\
b_{iu}^{A\Psi}(\mathbf{A},\boldsymbol{\Psi}) &= \sum_{k=1}^{m} \sigma_{ik}\sigma_{uk}^{\Psi} = \varepsilon\left\langle \int_{-\infty}^{\infty} \sum_{k,l=1}^{m}\sum_{r=1}^{n} \left(k_r^u \overline{G}_{ik}^{(1)}\Big|_t \overline{G}_{rl}^{(2)}\Big|_{t+\tau}\right) R_{kl}(\tau)\mathrm{d}\tau \right\rangle_t, \\
& i,j = 1,2,\ldots,n; \quad u,v = 1,2,\ldots,\alpha.
\end{aligned}
\tag{1.70}
$$

Due to the ergodicity of integrable and resonant Hamiltonian system on $(n-\alpha)$-dimensional torus, time averaging $\langle[\cdot\rangle_t$ in Eq. (1.70) can be replaced by averaging with respect to $\Phi_{\alpha+1}, \Phi_{\alpha+2}, \ldots, \Phi_n$. To complete the calculations in Eq. (1.70), it is needed to expand functions in Eq. (1.53) into Fourier series of the form (1.57). After averaging with respect to $\Phi_{\alpha+1}, \Phi_{\alpha+2}, \ldots, \Phi_n$ and integrating with respect to τ, the specific expressions for Eq. (1.70) can be obtained.

The averaged FPK equation associated with SDEs (1.69) is

$$
\begin{aligned}
\frac{\partial p}{\partial t} = &-\sum_{i=1}^{n} \frac{\partial}{\partial a_i}\left[m_i(\mathbf{a},\boldsymbol{\psi})p\right] - \sum_{u=1}^{\alpha} \frac{\partial}{\partial \psi_u}\left[m_u^{\psi}(\mathbf{a},\boldsymbol{\psi})p\right] \\
&+ \frac{1}{2}\sum_{i,j=1}^{n} \frac{\partial^2}{\partial a_i \partial a_j}\left[b_{ij}(\mathbf{a},\boldsymbol{\psi})p\right] + \sum_{i=1}^{n}\sum_{u=1}^{\alpha} \frac{\partial^2}{\partial a_i \partial \psi_u}\left[b_{iu}^{a\psi}(\mathbf{a},\boldsymbol{\psi})p\right] \\
&+ \frac{1}{2}\sum_{u,v=1}^{\alpha} \frac{\partial^2}{\partial \psi_u \partial \psi_v}\left[b_{uv}^{\psi}(\mathbf{a},\boldsymbol{\psi})p\right].
\end{aligned}
\tag{1.71}
$$

where

$$
\begin{aligned}
& m_i(\mathbf{a},\boldsymbol{\psi}) = m_i(\mathbf{A},\boldsymbol{\Psi})\big|_{\mathbf{A}=\mathbf{a},\boldsymbol{\Psi}=\boldsymbol{\psi}}, \quad m_u^{\psi}(\mathbf{a},\boldsymbol{\psi}) = m_u^{\Psi}(\mathbf{A},\boldsymbol{\Psi})\big|_{\mathbf{A}=\mathbf{a},\boldsymbol{\Psi}=\boldsymbol{\psi}}, \\
& b_{ij}(\mathbf{a},\boldsymbol{\psi}) = b_{ij}(\mathbf{A},\boldsymbol{\Psi})\big|_{\mathbf{A}=\mathbf{a},\boldsymbol{\Psi}=\boldsymbol{\psi}}, \quad b_{iu}^{a\psi}(\mathbf{a},\boldsymbol{\psi}) = b_{iu}^{A\Psi}(\mathbf{A},\boldsymbol{\Psi})\big|_{\mathbf{A}=\mathbf{a},\boldsymbol{\Psi}=\boldsymbol{\psi}}, \\
& b_{uv}^{\psi}(\mathbf{a},\boldsymbol{\psi}) = b_{uv}^{\Psi}(\mathbf{A},\boldsymbol{\Psi})\big|_{\mathbf{A}=\mathbf{a},\boldsymbol{\Psi}=\boldsymbol{\psi}}, \\
& \quad i,j = 1,2,\ldots,n; \quad u,v = 1,2,\ldots,\alpha.
\end{aligned}
\tag{1.72}
$$

In Eq. (1.71), $p = p(\mathbf{a},\boldsymbol{\psi},t|\mathbf{a}_0,\boldsymbol{\psi}_0)$ is the transition PDF of $[\mathbf{A}^{\mathrm{T}},\boldsymbol{\Psi}^{\mathrm{T}}]^{\mathrm{T}}$ with initial condition

$$
p(\mathbf{a},\boldsymbol{\psi},0|\mathbf{a}_0,\boldsymbol{\psi}_0) = \delta(\mathbf{a}-\mathbf{a}_0)\delta(\boldsymbol{\psi}-\boldsymbol{\psi}_0),
\tag{1.73}
$$

or $p = p(\mathbf{a},\boldsymbol{\psi},t)$ is the transient joint PDF of $[\mathbf{A}^{\mathrm{T}},\boldsymbol{\Psi}^{\mathrm{T}}]^{\mathrm{T}}$ with initial condition

$$p(\mathbf{a},\boldsymbol{\psi},0)=p(\mathbf{a}_0,\boldsymbol{\psi}_0). \tag{1.74}$$

When all V_i are entire phase planes, the boundary conditions are

$$\begin{aligned}
&p=\text{finite},\quad a_i=0,\\
&p,\ \partial p/\partial a_i\to 0,\ |\mathbf{a}|\to\infty,\\
&p(\mathbf{a},\psi_1,\psi_2,\ldots,\psi_u+2k\pi,\ldots,\psi_\alpha,t|\mathbf{a}_0,\boldsymbol{\psi}_0)\\
&\quad=p(\mathbf{a},\psi_1,\psi_2,\ldots,\psi_u,\ldots,\psi_\alpha,t|\mathbf{a}_0,\boldsymbol{\psi}_0),\\
&\quad i=1,2,\ldots,n,\quad u=1,2,\ldots,\alpha,\quad k\text{ is an integer.}
\end{aligned} \tag{1.75}$$

Letting $\partial p/\partial t=0$, Eq. (1.71) is degenerated to a reduced FPK equation. When system parameters satisfy certain compatibility conditions (see Example 1.4), an exact stationary PDF $p(\mathbf{a},\boldsymbol{\psi})$ may be obtained from solving the reduced FPK equation with boundary conditions in Eq. (1.75). Generally, $p(\mathbf{a},\boldsymbol{\psi})$ can be obtained only numerically.

Similar to Eq. (5.127) in vol. 1, once $p(\mathbf{a},\boldsymbol{\psi})$ is obtained, the approximate stationary joint PDF for the generalized displacements and generalized momenta of the system (1.4) is obtained as follows

$$\begin{aligned}
p(\mathbf{q},\mathbf{p})&=p(\mathbf{a},\boldsymbol{\psi},\boldsymbol{\phi}_1)\left|\frac{\partial(\mathbf{a},\boldsymbol{\psi},\boldsymbol{\phi}_1)}{\partial(\mathbf{q},\mathbf{p})}\right|=p(\boldsymbol{\phi}_1|\mathbf{a},\boldsymbol{\psi})p(\mathbf{a},\boldsymbol{\psi})\left|\frac{\partial(\mathbf{a},\boldsymbol{\psi},\boldsymbol{\phi}_1)}{\partial(\mathbf{q},\mathbf{p})}\right|\\
&=\frac{1}{(2\pi)^{n-\alpha}}p(\mathbf{a},\boldsymbol{\psi})\left|\frac{\partial(\mathbf{a},\boldsymbol{\psi},\boldsymbol{\phi}_1)}{\partial(\mathbf{q},\mathbf{p})}\right|.
\end{aligned} \tag{1.76}$$

Where $|\partial(\mathbf{a},\boldsymbol{\psi},\boldsymbol{\phi}_1)/\partial(\mathbf{q},\mathbf{p})|$ is Jacobi determinant for the transformation from $\mathbf{q}$, $\mathbf{p}$ to $\mathbf{a},\boldsymbol{\psi},\boldsymbol{\phi}_1$.

Using Eq. (1.51) and Itô differential rule, the averaged Itô SDEs for the sub-Hamiltonians $H_i(t)$ and combinations $\Psi_u(t)$ of angle variables can be derived from Eq. (1.69) as follows

$$\begin{aligned}
&\mathrm{d}H_i=\overline{m}_i(\mathbf{H},\boldsymbol{\Psi})\mathrm{d}t+\sum_{k=1}^{m}\overline{\sigma}_{ik}(\mathbf{H},\boldsymbol{\Psi})\mathrm{d}B_k(t),\\
&\mathrm{d}\Psi_u=\overline{m}_u^{\Psi}(\mathbf{H},\boldsymbol{\Psi})\mathrm{d}t+\sum_{k=1}^{m}\overline{\sigma}_{uk}^{\Psi}(\mathbf{H},\boldsymbol{\Psi})\mathrm{d}B_k(t),\\
&i=1,2,\ldots,n;\ u=1,2,\ldots,\alpha.
\end{aligned} \tag{1.77}$$

where $\mathbf{H}=[H_1,H_2,\ldots,H_n]^{\mathrm{T}}$, and the drift and diffusion coefficients are

$$\overline{m}_i(\mathbf{H},\boldsymbol{\Psi})=\{g_i(A_i+B_i)(1+d_i)m_i$$

$$
\begin{aligned}
&\left.\left. + \frac{1}{2}\frac{\mathrm{d}\left[g_i(A_i+B_i)(1+d_i)\right]}{\mathrm{d}A_i}\sum_{k=1}^{m}\sigma_{ik}\sigma_{ik}\right\}\right|_{A_i=U_i^{-1}(H_i)-B_i}, \\
&\overline{m}_u^{\Psi} = m_u^{\Psi}\Big|_{A_i=U_i^{-1}(H_i)-B_i}, \\
&\sum_{k=1}^{m}\left[\overline{\sigma}_{ik}\overline{\sigma}_{jk}(\mathbf{H},\mathbf{\Psi})\right] \\
&\quad = g_i(A_i+B_i)g_j(A_j+B_j)(1+d_i)(1+d_j)\sum_{k=1}^{m}\sigma_{ik}\sigma_{ik}\Big|_{A_i=U_i^{-1}(H_i)-B_i}, \\
&\sum_{k=1}^{m}\left[\overline{\sigma}_{uk}^{\Psi}\overline{\sigma}_{vk}^{\Psi}(\mathbf{H},\mathbf{\Psi})\right] = \sum_{k=1}^{m}\sigma_{uk}^{\Psi}\sigma_{vk}^{\Psi}\Big|_{A_i=U_i^{-1}(H_i)-B_i}, \\
&\sum_{k=1}^{m}\left[\overline{\sigma}_{ik}\overline{\sigma}_{uk}^{\Psi}(\mathbf{H},\mathbf{\Psi})\right] = g_i(A_i+B_i)(1+d_i)\sum_{k=1}^{m}\sigma_{ik}\sigma_{uk}^{\Psi}\Big|_{A_i=U_i^{-1}(H_i)-B_i}, \\
&i,j=1,2,\ldots,n;\quad u,v=1,2,\ldots,\alpha.
\end{aligned} \tag{1.78}
$$

Similarly, solving the averaged FPK equation associated with Itô Eq. (1.77) yields transition PDF $p(\mathbf{h},\boldsymbol{\psi},t|\mathbf{h}_0,\boldsymbol{\psi}_0)$ or PDF $p(\mathbf{h},\boldsymbol{\psi},t)$. Similar to Eq. (1.63), the stationary PDF $p(\mathbf{h},\boldsymbol{\psi})$ can be derived from $p(\mathbf{a},\boldsymbol{\psi})$ by using the relationship between A_i and H_i given in Eq. (1.51).

Example 1.3 Consider two Duffing-van der Pol oscillators coupled by nonlinear dampings and excited externally and parametrically by stationary rational noises. The motion equations of the system are

$$
\begin{aligned}
\dot{Q}_1 &= P_1, \\
\dot{Q}_2 &= P_2, \\
\dot{P}_1 &= -\omega_1^2 Q_1 - \alpha_1 Q_1^3 - \left(\beta_{10} + \beta_{11}Q_1^2 + \beta_{12}Q_2^2\right)P_1 \\
&\quad + Q_1\xi_1(t) + \xi_2(t), \\
\dot{P}_2 &= -\omega_2^2 Q_2 - \alpha_2 Q_2^3 - \left(\beta_{20} + \beta_{21}Q_1^2 + \beta_{22}Q_2^2\right)P_2 \\
&\quad + Q_2\xi_3(t) + \xi_4(t),
\end{aligned} \tag{1.79}
$$

where $\omega_i, \alpha_i, \beta_{ij}$ $(i,j=1,2)$ are constants; $\xi_k(t)$ $(k=1,2,3,4)$ are independent stationary rational noises with PSDs

$$
S_k(\omega) = \frac{D_k}{\pi(\omega^2+\Omega_k^2)},\quad k=1,2,3,4, \tag{1.80}
$$

where D_k, Ω_k are constants. Assume that $\alpha_i > 0$ and both β_{ij} and D_k are small quantities of ε order. Similar to Eqs. (1.38)–(1.41), the following expressions can be derived for the two Hamiltonian subsystems.

$$
\begin{aligned}
& g_i(q_i) = \omega_i^2 q_i + \alpha_i q_i^3, \quad U_i(q_i) = \omega_i^2 q_i^2/2 + \alpha_i q_i^4/4, \\
& b_i = d_i = 0, \\
& \nu_i(a_i, \phi_i) = \left[(\omega_i^2 + 3\alpha_i a_i^2/4)(1 + \eta_i \cos 2\phi_i)\right]^{1/2}, \\
& \eta_i = (\alpha_i a_i^2/4)/(\omega_i^2 + 3\alpha_i a_i^2/4) \le 1/3,
\end{aligned}
$$

$$
\begin{aligned}
\nu_i(a_i, \phi_i) &= b_{0i}(a_i)/2 + b_{2i}(a_i)\cos 2\phi_i + b_{4i}(a_i)\cos 4\phi_i \\
&\quad + b_{6i}(a_i)\cos 6\phi_i, \\
b_{0i}(a_i) &= (\omega_i^2 + 3\alpha_i a_i^2/4)^{1/2}(1 - \eta_i^2/16), \\
b_{2i}(a_i) &= (\omega_i^2 + 3\alpha_i a_i^2/4)^{1/2}(\eta_i/2 + 3\eta_i^3/64), \\
b_{4i}(a_i) &= (\omega_i^2 + 3\alpha_i a_i^2/4)^{1/2}(-\eta_i^2/16), \\
b_{6i}(a_i) &= (\omega_i^2 + 3\alpha_i a_i^2/4)^{1/2}(\eta_i^3/64), \\
\overline{\omega}_i(a_i) &= b_{0i}(a_i)/2, \quad i = 1, 2.
\end{aligned} \tag{1.81}
$$

Equation (1.52) can be derived from system (1.79) by applying the transformation (1.49). The coefficients can be obtained according to Eq. (1.53) as follows

$$
\begin{aligned}
\varepsilon F_i^{(1)}(\mathbf{A}, \mathbf{\Phi}) &= -\frac{A_i^2}{g_i(A_i)}\big(\beta_{i0} + \beta_{i1}A_1^2\cos^2\Phi_1 \\
&\quad + \beta_{i2}A_2^2\cos^2\Phi_2\big)\nu_i^2(A_i, \Phi_i)\sin^2\Phi_i, \\
\varepsilon F_i^{(2)}(\mathbf{A}, \mathbf{\Phi}) &= -\frac{A_i}{g_i(A_i)}\big(\beta_{i0} + \beta_{i1}A_1^2\cos^2\Phi_1 \\
&\quad + \beta_{i2}A_2^2\cos^2\Phi_2\big)\nu_i^2(A_i, \Phi_i)\sin\Phi_i\cos\Phi_i,
\end{aligned} \tag{1.82}
$$

$$
\begin{aligned}
\varepsilon^{1/2}G_{i1}^{(1)}(\mathbf{A},\mathbf{\Phi}) &= -\frac{A_i^2}{g_i(A_i)}\nu_i^2(A_i, \Phi_i)\sin\Phi_i\cos\Phi_i, \\
\varepsilon^{1/2}G_{i2}^{(1)}(\mathbf{A},\mathbf{\Phi}) &= -\frac{A_i}{g_i(A_i)}\nu_i(A_i, \Phi_i)\sin\Phi_i, \\
\varepsilon^{1/2}G_{i1}^{(2)}(\mathbf{A},\mathbf{\Phi}) &= -\frac{A_i}{g_i(A_i)}\nu_i^2(A_i, \Phi_i)\cos^2\Phi_i, \\
\varepsilon^{1/2}G_{i2}^{(2)}(\mathbf{A},\mathbf{\Phi}) &= -\frac{1}{g_i(A_i)}\nu_i(A_i, \Phi_i)\cos\Phi_i.
\end{aligned}
$$

When α_1 and α_2 are large, two Hamiltonian subsystems are strongly nonlinear and the probability of internal resonance between the two Hamiltonian subsystems is very small. So, only non-internal resonant case would be considered. The averaged Itô SDEs are of the form of Eq. (1.55) and the drift and diffusion coefficients can be obtained according to Eq. (1.56) as follows

$$
\begin{aligned}
m_i(\mathbf{A}) = {} & \frac{-A_i^2}{8g_i(A_i)}\beta_{i0}\left(4\omega_i^2 + \frac{5}{2}\alpha_i A_i^2\right) + \beta_{ii}A_i^2(\omega_i^2 + \frac{3}{4}\alpha_i A_i^2) \\
& + \beta_{i,3-i}A_{3-i}^2(2\omega_i^2 + \frac{5}{4}\alpha_i A_i^2)] \\
& \frac{\pi A_i}{32 g_i(A_i)}\sum_{n=2,4,6,\ldots}^{\infty}\left\{A_i\big(b_{i,n-2} - b_{i,n+2}\big)\left[\frac{\mathrm{d}}{\mathrm{d}A_i}\left(\frac{A_i^2\big(b_{i,n-2} - b_{i,n+2}\big)}{g_i(A_i)}\right)\right.\right. \\
& \left. + \frac{nA_i}{g_i(A_i)}\big(b_{i,n-2} + 2b_{i,n} + b_{i,n+2}\big)\right]S_{2i-1}(n\overline{\omega}_i) \\
& + 4\big(b_{i,n-2} - b_{i,n}\big)\left[\frac{\mathrm{d}}{\mathrm{d}A_i}\left(\frac{A_i\big(b_{i,n-2} - b_{i,n}\big)}{g_i(A_i)}\right) + \frac{n}{g_i(A_i)}\big(b_{i,n-2} + b_{i,n}\big)\right] \\
& \times S_{2i}((n-1)\overline{\omega}_i)\},
\end{aligned}
$$

$$
\begin{aligned}
b_{ii}(\mathbf{A}) &= \frac{\pi A_i^2}{16 g_i^2(A_i)}\sum_{n=2,4,\ldots}^{\infty}\{A_i^2(b_{i,n-2} - b_{i,n+2})^2 S_{2i-1}(n\overline{\omega}_i) \\
&\quad + 4(c_{i,n-2} - c_{i,n})^2 S_{2i}((n-1)\overline{\omega}_i)\}, \\
b_{12}(\mathbf{A}) &= b_{21}(\mathbf{A}) = 0, \quad i = 1, 2.
\end{aligned} \tag{1.83}
$$

The averaged FPK equation is of the form of Eq. (1.58) and boundary conditions are of the form of Eq. (1.62). In general, numerical method is needed to solve Eq. (1.58) to obtain stationary PDF $p(a_1, a_2)$. $p(h_1, h_2)$ and $p(q_1, q_2, p_1, p_2)$ are then calculated according to Eqs. (1.63) and (1.64), respectively.

In numerical calculation, the system parameters are $\omega_1 = 1$, $\omega_2 = 1.414$, $\alpha_1 = 0.3$, $\alpha_2 = 0.5$, $\beta_{10} = 0.1$, $\beta_{11} = 0.1$, $\beta_{12} = 0.1$, $\beta_{20} = 0.2$, $\beta_{21} = 0.1$, $\beta_{22} = 0.1$, $\Omega_1 = \Omega_2 = \Omega_3 = \Omega_4 = 5$, $D_1 = 2.7$, $D_2 = 5.5$, $D_3 = 2.7$, $D_4 = 5.5$. Comparisons between the results obtained by using the stochastic averaging method and from Monte Carlo simulation for system (1.79) are shown in Figs. 1.5, 1.6, 1.7 and 1.8. It is seen that the two results are in good agreement.

Example 1.4 Consider a system consisting of two coupled linear oscillators and two van der Pol oscillators excited by stationary wideband noises (Huang and Zhu 1997). The motion equations of the system are

$$
\begin{aligned}
\dot{Q}_1 &= P_1, \\
\dot{P}_1 &= -\omega_1^2 Q_1 - \beta_1 P_1 + \mu_1 Q_3 + \alpha_1 P_3 + \xi_1(t), \\
\dot{Q}_2 &= P_2,
\end{aligned}
$$

$$
\begin{aligned}
\dot{P}_2 &= -\omega_2^2 Q_2 - \beta_2 P_2 + \mu_2 Q_4 + \alpha_2 P_4 + \xi_2(t), \\
\dot{Q}_3 &= P_3,
\end{aligned}
$$

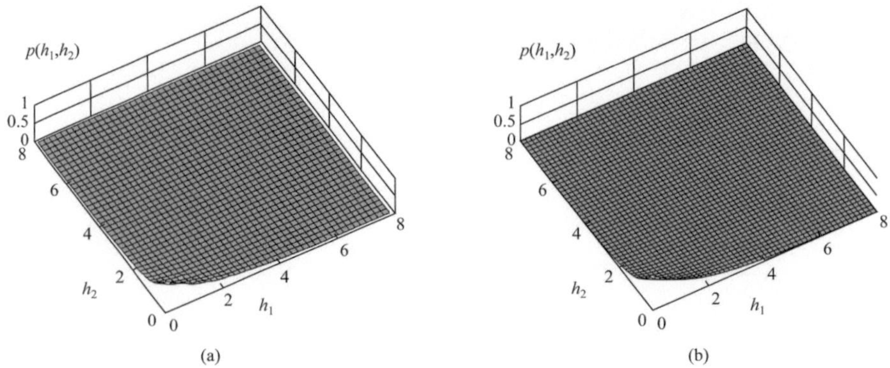

Fig. 1.5 Stationary joint PDF of Hamiltonian H_1, H_2 of system (1.79), **a** result from Monte Carlo simulation; **b** result from the stochastic averaging method

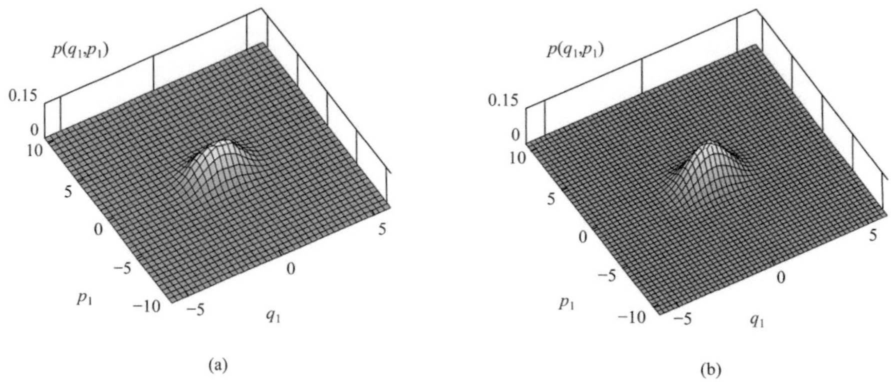

Fig. 1.6 Stationary joint PDF of generalized displacement Q_1 and generalized momentum P_1 of system (1.79). **a** Result from Monte Carlo simulation; **b** result from the stochastic averaging method

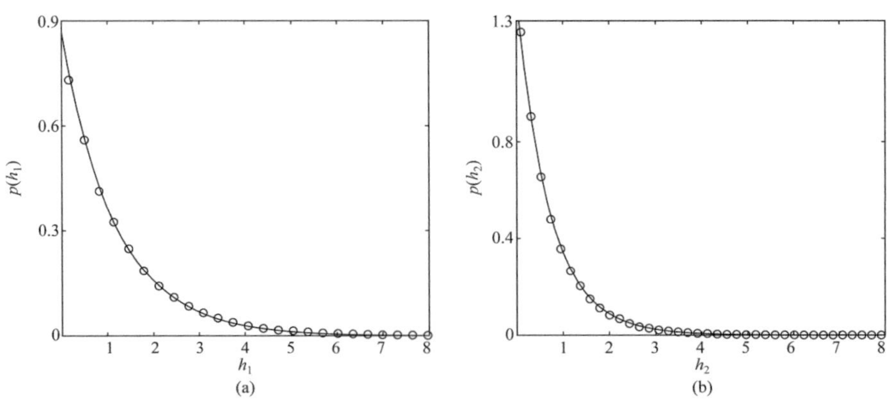

Fig. 1.7 Stationary marginal PDFs of Hamiltonian H_1 and H_2 of system (1.79). **a** $p(h_1)$; **b** $p(h_2)$; ——, results from the stochastic averaging method; ○, results from Monte Carlo simulation

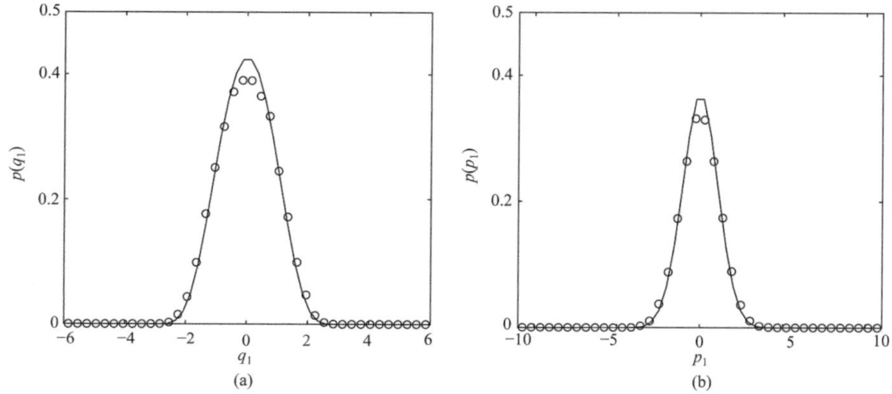

Fig. 1.8 Stationary marginal PDF of generalized displacement Q_1 and generalized momentum P_1 of system (1.79). **a** $p(q_1)$; **b** $p(p_1)$; ——, results from the stochastic averaging method; ○, results from Monte Carlo simulation

$$\begin{aligned}\dot{P}_3 &= -\omega_3^2 Q_3 - (-\beta_3 + \alpha_{33}P_3^2 + \alpha_{34}P_4^2)P_3 + \mu_3 Q_1 + \alpha_3 P_1 + \xi_3(t),\\ \dot{Q}_4 &= P_4,\\ \dot{P}_4 &= -\omega_4^2 Q_4 - (-\beta_4 + \alpha_{43}P_3^2 + \alpha_{44}P_4^2)P_4 + \mu_4 Q_2 + \alpha_4 P_2 + \xi_4(t).\end{aligned} \quad (1.84)$$

The first two oscillators represent the two adjacent modes of a structure while the last two oscillators represent the influence of shedding vortices. $\omega_3 = \omega_4 = 1$, $\omega_1 = 1 - \sigma_1$, $\omega_2 = 1 - \sigma_2$. Assume that α_i, β_i, μ_i, α_{ij}, σ_i are small quantities of ε order and that $\xi_k(t)$ $(k = 1, 2, 3, 4)$ are independent stationary wideband noises with zero mean. PSDs $S_k(\omega)$ are small quantities of ε order.

Generally, two oscillators in a system may occur internal resonance if the two natural frequencies of the two oscillators satisfy Eq. (5.118) in vol. 1 and at the same time the two equations of the two oscillators are coupled linearly or nonlinearly. For system (1.84), the four natural frequencies have relations $\omega_3 - \omega_1 = \sigma_1$, $\omega_4 - \omega_2 = \sigma_2$, $\omega_3 - \omega_4 = 0$, and at the same time the equations for oscillators 1 and 3, 2 and 4, and 3 and 4 are linear and nonlinear coupled. Thus, three internal resonances between oscillators 1 and 3, 2 and 4, and 3 and 4 may take place. Applying the stochastic averaging method described in Sect. 1.1.2(ii) yields the following averaged Itô SDEs

$$\begin{aligned}\mathrm{d}A_i &= (1/2)(-\beta_i A_i + \alpha_i A_{i+2}\cos\psi_i - \mu_i A_{i+2}\sin\psi_i + \pi K_i/A_i)\mathrm{d}t\\ &\quad + \sqrt{2\pi K_i}\mathrm{d}B_i(t),\\ \mathrm{d}A_{i+2} &= (1/2)\{-\beta_{i+2}A_{i+2} + \alpha_{i+2}A_i\cos\psi_i + \mu_{i+2}A_i\sin\psi_i\\ &\quad - (3\alpha_{i+2,i+2}/4)A_{i+2}^3 - (\alpha_{i+2,j+2}/4)[2 + \cos 2\psi_3]\\ &\quad A_3^2A_4^2/A_{i+2} + \pi K_{i+2}/A_{i+2}\}\mathrm{d}t + \sqrt{2\pi K_{i+2}}\mathrm{d}B_{i+2}(t).\end{aligned}$$

$$\mathrm{d}\psi_i = (1/2)\big[-\sigma_i - \big(\alpha_i A_{i+2}/A_i + \alpha_{i+2}A_i/A_{i+2}\big)\sin\psi_i - \big(\mu_i A_{i+2}/A_i$$

$$
\begin{aligned}
&-\mu_{i+2}A_i/A_{i+2}\big)\sin\psi_i + (-1)^{i+1}\big(\alpha_{i+2,j+2}/4\big)A_{5-i}^2\sin 2\psi_3]\mathrm{d}t \\
&+\Big(\sqrt{2\pi K_i}\Big/A_i\Big)\mathrm{d}B_{i+4}(t) - \Big(\sqrt{2\pi K_{i+2}}\Big/A_{i+2}\Big)\mathrm{d}B_{i+6}(t) \qquad (1.85)
\end{aligned}
$$

$$
\begin{aligned}
d\psi_3 =\ & (1/2)\big[\big(A_1/A_3\big)(-\alpha_3\sin\psi_1 + \mu_3\cos\psi_1) + \big(A_2/A_4\big) \\
& \times(-\alpha_4\sin\psi_2 + \mu_4\cos\psi_2) + (1/4)\big(\alpha_{34}A_4^2 + \alpha_{43}A_3^2\big)\cos 2\psi_3]\mathrm{d}t \\
& - \Big(\sqrt{2\pi K_3}\Big/A_3\Big)\mathrm{d}B_7(t) + \Big(\sqrt{2\pi K_4}\Big/A_4\Big)\mathrm{d}B_8(t) \\
& i,j = 1,2,\ \ j \neq i
\end{aligned}
$$

where

$$
\begin{aligned}
&\psi_i = \Phi_i - \Phi_{i+2},\ \ i = 1,2;\ \ \psi_3 = \Phi_4 - \Phi_3 \\
&K_k = S_k(1)\big/2,\ \ k = 1,2,3,4 \qquad (1.86)
\end{aligned}
$$

The reduced averaged FPK equation associated with averaged Itô equations (1.85) is

$$
\begin{aligned}
0 =\ & -\sum_{i=1}^{4}\frac{\partial}{\partial a_i}(m_i p) - \sum_{u=1}^{3}\frac{\partial}{\partial\psi_u}(m_u^{\psi}p) + \frac{1}{2}\sum_{i,j=1}^{4}\frac{\partial^2}{\partial a_i\partial a_j}(b_{ij}p) \\
& + \sum_{i=1}^{4}\sum_{u=1}^{3}\frac{\partial^2}{\partial a_i\partial\psi_u}(b_{iu}^{a\psi}p) + \frac{1}{2}\sum_{u,v=1}^{3}\frac{\partial^2}{\partial\psi_u\partial\psi_v}(b_{uv}^{\psi}p). \qquad (1.87)
\end{aligned}
$$

where the coefficients may be obtained from Eq. (1.85) by using Eq. (1.72). Suppose that $p(\mathbf{a},\boldsymbol{\psi})$ is of the form

$$
p(\mathbf{a},\boldsymbol{\psi}) = C\exp[-\lambda(a_1,a_2,a_3,a_4,\psi_1,\psi_2,\psi_3)], \qquad (1.88)
$$

where C is a normalization constant. $\lambda(\mathbf{a},\boldsymbol{\psi})$ is known as the probability potential function. Expand $\lambda(\mathbf{a},\boldsymbol{\psi})$ into a truncated Fourier series with respect to ψ_1, ψ_2, ψ_3, i.e.,

$$
\begin{aligned}
\lambda(\mathbf{a},\boldsymbol{\psi}) =\ & \lambda_0(\mathbf{a}) + \sum_{i=1}^{2}[\lambda_{1i}(\mathbf{a})\cos\psi_i + \overline{\lambda}_{1i}(\mathbf{a})\cos\psi_i] + \lambda_{23}(\mathbf{a})\cos 2\psi_3 \\
& + \overline{\lambda}_{23}(\mathbf{a})\sin 2\psi_3. \qquad (1.89)
\end{aligned}
$$

Also, expand the coefficient functions $m_i, m_u^{\psi}, b_{ij}, b_{uv}^{\psi}, b_{iu}^{a\psi}$ in the reduced FPK equation (1.87) into truncated Fourier series and substitute all these Fourier series and Eqs. (1.88), (1.89) into (1.87). An analytical solution of $\lambda(\mathbf{a},\boldsymbol{\psi})$ in Eq. (1.89) can be obtained if the system parameters satisfy the following conditions

$$\frac{\alpha_{34}}{K_3} = \frac{\alpha_{43}}{K_4}, \quad \frac{\alpha_{i+2}}{\pi K_{i+2}} = \frac{\pi K_i \alpha_i + d_i \mu_i}{\pi^2 K_i^2 + d_i^2}, \quad \frac{-\mu_{i+2}}{\pi K_{i+2}} = \frac{\pi K_i \mu_i + d_i \alpha_i}{\pi^2 K_i^2 + d_i^2}, \tag{1.90}$$

and $d_i = 2\pi K_i \sigma_i / \beta_i$. Substituting the exact solution of $\lambda(\mathbf{a}, \boldsymbol{\psi})$ into Eq. (1.88) results in the stationary solution of $p(\mathbf{a}, \boldsymbol{\psi})$. The stationary PDF in terms of action variables $I_i = a_i^2/2$, is then

$$\begin{aligned} p(I_1, I_2, I_3, I_4, \psi_1, \psi_2, \psi_3) = C \exp[-\zeta_1 I_1 - \zeta_2 I_2 + \zeta_3 I_3 \\ + \zeta_4 I_4 + \zeta_5 \sqrt{I_1 I_2} \cos(\psi_1 + \psi_{10}) + \zeta_6 \sqrt{I_2 I_4} \cos(\psi_2 + \psi_{20}) \\ - \zeta_7 I_3^2 - \zeta_8 I_4^2 - 2\zeta_9 I_3 I_4 (2 + \cos 2\psi_3)], \end{aligned} \tag{1.91}$$

where

$$\begin{aligned} &\zeta_i = \beta_i / 2\pi K_i, \ (i = 1, 2, 3, 4), \quad \zeta_{5,6} = \sqrt{\alpha_{3,4}^2 + \mu_{3,4}^2} \Big/ 2\pi K_{3,4}, \\ &\zeta_{7,8} = 3\alpha_{33,44} / 8\pi K_{33,44}, \quad \zeta_9 = \alpha_{34} / 8\pi K_3, \\ &\cos \psi_{u0} = \alpha_{u+2} \Big/ \sqrt{\alpha_{u+2}^2 + \mu_{u+2}^2}, \\ &\sin \psi_{u0} = -\mu_{u+2} \Big/ \sqrt{\alpha_{u+2}^2 + \mu_{u+2}^2}, \quad u = 1, 2. \end{aligned} \tag{1.92}$$

Taking the parameter values $\omega_i = 1.0$, $\pi k_i = 0.01$ $(i = 1, 2, 3, 4)$, $\alpha_i = \mu_i = 0.04$, $\beta_i = 0.06$ $(i = 1, 2)$, $\alpha_3 = \alpha_4 = -\mu_3 = -\mu_4 = 0.04$, $\beta_3 = \beta_4 = \alpha_{33} = \alpha_{44} = 0.05$, and $\alpha_{34} = \alpha_{43} = 0.02$, the stationary PDF $p(I_3, I_4)$ are calculated and shown in Fig. 1.9. It is seen that the result obtained by using the stochastic averaging method is in good agreement with that from Monte Carlo simulation of the original system (1.84).

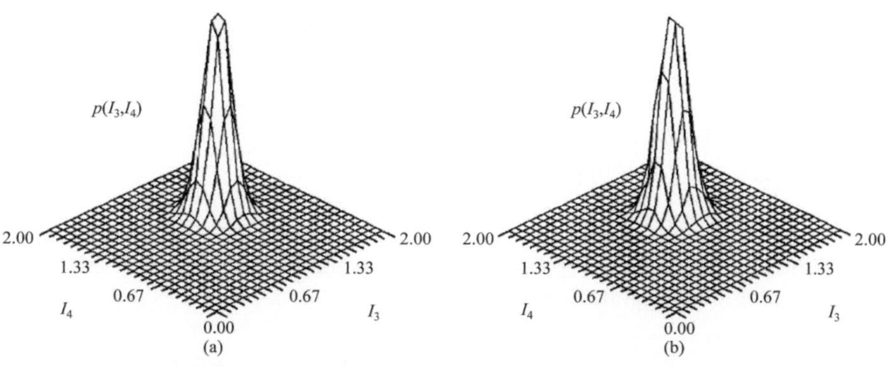

Fig. 1.9 Stationary PDF $p(I_3, I_4)$ of system (1.84). **a** Analytical result from integrating Eq. (1.91) with respect to $I_1, I_2, \psi_1, \psi_2, \psi_3$; **b** result from Monte Carlo simulation of original system (1.84) (Huang and Zhu 1997)

1.2 Fractional Gaussian Noise Excitation

In Chap. 7 of vol. 1, the stochastic averaging methods of quasi-Hamiltonian systems excited by fractional Gaussian noise (fGn) have been presented. Since the system response is not Markov process, no FPK equation can be used to obtain analytical solution. The only benefit of using the stochastic averaging method is to reduce system dimension and decrease computation time in Monte Carlo simulation. It is seen from Fig. 2.8, vol. 1 that the PSD curve of fGn change very slowly with frequency after a certain frequency. That is, fGns in frequency band higher than this frequency can be treated as wideband random processes. Therefore, if the natural frequencies of a quasi-integrable Hamiltonian system fall in this frequency band, fGns can be approximated as stationary wideband Gaussian noises and the stochastic averaging method of quasi-integrable Hamiltonian systems subjected to stationary wideband noise described in Sect. 1.1 can be applied. This idea is realized in this section and the applicability is examined (Deng and Zhu 2022; Lü et al. 2020a, b, 2022).

Consider a quasi-integrable Hamiltonian system excited by fGns and governed by the motion equations

$$\begin{aligned}&\dot{Q}_i = P_i,\\&\dot{P}_i = -g_i(Q_i) - \varepsilon\sum_{j=1}^{n} c_{ij}(\mathbf{Q},\mathbf{P})P_j + \varepsilon^{1/2}\sum_{k=1}^{m} f_{ik}(\mathbf{Q},\mathbf{P})W_{Hk}(t),\\&i = 1, 2, \ldots, n,\end{aligned} \tag{1.93}$$

where $\mathbf{Q}$, $\mathbf{P}$, c_{ij}, f_{ik}, ε are the same as those in system (1.4), $W_{Hk}(t)$ $(k = 1, 2, \ldots, m)$ are independent unit fGns with Hurst index $\mathcal{H}$ in the range of $0.5 < \mathcal{H} < 1.0$, and with PSD functions given by Eq. (2.219) in vol. 1. Suppose that the PSDs of m $W_{Hk}(t)$ at n natural frequencies of Hamiltonian system associated with Eq. (1.93) change very slowly with frequency and can be treated as stationary wideband noises. $g_i(Q_i)$ in system (1.93) are restoring forces and the corresponding potential energy functions are

$$U_i(Q_i) = \int_0^{Q_i} g_i(u)\mathrm{d}u, \quad i = 1, 2, \ldots, n. \tag{1.94}$$

Assume all $g_i(Q_i)$ and $U_i(Q_i)$ satisfy the four conditions given below Eq. (1.5). Introduce the following transformation

$$\begin{aligned}&Q_i(t) = A_i\cos\Phi_i(t) + B_i, \quad P_i(t) = -A_i\nu_i(A_i,\Phi_i)\sin\Phi_i(t),\\&\Phi_i(t) = \Gamma_i(t) + \Theta_i(t), \quad i = 1, 2, \ldots, n.\end{aligned} \tag{1.95}$$

where A_i, B_i, Φ_i, Γ_i, Θ_i are random processes, and

$$\nu_i(A_i, \Phi_i) = \frac{\sqrt{2[U_i(A_i + B_i) - U_i(A_i \cos\Phi_i + B_i)]}}{|A_i \sin\Phi_i|}. \tag{1.96}$$

A_i and $\nu_i(A_i, \Phi_i)$ are the amplitude and instantaneous frequency of the ith DOF subsystem. Similar to Eq. (1.10), the mean frequency of the ith DOF subsystem is

$$\overline{\omega}_i(A_i) = \frac{1}{2\pi}\int_0^{2\pi} \nu_i(A_i, \Phi_i)\mathrm{d}\Phi_i, \quad i = 1, 2, \ldots, n. \tag{1.97}$$

The following approximation will be used during stochastic averaging

$$\Phi_i(t) = \overline{\omega}_i(A_i)t + \Theta_i(t), \quad i = 1, 2, \ldots, n. \tag{1.98}$$

After using Eq. (1.95) to transform Q_i, P_i to A_i, Φ_i, system (1.93) becomes

$$\begin{aligned}
&\dot{A}_i = \varepsilon F_i^A(\mathbf{A}, \boldsymbol{\Phi}) + \varepsilon^{1/2}\sum_{k=1}^{m} G_{ik}^A(\mathbf{A}, \boldsymbol{\Phi})W_{Hk}(t),\\
&\dot{\Phi} = \nu_i(A_i, \Phi_i) + \varepsilon F_i^{\Phi}(\mathbf{A}, \boldsymbol{\Phi}) + \varepsilon^{1/2}\sum_{k=1}^{m} G_{ik}^{\Phi}(\mathbf{A}, \boldsymbol{\Phi})W_{Hk}(t),\\
&i = 1, 2, \ldots, n,
\end{aligned} \tag{1.99}$$

where

$$\begin{aligned}
F_i^A &= \frac{-A_i}{g_i(A_i + B_i)(1 + r_i)}\left[c_{ij}(\mathbf{A}, \Phi)A_j\nu_j(A_j, \Phi_j)\sin\Phi_j\right]\nu_i(A_i, \Phi_i)\sin\Phi_i,\\
F_i^{\Phi} &= \frac{-1}{g_i(A_i + B_i)(1 + r_i)}\left[c_{ij}(\mathbf{A}, \Phi)A_j\nu_j(A_j, \Phi_j)\sin\Phi_j\right]\nu_i(A_i, \Phi_i)(\cos\Phi_i + r_i),
\end{aligned}$$

$$\begin{aligned}
G_{ik}^A &= \frac{-A_i}{g_i(A_i + B_i)(1 + r_i)} f_{ik}(\mathbf{A}, \Phi)\nu_i(A_i, \Phi_i)\sin\Phi_i,\\
G_{ik}^{\Phi} &= \frac{-1}{g_i(A_i + B_i)(1 + r_i)} f_{ik}(\mathbf{A}, \Phi)\nu_i(A_i, \Phi_i)(\cos\Phi_i + r_i),\\
r_i &= \frac{g(-A_i + B_i) + g(A_i + B_i)}{g(-A_i + B_i) - g(A_i + B_i)}.
\end{aligned} \tag{1.100}$$

The dimension and form of the averaged Itô equations depend on whether system (1.93) has internal resonance.

1.2.1 Non-internal Resonant Case

In this case, $\mathbf{A} = [A_1, A_2, \ldots, A_n]^{\mathrm{T}}$ converges weakly to a n-dimensional vector Markov diffusion process as $\varepsilon \to 0$. The averaged Itô SDEs are of the form of Eq. (1.55), i.e.,

$$\mathrm{d}A_i = m_i(\mathbf{A})\mathrm{d}t + \sum_{k=1}^{m} \sigma_{ik}(\mathbf{A})\mathrm{d}B_k(t), \quad i = 1, 2, \ldots, n, \tag{1.101}$$

where $B_k(t)$ $(k = 1, 2, \ldots, m)$ are independent unit Wiener processes. The drift and diffusion coefficients are of the form of Eq. (1.56), i.e.,

$$\begin{aligned}
m_i(\mathbf{A}) &= \varepsilon\left\langle F_i^A + \sum_{k=1}^{m} \int_{-\infty}^{0} \sum_{j=1}^{n} \left(\frac{\partial G_{ik}^A}{\partial A_j}\bigg|_t G_{jk}^A\bigg|_{t+\tau} + \frac{\partial G_{ik}^A}{\partial \Phi_j}\bigg|_t G_{jk}^{\Phi}\bigg|_{t+\tau} \right) R_k(\tau)\mathrm{d}\tau \right\rangle_t, \\
b_{ij}(\mathbf{A}) &= \sum_{k=1}^{m} \sigma_{ik}\sigma_{jk}(\mathbf{A}) = \varepsilon\left\langle \sum_{k=1}^{m} \int_{-\infty}^{\infty} \left(G_{ik}^A\big|_t G_{jk}^A\big|_{t+\tau} \right) R_k(\tau)\mathrm{d}\tau \right\rangle_t, \\
i, j &= 1, 2, \ldots, n.
\end{aligned} \tag{1.102}$$

Since the fast-varying terms in Eq. (1.102) are periodic functions of $\mathbf{\Phi}$, the time averaging can be replaced by averaging with respect to $\mathbf{\Phi}$.

The following averaged Itô SDEs governing n sub-Hamiltonians can be obtained from Eq. (1.101) by using Eq. (1.51) and Itô differential rule

$$\mathrm{d}H_i = \overline{m}_i(\mathbf{H})\mathrm{d}t + \overline{\sigma}_{ik}(\mathbf{H})\mathrm{d}B_k(t), \quad i = 1, 2, \ldots, n; \quad k = 1, 2, \ldots, m, \tag{1.103}$$

where

$$\begin{aligned}
\overline{m}_i(\mathbf{H}) &= \Big[g_i(A_i + B_i)(1 + r_i)m_i(\mathbf{A}) \\
&\quad + \frac{1}{2}\frac{\mathrm{d}}{\mathrm{d}A_i}[g_i(A_i + B_i)(1 + r_i)]b_{ii}(\mathbf{A})\Big]\Big|_{A_i = U_i^{-1}(H_i) - B_i}, \\
\sum_{k=1}^{m} \overline{\sigma}_{ik}\overline{\sigma}_{jk}(\mathbf{H}) &= [g_i(A_i + B_i)g_j(A_j + B_j)(1 + r_i)(1 \\
&\quad + r_j)b_{ij}(\mathbf{A})]\big|_{A_i = U_i^{-1}(H_i) - B_i}.
\end{aligned} \tag{1.104}$$

The averaged FPK equation associated with Itô Eq. (1.103) is

$$\frac{\partial p}{\partial t} = -\sum_{i=1}^{n} \frac{\partial}{\partial h_i}\left[\overline{m}_i(\mathbf{h})p\right] + \frac{1}{2}\sum_{i,j=1}^{n} \frac{\partial^2}{\partial h_i \partial h_j}\left[\overline{b}_{ij}(\mathbf{h})p\right]. \tag{1.105}$$

where

$$\overline{m}_i(\mathbf{h}) = \overline{m}_i(\mathbf{H})|_{\mathbf{H}=\mathbf{h}}, \quad \overline{b}_{ij}(\mathbf{h}) = \sum_{k=1}^{m} \overline{\sigma}_{ik}\overline{\sigma}_{jk}(\mathbf{H})\Bigg|_{\mathbf{H}=\mathbf{h}}. \tag{1.106}$$

In (1.95), $p = p(\mathbf{h}, t|\mathbf{h}_0)$ is transition PDF of $\mathbf{H}(t)$ with initial condition

$$p(\mathbf{h}, 0|\mathbf{h}_0) = \delta(\mathbf{h} - \mathbf{h}_0), \tag{1.107}$$

or $p = p(\mathbf{h}, t)$ is transient PDF of $\mathbf{H}(t)$ with initial condition

$$p(\mathbf{h}, 0) = p(\mathbf{h}_0). \tag{1.108}$$

The boundary conditions for Eq. (1.105) depend on the system characteristics, for example, they are similar to Eq. (1.62) when V_i are all the entire plane (q_i, p_i).

Similar to Eq. (1.64), after stationary PDF $p(h)$ is obtained from solving the reduced FPK equation (1.105) with boundary conditions, the approximate stationary joint PDF of displacements and momenta of the original system (1.93) can be obtained as follows

$$p(\mathbf{q}, \mathbf{p}) = \frac{p(\mathbf{h})}{\prod_{i=1}^{n} T_i(h_i)}\Bigg|_{h_i=H_i(q_i,p_i)}, \tag{1.109}$$

where $T_i(h_i) = 4\int_0^a [2h_i - 2U_i(q_i)]^{-1/2} \mathrm{d}q_i$ is mean period of the ith subsystem. The marginal PDF $p(q_i)$ and mean-square $E[Q_i^2]$ can then be derived from Eq. (1.109).

Example 1.5 Consider two coupled Duffing oscillators excited by 4 fGns (Lü et al. 2020a). The motion equations for the system are

$$\begin{aligned}
\ddot{X}_1 + \beta_{11}\dot{X}_1 + \beta_{12}\dot{X}_2 + \omega_1^2 X_1 + \alpha_1 X_1^3 &= X_1\sqrt{2D_1}W_{H1}(t) + \sqrt{2D_2}W_{H2}(t), \\
\ddot{X}_2 + \beta_{21}\dot{X}_1 + \beta_{22}\dot{X}_2 + \omega_2^2 X_2 + \alpha_2 X_2^3 &= X_2\sqrt{2D_3}W_{H3}(t) + \sqrt{2D_4}W_{H4}(t),
\end{aligned} \tag{1.110}$$

where $\beta_{ij}, \omega_i, \alpha_i$ $(i = 1, 2)$ are constants, $W_{Hk}(t)$ $(k = 1, 2, 3, 4)$ are the fGns with Hurst index $0.5 < \mathcal{H} < 1$ and PSD in Eq. (2.219) in vol. 1; noise intensities $2D_k$ and β_{ij} are of the same ε order. The PSDs of four fGns at ω_1, ω_2 and higher frequencies change slowly.

Letting $X_1 = Q_1, \dot{X}_1 = P_1, X_2 = Q_2, \dot{X}_2 = P_2$, system (1.110) can be transformed into the form of Eq. (1.93), i.e., quasi-integrable Hamiltonian system excited by fGn. The Hamiltonians of whole system and subsystems are

$$
\begin{aligned}
&H = H_1 + H_2, \quad H_i = \frac{1}{2}P_i^2 + U_i(Q_i), \\
&U_i(Q_i) = \frac{1}{2}\omega_i^2 Q_i^2 + \frac{1}{4}\alpha_i Q_i^4, \quad i = 1, 2.
\end{aligned}
\tag{1.111}
$$

When $\alpha_1, \alpha_2 > 0$ and without damping and excitations, both two Duffing oscillators in system (1.110) have families of periodic solutions revolving around origin points ($B = 0$) in entire phase planes. The instantaneous frequencies of the two subsystems have the following expressions

$$
\begin{aligned}
&\nu_i(A_i, \Phi_i) = [(\omega_i^2 + 3\alpha_i A_i^2/4)(1 + \lambda_i \cos 2\Phi_i)]^{1/2}, \\
&\lambda_i = (\alpha_i A_i^2/4) \big/ (\omega_i^2 + 3\alpha_i A_i^2/4) \le 1\big/3.
\end{aligned}
\tag{1.112}
$$

Expand ν_i in Eq. (1.112) into Fourier series with respect to Φ_i, and keep only the first several terms, i.e.,

$$
\nu_i(A_i, \Phi_i) \approx c_{i,0}(A_i)\big/2 + c_{i,2}(A_i)\cos 2\Phi_i + c_{i,4}(A_i)\cos 4\Phi_i + c_{i,6}(A_i)\cos 6\Phi_i,
\tag{1.113}
$$

where

$$
\begin{aligned}
c_{i,0} &= (\omega_i^2 + 3\alpha_i A_i^2/4)^{1/2}(1 - \lambda_i^2/16), \\
c_{i,2} &= (\omega_i^2 + 3\alpha_i A_i^2/4)^{1/2}(\lambda_i/2 + 3\lambda_i^3/64), \\
c_{i,4} &= (\omega_i^2 + 3\alpha_i A_i^2/4)^{1/2}(-\lambda_i^2/16), \\
c_{i,6} &= (\omega_i^2 + 3\alpha_i A_i^2/4)^{1/2}(\lambda_i^3/64).
\end{aligned}
\tag{1.114}
$$

The mean frequencies are then

$$
\overline{\omega}_i(A_i) = c_{i,0}(A_i)\big/2, \quad i = 1, 2, \ldots, n.
\tag{1.115}
$$

By using transformation (1.95), the equations for amplitudes A_i and phase angles Φ_i can be obtained from system (1.110) as follows

$$
\begin{aligned}
&\dot{A}_i = F_i^A + \sum_{k=1}^{4} G_{ik}^A W_{Hk}(t), \\
&\dot{\Phi}_i = \nu_i + F_i^\Phi + \sum_{k=1}^{4} G_{ik}^\Phi W_{Hk}(t), \\
&i = 1, 2,
\end{aligned}
\tag{1.116}
$$

where

$$
\begin{aligned}
F_i^A &= -A_i\nu_i \sin\Phi_i \sum_{j=1}^{2} (\beta_{ij}A_j\nu_j \sin\Phi_j) \Big/ g_i, \\
F_i^\Phi &= -\nu_i \cos\Phi_i \sum_{j=1}^{2} (\beta_{ij}A_j\nu_j \sin\Phi_j) \Big/ g_i, \\
G_{11}^A &= -\sqrt{2D_1}A_1^2\nu_1 \sin\Phi_1 \cos\Phi_1 \Big/ g_1, \\
G_{12}^A &= -\sqrt{2D_2}A_1\nu_1 \sin\Phi_1 \Big/ g_1, \quad G_{13}^A = 0, \quad G_{14}^A = 0, \\
G_{21}^A &= 0, \quad G_{22}^A = 0, \quad G_{23}^A = -\sqrt{2D_3}A_2^2\nu_2 \sin\Phi_2 \cos\Phi_2 \Big/ g_2, \\
G_{24}^A &= -\sqrt{2D_4}A_2\nu_2 \sin\Phi_2 \Big/ g_2, \\
G_{11}^\Phi &= -\sqrt{2D_1}A_1\nu_1 \cos^2\Phi_1 \Big/ g_1, \quad G_{12}^\Phi = -\sqrt{2D_2}\nu_1 \cos\Phi_1 \Big/ g_1, \\
G_{13}^\Phi &= 0, \quad G_{14}^\Phi = 0, \\
G_{21}^\Phi &= 0, \quad G_{22}^\Phi = 0, \quad G_{23}^\Phi = -\sqrt{2D_3}A_2\nu_2 \cos^2\Phi_2 \Big/ g_2, \\
G_{24}^\Phi &= -\sqrt{2D_4}\nu_2 \cos\Phi_2 \Big/ g_2.
\end{aligned} \tag{1.117}
$$

If α_1 and α_2 are finite, the two subsystems are strongly nonlinear and the two frequencies of system (1.110) change randomly with the amplitudes. The probability of internal resonance is very low. Therefore, only the non-internal resonant case is considered.

By applying the stochastic averaging method described in Sect. 1.1.2(i), the following averaged Itô SDEs for $\mathbf{A} = [A_1, A_2]^{\mathrm{T}}$ can be derived

$$
\mathrm{d}A_i = m_i(\mathbf{A})\mathrm{d}t + \sum_{k=1}^{4} \sigma_{ik}(\mathbf{A})\mathrm{d}B_k(t), \quad i = 1, 2. \tag{1.118}
$$

where $m_i(\mathbf{A})$ and $\sigma_{ik}(\mathbf{A})$ can be obtained from Eq. (1.117) according to Eq. (1.102). Replacing time averaging with averaging with respect to Φ_i, the following drift and diffusion coefficients of Eq. (1.118) can be obtained

$$
\begin{aligned}
m_i(\mathbf{A}) = & \frac{-A_i^2}{8g_i}\beta_{ii}\left(4\omega_i^2 + \frac{5}{2}\alpha_i A_i^2\right) + \frac{\pi A_i}{32g_i} \\
& \times \sum_{n=2}^{\infty} \left\{ A_i\big(c_{i,n-2} - c_{i,n+2}\big)\left[\frac{\mathrm{d}}{\mathrm{d}A_i}\left(\frac{A_i^2\big(c_{i,n-2} - c_{i,n+2}\big)}{g_i}\right)\right.\right. \\
& \left. + \frac{nA_i}{g_i}\big(c_{i,n-2} + 2c_{i,n} + c_{i,n+2}\big)\right] S_{2i-1}(n\overline{\omega}_i)
\end{aligned}
$$

$$
\begin{aligned}
&+4\left(c_{i,n-2}-c_{i,n}\right)\left[\frac{\mathrm{d}}{\mathrm{d}A_i}\left(\frac{A_i\left(c_{i,n-2}-c_{i,n}\right)}{g_i}\right)\right.\\
&\left.+\frac{n}{g_i}\left(c_{i,n-2}+c_{i,n}\right)\right]S_{2i}((n-1)\overline{\omega}_i)\bigg\},
\end{aligned}
$$

$$
\begin{aligned}
&\sum_{k=1}^{4}\sigma_{ik}\sigma_{ik}=\frac{\pi A_i^2}{16g_i^2}\sum_{n=2}^{\infty}\left\{A_i^2\left(c_{i,n-2}-c_{i,n+2}\right)^2S_{2i-1}(n\overline{\omega}_i)\right.\\
&\left.+4\left(c_{i,n-2}-c_{i,n}\right)^2S_{2i}((n-1)\overline{\omega}_i)\right\},\\
&\sum_{k=1}^{4}\sigma_{1k}\sigma_{2k}=0,\quad \sum_{k=1}^{4}\sigma_{2k}\sigma_{1k}=0,\quad i=1,2,\quad n=2,4,6,\ldots. \qquad (1.119)
\end{aligned}
$$

The averaged Itô SDEs governing Hamiltonians H_i ($i = 1, 2$) of two subsystems can be derived from Eq. (1.118) by using Eq. (1.111) and Itô differential rules as follows

$$
\mathrm{d}H_i=\overline{m}_i(\mathbf{H})\mathrm{d}t+\sum_{k=1}^{4}\overline{\sigma}_{ik}(\mathbf{H})\mathrm{d}B_k(t). \qquad (1.120)
$$

According to Eq. (1.104), the drift and diffusion coefficients in (1.120) are

$$
\begin{aligned}
&\overline{m}_i(\mathbf{H})=\left.[(\omega_i^2A_i+\alpha_iA_i^3)m_i(\mathbf{A})+\frac{1}{2}(\omega_i^2+3\alpha_iA_i^2)\sum_{k=1}^{4}\sigma_{ik}\sigma_{ik}(\mathbf{A})]\right|_{A_i=U_i^{-1}(H_i)},\\
&\sum_{k=1}^{4}\overline{\sigma}_{ik}\overline{\sigma}_{ik}(\mathbf{H})=\left.[(\omega_i^2A_i+\alpha_iA_i^3)^2\sum_{k=1}^{4}\sigma_{ik}\sigma_{ik}(\mathbf{A})]\right|_{A_i=U_i^{-1}(H_i)},\\
&\sum_{k=1}^{4}\overline{\sigma}_{1k}\overline{\sigma}_{2k}(\mathbf{H})=0,\quad \sum_{k=1}^{4}\overline{\sigma}_{2k}\overline{\sigma}_{1k}(\mathbf{H})=0,\quad i=1,2. \qquad (1.121)
\end{aligned}
$$

The reduced averaged FPK equation associated with averaged Itô Eq. (1.120) is

$$
-\sum_{i=1}^{2}\frac{\partial}{\partial h_i}\left[\overline{m}_i(\mathbf{h})p\right]+\frac{1}{2}\sum_{i=1}^{2}\frac{\partial^2}{\partial h_i^2}\left[\overline{b}_{ii}(\mathbf{h})p\right]=0, \qquad (1.122)
$$

where

$$
\overline{m}_i(\mathbf{h})=\overline{m}_i(\mathbf{H})|_{\mathbf{H}=\mathbf{h}},\quad \overline{b}_{ii}(\mathbf{h})=\left.\sum_{k=1}^{4}\overline{\sigma}_{ik}\overline{\sigma}_{ik}(\mathbf{H})\right|_{\mathbf{H}=\mathbf{h}},\quad i=1,2. \qquad (1.123)
$$

Since H_i $(i = 1, 2)$ can change in $[0, \infty)$, the boundary conditions of Eq. (1.122) are

$$p(h_1, h_2 \to 0) = \text{finite}, \quad p(h_1 \to 0, h_2) = \text{finite},$$
$$p(h_1, h_2 \to \infty) = p(h_1 \to \infty, h_2) = 0. \tag{1.124}$$

Once the stationary PDF $p(h_1, h_2)$ is obtained, the approximate stationary joint PDF of the original system (1.110) can be obtained according to Eq. (1.109) as follows

$$p(q_1, q_2, p_1, p_2) = \frac{\overline{\omega}_1\overline{\omega}_2}{4\pi^2} p(h_1, h_2)\bigg|_{h_i=\frac{1}{2}p_i^2+U_i(q_i)}. \tag{1.125}$$

where $\overline{\omega}_1, \overline{\omega}_2$ are the mean frequencies of the two subsystems of system (1.110).

The marginal PDFs $p(h_1), p(h_2), p(q_1), p(p_1), p(q_1, p_1)$, mean-values $E[H_1], E[H_2]$ and mean-square values $E[Q_1^2], E[P_1^2]$ can be obtained as follows

$$p(h_1) = \int_0^{H_{c2}} p(h_1, h_2)\mathrm{d}h_2, \quad p(h_2) = \int_0^{H_{c1}} p(h_1, h_2)\mathrm{d}h_1,$$

$$E[H_1] = \int_0^{H_{c1}} h_1 p(h_1)\mathrm{d}h_1, \quad E[H_2] = \int_0^{H_{c2}} h_2 p(h_2)\mathrm{d}h_2,$$

$$p(q_1) = \int_{-\infty}^{\infty}\int_{-\infty}^{\infty}\int_{-\infty}^{\infty} p(q_1, q_2, p_1, p_2)\mathrm{d}p_1\mathrm{d}p_2\mathrm{d}q_2,$$

$$E[Q_1^2] = \int_{-\infty}^{\infty} q_1^2 p(q_1)\mathrm{d}q_1,$$

$$p(p_1) = \int_{-\infty}^{\infty}\int_{-\infty}^{\infty}\int_{-\infty}^{\infty} p(q_1, q_2, p_1, p_2)\mathrm{d}q_1\mathrm{d}q_2\mathrm{d}p_2,$$

$$E[P_1^2] = \int_{-\infty}^{\infty} p_1^2 p(p_1)\mathrm{d}p_1,$$

$$p(q_1, p_1) = \int_{-\infty}^{\infty}\int_{-\infty}^{\infty} p(q_1, q_2, p_1, p_2)\mathrm{d}q_2\mathrm{d}p_2 \tag{1.126}$$

For system (1.110) with parameters $\omega_1 = 1.414$ and $\omega_2 = 1$, $E[H_1]$, $E[H_2]$, $E[Q_1^2]$, and $E[P_1^2]$ are calculated using both the stochastic averaging method and Monte Carlo simulation and depicted in Fig. 1.10 with respect to the Hurst index $\mathcal{H}$. It is evident that the stochastic averaging method provides good prediction over a wide range of $\mathcal{H}$ values. For $\mathcal{H}$= 0.7, $E[H_1]$, $E[H_2]$, $E[Q_1^2]$ and $E[P_1^2]$ as functions of ω_2 are shown in Fig. 1.11. It is observed that for small ω_2 ($\omega_1 = 1.414\omega_2$), the stochastic averaging method of quasi-integrable Hamiltonian systems excited by wideband noises is not applicable. When ω_2 exceeds a certain value (approximately 0.6 in the figure), the PSD changes relatively slowly with frequency (see Fig. 2.8 in vol. 1), the results of the stochastic averaging method closely match the simulation results. The stationary PDF $p(q_1, p_1)$ of the first oscillator shown in Fig. 1.12 further verifies that when $\omega_2 = 1$ ($\omega_1 = 1.414$) the stochastic averaging method of quasi-integrable Hamiltonian systems under wideband noise excitation can provides good response prediction of system (1.93). In order to better demonstrate the applicability of this stochastic averaging method, the relative errors of $E[H_1]$, $E[H_2]$, $E[Q_1^2]$, and $E[P_1^2]$ as function of $\mathcal{H} \in (0.5,1)$ and ω_2 are shown in Fig. 1.13, $\omega_1 = 1.414\omega_2$. It can be seen from the figure that the relative errors decrease as natural frequencies increase.

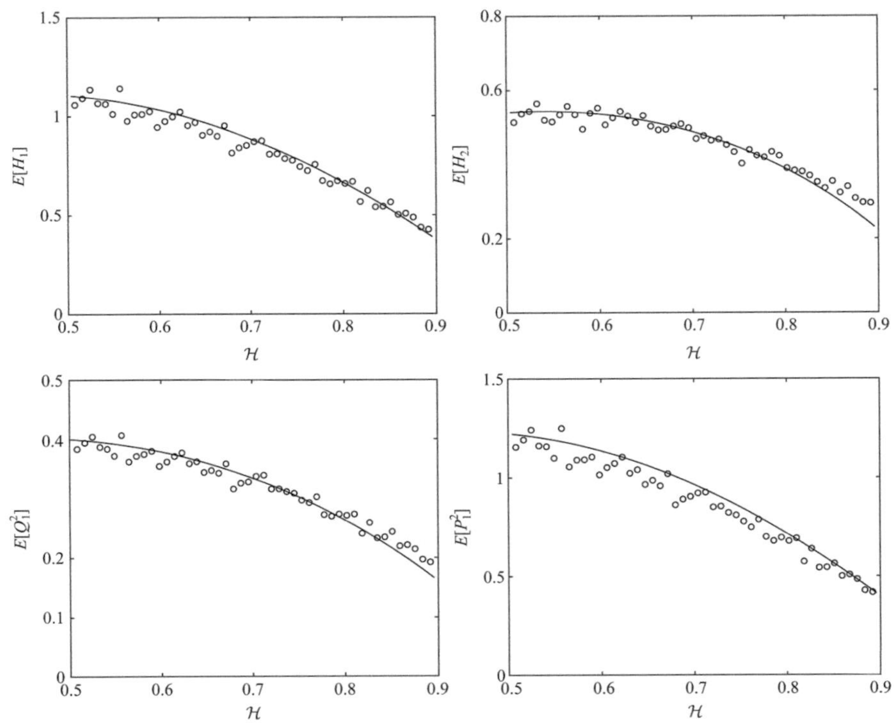

Fig. 1.10 Mean-values $E[H_1]$, $E[H_2]$ and mean-square values $E[Q_1^2]$, $E[P_1^2]$ versus Hurst index $\mathcal{H}$. ○, results from Monte Carlo simulation; ——, results from the stochastic averaging method (Lü et al. 2020a)

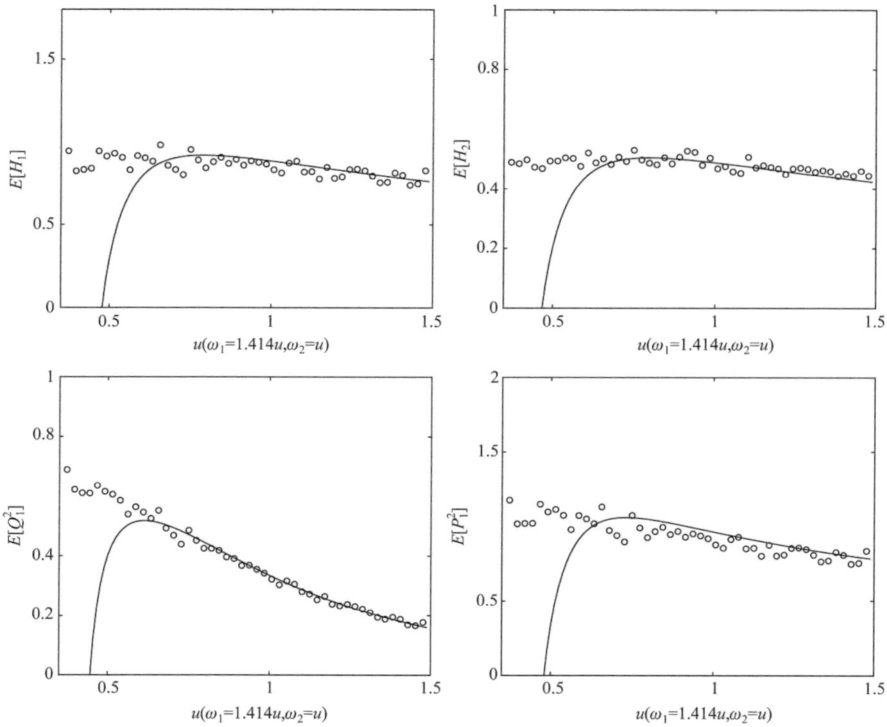

Fig. 1.11 Mean-values $E[H_1]$, $E[H_2]$ and mean-square value $E[Q_1^2]$, $E[P_1^2]$ versus natural frequencies ω_1, ω_2, $\mathcal{H} = 0.7$. ○, results from Monte Carlo simulation; ——, results from the stochastic averaging method (Lü et al. 2020a)

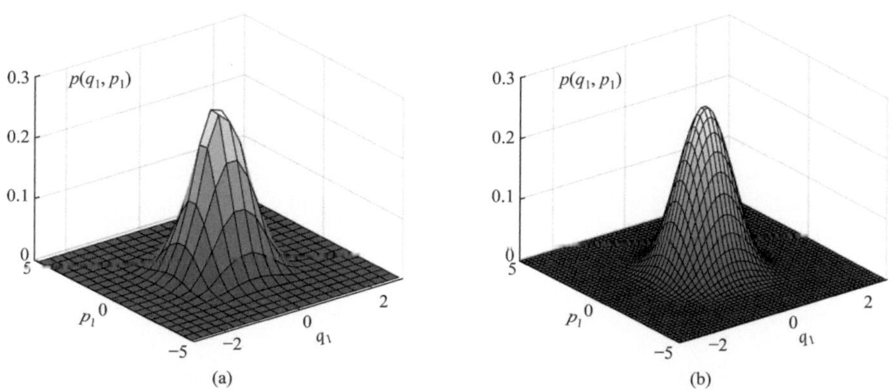

Fig. 1.12 Stationary PDF of $p(q_1, p_1)$ of system (1.110) with $\mathcal{H} = 0.7$, $\omega_1 = 1.414$, $\omega_2 = 1$. **a** result from Monte Carlo simulation; **b** result from the stochastic averaging method (Lü et al. 2020a)

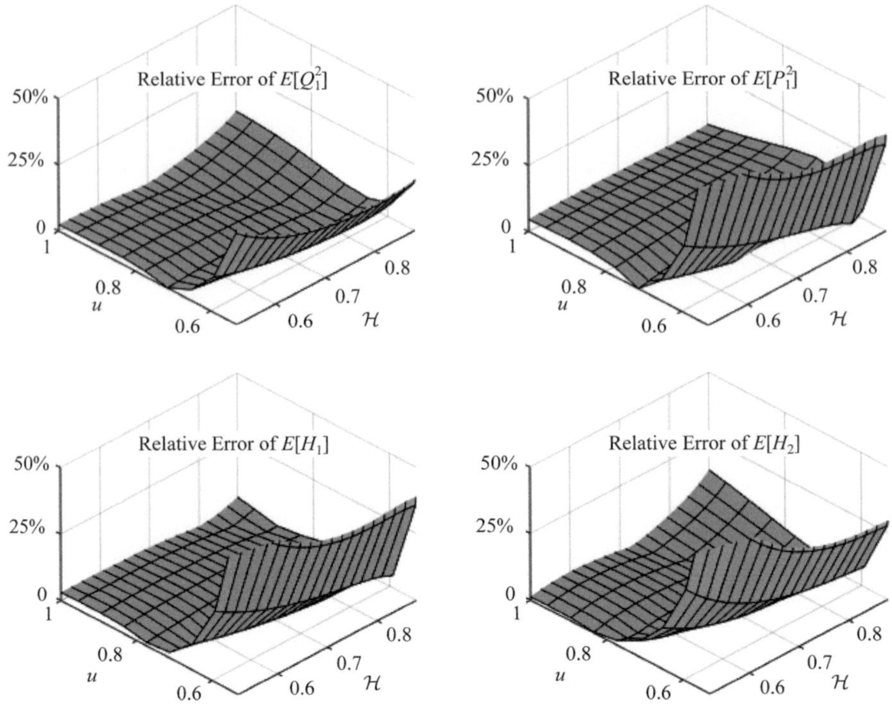

Fig. 1.13 Relative error of response statistics in Eq. (1.126), $\omega_1 = 1.414u$, $\omega_2 = u$ (Lü et al. 2020a)

1.2.2 Internal Resonant Case

It was pointed out in Sect. 3.2.5 in vol. 1 that non-degenerate nonlinear integrable Hamiltonian systems are hardly to occur internal resonance. For simplicity, to present the stochastic averaging method for system (1.93) subjected to fGns in internal resonant case, it is assumed that Eq. (1.93) is quasi-linear Hamiltonian systems. The Hamiltonian associated with system (1.93) is of the form

$$H(\mathbf{Q}, \mathbf{P}) = \sum_{i=1}^{n} H_i(Q_i, P_i), \quad H_i(Q_i, P_i) = \frac{1}{2}P_i^2 + \frac{1}{2}\omega_i^2 Q_i^2, \tag{1.127}$$

where H_i is the Hamiltonian of the ith subsystem. The relationship between H_i and the amplitude A_i is $H_i = \omega_i^2 A_i^2 / 2$. Equation (1.95) can be modified as

$$Q_i(t) = \frac{\sqrt{2H_i}}{\omega_i}\cos\Phi_i(t), \quad P_i(t) = -\sqrt{2H_i}\sin\Phi_i(t), \quad \Phi_i(t) = \omega_i t + \Theta_i(t). \tag{1.128}$$

Treating Eq. (1.128) as a transformation from $\boldsymbol{Q}$, $\mathbf{P}$ to $\mathbf{H}, \boldsymbol{\Phi}$, the equations for original system (1.93) become

$$\begin{aligned}
\dot{H}_i &= \varepsilon F_i^H(\mathbf{H}, \boldsymbol{\Phi}) + \varepsilon^{1/2} \sum_{k=1}^{m} G_{ik}^H(\mathbf{H}, \boldsymbol{\Phi}) W_{Hk}(t), \\
\dot{\Phi}_i &= \omega_i + \varepsilon F_i^{\Phi}(\mathbf{H}, \boldsymbol{\Phi}) + \varepsilon^{1/2} \sum_{k=1}^{m} G_{ik}^{\Phi}(\mathbf{H}, \boldsymbol{\Phi}) W_{Hk}(t), \\
i &= 1, 2, \ldots, n.
\end{aligned} \tag{1.129}$$

where $\boldsymbol{H} = [H_1, H_2, \ldots, H_n]^{\mathrm{T}}$, $\boldsymbol{\Phi} = [\Phi_1, \Phi_2, \ldots, \Phi_n]^{\mathrm{T}}$ and

$$\begin{aligned}
F_i^H &= -\sqrt{2H_i} \sin \Phi_i \sum_{j=1}^{n} c_{ij}(\mathbf{H}, \boldsymbol{\Phi}) \sqrt{2H_j} \sin \Phi_j, \\
F_i^{\Phi} &= \frac{-\cos \Phi_i}{\sqrt{2H_i}} \sum_{j=1}^{n} c_{ij}(\mathbf{H}, \boldsymbol{\Phi}) \sqrt{2H_j} \sin \Phi_j, \\
G_{ik}^H &= -\sqrt{2H_i} \sin \Phi_i f_{ik}(\mathbf{H}, \boldsymbol{\Phi}), \quad G_{ik}^{\Phi} = \frac{-\cos \Phi_i}{\sqrt{2H_i}} f_{ik}(\mathbf{H}, \boldsymbol{\Phi}).
\end{aligned} \tag{1.130}$$

Assume that the system (1.129) has α internal resonant relations $\sum_{r=1}^{n} k_r^u \omega_r = O_u(\varepsilon)$, $u = 1, 2, \ldots, \alpha$. Introduce the following α combinations of angle variables

$$\Psi_u = \sum_{i=1}^{n} k_i^u \Phi_i, \quad u = 1, 2, \ldots, \alpha. \tag{1.131}$$

Using Eq. (1.131) as a transformation and replacing $\Phi_1, \Phi_2, \ldots, \Phi_\alpha$ by $\Psi_1, \Psi_2, \ldots, \Psi_\alpha$, the motion equations for H_i $(i = 1, 2, \ldots, n)$, $\Psi_u (u = 1, 2, \ldots, \alpha)$, and $\Phi_r (r = \alpha + 1, \ldots, n)$ can be derived as

$$\begin{aligned}
\dot{H}_i &= \varepsilon \overline{F}_i^H(\mathbf{H}, \boldsymbol{\Psi}, \boldsymbol{\Phi}_1) + \varepsilon^{1/2} \sum_{k=1}^{m} \overline{G}_{ik}^H(\mathbf{H}, \boldsymbol{\Psi}, \boldsymbol{\Phi}_1) W_{Hk}(t), \\
\dot{\Psi}_u &= O_u(\varepsilon) + \varepsilon \sum_{r=1}^{n} k_r^u \overline{F}_r^{\Phi}(\mathbf{H}, \boldsymbol{\Psi}, \boldsymbol{\Phi}_1) \\
&\quad + \varepsilon^{1/2} \sum_{r=1}^{n} \sum_{k=1}^{m} k_r^u \overline{G}_{rk}^{\Phi}(\mathbf{H}, \boldsymbol{\Psi}, \boldsymbol{\Phi}_1) W_{Hk}(t), \\
\dot{\Phi}_s &= \omega_s + \varepsilon \overline{F}_s^{\Phi}(\mathbf{H}, \boldsymbol{\Psi}, \boldsymbol{\Phi}_1) + \varepsilon^{1/2} \sum_{k=1}^{m} \overline{G}_{sk}^{\Phi}(\mathbf{H}, \boldsymbol{\Psi}, \boldsymbol{\Phi}_1) W_{Hk}(t),
\end{aligned}$$

$$i = 1, 2, \ldots, n; \quad u = 1, 2, \ldots, \alpha; \quad s = \alpha + 1, \alpha + 2, \ldots, n, \tag{1.132}$$

where $\boldsymbol{\Psi} = [\Psi_1, \Psi_2, \cdots, \Psi_\alpha]^{\mathrm{T}}$, $\boldsymbol{\Phi}_1 = [\Phi_{\alpha+1}, \Phi_{\alpha+2}, \ldots, \Phi_n]^{\mathrm{T}}$, and $\overline{F}_i^H, \overline{F}_i^\Phi, \overline{G}_{ik}^H, \overline{G}_{ik}^\Phi$ can be obtained from $F_i^H, F_i^\Phi, G_{ik}^H, G_{ik}^\Phi$ with $\Phi_1, \Phi_2, \ldots, \Phi_\alpha$ replaced by $\boldsymbol{\Psi}$. In system (1.132), $\mathbf{H}(t)$ and $\boldsymbol{\Psi}(t)$ are slowly-varying vector process, while $\boldsymbol{\Phi}_1(t)$ is fast-varying vector process. Assume that PSDs of $W_{H1}, W_{H2}, \ldots, W_{Hm}$ at ω_i ($i = 1, 2, \ldots, n$) and in higher frequency bands change slowly with frequency. The stochastic averaging method of quasi-integrable Hamiltonian systems under wide-band noise excitation described in Sect. 1.12(ii) can be applied. Similar to Eq. (1.77), the following averaged SDEs for $(n+\alpha)$-dimensional vector process $[\mathbf{H}^{\mathrm{T}}, \boldsymbol{\Psi}^{\mathrm{T}}]^{\mathrm{T}}$ can be derived

$$\begin{aligned}
\mathrm{d}H_i &= m_i(\mathbf{H}, \boldsymbol{\Psi})\mathrm{d}t + \sum_{k=1}^{m} \sigma_{ik}(\mathbf{H}, \boldsymbol{\Psi})\mathrm{d}B_k(t), \\
\mathrm{d}\Psi_u &= [O_u(\varepsilon) + m_u^\Psi(\mathbf{H}, \boldsymbol{\Psi})]\mathrm{d}t + \sum_{k=1}^{m} \sigma_{uk}^\Psi(\mathbf{H}, \boldsymbol{\Psi})\mathrm{d}B_k(t), \\
i &= 1, 2, \ldots, n; \quad u = 1, 2, \ldots, \alpha.
\end{aligned} \tag{1.133}$$

where $B_1(t), B_2(t), \ldots, B_m(t)$ are independent unit Wiener processes. The drift and diffusion coefficients in Eq. (1.133) can be calculated as follows

$$\begin{aligned}
m_i(\mathbf{H}, \boldsymbol{\Psi}) &= \varepsilon \Bigg\langle \overline{F}_i^H + \int_{-\infty}^{0} \sum_{k=1}^{m} [\sum_{j=1}^{n} (\frac{\partial \overline{G}_{ik}^H}{\partial H_j}\bigg|_t \overline{G}_{jk}^H\bigg|_{t+\tau}) \\
&\quad + \sum_{u=1}^{\alpha} \sum_{r=1}^{n} (k_r^u \frac{\partial \overline{G}_{ik}^H}{\partial \Psi_u}\bigg|_t \overline{G}_{rk}^\Phi\bigg|_{t+\tau}) + \sum_{s=\alpha+1}^{n} (\frac{\partial \overline{G}_{ik}^H}{\partial \Phi_s}\bigg|_t \overline{G}_{sk}^\Phi\bigg|_{t+\tau})] R_k(\tau)\mathrm{d}\tau \Bigg\rangle_t, \\
m_u^\Psi(\mathbf{H}, \boldsymbol{\Psi}) &= \varepsilon \Bigg\langle \sum_{r=1}^{n} k_r^u \overline{F}_r^\Phi + \int_{-\infty}^{0} \sum_{k=1}^{m} [\sum_{j,r=1}^{n} (k_r^u \frac{\partial \overline{G}_{rk}^\Phi}{\partial H_j}\bigg|_t \overline{G}_{jk}^H\bigg|_{t+\tau}) \\
&\quad + \sum_{v=1}^{\alpha} \sum_{j,r=1}^{n} (k_r^u k_j^v \frac{\partial \overline{G}_{rk}^\Phi}{\partial \Psi_v}\bigg|_t \overline{G}_{jk}^\Phi\bigg|_{t+\tau}) + \sum_{s=\alpha+1}^{n} \sum_{r=1}^{n} (k_r^u \frac{\partial \overline{G}_{rk}^\Phi}{\partial \Phi_s}\bigg|_t \overline{G}_{sk}^\Phi\bigg|_{t+\tau})] R_k(\tau)\mathrm{d}\tau \Bigg\rangle_t, \\
b_{ij}(\mathbf{H}, \boldsymbol{\Psi}) &= \sum_{k=1}^{m} \sigma_{ik}\sigma_{jk} = \varepsilon \Bigg\langle \int_{-\infty}^{\infty} \sum_{k=1}^{m} (\overline{G}_{ik}^H\bigg|_t \overline{G}_{jk}^H\bigg|_{t+\tau}) R_k(\tau)\mathrm{d}\tau \Bigg\rangle_t, \\
b_{uv}^\Psi(\mathbf{H}, \boldsymbol{\Psi}) &= \sum_{k=1}^{m} \sigma_{uk}^\Psi \sigma_{vk}^\Psi = \varepsilon \Bigg\langle \int_{-\infty}^{\infty} \sum_{k=1}^{m} \sum_{j,r=1}^{n} (k_r^u k_j^v \overline{G}_{rk}^\Phi\bigg|_t \overline{G}_{jk}^\Phi\bigg|_{t+\tau}) R_k(\tau)\mathrm{d}\tau \Bigg\rangle_t,
\end{aligned}$$

$$b_{iu}^{H\Psi}(\mathbf{H},\boldsymbol{\Psi})=\sum_{k=1}^{m}\sigma_{ik}\sigma_{uk}^{\Psi}=\varepsilon\left\langle\int_{-\infty}^{\infty}\sum_{k=1}^{m}\sum_{r=1}^{n}(k_r^u\overline{G}_{ik}^{H}\Big|_t\overline{G}_{rk}^{\Phi}\Big|_{t+\tau})R_k(\tau)\mathrm{d}\tau\right\rangle_t,$$
$$i,j=1,2,\ldots,n;\quad u,v=1,2,\ldots,\alpha, \tag{1.134}$$

where $\Phi_i(t+\tau)$ can be replaced approximately by $\Phi_i(t)+\omega_i\tau$; time averaging $\langle[\cdot]\rangle_t$ can be replaced by averaging with respective to fast-varying phase angles

$$\langle[\cdot]\rangle_t=\frac{1}{(2\pi)^{n-\alpha}}\int_0^{2\pi}(\cdot)\mathrm{d}\boldsymbol{\Phi}_1. \tag{1.135}$$

The correlation functions of fGns is given by Eq. (2.217) in vol. 1. To obtain analytical expressions for $m_i, m_u^{\Psi}, b_{ij}, b_{uv}^{\Psi}, b_{iu}^{H\Psi}$ in Eq. (1.134), it is necessary to expand $\overline{F}_i^H, \overline{F}_i^{\Phi}, \overline{G}_{ik}^H, \overline{G}_{ik}^{\Phi}$ into Fourier series with respect to $\Phi_{\alpha+1}, \Phi_{\alpha+2}, \ldots, \Phi_n$, and convert the integrals with respect to τ into PSDs of fGns (see Eq. (1.23)). Finally, the analytical expressions for the drift and diffusion coefficients are obtained.

The averaged FPK equation associated with averaged Itô Eq. (1.133) is

$$\begin{aligned}\frac{\partial p}{\partial t}=&-\sum_{i=1}^{n}\frac{\partial}{\partial h_i}[m_i(\mathbf{h},\boldsymbol{\psi})p]-\sum_{u=1}^{\alpha}\frac{\partial}{\partial\psi_u}[m_u^{\psi}(\mathbf{h},\boldsymbol{\psi})p]\\&+\frac{1}{2}\sum_{i,j=1}^{n}\frac{\partial^2}{\partial h_i\partial h_j}[b_{ij}(\mathbf{h},\boldsymbol{\psi})p]+\sum_{i=1}^{n}\sum_{u=1}^{\alpha}\frac{\partial^2}{\partial h_i\partial\psi_u}\left[b_{iu}^{h\psi}(\mathbf{h},\boldsymbol{\psi})p\right]\\&+\frac{1}{2}\sum_{u,v=1}^{\alpha}\frac{\partial^2}{\partial\psi_u\partial\psi_v}[b_{uv}^{\psi}(\mathbf{h},\boldsymbol{\psi})p],\end{aligned} \tag{1.136}$$

where the first and second derivative moments are

$$\begin{aligned}&m_i(\mathbf{h},\ \boldsymbol{\psi})=m_i(\mathbf{H},\ \boldsymbol{\Psi})|_{\mathbf{H}=\mathbf{h},\ \boldsymbol{\Psi}=\boldsymbol{\psi}},\\&m_u^{\psi}(\mathbf{h},\ \boldsymbol{\psi})=m_u^{\Psi}(\mathbf{H},\ \boldsymbol{\Psi})\big|_{\mathbf{H}=\mathbf{h},\ \boldsymbol{\Psi}=\boldsymbol{\psi}},\\&b_{ij}(\mathbf{h},\ \boldsymbol{\psi})=b_{ij}(\mathbf{H},\ \boldsymbol{\Psi})\big|_{\mathbf{H}=\mathbf{h},\ \boldsymbol{\Psi}=\boldsymbol{\psi}},\\&b_{iu}^{h\psi}(\mathbf{h},\ \boldsymbol{\psi})=b_{iu}^{H\Psi}(\mathbf{H},\ \boldsymbol{\Psi})\big|_{\mathbf{H}=\mathbf{h},\ \boldsymbol{\Psi}=\boldsymbol{\psi}},\\&b_{uv}^{\psi}(\mathbf{h},\ \boldsymbol{\psi})=b_{uv}^{\Psi}(\mathbf{H},\ \boldsymbol{\Psi})\big|_{\mathbf{H}=\mathbf{h},\ \boldsymbol{\Psi}=\boldsymbol{\psi}},\\&i,j=1,2,\ldots,n;\quad u,v=1,2,\ldots,\alpha.\end{aligned} \tag{1.137}$$

In Eq. (1.136), $p=p(\mathbf{h},\psi,t|\mathbf{h}_0,\psi_0)$ is the transition PDF of $[\mathbf{H}^{\mathrm{T}},\boldsymbol{\Psi}^{\mathrm{T}}]^{\mathrm{T}}$ with initial condition

$$p(\mathbf{h},\boldsymbol{\psi},0|\mathbf{h}_0,\boldsymbol{\psi}_0)=\delta(\mathbf{h}-\mathbf{h}_0)\delta(\boldsymbol{\psi}-\boldsymbol{\psi}_0), \tag{1.138}$$

If $H_i \quad (i = 1, 2, \ldots, n)$ can change in $[0, \infty)$, then the boundary conditions of FPK Eq. (1.136) are

$$\begin{aligned}
&p = \text{finite}, \ h_i = 0, \\
&p, \quad \partial p/\partial h \to 0, \ |\mathbf{h}| \to \infty, \\
&p(\mathbf{h}, \psi_1, \psi_2, \ldots, \psi_u + 2k\pi, \ldots, \psi_\alpha, t|\mathbf{h}_0, \boldsymbol{\psi}_0) \\
&\quad = p(\mathbf{h}, \psi_1, \psi_2, \ldots, \psi_u, \ldots, \psi_\alpha, t|\mathbf{h}_0, \boldsymbol{\psi}_0), \\
&i = 1, 2, \ldots, n; \quad u = 1, 2, \ldots, \alpha; \quad k \text{ is an integer.}
\end{aligned} \tag{1.139}$$

The exact analytical solution for the reduced FPK Eq. (1.136) can be obtained under certain parameter conditions. In general, FPK Eq. (1.136) can be solved only numerically. Following a derivation similar to Eq. (1.76), the approximate stationary PDF $p(\mathbf{q}, \mathbf{p})$ of the original system (1.93) can be obtained as follows

$$p(\mathbf{q}, \mathbf{p}) = \frac{1}{(2\pi)^{n-\alpha}} p(\mathbf{h}, \boldsymbol{\psi}) \left| \frac{\partial(\mathbf{h}, \boldsymbol{\psi}, \boldsymbol{\phi}_1)}{\partial(\mathbf{q}, \mathbf{p})} \right|. \tag{1.140}$$

Other statistics such as marginal PDFs, mean-square values of the original state variables can then be obtained from $p(\mathbf{q}, \mathbf{p})$.

Example 1.6 Consider two coupled Rayleigh oscillators excited by fGns (Lü et al. 2022). The motion equations are

$$\begin{aligned}
&\ddot{X}_1 - \gamma_1 \dot{X}_1 + \gamma_2(\dot{X}_1^2 + \dot{X}_2^2)\dot{X}_1 + \omega^2 X_1 = \sqrt{2D_1} W_{H1}(t), \\
&\ddot{X}_2 - \gamma_1 \dot{X}_2 + \gamma_2(\dot{X}_1^2 + \dot{X}_2^2)\dot{X}_2 + \omega^2 X_2 = \sqrt{2D_2} W_{H2}(t),
\end{aligned} \tag{1.141}$$

where $W_{H1}(t)$, $W_{H2}(t)$ are independent fGns with Hurst index $1/2 < \mathcal{H} < 1$.

Letting $X_1 = Q_1, \dot{X}_1 = P_1, X_2 = Q_2, \dot{X}_2 = P_2$, system (1.141) can be transformed into a quasi-Hamiltonian system excited by fGns of the form of system (1.93). The Hamiltonian and sub-Hamiltonians are

$$H = H_1 + H_2, \quad H_i = P_i^2/2 + \omega^2 Q_i^2/2. \tag{1.142}$$

Due to the identical natural frequencies and coupling of the two oscillators in system (1.141), internal resonance may occurs. Introduce a combination of two angle variables

$$\Psi = \Phi_1 - \Phi_2 = \tan^{-1}\left(\frac{P_1}{\omega Q_1}\right) - \tan^{-1}\left(\frac{P_2}{\omega Q_2}\right). \tag{1.143}$$

First, applying the stochastic averaging method of quasi-integrable and resonant Hamiltonian systems excited by fGns presented in Sect. 7.3.2 in vol. 1 to system

(1.141) yields the following averaged fractional SDEs

$$
\begin{aligned}
\mathrm{d}H_1 &= m_1(\mathbf{H}, \Psi)\mathrm{d}t + \sum_{k=1}^{2} \sigma_{1k}(\mathbf{H}, \Psi)\mathrm{d}B_{Hk}(t), \\
\mathrm{d}H_2 &= m_2(\mathbf{H}, \Psi)\mathrm{d}t + \sum_{k=1}^{2} \sigma_{2k}(\mathbf{H}, \Psi)\mathrm{d}B_{Hk}(t), \\
\mathrm{d}\Psi &= m_3(\mathbf{H}, \Psi)\mathrm{d}t + \sum_{k=1}^{2} \sigma_{3k}(\mathbf{H}, \Psi)\mathrm{d}B_{Hk}(t).
\end{aligned} \tag{1.144}
$$

The drift and diffusion coefficients in Eq. (1.144) are

$$
\begin{aligned}
m_1 &= \gamma_1 H_1 - \frac{3}{2}\gamma_2 H_1^2 - \gamma_2(1 + \frac{1}{2}\cos 2\Psi)H_1 H_2, \\
m_2 &= \gamma_1 H_2 - \frac{3}{2}\gamma_2 H_2^2 - \gamma_2(1 + \frac{1}{2}\cos 2\Psi)H_1 H_2, \\
m_3 &= \frac{1}{4}\gamma_2 \sin(2\Psi)(H_1 + H_2), \\
b_{11} &= \sum_{k=1}^{2} \sigma_{1k}\sigma_{1k} = 2D_1 H_1, \quad b_{22} = \sum_{k=1}^{2} \sigma_{2k}\sigma_{2k} = 2D_2 H_2, \\
b_{33} &= \sum_{k=1}^{2} \sigma_{3k}\sigma_{3k} = \frac{1}{2}(\frac{D_1}{H_1} + \frac{D_2}{H_2}), \\
&\sum_{k=1}^{2} \sigma_{1k}\sigma_{2k} = \sum_{k=1}^{2} \sigma_{1k}\sigma_{3k} = \sum_{k=1}^{2} \sigma_{2k}\sigma_{1k} = \sum_{k=1}^{2} \sigma_{2k}\sigma_{3k} = \sum_{k=1}^{2} \sigma_{3k}\sigma_{1k} \\
&= \sum_{k=1}^{2} \sigma_{3k}\sigma_{2k} = 0.
\end{aligned} \tag{1.145}
$$

Next, treating fGns $W_{H1}(t)$ and $W_{H2}(t)$ in system (1.141) as stationary wideband noises and applying the stochastic averaging method described in Sect. 1.2.2 result in the following averaged Itô SDEs

$$
\begin{aligned}
\mathrm{d}H_1 &= \overline{m}_1(\mathbf{H}, \Psi)\mathrm{d}t + \sum_{k=1}^{2} \overline{\sigma}_{1k}(\mathbf{H}, \Psi)\mathrm{d}B_k(t), \\
\mathrm{d}H_2 &= \overline{m}_2(\mathbf{H}, \Psi)\mathrm{d}t + \sum_{k=1}^{2} \overline{\sigma}_{2k}(\mathbf{H}, \Psi)\mathrm{d}B_k(t), \\
\mathrm{d}\Psi &= \overline{m}_3(\mathbf{H}, \Psi)\mathrm{d}t + \sum_{k=1}^{2} \overline{\sigma}_{3k}(\mathbf{H}, \Psi)\mathrm{d}B_k(t).
\end{aligned} \tag{1.146}
$$

According to Eq. (1.134), the drift and diffusion coefficients in Eq. (1.146) can be obtained as follows

$$\begin{aligned}
&\overline{m}_1(H_1, H_2, \Psi) = \gamma_1 H_1 - \frac{3}{2}\gamma_2 H_1^2 - \gamma_2(1 + \frac{1}{2}\cos 2\Psi)H_1 H_2 \\
&\quad + 2\pi D_1 S(\omega), \\
&\overline{m}_2(H_1, H_2, \Psi) = \gamma_1 H_2 - \frac{3}{2}\gamma_2 H_2^2 - \gamma_2(1 + \frac{1}{2}\cos 2\Psi)H_1 H_2 \\
&\quad + 2\pi D_2 S(\omega), \\
&\overline{m}_3(H_1, H_2, \Psi) = \frac{1}{4}\gamma_2 \sin(2\Psi)(H_1 + H_2),
\end{aligned}$$

$$\begin{aligned}
&\overline{b}_{11}(H_1, H_2, \Psi) = \sum_{k=1}^{2} \overline{\sigma}_{1k}\overline{\sigma}_{1k} = 4\pi D_1 S(\omega) H_1, \\
&\overline{b}_{22} = \sum_{k=1}^{2} \overline{\sigma}_{2k}\overline{\sigma}_{2k} = 4\pi D_1 S(\omega) H_2, \\
&\overline{b}_{33}(H_1, H_2, \Psi) = \sum_{k=1}^{2} \overline{\sigma}_{3k}\overline{\sigma}_{3k} = \pi S(\omega)\left(\frac{D_1}{H_1} + \frac{D_2}{H_2}\right), \\
&\sum_{k=1}^{2} \overline{\sigma}_{1k}\overline{\sigma}_{2k} = \sum_{k=1}^{2} \overline{\sigma}_{1k}\overline{\sigma}_{3k} = \sum_{k=1}^{2} \overline{\sigma}_{2k}\overline{\sigma}_{1k} = \sum_{k=1}^{2} \overline{\sigma}_{2k}\overline{\sigma}_{3k} = \sum_{k=1}^{2} \overline{\sigma}_{3k}\overline{\sigma}_{1k} \\
&\quad = \sum_{k=1}^{2} \overline{\sigma}_{3k}\overline{\sigma}_{2k} = 0,
\end{aligned} \tag{1.147}$$

where $S(\omega)$ is the PSD of $W_{H1}(t)$ and $W_{H2}(t)$.

The reduced averaged FPK equation associated with Itô SDEs (1.146) is

$$-\frac{\partial(\overline{m}_1 p)}{\partial h_1} - \frac{\partial(\overline{m}_2 p)}{\partial h_2} - \frac{\partial(\overline{m}_3 p)}{\partial \psi} + \frac{1}{2}\frac{\partial^2(\overline{b}_{11} p)}{\partial h_1^2} + \frac{1}{2}\frac{\partial^2(\overline{b}_{22} p)}{\partial h_2^2} + \frac{1}{2}\frac{\partial^2(\overline{b}_{33} p)}{\partial \psi^2} = 0, \tag{1.148}$$

where

$$\begin{aligned}
&\overline{m}_i = \overline{m}_i(h_1, h_2, \psi) = \overline{m}_i(H_1, H_2, \Psi)|_{H_1=h_1,\, H_2=h_2,\, \Psi=\psi}, \\
&\overline{b}_{ii} = \overline{b}_{ii}(h_1, h_2, \psi) = \overline{b}_{ii}(H_1, H_2, \Psi)\Big|_{H_1=h_1,\, H_2=h_2,\, \Psi=\psi}, \\
&i = 1, 2, 3.
\end{aligned} \tag{1.149}$$

The boundary conditions of Eq. (1.148) are

$$
\begin{aligned}
&p(h_1, h_2 \to 0, \psi) = \text{finite}, \quad p(h_1 \to 0, h_2, \psi) = \text{finite}, \\
&p(h_1, h_2 \to \infty, \psi) = p(h_1 \to \infty, h_2, \psi) = 0, \\
&p(h_1, h_2, \psi = -\pi) = p(h_1, h_2, \psi = \pi).
\end{aligned} \tag{1.150}
$$

When $D_1 = D_2$, the reduced averaged FPK Eq. (1.148) has an exact solution

$$
\begin{aligned}
p(h_1, h_2, \psi) = C \exp\{&\frac{1}{2\pi D_1 S(\omega)}[\gamma_1(h_1 + h_2) \\
&- \frac{3}{4}\gamma_2(h_1^2 + h_2^2) - \gamma_2(1 + \frac{1}{2}\cos 2\psi)h_1 h_2]\}.
\end{aligned} \tag{1.151}
$$

The approximate joint PDF $p(q_1, q_2, p_1, p_2)$ for original system (1.141) can be obtained from Eq. (1.151) by using Eq. (1.140) as follows

$$
p(q_1, q_2, p_1, p_2) = C_2 p(h_1, h_2, \psi)|_{h_i = \frac{1}{2}p_i^2 + \frac{1}{2}\omega^2 q_i^2,\ \psi = \tan^{-1}(\frac{p_1}{\omega q_1}) - \tan^{-1}(\frac{p_2}{\omega q_2})}. \tag{1.152}
$$

The marginal PDFs and statistics for system (1.141), such as $p(h_1)$, $p(\psi)$, $p(q_1, p_1)$, $p(q_1)$, $E[H_1]$ and $E[Q_1^2]$, can be obtained as follows

$$
\begin{aligned}
p(\psi) &= \int_0^\infty \int_0^\infty p(h_1, h_2, \psi)\mathrm{d}h_2 \mathrm{d}h_1, \\
p(h_1) &= \int_0^\infty \int_{-\pi}^{\pi} p(h_1, h_2, \psi)\mathrm{d}h_2 \mathrm{d}\psi, \\
E[H_1] &= \int_0^\infty h_1 p(h_1)\mathrm{d}h_1, \\
p(q_1, p_1) &= \int_{-\infty}^\infty \int_{-\infty}^\infty p(q_1, q_2, p_1, p_2)\mathrm{d}q_2 \mathrm{d}p_2, \\
E[Q_1^2] &= \int_{-\infty}^\infty q_1^2 p(q_1)\mathrm{d}q_1, \\
p(q_1) &= \int_{-\infty}^\infty \int_{-\infty}^\infty \int_{-\infty}^\infty p(q_1, q_2, p_1, p_2)\mathrm{d}p_1 \mathrm{d}p_2 \mathrm{d}q_2.
\end{aligned} \tag{1.153}
$$

Given system parameters $\gamma_1 = 1$, $\gamma_2 = 0.1$, $D_1 = D_2 = 0.5$, 2000 sample points of system (1.141) in 3-dimensional space (H_1, H_2, Ψ) shown in Fig. 1.14

provide a rough observation for $p(h_1, h_2, \psi)$. A slice plot of $p(h_1, h_2, \psi)$ calculated from analytical solution (1.151) is given in Fig. 1.15. The stationary PDFs $p(h_1), p(h_2), p(\psi)$ of system (1.141) obtained from Monte Carlo simulation of original system (1.141), Monte Carlo simulation of averaged Itô Eq. (1.144) and analytical solution (1.153) are compared in Figs. 1.16 and 1.17. More comparisons between analytical solution (1.153) and simulation results of system (1.141) for various values of the Hurst index $\mathcal{H}$ and frequency ω are shown in Figs. 1.18, 1.19, 1.20, 1.21, and 1.22. It is seen that the stochastic averaging method of quasi-integrable Hamiltonian systems excited by stationary wideband noises can be applied to the quasi-integrable Hamiltonian system excited by fGns in a very large region of parameter plane $(\mathcal{H}, \omega)$.

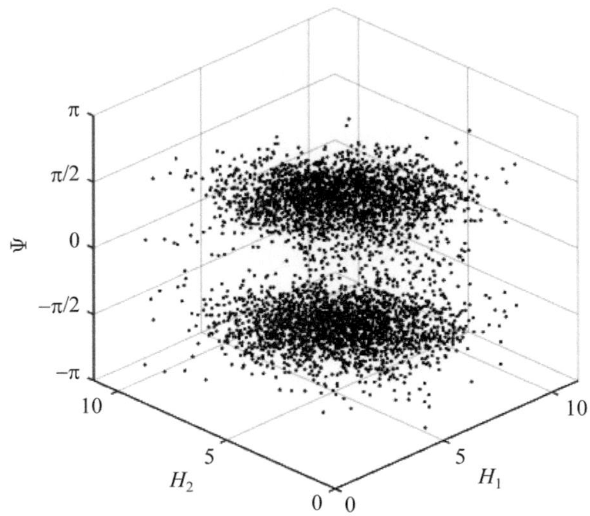

Fig. 1.14 2000 sample points of system (1.141) under stationary motion in three-dimensional space (H_1, H_2, Ψ), $\mathcal{H} = 0.7$, $\omega = 4$ (Lü et al. 2022)

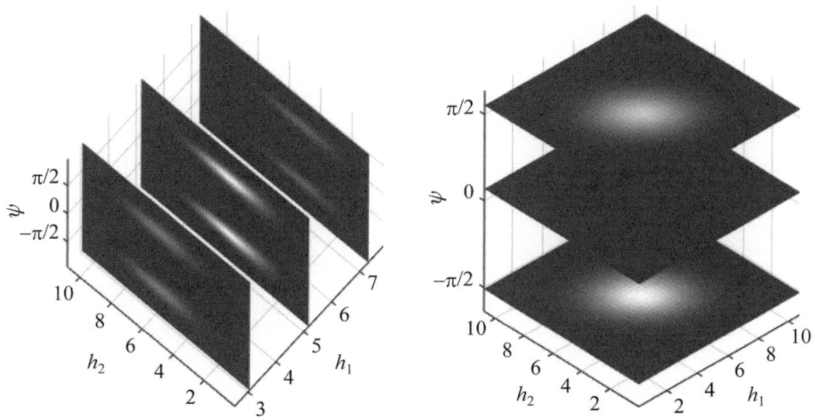

Fig. 1.15 Slices of stationary PDF $p(h_1, h_2, \psi)$ of system (1.141) in three-dimensional space (h_1, h_2, ψ) calculated according to analytical solution (1.151), $\mathcal{H} = 0.7$, $\omega = 4$ (Lü et al. 2022)

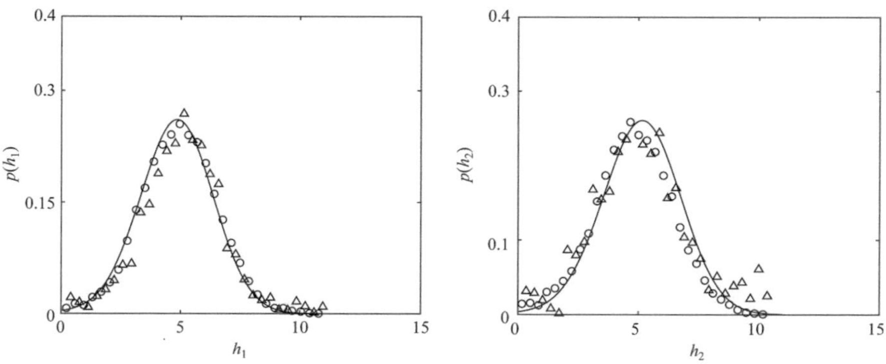

Fig. 1.16 Comparison of stationary PDF $p(h_1)$, $p(h_2)$ obtained from three methods. ○, result from Monte Carlo simulation of Eq. (1.141); △, result from Monte Carlo simulation of Eq. (1.144); ——, result from analytical solution (1.151) (Lü et al. 2022)

1.3 Combined Harmonic and Stationary Wideband Noise Excitations

1.3.1 Single-DOF System

The stochastic averaging method for single-DOF quasi-linear systems excited by combination of harmonic and stationary wideband noise has been presented in Sect. 4.7, vol. 1. In this section, the stochastic averaging method for strongly nonlinear systems excited by combination of harmonic and wideband noise will be presented. First, consider a SDOF quasi-Hamiltonian system subjected to harmonic

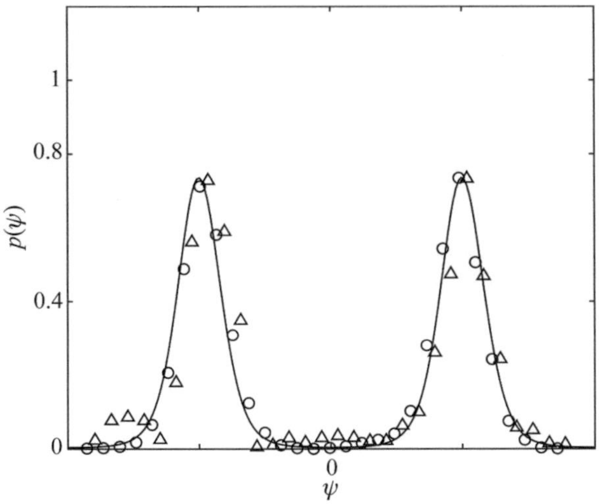

Fig. 1.17 Comparison of stationary PDF $p(\psi)$ obtained from three methods. ○, result from Monte Carlo simulation of Eq. (1.141); △, result from Monte Carlo simulation of Eq. (1.144); ——, result from analytical solution (1.151) (Lü et al. 2022)

and stationary wideband random excitations (Wu 2005; Wu and Zhu 2008). The motion equations are

$$\dot{Q} = P,$$
$$\dot{P} = -g(Q) - \varepsilon c(Q, P)P + \varepsilon h(Q, P)\cos \Omega t + \varepsilon^{1/2}\sum_{k=1}^{m} f_k(Q, P)\xi_k(t), \quad (1.154)$$

where ε is a small parameter, εh is the amplitude of a weak harmonic excitation with frequency Ω, $\xi_k(t)$ $(k = 1, 2, \ldots, m)$ are stationary wideband noises with correlation functions $R_{kl}(\tau)$ and PSDs $S_{kl}(\omega)$.

Assume that the functions $g(q)$ and $U(q) = \int_0^q g(u)\mathrm{d}u$ in system (1.154) satisfy the 4 conditions given below Eq. (1.5). When $\varepsilon = 0$, the system (1.154) has a family of periodic solutions (1.6) in the entire plane (q, p). When ε is small, the solution of the system (1.154) can be regarded as randomly diffused periodic solution in the phase plane. Introduce the transformation

$$Q(t) = A\cos \Phi(t) + B, \quad P(t) = -A\nu(A, \Phi)\sin \Phi(t), \quad \Phi(t) = \Gamma(t) + \Theta(t), \quad (1.155)$$

where A, Φ, Γ, Θ and ν are random processes, B is obtained from relation $U(A + B) = U(A - B) = H = P^2/2 + U(Q)$. Substituting Eq. (1.155) into Eq. (1.154) and deriving similar to that from Eq. (1.14) to Eq. (1.16) yield the following motion equations

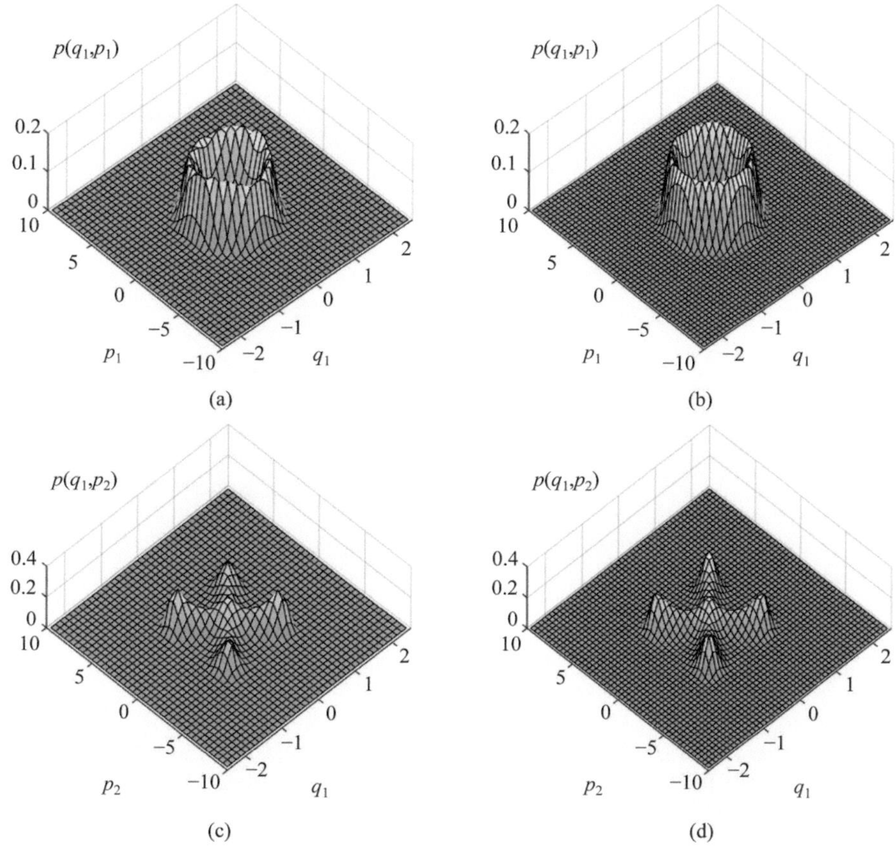

Fig. 1.18 Stationary joint PDF of system (1.141), $\mathcal{H} = 0.7, \omega = 4$. **a** $p(q_1, p_1)$, simulation result of Eq. (1.141); **b** $p(q_1, p_1)$, analytical results; **c** $p(q_1, p_2)$, simulation result of Eq. (1.141); **d** $p(q_1, p_2)$, analytical results (Lü et al. 2022)

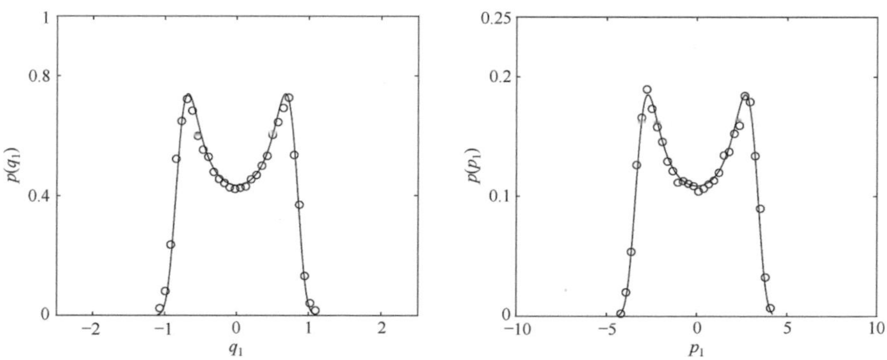

Fig. 1.19 Stationary marginal PDF $p(q_1)$ and $p(p_1)$ of system (1.141), $\mathcal{H} = 0.7, \omega = 4$. ○, results from Monte Carlo simulation of Eq. (1.141). ——, results from analytical solution (1.153) (Lü et al. 2022)

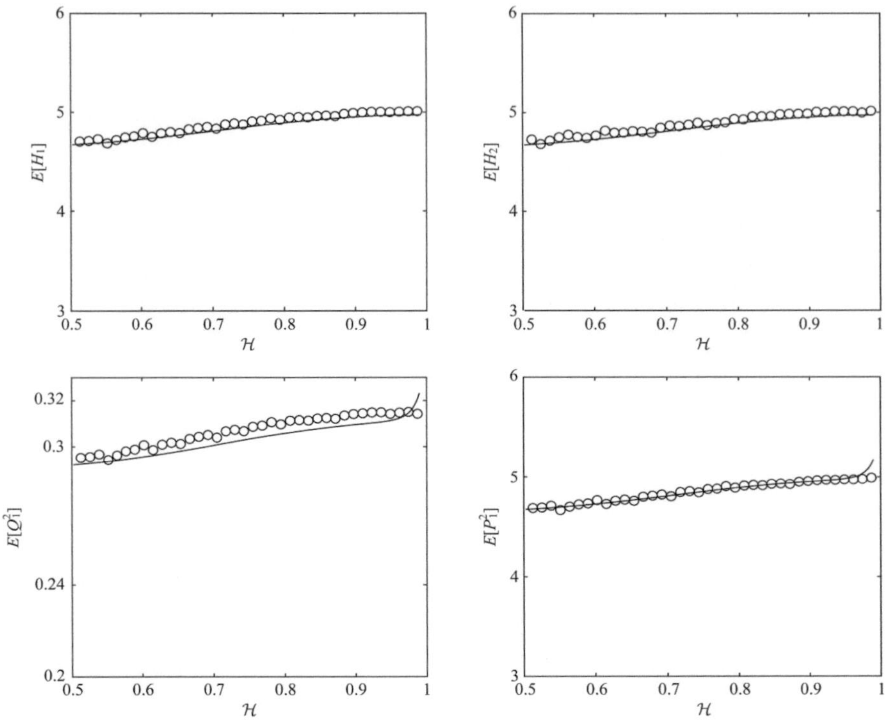

Fig. 1.20 Mean-value $E[H_1]$, $E[H_2]$ and Mean-square value $E[Q_1^2]$, $E[P_1^2]$ with Hurst index $\mathcal{H}$ changes. ○, results from Monte Carlo simulation of Eq. (1.141); ——, results from analytical solution (1.153) (Lü et al. 2022)

$$
\begin{aligned}
\frac{\mathrm{d}A}{\mathrm{d}t} &= \varepsilon F_1(A, \Phi, \Omega t) + \varepsilon^{1/2} \sum_{k=1}^{m} G_{1k}(A, \Phi)\xi_k(t), \\
\frac{\mathrm{d}\Phi}{\mathrm{d}t} &= \nu(A, \Phi) + \varepsilon F_2(A, \Phi, \Omega t) + \varepsilon^{1/2} \sum_{k=1}^{m} G_{2k}(A, \Phi)\xi_k(t),
\end{aligned} \tag{1.156}
$$

where

$$
\begin{aligned}
F_1 = {} & \frac{-A}{g(A+B)(1+d)}[c(A\cos\Phi + B, -A\nu\sin\Phi)A\nu\sin\Phi \\
& + h(A\cos\Phi + B, -A\nu\sin\Phi)\cos\Omega t]\nu\sin\Phi, \\
F_2 = {} & \frac{-1}{g(A+B)(1+d)}[c(A\cos\Phi + B, -A\nu\sin\Phi)A\nu\sin\Phi \\
& + h(A\cos\Phi + B, -A\nu\sin\Phi)\cos\Omega t]\nu(\cos\Phi + d),
\end{aligned}
$$

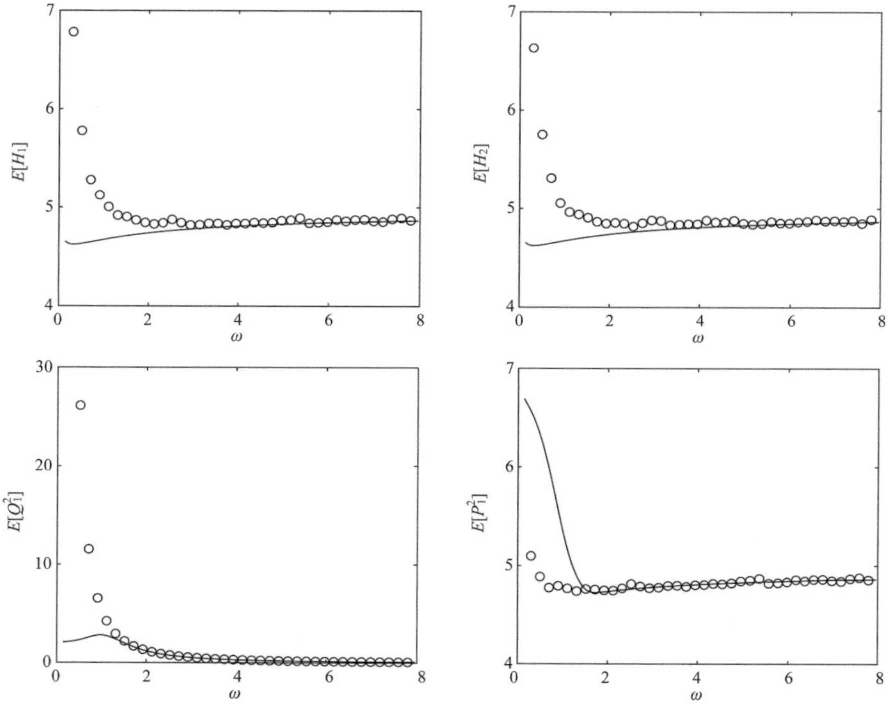

Fig. 1.21 Mean-value $E[H_1]$, $E[H_2]$ and Mean-square value $E[Q_1^2]$, $E[P_1^2]$ with natural frequency ω changes. ○, results from Monte Carlo simulation of Eq. (1.141); ——, results from analytical solution (1.153) (Lü et al. 2022)

$$
\begin{aligned}
G_{1k} &= \frac{-A}{g(A+B)(1+d)} f_k(A\cos\Phi + B, -A\nu\sin\Phi)\nu\sin\Phi, \\
G_{2k} &= \frac{-1}{g(A+B)(1+d)} f_k(A\cos\Phi + B, -A\nu\sin\Phi)\nu(\cos\Phi + d), \\
d &= \frac{\mathrm{d}B}{\mathrm{d}A} = \frac{g(-A+B)+g(A+B)}{g(-A+B)-g(A+B)}.
\end{aligned} \tag{1.157}
$$

The combination of harmonic and wideband noises is equivalent to a narrow-band noise. In the case of non-external resonance, the harmonic excitation has no effect on system response in the first approximation and could be neglected. Thus, only the external resonant case, i.e., the harmonic excitation frequency Ω near the mean frequency $\omega(A)$, is considered. Suppose that

$$
\frac{\Omega}{\omega(A)} = \frac{s}{r} + \varepsilon\sigma, \tag{1.158}
$$

where s, r are integers of mutual prime; $\varepsilon\sigma$ is a small detuning parameter. Multiplying both side of Eq. (1.158) by t, and utilizing the stochastic counterpart $\Phi = \Gamma(t) + \Theta \approx$

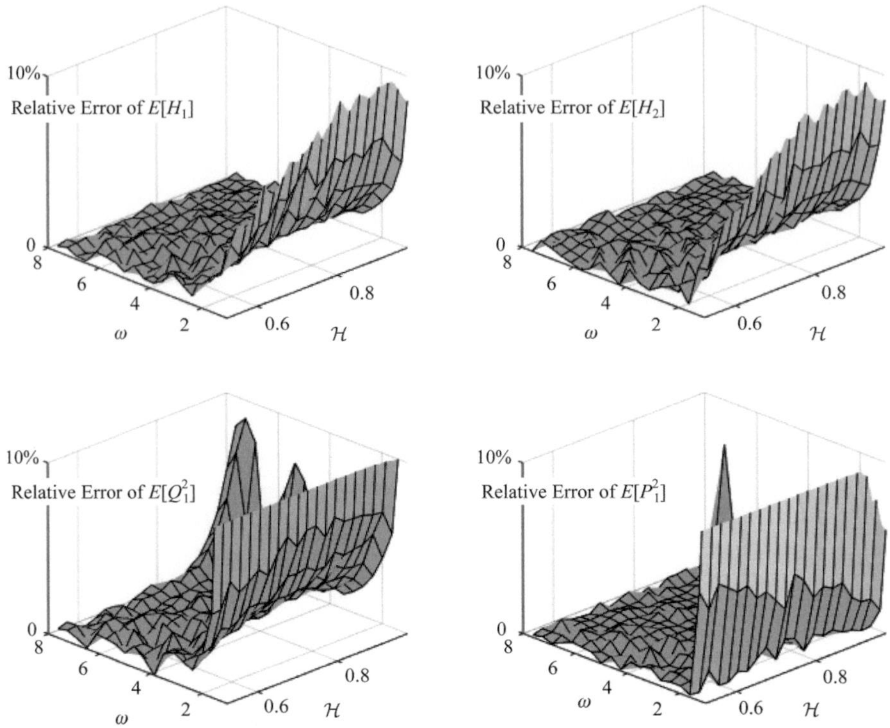

Fig. 1.22 Relative error between analytical solution (1.153) and simulation results of system (1.141) with varying $\mathcal{H}$, ω (Lü et al. 2022)

$\omega(A)t + \Theta$ of the approximation in Eq. (1.11) result in

$$\Omega t = \frac{s}{r}(\Phi - \Theta) + \varepsilon\sigma\Gamma. \tag{1.159}$$

Introduce a new variable

$$\Delta = \Omega t - \frac{s}{r}\Phi \tag{1.160}$$

which can be regarded as phase angle difference between the harmonic excitation phase angle Ωt and the multiple phase angle $\Phi s/r$. Replacing Φ with Δ and combining Eqs. (1.156), (1.157) and (1.160) yields the governing equations for A and Δ

$$\frac{\mathrm{d}A}{\mathrm{d}t} = \varepsilon F_1(A, \frac{r}{s}(\Omega t - \Delta), \Omega t) + \varepsilon^{1/2}\sum_{k=1}^{m} G_{1k}(A, \frac{r}{s}(\Omega t - \Delta))\xi_k(t),$$
$$\frac{\mathrm{d}\Delta}{\mathrm{d}t} = \Omega - \frac{s}{r}\nu(A, \frac{r}{s}(\Omega t - \Delta)) - \varepsilon\frac{s}{r}F_2(A, \frac{r}{s}(\Omega t - \Delta), \Omega t)$$

$$-\varepsilon^{1/2}\frac{s}{r}\sum_{k=1}^{m}G_{2k}(A,\frac{r}{s}(\Omega t-\Delta))\xi_k(t). \tag{1.161}$$

According to Khasminskii's theorem (Khasminskii 1966), when $\varepsilon \to 0$, A, Δ converge to 2-dimensional vector Markov diffusion process. Since the coefficients of Eq. (1.161) depend on Ωt, they need to be averaged with respect to Ωt. The resulting smoothed averaged Itô SDEs are

$$\begin{aligned}\mathrm{d}A &= m_1(A,\Delta)\mathrm{d}t+\sum_{k=1}^{m}\sigma_{1k}(A)\mathrm{d}B_k(t),\\ \mathrm{d}\Delta &= m_2(A,\Delta)\mathrm{d}t+\sum_{k=1}^{m}\sigma_{2k}(A)\mathrm{d}B_k(t),\end{aligned} \tag{1.162}$$

where

$$\begin{aligned}m_1(A,\Delta) &= \varepsilon\Big\langle F_1(A,\frac{r}{s}(\Omega t-\Delta),\Omega t)+\sum_{k,l=1}^{m}\int_{-\infty}^{0}(\frac{\partial G_{1k}}{\partial A}\Big|_t G_{1l}|_{t+\tau}\\ &\quad -\frac{s}{r}\frac{\partial G_{1k}}{\partial \Delta}\Big|_t G_{2l}|_{t+\tau})R_{kl}(\tau)\mathrm{d}\tau\Big\rangle_{\Omega t},\\ m_2(A,\Delta) &= \varepsilon\Big\langle \sigma\omega(A)-\frac{s}{r}F_2(A,\frac{r}{s}(\Omega t-\Delta),\Omega t)\\ &\quad -\sum_{k,l=1}^{m}\int_{-\infty}^{0}(\frac{s}{r}\frac{\partial G_{2k}}{\partial A}\Big|_t G_{1l}|_{t+\tau}+\frac{s^2}{r^2}\frac{\partial G_{2k}}{\partial \Delta}\Big|_t G_{2l}|_{t+\tau})R_{kl}(\tau)\mathrm{d}\tau\Big\rangle_{\Omega t},\\ \sum_{k=1}^{m}\sigma_{1k}\sigma_{1k} &= \varepsilon\Big\langle\sum_{k,l=1}^{m}\int_{-\infty}^{\infty}G_{1k}|_t G_{1l}|_{t+\tau}R_{kl}(\tau)\mathrm{d}\tau\Big\rangle_{\Omega t},\\ \sum_{k=1}^{m}\sigma_{2k}\sigma_{2k} &= \varepsilon\Big\langle\sum_{k,l=1}^{m}\frac{s^2}{r^2}\int_{-\infty}^{\infty}G_{2k}|_t G_{2l}|_{t+\tau}R_{kl}(\tau)\mathrm{d}\tau\Big\rangle_{\Omega t},\\ \sum_{k=1}^{m}\sigma_{1k}\sigma_{2k} &= \varepsilon\Big\langle-\sum_{k,l=1}^{m}\frac{s}{r}\int_{-\infty}^{\infty}G_{1k}|_t G_{2l}|_{t+\tau}R_{kl}(\tau)\mathrm{d}\tau\Big\rangle_{\Omega t},\\ \langle[\cdot]\rangle_{\Omega t} &= \frac{1}{2\pi}\int_{0}^{2\pi}[\cdot]\mathrm{d}(\Omega t).\end{aligned} \tag{1.163}$$

In order to obtain explicit expressions of Eq. (1.163), the functions in Eq. (1.157) are expanded into Fourier series with respect to Φ, and then $\Phi = r(\Omega t-\Delta)/s$ is used. In completing integration with respect to τ, the following Wiener-Hinchin relationship in Eq. (1.23) should be used.

$$S_{kl}(\omega) = \frac{1}{\pi}\int_{-\infty}^{0} R_{kl}(\tau)\cos\omega\tau\,\mathrm{d}\tau, \quad I_{kl}(\omega) = \frac{1}{\pi}\int_{-\infty}^{0} R_{kl}(\tau)\sin\omega\tau\,\mathrm{d}\tau. \tag{1.164}$$

The averaged FPK equation associated with the averaged Itô SDEs (1.162) is

$$\frac{\partial p}{\partial t} = -\frac{\partial}{\partial a}(a_1 p) - \frac{\partial}{\partial \delta}(a_2 p) + \frac{1}{2}\frac{\partial^2}{\partial a^2}(b_{11}p) + \frac{1}{2}\frac{\partial^2}{\partial \delta^2}(b_{22}p) \tag{1.165}$$

where the first and second derivative moments are

$$a_i = a_i(a,\delta) = m_i(A,\Delta)|_{A=a,\ \Delta=\delta}, \quad b_{ij} = b_{ij}(a,\delta) = \left.\sum_{k=1}^{m}\sigma_{ik}\sigma_{jk}\right|_{A=a,\ \Delta=\delta}. \tag{1.166}$$

In Eq. (1.165), $p = p(a, \delta, t)$ is joint PDF of $[\mathbf{A}, \boldsymbol{\Delta}]^{\mathrm{T}}$ with initial conditions

$$p(a,\delta,0) = p(a_0,\delta_0) \tag{1.167}$$

The boundary conditions are

$$\begin{aligned} &p = \text{finite}, \quad \text{when } a = 0, \\ &\partial p/\partial a \to 0, \quad \text{when } a \to \infty, \\ &p(a, \delta + 2n\pi, t) = p(a,\delta,t), \quad n = \pm 1, \pm 2, \ldots. \end{aligned} \tag{1.168}$$

Solving FPK Eq. (1.165) with $\partial p/\partial t = 0$ and boundary conditions in Eq. (1.168) yields the stationary PDF $p(a, \delta)$. Then the stationary PDF $p(q, p)$ for the original system (1.154) can be derived as follows

$$\begin{aligned} p(q,p) &= p(a,\delta)\left|\frac{\partial(a,\delta)}{\partial(q,p)}\right| \\ &= p(a,\delta)\left|\frac{\partial(a,\delta)}{\partial(a,\phi)}\right|\left|\frac{\partial(a,\phi)}{\partial(q,p)}\right| = \frac{s}{r}p(a,\delta)\left|\frac{\partial(a,\phi)}{\partial(q,p)}\right| \end{aligned} \tag{1.169}$$

The stationary marginal PDFs and statistics can be obtained from $p(q, p)$.

Example 1.7 Consider a Duffing oscillator with nonlinear damping and excited by combination of harmonic and wideband noises (Wu 2005; Wu and Zhu 2008). The motion equation is

$$\ddot{X} + (\beta_1 + \beta_2 X^2)\dot{X} + \omega_0^2 X + \alpha X^3 = E\cos\Omega t + \xi_1(t) + X\xi_2(t) \tag{1.170}$$

where $\beta_1, \beta_2, \omega_0, \alpha, E, \Omega$ are constants; $\xi_k(t)$ $(k = 1, 2)$ are two independent zero-mean stationary second-order rational noises with PSD

$$S_k(\omega) = \frac{D_k}{\pi}\frac{1}{\left(\omega^2 - \omega_k^2\right)^2 + 4\zeta_k^2\omega_k^2\omega^2}, \quad k = 1, 2 \tag{1.171}$$

where ζ_k, ω_k, D_k are constants.

Assume that $\alpha > 0$ and β_k, D_k, E are small quantities of ε order. Solution of system (1.170) is the form of Eq. (1.155) with $B = 0$. $\nu(a, \varphi)$ can be approximated by

$$\nu(a, \varphi) = b_0(a) + b_2(a)\cos 2\varphi + b_4(a)\cos 4\varphi + b_6(a)\cos 6\varphi \tag{1.172}$$

where

$$\begin{aligned}
b_0(a) &= \left(\omega_0^2 + 3\alpha a^2/4\right)^{1/2}\left(1 - \lambda^2/16\right),\\
b_2(a) &= \left(\omega_0^2 + 3\alpha a^2/4\right)^{1/2}\left(\lambda/2 + 3\lambda^3/64\right),\\
b_4(a) &= \left(\omega_0^2 + 3\alpha a^2/4\right)^{1/2}\left(-\lambda^2/16\right),\\
b_6(a) &= \left(\omega_0^2 + 3\alpha a^2/4\right)^{1/2}\left(\lambda^3/64\right),
\end{aligned} \tag{1.173}$$

The mean frequency is $\omega(a) = b_0(a)$. Using the transformation in Eq. (1.155), Eq. (1.170) becomes

$$\begin{aligned}
\frac{\mathrm{d}A}{\mathrm{d}t} &= F_1(A, \Phi, \Omega t) + G_{11}(A, \Phi)\xi_1(t) + G_{12}(A, \Phi)\xi_2(t),\\
\frac{\mathrm{d}\Phi}{\mathrm{d}t} &= \nu(A, \Phi) + F_2(A, \Phi, \Omega t) + G_{21}(A, \Phi)\xi_1(t) + G_{22}(A, \Phi)\xi_2(t),
\end{aligned} \tag{1.174}$$

where

$$\begin{aligned}
F_1 &= \frac{A}{g(A)}\left[\left(\beta_1 + \beta_2 A\cos^2\Phi\right)A\nu(A, \Phi)\sin\Phi + E\cos\Omega t\right]\nu(A, \Phi)\sin\Phi,\\
F_2 &= -\frac{1}{g(A)}\left[\left(\beta_1 + \beta_2 A\cos^2\Phi\right)A\nu(A, \Phi)\sin\Phi + E\cos\Omega t\right]\nu(A, \Phi)\cos\Phi,\\
G_{11} &= -\frac{A}{g(A)}\nu(A, \Phi)\sin\Phi, \quad G_{12} = -\frac{A^2}{g(A)}\nu(A, \Phi)\sin\Phi\cos\Phi,\\
G_{21} &= -\frac{1}{g(A)}\nu(A, \Phi)\cos\Phi, \quad G_{22} = -\frac{A}{g(A)}\nu(A, \Phi)\cos^2\Phi.
\end{aligned} \tag{1.175}$$

Consider the case of main external resonance. Suppose that the harmonic excitation frequency Ω is close to system mean frequency $\omega(a)$, i.e.,

$$\frac{\Omega}{\omega(a)} = 1 + \varepsilon\sigma \tag{1.176}$$

where $\varepsilon\sigma$ is a small detuning parameter. According to Eq. (1.160), introduce a new variable

$$\Delta = \Omega t - \Phi. \tag{1.177}$$

Completing the stochastic averaging and time averaging of Eq. (1.174) as described in Eqs. (1.162) and (1.163) yield the following averaged Itô SDEs

$$\begin{aligned} \mathrm{d}A &= m_1(A,\Delta)\mathrm{d}t + \sigma_{11}(A)\mathrm{d}B_1(t) + \sigma_{12}(A)\mathrm{d}B_2(t), \\ \mathrm{d}\Delta &= [\varepsilon\sigma\omega(A) + m_2(A,\Delta)]\mathrm{d}t + \sigma_{21}(A)\mathrm{d}B_1(t) + \sigma_{22}(A)\mathrm{d}B_2(t). \end{aligned} \tag{1.178}$$

where the drift and diffusion coefficients are

$$\begin{aligned} m_1 &= F_{10}(A,\Delta) + H_1(A), \\ m_2 &= F_{20}(A,\Delta), \\ F_{10}(A,\Delta) &= E\sin\Delta(2b_0(A) - \mathrm{b}_2(A))\big/4\left(\alpha A^2 + \omega_0^2\right) \\ &\quad - A\left[\beta_1\left(16\omega_0^2 + 10\alpha A^2\right) + A^2\beta_2\left(4\omega_0^2 + 3\alpha A^2\right)\right]\big/32\left(\alpha A^2 + \omega_0^2\right), \end{aligned}$$

$$\begin{aligned} F_{20}(A,\Delta) &= \Omega - b_0(A) + E\cos\Delta(2b_0(A) + \mathrm{b}_2(A))\big/4A\left(\alpha A^2 + \omega_0^2\right), \\ H_1(A) &= m_{11} + m_{12} + m_{13} + m_{14}, \\ m_{11} &= m_{111}S_1(\omega(A)) + m_{113}S_1(3\omega(A)) + m_{115}S_1(5\omega(A)) + m_{117}S_1(7\omega(A)), \\ m_{13} &= \overline{m}_{131}S_1(\omega(A)) + m_{133}S_1(3\omega(A)) + m_{135}S_1(5\omega(A)) + m_{137}S_1(7\omega(A)), \end{aligned}$$

$$\begin{aligned} m_{12} &= m_{122}S_2(2\omega(A)) + m_{124}S_2(4\omega(A)) + m_{126}S_2(6\omega(A)) \\ &\quad + m_{128}S_2(8\omega(A)), \\ m_{14} &= m_{142}S_2(2\omega(A)) + m_{144}S_2(4\omega(A)) + m_{146}S_2(6\omega(A)) \\ &\quad + m_{148}S_2(8\omega(A)), \end{aligned}$$

$$\begin{aligned} m_{111} &= \pi[b_2(A) - 2b_0(A)]\left\{2\alpha A[2b_0(A) - b_2(A)] + (A^2\alpha + \omega_0^2)\right. \\ &\quad \left.\times\left(\frac{\mathrm{d}b_2(A)}{\mathrm{d}A} - 2\frac{\mathrm{d}b_0(A)}{\mathrm{d}A}\right)\right\}\Big/\left[8\left(A^2\alpha + \omega_0^2\right)^3\right], \\ m_{113} &= \pi[b_2(A) - b_4(A)]\left\{2\alpha A[b_4(A) - b_2(A)] + (A^2\alpha + \omega_0^2)\right. \\ &\quad \left.\times\left(\frac{\mathrm{d}b_2(A)}{\mathrm{d}A} - \frac{\mathrm{d}b_4(A)}{\mathrm{d}A}\right)\right\}\Big/\left[8\left(A^2\alpha + \omega_0^2\right)^3\right], \end{aligned}$$

$$m_{115} = \pi[b_4(A) - b_6(A)] \times\left\{2\alpha A[b_6(A) - b_4(A)] + (A^2\alpha+\omega_0^2)\left(\frac{db_4(A)}{dA} - \frac{db_6(A)}{dA}\right)\right\}\Big/\left[8\left(A^2\alpha+\omega_0^2\right)^3\right],$$

$$m_{117} = \pi b_6(A)\left\{-2\alpha A b_6(A) + (A^2\alpha+\omega_0^2)\frac{db_6(A)}{dA}\right\}\Big/\left[8\left(A^2\alpha+\omega_0^2\right)^3\right],$$

$$m_{122} = \pi A[2b_0(A) - b_4(A)]\left\{[2b_0(A) - b_4(A)]\left(A^2\alpha-\omega_0^2\right) - A\left(A^2\alpha+\omega_0^2\right)\left(2\frac{db_0(A)}{dA} - \frac{db_4(A)}{dA}\right)\right\}\Big/\left[32\left(A^2\alpha+\omega_0^2\right)^3\right],$$

$$m_{124} = \pi A[b_2(A) - b_6(A)]\left\{[b_6(A) - b_2(A)]\left(A^2\alpha-\omega_0^2\right) + A\left(A^2\alpha+\omega_0^2\right)\left(\frac{db_2(A)}{dA} - \frac{db_6(A)}{dA}\right)\right\}\Big/\left[32\left(A^2\alpha+\omega_0^2\right)^3\right],$$

$$m_{126} = \pi A b_4(A)\left\{b_4(A)\left(A^2\alpha-\omega_0^2\right) + A\left(A^2\alpha+\omega_0^2\right)\frac{db_4(A)}{dA}\right\}\Big/\left[32\left(A^2\alpha+\omega_0^2\right)^3\right],$$

$$m_{128} = \pi A b_6(A)\left\{b_6(A)\left(A^2\alpha-\omega_0^2\right) + A\left(A^2\alpha+\omega_0^2\right)\frac{db_6(A)}{dA}\right\}\Big/\left[32\left(A^2\alpha+\omega_0^2\right)^3\right],$$

$$m_{131} = -\pi\left[b_2^2(A) - 4b_0^2(A)\right]\Big/\left[8A\left(A^2\alpha+\omega_0^2\right)^2\right],$$

$$m_{133} = -3\pi\left[b_4^2(A) - b_2^2(A)\right]\Big/\left[8A\left(A^2\alpha+\omega_0^2\right)^2\right],$$

$$m_{135} = -5\pi\left[b_6^2(A) - b_4^2(A)\right]\Big/\left[8A\left(A^2\alpha+\omega_0^2\right)^2\right],$$

$$m_{137} = 7\pi b_6^2(A)\Big/\left[8A\left(A^2\alpha+\omega_0^2\right)^2\right],$$

$$m_{142} = \pi A[2b_0(A) - b_4(A)][2b_0(A) + 2b_2(A) + b_4(A)]\Big/\left[16\left(A^2\alpha+\omega_0^2\right)^2\right],$$

$$m_{144} = \pi A[b_2(A) - b_6(A)][b_2(A) + 2b_4(A) + b_6(A)]\Big/\left[8\left(A^2\alpha+\omega_0^2\right)^2\right],$$

$$m_{146} = 3\pi A b_4(A)[b_4(A) + 2b_6(A)]\Big/\left[16\left(A^2\alpha+\omega_0^2\right)^2\right],$$

$$m_{148} = \pi A b_6^2(A)\Big/\left[4\left(A^2\alpha+\omega_0^2\right)^2\right],$$

$$\sigma_{11}(A)\sigma_{11}(A) + \sigma_{12}(A)\sigma_{12}(A) = b_{111} + b_{112},$$

$$\sigma_{21}(A)\sigma_{21}(A) + \sigma_{22}(A)\sigma_{22}(A) = b_{221} + b_{222},$$

$$\sigma_{i1}(A)\sigma_{j1}(A) + \sigma_{i2}(A)\sigma_{j2}(A) = 0, \quad i \neq j,$$

$$b_{111} = b_{1111}S_1(\omega(A)) + b_{1113}S_1(3\omega(A)) + b_{1115}S_1(5\omega(A)) + b_{1117}S_1(7\omega(A)),$$

$$
\begin{aligned}
b_{221} &= b_{2211}S_1(\omega(A)) + b_{2213}S_1(3\omega(A)) + b_{2215}S_1(5\omega(A)) \\
&\quad + b_{2217}S_1(7\omega(A)), \\
b_{112} &= b_{1122}S_2(2\omega(A)) + b_{1124}S_2(4\omega(A)) + b_{1126}S_2(6\omega(A)) \\
&\quad + b_{1128}S_2(8\omega(A)), \\
b_{222} &= b_{2220}S_2(0) + b_{2222}S_2(2\omega(A)) + b_{2224}S_2(4\omega(A)) + b_{2226}S_2(6\omega(A)) \\
&\quad + b_{2228}S_2(8\omega(A)),
\end{aligned}
$$

$$
\begin{aligned}
&b_{1111}=\pi[b_2(A)-2b_0(A)]^2\Big/4\left(\alpha A^2+\omega_0^2\right)^2,\\
&b_{1113}=\pi[b_2(A)-b_4(A)]^2\Big/4\left(\alpha A^2+\omega_0^2\right)^2,\\
&b_{1115}=\pi[b_4(A)-b_6(A)]^2\Big/4\left(\alpha A^2+\omega_0^2\right)^2,\\
&b_{1117}=\pi b_6^2(A)\Big/4\left(\alpha A^2+\omega_0^2\right)^2,\\
&b_{1122}=\pi A^2[b_4(A)-2b_0(A)]^2\Big/16\left(\alpha A^2+\omega_0^2\right)^2,\\
&b_{1124}=\pi A^2[b_2(A)-b_6(A)]^2\Big/16\left(\alpha A^2+\omega_0^2\right)^2,\\
&b_{1126}=\pi A^2 b_4^2(A)\Big/16\left(\alpha A^2+\omega_0^2\right)^2,\\
&b_{1128}=\pi A^2 b_6^2(A)\Big/16\left(\alpha A^2+\omega_0^2\right)^2,\\
&b_{2211}=\pi[2b_0(A)+b_2(A)]^2\Big/\left[4A^2\left(\alpha A^2+\omega_0^2\right)^2\right],\\
&b_{2213}=\pi[b_2(A)+b_4(A)]^2\Big/\left[4A^2\left(\alpha A^2+\omega_0^2\right)^2\right],\\
&b_{2215}=\pi[b_4(A)+b_6(A)]^2\Big/\left[4A^2\left(\alpha A^2+\omega_0^2\right)^2\right],\\
&b_{2217}=\pi b_6^2(A)\Big/\left[4A^2\left(\alpha A^2+\omega_0^2\right)^2\right],\\
&b_{2220}=\pi[2b_0(A)+b_2(A)]^2\Big/\left[8\left(\alpha A^2+\omega_0^2\right)^2\right],\\
&b_{2222}=\pi[2b_0(A)+2b_2(A)+b_4(A)]^2\Big/\left[16\left(\alpha A^2+\omega_0^2\right)^2\right],\\
&b_{2224}=\pi[b_2(A)+2b_4(A)+b_6(A)]^2\Big/\left[16\left(\alpha A^2+\omega_0^2\right)^2\right],\\
&b_{2226}=\pi[b_4(A)+2b_6(A)]^2\Big/\left[16\left(\alpha A^2+\omega_0^2\right)^2\right],\\
&b_{2228}=\pi b_6^2(A)\Big/\left[16\left(\alpha A^2+\omega_0^2\right)^2\right],
\end{aligned}
\tag{1.179}
$$

The reduced averaged FPK equation associated with averaged Itô SDE (1.178) is

$$0 = \left[-\frac{\partial}{\partial a}(a_1 p) - \frac{\partial}{\partial \delta}(a_2 p) + \frac{1}{2}\frac{\partial^2}{\partial a^2}(b_{11} p) + \frac{1}{2}\frac{\partial^2}{\partial \delta^2}(b_{22} p) \right] \tag{1.180}$$

where the first and second derivative moments are

$$\begin{aligned} a_1 &= a_1(a, \delta) = m_1(A, \Delta)|_{A=a,\ \Delta=\delta}, \quad a_2 = a_2(a, \delta) = m_2(A, \Delta)|_{A=a,\ \Delta=\delta}, \\ b_{11} &= b_{11}(a) = [b_{111}(A) + b_{112}(A)]|_{A=a}, \quad b_{22} = b_{22}(a) = [b_{221}(A) + b_{222}(A)]|_{A=a}. \end{aligned} \tag{1.181}$$

$p = p(a, \delta)$ in Eq. (1.180) is the stationary joint PDF of A, Δ. The boundary conditions for Eq. (1.180) are

$$\begin{aligned} &p = \text{finite}, \quad \text{when } a = 0, \\ &p, \quad \partial p / \partial a \to 0, \quad \text{when } a \to \infty, \\ &p(a, \delta + 2n\pi) = p(a, \delta), \quad n = \pm 1, \pm 2, \ldots. \end{aligned} \tag{1.182}$$

Equation (1.180) under boundary conditions in Eq. (1.182) can be solved by using the finite difference method.

In the numerical calculation, $S_1(\omega) = S_2(\omega) = S(\omega)$ and other parameters are (a) $\omega_k = 2.0$, $\zeta_k = 0.3$, $D_k = 0.1$ ($k = 1, 2$) and (b) $\omega_k = 5.0$, $\zeta_k = 0.5$, $D_k = 2.0$ ($k = 1, 2$). It is seen from Fig. 1.23 that case (a) is quite narrow band, and case (b) is typical wide band.

The numerical results for $p(a, \delta)$, $p(a)$ and $p(\delta)$ obtained by using the stochastic averaging method and from Monte Carlo simulation are shown in Fig. 1.24, 1.25,

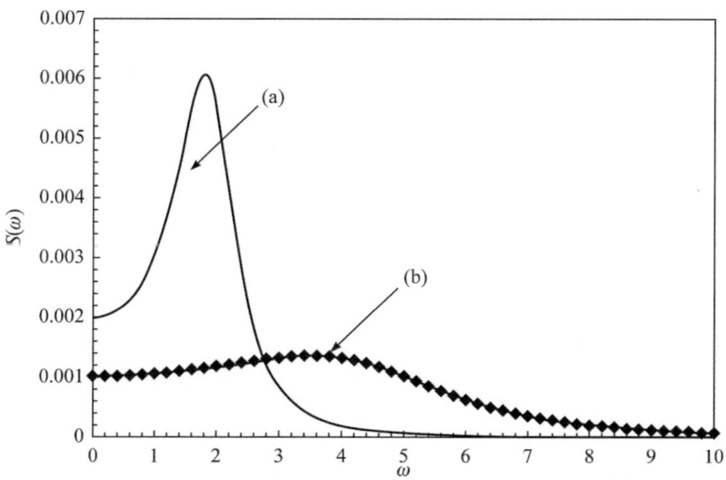

Fig. 1.23 PSD of second-order rational noise (1.171). **a** PSD used to calculate the results in Figs. 1.25 and 1.27, $\omega_k = 2.0$, $\zeta_k = 0.3$, $D_k = 0.1$ ($k = 1, 2$); **b** PSD used to calculate the results in Figs. 1.24 and 1.26, $\omega_k = 5.0$, $\zeta_k = 0.5$, $D_k = 2.0$ ($k = 1, 2$)

1.26 and 1.27. It can be observed from Figs. 1.24 and 1.25 that for the case of non-zero linear stiffness $\omega_0 \neq 0$, the stochastic averaging method provides good results for both (a) and (b) of excitation PSDs. When linear stiffness is zero, $\omega_0 = 0$, the stochastic averaging method can still produce quite satisfactory results for case (b) of excitation PSD, as shown in Fig. 1.26. For case (a) of excitation PSD, however, results from the stochastic averaging method deviate significantly from simulation results, as shown in Fig. 1.27.

It is known that Duffing oscillators under harmonic excitation can exhibit a phenomenon of amplitude jumps. Combination of harmonic and wideband noises is equivalent to a narrowband noise. Duffing oscillators excited by this combined excitation may undergo random jumps, as indicated by a bimodal PDF, meaning transition of system response from one peak (i.e., one more possible motion) to another peak (i.e., another more possible motion) or vice versa (Zhu 1998). The bimodal stationary joint PDF of displacement amplitude and phase angle under combination of harmonic and wideband noise is shown in Fig. 1.28, indicating random jumps. Samples of displacement and velocity shown in Fig. 1.29 confirm this conclusion.

Noise intensity, ratio of harmonic frequency to natural frequency, harmonic noise amplitude and intensity of nonlinearity, etc. can affect the occurrence of random

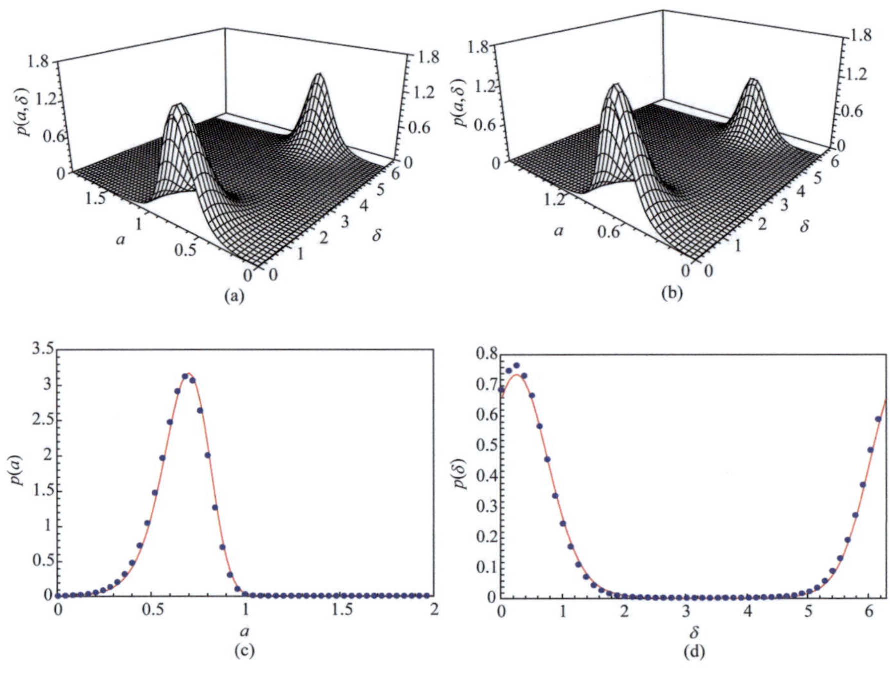

Fig. 1.24 Joint PDF $p(a, \delta)$ and marginal PDFs $p(a)$ and $p(\delta)$ of amplitude A and phase angle difference Δ. **a** Result from the stochastic averaging method; **b** result from Monte Carlo simulation; **c** $p(a)$; **d** $p(\delta)$. PSD of $\xi_1(t)$, $\xi_2(t)$ is shown in Fig. 1.23b. ——, results from the stochastic averaging method; ●, results from Monte Carlo simulation (Wu 2005)

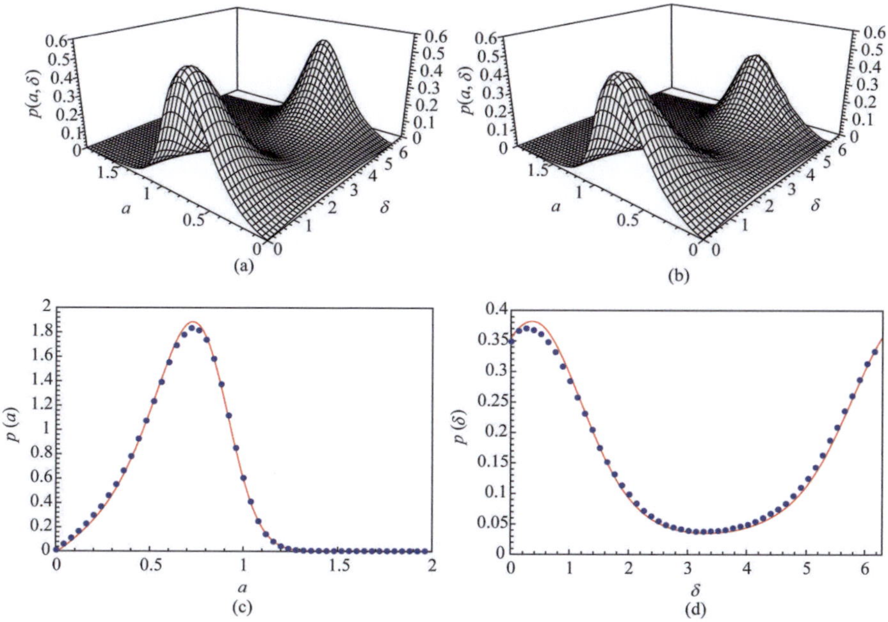

Fig. 1.25 Joint PDF $p(a, \delta)$ and marginal PDFs $p(a)$ and $p(\delta)$ of amplitude A and phase angle difference Δ. **a** Result from the stochastic averaging method; **b** result from Monte Carlo simulation; **c** $p(a)$; **d** $p(\delta)$. PSD of $\xi_1(t)$, $\xi_2(t)$ is shown in Fig. 1.23a. Other parameters are the same as those in Fig. 1.24. ——, results from the stochastic averaging method; ●, results from Monte Carlo simulation (Wu 2005)

jumps. The phenomena that random jumps appear or disappear due to parameter change is named by bifurcation of random jumps. Compared with Fig. 1.28, Fig. 1.30 shows that increasing noise intensity results in two higher and closer peaks in $p(a, \delta)$, indicating random jumps occur more likely. As the harmonic frequency decreases to 1.2 (Fig. 1.31) or increases to 1.8 (Fig. 1.32), harmonic excitation amplitude decreases to 0.1 (Fig. 1.33) or increases to 0.4 (Fig. 1.34), intensity of nonlinearity decreases to 0.6 (Fig. 1.35) or increases to 4.0 (Fig. 1.36), the joint PDF or marginal PDF $p(a)$ becomes unimodal, indicating random jumps disappear.

Example 1.8 Consider a nonlinear Duffing-Rayleigh oscillator excited by combination of harmonic and wideband noises (Wu and Zhu 2009). The motion equation is

$$\ddot{X} + \omega_0^2 X + \alpha X^3 + (\beta_1 + \beta_2 \dot{X}^2)\dot{X} = EX \cos \Omega\, t + \xi_1(t) + X^2 \xi_2(t) \tag{1.183}$$

where β_1, β_2, ω_0, α, E, Ω are constants, $\xi_1(t)$, $\xi_2(t)$ are wideband noises with rational PSD in Eq. (1.171), β_1, β_2, E, D_k are small quantities of ε order. Consider primary parametric resonance

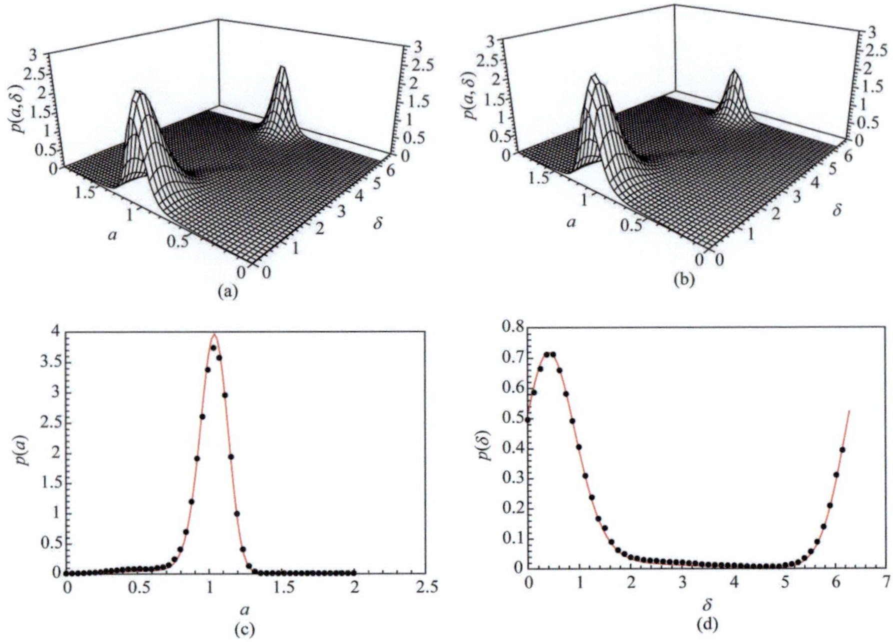

Fig. 1.26 Joint PDF $p(a, \delta)$ and marginal PDFs $p(a)$ and $p(\delta)$ of amplitude A and phase angle difference Δ. **a** Result from the stochastic averaging method; **b** result from Monte Carlo simulation; **c** $p(a)$; **d** $p(\delta)$. $\omega_0 = 0$. PSD of $\xi_1(t)$, $\xi_2(t)$ is shown in Fig. 1.23b. Other parameters are the same as those in Fig. 1.24. ——, results from the stochastic averaging method; ●, results from Monte Carlo simulation (Wu 2005)

$$\frac{\Omega}{\omega(a)} = 2 + \varepsilon\sigma \tag{1.184}$$

Introduce a new variable

$$\Delta = \Omega t - 2\Phi. \tag{1.185}$$

Applying stochastic averaging and time averaging according to Eq. (1.155) through Eq. (1.162) yields the following averaged Itô SDEs

$$\begin{aligned} \mathrm{d}A &= m_1(A, \Delta)\mathrm{d}t + \sigma_{11}(A)\mathrm{d}B_1(t) + \sigma_{12}(A)\mathrm{d}B_2(t), \\ \mathrm{d}\Delta &= [\varepsilon\sigma\omega(A) + m_2(A, \Delta)]\mathrm{d}t + \sigma_{21}(A)\mathrm{d}B_1(t) + \sigma_{22}(A)\mathrm{d}B_2(t), \end{aligned} \tag{1.186}$$

where drift and diffusion coefficients are

$$\begin{aligned} m_1(A, \Delta) &= F_{10}(A, \Delta) + H_1(A), \\ m_2(A, \Delta) &= F_{20}(A, \Delta). \end{aligned}$$

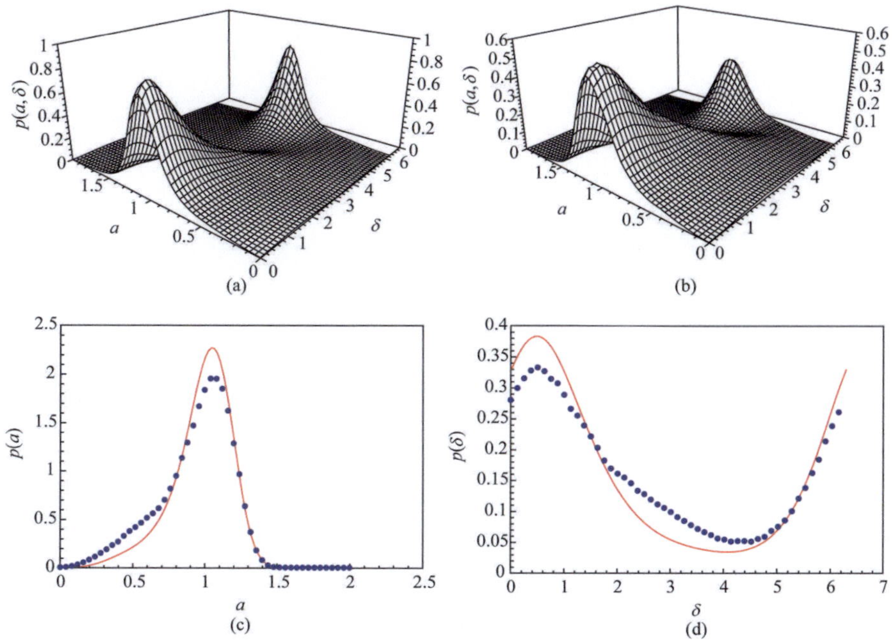

Fig. 1.27 Joint PDF $p(a, \delta)$ and marginal PDFs $p(a)$ and $p(\delta)$ of amplitude A and phase angle difference Δ. **a** Result from the stochastic averaging method; **b** result from Monte Carlo simulation; **c** $p(a)$; **d** $p(\delta)$. $\omega_0 = 0$. PSD of $\xi_1(t)$, $\xi_2(t)$ is shown in Fig. 1.23a. Other parameters are the same as those in Fig. 1.24. ——, results from the stochastic averaging method; ●, results from Monte Carlo simulation (Wu 2005)

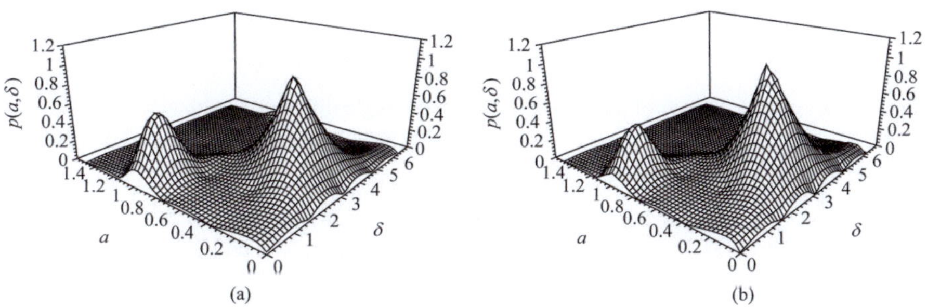

Fig. 1.28 Joint PDF $p(a, \delta)$ of amplitude A and phase angle difference Δ. **a** Results from the stochastic averaging method; **b** result from Monte Carlo simulation; system parameters $\omega_0 = 1.0$, $\alpha = 2.0$, $\beta_1 = 0.1$, $\beta_2 = 0$, $E = 0.2$, $\Omega = 1.5$; noise PSD parameters $\omega_k = 5.0$, $\zeta_k = 0.5$, $D_k = 2.0$ (k= 1, 2) (Wu and Zhu 2008)

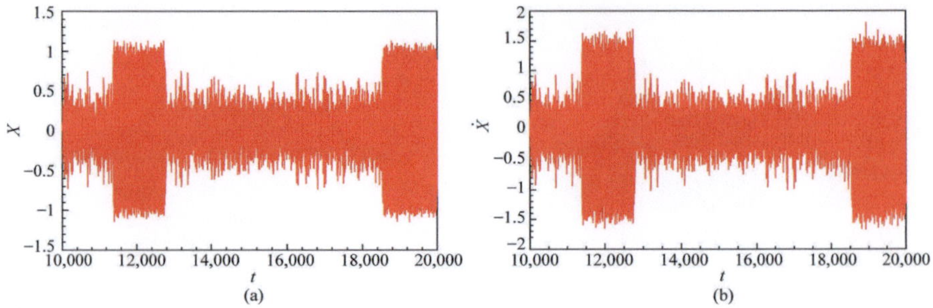

Fig. 1.29 Samples of random jumps. **a** displacement sample; **b** velocity sample; parameters are the same as those in Fig. 1.28 (Wu and Zhu 2008)

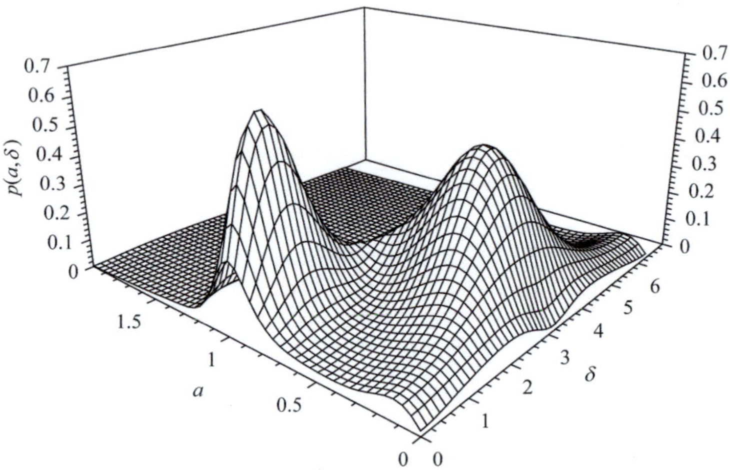

Fig. 1.30 Joint PDF $p(a, \delta)$ of amplitude A and phase angle difference Δ obtained by using the stochastic averaging method. Other parameters are the same as those in Fig. 1.28 (Wu and Zhu 2008)

$$
\begin{aligned}
F_{10}(A,\Delta) &= A\big\{67\alpha^2\beta_2A^6 + 256\beta_1\omega_0^2 + 224A^4\alpha\beta_2\omega_0^2 + 32A^2\big(5\alpha\beta_1 + 6\beta_2\omega_0^4\big) \\
&\quad -64[2b_0(A) - b_4(A)]E\sin\Delta\big\}\big/\,512\big(\alpha A^2 + \omega_0^2\big), \\
F_{20}(A,\Delta) &= \Omega - 2b_0(A) + [2b_0(A) + 2b_2(A) + b_4(A)]E\cos\Delta\big/\,4\big(\alpha A^2 + \omega_0^2\big), \\
H_1(A) &= m_{11} + m_{12} + m_{13} + m_{14},
\end{aligned}
$$

$$
\begin{aligned}
m_{11} &= m_{111}S_1(\omega(A)) + m_{113}S_1(3\omega(A)) + m_{115}S_1(5\omega(A)) \\
&\quad + m_{117}S_1(7\omega(A)), \\
m_{13} &= m_{131}S_1(\omega(A)) + m_{133}S_1(3\omega(A)) + m_{135}S_1(5\omega(A)) \\
&\quad + m_{137}S_1(7\omega(A)), \\
m_{12} &= m_{121}S_2(\omega(A)) + m_{123}S_2(3\omega(A)) + m_{125}S_2(5\omega(A))
\end{aligned}
$$

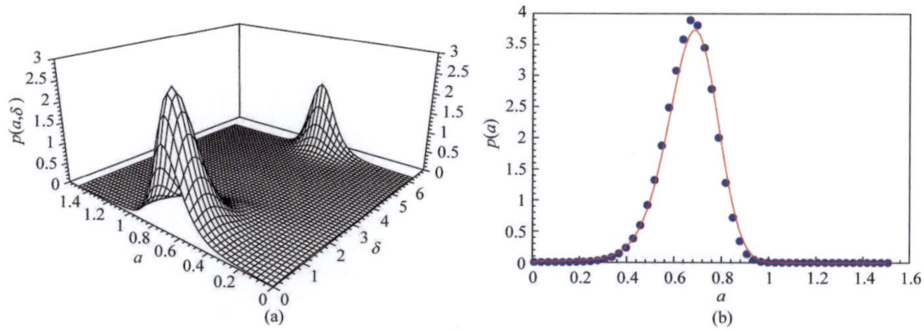

Fig. 1.31 **a** Joint PDF $p(a, \delta)$ of amplitude A and phase angle difference Δ obtained by using the stochastic averaging method; **b** Marginal PDF $p(a)$ of amplitude A, ——, result from the stochastic averaging method; ●, result from Monte Carlo simulation. $\Omega = 1.2$. Other parameters are the same as those in Fig. 1.28 (Wu 2005)

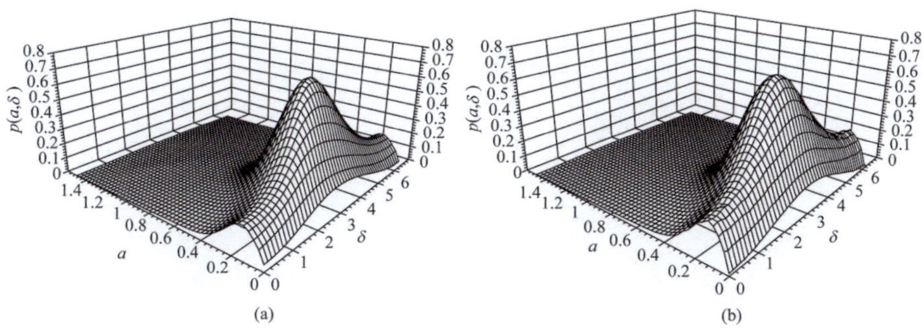

Fig. 1.32 Joint PDF $p(a, \delta)$ of amplitude A and phase angle difference Δ. **a** Result from the stochastic averaging method; **b** result from Monte Carlo simulation. $\Omega = 1.8$. Other parameters are the same as those in Fig. 1.28 (Wu 2005)

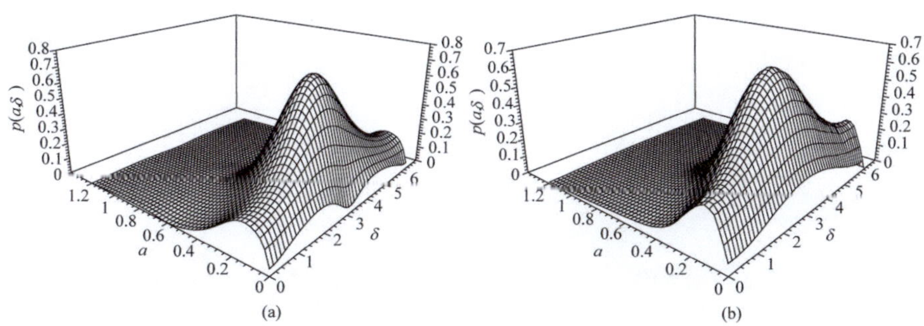

Fig. 1.33 Joint PDF $p(a, \delta)$ of amplitude A and phase angle difference Δ. **a** Result from the stochastic averaging method; **b** result from Monte Carlo simulation. $E = 0.1$. Other parameters are the same as those in Fig. 1.26 (Wu 2005)

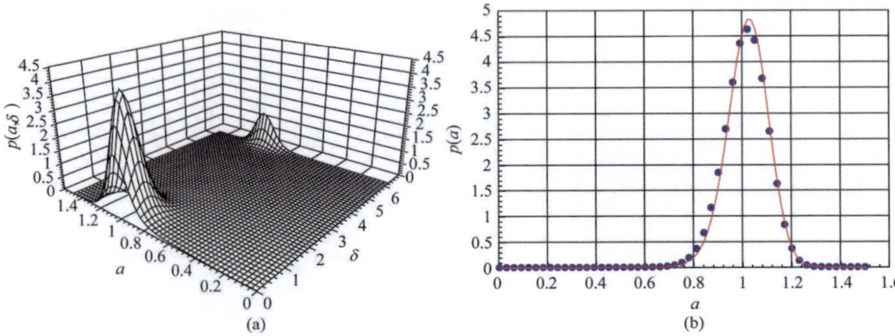

Fig. 1.34 **a** Joint PDF $p(a, \delta)$ of amplitude A and phase angle difference Δ obtained by using the stochastic averaging method; **b** Marginal PDF $p(a)$ of amplitude A, ——, result from the stochastic averaging method; ●, result from Monte Carlo simulation. $E = 0.4$. Other parameters are the same as those in Fig. 1.26 (Wu 2005)

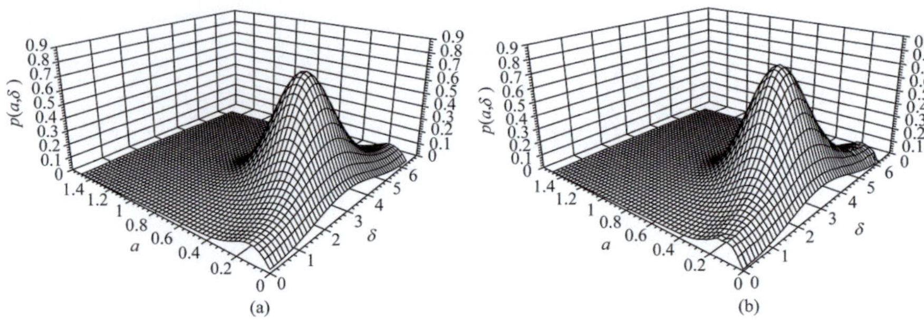

Fig. 1.35 Joint PDF $p(a, \delta)$ of amplitude A and phase angle difference Δ. **a** Result from the stochastic averaging method; **b** result from Monte Carlo simulation. $\alpha = 0.6$. Other parameters are the same as those in Fig. 1.26 (Wu 2005)

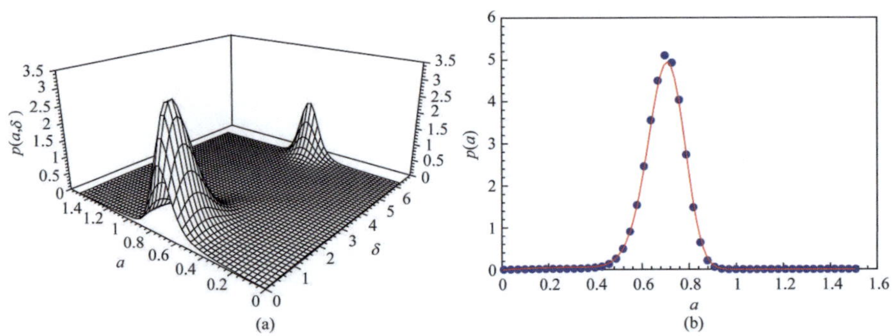

Fig. 1.36 **a** Joint PDF $p(a, \delta)$ of amplitude A and phase angle difference Δ obtained by the stochastic averaging method; **b** Marginal PDF $p(a)$ of amplitude A, ——, results from the stochastic averaging method; ●, result from Monte Carlo simulation. $\alpha = 4$. Other parameters are the same as those in Fig. 1.26 (Wu 2005)

$$
\begin{aligned}
&+ m_{127}S_2(7\omega(A)) + m_{129}S_2(9\omega(A)),\\
m_{14} &= m_{141}S_2(\omega(A)) + m_{143}S_2(3\omega(A)) + m_{145}S_2(5\omega(A))\\
&+ m_{147}S_2(7\omega(A)) + m_{149}S_2(9\omega(A)),
\end{aligned}
$$

$$
m_{121} = \pi A^3[2b_0(A) - b_4(A)]\Big\{ 2\omega_0^2[2b_0(A) - b_4(A)]
$$

$$
+A\left(A^2\alpha+\omega_0^2\right)\left(2\frac{db_0(A)}{dA} - \frac{db_4(A)}{dA}\right)\Big\}\Big/\left[128\left(A^2\alpha+\omega_0^2\right)^3\right],
$$

$$
m_{123} = \pi A^3[2b_0(A) + b_2(A) - b_4(A) - b_6(A)]\Big\{2\omega_0^2[2b_0(A) + b_2(A) - b_4(A) - b_6(A)]
$$

$$
+A\left(A^2\alpha+\omega_0^2\right)\left(2\frac{db_0(A)}{dA} + \frac{db_2(A)}{dA}\right)
$$

$$
-A\left(A^2\alpha+\omega_0^2\right)\left[\frac{db_4(A)}{dA} + \frac{db_6(A)}{dA}\right]\Big\}\Big/\left[128\left(A^2\alpha+\omega_0^2\right)^3\right]
$$

$$
m_{125} = \pi A^3[b_2(A) + b_4(A) - b_6(A)]\Big\{ 2\omega_0^2[b_2(A) + b_4(A) - b_6(A)],
$$

$$
+A\left(A^2\alpha+\omega_0^2\right)\left(\frac{db_2(A)}{dA} + \frac{db_4(A)}{dA}\right) - A\left(A^2\alpha+\omega_0^2\right)\frac{db_6(A)}{dA}\Big\}\Big/\left[128\left(A^2\alpha+\omega_0^2\right)^3\right],
$$

$$
m_{127} = \pi A^3[b_4(A) + b_6(A)]\Big\{ 2\omega_0^2[b_4(A) + b_6(A)]
$$

$$
+A\left(A^2\alpha+\omega_0^2\right)\left(\frac{db_4(A)}{dA} + \frac{db_6(A)}{dA}\right)\Big\}\Big/\left[128\left(A^2\alpha+\omega_0^2\right)^3\right],
$$

$$
m_{129} = \pi A^3 b_6(A)\Big\{ 2\omega_0^2 b_6(A) + A\left(A^2\alpha+\omega_0^2\right)\frac{db_6(A)}{dA}\Big\}\Big/\left[128\left(A^2\alpha+\omega_0^2\right)^3\right],
$$

$$
m_{141} = \pi A^3[2b_0(A) - b_4(A)][6b_0(A) + 4b_2(A) + b_4(A)]\Big/\left[128\left(A^2\alpha+\omega_0^2\right)^2\right],
$$

$$
m_{143} = 3\pi A^3[2b_0(A) + b_2(A) - b_4(A) - b_6(A)][2b_0(A) + 3b_2(A) + 3b_4(A)
$$

$$
+b_6(A)]\Big/\left[128\left(A^2\alpha+\omega_0^2\right)^2\right],
$$

$$
m_{145} = 5\pi A^3[b_2(A) + b_4(A) - b_6(A)][b_2(A) + 3b_4(A) + 3b_6(A)]\Big/\left[128\left(A^2\alpha+\omega_0^2\right)^2\right],
$$

$$
m_{147} = 7\pi A^3[b_4(A) + b_6(A)][b_4(A) + 3b_6(A)]\Big/\left[128\left(A^2\alpha+\omega_0^2\right)^2\right],
$$

$$
m_{149} = 9\pi A^3[b_6(A)]^2\Big/\left[128\left(A^2\alpha+\omega_0^2\right)^2\right],
$$

$$
\begin{aligned}
&\sigma_{11}(A)\sigma_{11}(A) + \sigma_{12}(A)\sigma_{12}(A) = b_{111} + b_{112},\\
&\sigma_{21}(A)\sigma_{21}(A) + \sigma_{22}(A)\sigma_{22}(A) = b_{221} + b_{222},\\
&\sigma_{i1}(A)\sigma_{j1}(A) + \sigma_{i2}(A)\sigma_{j2}(A) = 0, \quad i \neq j,\\
&b_{111} = b_{1111}S_1(\omega(A)) + b_{1113}S_1(3\omega(A)) + b_{1115}S_1(5\omega(A)) + b_{1117}S_1(7\omega(A)),\\
&b_{221} = b_{2211}S_1(\omega(A)) + b_{2213}S_1(3\omega(A)) + b_{2215}S_1(5\omega(A)) + b_{2217}S_1(7\omega(A)),\\
&b_{112} = b_{1121}S_2(\omega(A)) + b_{1123}S_2(3\omega(A)) + b_{1125}S_2(5\omega(A)) + b_{1127}S_2(7\omega(A))\\
&\quad + b_{1129}S_2(9\omega(A)),
\end{aligned}
$$

$$
\begin{aligned}
b_{222} = {} & b_{2221}S_2(\omega(A)) + b_{2223}S_2(3\omega(A)) + b_{2225}S_2(5\omega(A)) + b_{2227}S_2(7\omega(A)) \\
& + b_{2229}S_2(9\omega(A)),
\end{aligned}
$$

$$
\begin{aligned}
b_{1121} &= \pi A^4[-2b_0(A) + b_4(A)]^2 \Big/ \left[64\left(\alpha A^2 + \omega_0^2\right)^2\right], \\
b_{1123} &= \pi A^4[-2b_0(A) - b_2(A) + b_4(A) + b_6(A)]^2 \Big/ \left[64\left(\alpha A^2 + \omega_0^2\right)^2\right], \\
b_{1125} &= \pi A^4[b_2(A) + b_4(A) + b_6(A)]^2 \Big/ \left[64\left(\alpha A^2 + \omega_0^2\right)^2\right], \\
b_{1127} &= \pi A^4[b_4(A) + b_6(A)]^2 \Big/ \left[64\left(\alpha A^2 + \omega_0^2\right)^2\right], \\
b_{1129} &= \pi A^4[b_6(A)]^2 \Big/ \left[64\left(\alpha A^2 + \omega_0^2\right)^2\right], \\
b_{2221} &= \pi A^2[6b_0(A) + 4b_2(A) + b_4(A)]^2 \Big/ \left[64\left(\alpha A^2 + \omega_0^2\right)^2\right], \\
b_{2223} &= \pi A^2[2b_0(A) + 3b_2(A) + 3b_4(A) + b_6(A)]^2 \Big/ \left[64\left(\alpha A^2 + \omega_0^2\right)^2\right], \qquad (1.187)
\end{aligned}
$$

$$
\begin{aligned}
b_{2225} &= \pi A^2[b_2(A) + 3b_4(A) + 3b_6(A)]^2 \Big/ \left[64\left(\alpha A^2 + \omega_0^2\right)^2\right], \\
b_{2227} &= \pi A^2[b_4(A) + 3b_6(A)]^2 \Big/ \left[64\left(\alpha A^2 + \omega_0^2\right)^2\right], \\
b_{2229} &= \pi A^2[b_6(A)]^2 \Big/ \left[64\left(\alpha A^2 + \omega_0^2\right)^2\right],
\end{aligned}
$$

where m_{1jk} $(j = 1, 3;\ k = 1, 3, 5, 7)$ and b_{jj1k} $(j = 1, 2;\ k = 1, 3, 5, 7)$ are not affected by the parametric noise excitation term $X^2\xi_2(t)$. Their expressions are similar to those expressions for m_{1jk}, b_{jj1k} in Eq. (1.179).

Solving the reduced averaged FPK equation associated with averaged Itô Eq. (1.186) yields joint stationary PDF $p(a, \delta)$. Figure 1.37 provides a set of numerical results, indicating good agreement between the results obtained by using the stochastic averaging method and those from Monte Carlo simulation.

1.3.2 MDOF System

Consider the following n-DOF quasi-integrable Hamiltonian system excited by combination of harmonic and wideband noises (Huang and Zhu 2004a; Hu and Zhu 2015)

$$
\begin{aligned}
\dot{Q}_i &= P_i, \\
\dot{P}_i &= -g_i(Q_i) - \varepsilon \sum_{j=1}^{n} c_{ij}(\mathbf{Q}, \mathbf{P})P_j + \varepsilon \sum_{r=1}^{s} h_{ir}(\mathbf{Q}, \mathbf{P}) \cos(\Omega_r t
\end{aligned}
$$

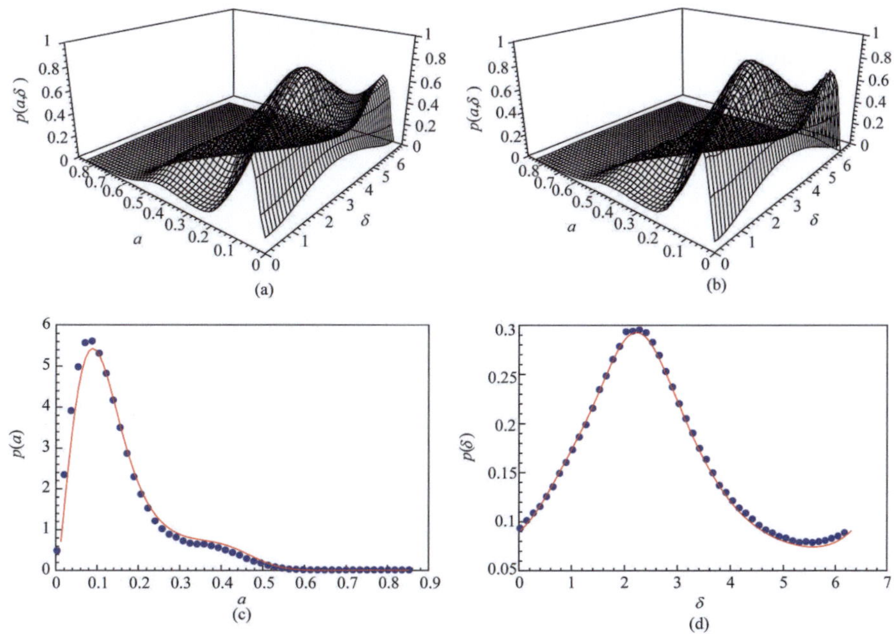

Fig. 1.37 Joint PDF $p(a, \delta)$ of amplitude A and phase angle difference Δ. **a** result from the stochastic averaging method; **b** result from Monte Carlo simulation; **c** $p(a)$; **d** $p(\delta)$. ——, results from the stochastic averaging method; ●, results from Monte Carlo simulation. System parameters $\omega_0 = 1.0, \alpha = 1.0, \beta_1 = \beta_2 = 0.05, E = 0.1, \Omega = 2.1$; noise PSD parameters $\omega_k = 5.0, \zeta_k = 0.5$, $D_k = 0.2$ ($k = 1, 2$) (Wu and Zhu 2009)

$$
\begin{aligned}
&+ \beta_r) + \varepsilon^{1/2} \sum_{k=1}^{m} f_{ik}(\mathbf{Q}, \mathbf{P}) \xi_k(t), \\
&i = 1, 2, \ldots, n,
\end{aligned} \tag{1.188}
$$

where $g_i(Q_i)$ are nonlinear restoring forces, ε is a small parameter, εc_{ij} are damping coefficients, εh_{ir} are amplitudes of harmonic excitations with frequencies Ω_r and phase angle β_r, $\xi_k(t)$ are wideband noises with cross correlation functions $R_{kl}(\tau)$ and cross PSDs $S_{kl}(\omega)$, $\varepsilon^{1/2} f_{ik}$ are stochastic excitation amplitudes. When h_{ir} or f_{ik} are constants, the excitations are external. When h_{ir} or f_{ik} are functions of state $\mathbf{Q}, \mathbf{P}$, the excitations are parametric.

The potential function of each subsystem in system (1.188) is

$$
U_i(Q_i) = \int_0^{Q_i} g_i(x) \mathrm{d}x. \tag{1.189}
$$

Assume that $g_i(Q_i)$ and $U_i(Q_i)$ $(i = 1, 2, \ldots, n)$ meet the four conditions below Eq. (1.5). Introduce the transformation

$$\begin{aligned}
&Q_i(t) = A_i \cos\Phi_i(t) + B_i, \quad P_i(t) = -A_i\nu_i(A_i, \Phi_i)\sin\Phi_i(t),\\
&\Phi_i(t) = \Gamma_i(t) + \Theta_i(t), \quad i = 1, 2, \ldots, n,
\end{aligned} \tag{1.190}$$

where the relations of B_i and A_i are $U_i(A_i+B_i) = U_i(A_i-B_i) = H_i$. The instantaneous frequencies $\nu_i(A_i, \Phi_i)$ are

$$\begin{aligned}
&\nu_i(A_i, \Phi_i) = \frac{\mathrm{d}\Gamma_i}{\mathrm{d}t} = \sqrt{\frac{2[U_i(A_i + B_i) - U_i(A_i\cos\Phi_i + B_i)]}{A_i^2\sin^2\Phi_i}},\\
&i = 1, 2, \ldots, n.
\end{aligned} \tag{1.191}$$

Defining mean frequency $\omega_i(A_i)$ of the ith subsystem as $\omega_i(A_i) = \int_0^{2\pi}\nu_i(A_i, \Phi_i)\mathrm{d}\Phi_i \Big/ 2\pi$, the following approximate relations will be used

$$\Phi_i(t) \approx \omega_i(A_i)t + \Theta_i(t). \tag{1.192}$$

Using Eq. (1.190) as a transformation from $\mathbf{Q}, \mathbf{P}$ to $\mathbf{A}, \boldsymbol{\Phi}$, the original system (1.188) becomes

$$\begin{aligned}
&\frac{\mathrm{d}A_i}{\mathrm{d}t} = \varepsilon F_i^{(1)}(\mathbf{A}, \boldsymbol{\Phi}, \boldsymbol{\Omega}) + \varepsilon^{1/2}\sum_{k=1}^{m} G_{ik}^{(1)}(\mathbf{A}, \boldsymbol{\Phi})\xi_k(t),\\
&\frac{\mathrm{d}\Phi_i}{\mathrm{d}t} = \nu_i(A_i, \Phi_i) + \varepsilon F_i^{(2)}(\mathbf{A}, \boldsymbol{\Phi}, \boldsymbol{\Omega}) + \varepsilon^{1/2}\sum_{k=1}^{m} G_{ik}^{(2)}(\mathbf{A}, \boldsymbol{\Phi})\xi_k(t),\\
&i = 1, 2, \ldots, n,
\end{aligned} \tag{1.193}$$

where

$$\begin{aligned}
&\mathbf{A} = [A_1, A_2, \ldots, A_n]^{\mathrm{T}}, \quad \boldsymbol{\Phi} = [\Phi_1, \Phi_2, \ldots, \Phi_n]^{\mathrm{T}}, \quad \boldsymbol{\Omega} = [\Omega_1, \Omega_2, \ldots, \Omega_s]^{\mathrm{T}},\\
&F_i^{(1)} = \frac{-A_i}{g_i(A_i + B_i)(1 + d_i)}[\sum_{j=1}^{n} c_{ij}(A_i\cos\Phi_i + B_i, -A_i\nu_i\sin\Phi_i)A_j\nu_j\sin\Phi_j\\
&\quad + \sum_{r=1}^{s} h_{ir}(A_i\cos\Phi_i + B_i, -A_i\nu_i\sin\Phi_i)\cos(\Omega_r t + \beta_r)]\nu_i\sin\Phi_i,\\
&F_i^{(2)} = \frac{-1}{g_i(A_i + B_i)(1 + d_i)}[\sum_{j=1}^{n} c_{ij}(A_i\cos\Phi_i + B_i, -A_i\nu_i\sin\Phi_i)A_j\nu_j\sin\Phi_j
\end{aligned}$$

$$+\sum_{r=1}^{s} h_{ir}(A_i\cos\Phi_i + B_i, -A_i\nu_i\sin\Phi_i)\cos(\Omega_r t+\beta_r)]\nu_i(\cos\Phi_i + d_i),$$

$$\begin{aligned}
G_{ik}^{(1)} &= \frac{-A_i}{g_i(A_i+B_i)(1+d_i)} f_{ik}(A_i\cos\Phi_i + B_i, -A_i\nu_i\sin\Phi_i)\nu_i\sin\Phi_i,\\
G_{ik}^{(2)} &= \frac{-1}{g_i(A_i+B_i)(1+d_i)} f_{ik}(A_i\cos\Phi_i + B_i, -A_i\nu_i\sin\Phi_i)\nu_i(\cos\Phi_i + d_i),\\
d_i &= \frac{\mathrm{d}B_i}{\mathrm{d}A_i} = \frac{g_i(-A_i+B_i)+g_i(A_i+B_i)}{g_i(-A_i+B_i)-g_i(A_i+B_i)}.
\end{aligned}\tag{1.194}$$

The quasi-integrable Hamiltonian system (1.188) may exhibit internal resonance between different subsystems and external resonance between the harmonic excitations and subsystems. In case without external resonance, the effect of harmonic excitation can be neglected in the first approximation. In this case, system (1.188) can be analyzed using the stochastic averaging method of quasi-integrable Hamiltonian systems under wideband noise excitation developed in Sect. 1.1. In the following, the case of both external resonance and internal resonance will be studied.

Assume that the n mean natural frequencies $\omega_i(a_i)$ $(i = 1, 2, \cdots, n)$ and the harmonic excitation frequencies $\Omega_1, \Omega_2, \ldots, \Omega_s$ satisfy the following resonant relations

$$\sum_{r=1}^{s} E_r^u\Omega_r + \sum_{i=1}^{n} I_i^u\omega_i = \varepsilon\chi_u, \quad u = 1,2,\ldots,\mu, \tag{1.195}$$

where E_r^u, I_i^u are integers and at least one of E_r^u and one of I_i^u are not zero; $\varepsilon\chi_u$ is a small detuning parameter. Note that Eq. (1.195) is applicable for only external resonant case or both internal and external resonant case. Introduce the following combinations of angle variables

$$\Delta_u = \sum_{r=1}^{s} E_r^u\Omega_r t + \sum_{i=1}^{n} I_i^u\Phi_i, \quad u = 1, 2, \ldots, \mu. \tag{1.196}$$

Deriving motion equations for $\Delta_1, \Delta_2, \ldots, \Delta_\mu$ from Eq. (1.193) using (1.196) and replacing $\Phi_1, \Phi_2, \ldots, \Phi_\mu$ with $\Delta_1, \Delta_2, \ldots, \Delta_\mu$ and $\Omega_1 t, \Omega_2 t, \ldots, \Omega_s t$ yield

$$\begin{aligned}
\frac{\mathrm{d}A_i}{\mathrm{d}t} &= \varepsilon\overline{F}_i^{(1)}(\mathbf{A}, \boldsymbol{\Delta}, \boldsymbol{\Phi}', \boldsymbol{\Omega}) + \varepsilon^{1/2}\sum_{k=1}^{m}\overline{G}_{ik}^{(1)}(\mathbf{A}, \boldsymbol{\Delta}, \boldsymbol{\Phi}')\xi_k(t),\\
\frac{\mathrm{d}\Delta_u}{\mathrm{d}t} &= \sum_{r=1}^{s} E_r^u\Omega_r + \sum_{i=1}^{n} I_i^u\nu_i(A_i, \Phi_i) + \varepsilon\sum_{i=1}^{n} I_i^u\overline{F}_i^{(2)}(\mathbf{A}, \boldsymbol{\Delta}, \boldsymbol{\Phi}', \boldsymbol{\Omega})\\
&\quad + \varepsilon^{1/2}\sum_{i=1}^{n}\sum_{k=1}^{m} I_i^u\overline{G}_{ik}^{(2)}(\mathbf{A}, \boldsymbol{\Delta}, \boldsymbol{\Phi}')\xi_k(t),
\end{aligned}$$

$$\frac{\mathrm{d}\Phi_w}{\mathrm{d}t} = \nu_w(A_w, \Phi_w) + \varepsilon \overline{F}_w^{(2)}(\mathbf{A}, \boldsymbol{\Delta}, \boldsymbol{\Phi}', \boldsymbol{\Omega}) + \varepsilon^{1/2} \sum_{k=1}^{m} \overline{G}_{wk}^{(2)}(\mathbf{A}, \boldsymbol{\Delta}, \boldsymbol{\Phi}')\xi_k(t),$$

$$i = 1, 2, \cdots, n; \quad u = 1, 2, \ldots, \mu; \quad w = \mu + 1, \mu + 2, \ldots, n, \tag{1.197}$$

where $\boldsymbol{\Delta} = [\Delta_1, \Delta_2, \ldots, \Delta_\mu]^{\mathrm{T}}$, $\boldsymbol{\Phi}' = [\Phi_{\mu+1}, \Phi_{\mu+2}, \ldots, \Phi_n]^{\mathrm{T}}$, $\boldsymbol{\Omega} = [\Omega_1, \Omega_2, \ldots, \Omega_s]^{\mathrm{T}}$, $\overline{F}_i^{(1)}, \overline{F}_i^{(2)}, \overline{G}_{ik}^{(1)}, \overline{G}_{ik}^{(2)}$ are obtained from $F_i^{(1)}, F_i^{(2)}, G_{ik}^{(1)}, G_{ik}^{(2)}$ with $\Phi_1, \Phi_2, \ldots, \Phi_\mu$ replaced by $\boldsymbol{\Delta}, \boldsymbol{\Phi}', \boldsymbol{\Omega}t$. The time averaging can be replaced by averaging with respect to $\boldsymbol{\Phi}', \boldsymbol{\Omega}t$. Finally, the averaged Itô SDEs are

$$\begin{aligned}
\mathrm{d}A_i &= V_i(\mathbf{A}, \boldsymbol{\Delta})\mathrm{d}t + \sum_{k=1}^{m} \sigma_{ik}^{(1)}(\mathbf{A})\mathrm{d}B_k(t), \\
\mathrm{d}\Delta_u &= W_u(\mathbf{A}, \boldsymbol{\Delta})\mathrm{d}t + \sum_{k=1}^{m} \sigma_{uk}^{(2)}(\mathbf{A})\mathrm{d}B_k(t), \\
i &= 1, 2, \ldots, n; \; u = 1, 2, \ldots, \mu,
\end{aligned} \tag{1.198}$$

where the drift and diffusion coefficients are obtained according to Eqs. (4.29) and (4.30) in vol. 1 as follows

$$\begin{aligned}
V_i = \frac{\varepsilon}{(2\pi)^{n-\mu+s}} \int_0^{2\pi}\int_0^{2\pi} \{\bar{F}_i^{(1)} &+ \sum_{k,l=1}^{m} \int_{-\infty}^{0} [\sum_{j=1}^{n} \left(\frac{\partial \bar{G}_{ik}^{(1)}}{\partial A_j}\bigg|_t \bar{G}_{jl}^{(1)}\bigg|_{t+\tau} \right) \\
&+ \sum_{v=1}^{\mu}\sum_{j=1}^{n} \left(I_j^u \frac{\partial \bar{G}_{ik}^{(1)}}{\partial \Delta_v}\bigg|_t \bar{G}_{jl}^{(2)}\bigg|_{t+\tau} \right) \\
&+ \sum_{w=\mu+1}^{n} (\frac{\partial \bar{G}_{ik}^{(1)}}{\partial \Phi_w}\bigg|_t \bar{G}_{wl}^{(2)}\bigg|_{t+\tau})] R_{kl}(\tau)\mathrm{d}\tau\} \mathrm{d}\boldsymbol{\Phi}'\mathrm{d}(\boldsymbol{\Omega}t), \\
W_u = \varepsilon\chi_u + \frac{\varepsilon}{(2\pi)^{n-\mu+s}} \int_0^{2\pi}\int_0^{2\pi} \{&\sum_{i=1}^{n} I_i^u \bar{F}_i^{(2)} \\
&+ \sum_{k,l=1}^{m} \int_{-\infty}^{0} [\sum_{i,j=1}^{n} \left(I_i^u \frac{\partial \bar{G}_{ik}^{(2)}}{\partial A_j}\bigg|_t \bar{G}_{jl}^{(1)}\bigg|_{t+\tau} \right) + \sum_{v=1}^{\mu}\sum_{i,j=1}^{n} (I_i^u I_j^v \frac{\partial \bar{G}_{ik}^{(2)}}{\partial \Delta_v}\bigg|_t \bar{G}_{jl}^{(2)}\bigg|_{t+\tau}) \\
&+ \sum_{w=\mu+1}^{n}\sum_{i=1}^{n} (I_i^u \frac{\partial \bar{G}_{ik}^{(2)}}{\partial \Phi_w}\bigg|_t \bar{G}_{wl}^{(2)}\bigg|_{t+\tau})] R_{kl}(\tau)\mathrm{d}\tau\} \mathrm{d}\boldsymbol{\Phi}'\mathrm{d}(\boldsymbol{\Omega}t),
\end{aligned}$$

$$\sum_{k=1}^{m} \sigma_{ik}^{(1)}\sigma_{jk}^{(1)} = \frac{\varepsilon}{(2\pi)^{n-\mu}} \int_0^{2\pi} \sum_{k=1}^{m} \int_{-\infty}^{\infty} (\overline{G}_{ik}^{(1)}\Big|_t \overline{G}_{jl}^{(1)}\Big|_{t+\tau} R_{kl}(\tau)\mathrm{d}\tau)\mathrm{d}\boldsymbol{\Phi}',$$

$$\sum_{k=1}^{m}\sigma_{uk}^{(2)}\sigma_{vk}^{(2)}=\frac{\varepsilon}{(2\pi)^{n-\mu}}\int_0^{2\pi}\sum_{k=1}^{m}\int_{-\infty}^{\infty}\sum_{i,j=1}^{n}I_i^uI_j^v\overline{G}_{ik}^{(2)}\Big|_t\overline{G}_{jl}^{(2)}\Big|_{t+\tau}R_{kl}(\tau)\mathrm{d}\tau)\mathrm{d}\boldsymbol{\Phi}',$$

$$\sum_{k=1}^{m}\sigma_{ik}^{(1)}\sigma_{uk}^{(2)}=\frac{\varepsilon}{(2\pi)^{n-\mu}}\int_0^{2\pi}\sum_{k=1}^{m}\int_{-\infty}^{\infty}\sum_{j=1}^{n}I_j^u\overline{G}_{ik}^{(1)}\Big|_t\overline{G}_{jl}^{(2)}\Big|_{t+\tau}R_{kl}(\tau)\mathrm{d}\tau)\mathrm{d}\boldsymbol{\Phi}',$$

$$i,j=1,2,\ldots,n;\quad u,v=1,2,\cdots,\mu. \tag{1.199}$$

The averaged FPK equation governing $p(\mathbf{a},\boldsymbol{\delta},t)$ can be derived from averaged Itô Eq. (1.198). Similar to Eq. (1.169), once the stationary PDF $p(\mathbf{a},\boldsymbol{\delta})$ is obtained, $p(\mathbf{q},\mathbf{p})$ can be determined by

$$p(\mathbf{q},\mathbf{p})=p(\mathbf{a},\boldsymbol{\delta})\left|\frac{\partial(\mathbf{a},\boldsymbol{\delta})}{\partial(\mathbf{q},\mathbf{p})}\right|=p(\mathbf{a},\boldsymbol{\delta})\left|\frac{\partial(\mathbf{a},\boldsymbol{\delta})}{\partial(\mathbf{a},\boldsymbol{\phi})}\right|\left|\frac{\partial(\mathbf{a},\boldsymbol{\phi})}{\partial(\mathbf{q},\mathbf{p})}\right|. \tag{1.200}$$

Example 1.9 Consider two coupled Duffing oscillators with nonlinear dampings and excited by one harmonic noise and four wideband noises (Wu 2005). The system motion equations are

$$\begin{aligned}&\ddot{X}_1+(\beta_{11}+\beta_{12}X_1^2+\beta_{13}X_2^2)\dot{X}_1+\omega_{01}^2X_1+\alpha_1X_1^3\\&\quad=E_1\cos\Omega t+\xi_1(t)+X_1\xi_2(t),\\&\ddot{X}_2+(\beta_{21}+\beta_{22}X_1^2+\beta_{23}X_2^2)\dot{X}_2+\omega_{02}^2X_2+\alpha_2X_2^3\\&\quad=E_2\cos\Omega t+\xi_3(t)+X_2\xi_4(t),\end{aligned} \tag{1.201}$$

where $\beta_{ij},\omega_{0i},\alpha_i,E_i,\Omega$ $(i=1,2;\ j=1,2,3)$ are constants, $\xi_k(t)$ $(k=1,\ldots,4)$ are wideband noises with PSD given in Eq. (1.171), β_{ij},E_i,D_k are small quantities of ε order.

Introduce the transformation

$$\begin{aligned}&X_i(t)=A_i\cos\Phi_i(t),\quad \dot{X}_i(t)=-A_i\nu_i(A_i,\Phi_i)\sin\Phi_i(t),\\&\Phi_i(t)=\Gamma_i(t)+\Theta_i(t),\quad i=1,2,\end{aligned} \tag{1.202}$$

where

$$\begin{aligned}&\nu_i(a_i,\phi_i)=\left[(\omega_{0i}^2+3\alpha_ia_i^2/4)(1+\lambda_i\cos 2\phi_i)\right]^{1/2},\\&\lambda_i=\alpha_ia_i^2/4(\omega_{0i}^2+3\alpha_ia_i^2/4).\end{aligned} \tag{1.203}$$

Expand $\nu_i(a_i,\phi_i)$ into a Fourier series and keep the first seven terms

$$\nu_i(a_i,\phi_i)\approx b_{i0}(a_i)+b_{i2}(a_i)\cos 2\phi_i+b_{i4}(a_i)\cos 4\phi_i+b_{i6}(a_i)\cos 6\phi_i, \tag{1.204}$$

where

$$
\begin{aligned}
b_{i0}(a_i) &= (\omega_{0i}^2 + 3\alpha_i a_i^2/4)^{1/2}(1 - \lambda_i^2/16),\\
b_{i2}(a_i) &= (\omega_{0i}^2 + 3\alpha_i a_i^2/4)^{1/2}(\lambda_i/2 + 3\lambda_i^3/64),\\
b_{i4}(a_i) &= (\omega_{0i}^2 + 3\alpha_i a_i^2/4)^{1/2}(-\lambda_i^2/16),\\
b_{i2}(a_i) &= (\omega_{0i}^2 + 3\alpha_i a_i^2/4)^{1/2}(\lambda_i^3/64).
\end{aligned} \tag{1.205}
$$

The mean frequency $\omega_i(a_i) = b_{i0}(a_i)$. Applying Eq. (1.202) to Eq. (1.201) yields

$$
\begin{aligned}
\frac{dA_1}{dt} &= F_1^{(1)}(\mathbf{A}, \boldsymbol{\Phi}, \Omega t) + G_{11}^{(1)}\xi_1(t) + G_{12}^{(1)}\xi_2(t),\\
\frac{dA_2}{dt} &= F_2^{(1)}(\mathbf{A}, \boldsymbol{\Phi}, \Omega t) + G_{23}^{(1)}\xi_3(t) + G_{24}^{(1)}\xi_4(t),\\
\frac{d\Phi_1}{dt} &= \nu_1(A_1, \Phi_1) + F_1^{(2)}(\mathbf{A}, \boldsymbol{\Phi}, \Omega t) + G_{11}^{(2)}\xi_1(t) + G_{12}^{(2)}\xi_2(t),\\
\frac{d\Phi_2}{dt} &= \nu_2(A_2, \Phi_2) + F_2^{(2)}(\mathbf{A}, \boldsymbol{\Phi}, \Omega t) + G_{23}^{(2)}\xi_3(t) + G_{24}^{(2)}\xi_4(t),
\end{aligned} \tag{1.206}
$$

where

$$
\begin{aligned}
F_i^{(1)} &= \frac{-A_i}{\omega_{0i}^2 A_i + \alpha_i A_i^3}[(\beta_{i1} + \beta_{i2}A_1^2\cos^2\Phi_1 + \beta_{i3}A_2^2\cos^2\Phi_2)A_i\nu_i(A_i, \Phi_i)\sin\Phi_i\\
&\quad + E_i\cos\Omega t]\nu_i(A_i, \Phi_i)\sin\Phi_i,\\
F_i^{(2)} &= \frac{-1}{\omega_{0i}^2 A_i + \alpha_i A_i^3}[(\beta_{i1} + \beta_{i2}A_1^2\cos^2\Phi_1 + \beta_{i3}A_2^2\cos^2\Phi_2)A_i\nu_i(A_i, \Phi_i)\sin\Phi_i\\
&\quad + E_i\cos\Omega t]\nu_i(A_i, \Phi_i)\cos\Phi_i,
\end{aligned}
$$

$$
\begin{aligned}
G_{i,2i-1}^{(1)} &= \frac{-A_i}{\omega_{0i}^2 A_i + \alpha_i A_i^3}\nu_i(A_i, \Phi_\iota)\sin\Phi_\iota,\\
G_{i,2i}^{(1)} &= \frac{-A_i^2}{\omega_{0i}^2 A_i + \alpha_i A_i^3}\nu_i(A_i, \Phi_\iota)\sin\Phi_\iota\cos\Phi_\iota,\\
G_{i,2i-1}^{(2)} &= \frac{-1}{\omega_{0i}^2 A_i + \alpha_i A_i^3}\nu_i(A_i, \Phi_\iota)\cos\Phi_\iota,\\
G_{i,2i}^{(2)} &= \frac{-A_i}{\omega_{0i}^2 A_i + \alpha_i A_i^3}\nu_i(A_i, \Phi_\iota)\cos^2\Phi_\iota,\\
i &= 1, 2.
\end{aligned} \tag{1.207}
$$

Considering the case that only the first oscillator is in external resonance with harmonic excitation, i.e.,

$$\frac{\Omega}{\omega_1(A_1)} = 1 + \varepsilon\sigma. \tag{1.208}$$

Introducing

$$\Delta = \Omega t - \Phi_1. \tag{1.209}$$

Completing the averaging with respect to Φ_2 and Ωt of equations for A_1, A_2, Δ results in the following averaged Itô SDEs

$$\begin{aligned}
\mathrm{d}A_1 &= m_1(A_1, A_2, \Delta)\mathrm{d}t + \sigma_{11}\mathrm{d}B_1(t) + \sigma_{12}\mathrm{d}B_2(t),\\
\mathrm{d}A_2 &= m_2(A_1, A_2, \Delta)\mathrm{d}t + \sigma_{23}\mathrm{d}B_3(t) + \sigma_{24}\mathrm{d}B_4(t),\\
\mathrm{d}\Delta &= m_3(A_1, A_2, \Delta)\mathrm{d}t + \sigma_{31}\mathrm{d}B_1(t) + \sigma_{32}\mathrm{d}B_2(t),
\end{aligned} \tag{1.210}$$

where drift and diffusion coefficients are obtained according to Eq. (1.199) as follows

$$\begin{aligned}
m_1 &= s_1 + m_{11} + m_{12} + m_{13} + m_{14},\\
m_2 &= s_2 + m_{21} + m_{22} + m_{23} + m_{24},\\
m_3 &= \Omega - b_{10}(A_1) + [b_{12}(A_1) + 2b_{10}(A_1)]E_1\cos\Delta\big/\left[4A_1\left(\omega_{01}^2 + \alpha_1 A_1^2\right)\right],\\
s_1 &= -\alpha_1 A_1^3\left(10\beta_{11} + 3\beta_{12}A_1^2 + 5\beta_{13}A_2^2\right)\big/\left[32\left(\alpha_1 A_1^2 + \omega_{01}^2\right)\right]\\
&\quad -\omega_{01}^2 A_1\left(4\beta_{11} + \beta_{12}A_1^2 + 2\beta_{13}A_2^2\right)\big/\left[8\left(\alpha_1 A_1^2 + \omega_{01}^2\right)\right]\\
&\quad + E_1(2b_{10} - b_{12})\sin\Delta\big/\left[4\left(\alpha_1 A_1^2 + \omega_{01}^2\right)\right],
\end{aligned}$$

$$\begin{aligned}
s_2 &= -\alpha_2 A_2^3\left(10\beta_{21} + 3\beta_{23}A_2^2 + 5\beta_{22}A_1^2\right)\big/\left[32\left(\alpha_2 A_2^2 + \omega_{02}^2\right)\right]\\
&\quad -\omega_{02}^2 A_2\left(4\beta_{21} + \beta_{22}A_1^2 + 2\beta_{23}A_2^2\right)\big/\left[8\left(\alpha_2 A_2^2 + \omega_{02}^2\right)\right],\\
m_{11} &= m_{11}^1 S_1(\omega(A_1)) + m_{11}^3 S_1(3\omega(A_1)) + m_{11}^5 S_1(5\omega(A_1)) + m_{11}^7 S_1(7\omega(A_1)),\\
m_{12} &= m_{12}^2 S_2(2\omega(A_1)) + m_{12}^4 S_2(4\omega(A_1)) + m_{12}^6 S_2(6\omega(A_1)) + m_{12}^8 S_2(8\omega(A_1)),\\
m_{13} &= m_{13}^1 S_1(\omega(A_1)) + m_{13}^3 S_1(3\omega(A_1)) + m_{13}^5 S_1(5\omega(A_1)) + m_{13}^7 S_1(7\omega(A_1)),
\end{aligned}$$

$$\begin{aligned}
m_{14} &= m_{14}^2 S_2(2\omega(A_1)) + m_{14}^4 S_2(4\omega(A_1)) + m_{14}^6 S_2(6\omega(A_1)) + m_{14}^8 S_2(8\omega(A_1)),\\
m_{21} &= m_{21}^1 S_3(\omega(A_2)) + m_{21}^3 S_3(3\omega(A_2)) + m_{21}^5 S_3(5\omega(A_2)) + m_{21}^7 S_3(7\omega(A_2)),\\
m_{22} &= m_{22}^2 S_4(2\omega(A_2)) + m_{22}^4 S_4(4\omega(A_2)) + m_{22}^6 S_4(6\omega(A_2)) + m_{22}^8 S_4(8\omega(A_2)),\\
m_{23} &= m_{23}^1 S_3(\omega(A_2)) + m_{23}^3 S_3(3\omega(A_2)) + m_{23}^5 S_3(5\omega(A_2)) + m_{23}^7 S_3(7\omega(A_2)),\\
m_{24} &= m_{24}^2 S_4(2\omega(A_2)) + m_{24}^4 S_4(4\omega(A_2)) + m_{24}^6 S_4(6\omega(A_2)) + m_{24}^8 S_4(8\omega(A_2)),
\end{aligned}$$

$$m_{i1}^1 = \pi[b_{i2}(A_i) - 2b_{i0}(A_i)]\Big\{ 2\alpha_i A_i[2b_{i0}(A_i) - b_{i2}(A_i)] - \left(A_i^2\alpha_i + \omega_{0i}^2\right)$$

$$\times\left(\frac{\mathrm{d}b_{i2}(A_i)}{\mathrm{d}A_i}-2\frac{\mathrm{d}b_{i0}(A_i)}{\mathrm{d}A_i}\right)\Big\}\Big/\Big[8\big(A_i^2\alpha_i+\omega_{0i}^2\big)^3\Big],$$

$$m_{i1}^3=\pi[b_{i2}(A_i)-b_{i4}(A_i)]\Big\{2\alpha_iA_i[b_{i4}(A_i)-b_{i2}(A_i)]+\big(A_i^2\alpha_i+\omega_{0i}^2\big)$$
$$\times\left(\frac{\mathrm{d}b_{i2}(A_i)}{\mathrm{d}A_i}-\frac{\mathrm{d}b_{i4}(A_i)}{\mathrm{d}A_i}\right)\Big\}\Big/\Big[8\big(A_i^2\alpha_i+\omega_{0i}^2\big)^3\Big],$$

$$m_{i1}^5=\pi[b_{i4}(A_i)-b_{i6}(A_i)]\Big\{2\alpha_iA_i[b_{i6}(A_i)-b_{14}(A_i)]+\big(A_i^2\alpha_1+\omega_{0i}^2\big)$$
$$\times\left(\frac{\mathrm{d}b_{i4}(A_i)}{\mathrm{d}A_i}-\frac{\mathrm{d}b_{i6}(A_i)}{\mathrm{d}A_i}\right)\Big\}\Big/\Big[8\big(A_i^2\alpha_i+\omega_{0i}^2\big)^3\Big],$$
$$m_{i1}^7=\pi b_{i6}(A_i)\Big\{-2\alpha_iA_ib_{i6}(A_i)+\big(A_i^2\alpha_i+\omega_{0i}^2\big)\frac{\mathrm{d}b_{i6}(A_i)}{\mathrm{d}A_i}\Big\}\Big/\Big[8\big(A_i^2\alpha_i+\omega_{0i}^2\big)^3\Big],$$

$$m_{i2}^2=\pi A_i[b_{i4}(A_i)-2b_{i0}(A_i)]\big\{[2b_{i0}(A_i)-b_{i4}(A_i)]\big(A_i^2\alpha_i-\omega_{0i}^2\big)+A_i\big(A_i^2\alpha_i+\omega_{0i}^2\big)$$
$$\times\left(\frac{\mathrm{d}b_{i4}(A_i)}{\mathrm{d}A_i}-\frac{2\mathrm{d}b_{i0}(A_i)}{\mathrm{d}A_i}\right)\Big\}\Big/\Big[32\big(A_i^2\alpha_1+\omega_{0i}^2\big)^3\Big],$$
$$m_{i2}^4=\pi A_i[b_{i2}(A_i)-b_{i6}(A_i)]\big\{[b_{i6}(A_i)-b_{i2}(A_i)]\big(A_i^2\alpha_i-\omega_{0i}^2\big)+A_i\big(A_i^2\alpha_i+\omega_{0i}^2\big)$$
$$\times\left(\frac{\mathrm{d}b_{i2}(A_i)}{\mathrm{d}A_i}-\frac{\mathrm{d}b_{i6}(A_i)}{\mathrm{d}A_i}\right)\Big\}\Big/\Big[32\big(A_i^2\alpha_i+\omega_{0i}^2\big)^3\Big],$$

$$m_{i2}^6=\pi A_ib_{i4}(A_i)\big\{b_{i4}(A_i)\big(A_i^2\alpha_i-\omega_{0i}^2\big)$$
$$+A_i\big(A_i^2\alpha_i+\omega_{0i}^2\big)\frac{\mathrm{d}b_{i4}(A_i)}{\mathrm{d}A_i}\Big\}\Big/\Big[32\big(A_i^2\alpha_i+\omega_{0i}^2\big)^3\Big],$$
$$m_{i2}^8=\pi A_ib_{i6}(A_i)\big\{b_{i6}(A_i)\big({A_i}^2\alpha-\omega_{0i}^2\big)$$
$$+A_i\big({A_i}^2\alpha+\omega_{0i}^2\big)\frac{\mathrm{d}b_{i6}(A_i)}{\mathrm{d}A_i}\Big\}\Big/\Big[32\big({A_i}^2\alpha_i+\omega_{0i}^2\big)^3\Big],$$
$$m_{i3}^1=\pi\big[4b_{i0}^2(A_i)-b_{i2}^2(A_i)\big]\Big/\Big[8A_i\big({A_i}^2\alpha_i+\omega_{0i}^2\big)^2\Big],$$
$$m_{i3}^3=3\pi\big[b_{i2}^2(A)-b_{i4}^2(A)\big]\Big/\Big[8A_i\big(A_i^2\alpha+\omega_{0i}^2\big)^2\Big],$$

$$m_{i3}^5=5\pi\big[b_{i4}^2(A_i)-b_{i6}^2(A_i)\big]\Big/\Big[8A_i\big(A_i^2\alpha+\omega_{0i}^2\big)^2\Big],$$
$$m_{i3}^7=7\pi b_{i6}^2(A_i)\Big/\Big[8A_i\big(A_i^2\alpha+\omega_{0i}^2\big)^2\Big],$$
$$m_{i4}^2=\pi A_i[2b_{i0}(A_i)-b_{i4}(A_i)][2b_{i0}(A_i)+2b_{i2}(A_i)+b_{i4}(A_i)]\Big/\Big[16\big(A_1^2\alpha_1+\omega_{0i}^2\big)^2\Big],$$
$$m_{i4}^4=\pi A_i[b_{i2}(A_i)-b_{i6}(A_i)][b_{i2}(A_i)+2b_{i4}(A_i)+b_{i6}(A_i)]\Big/\Big[8\big(A_1^2\alpha_i+\omega_{0i}^2\big)^2\Big],$$
$$m_{i4}^6=3\pi A_ib_{i4}(A_i)[b_{i4}(A_i)+2b_{i6}(A_i)]\Big/\Big[16\big(A_i^2\alpha_i+\omega_{0i}^2\big)^2\Big],$$

$$m_{i4}^{8} = \pi A_i b_{i6}^{2}(A_i) \Big/ \left[4\left(A_i^2\alpha_i+\omega_{0i}^2\right)^2\right], \quad (i = 1, 2),$$
$$\sigma_{11}(A_1)\sigma_{11}(A_1) + \sigma_{12}(A_1)\sigma_{12}(A_1) = b_{11}^{1} + b_{11}^{2},$$
$$\sigma_{23}(A_2)\sigma_{23}(A_2) + \sigma_{24}(A_2)\sigma_{24}(A_2) = b_{22}^{1} + b_{22}^{2},$$
$$\sigma_{31}(A_1)\sigma_{31}(A_1) + \sigma_{32}(A_1)\sigma_{32}(A_1) = b_{33}^{1} + b_{33}^{2},$$
$$\sum_{k=1}^{4}\sigma_{1k}\sigma_{2k} = \sum_{k=1}^{4}\sigma_{1k}\sigma_{3k} = \sum_{k=1}^{4}\sigma_{2k}\sigma_{3k} = 0,$$

$$b_{11}^{1} = b_{11}^{11}S_1(\omega(A_1)) + b_{11}^{13}S_1(3\omega(A_1)) + b_{11}^{15}S_1(5\omega(A_1)) + b_{11}^{17}S_1(7\omega(A_1)),$$
$$b_{11}^{2} = b_{11}^{22}S_2(2\omega(A_1)) + b_{11}^{24}S_2(4\omega(A_1)) + b_{11}^{26}S_2(6\omega(A_1)) + b_{11}^{28}S_2(8\omega(A_1)),$$
$$b_{22}^{1} = b_{22}^{11}S_3(\omega(A_2)) + b_{22}^{13}S_3(3\omega(A_2)) + b_{22}^{15}S_3(5\omega(A_2)) + b_{22}^{17}S_3(7\omega(A_2)),$$
$$b_{22}^{2} = b_{22}^{22}S_4(2\omega(A_2)) + b_{22}^{24}S_4(4\omega(A_2)) + b_{22}^{26}S_4(6\omega(A_2)) + b_{22}^{28}S_4(8\omega(A_2)),$$
$$b_{33}^{1} = b_{33}^{11}S_1(\omega(A_1)) + b_{33}^{13}S_1(3\omega(A_1)) + b_{33}^{15}S_1(5\omega(A_1)) + b_{33}^{17}S_1(7\omega(A_1)),$$
$$b_{33}^{2} = b_{33}^{20}S_2(0) + b_{33}^{22}S_2(2\omega(A_1)) + b_{33}^{24}S_2(4\omega(A_1)) + b_{33}^{26}S_2(6\omega(A_1)) + b_{33}^{28}S_2(8\omega(A_1)),$$

$$b_{ii}^{11} = \pi\left[b_{i2}(A_i) - 2b_{i0}(A_i)\right]^2 \Big/ 4\left(\alpha_i A_i^2 + \omega_{0i}^2\right)^2,$$
$$b_{ii}^{13} = \pi\left[b_{i2}(A_i) - b_{i4}(A_i)\right]^2 \Big/ 4\left(\alpha_i A_i^2 + \omega_{0i}^2\right)^2,$$
$$b_{ii}^{15} = \pi\left[b_{i4}(A_i) - b_{i6}(A_i)\right]^2 \Big/ 4\left(\alpha_i A_i^2 + \omega_{0i}^2\right)^2,$$
$$b_{ii}^{17} = \pi b_{i6}^{2}(A_i) \Big/ 4\left(\alpha_i A_i^2 + \omega_{0i}^2\right)^2,$$

$$b_{ii}^{22} = \pi A_i^2\left[b_{i4}(A_i) - 2b_{i0}(A_i)\right]^2 \Big/ 16\left(\alpha_i A_i^2 + \omega_{0i}^2\right)^2,$$
$$b_{ii}^{24} = \pi A_i^2\left[b_{i2}(A_i) - b_{i6}(A_i)\right]^2 \Big/ 16\left(\alpha_i A_i^2 + \omega_{0i}^2\right)^2,$$
$$b_{ii}^{26} = \pi A_i^2 b_{i4}^{2}(A_i) \Big/ 16\left(\alpha_i A_i^2 + \omega_{0i}^2\right)^2,$$
$$b_{ii}^{28} = \pi A_i^2 b_{i6}^{2}(A_i) \Big/ 16\left(\alpha_i A_i^2 + \omega_{0i}^2\right)^2, \tag{1.211}$$

$$(i = 1, 2)$$
$$b_{33}^{11} = \pi\left[2b_{10}(A_1) + b_{12}(A_1)\right]^2 \Big/ \left[4A_1^2\left(\alpha_1 A_1^2 + \omega_{01}^2\right)^2\right],$$
$$b_{33}^{13} = \pi\left[2b_{12}(A_1) + b_{14}(A_1)\right]^2 \Big/ \left[4A_1^2\left(\alpha_1 A_1^2 + \omega_{01}^2\right)^2\right],$$
$$b_{33}^{15} = \pi\left[2b_{14}(A_1) + b_{16}(A_1)\right]^2 \Big/ \left[4A_1^2\left(\alpha_1 A_1^2 + \omega_{01}^2\right)^2\right],$$

$$b_{33}^{17} = \pi\left[b_{16}(A_1)\right]^2 \Big/ \left[4A_1^2\left(\alpha_1 A_1^2 + \omega_{01}^2\right)^2\right],$$

$$b_{33}^{20}=\pi\left[2b_{10}(A_1)+b2(A_1)\right]^2\Big/\left[8\left(\alpha_1A_1^2+\omega_{01}^2\right)^2\right],$$
$$b_{33}^{22}=\pi\left[2b_{10}(A_1)+2b_{12}(A_1)+b_{14}(A_1)\right]^2\Big/\left[16\left(\alpha_1A_1^2+\omega_{01}^2\right)^2\right],$$
$$b_{33}^{24}=\pi\left[2b_{12}(A_1)+2b_{14}(A_1)+b_{16}(A_1)\right]^2\Big/\left[16\left(\alpha_1A_1^2+\omega_{01}^2\right)^2\right],$$
$$b_{33}^{26}=\pi\left[b_{14}(A_1)+2b_{16}(A_1)\right]^2\Big/\left[16\left(\alpha_1A_1^2+\omega_{01}^2\right)^2\right],$$
$$b_{33}^{28}=\pi\left[b_{16}(A_1)\right]^2\Big/\left[16\left(\alpha_1A_1^2+\omega_{01}^2\right)^2\right],$$

The reduced averaged FPK equation associated with averaged Itô Eq. (1.210) can be solved by using the finite difference method to obtain stationary joint PDF $p(a_1, a_2, \delta)$. Figure 1.38 provides a set of numerical results for PDFs $p(a_1, \delta)$, $p(a_1)$, $p(\delta)$ obtained by using the stochastic averaging method and from Monte Carlo simulation. It is seen the two results are in good agreement.

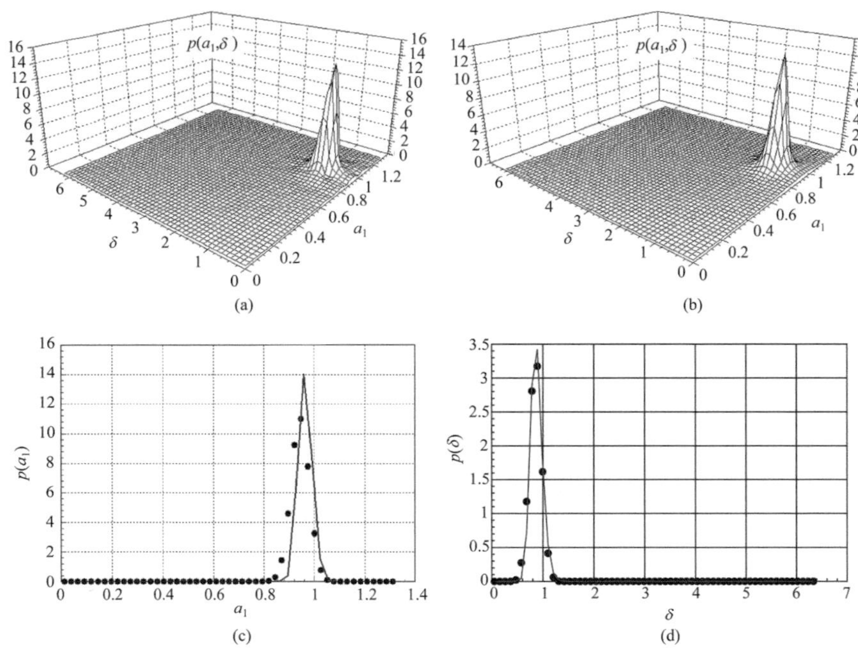

Fig. 1.38 **a** $p(a_1, \delta)$ obtained by using the stochastic averaging method; **b** $p(a_1, \delta)$ obtained from Monte Carlo simulation; **c** $p(a)$; **d** $p(\delta)$, ——, results from the stochastic averaging method; ●, results from Monte Carlo simulation. System parameters $\omega_{01} = 1.0, \omega_{02} = 2.0, \alpha_1 = 0.8, \alpha_2 = 0.5$, $\beta_{11} = \beta_{12} = 0.08$, $\beta_{13} = 0.05$, $\beta_{2i} = 0.05$, $E_1 = 0.15$, $E_2 = 0.1$, $\Omega = 1.2$, $\omega_j = 5.0$, $\zeta_j = 0.5$, $D_j = 0.2$ (i= 1, 2, 3; j= 1, 2, 3, 4) (Wu 2005)

Example 1.10 As an example of both internal and external resonances, consider the reliability of a two-DOF strongly nonlinear vibration system excited by combination of harmonic and wideband noises (Wu et al. 2013). The system motion equations are

$$
\begin{aligned}
&\dot{Q}_1 = P_1,\\
&\dot{P}_1 = -\omega_{01}^2 Q_1 - \alpha_1 Q_1^3 - (\beta_{10} + \beta_{11} Q_1^2 + \beta_{12} Q_2^2) P_1\\
&\quad - (\eta_{11} Q_2 + \eta_{12} P_2) + E_1 \cos \Omega t + \xi_{11}(t) + Q_1 \xi_{12}(t),\\
&\dot{Q}_2 = P_2,\\
&\dot{P}_2 = -\omega_{02}^2 Q_2 - \alpha_2 Q_2^3 - (\beta_{20} + \beta_{21} Q_1^2 + \beta_{22} Q_2^2) P_2\\
&\quad - (\eta_{21} Q_1 + \eta_{22} P_1) + E_2 \cos \Omega t + \xi_{21}(t) + Q_2 \xi_{22}(t),
\end{aligned}
\tag{1.212}
$$

where $\beta_{i0}, \beta_{ij}, \omega_{0i}, \alpha_i, \eta_{ij}, E_i\ (i, j = 1, 2)$ are constants; $\beta_{i0}, \beta_{ij}, \eta_{ij}, E_i, D_{ik}$ are small quantities of ε order; Ω is the frequency of harmonic excitation; $\xi_{ik}(t)$ are stationary wideband noises with PSD

$$
S_{ik}(\omega) = \frac{D_{ik}}{\pi} \frac{1}{\omega^2 + \omega_{ik}^2}, \quad i, k = 1, 2. \tag{1.213}
$$

In order to apply the stochastic averaging method to system (1.212), first introduce the following transformation

$$
\begin{aligned}
&Q_i(t) = A_i(t) \cos \Phi_i(t), \quad P_i(t) = -A_i(t) \nu_i(A_i, \Phi_i) \sin \Phi_i(t),\\
&\Phi_i(t) = \Gamma_i(t) + \Theta_i(t), \quad i = 1, 2,
\end{aligned}
\tag{1.214}
$$

Expand the instantaneous frequencies $\nu_i(A_i, \Phi_i)$ into Fourier series and keep the first several terms

$$
\nu_i(A_i, \Phi_i) \approx b_{i0}(A_i) + b_{i2}(A_i) \cos 2\Phi_i + b_{i4}(A_i) \cos 4\Phi_i + b_{i6}(A_i) \cos 6\Phi_i. \tag{1.215}
$$

Considering the case of both external and internal resonances, i.e., the first subsystem is in external resonance with harmonic excitation, and the first and second subsystems are in internal resonance. Introduce two combinations of angle variables

$$
\Delta_1 = \Omega t - \Phi_1, \quad \Delta_2 = \Phi_1 - \Phi_2. \tag{1.216}
$$

$[A_1, A_2, \Delta_1,\ \Delta_2]^{\mathrm{T}}$ is a vector of four slowly-varying processes. As $\varepsilon \to 0$, it converges weakly to a four-dimensional vector Markov diffusion process. The following averaged Itô SDEs can be obtained by applying the stochastic averaging method.

$$
\begin{aligned}
\mathrm{d}A_1 &= F_1(\boldsymbol{A},\boldsymbol{\Delta})\mathrm{d}t + f_{11}\mathrm{d}B_1(t) + f_{12}\mathrm{d}B_2(t),\\
\mathrm{d}A_2 &= F_2(\boldsymbol{A},\boldsymbol{\Delta})\mathrm{d}t + f_{21}\mathrm{d}B_1(t) + f_{22}\mathrm{d}B_2(t),\\
\mathrm{d}\Delta_1 &= G_1(\boldsymbol{A},\boldsymbol{\Delta})\mathrm{d}t + g_{11}\mathrm{d}B_1(t) + g_{12}\mathrm{d}B_2(t),\\
\mathrm{d}\Delta_2 &= G_2(\boldsymbol{A},\boldsymbol{\Delta})\mathrm{d}t + g_{21}\mathrm{d}B_1(t) + g_{22}\mathrm{d}B_2(t),
\end{aligned}
\tag{1.217}
$$

where $\boldsymbol{A}=[A_1,A_2]^{\mathrm{T}}$, $\boldsymbol{\Delta} = [\Delta_1,\Delta_2]^{\mathrm{T}}$. In Eq. (1.217), the drift and diffusion coefficients can be obtained by using Eq. (1.199) as follows

$$
\begin{aligned}
F_i &= s_i + m_{i1} + m_{i2} + m_{i3} + m_{i4},\\
G_1 &= \Omega - b_{10}(A_1) - c_1 - c_{11} - c_{12} - c_{13} - c_{14},\\
G_2 &= b_{10}(A_1) - b_{20}(A_2) + c_1 - c_2 + c_{11} + c_{12} + c_{13}\\
&\quad + c_{14} - (c_{21} + c_{22} + c_{23} + c_{24}),
\end{aligned}
$$

$$
\begin{aligned}
s_1 = \big[&-\alpha_1 A_1^3\big(10\beta_{10} + 3\beta_{11}A_1^2 + 5\beta_{12}A_2^2\big) - 4\omega_{01}^2 A_1\big(4\beta_{10} + \beta_{11}A_1^2 + 2\beta_{12}A_2^2\big)\\
&+ 8E_1(2b_{10} - b_{12})\sin\Delta_1 - 4A_2(2b_{10} - b_{12})(2b_{20} - b_{22})\eta_{12}\cos\Delta_2\\
&+ 2A_1A_2^2\beta_{12}\big(\alpha_1 A_1^2 + 2\omega_{01}^2\big)\cos(2\Delta_2) - 4A_2(b_{12} - b_{14})(b_{22} - b_{24})\eta_{12}\cos(3\Delta_2)\\
&- 4A_2(b_{14} - b_{16})(b_{24} - b_{26})\eta_{12}\cos(5\Delta_2)\\
&- 4A_2 b_{16}b_{26}\eta_{12}\cos(7\Delta_2) + 8A_2(2b_{10} - b_{12})\eta_{11}\sin\Delta_2\big]\Big/\big[32\big(\alpha_1 A_1^2 + \omega_{01}^2\big)\big],
\end{aligned}
$$

$$
\begin{aligned}
s_2 = \big[&-\alpha_2 A_2^3\big(10\beta_{20} + 5\beta_{21}A_1^2 + 3\beta_{22}A_2^2\big) - 4\omega_{02}^2 A_2\big(4\beta_{20} + 2\beta_{21}A_1^2 + \beta_{22}A_2^2\big)\\
&- 4A_1(2b_{10} - b_{12})(2b_{20} - b_{22})\eta_{22}\cos\Delta_2 + 2A_1^2A_2\beta_{21}\big(\alpha_2 A_2^2 + 2\omega_{02}^2\big)\cos(2\Delta_2)\\
&- 4A_1(b_{12} - b_{14})(b_{22} - b_{24})\eta_{22}\cos(3\Delta_2) - 4A_1(b_{14} - b_{16})(b_{24}\\
&- b_{26})\eta_{22}\cos(5\Delta_2) - 4A_1 b_{16}b_{26}\eta_{22}\cos(7\Delta_2) + 8A_1(b_{22} - 2b_{20})\eta_{21}\sin\Delta_2\\
&+8(2b_{20} - b_{22})E_2\sin(\Delta_1 + \Delta_2)\big]\Big/\big[32\big(\alpha_2 A_2^2 + \omega_{02}^2\big)\big],
\end{aligned}
\tag{1.218}
$$

$$
\begin{aligned}
c_1 = [&-8(b_{12} + 2b_{10})E_1\cos\Delta_1 + 8A_2(2b_{10} + b_{12})\eta_{11}\cos\Delta_2\\
&+ 4A_2(2b_{10} + b_{12})(2b_{20} - b_{22})\eta_{12}\sin\Delta_2 - A_1A_2^2\beta_{12}\left(3\alpha_1 A_1^2 + 4\omega_{01}^2\right)\sin(2\Delta_2)\\
&+ 4A_2(b_{12} + b_{14})(b_{22} - b_{24})\eta_{12}\sin(3\Delta_2) + 4A_2(b_{14} + b_{16})(b_{24} - b_{26})\eta_{12}\sin(5\Delta_2)\\
&+4A_2 b_{16}b_{26}\eta_{12}\sin(7\Delta_2)]\Big/\left[32A_1\left(\alpha_1 A_1^2 + \omega_{01}^2\right)\right],\\
c_2 = [&8A_1(2b_{20} + b_{22})\eta_{21}\cos\Delta_2 - 4A_1(2b_{10} - b_{12})(2b_{20} + b_{22})\eta_{22}\sin\Delta_2\\
&+ A_2A_1^2\beta_{21}\left(3\alpha_2 A_2^2 + 4\omega_{02}^2\right)\sin(2\Delta_2) - 4A_1(b_{12} - b_{14})(b_{22} + b_{24})\eta_{22}\sin(3\Delta_2)\\
&- 4A_1(b_{14} - b_{16})(b_{24} + b_{26})\eta_{22}\sin(5\Delta_2) - 4A_1 b_{16}b_{26}\eta_{22}\sin(7\Delta_2)\\
&-8(2b_{20} + b_{22})E_2\cos(\Delta_1 + \Delta_2)]\Big/\left[32A_2\left(\alpha_2 A_2^2 + \omega_{02}^2\right)\right],
\end{aligned}
$$

$$
\begin{aligned}
m_{i1} &= m_{i11}S_{i1}(\omega_i(A_i)) + m_{i13}S_{i1}(3\omega_i(A_i)) + m_{i15}S_{i1}(5\omega_i(A_i))\\
&\quad + m_{i17}S_{i1}(7\omega_i(A_i)),
\end{aligned}
$$

$$
\begin{aligned}
m_{i3} &= m_{i31}S_{i1}(\omega_i(A_i)) + m_{i33}S_{i1}(3\omega_i(A_i)) + m_{i35}S_{i1}(5\omega_i(A_i)) \\
&\quad + m_{i37}S_{i1}(7\omega_i(A_i)), \\
m_{i2} &= m_{i22}S_{i2}(2\omega_i(A_i)) + m_{i24}S_{i2}(4\omega_i(A_i)) + m_{i26}S_{i2}(6\omega_i(A_i)) \\
&\quad + m_{i28}S_{i2}(8\omega_i(A_i)), \\
m_{i4} &= m_{i42}S_{i2}(2\omega_i(A_i)) + m_{i44}S_{i2}(4\omega_i(A_i)) + m_{i46}S_{i2}(6\omega_i(A_i)) \\
&\quad + m_{i48}S_{i2}(8\omega_i(A_i)), \\
c_{i1} &= c_{i11}I_{i1}(\omega_i(A_i)) + c_{i13}I_{i1}(3\omega_i(A_i)) + c_{i15}I_{i1}(5\omega_i(A_i)) \\
&\quad + c_{i17}I_{i1}(7\omega_i(A_i)), \\
c_{i2} &= c_{i22}I_{i2}(2\omega_i(A_i)) + c_{i24}I_{i2}(4\omega_i(A_i)) + c_{i26}I_{i2}(6\omega_i(A_i)) \\
&\quad + c_{i28}I_{i2}(8\omega_i(A_i)), \\
c_{i3} &= c_{i31}I_{i1}(\omega_i(A_i)) + c_{i33}I_{i1}(3\omega_i(A_i)) + c_{i35}I_{i1}(5\omega_i(A_i)) \\
&\quad + c_{i37}I_{i1}(7\omega_i(A_i)), \\
c_{i4} &= c_{i42}I_{i2}(2\omega_i(A_i)) + c_{i44}I_{i2}(4\omega_i(A_i)) + c_{i46}I_{i2}(6\omega_i(A_i)) \\
&\quad + c_{i48}I_{i2}(8\omega_i(A_i)), \\
I_{ij}(\omega) &= \int_{-\infty}^{0} \sin(\omega\tau)C_{ij}(\tau)\mathrm{d}\tau = -\frac{D_{ij}}{\omega_{ij}}\frac{\omega}{\omega^2+\omega_{ij}^2}, \quad (i,j=1,2),
\end{aligned}
$$

$$
\begin{aligned}
m_{i11} &= \pi[2b_{i0}(A_i) - b_{i2}(A_1)]\frac{\mathrm{d}}{\mathrm{d}A_i}\left[\frac{2b_{i0}(A_i) - b_{i2}(A_i)}{A_i^2\alpha_i+\omega_{0i}^2}\right] \Big/ \left[8\left(A_i^2\alpha_i+\omega_{0i}^2\right)\right], \\
m_{i13} &= \pi[b_{i2}(A_i) - b_{i4}(A_i)]\frac{\mathrm{d}}{\mathrm{d}A_i}\left[\frac{b_{i2}(A_i) - b_{i4}(A_i)}{A_i^2\alpha_i+\omega_{0i}^2}\right] \Big/ \left[8\left(A_i^2\alpha_i+\omega_{0i}^2\right)\right], \\
m_{i15} &= \pi[b_{i4}(A_i) - b_{i6}(A_i)]\frac{\mathrm{d}}{\mathrm{d}A_i}\left[\frac{b_{i4}(A_i) - b_{i6}(A_i)}{A_i^2\alpha_i+\omega_{0i}^2}\right] \Big/ \left[8\left(A_i^2\alpha_i+\omega_{0i}^2\right)\right], \\
m_{i17} &= \pi b_{i6}(A_i)\frac{\mathrm{d}}{\mathrm{d}A_i}\left[\frac{b_{i6}(A_i)}{A_i^2\alpha_i+\omega_{0i}^2}\right] \Big/ \left[8\left(A_i^2\alpha_i+\omega_{0i}^2\right)\right],
\end{aligned}
$$

$$
\begin{aligned}
m_{i22} &= \pi A_i[2b_{i0}(A_i) - b_{i4}(A_i)]\frac{\mathrm{d}}{\mathrm{d}A_i}\left\{\frac{[2b_{i0}(A_i) - b_{i4}(A_i)]A_i}{{A_i}^2\alpha_i + \omega_{0i}^2}\right\} \Big/ \left[32\left({A_i}^2\alpha_i + \omega_{0i}^2\right)\right], \\
m_{i24} &= \pi A_i[b_{i2}(A_i) - b_{i6}(A_i)]\frac{\mathrm{d}}{\mathrm{d}A_i}\left\{\frac{[b_{i2}(A_i) - b_{i6}(A_i)]A_i}{{A_i}^2\alpha_i + \omega_{0i}^2}\right\} \Big/ \left[32\left({A_i}^2\alpha_i + \omega_{0i}^2\right)\right], \\
m_{i26} &= \pi A_i b_{i4}(A_i)\frac{\mathrm{d}}{\mathrm{d}A_i}\left[\frac{b_{i4}(A_i)A_i}{{A_i}^2\alpha_i + \omega_{0i}^2}\right] \Big/ \left[32\left({A_i}^2\alpha_i + \omega_{0i}^2\right)\right], \\
m_{i28} &= \pi A_i b_{i6}(A_i)\frac{\mathrm{d}}{\mathrm{d}A_i}\left[\frac{b_{i6}(A_i)A_i}{{A_i}^2\alpha_i + \omega_{0i}^2}\right] \Big/ \left[32\left({A_i}^2\alpha_i + \omega_{0i}^2\right)\right],
\end{aligned}
$$

$$
\begin{aligned}
m_{i31} &= \pi\left[4b_{i0}^2(A_i) - b_{i2}^2(A_i)\right] \Big/ \left[8A_i\left(A_i^2\alpha_i+\omega_{0i}^2\right)^2\right], \\
m_{i33} &= 3\pi\left[b_{i2}^2(A) - b_{i4}^2(A)\right] \Big/ \left[8A_i\left(A_i^2\alpha_i+\omega_{0i}^2\right)^2\right],
\end{aligned}
$$

$$m_{i35} = 5\pi\left[b_{i4}^2(A_i) - b_{i6}^2(A_i)\right]\Big/\left[8A_i\left(A_i^2\alpha_i + \omega_{0i}^2\right)^2\right],$$
$$m_{i37} = 7\pi b_{i6}^2(A_i)\Big/\left[8A_i\left(A_i^2\alpha_i + \omega_{0i}^2\right)^2\right],$$

$$m_{i42} = \pi A_i[2b_{i0}(A_i) - b_{i4}(A_i)][2b_{i0}(A_i) + 2b_{i2}(A_i) + b_{i4}(A_i)]\Big/\left[16\left(A_i^2\alpha_1 + \omega_{0i}^2\right)^2\right],$$
$$m_{i44} = \pi A_i[b_{i2}(A_i) - b_{i6}(A_i)][b_{i2}(A_i) + 2b_{i4}(A_i) + b_{i6}(A_i)]\Big/\left[8\left(A_i^2\alpha_i + \omega_{0i}^2\right)^2\right],$$
$$m_{i46} = 3\pi A_i b_{i4}(A_i)[b_{i4}(A_i) + 2b_{i6}(A_i)]\Big/\left[16\left(A_i^2\alpha_i + \omega_{0i}^2\right)^2\right],$$
$$m_{i48} = \pi A_i b_{i6}^2(A_i)\Big/\left[4\left(A_i^2\alpha_i + \omega_{0i}^2\right)^2\right],$$

$$c_{i11} = [2b_{i0}(A_i) + b_{i2}(A_i)]^2\Big/\left[8A_i^2\left(A_i^2\alpha_i + \omega_{0i}^2\right)^2\right],$$
$$c_{i13} = 3[b_{i2}(A_i) + b_{i4}(A_i)]^2\Big/\left[8A_i^2\left(A_i^2\alpha_i + \omega_{0i}^2\right)^2\right],$$
$$c_{i15} = 5[b_{i4}(A_i) + b_{i6}(A_i)]^2\Big/\left[8A_i^2\left(A_i^2\alpha_i + \omega_{0i}^2\right)^2\right],$$
$$c_{i17} = 7[b_{i6}(A_i)]^2\Big/\left[8A_i^2\left(A_i^2\alpha_i + \omega_{0i}^2\right)^2\right],$$

$$c_{i22} = [2b_{i0}(A_i) + 2b_{i2}(A_i) + b_{i4}(A_i)]^2\Big/\left[16\left(A_i^2\alpha_i + \omega_{0i}^2\right)^2\right],$$
$$c_{i24} = [b_{i2}(A_i) + 2b_{i4}(A_i) + b_{i6}(A_i)]^2\Big/\left[8\left(A_i^2\alpha_i + \omega_{0i}^2\right)^2\right],$$
$$c_{i26} = 3[b_{i4}(A_i) + 2b_{i6}(A_i)]^2\Big/\left[16\left(A_i^2\alpha_i + \omega_{0i}^2\right)^2\right],$$
$$c_{i28} = [b_{i6}(A_i)]^2\Big/\left[4\left(A_i^2\alpha_i + \omega_{0i}^2\right)^2\right],$$

$$c_{i31} = [2b_{i0}(A_i) - b_{i2}(A_i)]\frac{\mathrm{d}}{\mathrm{d}A_i}\left[\frac{2b_{i0}(A_i) + b_{i2}(A_i)}{A_i^3\alpha_i + \omega_{0i}^2 A_i}\right]\Big/\left[8\left(A_i^2\alpha_i + \omega_{0i}^2\right)\right],$$
$$c_{i33} = [b_{i2}(A_i) - b_{i4}(A_i)]\frac{\mathrm{d}}{\mathrm{d}A_i}\left[\frac{b_{i2}(A_i) + b_{i4}(A_i)}{A_i^3\alpha_i + \omega_{0i}^2 A_i}\right]\Big/\left[8\left(A_i^2\alpha_i + \omega_{0i}^2\right)\right],$$
$$c_{i35} = [b_{i4}(A_i) - b_{i6}(A_i)]\frac{\mathrm{d}}{\mathrm{d}A_i}\left[\frac{b_{i4}(A_i) + b_{i6}(A_i)}{A_i^3\alpha_i + \omega_{0i}^2 A_i}\right]\Big/\left[8\left(A_i^2\alpha_i + \omega_{0i}^2\right)\right],$$
$$c_{i37} = b_{i6}(A_i)\frac{\mathrm{d}}{\mathrm{d}A_i}\left[\frac{b_{i6}(A_i)}{A_i^3\alpha_i + \omega_{0i}^2 A_i}\right]\Big/\left[8\left(A_i^2\alpha_i + \omega_{0i}^2\right)\right],$$

$$c_{i42} = A_i[2b_{i0}(A_i) - b_{i4}(A_i)]\frac{\mathrm{d}}{\mathrm{d}A_i}\left[\frac{2b_{i0}(A_i) + 2b_{i2}(A_i) + b_{i4}(A_i)}{A_i^2\alpha_i + \omega_{0i}^2}\right]$$
$$/\left[32\left({A_i}^2\alpha_i + \omega_{0i}^2\right)\right],$$
$$c_{i44} = A_i[b_{i2}(A_i) - b_{i6}(A_i)]\frac{\mathrm{d}}{\mathrm{d}A_i}\left[\frac{b_{i2}(A_i) + 2b_{i4}(A_i) + b_{i6}(A_i)}{A_i^2\alpha_i + \omega_{0i}^2}\right]$$

$$
\begin{aligned}
&/\left[32\left({A_i}^2\alpha_i + \omega_{0i}^2\right)\right],\\
c_{i46} &= A_i b_{i4}(A_i)\frac{\mathrm{d}}{\mathrm{d}A_i}\left[\frac{b_{i4}(A_i) + 2b_{j6}(A_i)}{A_i^2\alpha_i + \omega_{0i}^2}\right]\Big/\left[32\left({A_i}^2\alpha_i + \omega_{0i}^2\right)\right],\\
c_{i48} &= A_i b_{i6}(A_i)\frac{\mathrm{d}}{\mathrm{d}A_i}\left[\frac{b_{j6}(A_i)}{A_i^2\alpha_i + \omega_{0i}^2}\right]\Big/\left[32\left({A_i}^2\alpha_i + \omega_{0i}^2\right)\right].
\end{aligned}
$$

and

$$
\begin{aligned}
B_{ii} &= f_{i1}(A_i)f_{i1}(A_i) + f_{i2}(A_i)f_{i2}(A_i) = b_{ii1} + b_{ii2},\\
K_{11} &= g_{11}^2 + g_{12}^2 = k_{111} + k_{112},\\
K_{22} &= g_{21}^2 + g_{22}^2 = K_{11} + k_{221} + k_{222},\\
K_{12} &= g_{11}g_{21} + g_{12}g_{22} = -K_{11},\\
B_{12} &= B_{21} = f_{11}f_{21} + f_{12}f_{22} = 0,
\end{aligned}
$$

$$
\begin{aligned}
&f_{i1}g_{j1} + f_{i2}g_{j2} = 0 \quad (i, j = 1, 2, i \neq j),\\
&b_{ii1} = b_{ii11}S_{i1}(\omega_i(A_i)) + b_{ii13}S_{i1}(3\omega_i(A_i)) + b_{ii15}S_{i1}(5\omega_i(A_i))\\
&\qquad + b_{ii17}S_{i1}(7\omega_i(A_i)),\\
&b_{ii2} = b_{ii22}S_{i2}(2\omega_i(A_i)) + b_{ii24}S_{i2}(4\omega_i(A_i)) + b_{ii26}S_{i2}(6\omega_i(A_i))\\
&\qquad + b_{ii28}S_{i2}(8\omega_i(A_i)),\\
&k_{ii1} = k_{ii11}S_{i1}(\omega_i(A_i)) + k_{ii13}S_{i1}(3\omega_i(A_i)) + k_{ii15}S_{i1}(5\omega_i(A_i))\\
&\qquad + k_{ii17}S_{i1}(7\omega_i(A_i)),\\
&k_{ii2} = k_{ii20}S_{i2}(0) + k_{ii22}S_{i2}(2\omega_i(A_i)) + k_{ii24}S_{i2}(4\omega_i(A_i))\\
&\qquad + k_{ii26}S_{i2}(6\omega_i(A_i)) + k_{ii28}S_{i2}(8\omega_i(A_i)),
\end{aligned}
$$

$$
\begin{aligned}
b_{ii11} &= \pi[b_{i2}(A_i) - 2b_{i0}(A_i)]^2\Big/\left[4\left(\alpha_i A_i^2 + \omega_{0i}^2\right)^2\right],\\
b_{ii13} &= \pi[b_{i2}(A_i) - b_{i4}(A_i)]^2\Big/\left[4\left(\alpha_i A_i^2 + \omega_{0i}^2\right)^2\right],\\
b_{ii15} &= \pi[b_{i4}(A_i) - b_{i6}(A_i)]^2\Big/\left[4\left(\alpha_i A_i^2 + \omega_{0i}^2\right)^2\right],\\
b_{ii17} &= \pi b_{i6}^2(A_i)\Big/\left[4\left(\alpha_i A_i^2 + \omega_{0i}^2\right)^2\right],
\end{aligned}
\tag{1.219}
$$

$$
\begin{aligned}
b_{ii22} &= \pi A_i^2[b_{i4}(A_i) - 2b_{i0}(A_i)]^2\Big/\left[16\left(\alpha_i A_i^2 + \omega_{0i}^2\right)^2\right],\\
b_{ii24} &= \pi A_i^2[b_{i2}(A_i) - b_{i6}(A_i)]^2\Big/\left[16\left(\alpha_i A_i^2 + \omega_{0i}^2\right)^2\right],\\
b_{ii26} &= \pi A_i^2 b_{i4}^2(A_i)\Big/\left[16\left(\alpha_i A_i^2 + \omega_{0i}^2\right)^2\right],\\
b_{ii28} &= \pi A_i^2 b_{i6}^2(A_i)\Big/\left[16\left(\alpha_i A_i^2 + \omega_{0i}^2\right)^2\right],
\end{aligned}
$$

$$
\begin{aligned}
k_{ii11} &= \pi\,[2b_{i0}(A_i) + b_{i2}(A_i)]^2 \Big/ \left[4A_i^2\left(\alpha_i A_i^2 + \omega_{0i}^2\right)^2\right],\\
k_{ii13} &= \pi\,[b_{i2}(A_i) + b_{i4}(A_i)]^2 \Big/ \left[4A_i^2\left(\alpha_i A_i^2 + \omega_{0i}^2\right)^2\right],\\
k_{ii15} &= \pi\,[b_{i4}(A_i) + b_{i6}(A_i)]^2 \Big/ \left[4A_i^2\left(\alpha_i A_i^2 + \omega_{0i}^2\right)^2\right],\\
k_{ii17} &= \pi\,[b_{i6}(A_i)]^2 \Big/ \left[4A_i^2\left(\alpha_i A_i^2 + \omega_{0i}^2\right)^2\right],
\end{aligned}
$$

$$
\begin{aligned}
k_{ii20} &= \pi\,[2b_{i0}(A_i) + b_{i2}(A_i)]^2 \Big/ \left[8\left(\alpha_i A_i^2 + \omega_{0i}^2\right)^2\right],\\
k_{ii22} &= \pi\,[2b_{i0}(A_i) + 2b_{i2}(A_i) + b_{i4}(A_i)]^2 \Big/ \left[16\left(\alpha_i A_i^2 + \omega_{0i}^2\right)^2\right],\\
k_{ii24} &= \pi\,[b_{i2}(A_i) + 2b_{i4}(A_i) + b_{i6}(A_i)]^2 \Big/ \left[16\left(\alpha_i A_i^2 + \omega_{0i}^2\right)^2\right],\\
k_{ii26} &= \pi\,[b_{i4}(A_i) + 2b_{i6}(A_i)]^2 \Big/ \left[16\left(\alpha_i A_i^2 + \omega_{0i}^2\right)^2\right],\\
k_{ii28} &= \pi\,[b_{i6}(A_i)]^2 \Big/ \left[16\left(\alpha_i A_i^2 + \omega_{0i}^2\right)^2\right], \quad (i, j = 1, 2).
\end{aligned}
$$

The averaged Itô Eq. (1.217) is used to estimate the reliability of the system (1.212). Define the system safety domain Π as $0 \le A_1 < A_{1c},\ \ 0 \le A_2 < A_{2c}$, where A_{1c}, A_{2c} are the boundaries of the safe domain. It is noted that the angle processes Δ_1 and Δ_2 have no effect on the reliability. The conditional reliability function is defined as

$$
\begin{aligned}
R(t|A_{10}, A_{20}, \Delta_{10}, \Delta_{20}) = \mathrm{Prob}[&A_1, A_2 \in \Pi,\ \ \forall\,\Delta_1 \text{ and } \Delta_2, \tau \in (0, t]\ |A_{10},\\
&A_{20} \in \Pi,\ \ \forall\,\Delta_{10} \text{ and } \Delta_{20}].
\end{aligned}
\tag{1.220}
$$

Equation (1.220) implies the probability that the process $A_1(\tau)$, $A_2(\tau)$ do not exceed safety domain Π within the time interval $\tau \in (0, t]$, given the initial values $A_1(\tau{=}\,0) = A_{10}$ and $A_2(\tau{=}\,0) = A_{20}$. The conditional reliability function $R(t|A_{10}, A_{20}, \Delta_{10}, \Delta_{20})$ satisfies the backward Kolmogorov equation

$$
\begin{aligned}
\frac{\partial R}{\partial t} &= F_{10}\frac{\partial R}{\partial a_{10}} + F_{20}\frac{\partial R}{\partial a_{20}} + G_{10}\frac{\partial R}{\partial \delta_{10}} + G_{20}\frac{\partial R}{\partial \delta_{20}}\\
&\quad + \frac{1}{2}B_{110}\frac{\partial^2 R}{\partial a_{10}^2} + \frac{1}{2}B_{220}\frac{\partial^2 R}{\partial a_{20}^2} + \frac{1}{2}K_{110}\frac{\partial^2 R}{\partial \delta_{10}^2} + \frac{1}{2}K_{220}\frac{\partial^2 R}{\partial \delta_{20}^2}\\
&\quad + K_{120}\frac{\partial^2 R}{\partial \delta_{10}\partial \delta_{20}}.
\end{aligned}
\tag{1.221}
$$

where the cross partial derivative terms vanished since the associated coefficient functions are zero.

The coefficients in Eq. (1.221) are obtained from Eq. (1.218) and Eq. (1.219) as follows

$$
\begin{aligned}
&F_{10} = F_1(a_{10}, a_{20}, \delta_{10}, \delta_{20}), \quad F_{20} = F_2(a_{10}, a_{20}, \delta_{10}, \delta_{20}),\\
&G_{10} = G_1(a_{10}, a_{20}, \delta_{10}, \delta_{20}), \quad G_{20} = G_2(a_{10}, a_{20}, \delta_{10}, \delta_{20}),\\
&B_{110} = B_{11}(a_{10}, a_{20}, \delta_{10}, \delta_{20}), \quad B_{220} = B_{22}(a_{10}, a_{20}, \delta_{10}, \delta_{20}),\\
&K_{110} = K_{11}(a_{10}, a_{20}, \delta_{10}, \delta_{20}), \quad K_{220} = K_{22}(a_{10}, a_{20}, \delta_{10}, \delta_{20}),\\
&K_{120} = K_{12}(a_{10}, a_{20}, \delta_{10}, \delta_{20}).
\end{aligned} \tag{1.222}
$$

The initial conditions are

$$
\begin{aligned}
&R(t = 0|A_{10}, A_{20}, \Delta_{10}, \Delta_{20}) = 1,\\
&\text{when } 0 \le A_{10} < A_{1c}, \ \ 0 \le A_{20} < A_{2c}, \ \ \forall\, \Delta_{10} \text{ and } \Delta_{20}.
\end{aligned} \tag{1.223}
$$

and the boundary conditions are

$$
\begin{aligned}
&R(t|A_{10}, A_{20}, \Delta_{10}, \Delta_{20}) = \text{finite}, \quad \text{when } A_{10} = 0 \ \text{ and/or } \ A_{20} = 0,\\
&R(t|A_{10}, A_{20}, \Delta_{10}, \Delta_{20}) = 0, \quad \text{when } A_{10} = A_{1c} \ \text{ and/or } \ A_{20} = A_{2c},\\
&R(t|A_{10}, A_{20}, \Delta_{10}, \Delta_{20}) = R(t|A_{10}, A_{20}, \Delta_{10} + 2\pi, \Delta_{20}),\\
&R(t|A_{10}, A_{20}, \Delta_{10}, \Delta_{20}) = R(t|A_{10}, A_{20}, \Delta_{10}, \Delta_{20} + 2\pi).
\end{aligned} \tag{1.224}
$$

Equation (1.224) implies that $A_{10}=A_{1c}$, $A_{20}=A_{2c}$ are absorbing boundaries, $A_{10}=0$, $A_{20}=0$ are reflecting boundaries, and Δ_{10}, Δ_{20} are periodic boundaries.

The first passage time T is defined as the time when $A_1(t)$ or $A_2(t)$ first reaches its critical value A_{1c} or A_{2c}, given the initial values $A_{10} < A_{1c}$ and $A_{20} < A_{2c}$. The conditional PDF of the first passage time T can be obtained from the conditional reliability function as

$$
p(T|A_{10}, A_{20}, \Delta_{10}, \Delta_{20}) = \left.\frac{-\partial R(t|A_{10}, A_{20}, \Delta_{10}, \Delta_{20})}{\partial t}\right|_{t=T}. \tag{1.225}
$$

The mean first passage time can be obtained from the conditional PDF as follow

$$
\mu(A_{10}, A_{20}, \Delta_{10}, \Delta_{20}) = E[T] = \int_0^\infty Tp(T|A_{10}, A_{20}, \Delta_{10}, \Delta_{20})\mathrm{d}T. \tag{1.226}
$$

The mean first passage time can also be obtained by solving the following Pontryagin equation

$$
\begin{aligned}
&F_{10}\frac{\partial \mu}{\partial a_{10}} + F_{20}\frac{\partial \mu}{\partial a_{20}} + G_{10}\frac{\partial \mu}{\partial \delta_{10}} + G_{20}\frac{\partial \mu}{\partial \delta_{20}}\\
&+ \frac{1}{2}B_{110}\frac{\partial^2 \mu}{\partial a_{10}^2} + \frac{1}{2}B_{220}\frac{\partial^2 \mu}{\partial a_{20}^2} + \frac{1}{2}K_{110}\frac{\partial^2 \mu}{\partial \delta_{10}^2}
\end{aligned}
$$

$$+\frac{1}{2}K_{220}\frac{\partial^2\mu}{\partial\delta_{20}^2}+K_{120}\frac{\partial^2\mu}{\partial\delta_{10}\partial\delta_{20}}=-1, \tag{1.227}$$

where boundary conditions are

$$\begin{aligned}
&\mu(A_{10},A_{20},\Delta_{10},\Delta_{20})=\text{finite}, \quad \text{when } A_{10}=0 \text{ or } A_{20}=0,\\
&\mu(A_{10},A_{20},\Delta_{10},\Delta_{20})=0, \quad \text{when } A_{10}=A_{1c} \text{ or } A_{20}=A_{2c},\\
&\mu(A_{10},A_{20},\Delta_{10},\Delta_{20})=\mu(A_{10},A_{20},\Delta_{10}+2\pi,\Delta_{20}),\\
&\mu(A_{10},A_{20},\Delta_{10},\Delta_{20})=\mu(A_{10},A_{20},\Delta_{10},\Delta_{20}+2\pi).
\end{aligned} \tag{1.228}$$

For system (1.212) with parameters $A_{ic}=0.5$, $\beta_{i0}=-0.01$, $\beta_{ij}=0.01$, $\eta_{ij}=0.01$, $\omega_{0i}=1.0$, $E_i=0.03$, $\Omega=1.1$, $\omega_{ij}=50$, $D_{i2}=10$ $(i,j=1,2)$, Figs. 1.39, 1.40 and 1.41 show a set of results for $R(t)$, $p(T)$ and μ obtained by using the stochastic averaging method and from Monte Carlo simulation. Figure 1.39 illustrates the conditional reliability function $R(t)$ for various excitation intensities $2D_{i1}$, $2D_{i2}$ and nonlinear stiffness parameters α_i. The values of the reliability functions start at $R(t=0)=1$ and decrease as time going on. Figure 1.40 shows PDF $p(T)$ of first passage time, which are calculated from conditional reliability functions $R(t)$ using Eq. (1.225). Figure 1.41 presents changes of the mean first passage time with varying initial values. As expected, the closer the initial value deviates from the boundary, the shorter the mean first passage time. The results by using the stochastic averaging method and those from Monte Carlo simulation of original system (1.212) are in quite good agreement.

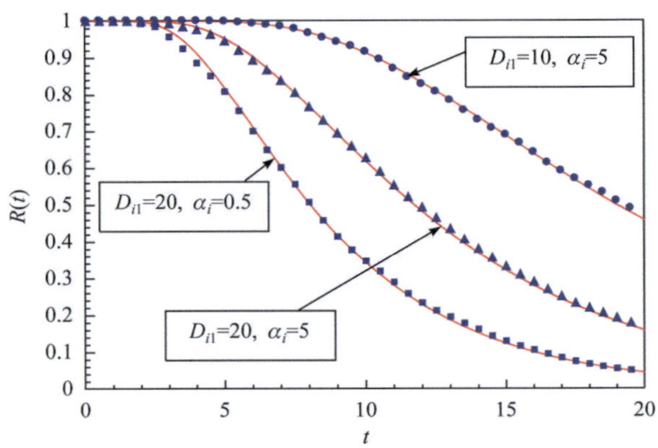

Fig. 1.39 Conditional reliability function of system (1.212). Initial conditions are $A_{10}=A_{20}=\Delta_{10}=\Delta_{20}=0$. ——, results from the stochastic averaging method; ■•▲, results from Monte Carlo simulation (Wu et al. 2013)

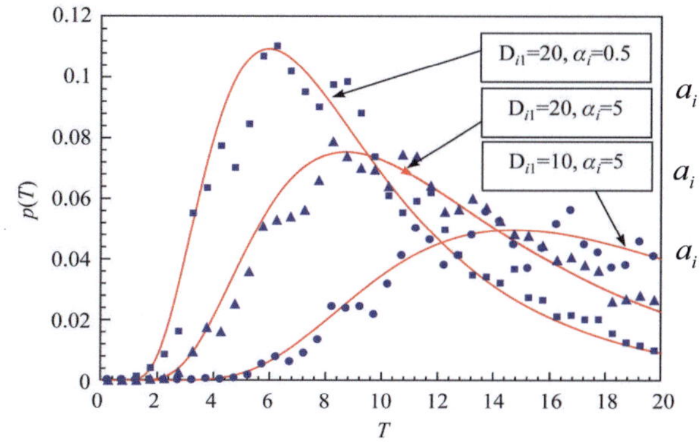

Fig. 1.40 PDF of first passage time of system (1.212) with initial conditions similar to those in Fig. 1.39. ——, results from the stochastic averaging method; ■•▲, results from Monte Carlo simulation (Wu et al. 2013)

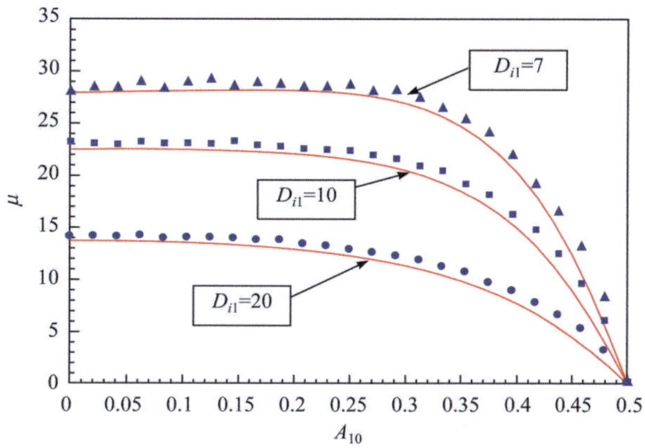

Fig. 1.41 Mean first passage time of system (1.212) with varying A_{10}. Other parameters are the same as those in Fig. 1.39. ——, results from the stochastic averaging method; ■•▲, results from Monte Carlo simulation (Wu et al. 2013)

1.4 Narrowband Randomized Harmonic Noise Excitation

Randomized harmonic noise, also known as bounded noise, was introduced in Sect. 2.6.3, vol. 1. This noise has bounded amplitude, and adjustable parameter to make the peak position and bandwidth of PSD aligned with those of actual noise. The randomized harmonic noise is defined as

$$\xi(t) = \sin(\Omega t + \sigma B(t) + \chi), \tag{1.229}$$

where Ω represents the center frequency; $B(t)$ is a unit Wiener process; χ is a random phase angle with uniform distribution on $[0, 2\pi)$; σ is a constant; $\sigma B(t) + \chi$ specifies the random perturbation of phase angle. According to Eqs. (2.309) and (2.310) in vol. 1, the autocorrelation function of $\xi(t)$ is

$$R_\xi(\tau) = \frac{1}{2}\exp\left(-\frac{1}{2}\sigma^2|\tau|\right)\cos\Omega\tau, \tag{1.230}$$

and PSD of $\xi(t)$ is

$$S_\xi(\omega) = \frac{\sigma^2}{4\pi} \frac{\omega^2 + \Omega^2 + \sigma^4/4}{(\omega^2 - \Omega^2 - \sigma^4/4)^2 + \sigma^4\omega^2}. \tag{1.231}$$

The bandwidth of $\xi(t)$ primarily depends on σ. When σ is small, $\xi(t)$ is a narrowband noise, when σ is large, $\xi(t)$ is a wideband noise, and when $\sigma \to \infty$, $\xi(t)$ becomes white noise. In the following, the stochastic averaging method of quasi-integrable Hamiltonian systems excited by narrowband randomized harmonic noises is presented.

1.4.1 SDOF System

Consider a SDOF quasi-Hamiltonian system excited by a narrowband randomized harmonic noise (Huang et al. 2002). The motion equations are

$$\begin{aligned} \dot{Q} &= P, \\ \dot{P} &= -g(Q) - \varepsilon c(Q,P)P + \varepsilon^{1/2} f(Q,P)\xi(t), \end{aligned} \tag{1.232}$$

where $\xi(t)$ is a randomized harmonic noise defined in Eq. (1.229), ε is a small parameter. Assume that the Hamiltonian system associated with system (1.232) has a family of periodic solutions of the form of Eq. (1.6) in domain V around equilibrium point $(b, 0)$. Applying the transformation in Eq. (1.12) and deriving according to Eqs. (1.14) to (1.18) lead to

$$\begin{aligned} \frac{\mathrm{d}A}{\mathrm{d}t} &= \varepsilon F_1(A, \Phi, \Omega t + \Lambda), \\ \frac{\mathrm{d}\Phi}{\mathrm{d}t} &= \nu(A, \Phi) + \varepsilon F_2(A, \Phi, \Omega t + \Lambda), \end{aligned} \tag{1.233}$$

where $\Lambda = \sigma B + \chi$ and

$$\begin{aligned} F_1 &= \frac{-A}{g(A+B)(1+d)}[c(A\cos\Phi + B, -A\nu(A,\Phi)\sin\Phi)A\nu(A,\Phi)\sin\Phi \\ &\quad + f(A\cos\Phi + B, -A\nu(A,\Phi)\sin\Phi)\sin(\Omega t + \Lambda)]\nu(A,\Phi)\sin\Phi, \\ F_2 &= \frac{-1}{g(A+B)(1+d)}[c(A\cos\Phi + B, -A\nu(A,\Phi)\sin\Phi)A\nu(A,\Phi)\sin\Phi \\ &\quad + f(A\cos\Phi + B, -A\nu(A,\Phi)\sin\Phi)\sin(\Omega t + \Lambda)]\nu(A,\Phi)(\cos\Phi + d), \\ d &= \frac{\mathrm{d}B}{\mathrm{d}A} = \frac{g(-A+B) + g(A+B)}{g(-A+B) - g(A+B)}. \end{aligned} \tag{1.234}$$

Assume that center frequency Ω of $\xi(t)$ and system mean frequency $\omega(A)$ satisfy the following resonant relation

$$\frac{\Omega}{\omega(A)} = \frac{s}{r} + \varepsilon\kappa, \tag{1.235}$$

where s, r are mutual prime numbers, $\varepsilon\kappa$ is a small detuning parameter.

Introducing a new variable

$$\Delta = \Omega t + \Lambda - \frac{s}{r}\Phi, \tag{1.236}$$

where Δ can be regarded as a measure of the difference between phase angle $\Omega t + \Lambda$ of the randomized harmonic noise and s/r phase angle $\Phi s/r$ of system response. Regarding Eq. (1.236) as a transformation from Φ to Δ, the following Itô SDE can be derived from Eq. (1.233)

$$\begin{aligned}
\mathrm{d}A &= \varepsilon F_1(A, \frac{r}{s}(\Omega t + \Lambda - \Delta), \Omega t + \Lambda)\mathrm{d}t,\\
\mathrm{d}\Delta &= \Big[\Omega - \frac{s}{r}\nu(A, \frac{r}{s}(\Omega t + \Lambda - \Delta)) - \frac{s}{r}\varepsilon F_2(A, \frac{r}{s}(\Omega t\\
&\quad + \Lambda - \Delta), \Omega t + \Lambda)\Big]\mathrm{d}t + \sigma \mathrm{d}B(t).
\end{aligned} \tag{1.237}$$

Both $A(t)$ and $\Delta(t)$ are slowly-varying processes. According to Khasminskii's theorem (Khasminskii 1968), $[A, \Delta]^{\mathrm{T}}$ converges weakly to a two-dimensional vector Markov diffusion process as $\varepsilon \to 0$. The averaged Itô SDEs for $[A, \Delta]^{\mathrm{T}}$ are obtained from time averaging for Eq. (1.237) as follows

$$\begin{aligned}
\mathrm{d}A &= m_1(A, \Delta)\mathrm{d}t,\\
\mathrm{d}\Delta &= m_2(A, \Delta)\mathrm{d}t + \sigma \mathrm{d}B(t),
\end{aligned} \tag{1.238}$$

where

$$\begin{aligned}
m_1(A, \Delta) &= \varepsilon\Big\langle F_1(A, \frac{r}{s}(\Omega t + \Lambda - \Delta))\Big\rangle_{\Omega t+\Lambda},\\
m_2(A, \Delta) &= \varepsilon\kappa\omega(A) - \varepsilon\Big\langle\frac{s}{r}F_2(A, \frac{r}{s}(\Omega t + \Lambda - \Delta))\Big\rangle_{\Omega t+\Lambda}.
\end{aligned} \tag{1.239}$$

The associated averaged FPK equation is

$$\frac{\partial p}{\partial t} = -\frac{\partial}{\partial a}(m_1 p) - \frac{\partial}{\partial \delta}(m_2 p) + \frac{\sigma^2}{2}\frac{\partial^2 p}{\partial \delta^2}, \tag{1.240}$$

where the first derivative moments are

$$m_1 = m_1(a, \delta) = m_1(A, \Delta)|_{A=a,\ \Delta=\delta}, \quad m_2 = m_2(a, \delta) = m_2(A, \Delta)|_{A=a,\ \Delta=\delta}. \tag{1.241}$$

In (1.240), $p = p(a, \delta, t|a_0, \delta_0)$ is the transition PSD of $[\mathbf{A}, \boldsymbol{\Delta}]^{\mathrm{T}}$ with the initial condition

$$p(a, \delta, 0|a_0, \delta_0) = \delta(a - a_0)\delta(\delta - \delta_0), \tag{1.242}$$

or $p = p(a, \delta, t)$ is the transient PSD of $[\mathbf{A}, \boldsymbol{\Delta}]^{\mathrm{T}}$ with the initial condition

$$p(a, \delta, 0) = p(a_0, \delta_0). \tag{1.243}$$

The boundary condition with respective to a depends on the domain V. When V is the entire plane (q, p), they have form of Eqs. (1.28) and (1.29). The periodic boundary condition with respective to δ is

$$p(a, \delta + 2k\pi, t|a_0, \delta_0) = p(a, \delta|a_0, \delta_0), \quad k = 0, \pm 1, \pm 2, \ldots \tag{1.244}$$

Solving the reduced averaged FPK equation of (1.240) yields the stationary PDF $p(a, \delta)$. The approximate stationary PDF of the original system (1.232) is them

$$p(q, p) = p(a, \delta)\left|\frac{\partial(a, \delta)}{\partial(q, p)}\right| = p(a, \delta)\left|\frac{\partial(a, \delta)}{\partial(a, \phi)}\right|\left|\frac{\partial(a, \phi)}{\partial(q, p)}\right| = \frac{s}{r}p(a, \delta)\left|\frac{\partial(a, \phi)}{\partial(q, p)}\right|. \tag{1.245}$$

Example 1.11 Consider a hard spring Duffing oscillator excited by a narrowband randomized harmonic noise (Huang et al. 2002). The motion equations are

$$\begin{aligned} \dot{Q} &= P, \\ \dot{P} &= -\omega_0^2 Q - \alpha Q^3 - \beta P + E\sin(\Omega t + \sigma B(t) + \chi). \end{aligned} \tag{1.246}$$

Equations (1.38) to (1.41) in Example 1.1 remain applicable to system (1.246). Considering the primary resonance ($s = r = 1$) of Eq. (1.235) and following the derivation from Eqs. (1.233) to (1.239) yield the averaged Itô Eq. (1.238) with the following drift coefficients

$$\begin{aligned} m_1 &= \frac{-\beta A(\omega_0^2 + 5\alpha A^2/8)}{2(\omega_0^2 + \alpha A^2)} - \frac{E\cos\Delta}{4(\omega_0^2 + \alpha A^2)}[2b_0(A) - b_2(A)], \\ m_2 &= \frac{E\sin\Delta}{4A(\omega_0^2 + \alpha A^2)}[2b_0(A) + b_2(A)] + \Omega - \omega(A). \end{aligned} \tag{1.247}$$

Solving the reduced averaged FPK Eq. (1.240) with first derivative moment obtained from Eq. (1.247) according to Eq. (1.241) yields stationary PDF $p(a, \delta)$. $p(a, \delta)$ with two peaks shown in Fig. 1.42 implies occurrence of random jumps. A displacement sample obtained from Monte Carlo simulation of system (1.246) is shown in Fig. 1.43 to confirm the occurrence of random jumps. It is noted that random jumps may occur bidirectionally. Similar to Example 1.7, the stochastic averaging method developed in this section can also provide an analysis of random jumps and their bifurcation during changing of various parameters.

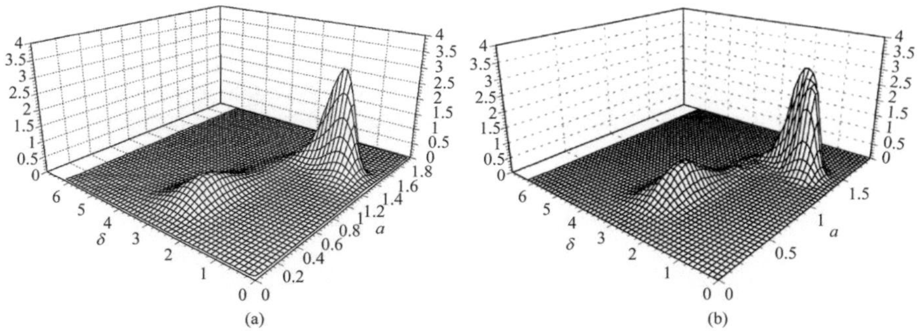

Fig. 1.42 Stationary PDF $p(a, \delta)$ of system (1.246), **a** results from the stochastic averaging method, **b** result from Monte Carlo simulation. System parameters $\omega_0 = 1.0$, $\alpha = 0.3$, $\beta = 0.1$, $\Omega = 1.2$, $E = 0.2$, $\sigma^2 = 0.02$ (Huang et al. 2002)

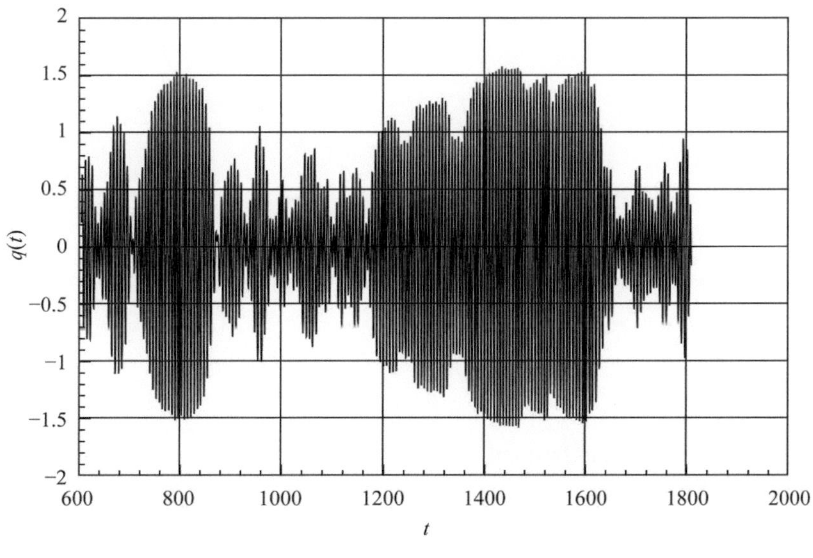

Fig. 1.43 Displacement samples $q(t)$ obtained from Monte Carlo simulation of system (1.246). System parameters are the same as those in Fig. 1.42 (Huang et al. 2002)

1.4.2 MDOF System

Consider a n-DOF quasi-integrable Hamiltonian system excited by weak narrowband randomized harmonic noises (Huang and Zhu 2004b). The motion equations are

$$\begin{aligned}
&\dot{Q}_i = P_i,\\
&\dot{P}_i = -g_i(Q_i) - \varepsilon \sum_{j=1}^{n} c_{ij}(\mathbf{Q},\mathbf{P})P_j + \varepsilon^{1/2}\sum_{k=1}^{m} f_{ik}(\mathbf{Q},\mathbf{P})\xi_k(t),\\
&i = 1, 2, \ldots, n,
\end{aligned} \tag{1.248}$$

where $g_i(Q_i)$ represent nonlinear restoring forces, ε is a small parameter, εc_{ij} are weak damping coefficients, $\xi_k(t)$ are narrowband randomized harmonic noises defined in Eq. (1.229), i.e.,

$$\xi_k(t) = \sin(\Omega_k t + \Lambda_k),\quad \Lambda_k = \sigma_k B_k(t) + \chi_k,\quad k = 1, 2, \ldots, m, \tag{1.249}$$

and $\varepsilon^{1/2} f_{ik}$ are small amplitudes of excitations.

Assume $g_i(Q_i)$ and $U_i(Q_i) = \int g_i(Q_i)\mathrm{d}Q_i$ satisfy the four conditions listed below Eq. (1.5). Introduce the transformation

$$\begin{aligned}
&Q_i(t) = A_i\cos\Phi_i(t) + B_i,\quad P_i(t) = -A_i\nu_i(A_i,\Phi_i)\sin\Phi_i(t),\\
&\Phi_i(t) = \Gamma_i(t) + \Theta_i(t).
\end{aligned} \tag{1.250}$$

where

$$\begin{aligned}
&\nu_i(A_i,\Phi_i) = \frac{\mathrm{d}\Gamma_i}{\mathrm{d}t} = \sqrt{\frac{2[U_i(A_i+B_i) - U_i(A_i\cos\Phi_i + B_i)]}{A_i^2\sin^2\Phi_i}},\\
&U_i(A_i+B_i) = U_i(A_i-B_i) = H_i.
\end{aligned} \tag{1.251}$$

in which H_i and $\nu_i(A_i,\Phi_i)$ are Hamiltonians and instantaneous frequencies of subsystem, respectively. Expanded $\nu_i(A_i,\Phi_i)$ into Fourier series

$$\nu_i(A_i,\Phi_i) = \omega_i(A_i) + \sum_{r=1}^{\infty}\omega_i^r(A_i)\cos r\Phi_i, \tag{1.252}$$

where $\omega_i(A_i)$ are the mean frequencies of subsystems. After transformation (1.250), system (1.248) becomes

$$\begin{aligned}
&\frac{\mathrm{d}A_i}{\mathrm{d}t} = \varepsilon F_i^{(1)}(\mathbf{A},\boldsymbol{\Phi},\boldsymbol{\Omega}t+\boldsymbol{\Lambda}),\\
&\frac{\mathrm{d}\Phi_i}{\mathrm{d}t} = \nu_i(A_i,\Phi_i) + \varepsilon F_i^{(2)}(\mathbf{A},\boldsymbol{\Phi},\boldsymbol{\Omega}t+\boldsymbol{\Lambda}),
\end{aligned}$$

$$i = 1, 2, \ldots, n. \tag{1.253}$$

where

$$\mathbf{A} = [A_1, A_2, \ldots, A_n]^{\mathrm{T}}, \quad \boldsymbol{\Phi} = [\Phi_1, \Phi_2, \ldots, \Phi_n]^{\mathrm{T}},$$
$$\boldsymbol{\Omega}t = [\Omega_1 t, \Omega_2 t, \ldots, \Omega_m t]^{\mathrm{T}}, \quad \boldsymbol{\Lambda} = [\Lambda_1, \Lambda_2, \ldots, \Lambda_m]^{\mathrm{T}},$$

$$F_i^{(1)} = \frac{-A_i}{g_i(A_i + B_i)(1 + d_i)} \Bigg[\sum_{j=1}^{n} c_{ij}(A_i \cos\Phi_i + B_i, -A_i \nu_i \sin\Phi_i A_j)\nu_j \sin\Phi_j$$
$$+ \sum_{k=1}^{m} f_{ik}(A_i \cos\Phi_i + B_i, -A_i \nu_i \sin\Phi_i)\sin(\Omega_k t + \Lambda_k) \Bigg] \nu_i \sin\Phi_i,$$

$$F_i^{(2)} = \frac{-1}{g_i(A_i + B_i)(1 + d_i)} \Bigg[\sum_{j=1}^{n} c_{ij}(A_i \cos\Phi_i + B_i, -A_i \nu_i \sin\Phi_i A_j)\nu_j \sin\Phi_j$$
$$+ \sum_{k=1}^{m} f_{ik}(A_i \cos\Phi_i + B_i, -A_i \nu_i \sin\Phi_i)\sin(\Omega_k t + \Lambda_k) \Bigg] \nu_i(\cos\Phi_i + d_i),$$
$$d_i = \frac{\mathrm{d}B_i}{\mathrm{d}A_i} = \frac{g_i(-A_i + B_i) + g_i(A_i + B_i)}{g_i(-A_i + B_i) - g_i(A_i + B_i)}, \quad i = 1, 2, \ldots, n. \tag{1.254}$$

System (1.248) may occur internal resonances among its subsystems. The narrowband randomized harmonic noise $\xi_k(t)$ may cause external resonances. If there is no external resonance, the impact of the narrowband noise on the system may be ignored in the first approximation. Thus, only the case of external resonance and the case of both internal and external resonances are studied in the following.

Suppose that n mean frequencies $\omega_i(A_i)$ $(i = 1, 2, \ldots, n)$ of subsystems of system (1.248) and $m$central frequencies $\Omega_1, \Omega_2, \ldots, \Omega_m$ of randomized harmonic noises satisfy the following μ resonance relations

$$\sum_{k=1}^{m} E_k^u \Omega_k + \sum_{i=1}^{n} I_i^u \omega_i = \varepsilon\kappa_u, \quad u = 1,2, \ldots, \mu, \tag{1.255}$$

where E_k^u, I_i^u are integers and at least one of E_k^u and one of I_i^u are not zero. $\varepsilon\kappa_u$ are small detuning quantities. It is noted that Eq. (1.255) is applicable for both cases of external resonance only and both internal and external resonance. Introducing μ combinations of angle variables

$$\Delta_u = \sum_{k=1}^{m} E_k^u(\Omega_k t + \Lambda_k) + \sum_{i=1}^{n} I_i^u \Phi_i, \quad u = 1,2, \ldots, \mu. \tag{1.256}$$

Regarding Eq. (1.256) as a transformation from $\boldsymbol{\Phi}$ to $\boldsymbol{\Delta} = [\Delta_1, \Delta_2, \ldots, \Delta_\mu]^{\mathrm{T}}$, the following equations can be obtained from Eq. (1.253)

$$\begin{aligned} \mathrm{d}\Delta_u = {} & [\sum_{k=1}^{m} E_k^u \Omega_k + \sum_{i=1}^{n} I_i^u v_i(A_i, \Phi_i) + \varepsilon \sum_{i=1}^{n} I_i^u F_i^{(2)}(\boldsymbol{A}, \overline{\Phi}, \boldsymbol{\Omega}t + \boldsymbol{\Lambda})]\mathrm{d}t \\ & + \sum_{k=1}^{m} E_k^u \sigma_k \mathrm{d}B_k(t), \quad u = 1, 2, \ldots, \mu. \end{aligned} \tag{1.257}$$

A_i in Eq. (1.253) and Δ_u in Eq. (1.257) constitute $(n + \mu)$–dimensional slowly-varying vector process. According to Khasminskii theorem (Khasminskii 1968), they converge to a $(n+\mu)$-dimensional vector Markov diffusion process. Replacing $\Phi_1, \Phi_2, \ldots, \Phi_\mu$ in $\boldsymbol{\Phi}$ with Δ_u $(u = 1, 2, \ldots, \mu)$ and $\Omega_k t + \Lambda_k$ $(k = 1, 2, \ldots, m)$ and replacing time averaging by averaging with respect to $\Phi_{\mu+1}, \Phi_{\mu+2}, \ldots, \Phi_n$ and $\Omega_k t + \Lambda_k$, the following averaged Itô SDEs can be obtained

$$\begin{aligned} & \mathrm{d}A_i = m_i^{(1)}(\boldsymbol{A}, \boldsymbol{\Delta})\mathrm{d}t, \\ & \mathrm{d}\Delta_u = m_u^{(2)}(\boldsymbol{A}, \boldsymbol{\Delta})\mathrm{d}t + \sum_{k=1}^{m} \sigma_{uk}^{(2)} \mathrm{d}B_k(t), \\ & i = 1,2, \ldots, n, \quad u = 1,2, \ldots, \mu, \end{aligned} \tag{1.258}$$

where the drift and diffusion coefficients are

$$\begin{aligned} & m_i^{(1)}(\mathbf{A}, \boldsymbol{\Delta}) = \varepsilon \left\langle F_i^{(1)}(\mathbf{A}, \overline{\Phi}, \boldsymbol{\Omega}t + \boldsymbol{\Lambda}) \right\rangle_{\overline{\Phi}, \boldsymbol{\Omega}t+\boldsymbol{\Lambda}}, \\ & m_u^{(2)}(\mathbf{A}, \boldsymbol{\Delta}) \\ & \quad = \varepsilon \left\langle \frac{1}{\varepsilon}[\sum_{k=1}^{m} E_k^u \Omega_k + \sum_{j=1}^{n} I_j^u v_j(A_j, \overline{\Phi}_j)] + \sum_{j=1}^{n} I_j^u F_j^{(2)}(\mathbf{A}, \overline{\Phi}, \boldsymbol{\Omega}t + \boldsymbol{\Lambda}) \right\rangle_{\overline{\Phi}, \boldsymbol{\Omega}t+\boldsymbol{\Lambda}}, \\ & \sum_{k=1}^{m} \sigma_{uk}^{(2)} \sigma_{vk}^{(2)} = \sum_{k=1}^{m} E_k^u E_k^v \sigma_k^2, \\ & i = 1, 2, \ldots, n, \quad u, v = 1, 2, \ldots, \mu. \end{aligned} \tag{1.259}$$

The averaging operator is defined as

$$\langle [\cdot] \rangle_{\overline{\Phi}, \boldsymbol{\Omega}t+\boldsymbol{\Lambda}} = \frac{1}{(2\pi)^{n-\mu+m}} \int_0^{2\pi} [\cdot]\mathrm{d}\Phi_{\mu+1} \cdots \mathrm{d}\Phi_n \mathrm{d}(\Omega_1 t + \Lambda_1) \cdots \mathrm{d}(\Omega_m t + \Lambda_m). \tag{1.260}$$

where $\boldsymbol{\Omega}t+\boldsymbol{\Lambda}=[\Omega_1 t+\Lambda_1, \Omega_2 t+\Lambda_2, \ldots, \Omega_m t+\Lambda_m]$, $\overline{\boldsymbol{\Phi}}=[\Phi_{\mu+1}, \Phi_{\mu+2}, \ldots, \Phi_n]$. The averaged FPK equation associated with Eq. (1.258) is

$$\frac{\partial p}{\partial t}=-\sum_{i=1}^{n}\frac{\partial}{\partial a_i}[a_i^{(1)}(\mathbf{a},\delta)p]-\sum_{u=1}^{\mu}\frac{\partial}{\partial \delta_u}[a_u^{(2)}(\mathbf{a},\delta)p]+\frac{1}{2}\sum_{u,v=1}^{\mu}b_{uv}\frac{\partial^2 p}{\partial\delta_u\partial\delta_v}, \tag{1.261}$$

where the first and second derivative moments are

$$a_i^{(1)}(\mathbf{a},\delta)=m_i^{(1)}(\boldsymbol{A},\boldsymbol{\Delta})\Big|_{\boldsymbol{A}=\boldsymbol{a},\ \boldsymbol{\Delta}=\boldsymbol{\delta}},\quad a_u^{(2)}(\mathbf{a},\delta)=m_u^{(2)}(\boldsymbol{A},\boldsymbol{\Delta})\big|_{\boldsymbol{A}=\boldsymbol{a},\ \boldsymbol{\Delta}=\boldsymbol{\delta}},$$
$$b_{uv}=\sum_{k=1}^{m}\sigma_{uk}^{(2)}\sigma_{vk}^{(2)}. \tag{1.262}$$

In Eq. (1.261), $p=p(\mathbf{a},\delta,t|\mathbf{a}_0,\delta_0)$ with initial condition

$$p(\mathbf{a},\delta,0|\mathbf{a}_0,\delta_0)=\delta(a_1-a_{10})\cdots\delta(a_n-a_{n0})\delta(\delta_1-\delta_{10})\cdots\delta(\delta_\mu-\delta_{\mu 0}), \tag{1.263}$$

where $\delta(\cdot)$ is the Dirac function. If $g_i(q_i)$, $U_i(q_i)$ satisfy the four conditions below Eq. (1.5) in all entire planes (q_i, p_i), the boundary conditions for Eq. (1.261) at $|\mathbf{a}|=0$ and $\mathbf{a}\to\infty$ are the same as those in Eq. (1.62). The periodic boundary conditions with respective to δ are

$$\begin{aligned}&p(\mathbf{a},\delta_1,\delta_2,\ldots,\delta_u+2k\pi,\ldots,\delta_\mu,t\big|\mathbf{a}_0,\delta_0)\\&\quad=p(\mathbf{a},\delta_1,\delta_2,\ldots,\delta_u,\ldots,\delta_\mu,t\big|\mathbf{a}_0,\delta_0),\\&k=0,\pm1,\pm2,\ldots.\end{aligned} \tag{1.264}$$

Once the stationary solution $p(\mathbf{a},\boldsymbol{\delta})$ is obtained from solving the reduced averaged FPK Eq. (1.261), $p(\mathbf{q},\mathbf{p})$ can be derived similar to Eq. (1.245) as follows

$$p(\mathbf{q},\mathbf{p})=p(\mathbf{a},\boldsymbol{\delta})\left|\frac{\partial(\mathbf{a},\boldsymbol{\delta})}{\partial(\mathbf{q},\mathbf{p})}\right|=p(\mathbf{a},\boldsymbol{\delta})\left|\frac{\partial(\mathbf{a},\boldsymbol{\delta})}{\partial(\mathbf{a},\boldsymbol{\phi})}\right|\left|\frac{\partial(\mathbf{a},\boldsymbol{\phi})}{\partial(\mathbf{q},\mathbf{p})}\right|. \tag{1.265}$$

Example 1.12 Consider a two-DOF nonlinear system excited by a narrowband randomized harmonic noise. The motion equations are

$$\begin{aligned}&\ddot{X}_1+\gamma_1\dot{X}_1+(\alpha_1+\beta_1\dot{X}_2)\dot{X}_2+\omega_1^2X_1=\xi(t),\\&\ddot{X}_2+(\alpha_2+\beta_2\dot{X}_1)\dot{X}_1+\gamma_2\dot{X}_2+\omega_2^2X_2=0.\end{aligned} \tag{1.266}$$

Letting $Q_1=X_1,\ P_1=\dot{X}_1, Q_2=X_2,\ P_2=\dot{X}_2$, the quasi-integrable Hamiltonian system associated with original system (1.266) is

$$
\begin{aligned}
\dot{Q}_1 &= P_1, \\
\dot{Q}_2 &= P_2, \\
\dot{P}_1 &= -\omega_1^2 Q_1 - \gamma_1 P_1 - (\alpha_1 P_2 + \beta_1 P_2^2) + \xi(t), \\
\dot{P}_2 &= -\omega_2^2 Q_2 - \gamma_2 P_2 - (\alpha_2 P_1 + \beta_2 P_1^2).
\end{aligned} \tag{1.267}
$$

The Hamiltonians associated with system (1.267) and its sub-systems are

$$
H = H_1 + H_2, \quad H_i = \frac{1}{2}P_i^2 + \frac{1}{2}\omega_i^2 Q_i^2, \quad i = 1, 2. \tag{1.268}
$$

Using transformation Eq. (1.250) with $B = 0$ to change state variables from Q_i, P_i to H_i, Φ_i, the following motion equations can be obtained from Eq. (1.267)

$$
\begin{aligned}
\dot{H}_1 &= \sqrt{2H_1}\sin\Phi_1(-\gamma_1\sqrt{2H_1}\sin\Phi_1 - \alpha_1\sqrt{2H_2}\sin\Phi_2 + 2\beta_1 H_2\sin^2\Phi_2) \\
&\quad - E\sqrt{2H_1}\sin\Phi_1\xi(t), \\
\dot{H}_2 &= \sqrt{2H_2}\sin\Phi_2(-\gamma_2\sqrt{2H_2}\sin\Phi_2 - \alpha_2\sqrt{2H_1}\sin\Phi_1 + 2\beta_2 H_1\sin^2\Phi_1),
\end{aligned}
$$

$$
\begin{aligned}
\dot{\Phi}_1 &= \omega_1 + \frac{\cos\Phi_1}{\sqrt{2H_1}}(-\gamma_1\sqrt{2H_1}\sin\Phi_1 - \alpha_1\sqrt{2H_2}\sin\Phi_2 \\
&\quad + 2\beta_1 H_2\sin^2\Phi_2) - \frac{E\cos\Phi_1}{\sqrt{2H_1}}\xi(t), \\
\dot{\Phi}_2 &= \omega_2 + \frac{\cos\Phi_2}{\sqrt{2H_2}}(-\gamma_2\sqrt{2H_2}\sin\Phi_2 - \alpha_2\sqrt{2H_1}\sin\Phi_1 \\
&\quad + 2\beta_2 H_1\sin^2\Phi_1).
\end{aligned} \tag{1.269}
$$

System (1.267) has a center frequency Ω of excitation and two natural frequencies ω_1, ω_2. In the following, the case of external resonance only and the case of both internal and external resonances will be studied.

(i) **Case of external resonance only**

Consider the case of external resonance only, i.e., $\Omega - \omega_1 = 0$. Introduce a combination of angle variable

$$
\Delta = \Omega t + \Lambda - \Phi_1. \tag{1.270}
$$

The following equations for Δ can be obtained from the equation for Φ_1 in Eq. (1.269).

$$
\mathrm{d}\Delta = \{O(\varepsilon) + \frac{\cos\Phi_1}{\sqrt{2H_1}}(\gamma_1\sqrt{2H_1}\sin\Phi_1 + \alpha_1\sqrt{2H_2}\sin\Phi_2
$$

$$- 2\beta_1 H_2 \sin^2 \Phi_2) + \frac{E \cos \Phi_1}{\sqrt{2H_1}} \sin(\Omega t + \Lambda)\}\mathrm{d}t + \sigma \mathrm{d}B(t). \tag{1.271}$$

Consider the system combining of equations for H_1, H_2, Φ_2 in Eq. (1.269) and equation for Δ in Eq. (1.271) where Φ_1 is replaced by Δ and $\Omega t + \Lambda$ according to Eq. (1.270). $[H_1, H_2, \Delta]^{\mathrm{T}}$ is a slowly-varying vector process while Φ_2 is fast-varying process. As $\gamma_i, \alpha_i, \beta_i$ $(i = 1, 2) \to 0$, $[H_1, H_2, \Delta]^{\mathrm{T}}$ converges weakly to a 3-dimensional vector Markov diffusion process. The averaged Itô SDEs for $[H_1, H_2, \Delta]^{\mathrm{T}}$ can be obtained according to Eq. (1.258) as follows

$$\begin{aligned} \mathrm{d}H_1 &= a_1^{(H)} \mathrm{d}t, \\ \mathrm{d}H_2 &= a_2^{(H)} \mathrm{d}t, \\ \mathrm{d}\Delta &= a^{(\Delta)} \mathrm{d}t + \sigma \mathrm{d}B(t). \end{aligned} \tag{1.272}$$

The associated reduced averaged FPK equation is

$$\frac{1}{2}\sigma^2 \frac{\partial^2 p}{\partial \delta^2} - \frac{\partial [a^{(\delta)}(h_1, h_2, \delta)p]}{\partial \delta} - \frac{\partial [a_1^{(h)}(h_1, h_2, \delta)p]}{\partial h_1} - \frac{\partial [a_2^{(h)}(h_1, h_2, \delta)p]}{\partial h_2} = 0, \tag{1.273}$$

where $p = p(h_1, h_2, \delta)$. According to Eq. (1.259), Eq. (1.260), Eq. (1.262), Eq. (1.269) and Eq. (1.271), the first derivative moments are obtained as follows

$$\begin{aligned} a_1^{(h)} &= -\gamma_1 h_1 - E\sqrt{h_1/2} \cos \delta, \\ a_2^{(h)} &= -\gamma_2 h_2, \\ a^{(\delta)} &= \Omega - \omega_1 + E \sin \delta \Big/ (2\sqrt{2h_1}). \end{aligned} \tag{1.274}$$

The boundary conditions for the FPK Eq. (1.273) are

$$\begin{aligned} &p = \text{finite}, \quad \text{when } h_1 = 0 \text{ or } h_2 = 0, \\ &p \to 0, \ \partial p / \partial h_1 \to 0, \quad \text{when } h_1 \to \infty, \\ &p \to 0, \ \partial p / \partial h_2 \to 0, \quad \text{when } h_2 \to \infty, \\ &p(h_1, h_2, \delta + 2k\pi) = p(h_1, h_2, \delta), \\ &k = 0, \pm 1, \pm 2, \ldots. \end{aligned} \tag{1.275}$$

It can be observed from Eq. (1.274) that the equations for $a_1^{(h)}, a^{(\delta)}$ do not contain h_2, and the equations for $a_2^{(h)}$ has only a h_2 term. Thus we may assume $p(h_1, h_2, \delta) = p(h_1, \delta)p(h_2)$ for the solution of Eq. (1.273). By substituting this solution into Eq. (1.273), $p(h_2) = \delta(h_2)$ is obtained, where $\delta(h_2)$ is Dirac function. It indicates that the stationary state of the second oscillator is static due to the damping and no random excitation. Given system (1.266) with parameters $\omega_1 = 1$, $\omega_2 = 1.414$, $\gamma_1 = \gamma_2 = 0.1$, $\alpha_1 = \alpha_2 = 0.04$, $\beta_1 = \beta_2 = 0.05$, $E = 0.2$, $\Omega = 1.05$, $\sigma^2 = 0.02$, the numerical result shown in Fig. 1.44 indicates that in this case the first oscillator of system (1.267) absorbs almost all noise energy due to external resonance. Comparisons between the results from Monte Carlo simulation and those obtained by using the stochastic averaging method shown in Figs. 1.45, 1.46 and 1.47 demonstrate that the two results are in good agreement.

(ii) **Case of both internal and external resonances**

Consider the case of both internal resonance $\omega_1 - \omega_2 = O_2(\varepsilon)$ and external resonance $\Omega - \omega_1 = O_1(\varepsilon)$ of system (1.267). Introduce two combinations of angle variable

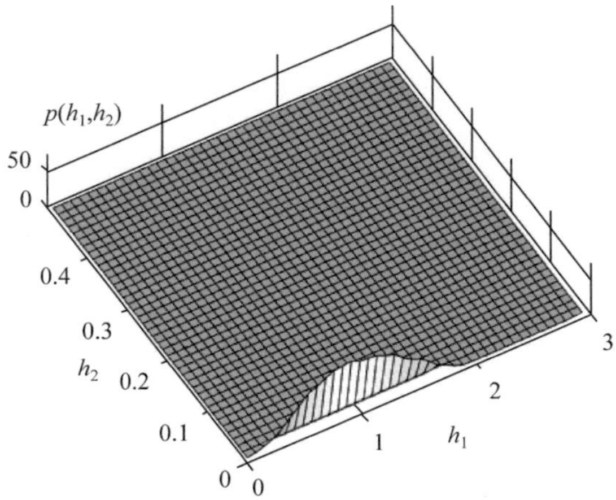

Fig. 1.44 Stationary PDF $p(h_1, h_2)$ of system (1.267) in the case of only external resonance

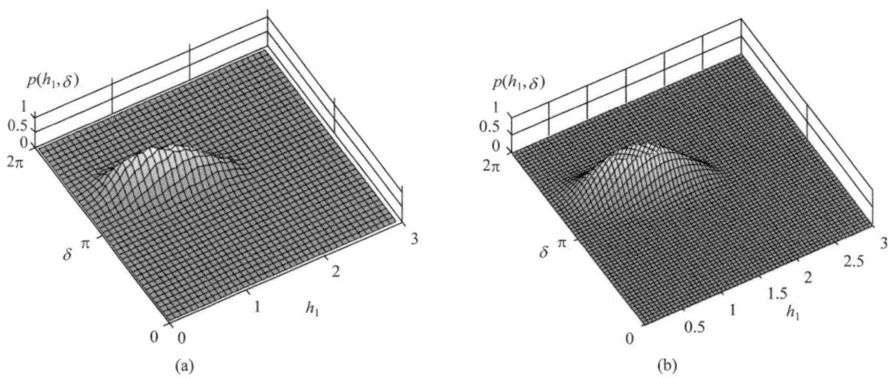

Fig. 1.45 Stationary PDF $p(h_1, \delta)$ of Hamiltonian H_1 and phase angle difference Δ of system (1.267) in the case of only external resonance. **a** result from Monte Carlo simulation; **b** result from the stochastic averaging method

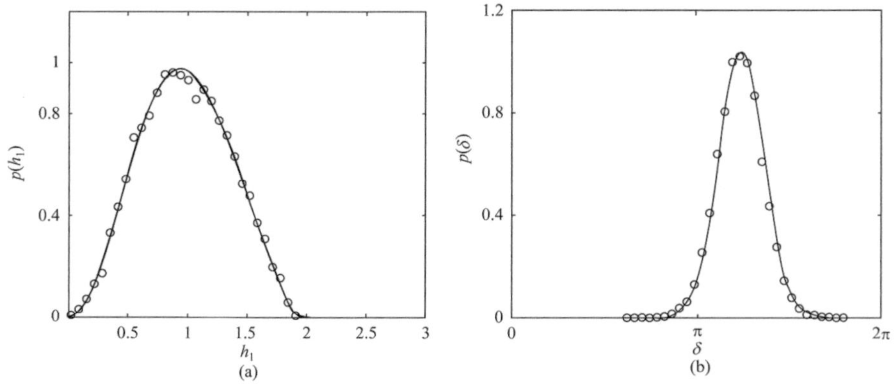

Fig. 1.46 In the case of only external resonance, **a** stationary PDF $p(h_1)$ of Hamiltonian H_1; **b** stationary PDF $p(\delta)$ of phase angle difference Δ. ——, result from the stochastic averaging method; ○, results from Monte Carlo simulation

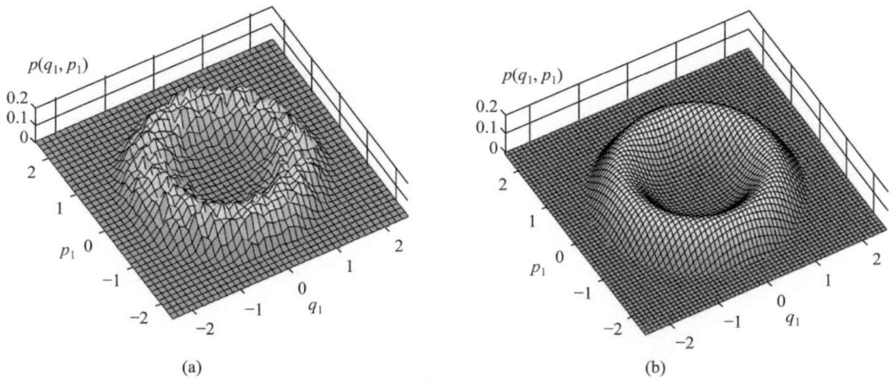

Fig. 1.47 Stationary PDF $p(q_1, p_1)$ of generalized displacement Q_1 and generalized momentum P_1 of system (1.267) in the case of only external resonance. **a** result from Monte Carlo simulation; **b** result from the stochastic averaging method

$$\Delta_1 = \Omega t + \Delta - \Phi_1, \quad \Delta_2 = \Phi_1 - \Phi_2. \tag{1.276}$$

The equations for $[H_1, H_2, \Delta_1, \Delta_2]^{\mathrm{T}}$ can be obtained from the motion Eq. (1.269) as follows

$$\begin{aligned}
\mathrm{d}H_1 = &\{\sqrt{2H_1}\sin\Phi_1(-\gamma_1\sqrt{2H_1}\sin\Phi_1 - \alpha_1\sqrt{2H_2}\sin\Phi_2 \\
&+ 2\beta_1 H_2 \sin^2\Phi_2) - E\sqrt{2H_1}\sin\Phi_1\sin(\Omega t + \Lambda)\}\mathrm{d}t, \\
\mathrm{d}H_2 = &\{\sqrt{2H_2}\sin\Phi_2(-\gamma_2\sqrt{2H_2}\sin\Phi_2 - \alpha_2\sqrt{2H_1}\sin\Phi_1 \\
&+ 2\beta_2 H_1 \sin^2\Phi_1)\}\mathrm{d}t,
\end{aligned}$$

$$
\begin{aligned}
\mathrm{d}\Delta_1 &= \{O(\varepsilon) + \frac{\cos\Phi_1}{\sqrt{2H_1}}(\gamma_1\sqrt{2H_1}\sin\Phi_1 + \alpha_1\sqrt{2H_2}\sin\Phi_2 \\
&\quad - 2\beta_1 H_2 \sin^2\Phi_2) + \frac{E\cos\Phi_1}{\sqrt{2H_1}}\sin(\Omega t + \Lambda)\}\mathrm{d}t + \sigma\mathrm{d}B(t), \\
\mathrm{d}\Delta_2 &= \{O(\varepsilon) + \frac{\cos\Phi_1}{\sqrt{2H_1}}(-\gamma_1\sqrt{2H_1}\sin\Phi_1 - \alpha_1\sqrt{2H_2}\sin\Phi_2 \\
&\quad + 2\beta_1 H_2 \sin^2\Phi_2) - \frac{E\cos\Phi_1}{\sqrt{2H_1}} \times \sin(\Omega t + \Lambda) \\
&\quad - \frac{\cos\Phi_2}{\sqrt{2H_2}}(-\gamma_2\sqrt{2H_2}\sin\Phi_2 - \alpha_2\sqrt{2H_1}\sin\Phi_1 + 2\beta_2 H_1\sin^2\Phi_1)\}\mathrm{d}t.
\end{aligned}
\tag{1.277}
$$

where Φ_1 and Φ_2 should be replaced by Δ_1 and Δ_2 according to Eq. (1.276). The following averaged Itô SDEs can be obtained

$$
\begin{aligned}
\mathrm{d}H_1 &= a_1^{(H)}\mathrm{d}t, \\
\mathrm{d}H_2 &= a_2^{(H)}\mathrm{d}t, \\
\mathrm{d}\Delta_1 &= a_1^{(\Delta)}\mathrm{d}t + \sigma\mathrm{d}B(t), \\
\mathrm{d}\Delta_2 &= a_2^{(\Delta)}\mathrm{d}t.
\end{aligned}
\tag{1.278}
$$

The associated reduced averaged FPK equation is

$$
\frac{1}{2}\sigma^2\frac{\partial^2 p}{\partial\delta_1^2} - \frac{\partial(a_1^{(\delta)}p)}{\partial\delta_1} - \frac{\partial(a_2^{(\delta)}p)}{\partial\delta_2} - \frac{\partial(a_1^{(h)}p)}{\partial h_1} - \frac{\partial(a_2^{(h)}p)}{\partial h_2} = 0. \tag{1.279}
$$

where $p = p(h_1, h_2, \delta_1, \delta_2)$. According to Eqs. (1.259), (1.260), (1.262) and (1.277), the first derivative momenta are derived as

$$
\begin{aligned}
a_1^{(h)} &= -\gamma_1 h_1 - \alpha_1\sqrt{h_1 h_2}\cos\delta_2 - E\sqrt{h_1/2}\cos\delta_1, \\
a_2^{(h)} &= -\gamma_2 h_2 - \alpha_2\sqrt{h_1 h_2}\cos\delta_2, \\
a_1^{(\delta)} &= \Omega - \omega_1 + \frac{E\sin\delta_1}{2\sqrt{2h_1}} - \alpha_1\sin\delta_2\sqrt{\frac{h_2}{4h_1}}, \\
a_2^{(\delta)} &= \omega_1 - \omega_2 - \frac{E\sin\delta_1}{2\sqrt{2h_1}} + \sin\delta_2\left(\alpha_1\sqrt{\frac{h_2}{4h_1}} + \alpha_2\sqrt{\frac{h_1}{4h_2}}\right).
\end{aligned}
\tag{1.280}
$$

The boundary conditions for FPK Eq. (1.279) are

$$
p = \text{finite}, \quad \text{when } h_1 = 0 \text{ or } h_2 \to 0,
$$

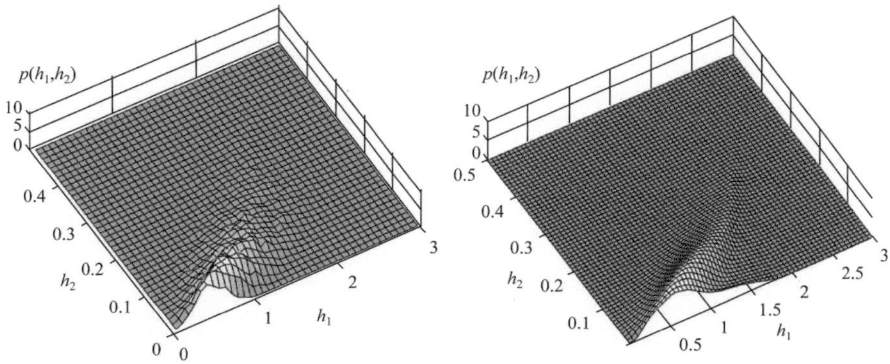

(a) (b)

Fig. 1.48 Stationary PDF $p(h_1, h_2)$ of Hamiltonian H_1, H_2 of system (1.267) in the case of both internal and external resonances. **a** result from Monte Carlo simulation; **b** result from the stochastic averaging method

$$
\begin{aligned}
&p \to 0,\ \partial p/\partial h_1 \to 0, \quad \text{when } h_1 \to \infty,\\
&p \to 0,\ \partial p/\partial h_2 \to 0, \quad \text{when } h_2 \to \infty,\\
&p(h_1, h_2, \delta_1 + 2k\pi, \delta_2) = p(h_1, h_2, \delta_1, \delta_2),\\
&p(h_1, h_2, \delta_1, \delta_2 + 2k\pi) = p(h_1, h_2, \delta_1, \delta_2),\\
&k = 0, \pm 1, \pm 2, \ldots.
\end{aligned}
\tag{1.281}
$$

Given $\omega_1 = \omega_2 = \Omega = 1$ and the other parameters the same as those in the previous case of only external resonance. The numerical results shown in Fig. 1.47 indicate that in the case of both internal and external resonance, the first oscillator of the system (1.267) absorbs most amount of noise energy due to external resonance. While the second oscillator absorbs small amount of energy from the first oscillator due to internal resonance. The numerical results obtained by using the stochastic averaging method and those from Monte Carlo simulation shown in Figs. 1.48 and 1.49 are in good agreement.

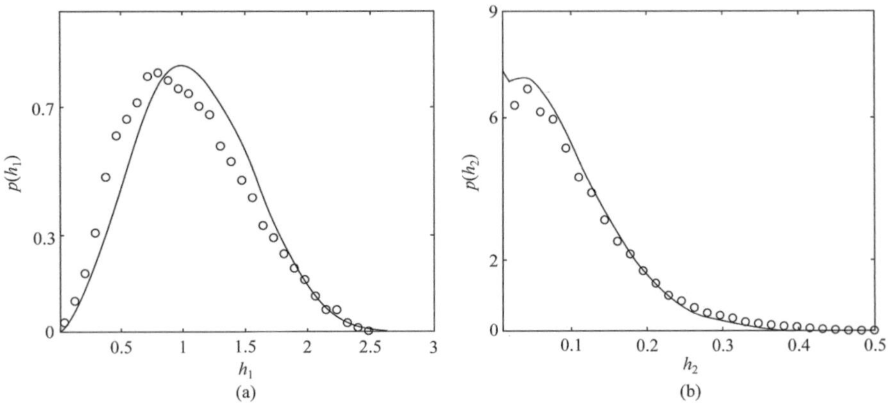

Fig. 1.49 Stationary PDFs of system (1.267) is in the case of both internal and external resonance, **a** stationary PDF $p(h_1)$ of Hamiltonian H_1, **b** stationary PDF $p(h_2)$ of Hamiltonian H_2. ——, results from the stochastic averaging method; ○, results from Monte Carlo simulation

References

Deng M L, Zhu W Q. 2007. Stochastic averaging of MDOF quasi-integrable Hamiltonian systems under wideband random excitation. Journal of Sound and Vibration, 305(4–5): 783–794.

Deng M L, Zhu W Q. 2022. Analysis of applicability of the approximate method for nonlinear systems excited by fGn. Journal of Vibration Engineering, 35(5): 1076–1083. (in Chinese)

Hu R C, Zhu W Q. 2015. Stochastic optimal control of MDOF nonlinear systems under combined harmonic and wide-band noise excitations. Nonlinear Dynamics, 179: 1115–1129.

Huang Z L, Gao L X, Zhu W Q. 2003. Stationary response of SDOF vibration-impact system excited by wide-band random process. Journal of Zhejiang University (Engineering Edition), 37(1): 94–97. (in Chinese)

Huang Z L, Zhu W Q. 1997. Exact stationary solutions of averaged equations of stochastically and harmonically excited MDOF quasi-linear systems with internal and/or external resonances. Journal of Sound and Vibration, 204: 249–258.

Huang Z L, Zhu W Q. 2004a. Stochastic averaging of quasi-integrable Hamiltonian systems under combined harmonic and white noise excitations. International Journal of Non-Linear Mechanics, 39: 1421–1434.

Huang Z L, Zhu W Q. 2004b. Stochastic averaging of quasi-integrable Hamiltonian systems under bounded noise excitations. Probabilistic Engineering Mechanics, 19: 219–228.

Huang Z L. Zhu W Q, Ni Y Q, Ko J M. 2002. Stochastic averaging of strongly non-linear oscillators under bounded noise excitation. Journal of Sound and Vibration, 254: 245–267.

Khasminskii R Z. 1966. A limit theorem for the solution of differential equations with random right hand sides. Theory of Probability and Application, 11(3): 390–405.

Khasminskii R Z. 1968. On the averaging principle for Itô stochastic differential equations. Kibernetika, 3(4): 260–279. (in Russian)

Lü Q F, Zhu W Q, Deng M L. 2020a. Response of quasi-integrable and non-resonant Hamiltonian systems to fractional Gaussian noise. IEEE Access, 8(1): 72372–72380.

Lü Q F, Zhu W Q, Deng M L. 2020b. Reliability of quasi integrable and non-resonant Hamiltonian systems under fractional Gaussian noise excitation. Acta Mechanica Sinica, 36(4): 902–909.

Lü Q F, Zhu W Q, Deng M L. 2022. Response of quasi-integrable and resonant Hamiltonian systems to fractional Gaussian noise. Journal of Vibration and Acoustic, 144: 011010.

Wu Y J. 2005. Stochastic dynamics and control of strongly nonlinear system under harmonic and white/wideband noise excitations. Ph.D. Dissertation. Hangzhou: Zhejiang University. (in Chinese)

Wu Y J, Gao Y Y, Zhang L. 2013. First-passage problem of strongly nonlinear stochastic oscillators with external and internal resonances. European Journal of Mechanics-A/Solids, 39: 60–68.

Wu Y J, Zhu W Q. 2008. Stochastic averaging of strongly nonlinear oscillators under combined harmonic and wide-band noise excitations. Journal of Vibration and Acoustics, 130: 051004.

Wu Y J, Zhu W Q. 2009. Stationary response of Duffing-Rayleigh-Mathieu system. Journal of Vibration Engineering, 22(2): 207–212. (in Chinese)

Xu Z, Chung Y K. 1994. Averaging method using generalized harmonic functions for strongly non-linear oscillators. Journal of Sound and Vibration, 174: 563–576.

Zhu W Q. 1998. Random Vibration. Beijing: Science Press. (in Chinese)

Zhu W Q, Huang Z L, Suzuki Y. 2001. Response and stability of strongly non-linear oscillators under wide-band random excitation. International Journal of Nonlinear Mechanics, 36(8): 1235–1250.

Zhu W Q. 2003. Nonlinear Stochastic Dynamics and Control - Hamilton Theoretical Framework. Beijing: Science Press. (in Chinese)

Chapter 2
Stochastic Averaging Methods of Quasi-integrable Hamiltonian Systems with Genetic Effective Forces

Genetic effective forces can be considered as nonlinear coupling of elastic restoring forces and viscous damping forces. For quasi-integrable Hamiltonian systems with genetic effective forces, the stochastic averaging method described in Chaps. 4–7 in vol. 1 and Chap. 1 in this volume cannot be applied directly. Instead, it is necessary first to decouple the genetic effective forces into equivalent elastic restoring forces and viscous damping forces using certain criteria and procedures. Such a way, a quasi-integrable Hamiltonian system with genetic effective forces can be transformed into an equivalent quasi-integrable Hamiltonian system without genetic effective force, and then the stochastic averaging method described in Chaps. Chaps. 4–7 in vol. 1 and Chap. 1 in this volume can be applied. In this chapter, the stochastic averaging method of quasi-integrable Hamiltonian systems with genetic effective forces, including hysteretic restoring forces, viscoelastic forces, fractional derivative damping forces, and time-delay forces will be developed.

2.1 Quasi-integrable Hamiltonian System with Hysteretic Forces

2.1.1 Equalization of Hysteretic Forces

The equalization of hysteretic restoring force is a procedure to decouple a hysteretic restoring force into equivalent elastic restoring force and viscous damping force. There are two common methods for the equalization. The first method is to use the generalized harmonic balance technique, and the second method is based on calculation of the potential and dissipated energies of the hysteretic restoring force.

Consider a quasi-integrable Hamiltonian system with hysteretic restoring forces

W. Zhu et al., *Stochastic Averaging*, https://doi.org/10.1007/978-981-96-3833-8_2

$$
\begin{aligned}
&\dot{Q}_i = P_i, \\
&\dot{P}_i = -g_i'(Q_i) - \varepsilon \sum_{j=1}^{n} c_{ij}(\boldsymbol{Q}, \boldsymbol{P})P_j - \varepsilon f_i(Q_i, P_i) + \varepsilon^{1/2} \sum_{k=1}^{m} f_{ik}(\boldsymbol{Q}, \boldsymbol{P})\xi_k(t), \\
&\quad i = 1, 2, \ldots, n,
\end{aligned} \tag{2.1}
$$

where $\boldsymbol{Q}, \boldsymbol{P}, g_i'(Q_i), \varepsilon, c_{ij}, f_{ik}$ are the same as those in Eq. (1.4), $\xi_k(t)$ can be Gaussian white noise, Poisson white noise or colored noise, $\varepsilon f_i(Q_i, P_i)$ is the hysteretic restoring force of the ith DOF.

(i) **First Equalization Method**

Assume that $g_i'(Q_i)$ $i = 1, 2, \ldots, n$ satisfy the conditions listed below Eq. (1.5). For simplicity, let $g_i'(0) = 0$. System (2.1) has a family of the following randomized periodic solutions

$$
Q_i(t) = A_i \cos \Phi_i(t), \quad P_i(t) = -A_i \nu_i(A_i, \Phi_i) \sin \Phi_i(t), \tag{2.2}
$$

where

$$
\begin{aligned}
&\nu_i(A_i, \Phi_i) = \frac{\mathrm{d}\Phi_i}{\mathrm{d}t} = \sqrt{\frac{2[U_i(A_i) - U_i(A_i \cos \Phi_i)]}{A_i^2 \sin^2 \Phi_i}}, \\
&U_i(A_i) = U_i(-A_i) = H_i
\end{aligned} \tag{2.3}
$$

where H_i, U_i, A_i and ν_i are the total energy, potential energy, amplitude and instantaneous frequency for the ith DOF, respectively. Let

$$
\begin{aligned}
&f_i(Q_i, P_i) = K_i(A_i)Q_i + C_i(A_i)P_i, \\
&i = 1, 2, \ldots, n.
\end{aligned} \tag{2.4}
$$

Equation (2.4) indicates that hysteretic restoring force $f_i(Q_i, P_i)$ is equivalised by the combination of restoring force $K_i(A_i)Q_i$ and damping force $C_i(A_i)P_i$. Multiplying both sides of Eq. (2.4) by $\cos \Phi_i(t)$ and $\sin \Phi_i(t)$, respectively, and integrating them with respect to Φ_i from 0 to 2π yield

$$
\begin{aligned}
&K_i(A_i) = \frac{1}{\pi a_i} \int_0^{2\pi} f_i(A_i \cos \Phi_i, -A_i \nu_i \sin \Phi_i) \cos \Phi_i \mathrm{d}\Phi_i, \\
&C_i(A_i) = \frac{-1}{\pi a_i \nu_i} \int_0^{2\pi} f_i(A_i \cos \Phi_i, -A_i \nu_i \sin \Phi_i) \sin \Phi_i \mathrm{d}\Phi_i.
\end{aligned} \tag{2.5}
$$

For bilinear hysteretic restoring forces, substituting Eq. (3.114) of vol. 1 into Eq. (2.5) and carrying out the integration yield

$$K_i(A_i) = \begin{cases} 1, & A_i \le 1 \\ \alpha + 2(1-\alpha)(2-A_i)\frac{\sqrt{A_i-1}}{\pi A_i^2} + \frac{1-\alpha}{\pi}\arccos\left(\frac{A_i-2}{A_i}\right), & A_i > 1 \end{cases},$$

$$C_i(A_i) = \begin{cases} 0, & A_i \le 1 \\ \frac{4(1-\alpha)(A_i-1)}{\pi\omega(A_i)A_i^2}, & A_i > 1 \end{cases}. \tag{2.6}$$

For Bouc-Wen hysteretic restoring forces, substituting Eqs. (3.116) and (3.118) of vol. 1 into Eq. (2.5) and completing the integration yield

$$\begin{aligned} K_i(A_i) &= \alpha + \frac{2(1-\alpha)}{\pi A_i}\int_0^{\arccos(q_{i0}/A_i)} \frac{\exp[(\gamma-\beta)(A_i\cos\Phi_i - q_{i0})]-1}{\gamma-\beta}\cos\Phi_i \mathrm{d}\Phi_i \\ &+ (1-\alpha)\frac{2}{\pi A_i}\int_{\arccos(q_{i0}/A_i)}^{\pi} \frac{\exp[(\gamma+\beta)(A_i\cos\Phi_i - q_{i0})]-1}{\gamma+\beta}\cos\Phi_i \mathrm{d}\Phi_i, \\ C_i(A_i) &= \frac{2(1-\alpha)}{\pi A_i^2\omega(A_i)}\left[\frac{A_i+q_{i0}}{\beta+\gamma} + \frac{\exp[-(A_i+q_{i0})(\beta+\gamma)]-1}{(\beta+\gamma)^2}\right. \\ &\left. + \frac{1-\exp[-(A_i-q_{i0})(\beta-\gamma)]-(A_i-q_{i0})(\beta-\gamma)}{(\beta-\gamma)^2}\right]. \end{aligned} \tag{2.7}$$

For Duhem hysteretic restoring forces, substituting Eq. (3.123) of vol. 1 into Eq. (2.5) and completing the integration result in

$$\begin{aligned} K_i(A_i) &= k_1 + \frac{3}{4}k_3A_i^2 + \frac{2}{\beta A_i}\exp(-\gamma q_{i0})\text{besseli}(1, \gamma A_i), \\ C_i(A_i) &= \frac{4\gamma A_i - 4\exp(-\gamma q_{i0})\sinh(\gamma A_i)}{\pi\beta\gamma\omega_i(A_i)A_i^2}. \end{aligned} \tag{2.8}$$

where besseli$(\cdot,\cdot)$ is the modified first kind Bessel function.

For the Preisach hysteretic restoring forces, substituting Eqs. (3.133) and (3.134) of vol. 1 into Eq. (2.5) and completing the integration yield $K_i(A_i)$ and $C_i(A_i)$. Please refer to Example 2.1 for more details.

After $K_i(A_i)$ and $C_i(A_i)$ are found and $f_i(Q_i, P_i)$ in Eq. (2.1) are replaced by the right hand side of Eq. (2.4), the following equivalent quasi-integrable Hamiltonian system without hysteretic restoring force is obtained

$$\begin{aligned} \dot{Q}_i &= P_i, \\ \dot{P}_i &= -[g_i'(Q_i) + \varepsilon K_i(A_i)Q_i] - \varepsilon\left[\sum_{j=1}^{n} c_{ij}'(\boldsymbol{Q},\boldsymbol{P})P_j + C_i(A_i)P_i\right] \\ &+ \varepsilon^{1/2}\sum_{k=1}^{m} f_{ik}(\boldsymbol{Q},\boldsymbol{P})\xi_k(t), \ i = 1, 2, \ldots, n. \end{aligned} \tag{2.9}$$

(ii) **Second Equalization Method**

$f_{i1}(q_i)$ and $f_{i2}(q_i)$ in Fig. 2.1 are the rising and decline branches of the hysteretic loop, and the shaded areas are the potential energy of the rising branch $f_{i1}(q_i)$ $(\dot{q}_i > 0)$ which can be expressed as

$$U(q_i) = \begin{cases} \int_{-q_{i10}}^{q_i} f_{i1}(u)\mathrm{d}u, & 当\dot{q}_i > 0, \quad -A_{i1} \le q_i \le -q_{i10}, \\ \int_{q_{i20}}^{f_{i2}^{-1}[f_{i1}(q_i)]} f_{i2}(u)\mathrm{d}u, & 当\dot{q}_i > 0, \quad -q_{i10} \le q_i \le A_{i2}, \end{cases} \tag{2.10}$$

where $-A_{i1}$ and A_{i2} are negative and positive displacement amplitudes, $-q_{i10}$ and q_{i20} are residual displacements. The potential energy $U(q_i)$ for the decline branch $f_{i2}(q_i)$ $(\dot{q}_i < 0)$ can be obtained similar to Eq. (2.10). For most hysteretic models, hysteretic loop is antisymmetric, i.e., $f_{i2}(q_i) = -f_{i1}(-q_i)$, $A_{i1} = A_{i2} = A_i$, $q_{i10} = q_{i20} = q_{i0}$. In this case, the equivalent restoring force for hysteretic restoring force is $\mathrm{d}U_i(q_i)/\mathrm{d}q_i$.

The energy dissipated by one cycle of vibration is equal to the area enclosed by the hysteretic loop

$$A_{ri} = \int_{-A_{i1}}^{A_{i2}} f_{i1}(q_i)\mathrm{d}q_i + \int_{A_{i2}}^{-A_{i1}} f_{i2}(q_i)\mathrm{d}q_i. \tag{2.11}$$

The equivalent viscous damping coefficient for hysteretic restoring forces is

$$2\zeta_i(H_i) = \frac{A_{ri}}{2\int_{-A_{i1}}^{A_{i2}} \sqrt{2H_i - 2U_i(q_i)}\mathrm{d}q_i}. \tag{2.12}$$

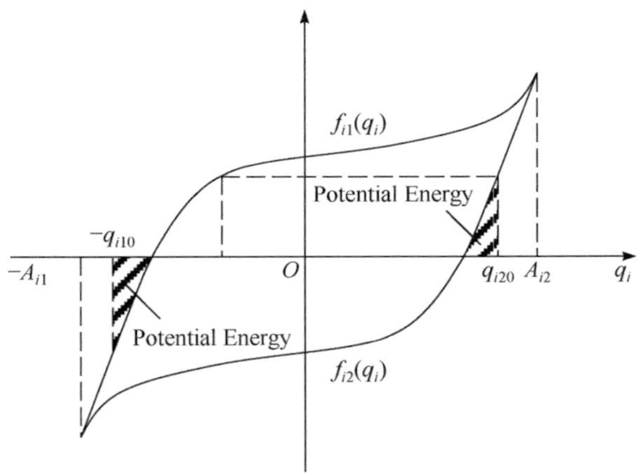

Fig. 2.1 Hysteretic loop and potential energy

where H_i is the energy of ith degree-of-freedom.

For the Bouc-Wen hysteretic restoring force model with $n = 1, A_1 = 1, \beta \neq \gamma$, the potential and dissipated energies can be obtained using Eqs. (2.10), (2.11) and Eq. (3.119) of vol. 1 as follows

$$U_i(q_i) = \begin{cases} (\alpha - k_1)q_i^2/2 + (1-\alpha)\{q_i + q_{i0} + [e^{-(\gamma-\beta)(q_i+q_{i0})} - 1] \\ \quad /(\gamma-\beta)\}/(\gamma-\beta), \quad -A_i \le q_i \le -q_{i0}, \quad \gamma \neq \pm\beta \\ (\alpha - k_1)q_i^2/2 + (1-\alpha)\{1 - e^{-(\gamma+\beta)(q_i+q_{i0})} - [(\gamma+\beta) \\ \quad /(\gamma-\beta)]\ln[1 + (\gamma+\beta)(1 - e^{-(\gamma+\beta)(q_i+q_{i0})})/(\gamma-\beta)]\} \\ \quad /(\gamma^2-\beta^2), \quad -q_{i0} \le q_i \le A_i, \quad \gamma \neq \pm\beta \end{cases}$$
$$A_{ri} = [4(1-\alpha)/(\gamma^2-\beta^2)]\{\gamma A_i - \beta q_{i0} + \gamma[e^{-(\gamma+\beta)(q_i+q_{i0})} - 1]/(\gamma+\beta)\},$$
$$\gamma \neq \pm\beta, \tag{2.13}$$

where q_{i0} are the residual hysteretic displacements. A_i and q_{i0} are determined from H_i using the following equations

$$(\gamma+\beta)e^{(\gamma-\beta)(A_i-q_{i0})} + (\gamma-\beta)e^{-(\gamma+\beta)(A_i+q_{i0})} = 2\gamma,$$
$$2H_i - (\alpha-k_1)A_i^2 = (1-\alpha)\left[-A_i + q_{i0} + \left(e^{-(\gamma-\beta)(A_i-q_{i0})} - 1\right)/(\gamma-\beta)\right]/(\gamma-\beta),$$
$$\gamma \neq \pm\beta. \tag{2.14}$$

For the Duhem hysteretic restoring force model, the potential and dissipated energies can be obtained using Eqs. (2.10), (2.11) and (3.121) of vol. 1 as follows

$$U_i(q_i) = k_1 q_i^2/2 + k_3 q_i^4/4 + (q_i + q_{i0})/\beta + [e^{-\gamma(q_i+q_{i0})} - 1]/(\beta\gamma),$$
$$-A_i \le q_i \le -q_{i0},$$
$$U_i(q_i) = k_1 q_i^2/2 + k_3 q_i^4/4 + [1 - e^{-\gamma(q_i+q_{i0})}]/(\beta\gamma) - \ln[2 - e^{-\gamma(q_i+q_{i0})}]/(\beta\gamma),$$
$$-q_{i0} \le q_i \le A_i,$$
$$A_{ri} = 4[(1 + A_i\gamma) - e^{\gamma(A_i-q_{i0})}]/(\beta\gamma), \tag{2.15}$$

where q_{i0} and A_i are determined from H_i using the following equations

$$q_{i0} = -A_i + \frac{1}{\gamma}\ln\frac{1+\exp(2A_i\gamma)}{2},$$
$$H_i = k_1 A_i^2/2 + k_3 A_i^4/4 - (A_i - q_{i0})/\beta + [e^{\gamma(A_i-q_{i0})} - 1]/(\beta\gamma). \tag{2.16}$$

After determining $U_i(q_i)$ and $2\zeta_i(H_i)$, the equivalent quasi-integrable Hamiltonian system without hysteretic restoring forces for system (2.1) is obtained as follows

$$\begin{aligned}
\dot{Q}_i &= P_i, \\
\dot{P}_i &= -[g_i'(Q_i) + \varepsilon dU_i/dQ_i] - \varepsilon\left[\sum_{j=1}^{n} c_{ij}'(\boldsymbol{Q},\boldsymbol{P})P_j + 2\zeta_i(H_i)P_i\right] \\
&\quad + \varepsilon^{1/2}\sum_{k=1}^{m} f_{ik}(\boldsymbol{Q},\boldsymbol{P})\xi_k(t), \\
i &= 1, 2, \ldots, n.
\end{aligned} \tag{2.17}$$

2.1.2 *Stochastic Averaging for the Equivalent Quasi-Integrable Hamiltonian Systems*

Once the equivalent quasi-integrable Hamiltonian system (2.9) or system (2.17) is obtained using the methods described in Sect. 2.1.1, the stochastic averaging method for Gaussian white noises, combination of Gaussian and Poisson white noises, fractional Gaussian noises, and colored noises, as described in Sects. 5.2, 6.3, 7.3, vol. 1 and Chap. 1 in this volume, can be applied, respectively. Note that possible Wong-Zakai correction terms due to Gaussian white noise excitations and/or Poisson white noise excitations should be added. Below are two examples to illustrate the applications.

Example 2.1 Consider a SDOF hysteretic system excited by Gaussian white noise as shown in Fig. 2.2 (Wang 2008; Wang et al. 2009). The motion equation of system is given by

$$m\ddot{X}(t) + c\dot{X}(t) + kX(t) + f(t) = W_g(t), \tag{2.18}$$

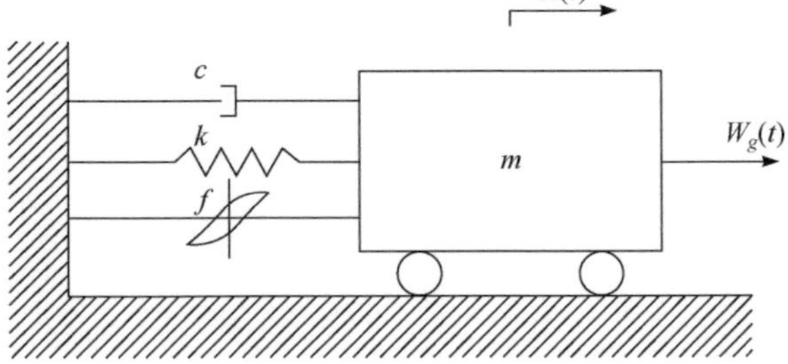

Fig. 2.2 Schematic diagram of single DOF hysteretic system (Spanos et al. 2004)

where $W_g(t)$ is a zero-mean Gaussian white noise with power spectral density S_W. $f(t)$ is the Preisach model of hysteretic restoring force described in Sect. 3.4.1 of vol. 1, and the rising branch and decline branch are given by Eqs. (3.130)–(3.134) of vol. 1, respectively.

Introducing dimensionless parameters $\mu = k/k_J$, where k_J represents the stiffness coefficient of the linear part of the hysteretic restoring force. System (2.18) can be written as

$$\ddot{X}(t) + 2\zeta\overline{\omega}\dot{X}(t) + \overline{\omega}^2 X(t) + \frac{f_H(t)}{m} = \frac{W_g(t)}{m}, \tag{2.19}$$

where $\overline{\omega} = \sqrt{(k + k_J)/m} = \omega_J\sqrt{1+\mu}$, $\omega_J = \sqrt{k_j/m}$, $\mu = \omega^2/\omega_J^2$, $\omega = \sqrt{k/m}$, $2\zeta\overline{\omega} = c/m$. $f_H(t)$ is the nonlinear part of the hysteretic force $f(t)$. Introducing dimensionless displacement

$$Y = \frac{X\sqrt{2\zeta\overline{\omega}^3 m^2}}{\sqrt{\pi S_w}} \tag{2.20}$$

system (2.19) can be written in the following form

$$\ddot{Y}(t) + 2\zeta\overline{\omega}\dot{Y}(t) + \overline{\omega}^2[Y(t) + \lambda p_H(t)] = \widehat{W}(t), \tag{2.21}$$

where

$$\begin{gathered}\widehat{W}(t) = W_g(t)\sqrt{2\zeta\overline{\omega}^3}\Big/\sqrt{\pi S_w},\quad S_{\widehat{W}(t)} = 2\zeta\overline{\omega}^3/\pi,\quad \lambda = \overline{\omega}^2\frac{\sqrt{\pi S_w}}{f_y^*\sqrt{2\zeta\overline{\omega}^3}},\\ p_H(t) = \frac{f_y^* 2\zeta}{\pi S_w\overline{\omega}} f_H(t),\quad f_y^* = \frac{f_{y,\max} + f_{y,\min}}{2}.\end{gathered} \tag{2.22}$$

Furthermore, introduce dimensionless parameters ν and Δ

$$\nu = \frac{f_{y,\max} - f_{y,\min}}{2f_y^*},\quad \frac{2f_{y,\min}}{k_J}\frac{\sqrt{2\zeta\overline{\omega}^3 m^2}}{\sqrt{\pi S_w}} = 2\frac{(1+\mu)(1-\nu)}{\lambda} = \Delta. \tag{2.23}$$

The parameter ν characterizes the distribution of Jenkins units. When $\nu \to 0$, the domain A (see Fig. 3.10 in vol. 1) degenerates into a straight line, representing the definition domain of a single Jenkins model. When $\nu < 1$, the system has a scaling limit. However, when $\nu = 1$, the scaling limit disappears, and the system exhibits strong nonlinear behavior. Obviously, when $\nu = 1$, the domain of the weight function is the entire D domain. Similar to Eqs. (3.133)–(3.135) of vol. 1, when the dimensionless amplitude $A = \frac{\overline{a}\sqrt{2\zeta\overline{\omega}^3 m^2}}{\sqrt{\pi S_w}} > \frac{\Delta}{2}$, the dimensionless expression for the nonlinear part of the hysteretic restoring force is

$$p_H(t)=\frac{1}{4\nu(1+\mu)^2}\{-[2(\tilde{F}(\frac{\Delta}{2},\beta'_P)-\tilde{F}(\frac{\Delta}{2},\beta'_1))+H(Y(t)-\beta'_{n-1}$$
$$-\Delta)\tilde{F}(Y,\beta'_{n-1})]-\frac{1}{2}[2\sum_{j=2}^{n-1}(\tilde{F}(\alpha'_j,\beta'_{j-1})-\tilde{F}(\alpha'_j,\beta'_j))-\tilde{F}(\alpha'_P,\beta'_P)]\}. \tag{2.24}$$

for the rising branch and

$$p_H(t)=\frac{1}{4\nu(1+\mu)^2}\{-[2(\tilde{F}(\frac{\Delta}{2},\beta'_P)-\tilde{F}(\frac{\Delta}{2},\beta'_1))+\tilde{F}(\alpha'_n,\beta'_{n-1})$$
$$-H(\alpha'_n-\Delta-Y(t))\tilde{F}(\alpha'_n,Y(t))]-\frac{1}{2}[2\sum_{j=2}^{n-1}(\tilde{F}(\alpha'_j,\beta'_{j-1})$$
$$-\tilde{F}(\alpha'_j,\beta'_j))-\tilde{F}(\alpha'_P,\beta'_P)]\}. \tag{2.25}$$

for the decline branch. When the dimensionless amplitude $A<\Delta/2$, the dimensionless expression for the nonlinear part of the hysteretic restoring force in both the rising and decline branch is

$$p_H(t)=\frac{1}{4\nu(1+\mu)^2}\{-2[\tilde{F}(\frac{\Delta}{2},\beta'_P)-\tilde{F}(\frac{\Delta}{2},\beta'_1)]$$
$$-\frac{1}{2}[2\sum_{j=2}^{s}(\tilde{F}(\alpha'_j,\beta'_{j-1})-\tilde{F}(\alpha'_j,\beta'_j))-\tilde{F}(\alpha'_P,\beta'_P)]\}, \tag{2.26}$$

where $\alpha'_j=\frac{\alpha_j\sqrt{2\zeta\overline{\omega}^3m^2}}{\sqrt{\pi S_w}}$ and $\beta'_j=\frac{\beta_j\sqrt{2\zeta\overline{\omega}^3m^2}}{\sqrt{\pi S_w}}$ are dimensionless local maxima and minima, respectively, and $\tilde{F}(\alpha'_j,\beta'_j)=\frac{1}{2}[\Delta+(\beta'_j-\alpha'_j)]^2$.

The nonlinear part of the hysteretic restoring force is coupled nonlinear elastic restoring force and nonlinear dissipated forces. To apply the stochastic averaging method of energy envelope, it must be first decoupled. By using the generalized harmonic balance technique described in Sect. 2.1.1, the nonlinear part of the hysteretic restoring force $\overline{\omega}^2p_H(Y)$ can be decoupled as

$$\overline{\omega}^2p_H(Y)=C(A)\dot{Y}+K(A)Y, \tag{2.27}$$

where

$$C(A)=-\frac{\overline{\omega}^2}{A\pi\omega(A)}\int_0^{2\pi}p_H(A\cos\phi)\sin\phi\mathrm{d}\phi,$$

$$K(A) = \frac{\overline{\omega}^2}{A\pi}\int_0^{2\pi} p_H(A\cos\phi)\cos\phi \mathrm{d}\phi. \tag{2.28}$$

When dimensionless amplitude $A > \Delta/2$,

$$\begin{aligned} C(A) = {} & \frac{\overline{\omega}^2}{A\pi\omega(A)}\frac{1}{4\nu(1+\mu)^2} \\ & \times \{\int_0^{\pi} [\tilde{F}(\alpha_n', \beta_{n-1}') - H(\alpha_n' - \Delta - A\cos\phi)\tilde{F}(\alpha_n', A\cos\phi)] \\ & \times \sin\phi \mathrm{d}\phi + \int_{\pi}^{2\pi} [H(A\cos\phi - \beta_{n-1}' - \Delta)\tilde{F}(A\cos\phi, \beta_{n-1}')]\sin\phi \mathrm{d}\phi\}, \end{aligned} \tag{2.29}$$

$$\begin{aligned} K(A) = {} & \frac{\overline{\omega}^2}{A\pi}\frac{1}{4\nu(1+\mu)^2} \\ & \times \{\int_0^{\pi} [H(\alpha_n' - \Delta - A\cos\phi)\tilde{F}(\alpha_n', A\cos\phi)]\cos\phi \mathrm{d}\phi \\ & - \int_{\pi}^{2\pi} [H(A\cos\phi - \beta_{n-1}' - \Delta)\tilde{F}(A\cos\phi, \beta_{n-1}')]\cos\phi \mathrm{d}\phi\}. \end{aligned}$$

Since the response of undamped and unexcited system associated with Eq. (2.21) is harmonic vibration, the dimensionless local maxima and minima α_n', β_{n-1}' are approximately equal to the dimensionless amplitude A and $-A$ respectively, i.e.,

$$\alpha_n' = A, \quad \beta_{n-1}' = -A. \tag{2.30}$$

Thus, Eq. (2.29) can be written as

$$\begin{aligned} C(A) = {} & \frac{\overline{\omega}^2}{A\pi\omega(A)}\frac{1}{4\nu(1+\mu)^2} \\ & \times \{\int_0^{\pi} [\tilde{F}(A, -A) - H(A - \Delta - A\cos\phi)\tilde{F}(A, A\cos\phi)]\sin\phi \mathrm{d}\phi \\ & + \int_{\pi}^{2\pi} [H(A\cos\phi + A - \Delta)\tilde{F}(A\cos\phi, -A)]\sin\phi \mathrm{d}\phi\}, \end{aligned}$$

$$K(A)=\frac{\overline{\omega}^2}{A\pi}\frac{1}{4\nu(1+\mu)^2}\{\int_0^{\pi}[H(A-\Delta-A\cos\phi)\tilde{F}(A,A\cos\phi)]\cos\phi\mathrm{d}\phi$$
$$-\int_{\pi}^{2\pi}[H(A\cos\phi+A-\Delta)\tilde{F}(A\cos\phi,-A)]\cos\phi\mathrm{d}\phi\}. \tag{2.31}$$

Rearranging the above equation leads to

$$C(A)=\frac{\overline{\omega}^2}{A\pi\omega(A)}\frac{1}{4\nu(1+\mu)^2}\{(\Delta-2A)^2$$
$$-\int_{\arccos\frac{A-\Delta}{A}}^{\pi}\frac{1}{2}[\Delta+A(\cos\phi-1)]^2\sin\phi\mathrm{d}\phi$$
$$+\int_{2\pi-\arccos\frac{\Delta-A}{A}}^{2\pi}\frac{1}{2}[\Delta-A(\cos\phi+1)]^2\sin\phi\mathrm{d}\phi\},$$
$$K(A)=\frac{\overline{\omega}^2}{A\pi}\frac{1}{4\nu(1+\mu)^2}\{\int_{\arccos\frac{A-\Delta}{A}}^{\pi}\frac{1}{2}[\Delta+A(\cos\phi-1)]^2\cos\phi\mathrm{d}\phi$$
$$-\int_{2\pi-\arccos\frac{\Delta-A}{A}}^{2\pi}\frac{1}{2}[\Delta-A(\cos\phi+1)]^2\cos\phi\mathrm{d}\phi\}. \tag{2.32}$$

When the dimensionless amplitude $A<\Delta/2$,

$$C(A)=0,\quad K(A)=0. \tag{2.33}$$

Therefore, the equivalent quasi-linear stochastic system of Eq. (2.21) is

$$\ddot{Y}(t)+[2\zeta\overline{\omega}+\lambda C(A)]\dot{Y}(t)+(\overline{\omega}^2+\lambda K(A))Y(t)=\widehat{W}(t). \tag{2.34}$$

It can also be rewritten as

$$\begin{cases}\ddot{Y}(t)+2\zeta\overline{\omega}\dot{Y}(t)+\overline{\omega}^2Y(t)=\widehat{W}(t), & \text{when } A<\Delta/2,\\ \ddot{Y}(t)+[2\zeta\overline{\omega}+\lambda C(A)]\dot{Y}(t)+[\overline{\omega}^2+\lambda K(A)]Y(t)=\widehat{W}(t), & \text{when } A>\Delta/2.\end{cases} \tag{2.35}$$

Equation (2.35) shows that system motion is linear when the dimensionless amplitude $A<\Delta/2$ and the nonlinear part of the hysteretic restoring force has no effect on the response of system. When the dimensionless amplitude $A > \Delta/2$, system motion is quasi-linear with damping and stiffness coefficients depending on amplitude or energy.

The total energy of system is

$$H = \frac{1}{2}\dot{Y}^2 + \frac{1}{2}[\overline{\omega}^2 + \lambda K(A)]Y^2. \tag{2.36}$$

When $\dot{Y} = 0,\ Y = A, H = \frac{1}{2}(\overline{\omega}^2+\lambda K(A))A^2$, which is the relationship $A = A(H)$ between amplitude A and energy H.

Equation (2.34) can be rewritten as

$$\ddot{Y}(t) + \zeta'(H)\dot{Y}(t) + \overline{\overline{\omega}}^2(H)Y(t) = \widehat{W}(t), \tag{2.37}$$

where

$$\zeta'(H) = 2\zeta\overline{\omega} + \lambda C(A(H)), \quad \overline{\overline{\omega}}^2(H) = \overline{\omega}^2 + \lambda K(A(H)). \tag{2.38}$$

Letting $Q_1 = Y,\ P = \dot{Y}$, system (2.37) can be further rewritten in the following quasi-Hamiltonian system form

$$\dot{Q} = P, \quad \dot{P} = -\overline{\overline{\omega}}^2(H)Q - \zeta'(H)P + \widehat{W}(t), \tag{2.39}$$

where the intensity of $\widehat{W}(t)$ is $2D = 2\pi S_{\widehat{w}} = 4\zeta\overline{\omega}^3$. The equivalent Itô stochastic differential equation of Eq. (2.39) is

$$\mathrm{d}Q = P\mathrm{d}t, \quad \mathrm{d}P = [-\overline{\overline{\omega}}^2(H)Q - \zeta'(H)P]\mathrm{d}t + \sqrt{2D}\mathrm{d}B(t). \tag{2.40}$$

where $B(t)$ is unit Wiener process. The system energy is

$$H = \frac{1}{2}P^2 + \frac{1}{2}\overline{\overline{\omega}}^2(H)Q^2 = H(Q, P). \tag{2.41}$$

Using Eq. (2.41) as a transformation from P to H and using the Itô differential rule, an Itô stochastic differential equation for H is obtained from Eq. (2.40). Then Eq. (2.40) is replaced by the following equations

$$\mathrm{d}Q = \pm\sqrt{2H - \overline{\overline{\omega}}^2(H)Q^2}\mathrm{d}t,$$

$$\mathrm{d}H = \left\{\begin{bmatrix} \frac{\partial H}{\partial Q} & \frac{\partial H}{\partial P} \end{bmatrix}\begin{bmatrix} P \\ -\zeta'(H)P - \overline{\overline{\omega}}^2(H)Q \end{bmatrix}\right.$$

$$+\frac{1}{2}\mathrm{tr}\left(\begin{bmatrix}\frac{\partial^2 H}{\partial Q^2} & \frac{\partial^2 H}{\partial Q\partial P}\\ \frac{\partial^2 H}{\partial P\partial Q} & \frac{\partial^2 H}{\partial P^2}\end{bmatrix}\begin{bmatrix}0\\ \sqrt{2D}\end{bmatrix}[0\ \sqrt{2D}]\right)\mathrm{d}t\Bigg\}$$

$$+\begin{bmatrix}\frac{\partial H}{\partial Q} & \frac{\partial H}{\partial P}\end{bmatrix}\begin{bmatrix}0\\ \sqrt{2D}\end{bmatrix}\mathrm{d}B(t), \tag{2.42}$$

where tr(·) represents the trace of a matrix. Note that

$$\frac{\partial H}{\partial Q}=\frac{\overline{\overline{\omega}}^2(H)Q}{1-\frac{1}{2}\frac{\partial\overline{\overline{\omega}}^2(H)}{\partial H}Q^2},\quad \frac{\partial H}{\partial P}=\frac{P}{1-\frac{1}{2}\frac{\partial\overline{\overline{\omega}}^2(H)}{\partial H}Q^2},$$

$$\frac{\partial^2 H}{\partial P^2}=\frac{\left(1-\frac{1}{2}\frac{\partial\overline{\overline{\omega}}^2(H)}{\partial H}Q^2\right)-P\left(-\frac{1}{2}\frac{\partial^2\overline{\overline{\omega}}^2(H)}{\partial H^2}Q^2\frac{\partial H}{\partial P}\right)}{\left(1-\frac{1}{2}\frac{\partial\overline{\overline{\omega}}^2(H)}{\partial H}Q^2\right)^2}. \tag{2.43}$$

Equation (2.42) becomes

$$\mathrm{d}Q=\pm\sqrt{2H-\overline{\overline{\omega}}^2(H)Q^2}\mathrm{d}t,$$

$$\mathrm{d}H=\Bigg\{-\zeta'(H)\frac{2H-\overline{\overline{\omega}}^2(H)Q^2}{1-\frac{1}{2}\frac{\partial\overline{\overline{\omega}}^2(H)}{\partial H}Q^2}+\frac{D}{1-\frac{1}{2}\frac{\partial\overline{\overline{\omega}}^2(H)}{\partial H}Q^2}$$

$$+D\left(1-\frac{1}{2}\frac{\partial\overline{\overline{\omega}}^2(H)}{\partial H}Q^2\right)^{-3}\times\frac{1}{2}\frac{\partial^2\overline{\overline{\omega}}^2(H)}{\partial H^2}Q^2(2H-\overline{\overline{\omega}}^2(H)Q^2)\Bigg\}\mathrm{d}t$$

$$+\sqrt{2D}\frac{\pm\sqrt{2H-\overline{\overline{\omega}}^2(H)Q^2}}{1-\frac{1}{2}\frac{\partial\overline{\overline{\omega}}^2(H)}{\partial H}Q^2}\mathrm{d}B(t). \tag{2.44}$$

In system (2.44), energy $H(t)$ is a slowly-varying process and $Q(t)$ is a fast-varying process. According to Khasminskii theorem (Khasminskii 1968), the energy process $H(t)$ approaches a Markov diffusion process as $\zeta', D\to 0$. The averaged Itô stochastic differential equation can be derived from the second equation of Eq. (2.44) by carrying out the time averaging which can be replaced by spatial averaging with respect to the fast-varying process Q (Zhu and Lin 1991). The result is

$$\mathrm{d}H=m(H)\mathrm{d}t+\sigma(H)\mathrm{d}B(t) \tag{2.45}$$

where

$$m(H)=\frac{1}{T(H)}2\int_{-A(H)}^{A(H)}\Bigg\{-\zeta'(H)\frac{\sqrt{2H-\overline{\overline{\omega}}^2(H)Q^2}}{1-\frac{1}{2}\frac{\partial\overline{\overline{\omega}}^2(H)}{\partial H}Q^2}$$

$$+\frac{D}{\left(1-\frac{1}{2}\frac{\partial\overline{\overline{\omega}}^2(H)}{\partial H}Q^2\right)\sqrt{2H-\overline{\overline{\omega}}^2(H)Q^2}}$$

$$\left.+D\frac{\frac{1}{2}\frac{\partial^2\overline{\overline{\omega}}^2(H)}{\partial H^2}Q^2\sqrt{2H-\overline{\overline{\omega}}^2(H)Q^2}}{(1-\frac{1}{2}\frac{\partial\overline{\overline{\omega}}^2(H)}{\partial H}Q^2)^3}\right\}\mathrm{d}Q,$$

$$\sigma^2(H)=\frac{4D}{T(H)}\int_{-A(H)}^{A(H)}\frac{\sqrt{2H-\overline{\overline{\omega}}^2(H)Q^2}}{\left(1-\frac{1}{2}\frac{\partial\overline{\overline{\omega}}^2(H)}{\partial H}Q^2\right)^2}\mathrm{d}Q,$$

$$T(H)=2\int_{-A(H)}^{A(H)}\frac{\mathrm{d}Q}{\sqrt{2H-\overline{\overline{\omega}}^2(H)Q^2}}. \tag{2.46}$$

The FPK equation associated with averaged Itô Eq. (2.45) is

$$\frac{\partial p}{\partial t}=-\frac{\partial}{\partial H}[m(H)p]+\frac{1}{2}\frac{\partial^2}{\partial H^2}[\sigma^2(H)p], \tag{2.47}$$

where $p=p(H,t|H_0)$ is the transition PDF of energy H. The following stationary PDF $p(h)$ can be obtained by solving the reduced FPK equation, i.e., Eq. (2.47) without the time derivative term as follows

$$p(h)=\tilde{C}\exp\left\{-\int_0^h\left[\frac{\mathrm{d}\sigma^2(u)}{\mathrm{d}u}-2m(u)\right]/\sigma^2(u)\mathrm{d}u\right\}, \tag{2.48}$$

where $\tilde{C}$ is a normalization constant. According to Eq. (5.18) in vol. 1, the joint stationary PDF of displacement Q and momentum P is

$$p(q,p)=\frac{p(h)}{T(h)}|_{h=H(q,p)}. \tag{2.49}$$

The stationary PDF of amplitude A can be obtained from Eq. (2.48) as follows

$$p(a)=p(h)\left|\frac{\partial H(a)}{\partial a}\right|\Bigg|_{h=H(a)}. \tag{2.50}$$

Numerical calculations were carried out for the system (2.18) with parameters $\zeta=0.01,\overline{\omega}=4\pi$ using the present stochastic averaging method (Wang et al. 2009). The stationary PDF of displacement amplitude is shown in Fig. 2.3 for the cases of $\nu=1$, $\mu=0.5,1,2$ and different λ values. For comparison, results from Monte Carlo simulation and results from the other stochastic averaging method (Spanos et al. 2004) are also shown in the figure. It can be observed that for larger λ, the results from the present stochastic averaging method are more accurate than those from the other stochastic averaging method (Spanos et al. 2004). Figs. 2.4 and 2.5 show the stationary PDFs of amplitude for the cases of $\nu = 0.5$ and 0.2, respectively. Since the other stochastic averaging method (Spanos et al. 2004) can be applied only for the case of $\nu = 1$, only the results from Monte Carlo simulations are shown in Figs. 2.4 and 2.5 for comparison. It is seen from Eq. (2.21) that λ is a nonlinear coefficient. As λ decreases, the effect of nonlinear part $p_H(t)$ of hysteretic restoring force on the response becomes weaker, the system approaches a linear one, and the amplitude tends to a Rayleigh distribution, as shown in all three figures. As μ decreases, the peak of $p(a)$ becomes higher, which indicates the nonlinear characteristic of system response become more significant. Figs. 2.6 and 2.7 depict the stationary PDFs of energy for cases of $\nu = 1$ and $\nu = 0.5$, respectively, obtained by using the present stochastic averaging method.

Example 2.2 Consider a two-DOF nonlinear system with hysteretic restoring forces and excited by Gaussian white noises

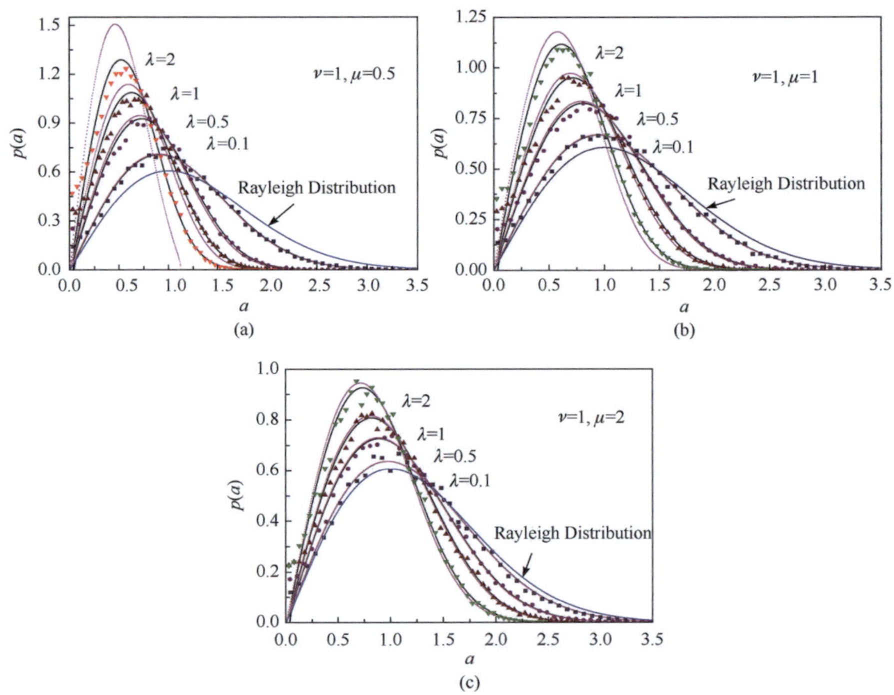

Fig. 2.3 Stationary PDF of displacement amplitude of system (2.18) with $\nu = 1.0$. Lines —, result from the present stochastic averaging method (Wang et al. 2009); lines , result from reference (Spanos et al. 2004); symbols ▼▲●■, results from Monte Carlo simulation

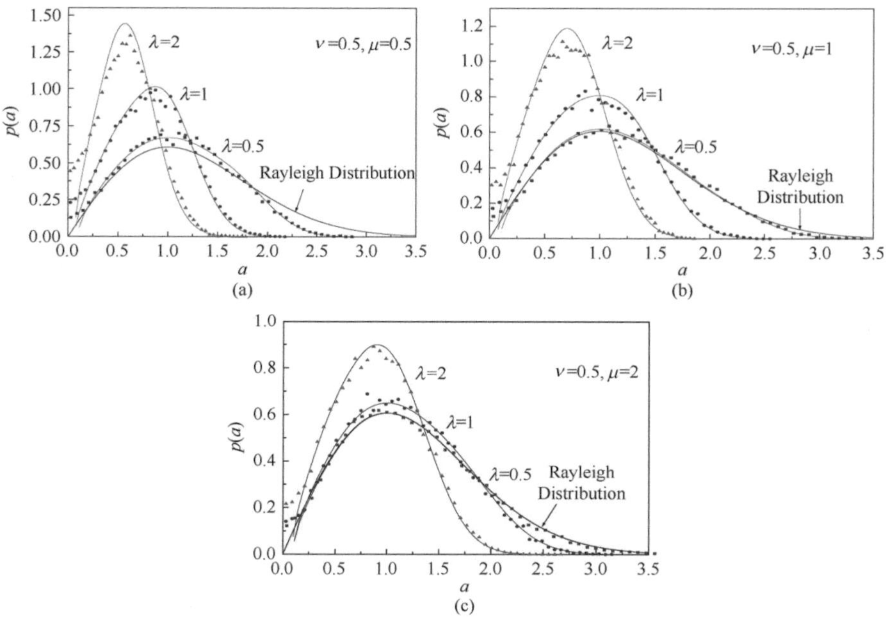

Fig. 2.4 Stationary PDF of displacement amplitude of system (2.18) with $\nu = 0.5$. Lines —, results from the stochastic averaging method; symbols ▲●■, results from Monte Carlo simulation (Wang et al. 2009)

$$\begin{aligned} &\ddot{X}_1 + (\lambda_1 + \eta_1 \dot{X}_2^2)\dot{X}_1 + \mu_1 f_1(X_1) + \omega_1^2 X_1 = W_{g1}(t), \\ &\ddot{X}_2 + (\lambda_2 + \eta_2 X_1^2)\dot{X}_2 + \mu_2 f_2(X_2) + \omega_2^2 X_2 = W_{g2}(t). \end{aligned} \tag{2.51}$$

where $W_{g1}(t)$ and $W_{g2}(t)$ are independent Gaussian white noises with intensities $2D_1$ and $2D_2$, respectively; $f_1(X_1)$ and $f_2(X_2)$ are the Bouc-Wen model of hysteretic restoring forces introduced in Eqs. (3.116) and (3.118) of vol. 1; $\lambda_i, \eta_i, \mu_i, D_i, \ i = 1, 2$ are small quantities of order ε. Using Eqs. (2.4) and (2.7), the hysteretic restoring forces can be decoupled into elastic restoring forces and viscous damping forces, which are combined with the original elastic and damping forces to form the following equivalent system

$$\begin{aligned} &\ddot{X}_1 + (\lambda_1 + \mu_1 C_1(A_1) + \eta_1 \dot{X}_2^2)\dot{X}_1 + (\omega_1^2 + \mu_1 K_1(A_1))X_1 = W_{g1}(t), \\ &\ddot{X}_2 + (\lambda_2 + \mu_2 C_2(A_2) + \eta_2 X_1^2)\dot{X}_2 + (\omega_2^2 + \mu_2 K_2(A_2))X_2 = W_{g2}(t), \end{aligned} \tag{2.52}$$

where A_1, A_2 are the amplitudes of the two equivalent oscillators in the system, and $C_1(A_1)$, $C_2(A_2)$, $K_1(A_1)$, $K_2(A_2)$ are obtained from Eq. (2.7).

Letting $Q_1 = X_1, \ P_1 = \dot{X}_1, Q_2 = X_2, \ P_2 = \dot{X}_2$, Eq. (2.52) can be rewritten as the following quasi-integrable Hamiltonian system

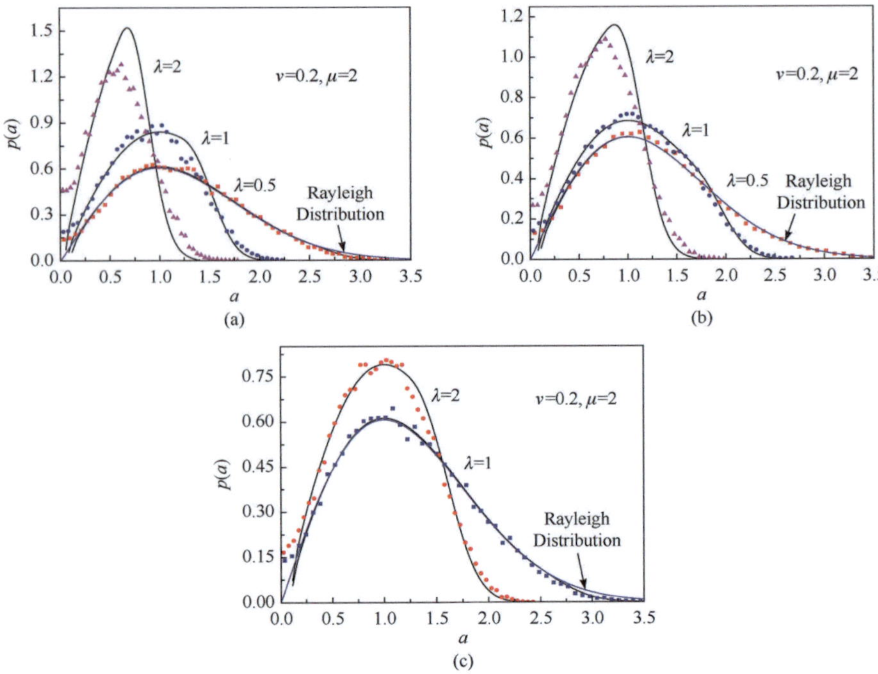

Fig. 2.5 Stationary PDF of displacement amplitude of system (2.18) with $\nu = 0.2$ and $\mu = 2$. Lines —, results from the stochastic averaging method; symbols ▲●■, results from Monte Carlo simulation (Wang et al. 2009)

$$
\begin{aligned}
\dot{Q}_1 &= \frac{\partial H}{\partial P_1}, \\
\dot{Q}_2 &= \frac{\partial H}{\partial P_2}, \\
\dot{P}_1 &= -\frac{\partial H}{\partial Q_1} - (\lambda_1 + \mu_1 \overline{C}_1(H_1) + \eta_1 P_2^2) P_1 + W_{g1}(t), \\
\dot{P}_2 &= -\frac{\partial H}{\partial Q_2} - (\lambda_2 + \mu_2 \overline{C}_2(H_2) + \eta_2 Q_1^2) P_2 + W_{g2}(t).
\end{aligned} \tag{2.53}
$$

The Hamiltonian associated with the quasi-integrable Hamiltonian system (2.53) is

$$
\begin{aligned}
H &= H_1 + H_2, \\
H_1 &= \frac{1}{2} P_1^2 + \frac{1}{2}(\omega_1^2 + \mu_1 \overline{K}_1(H_1)) Q_1^2, \\
H_2 &= \frac{1}{2} P_2^2 + \frac{1}{2}(\omega_2^2 + \mu_2 \overline{K}_2(H_2)) Q_2^2.
\end{aligned} \tag{2.54}
$$

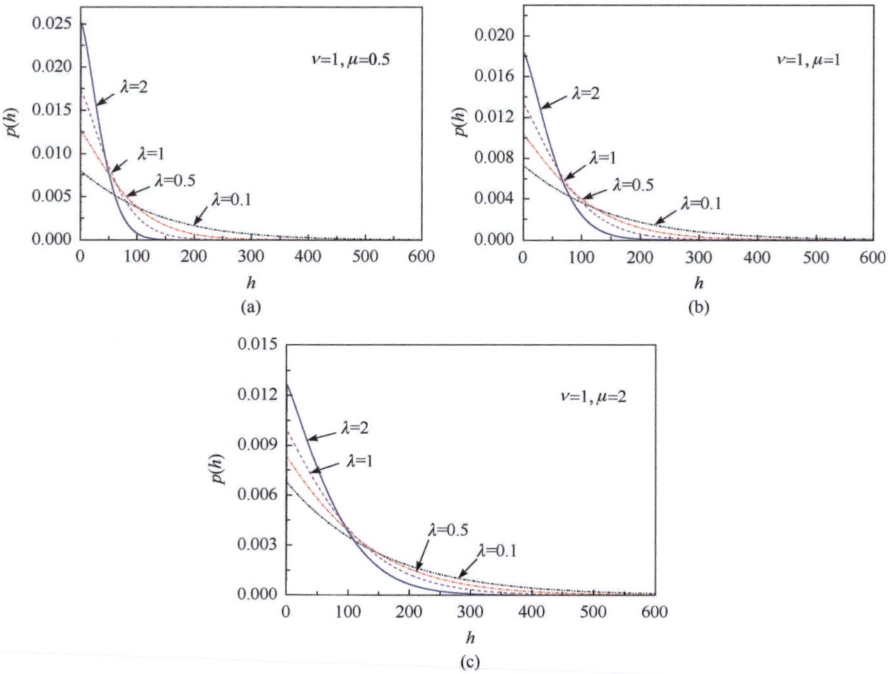

Fig. 2.6 Stationary PDF of system energy of system (2.18) with $\nu = 1.0$. Results are obtained from the present stochastic averaging method (Wang et al. 2009)

The values of $\overline{C}_i(H_i), \overline{K}_i(H_i)$ $i = 1, 2$ in Eqs. (2.53) and (2.54) are obtained from following equations

$$\overline{C}_i(H_i) = C_i(A_i)|_{A_i=\sqrt{\frac{2H_i}{\omega_i^2+\overline{K}_i(H_i)}}}, \quad \overline{K}_i(H_i) = K_i(A_i)|_{A_i=\sqrt{\frac{2H_i}{\omega_i^2+\overline{K}_i(H_i)}}}. \tag{2.55}$$

It is difficult to obtain analytical expressions for $\overline{C}_i(H_i), \overline{K}_i(H_i)$. However, they can be calculated by using numerical method.

If the equivalent frequencies $\sqrt{\omega_1^2 + \mu_1\overline{K}_1(H_1)}$ and $\sqrt{\omega_2^2 + \mu_2\overline{K}_2(H_2)}$ of the two oscillators in system (2.53) do not satisfy resonant relation, the stochastic averaging method of quasi-integrable and non-resonant Hamiltonian systems excited by Gaussian white noises described in Sect. 5.2.1, vol. 1 can be applied. Since $[H_1(t), H_2(t)]^{\mathrm{T}}$ converges weakly to a two-dimensional vector Markov diffusion process as $\lambda_i, \mu_i, \eta_i, D_i \to 0$, the averaged Itô stochastic differential equations are of the form

$$\begin{aligned} \mathrm{d}H_1 &= m_1(H_1, H_2)\mathrm{d}t + \sigma_{11}(H_1, H_2)\mathrm{d}B_1(t) + \sigma_{12}(H_1, H_2)\mathrm{d}B_2(t), \\ \mathrm{d}H_2 &= m_2(H_1, H_2)\mathrm{d}t + \sigma_{21}(H_1, H_2)\mathrm{d}B_1(t) + \sigma_{22}(H_1, H_2)\mathrm{d}B_2(t). \end{aligned} \tag{2.56}$$

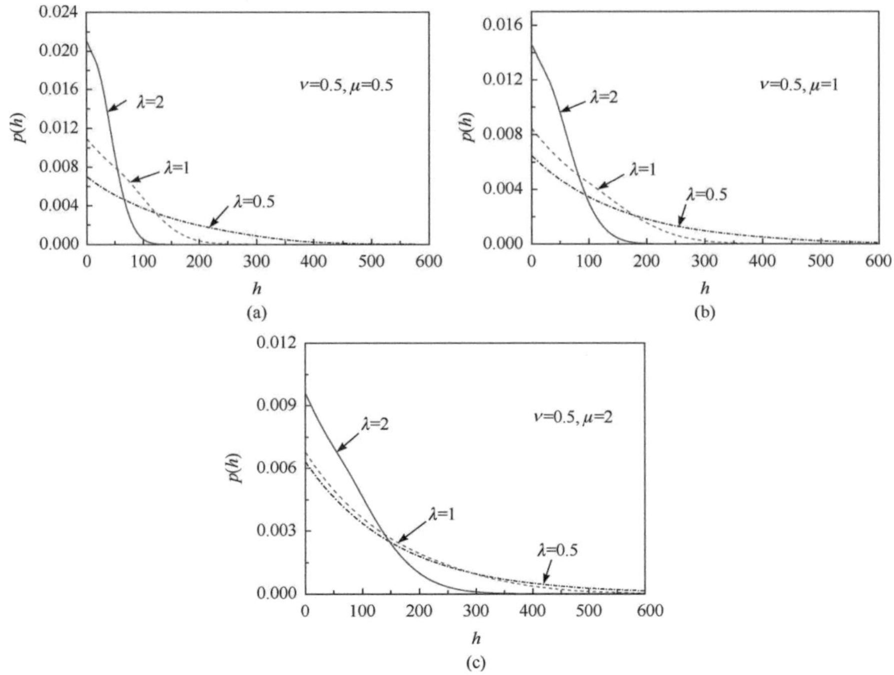

Fig. 2.7 Stationary PDF of system energy of system (2.18) with $\nu = 0.5$. Results are obtained by using the present stochastic averaging method (Wang et al. 2009)

The drift and diffusion coefficients can be derived according to Eq. (2.25) as follows

$$\begin{aligned}
m_1(H_1, H_2) &= D_1 - (\lambda_1 + \mu_1 \overline{C}_1(H_1))H_1 - \eta_1 H_1 H_2, \\
m_2(H_1, H_2) &= D_2 - (\lambda_2 + \mu_2 \overline{C}_2(H_2))H_2 - \eta_2 \frac{H_1 H_2}{\omega_1^2 + \mu_1 \overline{K}_1(H_1)}, \\
b_{11}(H_1, H_2) &= \sigma_{11}\sigma_{11} + \sigma_{12}\sigma_{12} = 2D_1 H_1, \\
b_{22}(H_1, H_2) &= \sigma_{21}\sigma_{21} + \sigma_{22}\sigma_{22} = 2D_2 H_2, \\
b_{12} &= b_{21} = 0.
\end{aligned} \tag{2.57}$$

The reduced averaged FPK equation associated with averaged Itô equation (2.56) is

$$-\frac{\partial[m_1(h_1, h_2)p]}{\partial h_1} - \frac{\partial[m_2(h_1, h_2)p]}{\partial h_2} + \frac{1}{2}\frac{\partial^2[b_{11}(h_1, h_2)p]}{\partial h_1^2} + \frac{1}{2}\frac{\partial^2[b_{22}(h_1, h_2)p]}{\partial h_2^2} = 0, \tag{2.58}$$

where $p = p(h_1, h_2)$ is the stationary PDF and the first and second derivative moments are given by

$$m_1(h_1,h_2)=m_1(H_1,H_2)|_{H_1=h_1,\,H_2=h_2},\quad m_2(h_1,h_2)=m_2(H_1,H_2)|_{H_1=h_1,\,H_2=h_2},$$
$$b_{11}(h_1,h_2)=b_{11}(H_1,H_2)|_{H_1=h_1,\,H_2=h_2},\quad b_{22}(h_1,h_2)=b_{22}(H_1,H_2)|_{H_1=h_1,\,H_2=h_2}. \tag{2.59}$$

The boundary conditions for the FPK Eq. (2.58) are

$$\begin{aligned}
&p=\text{finite},\quad \text{when } h_1=0 \text{ or } h_2=0,\\
&p\to 0,\ \partial p/\partial h_1\to 0,\quad \text{when } h_1\to\infty,\\
&p\to 0,\ \partial p/\partial h_2\to 0,\quad \text{when } h_2\to\infty.
\end{aligned} \tag{2.60}$$

Once $p(h_1,h_2)$ is obtained, the joint PDFs $p(q_1,q_2,p_1,p_2)$ and $p(q_1,p_1)$ can be obtained according to Eq. (2.36) as follows

$$\begin{aligned}
&p(q_1,q_2,p_1,p_2)\\
&=\frac{\sqrt{[\omega_1^2+\mu_1\overline{K}_1(h_1)][\omega_2^2+\mu_2\overline{K}_2(h_2)]}}{4\pi^2}p(h_1,h_2)\Bigg|_{\substack{h_1=p_1^2/2+[\omega_1^2+\mu_1\overline{K}_1(h_1)]q_1^2/2,\\ h_2=p_2^2/2+[\omega_2^2+\mu_2\overline{K}_2(h_2)]q_2^2/2}}\\
&p(q_1,p_1)=\int_{-\infty}^{\infty}\int_{-\infty}^{\infty}p(q_1,q_2,p_1,p_2)\mathrm{d}q_2\mathrm{d}p_2.
\end{aligned} \tag{2.61}$$

The marginal PDFs and statistics can be calculated from Eq. (2.61).

In numerical calculations for system (2.51), parameters $\omega_1=1$, $\omega_2=1.414$, $\lambda_1=0.1$, $\lambda_2=0.1$, $\eta_1=0.02$, $\eta_2=0.02$, $D_1=0.1$, $D_2=0.3$, $\mu_1=\mu_2=0.6$, $\beta=0.8$, $\gamma=1.1$, and $\alpha=0.1$ are used. Figures 2.8 and 2.9 show the results from the stochastic averaging method and from Monte Carlo simulations of system (2.51). It is seen that the two results agree well.

2.2 Quasi-integrable Hamiltonian Systems with Viscoelastic Forces

The stochastic averaging method for single-DOF systems with viscoelastic forces described in Sect. 4.5 of vol. 1 can be extended to the following MDOF quasi-integrable Hamiltonian systems with coupled viscoelastic forces. Consider the following quasi-integrable Hamiltonian system with viscoelastic forces

$$\begin{aligned}
&\dot{Q}_i=P_i,\\
&\dot{P}_i=-g_i'(Q_i)-\varepsilon\sum_{j=1}^{n}c_{ij}'(\boldsymbol{Q},\boldsymbol{P})P_j-\varepsilon\mu_{ii}Z_i(P_i)-\varepsilon\sum_{j=1,j\neq i}^{n}\mu_{ij}Z_i(P_i-P_j)
\end{aligned}$$

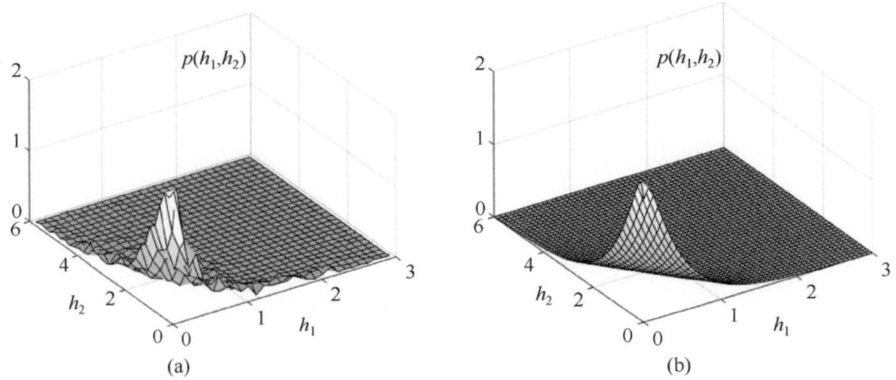

Fig. 2.8 Stationary joint PDF of Hamiltonians of system (2.51). **a** Result from Monte Carlo simulation; **b** result from the stochastic averaging method

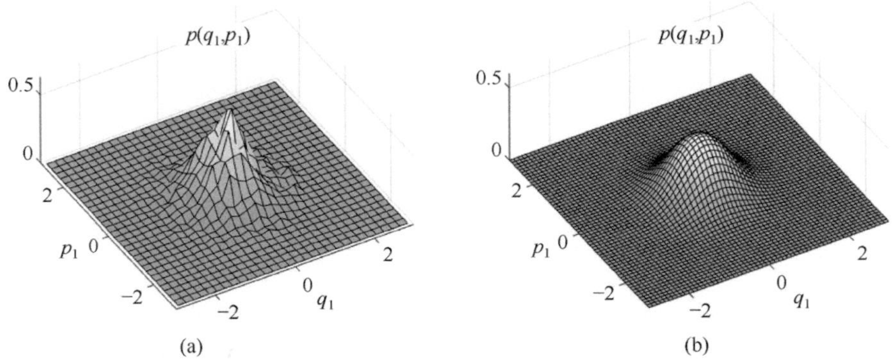

Fig. 2.9 Stationary joint PDF of generalized displacements and generalized momenta of system (2.51). **a** Result from Monte Carlo simulation; **b** result from the stochastic averaging method

$$+ \varepsilon^{1/2} \sum_{k=1}^{m} f_{ik}(\boldsymbol{Q}, \boldsymbol{P}) \xi_k(t),$$
$$i = 1, 2, \ldots, n. \tag{2.62}$$

where $\varepsilon\mu_{ii}Z_i(P_i), i = 1, 2, \ldots, n$ are the viscoelastic force of ith oscillator; $\varepsilon \sum_{j=1, j \neq i}^{n} \mu_{ij} Z_i(P_i - P_j)$ are the sum of the coupled viscoelastic forces between ith and ith $(j \neq i)$ oscillator; $\xi_k(t) \quad (k = 1, 2, \ldots, m)$ are stationary wideband processes with correlation functions $R_{kl}(\tau)$ $(k, l = 1, 2, \ldots, m)$. The other terms in the system (2.62) are the same as those in system (2.1). According to Eq. (3.150) of vol. 1,

$$Z_i(P_i) = \int_0^t G_i(t-\tau)P_i(\tau)\mathrm{d}\tau, \tag{2.63}$$

where $G_i(t)$ is the relaxation modulus. According to Eq. (3.151) of vol. 1, the following approximation is used (Zhang 1994; Christensen 1982; Drozdov 1998)

$$G_i(t) = \sum_{s=1}^{M} \beta_{is} \exp(-t/\lambda_{is}), \quad \lambda_{is} > 0. \tag{2.64}$$

For simplicity, only the first term in Eq. (2.64) is considered. Substituting the first term in Eq. (2.64) into Eq. (2.63) yields

$$Z_i(P_i) = \beta_i \int_0^t \exp[-(t-\tau)/\lambda_i]P_i(\tau)\mathrm{d}\tau. \tag{2.65}$$

Equation (2.65) indicates that $Z_i(P_i)$ is a linear operator, thus

$$Z_i(P_i - P_j) = Z_i(P_i) - Z_i(P_j) \tag{2.66}$$

In order to decouple a viscoelastic force into elastic restoring force and viscous damping force, the generalized harmonic balance technique is applied. Similar to Eq. (2.1), assume that system (2.62) has the following randomized periodic solution

$$Q_i(t) = A_i \cos\Phi_i(t), \quad P_i(t) = -A_i\overline{\omega}_i \sin\Phi_i(t). \tag{2.67}$$

where $\overline{\omega}_i$ represents the mean frequency of the i-th DOF oscillator. Note that the relationship between amplitude A_i and the Hamiltonian function H_i is given by $A_i = \sqrt{2H_i}/\overline{\omega}_i$. Substituting Eq. (2.67) into Eqs. (2.65) and (2.66), completing the integration and neglecting transient terms yield

$$\begin{aligned}
Z_i(P_i) &= C_{ii}(H_i)P_i + K_{ii}(H_i)Q_i,\\
Z_i(P_i - P_j) &= C_{ii}(H_i)P_i - C_{ij}(H_j)P_j + K_{ii}(H_i)Q_i - K_{ij}(H_j)Q_j,\\
C_{ii}(H_i) &= \frac{\beta_i\lambda_i}{1+\overline{\omega}_i^2\lambda_i^2}, \quad K_{ii}(H_i) = \frac{\beta_i\lambda_i^2\overline{\omega}_i^2}{1+\overline{\omega}_i^2\lambda_i^2},\\
C_{ij}(H_j) &= \frac{\beta_i\lambda_i}{1+\overline{\omega}_j^2\lambda_i^2}, \quad K_{ij}(H_j) = \frac{\beta_i\lambda_i^2\overline{\omega}_j^2}{1+\overline{\omega}_j^2\lambda_i^2}.
\end{aligned} \tag{2.68}$$

As a result, the equivalent quasi-Hamiltonian system can be expressed as

$$\dot{Q}_i = P_i,$$

$$\dot{P}_i = -g_i(Q_i) - \varepsilon \sum_{j=1}^{n} c_{ij}(\boldsymbol{Q}, \boldsymbol{P})P_j + \varepsilon \sum_{j=1,\, j\neq i}^{n} \mu_{ij}K_{ij}(H_j)Q_j$$
$$+ \varepsilon^{1/2} \sum_{k=1}^{m} f_{ik}(\boldsymbol{Q}, \boldsymbol{P})\xi_k(t), \quad i = 1, 2, \ldots, n, \tag{2.69}$$

where

$$g_i(Q_i) = g_i'(Q_i) + \varepsilon(\sum_{r=1}^{n} \mu_{ir})K_{ii}(H_i)Q_i,$$
$$c_{ij}(\boldsymbol{Q}, \boldsymbol{P}) = \begin{cases} c_{ii}'(\boldsymbol{Q}, \boldsymbol{P}) + (\sum\limits_{r=1}^{n} \mu_{ir})C_{ii}(H_i), & j = i, \\ c_{ij}'(\boldsymbol{Q}, \boldsymbol{P}) - \mu_{ij}C_{ij}(H_j), & j \neq i. \end{cases} \tag{2.70}$$

System (2.69) is a quasi-non-integrable Hamiltonian system due to the terms $\varepsilon \sum\limits_{j=1,\, j\neq i}^{n} \mu_{ij}K_{ij}(H_j)Q_j$. However, these terms are small compared to $g_i(Q_i)$. Based on the analysis of Sect. 5.4.3 in vol. 1, the stochastic averaging method of quasi-integrable Hamiltonian systems can be applied to system (2.69). Based on whether the mean frequencies $\overline{\omega}_i$ $(i = 1, 2, \ldots, n)$ in the system (2.69) satisfy the following resonance conditions

$$\sum_{i=1}^{n} k_i^u \overline{\omega}_i = O_u(\varepsilon), \quad u = 1, 2, \ldots, \alpha. \tag{2.71}$$

where k_i^u are integers not all equal to zero, two cases of non-resonance and resonance will be considered in applying the stochastic averaging method.

(i) **Non-resonant case**

In the case where the mean frequencies $\overline{\omega}_i$ $(i = 1, 2, \ldots, n)$ of system (2.69) do not satisfy Eq. (2.71) based on Eq. (2.67) and the relationship between A_i and H_i, the stochastic differential equations for the Hamiltonian function H_i $(i = 1, 2, \ldots, n)$ and the phase angles Φ_i $(i = 1, 2, \ldots, n)$ can be derived from Eq. (2.69) (see Sect. 2.1.2)

$$\dot{H}_i = \varepsilon F_i^H(\boldsymbol{H}, \boldsymbol{\Phi}) + \varepsilon^{1/2} \sum_{k=1}^{m} G_{ik}^H(\boldsymbol{H}, \boldsymbol{\Phi})\xi_k(t),$$
$$\dot{\Phi}_i = \overline{\omega}_i(H_i) + \varepsilon F_i^{\Phi}(\boldsymbol{H}, \boldsymbol{\Phi}) + \varepsilon^{1/2} \sum_{k=1}^{m} G_{ik}^{\Phi}(\boldsymbol{H}, \boldsymbol{\Phi})\xi_k(t), \tag{2.72}$$
$$i = 1, 2, \ldots, n,$$

where $\boldsymbol{H} = [H_1, H_2, \ldots, H_n]^{\mathrm{T}}$, $\boldsymbol{\Phi} = [\Phi_1, \Phi_2, \ldots, \Phi_n]^{\mathrm{T}}$,

$$
\begin{aligned}
F_i^H &= \sqrt{2H_i}\sin\Phi_i[\sum_{j=1}^{n} c_{ij}(\boldsymbol{Q},\boldsymbol{P})P_j - \sum_{j=1,j\neq i}^{n}\mu_{ij}K_{ij}(H_j)Q_j],\\
F_i^{\Phi} &= \frac{\cos\Phi_i}{\sqrt{2H_i}}[\sum_{j=1}^{n} c_{ij}(\boldsymbol{Q},\boldsymbol{P})P_j - \sum_{j=1,j\neq i}^{n}\mu_{ij}K_{ij}(H_j)Q_j],\\
G_{ik}^H &= -\sqrt{2H_i}\sin\Phi_i f_{ik}(\boldsymbol{Q},\boldsymbol{P}),\quad G_{ik}^{\Phi} = -\frac{\cos\Phi_i}{\sqrt{2H_i}} f_{ik}(\boldsymbol{Q},\boldsymbol{P}).
\end{aligned}
\tag{2.73}
$$

In Eq. (2.73), Q_i, P_i should be replaced by H_i, Φ_i according to Eq. (2.67) and the relationship between A_i and H_i.

The variable $\boldsymbol{\Phi}(t)$ in the system (2.72) is fast-varying random process and $\boldsymbol{H}(t)$ is slowly-varying random process. According to the Khasminskii theorem (Khasminskii, 1966, 1968), when $\varepsilon \to 0$, $\boldsymbol{H}(t)$ converges weakly to a n-dimensional vector Markov diffusion process, which is governed by the following averaged Itô stochastic differential equations

$$
\begin{aligned}
\mathrm{d}H_i &= m_i(\boldsymbol{H})\mathrm{d}t + \sum_{k=1}^{m}\sigma_{ik}(\boldsymbol{H})\mathrm{d}B_k(t),\\
i &= 1,2,\ldots,n,
\end{aligned}
\tag{2.74}
$$

where the drift and diffusion coefficients can be obtained similar to Eqs. (4.27) and (4.28) in vol. 1 as follows

$$
\begin{aligned}
m_i(\boldsymbol{H}) &= \varepsilon\left\langle F_i^H + \sum_{k,l=1}^{m}\int_{-\infty}^{0}\sum_{j=1}^{n}\left(\frac{\partial G_{ik}^H}{\partial H_j}\bigg|_t G_{jl}^H\bigg|_{t+\tau} + \frac{\partial G_{ik}^H}{\partial \Phi_j}\bigg|_t G_{jl}^{\Phi}\bigg|_{t+\tau}\right)R_{kl}(\tau)\mathrm{d}\tau\right\rangle_t,\\
b_{ij}(\boldsymbol{H}) &= \sum_{k=1}^{m}\sigma_{ik}(\boldsymbol{H})\sigma_{jk}(\boldsymbol{H}) = \varepsilon\left\langle\sum_{k,l=1}^{m}\int_{-\infty}^{\infty}\left(G_{ik}^H\big|_t G_{jl}^H\big|_{t+\tau}\right)R_{kl}(\tau)\mathrm{d}\tau\right\rangle_t.
\end{aligned}
\tag{2.75}
$$

where the time averaging can be replaced by spatial averaging with respect to $\boldsymbol{\Phi}$, i.e.,

$$
\langle[\cdot]\rangle_t = \frac{1}{(2\pi)^n}\int_0^{2\pi}[\cdot]\mathrm{d}\Phi_1\mathrm{d}\Phi_2\cdots\mathrm{d}\Phi_n,\quad i,j=1,2,\ldots,n,
\tag{2.76}
$$

The averaged FPK equation associated with Eq. (2.74) can be established as

$$
\frac{\partial p}{\partial t} = \sum_{i=1}^{n}\frac{\partial(m_i p)}{\partial h_i} + \frac{1}{2}\sum_{i,j=1}^{n}\frac{\partial^2(b_{ij}p)}{\partial h_i\partial h_j}.
\tag{2.77}
$$

where the first derivative moments $m_i(\boldsymbol{h}) = m_i(\boldsymbol{H})|_{\boldsymbol{H}=\boldsymbol{h}}$, and the second derivative moments $b_{ij}(\boldsymbol{h}) = b_{ij}(\boldsymbol{H})\big|_{\boldsymbol{H}=\boldsymbol{h}}$. In Eq. (2.77), $p = p(\boldsymbol{h},t|\boldsymbol{h}_0)$ is the transition PDF of

$\boldsymbol{H}(t)$. The initial conditions and boundary conditions for Eq. (2.77) are

$$\begin{aligned} &p(\boldsymbol{h}, 0|\boldsymbol{h}_0) = \delta(\boldsymbol{h} - \boldsymbol{h}_0), \\ &p = \text{finite}, \quad \text{when } |\boldsymbol{h}| = 0, \\ &p = 0, \quad \partial p/\partial h_i = 0, \quad \text{when } |\boldsymbol{h}| \to \infty, \end{aligned} \tag{2.78}$$

and the normalization condition is $\int_0^\infty p(\boldsymbol{h})\mathrm{d}\boldsymbol{h} = 1$.

(ii) **Resonant case**

In the case where the mean frequencies $\overline{\omega}_i$ $(i = 1, 2, \ldots, n)$ of system (2.69) satisfy Eq. (2.71), the following combinations of angle variables are introduced

$$\Psi_u = \sum_{i=1}^{n} k_i^u \Phi_i, \quad u = 1, 2, \ldots, \alpha. \tag{2.79}$$

By replacing α equations for $\Phi_1, \Phi_2, \ldots, \Phi_\alpha$ in Eq. (2.72) by α equations for $\Psi_1, \Psi_2, \ldots, \Psi_\alpha$ according to Eq. (2.79), the combinations Ψ_u of angle variables and the phase angles $\Phi_{\alpha+1}, \Phi_{\alpha+2}, \ldots, \Phi_n$ can be established from system (2.72) as follows

$$\begin{aligned} &\dot{H}_i = \varepsilon \overline{F}_i^H(\boldsymbol{H}, \boldsymbol{\Psi}, \Phi') + \varepsilon^{1/2} \sum_{k=1}^{m} \overline{G}_{ik}^H(\boldsymbol{H}, \boldsymbol{\Psi}, \Phi')\xi_k(t), \\ &\dot{\Psi}_u = O(\varepsilon) + \varepsilon \sum_{i=1}^{n} k_i^u \overline{F}_i^\Phi(\boldsymbol{H}, \boldsymbol{\Psi}, \Phi') + \varepsilon^{1/2} \sum_{k=1}^{m} \sum_{i=1}^{n} k_i^u \overline{G}_{ik}^\Phi(\boldsymbol{H}, \boldsymbol{\Psi}, \Phi')\xi_k(t), \\ &\dot{\Phi}_r = \overline{\omega}_r(H_r) + \varepsilon \overline{F}_r^\Phi(\boldsymbol{H}, \boldsymbol{\Psi}, \Phi') + \varepsilon^{1/2} \sum_{k=1}^{m} \overline{G}_{rk}^\Phi(\boldsymbol{H}, \boldsymbol{\Psi}, \Phi')\xi_k(t), \\ &\quad i = 1, 2, \ldots, n; \quad u = 1, 2, \ldots, \alpha; \quad r = \alpha + 1, \alpha + 2, \ldots, n, \end{aligned} \tag{2.80}$$

where $\boldsymbol{\Psi} = [\Psi_1, \Psi_2, \ldots, \Psi_\alpha]^{\mathrm{T}}$, $\Phi' = [\Phi_{\alpha+1}, \Phi_{\alpha+2}, \ldots, \Phi_n]^{\mathrm{T}}$. $\overline{F}_i^H, \overline{F}_i^\Phi, \overline{G}_{ik}^H, \overline{G}_{ik}^\Phi$ are obtained from $F_i^H, F_i^\Phi, G_{ik}^H, G_{ik}^\Phi$ in Eq. (2.73) by replacing $\Phi_1, \Phi_2, \ldots, \Phi_\alpha$ with $\boldsymbol{\Psi}$.

In system (2.80), $\Phi'(t)$ is fast-varying random process and both $\boldsymbol{H}(t)$ and $\boldsymbol{\Psi}(t)$ are slowly-varying random processes. According to the Khasminskii theorems (Khasminskii 1966, 1968), when $\varepsilon \to 0$, $[\boldsymbol{H}^{\mathrm{T}}(t), \boldsymbol{\Psi}^{\mathrm{T}}(t)]^{\mathrm{T}}$ converges to $(n+\alpha)$-dimensional vector Markov diffusion process governed by the following averaged Itô stochastic differential equations

$$\begin{aligned} &\mathrm{d}H_i = m_i^H(\boldsymbol{H}, \boldsymbol{\Psi})\mathrm{d}t + \sum_{k=1}^{m} \sigma_{ik}^H(\boldsymbol{H}, \boldsymbol{\Psi})\mathrm{d}B_k(t), \\ &\mathrm{d}\Psi_u = m_u^\Psi(\boldsymbol{H}, \boldsymbol{\Psi})\mathrm{d}t + \sum_{k=1}^{m} \sigma_{uk}^\Psi(\boldsymbol{H}, \boldsymbol{\Psi})\mathrm{d}B_k(t), \\ &i = 1, 2, \ldots, n; \quad u = 1, 2, \ldots, \alpha, \end{aligned} \tag{2.81}$$

where the drift and diffusion coefficients can be obtained according to Eqs. (4.27) and (4.28) in vol. 1 as follows

$$
\begin{aligned}
m_i^H &= \varepsilon\Bigg\langle \overline{F}_i^H + \sum_{k,l=1}^{m}\int_{-\infty}^{0}\Bigg(\sum_{j=1}^{n}\frac{\partial \overline{G}_{ik}^H}{\partial H_j}\bigg|_t \overline{G}_{jl}^H\bigg|_{t+\tau} + \sum_{u=1}^{\alpha}\sum_{j=1}^{n} k_j^u \frac{\partial \overline{G}_{ik}^H}{\partial \Psi_u}\bigg|_t \overline{G}_{jl}^\Phi\bigg|_{t+\tau} \\
&\quad + \sum_{r=\alpha+1}^{n}\frac{\partial \overline{G}_{ik}^H}{\partial \Phi_r}\bigg|_t \overline{G}_{rl}^\Phi\bigg|_{t+\tau}\Bigg) R_{kl}(\tau)\mathrm{d}\tau\Bigg\rangle_t, \\
m_u^\Psi &= O_u(\varepsilon) + \varepsilon\Bigg\langle \sum_{j=1}^{n} k_j^u \overline{F}_j^\Phi + \sum_{k,l=1}^{m}\int_{-\infty}^{0}\Bigg(\sum_{i,j=1}^{n} k_i^u \frac{\partial \overline{G}_{ik}^\Phi}{\partial A_j}\bigg|_t \overline{G}_{jl}^H\bigg|_{t+\tau} \\
&\quad + \sum_{v=1}^{\alpha}\sum_{i,j=1}^{n} k_i^u k_j^v \frac{\partial \overline{G}_{ik}^\Phi}{\partial \Psi_v}\bigg|_t \overline{G}_{jl}^\Phi\bigg|_{t+\tau} + \sum_{r=\alpha+1}^{n}\sum_{i=1}^{n} k_i^u \frac{\partial \overline{G}_{ik}^\Phi}{\partial \Phi_r}\bigg|_t \overline{G}_{rl}^\Phi\bigg|_{t+\tau}\Bigg) R_{kl}(\tau)\mathrm{d}\tau\Bigg\rangle_t,
\end{aligned}
\tag{2.82}
$$

$$
\begin{aligned}
b_{ij}^H &= \sum_{k=1}^{m}\sigma_{ik}^H\sigma_{jk}^H = \varepsilon\Bigg\langle \sum_{k,l=1}^{m}\int_{-\infty}^{\infty} \overline{G}_{ik}^H\Big|_t \overline{G}_{jl}^H\Big|_{t+\tau} R_{kl}(\tau)\mathrm{d}\tau\Bigg\rangle_t, \\
b_{uv}^\Psi &= \sum_{k=1}^{m}\sigma_{uk}^\Psi\sigma_{vk}^\Psi = \varepsilon\Bigg\langle \sum_{k,l=1}^{m}\int_{-\infty}^{\infty}\sum_{i,j=1}^{n} k_i^u k_j^v \overline{G}_{ik}^\Phi\Big|_t \overline{G}_{jl}^\Phi\Big|_{t+\tau} R_{kl}(\tau)\mathrm{d}\tau\Bigg\rangle_t, \\
b_{iu}^{H\Psi} &= \sum_{k=1}^{m}\sigma_{ik}^H\sigma_{uk}^\Psi = \varepsilon\Bigg\langle \sum_{k,l=1}^{m}\int_{-\infty}^{\infty}\sum_{j=1}^{n} k_j^u \overline{G}_{ik}^H\Big|_t \overline{G}_{jl}^\Phi\Big|_{t+\tau} R_{kl}(\tau)\mathrm{d}\tau\Bigg\rangle_t, \\
\langle[\cdot]\rangle_t &= \frac{\varepsilon}{(2\pi)^{n-\alpha}}\int_0^{2\pi}[\cdot]\mathrm{d}\Phi_{\alpha+1}\mathrm{d}\Phi_{\alpha+2}\cdots\mathrm{d}\Phi_n, \\
& i,j = 1,2,\ldots,n; \quad u,v = 1,2,\ldots,\alpha.
\end{aligned}
$$

where $\overline{F}_i^H$, $\overline{F}_i^\Phi$, $\overline{G}_{ik}^H$, $\overline{G}_{ik}^\Phi$, are the F_i^H, F_i^Φ, G_{ik}^H, G_{ik}^Φ, in Eq. (2.73), respectively, with $\Phi_1, \Phi_2, \ldots, \Phi_\alpha$ replaced by $\Psi_1, \Psi_2, \ldots, \Psi_\alpha$ according to Eq. (2.79).

The averaged FPK equation associated with Itô Eq. (2.81) is given by

$$
\begin{aligned}
\frac{\partial p}{\partial t} &= -\sum_{i=1}^{n}\frac{\partial(m_i^h p)}{\partial h_i} - \sum_{u=1}^{\alpha}\frac{\partial(m_u^\psi p)}{\partial \psi_u} + \frac{1}{2}\sum_{i,j=1}^{n}\frac{\partial^2(b_{ij}^h p)}{\partial h_i\partial h_j} \\
&\quad + \frac{1}{2}\sum_{u,v=1}^{\alpha}\frac{\partial^2(b_{uv}^\psi p)}{\partial\psi_u\partial\psi_v} + \sum_{u=1}^{\alpha}\sum_{i=1}^{n}\frac{\partial^2(b_{iu}^{h\psi} p)}{\partial h_i\partial\psi_u}.
\end{aligned}
\tag{2.83}
$$

The first and second derivative moments for this FPK equation are

$$
\begin{aligned}
&m_i^h = m_i^H\big|_{H=h,\ \Psi=\psi}, \quad m_u^\psi = m_u^\Psi\big|_{H=h,\ \Psi=\psi},\\
&b_{ij}^h = b_{ij}^H\Big|_{H=h,\ \Psi=\psi}, \quad b_{uv}^\psi = b_{uv}^\Psi\big|_{H=h,\ \Psi=\psi}, \quad b_{iu}^{h\psi} = b_{iu}^{H\Psi}\big|_{H=h,\ \Psi=\psi},\\
&i,j = 1,2,\ldots,n; \quad u,v = 1,2,\ldots,\alpha,
\end{aligned}
\tag{2.84}
$$

where $p = p(\boldsymbol{h}, \boldsymbol{\psi}, t|\boldsymbol{h}_0, \boldsymbol{\psi}_0)$ represents the transition PDF of process $[\boldsymbol{H}^{\mathrm{T}}, \boldsymbol{\Psi}^{\mathrm{T}}]^{\mathrm{T}}$. The initial conditions and boundary conditions are

$$
\begin{aligned}
&p(\boldsymbol{h}, \boldsymbol{\psi}, 0|\boldsymbol{h}_0, \boldsymbol{\psi}_0) = \delta(\boldsymbol{h} - \boldsymbol{h}_0)\delta(\boldsymbol{\psi} - \boldsymbol{\psi}_0),\\
&p = \text{finite}, \quad \text{when } |\boldsymbol{h}| = 0,\\
&p = 0, \quad \partial p/\partial h_i = 0, \quad \text{when } |\boldsymbol{h}| \to \infty.
\end{aligned}
\tag{2.85}
$$

The periodic condition is

$$
p(\mathbf{h}, \boldsymbol{\psi}+2\pi, t|\mathbf{h}_0, \boldsymbol{\psi}_0) = p(\mathbf{h}, \boldsymbol{\psi}, t|\mathbf{h}_0, \boldsymbol{\psi}_0),
\tag{2.86}
$$

and the normalization condition is $\int_0^{2\pi}\int_0^{\infty} p(\boldsymbol{h}, \boldsymbol{\psi})\mathrm{d}\boldsymbol{h}\mathrm{d}\boldsymbol{\psi} = 1$.

Example 2.3 Consider the following viscoelastic system excited by Gaussian white noise

$$
\ddot{X} + \gamma'\dot{X} + \omega'^2 X + kX^3 + \varepsilon Z = W_g(t).
\tag{2.87}
$$

where $W_g(t)$ is a zero-mean Gaussian white noise with intensity $2D$, γ', D are small quantities of order ε. For simplicity, the simplified viscoelastic model in Eq. (2.65) is used, i.e.,

$$
Z = \beta \int_0^t \exp\left(-\frac{t-\tau}{\lambda}\right)\dot{X}(\tau)\mathrm{d}\tau.
\tag{2.88}
$$

Using Eqs. (2.68)–(2.70) to decouple the viscoelastic force into an elastic restoring force and a damping force, system (2.87) is equivalent to

$$
\ddot{X} + \gamma\dot{X} + \omega^2 X + kX^3 = W_g(t),
\tag{2.89}
$$

where

$$
\omega^2 = \omega'^2 + \frac{\varepsilon\beta\lambda^2\overline{\omega}^2}{1 + \overline{\omega}^2\lambda^2}, \quad \gamma = \gamma' + \frac{\varepsilon\beta\lambda}{1 + \overline{\omega}^2\lambda^2}.
\tag{2.90}
$$

and where the mean frequency $\overline{\omega} = \overline{\omega}(H)$ can be obtained by solving the following transcendental equation

$$\frac{\pi}{\overline{\omega}} = 2\int_{0}^{U^{-1}(H)} \frac{\mathrm{d}x}{\sqrt{2H - 2U(x)}}, \quad U(x) = \frac{1}{2}\omega^2 x^2 + \frac{1}{4}kx^4. \tag{2.91}$$

Letting $X_1 = X, \quad X_2 = \dot{X}$, system Eq. (2.89) can be transformed into the following Itô stochastic differential equation

$$\begin{aligned} \mathrm{d}X_1 &= X_2\mathrm{d}t, \\ \mathrm{d}X_2 &= [-\gamma X_2 - \omega^2 X_1 - kX_1^3]\mathrm{d}t + \sqrt{2D}\mathrm{d}B(t), \end{aligned} \tag{2.92}$$

where $B(t)$ is a unit Wiener process. The Hamiltonian function or total energy of system (2.92) is

$$H(X_1, X_2) = \frac{1}{2}X_2^2 + \frac{1}{2}\omega^2 X_1^2 + \frac{1}{4}kX_1^4. \tag{2.93}$$

By applying the stochastic averaging method of quasi-non-integrable Hamiltonian systems, the averaged Itô stochastic differential equation for the energy is derived as

$$\mathrm{d}H = m(H)\mathrm{d}t + \sigma(H)\mathrm{d}B(t), \tag{2.94}$$

where the drift and diffusion coefficients are

$$\begin{aligned} m(H) &= D - \frac{2\gamma}{T(H)}\int_{X_1'}^{X_1''} \sqrt{2H - \omega^2 X_1^2 - kX_1^4/2}\mathrm{d}X_1, \\ \sigma^2(H) &= \frac{4D}{T(H)}\int_{X_1'}^{X_1''} \sqrt{2H - \omega^2 X_1^2 - kX_1^4/2}\mathrm{d}X_1, \\ T(H) &= 2\int_{X_1'}^{X_1''} \frac{\mathrm{d}X_1}{\sqrt{2H - \omega^2 X_1^2 - kX_1^4/2}}. \end{aligned} \tag{2.95}$$

and the integral limits $X_1'' = X_1''(H)$, $X_1' = X_1'(H)$ are the positive and negative roots of the equation $-w^2X_1^2 - kX_1^4/2 = 0$.

To obtain the probability and statistics of system response, the following averaged FPK equation associated with Eq. (2.94) is derived

$$\frac{\partial p}{\partial t} = -\frac{\partial}{\partial h}[m(h)p] + \frac{1}{2}\frac{\partial}{\partial h^2}[\sigma^2(h)p], \tag{2.96}$$

where $p = p(h, t)$ is the PDF of energy. The first and second derivative moments are

$$m(h) = m(H)|_{H=h}, \quad \sigma^2(h) = \sigma^2(H)\big|_{H=h}. \tag{2.97}$$

The initial condition is

$$p(h, 0) = p(h_0). \tag{2.98}$$

The boundary conditions are

$$\begin{aligned} &p(0, t) = \text{finite}, \quad \partial p(0, t)/\partial h = \text{finite}, \\ &p(\infty, t) = 0, \quad \partial p(\infty, t)/\partial h = \text{finite}. \end{aligned} \tag{2.99}$$

Solving Eq. (2.96) with $\partial p/\partial t = 0$ yields the stationary PDF of energy

$$p(h) = \frac{C}{\overline{\sigma}^2(h)} \exp\left[2\int_0^h \frac{m(u)}{\sigma^2(u)} du\right], \tag{2.100}$$

where C is a normalization constant.

Once $p(h)$ is obtained, the joint PDF $p(x_1, x_2)$ of displacement X_1 and velocity X_2, can be obtained according to Eq. (5.18) of vol. 1 as follows

$$p(x_1, x_2) = \frac{\overline{\omega}(h)}{2\pi} p(h)\Big|_{h=x_2^2/2+\omega^2 x_1^2/2+kx_1^4/4}. \tag{2.101}$$

For system (2.87) with parameters $\gamma' = 0.2$, $\omega' = 1$, $k = 2$, $D = 0.2$, $\beta = 0.2$, $\lambda = 1$, Figs. 2.10 and 2.11 show the results obtained by using the stochastic averaging method and Monte Carlo simulation. It is seen that the two results are in good agreement.

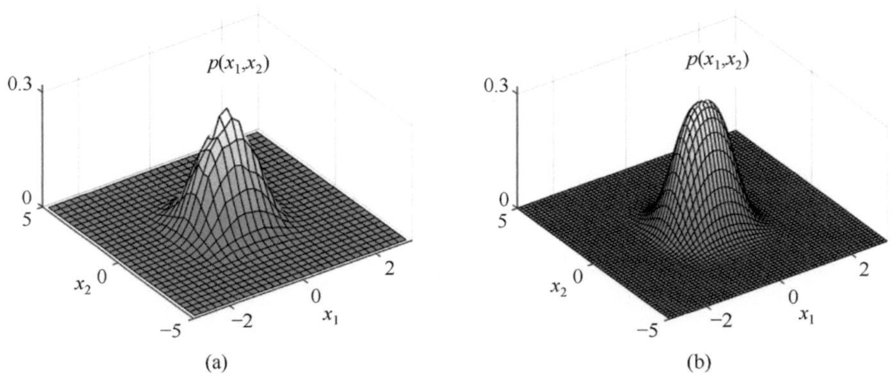

Fig. 2.10 Stationary joint PDF of displacement and velocity of system (2.87). **a** Result from Monte Carlo simulation; **b** result from the stochastic averaging method

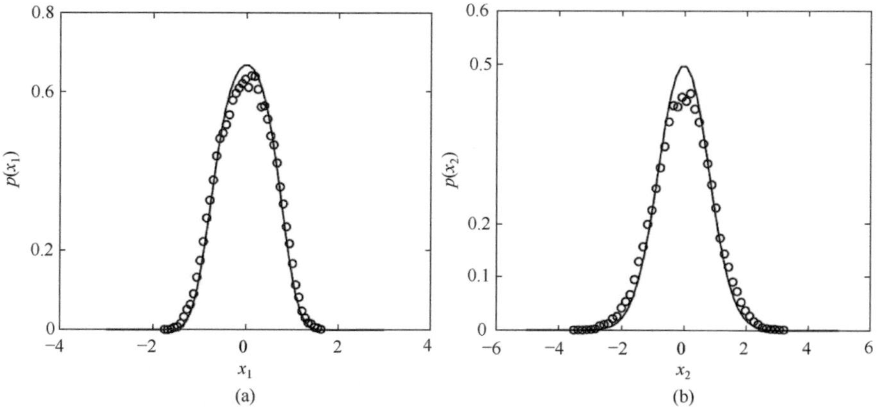

Fig. 2.11 Stationary marginal PDF of system (2.87). **a** $p(x_1)$; **b** $p(x_2)$, Lines —, results from the stochastic averaging method; symbols ○, results from Monte Carlo simulation

Example 2.4 Consider a two-DOF viscoelastic system excited by Gaussian white noises

$$\begin{aligned}&\ddot{X}_1+[\gamma_1+\eta_1(X_1^2+X_2^2)]\dot{X}_1+\omega_1^2X_1+\mu_1Z_1(\dot{X}_1-\dot{X}_2)=W_{g1}(t),\\&\ddot{X}_2+[\gamma_2+\eta_2(X_1^2+X_2^2)]\dot{X}_2+\omega_2^2X_2+\mu_2Z_2(\dot{X}_2-\dot{X}_1)=W_{g2}(t),\end{aligned}\tag{2.102}$$

where $W_{g1}(t)$ and $W_{g2}(t)$ are independent Gaussian white noises with intensities $2D_1$ and $2D_2$, respectively, and parameters $\gamma_i, \eta_i, \mu_i, D_i,\ (i=1,2)$ are all small quantities of order ε. The viscoelastic forces $Z_1(\dot{X}_1-\dot{X}_2)$, $Z_2(\dot{X}_2-\dot{X}_1)$ in Eq. (2.102) are

$$\begin{aligned}Z_1(\dot{X}_1-\dot{X}_2)&=\beta_1\int_0^t\exp\left(-\frac{t-\tau}{\lambda_1}\right)[\dot{X}_1(\tau)-\dot{X}_2(\tau)]\mathrm{d}\tau,\\Z_2(\dot{X}_2-\dot{X}_1)&=\beta_2\int_0^t\exp\left(-\frac{t-\tau}{\lambda_2}\right)[\dot{X}_2(\tau)-\dot{X}_1(\tau)]\mathrm{d}\tau.\end{aligned}\tag{2.103}$$

Using Eq. (2.68), the equivalent restoring forces and damping forces of the viscoelastic forces in Eq. (2.103) are obtained as follows

$$\begin{aligned}&Z_1(\dot{X}_1-\dot{X}_2)=C_{11}(H_1)\dot{X}_1-C_{12}(H_2)\dot{X}_2+K_{11}(H_1)X_1-K_{12}(H_2)X_2,\\&Z_2(\dot{X}_2-\dot{X}_1)=C_{22}(H_2)\dot{X}_2-C_{21}(H_1)\dot{X}_1+K_{22}(H_2)X_2-K_{21}(H_1)X_1,\\&C_{11}(H_1)=\frac{\beta_1\lambda_1}{1+\overline{\omega}_1^2\lambda_1^2},\quad K_{11}(H_1)=\frac{\beta_1\lambda_1^2\overline{\omega}_1^2}{1+\overline{\omega}_1^2\lambda_1^2},\end{aligned}$$

$$
\begin{aligned}
C_{22}(H_2) &= \frac{\beta_2\lambda_2}{1+\overline{\omega}_2^2\lambda_2^2}, \quad K_{22}(H_2) = \frac{\beta_2\lambda_2^2\overline{\omega}_2^2}{1+\overline{\omega}_2^2\lambda_2^2},\\
C_{12}(H_2) &= \frac{\beta_1\lambda_1}{1+\overline{\omega}_2^2\lambda_1^2}, \quad K_{12}(H_2) = \frac{\beta_1\lambda_1^2\overline{\omega}_2^2}{1+\overline{\omega}_2^2\lambda_1^2},\\
C_{21}(H_1) &= \frac{\beta_2\lambda_2}{1+\overline{\omega}_1^2\lambda_2^2}, \quad K_{21}(H_1) = \frac{\beta_2\lambda_2^2\overline{\omega}_1^2}{1+\overline{\omega}_1^2\lambda_2^2},
\end{aligned} \tag{2.104}
$$

where $\overline{\omega}_1, \overline{\omega}_2$ are the mean frequencies of the first and second oscillators.

Combining the equivalent elastic restoring forces and viscous damping forces in Eq. (2.104) with the original restoring forces and damping forces in system (2.102) respectively, and letting $Q_1 = X_1,\ P_1 = \dot{X}_1,\ Q_2 = X_2,\ P_2 = \dot{X}_2$, an equivalent quasi-Hamiltonian system is obtained as follows

$$
\begin{aligned}
\dot{Q}_1 &= \frac{\partial H}{\partial P_1},\\
\dot{Q}_2 &= \frac{\partial H}{\partial P_2},\\
\dot{P}_1 &= -\frac{\partial H}{\partial Q_1} - [\gamma_1 + \mu_1 C_{11}(H_1) + \eta_1(Q_1^2 + Q_2^2)]P_1\\
&\quad - \mu_1 C_{12}(H_2)P_2 - \mu_1 K_{12}(H_2)Q_2 + W_{g1}(t),\\
\dot{P}_2 &= -\frac{\partial H}{\partial Q_2} - [\gamma_2 + \mu_2 C_{22}(H_2) + \eta_2(Q_1^2 + Q_2^2)]P_2\\
&\quad - \mu_2 C_{21}(H_1)P_1 - \mu_2 K_{21}(H_1)Q_1 + W_{g2}(t).
\end{aligned} \tag{2.105}
$$

The Hamiltonian function is

$$
\begin{aligned}
&H = H_1 + H_2, \quad H_i = \frac{1}{2}P_i^2 + \frac{1}{2}\overline{\omega}_i^2 Q_i^2,\\
&\overline{\omega}_i^2 = \omega_i^2 + \mu_i K_{ii}(H_i), \quad i = 1, 2.
\end{aligned} \tag{2.106}
$$

The terms $\mu_2 K_{21}(H_1)Q_1$ and $\mu_1 K_{12}(H_2)Q_2$ in system (2.105) are small compared with $\overline{\omega}_1^2 Q_1$ and $\overline{\omega}_2^2 Q_2$, respectively. Based on the analysis of Sect. 5.4.3 in vol. 1, the stochastic averaging method of quasi-integrable Hamiltonian systems can be applied to system (2.105). To proceed further, two cases should be treated separately.

(i) **Non-resonant case**

Suppose that the mean frequencies $\overline{\omega}_1, \overline{\omega}_2$ of the two oscillators in system (2.105) do not satisfy weak internal the resonant relation. $[H_1(t), H_2(t)]^{\mathrm{T}}$ in Eq. (2.105) converges to 2-dimensional vector Markov diffusion process. The averaged Itô stochastic differential equations are obtained by using the stochastic averaging method of quasi-integrable and non-resonant Hamiltonian systems under Gaussian white noises described in Sect. 5.2.1 in vol. 1 as follows

$$
\begin{aligned}
\mathrm{d}H_1 &= m_1(H_1, H_2)\mathrm{d}t + \sigma_{11}(H_1, H_2)\mathrm{d}B_1(t) + \sigma_{12}(H_1, H_2)\mathrm{d}B_2(t), \\
\mathrm{d}H_2 &= m_2(H_1, H_2)\mathrm{d}t + \sigma_{21}(H_1, H_2)\mathrm{d}B_1(t) + \sigma_{22}(H_1, H_2)\mathrm{d}B_2(t).
\end{aligned} \tag{2.107}
$$

The drift and diffusion coefficients are derived by using Eq. (5.95) in vol. 1 as follows

$$
\begin{aligned}
m_1(H_1, H_2) &= -\left(\gamma_1 + \frac{\mu_1\beta_1\lambda_1}{1+\overline{\omega}_1^2\lambda_1^2}\right)H_1 - \frac{\eta_1}{2\overline{\omega}_1^2}H_1^2 - \frac{\eta_1}{\overline{\omega}_2^2}H_1H_2 + D_1, \\
m_2(H_1, H_2) &= -\left(\gamma_2 + \frac{\mu_2\beta_2\lambda_2}{1+\overline{\omega}_2^2\lambda_2^2}\right)H_2 - \frac{\eta_2}{2\overline{\omega}_2^2}H_2^2 - \frac{\eta_2}{\overline{\omega}_1^2}H_1H_2 + D_2, \\
b_{11}(H_1, H_2) &= \sum_{k=1}^{2} \sigma_{1k}\sigma_{1k} = 2D_1H_1, \\
b_{22}(H_1, H_2) &= \sum_{k=1}^{2} \sigma_{2k}\sigma_{2k} = 2D_2H_2, \\
b_{12}(H_1, H_2) &= b_{21}(H_1, H_2) = \sum_{k=1}^{2} \sigma_{1k}\sigma_{2k} = 0.
\end{aligned} \tag{2.108}
$$

The reduced averaged FPK equation associated with Itô equation (2.107) is

$$
-\frac{\partial[m_1(h_1,h_2)p]}{\partial h_1} - \frac{\partial[m_2(h_1,h_2)p]}{\partial h_2} + \frac{1}{2}\frac{\partial^2[b_{11}(h_1,h_2)p]}{\partial h_1^2} + \frac{1}{2}\frac{\partial^2[b_{22}(h_1,h_2)p]}{\partial h_2^2} = 0, \tag{2.109}
$$

where $p = p(h_1, h_2)$ represents the stationary PDF, and the first and second derivative moments are given by

$$
\begin{aligned}
m_1(h_1, h_2) &= m_1(H_1, H_2)|_{H_1=h_1,\, H_2=h_2}, \quad m_2(h_1, h_2) = m_2(H_1, H_2)|_{H_1=h_1,\, H_2=h_2}, \\
b_{11}(h_1, h_2) &= b_{11}(H_1, H_2)|_{H_1=h_1,\, H_2=h_2}, \quad b_{22}(h_1, h_2) = b_{22}(H_1, H_2)|_{H_1=h_1,\, H_2=h_2}.
\end{aligned} \tag{2.110}
$$

The boundary conditions for the FPK Eq. (2.109) are

$$
\begin{aligned}
&p = \text{finite}, \quad \text{when } h_1 = 0 \text{ or } h_2 = 0, \\
&p \to 0,\ \partial p/\partial h_1 \to 0, \quad \text{when } h_1 \to \infty, \\
&p \to 0,\ \partial p/\partial h_2 \to 0, \quad \text{when } h_2 \to \infty.
\end{aligned} \tag{2.111}
$$

After obtaining $p(h_1, h_2)$, the joint PDF $p(q_1, q_2, p_1, p_2)$ of generalized displacements and generalized momenta can be calculated using Eq. (5.106) in vol. 1 as follows

$$
p(q_1, q_2, p_1, p_2) = \frac{\overline{\omega}_1\overline{\omega}_2}{4\pi^2} p(h_1, h_2)\bigg|_{\substack{h_1=p_1^2/2+\overline{\omega}_1^2q_1^2/2,\\ h_2=p_2^2/2+\overline{\omega}_2^2q_2^2/2}} \tag{2.112}
$$

The marginal PDFs can be obtained from Eq. (2.112) by integration. The numerical calculations were completed for system (2.102) with parameters $\omega_1 = 1$, $\omega_2 = 1.414$, $\gamma_1 = \gamma_2 = 0.2$, $\eta_1 = \eta_2 = 0.1$, $\mu_1 = \mu_2 = 0.1$, $D_1 = 0.2$, $D_2 = 0.02$, $\beta_1 = \beta_2 = 2$ and $\lambda_1 = \lambda_2 = 1$. Figures 2.12 and 2.13 show the results from the stochastic averaging method and Monte Carlo simulations of system (2.102). It is seen that the two results agree well.

(ii) **Resonant case**

When the mean frequencies $\overline{\omega}_1, \overline{\omega}_2$ of the two oscillators in system (2.105) satisfy the internal resonant relation $\overline{\omega}_1 = \overline{\omega}_2$, the stochastic averaging method of quasi-integrable and internal resonant Hamiltonian systems under Gaussian white noises

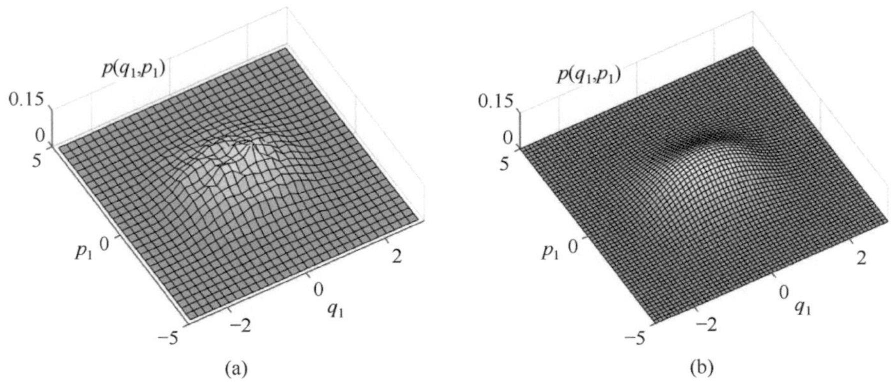

Fig. 2.12 Stationary joint PDF of generalized displacement and generalized momentum of system (2.102) in non-resonant case. **a** result from Monte Carlo simulation; **b** result from the stochastic averaging method

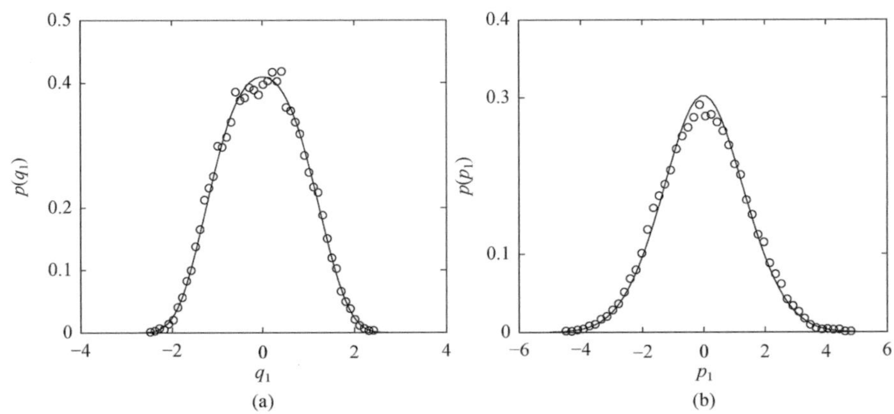

Fig. 2.13 Stationary marginal PDF of system (2.102) in non-resonant case. **a** $p(q_1)$; **b** $p(p_1)$. Lines —, results from the stochastic averaging method; symbols ○, results from Monte Carlo simulation

described in Sect. 5.2.2 in vol. 1 is applicable. In this case, $H_1(t), H_2(t)$ and the phase difference $\Psi(t) = \Phi_1(t) - \Phi_2(t)$ weakly converge to a 3-dimensional vector Markov diffusion process. The averaged Itô stochastic differential equations for $[H_1(t), H_2(t), \Psi(t)]^{\mathrm{T}}$ can be obtained from Eq. (5.119) in vol. 1 as follows

$$\begin{aligned}
\mathrm{d}H_1 &= a_1(H_1, H_2, \Psi)\mathrm{d}t + \sigma_{11}(H_1, H_2, \Psi)\mathrm{d}B_1(t) + \sigma_{12}(H_1, H_2, \Psi)\mathrm{d}B_2(t), \\
\mathrm{d}H_2 &= a_2(H_1, H_2, \Psi)\mathrm{d}t + \sigma_{21}(H_1, H_2, \Psi)\mathrm{d}B_1(t) + \sigma_{22}(H_1, H_2, \Psi)\mathrm{d}B_2(t), \\
\mathrm{d}\Psi &= a_\Psi(H_1, H_2, \Psi)\mathrm{d}t + \sigma_{\Psi 1}(H_1, H_2, \Psi)\mathrm{d}B_1(t) + \sigma_{\Psi 2}(H_1, H_2, \Psi)\mathrm{d}B_2(t).
\end{aligned} \tag{2.113}$$

The following drift and diffusion coefficients can be derived from Eq. (2.105) by using Eq. (5.121) in vol. 1

$$\begin{aligned}
& m_1(H_1, H_2, \Psi) \\
&= -(\gamma_1 + \mu_1 C_{11})H_1 - \frac{\eta_1}{2\overline{\omega}_1^2}H_1^2 - \frac{\eta_1}{\overline{\omega}_2^2}\left(1 - \frac{1}{2}\cos 2\Psi\right)H_1 H_2 \\
&\quad + \mu_1\left(C_{12}\cos\Psi - K_{12}\frac{\sin\Psi}{\overline{\omega}_2}\right)\sqrt{H_1 H_2} + D_1, \\
& m_2(H_1, H_2, \Psi) \\
&= -(\gamma_2 + \mu_2 C_{22})H_2 - \frac{\eta_2}{2\overline{\omega}_2^2}H_2^2 - \frac{\eta_2}{\overline{\omega}_1^2}\left(1 - \frac{1}{2}\cos 2\Psi\right)H_1 H_2 \\
&\quad + \mu_2\left(C_{21}\cos\Psi + K_{21}\frac{\sin\Psi}{\overline{\omega}_1}\right)\sqrt{H_1 H_2} + D_2, \\
& m_\Psi(H_1, H_2, \Psi) = -\frac{\sin 2\Psi}{4}\left(\frac{\eta_1 H_2}{\overline{\omega}_2^2} + \frac{\eta_2 H_1}{\overline{\omega}_1^2}\right) \\
&\qquad - \frac{\mu_1}{2}\sqrt{\frac{H_2}{H_1}}\left(C_{12}\sin\Psi + K_{12}\frac{\cos\Psi}{\overline{\omega}_2}\right) \\
&\qquad - \frac{\mu_2}{2}\sqrt{\frac{H_1}{H_2}}\left(C_{21}\sin\Psi - K_{21}\frac{\cos\Psi}{\overline{\omega}_1}\right), \\
& b_{11}(H_1, H_2, \Psi) = 2D_1 H_1, \quad b_{22}(H_1, H_2, \Psi) = 2D_2 H_2, \\
& b_{\Psi\Psi}(H_1, H_2, \Psi) = \frac{1}{2}\left(\frac{D_1}{H_1} + \frac{D_2}{H_2}\right), \\
& b_{12} = b_{21} = b_{1\psi} = b_{\psi 1} = b_{2\psi} = b_{\psi 2} = 0.
\end{aligned} \tag{2.114}$$

The reduced averaged FPK equation associated with Itô Eq. (2.113) is

$$-\frac{\partial(m_1 p)}{\partial h_1} - \frac{\partial(m_2 p)}{\partial h_2} - \frac{\partial(m_\Psi p)}{\partial \psi} + \frac{1}{2}\frac{\partial^2(b_{11}p)}{\partial h_1^2} + \frac{1}{2}\frac{\partial^2(b_{22}p)}{\partial h_2^2} + \frac{1}{2}\frac{\partial^2(b_{\Psi\Psi}p)}{\partial \psi^2} = 0, \tag{2.115}$$

where $p = p(h_1, h_2, \psi)$ is stationary PDF. The first and second derivative moments are

$$\begin{aligned}
m_1(h_1, h_2, \psi) &= m_1(H_1, H_2, \Psi)|_{\substack{H_1=h_1,\, H_2=h_2,\\ \Psi=\psi}}\\
m_2(h_1, h_2, \psi) &= m_2(H_1, H_2, \Psi)|_{\substack{H_1=h_1,\, H_2=h_2,\\ \Psi=\psi}}\\
m_\Psi(h_1, h_2, \psi) &= m_\Psi(H_1, H_2, \Psi)|_{\substack{H_1=h_1,\, H_2=h_2,\\ \Psi=\psi}}\\
b_{11}(h_1, h_2, \psi) &= b_{11}(H_1, H_2, \Psi)|_{\substack{H_1=h_1,\, H_2=h_2,\\ \Psi=\psi}}\\
b_{22}(h_1, h_2, \psi) &= b_{22}(H_1, H_2, \Psi)|_{\substack{H_1=h_1,\, H_2=h_2,\\ \Psi=\psi}}\\
b_{\Psi\Psi}(h_1, h_2, \psi) &= b_{\Psi\Psi}(H_1, H_2, \Psi)|_{\substack{H_1=h_1,\, H_2=h_2.\\ \Psi=\psi}}
\end{aligned} \tag{2.116}$$

The boundary conditions for FPK Eq. (2.115) are

$$\begin{aligned}
&p = \text{finite}, \quad \text{when } h_1 = 0 \text{ or } h_2 = 0,\\
&p \to 0,\ \partial p/\partial h_1 \to 0, \quad \text{when } h_1 \to \infty,\\
&p \to 0,\ \partial p/\partial h_2 \to 0, \quad \text{when } h_2 \to \infty,\\
&p(h_1, h_2, \psi + 2\pi) = p(h_1, h_2, \psi).
\end{aligned} \tag{2.117}$$

After obtaining $p(h_1, h_2, \psi)$, the joint PDF $p(q_1, q_2, p_1, p_2)$ of generalized displacements and generalized momenta can be calculated using Eq. (5.127) of vol. 1 as follows

$$p(q_1, q_2, p_1, p_2) = Cp(h_1, h_2, \psi)|_{\substack{h_1=p_1^2/2+\overline{\omega}_1^2 q_1^2/2,\\ h_2=p_2^2/2+\overline{\omega}_2^2 q_2^2/2,\\ \psi=\tan^{-1}(p_1/\overline{\omega}_1 q_1)-\tan^{-1}(p_2/\overline{\omega}_2 q_2)}}. \tag{2.118}$$

The marginal PDFs can be obtained from Eq. (2.118) by integration.

Numerical calculations were carried out for the system (2.102) with parameters $\omega_1 = \omega_2 = 1$, $\gamma_1 = \gamma_2 = 0.2$, $\eta_1 = \eta_2 = 0.1$, $\mu_1 = \mu_2 = 0.1$, $D_1 = 0.2$, $D_2 = 0.02$, $\beta_1 = \beta_2 = 2$, and $\lambda_1 = \lambda_2 = 1$. Note that the excitation intensity of the first oscillator is much large than that of the second one ($D_1 \gg D_2$). Figures 2.14 and 2.15 show that the results from the stochastic averaging method agree well with the results from Monte Carlo simulation of system (2.102). By comparing the stationary PDFs $p(h_2)$ between the two cases, shown in Fig. 2.16a and b, respectively, it is revealed that $E[H_2]$ would be lower in non-resonant case than $E[H_2]$ in resonant case since more energy is injected from the first oscillator to the second oscillator in resonant case.

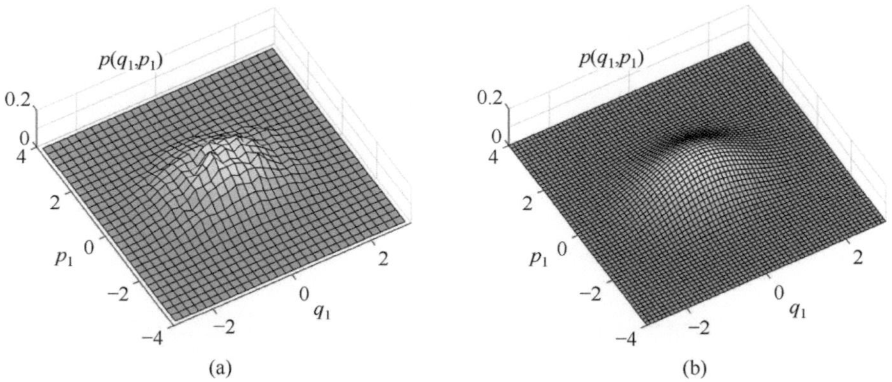

Fig. 2.14 Stationary joint PDF of generalized displacement and generalized momentum of system (2.102) in internal resonant case. **a** result from Monte Carlo simulation; **b** result from the stochastic averaging method

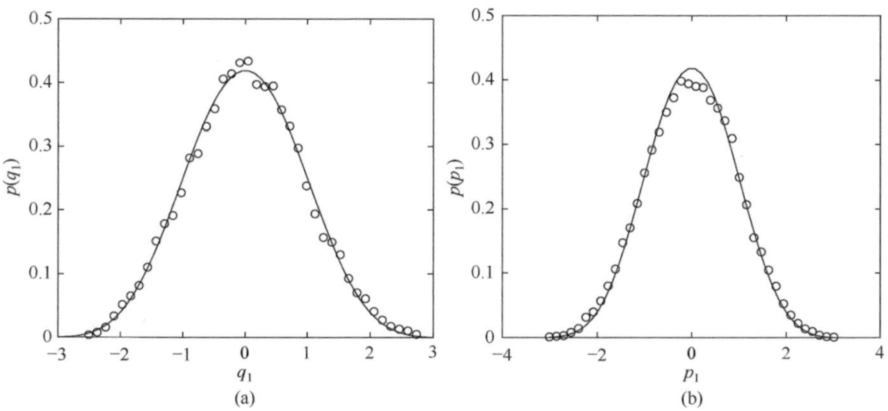

Fig. 2.15 Stationary marginal PDF of system (2.102) in internal resonant case. **a** $p(q_1)$; **b** $p(p_1)$. Lines —, results from the stochastic averaging method; symbols ○, results from Monte Carlo simulation

2.3 Quasi-integrable Hamiltonian Systems with Fractional Derivative Damping Forces

Consider a quasi-integrable Hamiltonian system with fractional derivative damping forces. The governing equations are given by

$$
\begin{aligned}
\dot{Q}_i &= P_i, \\
\dot{P}_i &= -g_i'(Q_i) - \varepsilon \sum_{j=1}^{n} c_{ij}'(\boldsymbol{Q}, \boldsymbol{P}) P_j - \varepsilon \mu_{ii} D^{\alpha_i} Q_i
\end{aligned}
$$

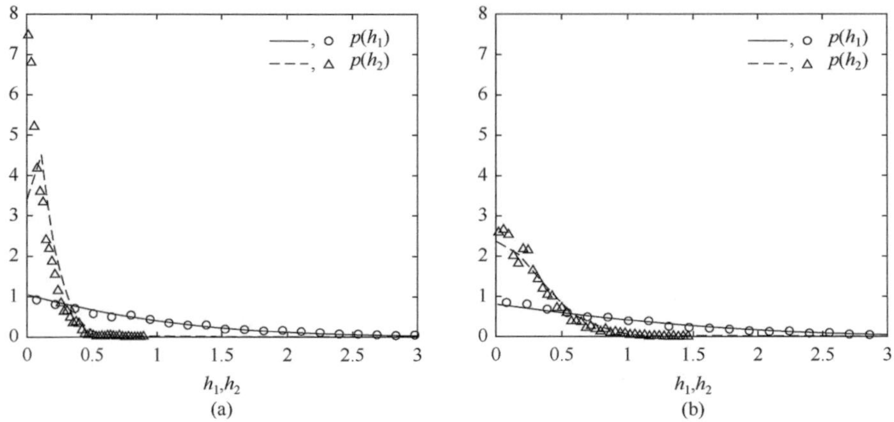

Fig. 2.16 Stationary PDF of Hamiltonians H_1, H_2 of system (2.102). **a** Non-resonant case; **b** internal resonant case. Lines —, lines —, results from the stochastic averaging method; symbols ○△, results from Monte Carlo numerical simulation

$$-\varepsilon \sum_{j=1,\, j\neq i}^{n} \mu_{ij} D^{\alpha_i}(Q_i - Q_j) + \varepsilon^{1/2} \sum_{k=1}^{m} f_{ik}(\boldsymbol{Q}, \boldsymbol{P})\xi_k(t),$$
$$i = 1, 2, \ldots, n. \tag{2.119}$$

In system (2.119), the ith oscillator is subjected to a fractional derivative damping force $\varepsilon\mu_{ii}D^{\alpha_i}Q_i$, and $n-1$ coupled fractional derivative damping forces $\varepsilon \sum\limits_{j=1,j\neq i}^{n} \mu_{ij}D^{\alpha_i}(Q_i - Q_j)$. Here, $D^{\alpha_i}Q$ is fractional derivative defined by Riemann–Liouville as

$$D^{\alpha_i}Q = D^{\alpha_i}Q(t) = \frac{1}{\Gamma(l-\alpha_i)}\left(\frac{\mathrm{d}}{\mathrm{d}t}\right)^l \int_0^t \frac{Q(\tau)}{(t-\tau)^{\alpha_i - l + 1}}\mathrm{d}\tau, \quad (l-1) \le \alpha_i < l \tag{2.120}$$

where l is integer, and $\Gamma(\cdot)$ is the Gamma function. In order to ensure that fractional derivative damping force lie between elastic force and Newtonian viscous force, α_i is restricted to be $0 < \alpha_i < 1$. According to Eq. (2.120), l must be 1 and $D^{\alpha_i}Q$ is a linear operator. Thus,

$$D^{\alpha_i}(Q_1 - Q_2) = D^{\alpha_i}(Q_1) - D^{\alpha_i}(Q_2) \tag{2.121}$$

Assume that $g_i'(Q_i)$ satisfy the four conditions stated below Eq. (1.5), $\xi_k(t)$ are wideband stochastic processes and Eq. (2.119) has the following randomized periodic solution

$$Q_i(t) = A_i(t)\cos\Phi_i(t) + B_i, \quad P_i(t) = -A_i(t)\nu_i(A_i, \Phi_i)\sin\Phi_i(t),$$
$$\Phi_i(t) = \Gamma_i(t) + \Theta_i(t), \quad i = 1, 2, \ldots, n. \tag{2.122}$$

where A_i, Φ_i, Γ_i, Θ_i are random processes, A_i, Φ_i are the displacement amplitude and phase angle of the ith DOF, ν_i is instantaneous frequency of the ith DOF given by

$$\nu_i(A_i, \Phi_i) = \frac{\sqrt{2[U_i(A_i + B_i) - U_i(A_i\cos\Phi_i + B_i)]}}{|A_i \sin\Phi_i|}. \tag{2.123}$$

The Hamiltonian functions and the potentials for sub-systems associated with system (2.119) are

$$H_i(Q_i, P_i) = P_i^2/2 + U_i(Q_i), \quad U_i(Q_i) = \int_0^{Q_i} g_i(u)\mathrm{d}u. \tag{2.124}$$

where g_i is the sum of g_i' plus the elastic restoring forces decoupled from fractional derivative dampings of ith oscillator. Equation (2.123) shows that instantaneous frequency ν_i is an even function of Φ_i and can be expanded into the following Fourier series

$$\nu_i(A_i, \Phi_i) = \omega_i(A_i) + \sum_{r=1}^{\infty} \omega_{ir}(A_i)\cos r\Phi_i, \tag{2.125}$$

where $\omega_i(A_i)$ is mean value of ν_i over one period, i.e., the mean frequency. In the following, the following approximation will be used

$$\Phi_i(t) = \omega_i(A_i)t + \Theta_i(t). \tag{2.126}$$

Assume that $D^{\alpha_i}Q_i$ can be decoupled as

$$D^{\alpha_i}Q_i = C_{ii}(A_i)P_i + K_{ii}(A_i)Q_i. \tag{2.127}$$

Multiplying Eq. (2.123) by $\sin\Phi_i(t)$ and integrating with respect to $\Phi_i(t)$ over $(0, 2\pi]$ yields

$$\begin{aligned} C_i(A_i) &= \frac{-1}{\pi A_i \omega_i(A_i)} \int_0^{2\pi} D^{\alpha_i}(A_i\cos\Phi_i)\sin\Phi_i \mathrm{d}\Phi_i \\ &= \frac{-2}{\Gamma(1-\alpha_i)A_i\omega_i(A_i)} \lim_{T\to\infty} \frac{1}{T} \int_0^T [\frac{\mathrm{d}}{\mathrm{d}t} \int_0^t \frac{Q_i(t-\tau)}{\tau^{\alpha_i}} \mathrm{d}\tau] \sin\Phi_i \mathrm{d}t \end{aligned}$$

$$
\begin{aligned}
&= \frac{-2}{\Gamma(1-\alpha_i)A_i\omega_i(A_i)} \lim_{T\to\infty}\frac{1}{T}\left\{ \sin\Phi_i \int_0^t \frac{A_i\cos\Phi_i(t-\tau)}{\tau^\alpha}\mathrm{d}\tau)\Bigg|_0^T \right. \qquad (2.128)\\
&\left. + \int_0^T v_i\cos\Phi_i\left[\left(\left(\int_0^t \frac{A_i\cos\Phi_i(t-\tau)}{\tau^\alpha}\mathrm{d}\tau\right)\right)\right]\mathrm{d}t \right\}\\
&= \frac{2}{\Gamma(1-\alpha_i)\omega_i(A_i)} \lim_{T\to\infty}\frac{1}{T}\int_0^T\left[v_i\cos\Phi_i\int_0^t\frac{\cos\Phi_i(t-\tau)}{\tau^\alpha}\mathrm{d}\tau\right]\mathrm{d}t\\
&\approx \frac{2}{\Gamma(1-\alpha_i)\omega_i(A_i)} \lim_{T\to\infty}\frac{1}{T}\int_0^T\left[v_i\cos\Phi_i\left(\cos\Phi_i\int_0^t\frac{\cos\omega_i\tau}{\tau^\alpha}\mathrm{d}\tau\right.\right.\\
&\left.\left. +\sin\Phi_i\int_0^t\frac{\sin\omega_i\tau}{\tau^\alpha}\mathrm{d}\tau\right)\right]\mathrm{d}t\,.
\end{aligned}
$$

Similarly, multiplying Eq. (2.123) by $\cos\Phi_i(t)$ and integrating with respect to $\Phi_i(t)$ over $(0, 2\pi]$ yields

$$
\begin{aligned}
K_i(A_i) &= \frac{1}{\pi A_i}\int_0^{2\pi} D^{\alpha_i}(A_i\cos\Phi_i)\cos\Phi_i\mathrm{d}\Phi_i\\
&= \frac{2}{A_i\Gamma(1-\alpha)} \lim_{T\to\infty}\frac{1}{T}\int_0^T[\frac{\mathrm{d}}{\mathrm{d}t}\int_0^t\frac{Q_i(t-\tau)}{\tau^\alpha}\mathrm{d}\tau]\cos\Phi_i\mathrm{d}t\\
&= \frac{2}{A_i\Gamma(1-\alpha_i)} \lim_{T\to\infty}\frac{1}{T}\left\{\cos\Phi_i\int_0^t\frac{A_i\cos\Phi_i(t-\tau)}{\tau^{\alpha_i}}\mathrm{d}\tau)\Bigg|_0^T\right. \qquad (2.129)\\
&\left. + \int_0^T\left[(\int_0^t\frac{A_i\cos\Phi_i(t-\tau)}{\tau^{\alpha_i}}\mathrm{d}\tau)v_i\sin\Phi_i\right]\mathrm{d}t\right\}\\
&\approx \frac{2}{\Gamma(1-\alpha_i)} \lim_{T\to\infty}\frac{1}{T}\int_0^T\left[v_i\sin\Phi_i\left(\cos\Phi_i\int_0^t\frac{\cos\omega_i\tau}{\tau^{\alpha_i}}\mathrm{d}\tau + \sin\Phi_i\int_0^t\frac{\sin\omega_i\tau}{\tau^{\alpha_i}}\mathrm{d}\tau\right)\right]\mathrm{d}t.
\end{aligned}
$$

To complete the integrations in Eqs. (2.128) and (2.129), the following two asymptotic series are introduced

$$
\int_0^t\frac{\cos(\omega\tau)}{\tau^q}\mathrm{d}\tau = \omega^{q-1}\int_0^{\omega t}\frac{\cos(u)}{u^q}\mathrm{d}u \approx \omega^{q-1}[\Gamma(1-q)\sin(\frac{q\pi}{2}) + \frac{\sin(\omega t)}{(\omega t)^q}
$$

$$+ O((\omega t)^{-q-1})],$$

$$\int_0^t \frac{\sin(\omega\tau)}{\tau^q} d\tau = \omega^{q-1} \int_0^{\omega t} \frac{\sin(u)}{u^q} du \approx \omega^{q-1}[\Gamma(1-q)\cos(\frac{q\pi}{2}) - \frac{\cos(\omega t)}{(\omega t)^q}$$
$$+ O((\omega t)^{-q-1})]. \tag{2.130}$$

Substituting the asymptotic Eq. (2.130) into Eq. (2.128) and Eq. (2.129), the following simple expressions can be obtained (Huang and Jin 2009; Chen et al. 2013a, b)

$$C_{ii}(A_i) = \omega_i^{\alpha_i - 2}(\omega_i + \frac{\omega_{i2}}{2}) \sin\left(\frac{\alpha_i \pi}{2}\right),$$
$$K_{ii}(A_i) = \omega_i^{\alpha_i - 1}\left(\omega_i - \frac{\omega_{i2}}{2}\right) \cos\left(\frac{\alpha_i \pi}{2}\right). \tag{2.131}$$

where $\omega_i(A_i), \omega_{i2}(A_i)$ are coefficients in Eq. (2.125). Similarly, the fractional derivative damping forces of the displacement difference can be decoupled as

$$D^{\alpha_i}(Q_i - Q_j) = C_{ii}(A_i)P_i - C_{ij}(A_j)P_j + K_{ii}(A_i)Q_i - K_{ij}(A_j)Q_j,$$
$$C_{ij}(A_j) = \omega_j^{\alpha_i - 2}(\omega_j + \frac{\omega_{j2}}{2}) \sin(\frac{\alpha_i \pi}{2}), \quad K_{ij}(A_j) = \omega_j^{\alpha_i - 1}(\omega_j - \frac{\omega_{j2}}{2}) \cos(\frac{\alpha_i \pi}{2}). \tag{2.132}$$

Substituting the equivalent relations in Eqs. (2.127) and (2.132) into the original quasi-Hamiltonian system (2.119), the following equivalent quasi-Hamiltonian system without fractional derivative dampings is obtained

$$\dot{Q}_i = P_i,$$
$$\dot{P}_i = -g_i(Q_i) - \varepsilon \sum_{j=1}^{n} c_{ij}(\boldsymbol{Q},\boldsymbol{P})P_j + \varepsilon \sum_{j=1,\, j\neq i}^{n} \mu_{ij} K_{ij}(A_j) Q_j$$
$$+ \varepsilon^{1/2} \sum_{k=1}^{m} f_{ik}(\boldsymbol{Q},\boldsymbol{P}) \xi_k(t),$$
$$i = 1, 2, \ldots, n. \tag{2.133}$$

In this system, the new damping coefficients and restoring force functions are given as

$$g_i(Q_i) = g_i'(Q_i) + \varepsilon \left(\sum_{r=1}^{n} \mu_{ir} \right) K_{ii}(A_i) Q_i,$$
$$c_{ij}(\boldsymbol{Q},\boldsymbol{P}) = \begin{cases} c_{ii}'(\boldsymbol{Q},\boldsymbol{P}) + \left(\sum_{r=1}^{n} \mu_{ir} \right) C_{ii}(A_i), & j = i, \\ c_{ij}'(\boldsymbol{Q},\boldsymbol{P}) - \mu_{ij} C_{ij}(A_j), & j \neq i. \end{cases} \tag{2.134}$$

The terms $\varepsilon \sum_{j=1,\, j\neq i}^{n} \mu_{ij} K_{ij}(A_j) Q_j$ in Eq. (2.133) are much smaller than $g_i(Q_i)$. Based on the analysis in Sect. 5.4.2 in vol. 1, system (2.133) can be considered as a quasi-integrable Hamiltonian system and the stochastic averaging method of quasi-integrable Hamiltonian systems described in Sect. 1.1.2 can be applied to system (2.133).

Using Eq. (2.122), system (2.133) can be transformed into

$$\begin{aligned}
\dot{A}_i &= \varepsilon F_i^A(\boldsymbol{A}, \boldsymbol{\Phi}) + \varepsilon^{1/2} \sum_{k=1}^{m} G_{ik}^A(\boldsymbol{A}, \boldsymbol{\Phi}) \xi_k(t), \\
\dot{\Phi}_i &= \nu_i(A_i, \Phi_i) + \varepsilon F_i^{\Phi}(\boldsymbol{A}, \boldsymbol{\Phi}) + \varepsilon^{1/2} \sum_{k=1}^{m} G_{ik}^{\Phi}(\boldsymbol{A}, \boldsymbol{\Phi}) \xi_k(t), \\
i &= 1, 2, \ldots, n,
\end{aligned} \tag{2.135}$$

where $\boldsymbol{A} = [A_1, A_2, \ldots, A_n]^{\mathrm{T}}$, $\Phi = [\Phi_1, \Phi_2, \ldots, \Phi_n]^{\mathrm{T}}$ and

$$\begin{aligned}
F_i^A &= \frac{-A_i \nu_i \sin \Phi_i}{g(A_i + B_i)(1 + r_i)} [\sum_{j=1}^{n} c_{ij} A_j \nu_j \sin \Phi_j \\
&\quad + \sum_{j=1,\, j\neq i}^{n} \mu_{ij} K_{ij}(A_j \cos \Phi_j + B_j)], \\
F_i^{\Phi} &= \frac{-\nu_i(\cos \Phi_i + r_i)}{g(A_i + B_i)(1 + r_i)} [\sum_{j=1}^{n} c_{ij} A_j \nu_j \sin \Phi_j \\
&\quad + \sum_{j=1,\, j\neq i}^{n} \mu_{ij} K_{ij}(A_j \cos \Phi_j + B_j)],
\end{aligned} \tag{2.136}$$

$$\begin{aligned}
G_{ik}^A &= \frac{-A_i \nu_i \sin \Phi_i}{g(A_i + B_i)(1 + r_i)} f_{ik}, \quad G_{ik}^{\Phi} = \frac{-\nu_i(\cos \Phi_i + r_i)}{g(A_i + B_i)(1 + r_i)} f_{ik}, \\
r_i &= \frac{\mathrm{d}B_i}{\mathrm{d}A_i} = \frac{g(-A_i + B_i) + g(A_i + B_i)}{g(-A_i + B_i) - g(A_i + B_i)}.
\end{aligned}$$

In system (2.135), the amplitude $\boldsymbol{A}$ is a slowly-varying vector process while the phase angle $\boldsymbol{\Phi}$ is a fast-varying vector process. The dimension and form of averaged Itô equations and FPK equation depend on the resonance properties of the associated Hamiltonian system of Eq. (2.133).

(i) **Non-resonant case**

In the case of non-resonance, based on the Khasminskii theorems (Khasminskii 1966, 1968), $\boldsymbol{A}(t)$ in system (2.135) weakly converges to a n-dimensional Markov diffusion process as the damping forces and intensities of random excitations approaches to zero. The averaged Itô stochastic differential equations governing $\boldsymbol{A}(t)$ are given by

$$\mathrm{d}A_i = m_i(\boldsymbol{A})\mathrm{d}t + \sum_{k=1}^{m} \sigma_{ik}(\boldsymbol{A})\mathrm{d}B_k(t),$$
$$i = 1, 2, \ldots, n. \tag{2.137}$$

According to Eq. (1.56), the drift and diffusion coefficients are derived as

$$m_i(\boldsymbol{A}) = \varepsilon\left\langle F_i^A + \sum_{k,l=1}^{m} \int_{-\infty}^{0} \sum_{j=1}^{n} \left(\frac{\partial G_{ik}^A}{\partial A_j}\bigg|_t G_{jl}^A\bigg|_{t+\tau} + \frac{\partial G_{ik}^A}{\partial \Phi_j}\bigg|_t G_{jl}^{\Phi}\bigg|_{t+\tau} \right) R_{kl}(\tau)\mathrm{d}\tau \right]\right\rangle_t,$$
$$b_{ij}(\boldsymbol{A}) = \sum_{k=1}^{m} \sigma_{ik}(\boldsymbol{A})\sigma_{jk}(\boldsymbol{A}) = \varepsilon\left\langle \sum_{k,l=1}^{m} \int_{-\infty}^{\infty} \left(G_{ik}^A\big|_t G_{jl}^A\big|_{t+\tau} \right) R_{kl}(\tau)\mathrm{d}\tau \right\rangle_t,$$
$$i, j = 1, 2, \ldots, n. \tag{2.138}$$

The time averaging in Eq. (2.138) can be replaced by averaging with respect to $\Phi_1, \Phi_2, \ldots, \Phi_n$, i.e.,

$$\langle[\bullet]\rangle_t = \lim_{T\to\infty} \frac{1}{T} \int_0^T [\bullet]\mathrm{d}t = \frac{1}{(2\pi)^n} \int_0^{2\pi} [\,\bullet\,]\mathrm{d}\boldsymbol{\Phi}. \tag{2.139}$$

In calculations of Eq. (2.138), the following relations will be used

$$S_{kl}(\omega) = \frac{1}{\pi} \int_{-\infty}^{0} R_{kl}(\tau) \cos\omega\tau\mathrm{d}\tau, \quad I_{kl}(\omega) = \frac{1}{\pi} \int_{-\infty}^{0} R_{kl}(\tau) \sin\omega\tau\mathrm{d}\tau,$$
$$k, l = 1, 2, \ldots, m. \tag{2.140}$$

where $S_{kl}(\omega)$ and $I_{kl}(\omega)$ are the real and imaginary parts of the cross-power spectral density of wideband noise $\xi_k(t)$ and $\xi_l(t)$. Firstly, carry out Fourier expansion of each term in Eq. (2.136) with respect to Φ_i. Then, substitute these Fourier series into Eq. (2.20) and complete the integral with respect to τ and averaging with respect to Φ_i to yield the final expressions for $m_i(\boldsymbol{A})$ and $b_{ij}(\boldsymbol{A})$.

Based on the relationship $H_i = U_i(A_i + B_i)$ or $A_i = U_i^{-1}(H_i) - B_i$ between A_i and H_i, and using the Itô differential rule, the following averaged Itô stochastic differential equations governing $\boldsymbol{H} = [H_1, H_2, \ldots, H_n]^{\mathrm{T}}$ can be derived from Eq. (2.137)

$$\mathrm{d}H_i = \varepsilon\overline{m}_i(\boldsymbol{H})\mathrm{d}t + \varepsilon^{1/2} \sum_{k=1}^{m} \overline{\sigma}_{ik}(\boldsymbol{H})\mathrm{d}B_k(t),$$
$$i = 1, 2, \ldots, n. \tag{2.141}$$

The drift and diffusion coefficients are

$$\overline{m}_i = \left[g_i(A_i + B_i)(1 + r_i)m_i(\boldsymbol{A}) + \frac{1}{2}\frac{\mathrm{d}[g_i(A_i + B_i)(1 + r_i)]}{\mathrm{d}A_i}b_{ii}(\boldsymbol{A})\right]\Bigg|_{A_i = U_i^{-1}(H_i) - B_i},$$

$$\overline{b}_{ij} = \sum_{k=1}^{m}\overline{\sigma}_{ik}\overline{\sigma}_{jk} = [g_i(A_i + B_i)g_j(A_j + B_j)(1 + r_i)(1 + r_j)b_{ij}(\boldsymbol{A})]\Big|_{\substack{A_i = U_i^{-1}(H_i) - B_i,\\ A_j = U_j^{-1}(H_j) - B_j}},$$

$$i, j = 1, 2, \ldots, n. \tag{2.142}$$

The averaged FPK equation associated with Itô Eq. (2.141) can be established as

$$\frac{\partial p}{\partial t} = \sum_{i=1}^{n}\frac{\partial(\overline{m}_i p)}{\partial h_i} + \frac{1}{2}\sum_{i,j=1}^{n}\frac{\partial^2(\overline{b}_{ij}p)}{\partial h_i \partial h_j}, \tag{2.143}$$

where $\overline{m}_i(\boldsymbol{h}) = \overline{m}_i(\boldsymbol{H})|_{\boldsymbol{H}=\boldsymbol{h}}, \overline{b}_{ij}(\boldsymbol{h}) = \overline{b}_{ij}(\boldsymbol{H})\Big|_{\boldsymbol{H}=\boldsymbol{h}}$, and $p = p(\boldsymbol{h}, t|\boldsymbol{h}_0)$ is the transition PDF of $\boldsymbol{H}(t)$. The initial conditions and boundary conditions for Eq. (2.143) are

$$\begin{aligned} &p(\boldsymbol{h}, 0|\boldsymbol{h}_0) = \delta(\boldsymbol{h} - \boldsymbol{h}_0),\\ &\qquad p = \text{finite}, \quad \partial p/\partial h_i = \text{finite}, \quad \text{when } |\boldsymbol{h}| = 0,\\ &\qquad p = 0, \quad \partial p/\partial h_i = 0, \quad \text{when } |\boldsymbol{h}| \to \infty. \end{aligned} \tag{2.144}$$

The normalization condition of stationary PDF $p(\boldsymbol{h})$ is

$$\int_0^\infty p(\boldsymbol{h})\mathrm{d}\boldsymbol{h} = 1. \tag{2.145}$$

(ii) **Resonant case**

Assuming that the mean frequencies ω_i $(i = 1, 2, \ldots, n)$ of the i th subsystems in the equivalent quasi-Hamiltonian system (2.133) satisfy the α $(1 \le \alpha \le n - 1)$ internal resonance conditions

$$\sum_{i=1}^{n} k_i^u \omega_i = O_u(\varepsilon), \quad u = 1, 2, \ldots, \alpha. \tag{2.146}$$

where k_i^u are integers that are not all zero. Introduce α combinations $\Psi_u = \sum_{i=1}^{n} k_i^u \Phi_i$ of angle variables. Equations for Φ_i $(i = 1, 2, \ldots, \alpha)$ in Eq. (2.135) are replaced by equations for Ψ_i $(i = 1, 2, \ldots, \alpha)$ and the new set of stochastic differential equations are

$$\dot{A}_i = \varepsilon\overline{F}_i^A(\boldsymbol{A}, \boldsymbol{\Psi}, \Phi') + \varepsilon^{1/2}\sum_{k=1}^{m}\overline{G}_{ik}^A(\boldsymbol{A}, \boldsymbol{\Psi}, \Phi')\xi_k(t),$$

$$\dot{\Psi}_u = \sum_{i=1}^{n} k_i^u \nu_i(A_i, \Phi_i) + \varepsilon\sum_{i=1}^{n} k_i^u\overline{F}_i^\Phi(\boldsymbol{A}, \boldsymbol{\Psi}, \Phi') + \varepsilon^{1/2}\sum_{k=1}^{m}\sum_{i=1}^{n} k_i^u\overline{G}_{ik}^\Phi(\boldsymbol{A}, \boldsymbol{\Psi}, \Phi')\xi_k(t),$$

$$\dot{\Phi}_r = \nu_r(A_r, \Phi_r) + \varepsilon\overline{F}_r^\Phi(\boldsymbol{A}, \boldsymbol{\Psi}, \Phi') + \varepsilon^{1/2}\sum_{k=1}^{m}\overline{G}_{rk}^\Phi(\boldsymbol{A}, \boldsymbol{\Psi}, \Phi')\xi_k(t),$$

$$i = 1, 2, \ldots, n, \quad u = 1, 2, \ldots, \alpha, \quad r = \alpha+1, \alpha+2, \ldots, n, \tag{2.147}$$

where $\boldsymbol{\Psi} = [\Psi_1, \Psi_2, \ldots, \Psi_\alpha]^{\mathrm{T}}$, $\Phi' = [\Phi_{\alpha+1}, \Phi_{\alpha+2}, \ldots, \Phi_n]^{\mathrm{T}}$, $\overline{F}_i^A, \overline{F}_i^\Phi, \overline{G}_{ik}^A, \overline{G}_{ik}^\Phi$ are $F_i^A, F_i^\Phi, G_{ik}^A, G_{ik}^\Phi$, respectively with $\Phi_1, \Phi_2, \ldots, \Phi_\alpha$ replaced by $\Psi_1, \Psi_2, \ldots, \Psi_\alpha$ according to $\Psi_u = \sum_{i=1}^{n} k_i^u\Phi_i$.

In system (2.147), both $\boldsymbol{A}(t)$ and $\boldsymbol{\Psi}(t)$ are slowly-varying vector processes while $\Phi'(t)$ is fast-varying vector process. According to the Khasminskii theorems (Khasminskii, 1966; 1968), when $\varepsilon \to 0$, $[\boldsymbol{A}^{\mathrm{T}}, \boldsymbol{\Psi}^{\mathrm{T}}]^{\mathrm{T}}$ converges weakly to a $(n+\alpha)$-dimensional vector Markov diffusion vector process. The averaged Itô stochastic differential equations for $[\boldsymbol{A}^{\mathrm{T}}, \boldsymbol{\Psi}^{\mathrm{T}}]^{\mathrm{T}}$ are

$$\mathrm{d}A_i = m_i^A(\mathbf{A}, \boldsymbol{\Psi})\mathrm{d}t + \sum_{k=1}^{m}\sigma_{ik}^A(\mathbf{A}, \boldsymbol{\Psi})\mathrm{d}B_k(t),$$

$$\mathrm{d}\Psi_u = m_u^\Psi(\mathbf{A}, \boldsymbol{\Psi})\mathrm{d}t + \sum_{k=1}^{m}\sigma_{uk}^\Psi(\mathbf{A}, \boldsymbol{\Psi})\mathrm{d}B_k(t),$$

$$i = 1, 2, \ldots, n; \quad u = 1, 2, \ldots, \alpha, \tag{2.148}$$

According to Eq. (1.70), the drift and diffusion coefficients are derived as

$$m_i^A = \frac{\varepsilon}{(2\pi)^{n-\alpha}}\int_0^{2\pi}[\overline{F}_i^A + \sum_{k,l=1}^{m}\int_{-\infty}^{0}(\sum_{j=1}^{n}\frac{\partial\overline{G}_{ik}^A}{\partial A_j}\bigg|_t \overline{G}_{jl}^A\bigg|_{t+\tau} + \sum_{u=1}^{\alpha}\sum_{j=1}^{n}k_j^u\frac{\partial\overline{G}_{ik}^A}{\partial\Psi_u}\bigg|_t \overline{G}_{jl}^\Phi\bigg|_{t+\tau}$$
$$+ \sum_{r=\alpha+1}^{n}\frac{\partial\overline{G}_{ik}^A}{\partial\Phi_r}\bigg|_t \overline{G}_{rl}^\Phi\bigg|_{t+\tau})R_{kl}(\tau)\mathrm{d}\tau]\mathrm{d}\Phi',$$

$$m_u^\Psi = O_u(\varepsilon) + \frac{\varepsilon}{(2\pi)^{n-\alpha}}\int_0^{2\pi}[\sum_{j=1}^{n}k_j^u\overline{F}_j^\Phi + \sum_{k,l=1}^{m}\int_{-\infty}^{0}(\sum_{i,j=1}^{n}k_i^u\frac{\partial\overline{G}_{ik}^\Phi}{\partial A_j}\bigg|_t \overline{G}_{jl}^A\bigg|_{t+\tau}$$
$$+\sum_{v=1}^{\alpha}\sum_{i,j=1}^{n}k_i^u k_j^v\frac{\partial\overline{G}_{ik}^\Phi}{\partial\Psi_v}\bigg|_t \overline{G}_{jl}^\Phi\bigg|_{t+\tau} + \sum_{r=\alpha+1}^{n}\sum_{i=1}^{n}k_i^u\frac{\partial\overline{G}_{ik}^\Phi}{\partial\Phi_r}\bigg|_t \overline{G}_{rl}^\Phi\bigg|_{t+\tau})R_{kl}(\tau)\mathrm{d}\tau]\mathrm{d}\Phi', \tag{2.149}$$

$$
\begin{aligned}
b_{ij}^{A} &= \sum_{k=1}^{m}\sigma_{ik}^{A}\sigma_{jk}^{A} = \frac{\varepsilon}{(2\pi)^{n-\alpha}}\sum_{k,l=1}^{m}\int_{0}^{2\pi}\int_{-\infty}^{\infty}\overline{G}_{ik}^{A}\Big|_{t}\overline{G}_{jl}^{A}\Big|_{t+\tau}R_{kl}(\tau)\mathrm{d}\tau\mathrm{d}\Phi',\\
b_{uv}^{\Psi} &= \sum_{k=1}^{m}\sigma_{uk}^{\Psi}\sigma_{vk}^{\Psi} = \frac{\varepsilon}{(2\pi)^{n-\alpha}}\sum_{k,l=1}^{m}\int_{0}^{2\pi}\int_{-\infty}^{\infty}\sum_{i,j=1}^{n}k_i^{u}k_j^{v}\overline{G}_{ik}^{\Phi}\Big|_{t}\overline{G}_{jl}^{\Phi}\Big|_{t+\tau}R_{kl}(\tau)\mathrm{d}\tau\mathrm{d}\Phi',\\
b_{iu}^{A\Psi} &= \sum_{k=1}^{m}\sigma_{ik}^{A}\sigma_{uk}^{\Psi} = \frac{\varepsilon}{(2\pi)^{n-\alpha}}\sum_{k,l=1}^{m}\int_{0}^{2\pi}\int_{-\infty}^{\infty}\sum_{j=1}^{n}k_j^{u}\overline{G}_{ik}^{A}\Big|_{t}\overline{G}_{jl}^{\Phi}\Big|_{t+\tau}R_{kl}(\tau)\mathrm{d}\tau\mathrm{d}\Phi',\\
& i,j = 1,2,\ldots,n;\quad u,v = 1,2,\ldots,\alpha.
\end{aligned}
$$

In Eq. (2.149), time averaging has been replaced by averaging with respect to $\Phi' = [\Phi_\alpha, \Phi_{\alpha+1}, \ldots, \Phi_n]^{\mathrm{T}}$. Applying the relationship between A_i and H_i, and using the Itô stochastic differential rule, the following averaged Itô stochastic differential equations for $[\boldsymbol{H}^{\mathrm{T}}, \boldsymbol{\Psi}^{\mathrm{T}}]^{\mathrm{T}}$ can be derived from Eq. (2.148)

$$
\begin{aligned}
\mathrm{d}H_i &= \overline{m}_i^{H}(\boldsymbol{H},\boldsymbol{\Psi})\mathrm{d}t + \sum_{k=1}^{m}\overline{\sigma}_{ik}^{H}(\boldsymbol{H},\boldsymbol{\Psi})\mathrm{d}B_k(t),\\
\mathrm{d}\Psi_u &= \overline{m}_u^{\Psi}(\boldsymbol{H},\boldsymbol{\Psi})\mathrm{d}t + \sum_{k=1}^{m}\overline{\sigma}_{uk}^{\Psi}(\boldsymbol{H},\boldsymbol{\Psi})\mathrm{d}B_k(t),\\
& i = 1,2,\ldots,n;\quad u = 1,2,\ldots,\alpha.
\end{aligned}
\tag{2.150}
$$

where

$$
\begin{aligned}
\overline{m}_i^{H} &= \{m_i^{A}g_i(A_i+B_i)(1+r_i) + \frac{1}{2}b_{ii}^{A}\frac{\mathrm{d}[g_i(A_i+B_i)(1+r_i)]}{\mathrm{d}A_i}\}\Big|_{A_i=U_i^{-1}(H_i)-B_i},\\
\overline{m}_u^{\Psi} &= m_u^{\Psi}\big|_{A_i=U_i^{-1}(H_i)-B_i},
\end{aligned}
\tag{2.151}
$$

$$
\begin{aligned}
\overline{b}_{ij}^{H} &= \sum_{k=1}^{m}\overline{\sigma}_{ik}^{H}\overline{\sigma}_{jk}^{H} = [g_i(A_i+B_i)g_j(A_j+B_j)(1+r_i)(1+r_j)b_{ij}^{A}]\Big|_{\substack{A_i=U_i^{-1}(H_i)-B_i,\\ A_j=U_j^{-1}(H_j)-B_j}},\\
\overline{b}_{uv}^{\Psi} &= \sum_{k=1}^{m}\overline{\sigma}_{uk}^{\Psi}\overline{\sigma}_{vk}^{\Psi} = b_{uv}^{\Psi}\big|_{A_i=U_i^{-1}(H_i)-B_i},\\
\overline{b}_{iu}^{H\Psi} &= \sum_{k=1}^{m}\overline{\sigma}_{ik}^{H}\overline{\sigma}_{uk}^{\Psi} = [g_i(A_i+B_i)(1+r_i)b_{iu}^{A\Psi}]\big|_{A_i=U_i^{-1}(H_i)-B_i}.\\
& i,j = 1,2,\ldots,n;\quad u,v = 1,2,\ldots,\alpha.
\end{aligned}
$$

The averaged FPK equation associated with Itô Eq. (2.150) is

$$\frac{\partial p}{\partial t} = -\sum_{i=1}^{n} \frac{\partial(\overline{m}_i^h p)}{\partial h_i} - \sum_{u=1}^{\alpha} \frac{\partial(\overline{m}_u^\psi p)}{\partial \psi_u} + \frac{1}{2}\sum_{i,j=1}^{n} \frac{\partial^2(\overline{b}_{ij}^h p)}{\partial h_i \partial h_j}$$
$$+ \frac{1}{2}\sum_{u,v=1}^{\alpha} \frac{\partial^2(\overline{b}_{uv}^\psi p)}{\partial \psi_u \partial \psi_v} + \sum_{u=1}^{\alpha}\sum_{i=1}^{n} \frac{\partial^2(\overline{b}_{iu}^{h\psi} p)}{\partial h_i \partial \psi_u}, \tag{2.152}$$

where the first and second derivative moments are

$$\overline{m}_i^h = \overline{m}_i^H\Big|_{\boldsymbol{H}=\boldsymbol{h},\ \boldsymbol{\Psi}=\boldsymbol{\psi}}, \quad \overline{m}_u^\psi = \overline{m}_u^\Psi\Big|_{\boldsymbol{H}=\boldsymbol{h},\ \boldsymbol{\Psi}=\boldsymbol{\psi}},$$
$$\overline{b}_{ij}^h = \overline{b}_{ij}^H\Big|_{\boldsymbol{H}=\boldsymbol{h},\ \boldsymbol{\Psi}=\boldsymbol{\psi}}, \quad \overline{b}_{uv}^\psi = \overline{b}_{uv}^\Psi\Big|_{\boldsymbol{H}=\boldsymbol{h},\ \boldsymbol{\Psi}=\boldsymbol{\psi}}, \quad \overline{b}_{iu}^{h\psi} = \overline{b}_{iu}^{H\Psi}\Big|_{\boldsymbol{H}=\boldsymbol{h},\ \boldsymbol{\Psi}=\boldsymbol{\psi}},$$
$$i,j = 1,2,\ldots,n; \quad u,v = 1,2,\ldots,\alpha. \tag{2.153}$$

$p = p(\boldsymbol{h}, \boldsymbol{\psi}, t|\boldsymbol{h}_0, \boldsymbol{\psi}_0)$ is the transition PDF of vector process $[\boldsymbol{H}^{\mathrm{T}}, \boldsymbol{\Psi}^{\mathrm{T}}]^{\mathrm{T}}$. The initial conditions and boundary conditions for Eq. (2.152) are

$$p(\boldsymbol{h}, \boldsymbol{\psi}, 0|\boldsymbol{h}_0, \boldsymbol{\psi}_0) = \delta(\boldsymbol{h} - \boldsymbol{h}_0)\delta(\boldsymbol{\psi} - \boldsymbol{\psi}_0),$$
$$p = \text{finite}, \quad \partial p/\partial h_i = \text{finite}, \quad \text{when } \boldsymbol{h} = 0,$$
$$p = 0, \quad \partial p/\partial h_i = 0, \quad \text{when } \boldsymbol{h} \to \infty,$$
$$i = 1,2,\ldots,n. \tag{2.154}$$

The periodic condition are

$$p(\mathbf{h}, \boldsymbol{\psi}+2\overline{n}\pi, t|\mathbf{h}_0, \boldsymbol{\psi}_0) = p(\mathbf{h}, \boldsymbol{\psi}, t|\mathbf{h}_0, \boldsymbol{\psi}_0),$$
$$\overline{n} = \pm 1, 2, \ldots, \tag{2.155}$$

and the normalization condition of stationary PDF $p(\boldsymbol{h}, \boldsymbol{\psi})$ is

$$\int_0^{2\pi}\int_0^{\infty} p\mathrm{d}\boldsymbol{h}\mathrm{d}\boldsymbol{\psi} = 1. \tag{2.156}$$

Example 2.5 Consider a Duffing-van der Pol oscillator with fractional derivative damping and excited by weak wideband noises (Chen et al., 2014). The motion equation is

$$\ddot{X} + (\beta_0 + \beta_1 X^2)\dot{X} + \omega_0^2 X + \alpha_0 X^3 + \chi D^\alpha X = \xi_1(t) + X\xi_2(t), \tag{2.157}$$

where $\omega_0, \beta_0, \beta_1, \alpha_0$ and χ are constants, $\xi_1(t), \xi_2(t)$ are independent wideband noises with the following rational power spectral densities

$$S_i(\omega) = \frac{D_i}{\pi} \frac{1}{(\omega^2 - \omega_i^2)^2 + 4\xi_i^2\omega^2\omega_i^2}, \quad i = 1, 2. \tag{2.158}$$

In Eqs. (2.157) and (2.158), $\beta_0, \beta_1, \chi, D_i$ are small quantities of order ε, $D^\alpha X$ is a fractional derivative defined by Riemann–Liouville as

$$D^\alpha X = D^\alpha X(t) = \frac{1}{\Gamma(1-\alpha)} \frac{\mathrm{d}}{\mathrm{d}t} \int_0^t \frac{x(\tau)}{(t-\tau)^\alpha} \mathrm{d}\tau, \quad 0 < \alpha < 1. \tag{2.159}$$

Applying the generalized harmonic balance technique, $\chi D^\alpha X(t)$ can be decoupled into equivalent quasi-linear damping force and quasi-linear restoring force, i.e.,

$$\chi D^\alpha X(t) = C(A)\dot{X}(t) + K(A)X(t), \tag{2.160}$$

where

$$\begin{aligned} C(A) &= -\frac{\chi}{\pi A\omega(A)} \int_0^{2\pi} D^\alpha(A\cos\phi)\sin\phi \mathrm{d}\phi \\ &= \chi b_0^{\alpha-2}(A)[b_0(A) + b_2(A)/2]\sin\left(\frac{\alpha\pi}{2}\right), \\ K(A) &= \frac{\chi}{\pi A} \int_0^{2\pi} D^\alpha(A\cos\phi)\cos\phi \mathrm{d}\phi = \chi b_0^{\alpha-1}(A)[b_0(A) - b_2(A)/2]\cos\left(\frac{\alpha\pi}{2}\right). \end{aligned} \tag{2.161}$$

The detailed derivation process can be found in the appendix of the reference (Chen et al. 2014).

Substituting Eq. (2.160) into Eq. (2.157) yields the equivalent system which can be written as the following quasi Hamiltonian system by letting $Q = X,\ P = \dot{X}$

$$\dot{Q} = \frac{\partial H}{\partial P}, \quad \dot{P} = -\frac{\partial H}{\partial Q} - (\beta + \beta_1 Q^2)P + \xi_1(t) + Q\xi_2(t), \tag{2.162}$$

where $\beta = \beta_0 + C(A)$, system potential function $U(Q)$ and Hamiltonian function H are

$$H = \frac{1}{2}P^2 + U(Q), \quad U(Q) = \frac{1}{2}[K(A) + \omega_0^2]Q^2 + \frac{1}{4}\alpha_0 Q^4. \tag{2.163}$$

The relationship between energy H and amplitude A is $H = U(A)$.

If $\alpha_0 > 0$, system (2.162) has the following randomized periodic solution

$$Q(t) = A\cos\Phi(t), \quad P(t) = -A\nu(A,\Phi)\sin\Phi(t), \quad \Phi(t) = \Gamma(t) + \Theta(t). \tag{2.164}$$

where

$$\begin{aligned}\nu(A,\Phi) &= \frac{\mathrm{d}\Phi}{\mathrm{d}t} = \sqrt{\frac{2[U(A)-U(A\cos\Phi)]}{A^2\sin^2\Phi}} \\ &= [(\omega_0^2 + K(A) + \frac{3}{4}\alpha_0 A^2)(1+\eta\cos 2\Phi)]^{1/2}, \\ \eta &= \alpha_0 A^2 \big/ (4\omega_0^2 + 4K(A) + 3\alpha_0 A^2).\end{aligned} \tag{2.165}$$

The instantaneous frequency $\nu(A, \Phi)$ is an even function of Φ, which can be expanded into a Fourier series. The approximation with first four terms is

$$\nu(A,\Phi) = \sum_{r=0}^{3} b_{2r}(A)\cos 2r\Phi, \tag{2.166}$$

where

$$\begin{aligned} b_0 &= (\omega_0^2 + K(A) + \frac{3}{4}\alpha_0 A^2)^{1/2}(1 - \frac{\eta^2}{16}), \\ b_2 &= (\omega_0^2 + K(A) + \frac{3}{4}\alpha_0 A^2)^{1/2}(\frac{\eta}{2} + \frac{3\eta^3}{64}), \\ b_4 &= (\omega_0^2 + K(A) + \frac{3}{4}\alpha_0 A^2)^{1/2}(\frac{-\eta^2}{16}), \quad b_6 = (\omega_0^2 + K(A) + \frac{3}{4}\alpha_0 A^2)^{1/2}(\frac{\eta^3}{64}). \end{aligned} \tag{2.167}$$

$b_0(A)$ and $b_2(A)$ can be obtained through numerical iteration, and then $C(A)$, $K(A)$, $b_4(A)$ and $b_6(A)$ can be determined.

Using Eq. (2.164), the following stochastic differential equation for A, Φ can be derived from Eq. (2.162)

$$\begin{aligned} \frac{\mathrm{d}A}{\mathrm{d}t} &= F^A(A,\Phi) + G_1^A(A,\Phi)\xi_1(t) + G_2^A(A,\Phi)\xi_2(t), \\ \frac{\mathrm{d}\Phi}{\mathrm{d}t} &= \nu(A,\Phi) + F^\Phi(A,\Phi) + G_1^\Phi(A,\Phi)\xi_1(t) + G_2^\Phi(A,\Phi)\xi_2(t), \end{aligned} \tag{2.168}$$

where

$$\begin{aligned} F^A &= \frac{-A\nu^2(A,\Phi)[\beta + \beta_1 A^2\cos^2\Phi]\sin^2\Phi}{K(A) + \omega_0^2 + \alpha_0 A^2}, \\ F^\Phi &= \frac{-\nu(A,\Phi)[\beta + \beta_1 A^2\cos^2\Phi]\sin\Phi\cos\Phi}{K(A) + \omega_0^2 + \alpha_0 A^2}, \\ G_1^A &= \frac{-\nu(A,\Phi)\sin\Phi}{K(A) + \omega_0^2 + \alpha_0 A^2}, \quad G_2^A = \frac{-A\nu(A,\Phi)\sin\Phi\cos\Phi}{K(A) + \omega_0^2 + \alpha_0 A^2}, \end{aligned}$$

$$G_1^{\Phi} = \frac{-\nu(A, \Phi)\cos\Phi}{A(K(A) + \omega_0^2 + \alpha_0 A^2)}, \quad G_2^{\Phi} = \frac{-\nu(A, \Phi)\cos^2\Phi}{K(A) + \omega_0^2 + \alpha_0 A^2}. \tag{2.169}$$

In Eq. (2.168), $A(t)$ is a slowly-varying process while $\Phi(t)$ is a fast-varying stochastic process. By applying the stochastic averaging method described in Sect. 1, an averaged Itô stochastic differential equation for $A(t)$ can be obtained as follows

$$\mathrm{d}A = m(A)\mathrm{d}t + \sigma(A)\mathrm{d}B(t), \tag{2.170}$$

According to Eq. (1.20), the drift and diffusion coefficients are

$$m(A) = \langle F^A\rangle_t + \left\langle \sum_{k,l=1}^{2} \int_{-\infty}^{0} \left(\left.\frac{\partial G_k^A}{\partial A}\right|_t G_l^A\Big|_{t+\tau} + \left.\frac{\partial G_k^A}{\partial \Phi}\right|_t G_l^{\Phi}\Big|_{t+\tau}\right) R_{kl}(\tau)\mathrm{d}\tau \right\rangle_t,$$

$$\sigma^2(A) = \left\langle \sum_{k,l=1}^{2} \int_{-\infty}^{\infty} G_k^A\Big|_t G_l^A\Big|_{t+\tau} R_{kl}(\tau)\mathrm{d}\tau \right\rangle_t. \tag{2.171}$$

where $R_{kl}(\tau)$ represents the cross-correlation function of the excitations for cases $k \neq l$, and the auto-correlation function for cases $k = l$. For the present example, the cross-correlation functions are zero.

Expanding the functions in Eq. (2.169) into Fourier series and substituting them into Eq. (2.171), completing the integration with respect to τ, and replacing time averaging with averaging with respect to Φ result in

$$\begin{aligned}
m(A) = &-\frac{\beta(4\omega_0^2 A + 4K(A)A + 2.5\alpha_0 A^3)}{8(\omega_0^2 + K(A)A + \alpha_0 A^2)} - \frac{\beta_1(K(A)A^3 + \omega_0^2 A^3 + 0.75\alpha_0 A^5)}{8(\omega_0^2 + K(A)A + \alpha_0 A^2)} \\
&+ \frac{\pi A}{32(\omega_0^2 + K(A)A + \alpha_0 A^2)}\{(2b_0 - b_4)S_2(2\omega)\frac{\mathrm{d}}{\mathrm{d}A}[\frac{A(2b_0 - b_4)}{\omega_0^2 + K(A)A + \alpha_0 A^2}] \\
&+ (b_2 - b_6)S_2(4\omega)\frac{\mathrm{d}}{\mathrm{d}A}[\frac{A(b_2 - b_6)}{\omega_0^2 + K(A)A + \alpha_0 A^2}] + b_4 S_2(6\omega)\frac{\mathrm{d}}{\mathrm{d}A}[\frac{Ab_4}{\omega_0^2 + K(A)A + \alpha_0 A^2}] \\
&+ (b_2 - b_4)S_1(3\omega)\frac{\mathrm{d}}{\mathrm{d}A}[\frac{b_2 - b_4}{\omega_0^2 + K(A)A + \alpha_0 A^2}] \\
&+ (b_4 - b_6)S_1(5\omega)\frac{\mathrm{d}}{\mathrm{d}A}[\frac{b_4 - b_6}{\omega_0^2 + K(A)A + \alpha_0 A^2}] + b_6 S_1(7\omega)\frac{\mathrm{d}}{\mathrm{d}A}[\frac{b_6}{\omega_0^2 + K(A)A + \alpha_0 A^2}]\} \\
&+ \frac{\pi}{8A(\omega_0^2 + K(A)A + \alpha_0 A^2)^2}\{(4b_0^2 - b_2^2)S_1(\omega) + 3(b_2^2 - b_4^2)S_1(3\omega) + 5(b_4^2 - b_6^2)S_1(5\omega) \\
&+ 7b_6^2 S_1(7\omega)\}, \\
&+ b_6 S_2(8\omega)\frac{\mathrm{d}}{\mathrm{d}A}[\frac{Ab_6}{\omega_0^2 + K(A)A + \alpha_0 A^2}]\}
\end{aligned} \tag{2.172}$$

$$+ \frac{\pi A}{16(\omega_0^2 + K(A)A + \alpha_0 A^2)^2}\{(2b_0 - b_4)(2b_0 + 2b_2 + b_4)$$

$$\times S_2(2\omega)+2(b_2-b_6)(b_2+2b_4+b_6)S_2(4\omega)+3b_4(b_4+2b_6)S_2(6\omega)+4b_6^2S_2(8\omega)\}$$
$$+\frac{\pi}{8(\omega_0^2+K(A)A+\alpha_0A^2)}\{(2b_0-b_2)S_1(\omega)\frac{\mathrm{d}}{\mathrm{d}A}[\frac{2b_0-b_2}{\omega_0^2+K(A)A+\alpha_0A^2}],$$

$$\sigma^2(A)=\frac{\pi A^2}{16(\omega_0^2+K(A)A+\alpha_0A^2)^2}\{(2b_0-b_4)^2S_2(2\omega)+(b_2-b_6)^2S_2(4\omega)$$
$$+b_4^2S_2(6\omega)+b_6^2S_2(8\omega)\}+\frac{\pi}{4(\omega_0^2+K(A)A+\alpha_0A^2)^2}$$
$$\times\{(2b_0-b_2)^2S_1(\omega)+(b_2-b_4)^2S_1(3\omega)+(b_4-b_6)^2S(5\omega)+b_6^2S(7\omega)\}.$$

According to relationship $H=U(A)$ and Itô differential rules, an averaged Itô equation for energy H can be obtained as

$$\mathrm{d}H=\overline{m}(H)\mathrm{d}t+\overline{\sigma}(H)\mathrm{d}B(t) \tag{2.173}$$

where

$$\overline{m}(H)=[m(A)\frac{\mathrm{d}H}{\mathrm{d}A}+\frac{\sigma^2}{2}\frac{\mathrm{d}^2H}{\mathrm{d}A^2}]\bigg|_{A=U^{-1}(H)},\quad \overline{\sigma}^2(H)=\sigma^2(A)(\frac{\mathrm{d}H}{\mathrm{d}A})^2\bigg\|_{A=U^{-1}(H)}. \tag{2.174}$$

Consider reliability of system (2.170). Let h_c be the threshold and $\Omega_s=[0,h_c)$ as the safe domain. The conditional reliability function $R(t|h_0)=\mathrm{Prob}[H(t)\in\Omega_s,\tau\in(0,t]|h_0\in\Omega_s]$ represents the probability that $H(\tau)$ does not exceed the safe domain Ω_s within the time interval $\tau\in(0,t]$, given the initial value $H(\tau=0)=h_0\in\Omega_s$. $R(t|h_0)$ satisfies the following backward Kolmogorov equation

$$\frac{\partial R}{\partial t}=\overline{m}(h_0)\frac{\partial R}{\partial h_0}+\frac{1}{2}\overline{\sigma}^2(h_0)\frac{\partial^2R}{\partial h_0^2}, \tag{2.175}$$

where the first and second derivative moments are equal to the drift and diffusion coefficients in Eq. (2.174) with H replaced by h_0. The initial condition of Eq. (2.175) is

$$R(0|h_0)=1,\quad h_0\in\Omega_s. \tag{2.176}$$

The safe domain has two boundaries. The left boundary $h_0=0$ is reflecting boundary and right boundary $h_0=h_c$ is absorbing boundary, i.e.,

$$R(t|0)=\text{finite},$$
$$R(t|h_c)=0. \tag{2.177}$$

Solving Eq. (2.175) with initial condition in Eq. (2.176) and boundary conditions in Eq. (2.177) numerically yields the reliability function $R(t|h_0)$. Then, the mean first passage time $\mu(h_0)$ can be obtained from

$$\mu(h_0) = \int_0^\infty R(t|h_0)\mathrm{d}t \tag{2.178}$$

For system (2.157) with parameters $\beta_0 = 0.05$, $\beta_1 = 0.05$, $\omega_0 = 1$, $\alpha_0 = 1$, $\chi = 0.1$, $H_0 = 0$, $h_c = 0.001$, $\omega_1 = \omega_2 = 10$, $\xi_1 = \xi_2 = 0.3$, $D_1 = D_2 = 1$, Figs. 2.17 and 2.18 show the reliability functions and the means of the first passage time for different order of fractional derivatives. Both the results from the stochastic averaging method and from Monte Carlo simulation of the original system (2.162) are presented. It can be seen from these figures that the two results match very well. The reliability and mean first passage time of the system increase with the increasing of the order α of fractional derivative due to a larger value of α associated with stronger damping force.

Example 2.6 Consider a two-DOF nonlinear system with fractional derivative damping forces and excited by Gaussian white noises. The motion equations are given by

$$\begin{aligned}
&\ddot{X}_1 + \left[-\gamma_1' + \beta_1\left(\dot{X}_1^2 + \dot{X}_2^2\right)\right]\dot{X}_1 + \omega_1'^2 X_1 + \mu_1 D^{\alpha_1}(X_1 - X_2) = W_{g1}(t),\\
&\ddot{X}_2 + \left[-\gamma_2' + \beta_2\left(\dot{X}_1^2 + \dot{X}_2^2\right)\right]\dot{X}_2 + \omega_2'^2 X_2 + \mu_2 D^{\alpha_2}(X_2 - X_1) = W_{g2}(t),
\end{aligned} \tag{2.179}$$

where $\gamma_1', \gamma_2', \beta_1, \beta_2, \omega_1', \omega_2', \mu_1, \mu_2$ are positive constants, $\mu_1 D^{\alpha_1}(X_1 - X_2)$, $(0 < \alpha_1 < 1)$ and $\mu_2 D^{\alpha_2}(X_2 - X_1)$, $(0 < \alpha_2 < 1)$ are fractional derivative damping forces, $W_{g1}(t), W_{g2}(t)$ are independent Gaussian white noises with excitation intensities $2\pi K_1$ and $2\pi K_2$, respectively, $\gamma_i', \beta_i, \mu_i, K_i,\ i = 1, 2$ are small parameters of ε

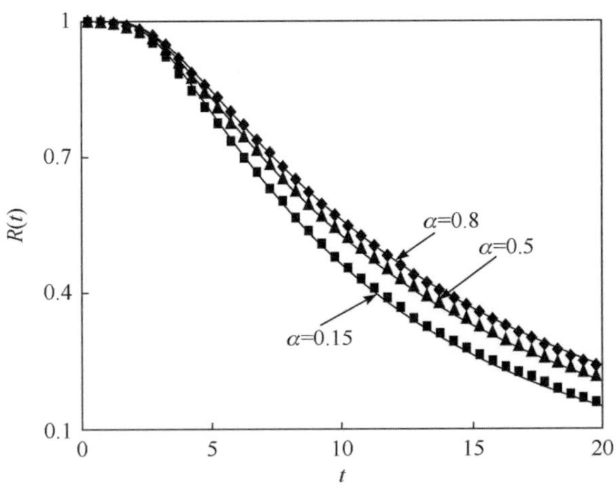

Fig. 2.17 Reliability function of system (2.157) for different fractional derivative order α. Lines —, results from the stochastic averaging method; symbols ■●▼, results from Monte Carlo simulation (Chen et al. 2014)

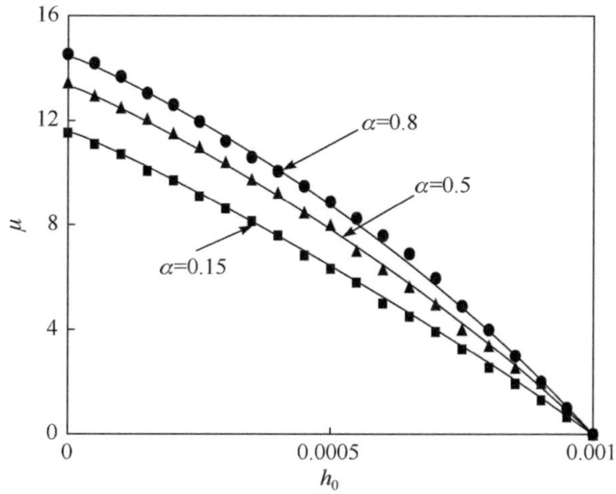

Fig. 2.18 Mean first passage time of system (2.157) for different fractional derivative order α. Lines —, results from the stochastic averaging method; symbols ■●▼, results from Monte Carlo simulation (Chen et al. 2014)

order. System (2.179) without fractional derivative damping has two diffused limit cycles in the 4-dimensional phase space (Deng and Zhu 2004).

Applying the generalized harmonic balance technique in Eq. (2.132) to the fractional derivative damping forces $D^{\alpha_1}(X_1 - X_2)$ and $D^{\alpha_2}(X_2 - X_1)$ yields

$$\begin{aligned}
D^{\alpha_1}(X_1 - X_2) &= C_{11}(A_1)\dot{X}_1 - C_{12}(A_2)\dot{X}_2 + K_{11}(A_1)X_1 - K_{12}(A_2)X_2,\\
D^{\alpha_2}(X_2 - X_1) &= C_{22}(A_2)\dot{X}_2 - C_{21}(A_1)\dot{X}_1 + K_{22}(A_2)X_2 - K_{21}(A_1)X_1,\\
C_{11}(A_1) &= \omega_1^{\alpha_1 - 1}\sin\frac{\alpha_1\pi}{2},\quad K_{11}(A_1) = \omega_1^{\alpha_1}\cos\frac{\alpha_1\pi}{2},\\
C_{22}(A_2) &= \omega_2^{\alpha_2 - 1}\sin\frac{\alpha_2\pi}{2},\quad K_{22}(A_2) = \omega_2^{\alpha_2}\cos\frac{\alpha_2\pi}{2},\\
C_{12}(A_2) &= \omega_2^{\alpha_1 - 1}\sin\frac{\alpha_1\pi}{2},\quad K_{12}(A_2) = \omega_2^{\alpha_1}\cos\frac{\alpha_1\pi}{2},\\
C_{21}(A_1) &= \omega_1^{\alpha_2 - 1}\sin\frac{\alpha_2\pi}{2},\quad K_{21}(A_1) = \omega_1^{\alpha_2}\cos\frac{\alpha_2\pi}{2},
\end{aligned} \tag{2.180}$$

where ω_1, ω_2 are the mean frequencies of equivalent system to be determined in Eq. (2.184).

Substituting Eq. (2.180) into the original system (2.179) and letting $Q_1 = X_1,\ P_1 = \dot{X}_1, Q_2 = X_2,\ P_2 = \dot{X}_2$ lead to the following equivalent quasi-Hamiltonian system

$$\begin{aligned}
\dot{Q}_1 &= \frac{\partial H}{\partial P_1},\\
\dot{Q}_2 &= \frac{\partial H}{\partial P_2},\\
\dot{P}_1 &= -\frac{\partial H}{\partial Q_1} - [-\gamma_1 + \beta_1(P_1^2 + P_2^2)]P_1 - \mu_1 C_{12}(A_2)P_2
\end{aligned}$$

$$
\begin{aligned}
&- \mu_1 K_{12}(A_2) Q_2 + W_{g1}(t), \\
\dot{P}_2 &= -\frac{\partial H}{\partial Q_2} - [-\gamma_2 + \beta_2(P_1^2 + P_2^2)]P_2 - \mu_2 C_{21}(A_1)P_1 \\
&- \mu_2 K_{21}(A_1) Q_1 + W_{g2}(t).
\end{aligned}
\tag{2.181}
$$

where the parameters γ_1, γ_2 are given by

$$
\gamma_1 = \gamma_1' - \mu_1 C_{11}(A_1), \quad \gamma_2 = \gamma_2' - \mu_2 C_{22}(A_2),
\tag{2.182}
$$

and the associated Hamiltonian function is given by

$$
H = H_1 + H_2, \quad H_i = \frac{1}{2} P_i^2 + \frac{1}{2}\omega_i^2 Q_i^2, \quad i = 1, 2.
\tag{2.183}
$$

where

$$
\omega_1^2 = \omega_1'^2 + \mu_1 K_{11}(A_1), \quad \omega_2^2 = \omega_2'^2 + \mu_2 K_{22}(A_2)
\tag{2.184}
$$

Note that to obtain the frequencies ω_1, ω_2 of equivalent system require solving transcendental equations in Eq. (2.184). $\mu_1 K_{12}(A_2) Q_2$ and $\mu_2 K_{21}(A_1) Q_1$ in system (2.181) are small compared with $\omega_1^2 Q_1$ and $\omega_2^2 Q_2$, respectively. Based on the analysis in Sect. 5.4.2 in vol. 1, system (2.181) can be treated as a quasi-integrable Hamiltonian system, and the stochastic averaging method described in Sect. 5.2 in vol. 1 can be applied.

Applying transformation (2.122) and the relation $H_i = H_i(A_i)$, the following equations for the Hamiltonian functions H_i and the phase angles Φ_i can be obtained from Eq. (2.181)

$$
\begin{aligned}
\dot{H}_1 &= -2H_1 \sin^2\Phi_1[-\gamma_1 + 2\beta_1(H_1 \sin^2\Phi_1 + H_2 \sin^2\Phi_2)] \\
&+ 2\mu_1 \sin\Phi_1 \sqrt{H_1 H_2}\left(C_{12} \sin\Phi_2 - \frac{K_{12}}{\omega_2}\cos\Phi_2\right) - \sqrt{2H_1}\sin\Phi_1 W_{g1}(t), \\
\dot{H}_2 &= -2H_2 \sin^2\Phi_2[-\gamma_2 + 2\beta_2(H_1 \sin^2\Phi_1 + H_2 \sin^2\Phi_2)] \\
&+ 2\mu_2 \sin\Phi_2 \sqrt{H_1 H_2}\left(C_{21} \sin\Phi_1 - \frac{K_{21}}{\omega_1}\cos\Phi_1\right) - \sqrt{2H_2}\sin\Phi_2 W_{g2}(t),
\end{aligned}
\tag{2.185}
$$

$$
\begin{aligned}
\dot{\Phi}_1 &= \omega_1 - \sin\Phi_1\cos\Phi_1[-\gamma_1 + 2\beta_1(H_1 \sin^2\Phi_1 + H_2 \sin^2\Phi_2)] \\
&+ \mu_1 \cos\Phi_1 \sqrt{\frac{H_2}{H_1}}\left(C_{12}\sin\Phi_2 - \frac{K_{12}}{\omega_2}\cos\Phi_2\right) - \frac{\cos\Phi_1}{\sqrt{2H_1}} W_{g1}(t), \\
\dot{\Phi}_2 &= \omega_2 - \sin\Phi_2\cos\Phi_2[-\gamma_2 + 2\beta_2(H_1 \sin^2\Phi_1 + H_2 \sin^2\Phi_2)] \\
&+ \mu_2 \cos\Phi_2 \sqrt{\frac{H_1}{H_2}}\left(C_{21}\sin\Phi_1 - \frac{K_{21}}{\omega_1}\cos\Phi_1\right) - \frac{\cos\Phi_2}{\sqrt{2H_2}} W_{g2}(t).
\end{aligned}
$$

Based on whether ω_1 and ω_2 satisfy the internal resonance relation (2.146), system (2.185) should be treated separately for the following two cases.

(i) **Non-resonant case**

In the non-resonant case, ω_1, ω_2 do not satisfy the internal resonance relation in Eq. (2.146). $[H_1(t), H_2(t)]^{\mathrm{T}}$ weakly converges to a 2-dimensional vector Markov diffusion process. Applying the stochastic averaging method of quasi-integrable and non-resonant Hamiltonian systems under Gaussian white noise described in Sect. 5.2.1 in vol. 1, the averaged Itô stochastic differential equations for $[H_1(t), H_2(t)]^{\mathrm{T}}$ and its associated reduced averaged FPK equation can be derived from Eq. (2.185)

$$\frac{1}{2}\frac{\partial^2}{\partial h_1^2}(b_{11}p) + \frac{1}{2}\frac{\partial^2}{\partial h_2^2}(b_{22}p) - \frac{\partial}{\partial h_1}(a_1 p) - \frac{\partial}{\partial h_2}(a_2 p) = 0, \tag{2.186}$$

where the first and second derivative moments can be obtained according to Eqs. (5.95)–(5.99) in vol. 1 as follows

$$\begin{aligned}
a_1 &= \gamma_1 h_1 - \frac{3}{2}\beta_1 h_1^2 - \beta_1 h_1 h_2 + \pi K_1,\\
a_2 &= \gamma_2 h_2 - \frac{3}{2}\beta_2 h_2^2 - \beta_2 h_2 h_1 + \pi K_2,\\
b_{11} &= 2\pi K_1 h_1, \quad b_{22} = 2\pi K_2 h_2.
\end{aligned} \tag{2.187}$$

In Eq. (2.186) $p = p(h_1, h_2)$ is the stationary PDF and the boundary conditions are

$$\begin{aligned}
p(h_1, h_2) &= \text{finite}, \quad \text{when } h_1 = 0 \text{ or } h_2 = 0,\\
p(h_1, h_2) &= 0, \quad \text{when } h_1 \to \infty \text{ or } h_2 \to \infty,\\
\partial p/\partial h_1 &= 0, \quad \text{when } h_1 \to \infty,\\
\partial p/\partial h_2 &= 0, \quad \text{when } h_2 \to \infty.
\end{aligned} \tag{2.188}$$

Generally, the FPK Eq. (2.186) with boundary conditions in Eq. (2.188) can be solved only numerically. However, when system parameters satisfy the condition $\beta_1/K_1 = \beta_2/K_2$ (Zhu et al. 1997), Eq. (2.186) has exact analytical solution

$$p(h_1, h_2) = C\exp\{\frac{\beta_1}{\pi K_1}[\frac{\gamma_1}{\beta_1}h_1 + \frac{\gamma_2}{\beta_2}h_2 - \frac{3}{4}(h_1^2 + h_2^2) - h_1 h_2]\}, \tag{2.189}$$

where C is a normalization constant. Once $p(h_1, h_2)$ is obtained, the stationary joint PDF $p(q_1, q_2, p_1, p_2)$ can be obtained according to Eq. (5.106) in vol. 1 as follows

$$p(q_1, q_2, p_1, p_2) = \frac{\omega_1\omega_2}{4\pi^2}p(h_1, h_2)\Big|_{\substack{h_1=(p_1^2+\omega_1^2 q_1^2)/2 \\ h_2=(p_2^2+\omega_2^2 q_2^2)/2}}. \tag{2.190}$$

Interested marginal PDFs and statistics can then be derived from $p(q_1, q_2, p_1, p_2)$.

For system (2.179) with parameters $\gamma_1' = 0.2$, $\gamma_2' = 0.2$, $\beta_1 = 0.05$, $\beta_2 = 0.05$, $\omega_1' = 1.414$, $\omega_2' = 2$, $\mu_1 = 0.05$, $\mu_2 = 0.05$, $D_1 = 0.01$, $\alpha_1 = 0.5$, $\alpha_2 = 0.5$, $\pi K_1 = 0.01$, $\pi K_2 = 0.01$, Figs. 2.19, 2.20 and 2.21 show the results from the stochastic averaging method and from Monte Carlo simulation of original system (2.179). It can be seen that the two results match well.

(ii) **Resonant case**

Assume that the frequencies of equivalent system satisfy the internal resonance condition $\omega_1 - \omega_2 = 0$. Introduce phase angle difference $\Psi = \Phi_1 - \Phi_2$. $[H_1(t), H_2(t), \Psi(t)]^T$ converges weakly to a 3-dimensional vector Markov diffusion process. Applying the stochastic averaging method of quasi-integrable and resonant Hamiltonian systems excited by Gaussian white noise in Sect. 5.2.2 in vol. 1, the averaged Itô stochastic differential equation for $[H_1(t), H_2(t), \Psi(t)]^T$ and their

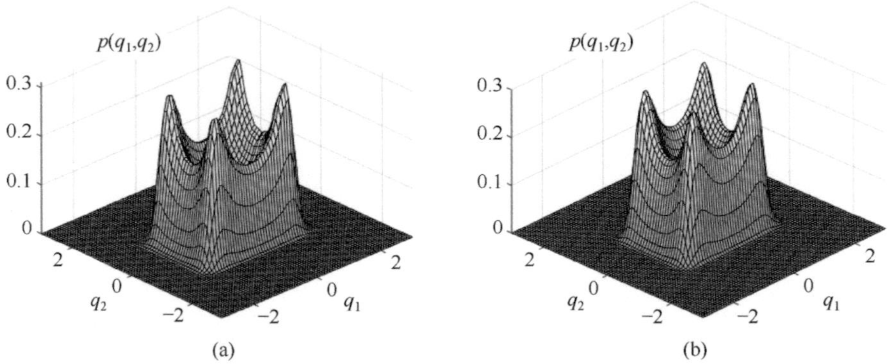

Fig. 2.19 Stationary joint PDF $p(q_1, q_2)$ of system (2.179) in non-resonant case. **a** Result from Monte Carlo simulation; **b** result from the stochastic averaging method

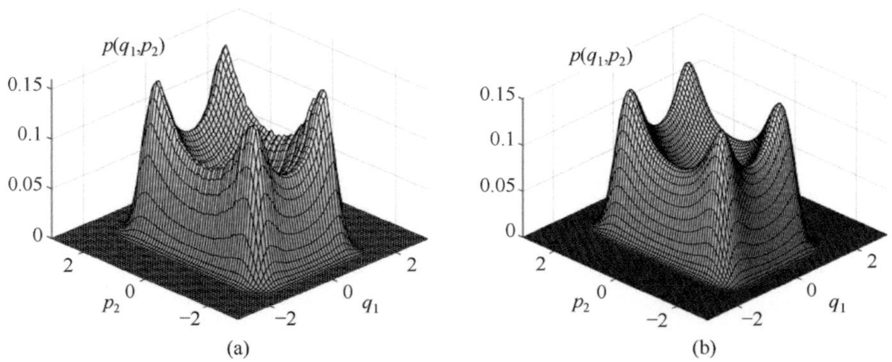

Fig. 2.20 Stationary joint PDF $p(q_1, p_2)$ of system (2.179) in non-resonant case. **a** Result from Monte Carlo simulation; **b** result from the stochastic averaging method

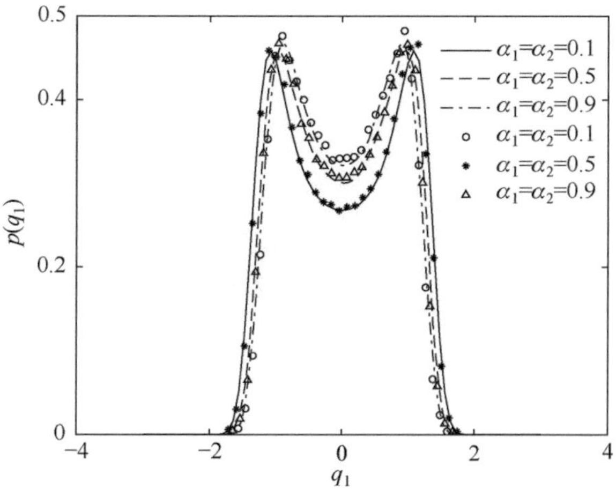

Fig. 2.21 Stationary PDF $p(q_1)$ of system (2.179) in non-resonant case with different fractional derivative orders α_1, α_2. Line —, line —, line -••, results from the stochastic averaging method; symbols ○•△, results from Monte Carlo simulation

associated reduced averaged FPK equation can be obtained from Eq. (2.185)

$$\frac{1}{2}\frac{\partial^2}{\partial h_1^2}(b_{11}p) + \frac{1}{2}\frac{\partial^2}{\partial h_2^2}(b_{22}p) + \frac{1}{2}\frac{\partial^2}{\partial \psi^2}(b^{\Psi}p)$$
$$-\frac{\partial}{\partial h_1}(a_1 p) - \frac{\partial}{\partial h_2}(a_2 p) - \frac{\partial}{\partial \psi}(a^{\Psi}p) = 0, \tag{2.191}$$

where the first and second derivative moments can be obtained similar to Eq. (5.121) in vol. 1

$$a_1 = \gamma_1 h_1 - \frac{3}{2}\beta_1 h_1^2 - \beta_1(1 + \frac{1}{2}\cos 2\psi)h_1 h_2$$
$$+ \mu_1[C_{12}(\frac{\sqrt{2h_2}}{\omega_2})\cos\psi - K_{12}(\frac{\sqrt{2h_2}}{\omega_2})\frac{\sin\psi}{\omega_2}]\sqrt{h_1 h_2} + \pi K_1,$$
$$a_2 = \gamma_2 h_2 - \frac{3}{2}\beta_2 h_2^2 - \beta_2(1 + \frac{1}{2}\cos 2\psi)h_1 h_2$$
$$+ \mu_2[C_{21}(\frac{\sqrt{2h_1}}{\omega_1})\cos\psi + K_{21}(\frac{\sqrt{2h_1}}{\omega_1})\frac{\sin\psi}{\omega_1}]\sqrt{h_1 h_2} + \pi K_2, \tag{2.192}$$

$$a^{\Psi} = \frac{1}{4}\sin 2\psi(\beta_1 h_2 + \beta_2 h_1) - \frac{\mu_1}{2}\sqrt{\frac{h_2}{h_1}}[C_{12}(\frac{\sqrt{2h_2}}{\omega_2})\sin\psi + K_{12}(\frac{\sqrt{2h_2}}{\omega_2})\frac{\cos\psi}{\omega_2}]$$
$$-\frac{\mu_2}{2}\sqrt{\frac{h_1}{h_2}}[C_{21}(\frac{\sqrt{2h_1}}{\omega_1})\sin\psi - K_{21}(\frac{\sqrt{2h_1}}{\omega_1})\frac{\cos\psi}{\omega_1}],$$
$$b_{11} = 2\pi K_1 h_1, \quad b_{22} = 2\pi K_2 h_2, \quad b^{\Psi} = \frac{\pi}{2}(K_1/h_1 + K_2/h_2).$$

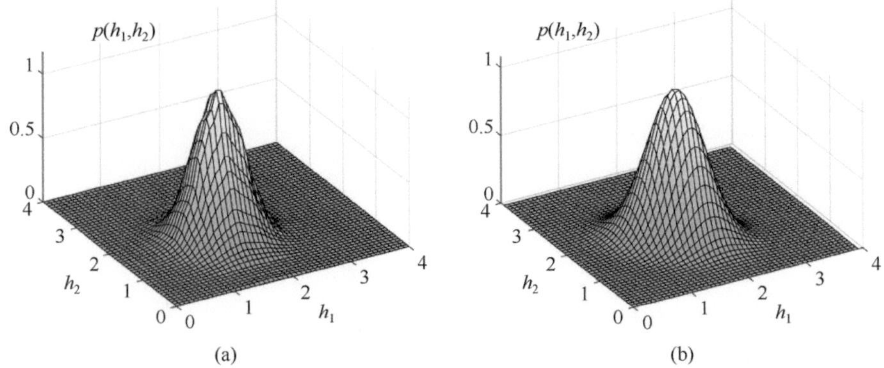

Fig. 2.22 Stationary joint PDF $p(h_1, h_2)$ of system (2.179) in internal resonant case, **a** result from Monte Carlo simulation; **b** result from the stochastic averaging method

In Eq. (2.191), $p = p(h_1, h_2, \psi)$is stationary PDF. Besides the boundary condition (2.188), Eq. (2.191) has the following periodic boundary condition

$$p(h_1, h_2, \psi + 2k\pi) = p(h_1, h_2, \psi), \quad k = \pm 1, 2, \ldots. \tag{2.193}$$

Generally, the reduced averaged FPK Eq. (2.191) can be solved only numerically. Once $p(h_1, h_2, \psi)$ is obtained, the stationary joint PDF of generalized displacements and generalized momenta can be obtained according to Eq. (5.178) in vol. 1 as follows

$$p(q_1, q_2, p_1, p_2) = \frac{\omega_1\omega_2}{2\pi} p(h_1, h_2, \psi)\Big|_{\substack{h_1=(p_1^2+\omega_1^2 q_1^2)/2 \\ h_2=(p_2^2+\omega_2^2 q_2^2)/2 \\ \psi=\tan^{-1}(p_1/\omega_1 q_1)-\tan^{-1}(p_2/\omega_2 q_2)}}. \tag{2.194}$$

The marginal PDFs and statistics can then be derived from $p(q_1, q_2, p_1, p_2)$.

For system (2.179) with parameters $\gamma_1' = 0.2$, $\gamma_2' = 0.2$, $\beta_1 = 0.05$, $\beta_2 = 0.05$, $\omega_1' = 2$, $\omega_2' = 2$, $\mu_1 = 0.05$, $\mu_2 = 0.05$, $\alpha_1 = 0.5$, $\alpha_2 = 0.5$, $\pi K_1 = 0.01$, $\pi K_2 = 0.01$, Figs. 2.22, 2.23 and 2.24 show the results from the stochastic averaging method and from Monte Carlo simulation of original system (2.179). It can be seen that the two results match well.

2.4 Quasi-integrable Hamiltonian Systems with Time-Delay Forces

Currently, there is limited research on stochastic systems with time-delay forces. Grigoriu (1997) studied time-delay problem in deterministic linear control systems under Gaussian white noise excitation. Di Paola et al. (2001) used Taylor series to

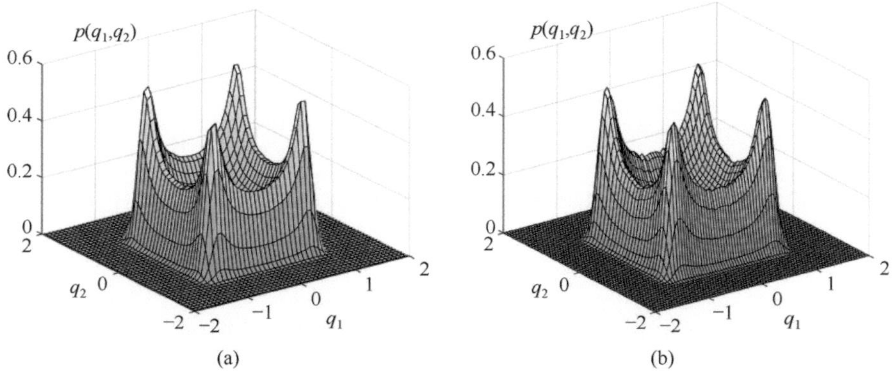

Fig. 2.23 Stationary joint PDF $p(q_1, q_2)$ of system (2.179) in internal resonant case, **a** result from Monte Carlo simulation; **b** result from the stochastic averaging method

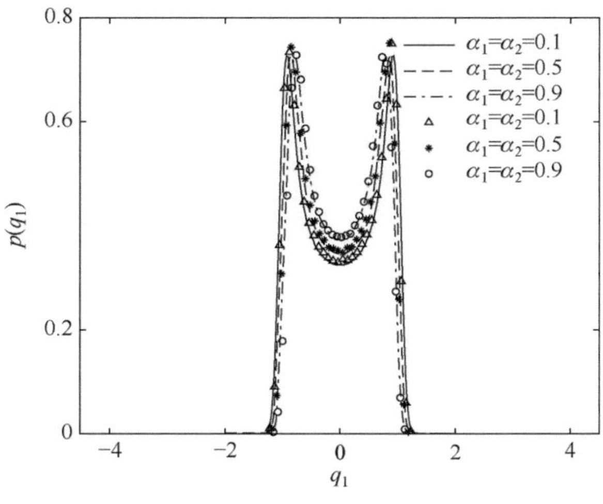

Fig. 2.24 Stationary PDF $p(q_1)$ of system (2.179) in internal resonant case with different fractional derivative orders α_1, α_2. Line —, line —, line -••, results from the stochastic averaging method; symbols ○◆△, results from Monte Carlo simulation

investigate the influence of time-delay on the response of controlled linear systems under Gaussian white noise excitation. In this section, the stochastic averaging method of quasi-integrable Hamiltonian systems with time-delay forces (Liu 2007; Zhu and Liu 2007) is presented.

Consider a quasi-integrable Hamiltonian system with time-delay control forces governed by following motion equations

$$\dot{Q}_i = P_i,$$
$$\dot{P}_i = -g_i'(Q_i) - \varepsilon \sum_{j=1}^{n} c_{ij}'(\boldsymbol{Q}, \boldsymbol{P})P_j - \varepsilon F_i(Q_{i\tau}, P_{i\tau})$$

$$+\varepsilon^{1/2}\sum_{k=1}^{m} f_{ik}(\boldsymbol{Q},\boldsymbol{P})\xi_k(t),$$
$$i = 1, 2, \ldots, n, \tag{2.195}$$

where $\varepsilon F_i(Q_{i\tau}, P_{i\tau})$ represent the time-delay forces, $Q_{i\tau} = Q_i(t-\tau), P_{i\tau} = P_i(t-\tau)$, τ is the delay time, and the other terms are the same as those in system (2.1). Assume that $g_i'(0) = 0$ and $g_i'(q_i)$ satisfy the four conditions listed below Eq. (1.5). Apply the following transformation to system (2.195)

$$\begin{aligned}&Q_i(t) = A_i \cos\Phi_i(t), \quad P_i(t) = -A_i\omega_i(A_i)\sin\Phi_i(t),\\ &\Phi_i(t) = \Gamma_i(t) + \Theta_i(t), \quad i = 1, 2, \ldots, n,\end{aligned} \tag{2.196}$$

where $\omega_i(A_i)$ are the mean value of $d\Phi_i/dt$. For small delay time τ and due to the slowly-varying nature of $A_i(t)$, $\Theta_i(t)$, the following approximations can be made

$$\begin{aligned}Q_i(t-\tau) &= A_i(t-\tau)\cos\Phi_i(t-\tau)\\ &\approx A_i(t)\cos[\omega_i(A_i)(t-\tau) + \Theta_i(t)]\\ &= A_i\{\cos[\omega_i(A_i)t + \Theta_i(t)]\cos[\omega_i(A_i)\tau]\\ &\quad + \sin[\omega_i(A_i)t + \Theta_i(t)]\sin[\omega_i(A_i)\tau]\}\\ &\approx Q_i(t)\cos[\omega_i(A_i)\tau] - \frac{P_i(t)}{\omega_i(A_i)}\sin[\omega_i(A_i)\tau],\end{aligned} \tag{2.197}$$

$$\begin{aligned}P_i(t-\tau) &= -A_i(t-\tau)\omega_i(t-\tau)\sin\Phi_i(t-\tau)\\ &\approx -A_i(t)\omega_i(A_i)\sin[\omega_i(A_i)(t-\tau) + \Theta_i(t)]\\ &= -A_i\omega_i(A_i)\{\sin[\omega_i(A_i)t + \Theta_i(t)]\cos[\omega_i(A_i)\tau]\\ &\quad - \cos[\omega_i(A_i)t + \Theta_i(t)]\sin[\omega_i(A_i)\tau]\}\\ &\approx P_i(t)\cos[\omega_i(A_i)\tau] + Q_i(t)\omega_i(A_i)\sin[\omega_i(A_i)\tau].\end{aligned}$$

Therefore, the time-delay forces $\varepsilon F_i(Q_{i\tau}, P_{i\tau})$ in system (2.195) can be represented approximately in terms of the Q_i, P_i at time t. These forces can be split into conservative and dissipated components and then combine them with $g_i'(Q_i)$ and $\sum_{j=1}^{n} c_{ij}'(\boldsymbol{Q},\boldsymbol{P})P_j$, respectively. Thus, the following equivalent quasi-integrable Hamiltonian system without time delay can be obtained

$$\begin{aligned}&\dot{Q}_i = P_i,\\ &\dot{P}_i = -g_i(Q_i) - \varepsilon\sum_{j=1}^{n} c_{ij}(\boldsymbol{Q},\boldsymbol{P})P_j + \varepsilon^{1/2}\sum_{k=1}^{m} f_{ik}(\boldsymbol{Q},\boldsymbol{P})\xi_k(t),\\ &i = 1, 2, \ldots, n.\end{aligned} \tag{2.198}$$

If $\xi_k(t)$ $(k = 1, 2, \ldots, m)$ are stationary wideband noises, the stochastic averaging method described in Sect. 1.1 can be applied to Eq. (2.198). If $\xi_k(t)$ are Gaussian white noises, possible Wong-Zakai correction terms should also be split into conservative and dissipated components and then combined them with $g_i(Q_i)$ and $\sum_{j=1}^{n} c_{ij}(\boldsymbol{Q}, \boldsymbol{P})P_j$, respectively. Equation (2.198) can then be transformed into the following Itô stochastic differential equations

$$
\begin{aligned}
&\mathrm{d}Q_i = P_i \mathrm{d}t, \\
&\mathrm{d}P_i = [-\overline{g}_i(Q_i) - \varepsilon \sum_{j=1}^{n} \overline{c}_{ij}(\boldsymbol{Q}, \boldsymbol{P})P_j]\mathrm{d}t + \varepsilon^{1/2} \sum_{k=1}^{m} \sigma_{ik} \mathrm{d}B_k(t), \\
&\quad i = 1, 2, \ldots, n.
\end{aligned}
\tag{2.199}
$$

where $B_k(t),\ k = 1, 2, \ldots, m$ are independent unit Wiener processes, $\sigma\sigma^{\mathrm{T}} = 2\boldsymbol{fDf}^{\mathrm{T}}$, $\boldsymbol{f} = [f_{ik}]$, $2\boldsymbol{D} = [2D_{kl}]$ is an intensity matrix of Gaussian white noises. The stochastic averaging method described in Sect. 5.2 (vol. 1) can be applied to Eq. (2.199). If $\xi_k(t)$ are combined Gaussian and Poisson white noises or fractional Gaussian noises, Eq. (2.195) can be treated similarly.

Consider the case that the time-delay forces in (2.195) are the following time-delay Bang-Bang control forces

$$
\varepsilon F_i(Q_{i\tau}, P_{i\tau}) = \varepsilon u_i(P_{i\tau}) = \varepsilon b_i \mathrm{sgn}[P_i(t - \tau)], \quad i = 1, 2, \ldots, n,
\tag{2.200}
$$

where sgn(·) is the sign function. Equation (2.200) shows that the amplitude of time-delay force is εb_i with the same direction as $P_{i\tau}$. To find an equivalent non-time-delay Bang-Bang control force $K_i u_i[P_i(t)]$, suppose that a time-delayed force and its equivalent force without time delay dissipate the same energy in a cycle, i.e.,

$$
\int_0^{2\pi/\omega_i} u_i(P_{i\tau})P_i(t)\mathrm{d}t = \int_0^{2\pi/\omega_i} K_i u_i[P_i(t)]P_i(t)\mathrm{d}t.
\tag{2.201}
$$

Note that in Eq. (2.196), both $A_i(t)$ and $\Theta_i(t)$ are slowly-varying processes and can be approximated as constants within $t \in [0, 2\pi/\omega_i]$. For simplification, letting $\Theta_i = 0$,

$$
\begin{aligned}
&\int_0^{2\pi/\omega_i} u_i[P_i(t-\tau)]P_i(t)\mathrm{d}t = \int_0^{2\pi/\omega_i} b_i \mathrm{sgn}[P_i(t-\tau)]P_i(t)\mathrm{d}t \\
&= -\int_0^{\tau} b_i A_i \omega_i \sin\omega_i t \mathrm{d}t + \int_{\tau}^{\pi/\omega_i+\tau} b_i A_i \omega_i \sin\omega_i t \mathrm{d}t
\end{aligned}
$$

$$-\int_{\pi/\omega_i+\tau}^{2\pi/\omega_i} b_i A_i \omega_i \sin \omega_i t \mathrm{d}t = 4 b_i A_i \cos \omega_i \tau. \tag{2.202}$$

$$\begin{aligned}
\int_0^{2\pi/\omega_i} K_i u_i[P_i(t)] P_i(t) \mathrm{d}t &= \int_0^{2\pi/\omega_i} K_i b_i \mathrm{sgn}[P_i(t)] P_i(t) \mathrm{d}t \\
&= \int_0^{\pi/\omega_i} K_i b_i A_i \omega_i \sin \omega_i t \mathrm{d}t - \int_{\pi/\omega_i}^{2\pi/\omega_i} K_i b_i A_i \omega_i \sin \omega_i t \mathrm{d}t \\
&= 4 K_i b_i A_i.
\end{aligned} \tag{2.203}$$

Substituting Eqs. (2.202) and (2.203) into Eq. (2.201) yields

$$K_i = \cos \omega_i \tau. \tag{2.204}$$

Hence,

$$\varepsilon F_i(Q_{i\tau}, P_{i\tau}) = \varepsilon u_i(P_{i\tau}) = \varepsilon b_i \cos \omega_i \tau \mathrm{sgn}[P_i(t)]. \tag{2.205}$$

Substituting Eq. (2.205) into system (2.195) yields an equivalent quasi-integrable Hamiltonian system without time delay

$$\begin{aligned}
\dot{Q}_i &= P_i, \\
\dot{P}_i &= -g_i'(Q_i) - \varepsilon \sum_{j=1}^{n} c_{ij}'(\boldsymbol{Q}, \boldsymbol{P}) P_j - \varepsilon b_i \mathrm{sgn}[P_i(t)] \cos \omega_i \tau \\
&\quad + \varepsilon^{1/2} \sum_{k=1}^{m} f_{ik}(\boldsymbol{Q}, \boldsymbol{P}) \xi_k(t), \quad i = 1, 2, \ldots, n.
\end{aligned} \tag{2.206}$$

The stochastic averaging method of quasi-integrable Hamiltonian systems presented in Sects. 5.2, 6.2, 7.2 in vol. 1 or 1.2 in this volume may be applied to Eq. (2.206) depending on the excited noises.

Example 2.7 Consider a Duffing-van der Pol oscillator with time-delay Bang-Bang control force and excited by Gaussian white noise (Zhu and Liu 2007). The motion equation is

$$\ddot{X} + \omega_0^2 X + \alpha X^3 - \varepsilon(\beta - X^2)\dot{X} = W_g(t) + \varepsilon u_\tau, \tag{2.207}$$

where ε is a small positive parameter, ω_0, α and β are positive constants, $W_g(t)$ is a Gaussian white noise with intensity $2D$, $u_\tau = -b\mathrm{sgn}[\dot{X}(t-\tau)]$ is the time-delay Bang-Bang control force. Letting $X = Q$, $\dot{X} = P$, system (2.207) can be written as

$$\dot{Q} = P,$$
$$\dot{P} = -\omega_0^2 Q - \alpha Q^3 + \varepsilon(\beta - Q^2)P + \varepsilon u_\tau + W_g(t). \tag{2.208}$$

For this example there is no Wong-Zakai correction term. The associated Hamiltonian function is

$$H = \frac{1}{2}P^2 + \frac{1}{2}\omega_0^2 Q^2 + \frac{1}{4}\alpha Q^4. \tag{2.209}$$

Based on Eq. (2.204), the time-delay Bang-Bang control force is equivalent to

$$u_\tau = -b\cos\omega\tau\,\mathrm{sgn}(P). \tag{2.210}$$

The mean frequency is

$$\omega(H) = \frac{\pi\sqrt{\alpha}}{2\sqrt{2}}\frac{\sqrt{A^2+B^2}}{K(r)}, \tag{2.211}$$

where $K(r)$ is the hyperbolic function of the first kind, $r = A\Big/\sqrt{A^2+B^2}$, $A^2 = \frac{\omega_0^2}{\alpha}\left(\sqrt{1+\frac{4\alpha H}{\omega_0^4}} - 1\right)$, and $B^2 = \frac{\omega_0^2}{\alpha}\left(\sqrt{1+\frac{4\alpha H}{\omega_0^4}} + 1\right)$.

Applying Itô differential rule to the function in Eq. (2.209), the following Itô stochastic differential equation can be obtained from Eqs. (2.208), (2.210) and (2.211)

$$\mathrm{d}H = \left\{[\varepsilon(\beta - Q^2)P - \varepsilon b\cos\omega\tau\,\mathrm{sgn}(P)]P + D\right\}\mathrm{d}t + P\mathrm{d}B(t). \tag{2.212}$$

Applying the stochastic averaging method in Sect. 5.1, vol. 1 to Eq. (2.212) yields the averaged FPK equation

$$\frac{\partial p}{\partial t} = -\frac{\partial}{\partial H}[a(H)p] + \frac{1}{2}\frac{\partial^2}{\partial H^2}[b(H)p]. \tag{2.213}$$

where the first and second derivate moments are given by

$$a(H) = \frac{1}{T(H)}\oint_\Omega \left[\varepsilon[(\beta - q^2)p - b\cos\omega\tau]p + D\right]/\,(p)\mathrm{d}q,$$
$$b(H) = \frac{1}{T(H)}\oint_\Omega \left[[2Dp^2]/\,(p)\mathrm{d}q,\right.$$
$$T(H) = \oint_\Omega \left({}^1\!/_p\right)\mathrm{d}q, \quad \Omega = \{q|H = \omega_0^2 q^2/2 + \alpha q^4/4\}. \tag{2.214}$$

where C is a normalization constant. The stationary solution of the FPK Eq. (2.213) is

$$p(H) = C\exp\{-\int_0^H \frac{1}{b(h)}[\frac{\mathrm{d}b(h)}{\mathrm{d}h} - 2a(h)]\mathrm{d}h\}. \tag{2.215}$$

The stationary PDF for system (2.208) is

$$p(q, p) = C'p(H)|_{H=\frac{1}{2}p^2+\frac{1}{2}\omega_0^2q^2+\frac{1}{4}\alpha q^4}. \tag{2.216}$$

where C' is a normalization constant. The marginal stationary PDF and statistics, such as $p(q)$, $E[Q^2]$, can be calculated from $p(q, p)$.

For different delay times, PDF $p(q)$ and the mean-square displacement $E[Q^2]$ obtained by using the stochastic averaging method and from Monte Carlo simulation are shown in Figs. 2.25 and 2.26, respectively. The figures illustrate the significant impact of time delay on control effectiveness. When $\tau = 0$, the Bang-Bang control force has a significant effect on reducing the system response. At $\tau = 1.0$, the control force has a mild effect. When $\tau = 2.0$, the control force has the opposite effect, i.e., the system response is larger than that in the uncontrolled case.

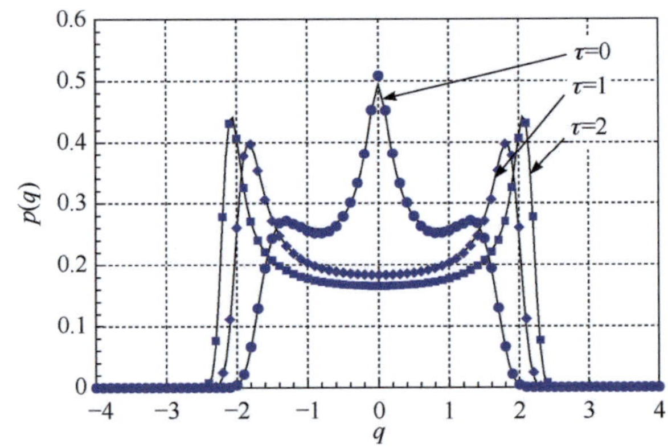

Fig. 2.25 Stationary PDF $p(q)$ of system (2.208). Lines —, results from the stochastic averaging method; symbols ●◆■, results from Monte Carlo simulation. System parameters are $\alpha = 0.5$, $\beta = 1$, $\omega_0 = 1$, $D = 0.001$, $b = 0.6$, $\varepsilon = 0.01$ (Zhu and Liu 2007)

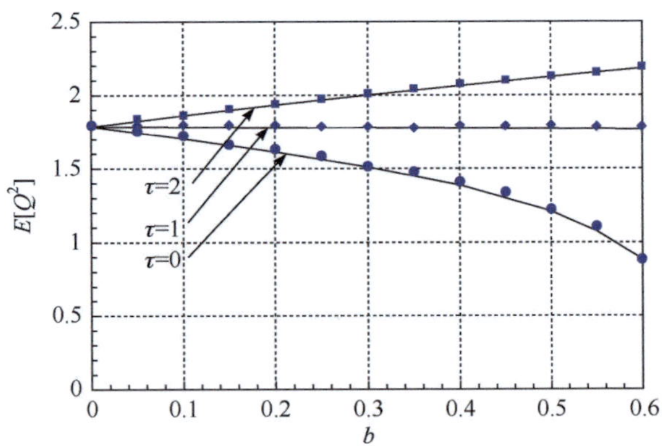

Fig. 2.26 Mean-square $E[Q^2]$ of system (2.208). Lines —, results from the stochastic averaging method; symbols ●◆■, results from Monte Carlo simulation. System parameters are $\alpha = 0.5$, $\beta = 1$, $\omega_0 = 1$, $D = 0.001$, $b = 0.6$, $\varepsilon = 0.01$ (Zhu and Liu 2007)

Example 2.8 Consider two linear oscillators coupled by linear and nonlinear damping forces under time-delay Bang-Bang control and Gaussian white noise excitations (Zhu and Liu 2007). The motion equations of the system are

$$\begin{aligned}
&\ddot{X}_1 + \alpha_{11}\dot{X}_1 + \alpha_{12}\dot{X}_2 + \beta_1\dot{X}_1(X_1^2 + X_2^2) + \omega_1^2 X_1 = u_{1\tau} + W_{g1}(t),\\
&\ddot{X}_2 + \alpha_{21}\dot{X}_1 + \alpha_{22}\dot{X}_2 + \beta_2\dot{X}_2(X_1^2 + X_2^2) + \omega_2^2 X_2 = u_{2\tau} + W_{g2}(t),
\end{aligned} \tag{2.217}$$

where $\alpha_{ij}, \beta_i, \omega_i (i, j = 1, 2)$ are constants, $W_{gi}(t) (i = 1, 2)$ are independent Gaussian white noises with intensities $2D_{ii}$, $u_{i\tau} = -b_i \mathrm{sgn}\big(\dot{X}_i(t - \tau)\big)$ $(i = 1, 2)$ are time-delay Bang-Bang control forces with delay time τ. Assuming that $\alpha_{ij}, \beta_i, D_{ii}$ and $u_{i\tau}$ are small quantities of the same order of ε. Letting $Q_1 = X_1,\ Q_2 = X_2, P_1 = \dot{X}_1,\ P_2 = \dot{X}_2$, the following Itô stochastic differential equation can be obtained from Eq. (2.217)

$$\begin{aligned}
\mathrm{d}Q_1 &= P_1 \mathrm{d}t,\\
\mathrm{d}P_1 &= \{-\omega_1^2 Q_1 - [\alpha_{11} + \beta_1(Q_1^2 + Q_2^2)]P_1 - \alpha_{12}P_2 + u_{1\tau}\}\mathrm{d}t + \sqrt{2D_1}\mathrm{d}B_1(t),\\
\mathrm{d}Q_2 &= P_2 \mathrm{d}t,\\
\mathrm{d}P_2 &= \{-\omega_2^2 Q_2 - [\alpha_{22} + \beta_2(Q_1^2 + Q_2^2)]P_2 - \alpha_{21}P_1 + u_{2\tau}\}\mathrm{d}t + \sqrt{2D_2}\mathrm{d}B_2(t).
\end{aligned} \tag{2.218}$$

The associated Hamiltonian function is

$$\begin{aligned}
&H = \sum_{i=1}^{2} \omega_i I_i,\\
&I_i = \frac{1}{2\omega_i}(P_i^2 + \omega_i^2 Q_i^2), \quad \theta_i = -\tan^{-1}\left(\frac{P_i}{\omega_i Q_i}\right), \quad i = 1, 2.
\end{aligned} \tag{2.219}$$

According to Eq. (2.204), the equivalent control forces are

$$u_{i\tau} = -b_i \cos\omega_i\tau\,\mathrm{sgn}(P_i), \quad i = 1, 2. \tag{2.220}$$

Using Itô differential rule to functions I_i, θ_i in Eq. (2.219), the Itô stochastic differential equations for I_i, θ_i are derived from Eq. (2.218) as follows

$$\begin{aligned}
\mathrm{d}I_i &= \left\{-[(\alpha_{i1}P_1 + \alpha_{i2}P_2 + \beta_i(Q_1^2 + Q_2^2)P_i + b_i\cos\omega_i\tau\,\mathrm{sgn}(P_i)]\frac{P_i}{\omega_i} + \frac{D_{ii}}{\omega_i}\right\}\mathrm{d}t\\
&\quad + \sqrt{2D_{ii}}\frac{P_i}{\omega_i}\mathrm{d}B_i(t),\\
\mathrm{d}\theta_i &= \left\{\omega_i + [(\alpha_{i1}P_1 + \alpha_{i2}P_2 + \beta_i(Q_1^2 + Q_2^2)P_i + b_i\cos\omega_i\tau\,\mathrm{sgn}(P_i)]\frac{\omega_i Q_i}{\omega_i^2 Q_i^2 + P_i^2}\right.\\
&\quad \left. + D_{ii}\frac{2\omega_i Q_i P_i}{(\omega_i^2 Q_i^2 + P_i^2)^2}\right\}\mathrm{d}t - \sqrt{2D_{ii}}\frac{\omega_i Q_i}{\omega_i^2 Q_i^2 + P_i^2}\mathrm{d}B_i(t), \quad i = 1, 2,
\end{aligned} \tag{2.221}$$

where $B_i(t)$ are independent unit Wiener processes. In the following, the stochastic averaging method described in Sect. 5.2, vol. 1 for non-resonant case and internal resonant case is applied to Eq. (2.221), respectively.

(i) **Non-resonant case**

In this case, $r\omega_1 + s\omega_2 \neq 0$, r, s are integers, the averaged FPK equation has the form given in Eq. (5.81) of vol. 1. The first and second derivative moment are

$$
\begin{aligned}
a_1(I_1, I_2) &= -\alpha_{11}I_1 - \frac{\beta_1}{2\omega_1}I_1^2 - \frac{\beta_1}{\omega_2}I_1I_2 - \frac{2b_1\cos\omega_1\tau}{\pi}\sqrt{\frac{2I_1}{\omega_1}} + \frac{D_{11}}{\omega_1},\\
a_2(I_1, I_2) &= -\alpha_{22}I_2 - \frac{\beta_2}{2\omega_2}I_2^2 - \frac{\beta_2}{\omega_1}I_1I_2 - \frac{2b_2\cos\omega_2\tau}{\pi}\sqrt{\frac{2I_2}{\omega_2}} + \frac{D_{22}}{\omega_2},\\
b_{11}(I_1, I_2) &= \frac{2}{\omega_1}D_{11}I_1,\quad b_{22}(I_1, I_2) = \frac{2}{\omega_2}D_{22}I_2,\quad b_{12} = b_{21} = 0.
\end{aligned}
\tag{2.222}
$$

When system parameters satisfy the compatibility condition

$$
\frac{\beta_1\omega_1}{D_{11}\omega_2} = \frac{\beta_2\omega_2}{D_{22}\omega_1} = \gamma. \tag{2.223}
$$

the following exact stationary solution of averaged FPK equation can be obtained

$$
\begin{aligned}
p(I_1, I_2) &= C\exp[-\lambda(I_1, I_2)],\\
\lambda(I_1, I_2) &= \frac{1}{D_{11}}\left(\alpha_{11}\omega_1I_1 + \frac{\beta_1}{4}I_1^2 + \frac{4b_1\cos\omega_1\tau}{\pi}\sqrt{2\omega_1I_1}\right)\\
&\quad + \frac{1}{D_{22}}\left(\alpha_{22}\omega_2I_2 + \frac{\beta_2}{4}I_2^2 + \frac{4b_2\cos\omega_2\tau}{\pi}\sqrt{2\omega_2I_2}\right) + \gamma I_1I_2,
\end{aligned}
\tag{2.224}
$$

where C is a normalization constant. According to Eq. (5.104) in vol. 1, the joint stationary PDF for the generalized displacements and momenta of the system (2.218) is

$$
p(q_1, p_1, q_2, p_2) = C'p(I_1, I_2)|_{I_i=(q_i^2+\omega_i^2p_i^2)/2\omega_i}, \tag{2.225}
$$

where C' is another normalization constant. If system parameters do not satisfy the condition in Eq. (2.223), numerical methods are required to solve the reduced averaged FPK equation.

(ii) **Internal resonant case**

Assuming system is in primary resonance, i.e., $\omega_1 = \omega_2 = \omega$, and letting $\psi = \theta_1 - \theta_2$, the averaged FPK equation is of the form of Eq. (5.122) in vol. 1 with first and second derivative moments given by

$$
\begin{aligned}
a_1 &= -\alpha_{11}I_1 - \alpha_{12}\sqrt{I_1I_2}\cos\psi - \frac{\beta_1}{2\omega}I_1^2 - \frac{\beta_1}{\omega}I_1I_2\left(1-\frac{1}{2}\cos 2\psi\right) \\
&\quad - \frac{2b_1\cos\omega\tau}{\pi}\sqrt{\frac{2I_1}{\omega}} + \frac{D_{11}}{\omega}, \\
a_2 &= -\alpha_{22}I_2 - \alpha_{21}\sqrt{I_1I_2}\cos\psi - \frac{\beta_2}{2\omega}I_2^2 - \frac{\beta_2}{\omega}I_1I_2\left(1-\frac{1}{2}\cos 2\psi\right) \\
&\quad - \frac{2b_2\cos\omega\tau}{\pi}\sqrt{\frac{2I_2}{\omega}} + \frac{D_{22}}{\omega}, \\
a_3 &= \frac{1}{2}\left(\alpha_{12}\sqrt{\frac{I_2}{I_1}} + \alpha_{21}\sqrt{\frac{I_1}{I_2}}\right)\sin\psi - \frac{1}{4\omega}(\beta_1I_2+\beta_2I_1)\sin 2\psi, \\
b_{11} &= \frac{2}{\omega}D_{11}I_1, \quad b_{22} = \frac{2}{\omega}D_{22}I_2, \quad b_{33} = \frac{1}{2\omega}\left(\frac{D_{11}}{I_1}+\frac{D_{22}}{I_2}\right), \\
b_{12} &= b_{21} = b_{13} = b_{31} = b_{23} = b_{32} = 0.
\end{aligned} \tag{2.226}
$$

When system parameters satisfy the following compatibility condition

$$
\beta_1/D_{11} = \beta_2/D_{22} = \gamma_1, \quad \alpha_{12}/D_{11} = \alpha_{21}/D_{22} = \gamma_2, \tag{2.227}
$$

the following exact stationary solution $p(I_1, I_2, \psi)$ for the averaged FPK equation can be obtained

$$
\begin{aligned}
p(I_1, I_2, \psi) &= C\exp[-\lambda(I_1, I_2, \psi)], \\
\lambda(I_1, I_2, \psi) &= \frac{\alpha_{11}\omega}{D_{11}}I_1 + \frac{\alpha_{22}\omega}{D_{22}}I_2 + \frac{\beta_1}{4D_{11}}I_1^2 + \frac{\beta_2}{4D_{22}}I_2^2 \\
&\quad + \frac{4b_1\cos\omega_1\tau}{\pi D_{11}}\sqrt{2\omega_1I_1} + \frac{4b_2\cos\omega_2\tau}{\pi D_{22}}\sqrt{2\omega_2I_2} \\
&\quad + \gamma_1I_1I_2 - \frac{\gamma_1}{2}I_1I_2\cos 2\psi + 2\gamma_2\omega\sqrt{I_1I_2}\cos\psi,
\end{aligned} \tag{2.228}
$$

where C is a normalization constant. According to Eq. (5.127) in vol. 1, the joint stationary PDF for generalized displacements and momenta of system (2.218) is

$$
p(q_1, p_1, q_2, p_2) = C'p(I_1, I_2, \psi)\big|_{I_i=I_i(\boldsymbol{q},\boldsymbol{p}),\ \psi=\psi(\boldsymbol{q},\boldsymbol{p})}, \tag{2.229}
$$

where C' is another normalization constant, $I_i = I_i(\boldsymbol{q},\boldsymbol{p})$ and $\psi = \theta_1(\boldsymbol{q},\boldsymbol{p}) - \theta_2(\boldsymbol{q},\boldsymbol{p})$ are obtained from Eq. (2.219). If system parameters do not satisfy the compatibility condition in Eq. (2.227), numerical methods are needed to solve the FPK equation.

The stationary PDFs $p(q_1, q_2)$ and $p(q_1)$, and the mean-square displacement $E[Q_1^2]$ can then be calculated from $p(q_1, p_1, q_2, p_2)$ as follows

$$p(q_1, q_2) = \int_{-\infty}^{\infty}\int_{-\infty}^{\infty} p(q_1, p_1, q_2, p_2)\mathrm{d}p_1\mathrm{d}p_2,$$

$$p(q_1) = \int_{-\infty}^{\infty}\int_{-\infty}^{\infty}\int_{-\infty}^{\infty} p(q_1, p_1, q_2, p_2)\mathrm{d}p_1\mathrm{d}p_2\mathrm{d}q_2,$$

$$E[Q_1^2] = \int_{-\infty}^{\infty}\int_{-\infty}^{\infty}\int_{-\infty}^{\infty}\int_{-\infty}^{\infty} q_1^2 p(q_1, p_1, q_2, p_2)\mathrm{d}q_1\mathrm{d}q_2\mathrm{d}p_1\mathrm{d}p_2. \tag{2.230}$$

Numerical calculations were carried out for system (2.218) and the results are shown in Figs. 2.27, 2.28 and 2.29 for the non-resonant case and in Figs. 2.30, 2.31 and 2.32 for resonant case. Figure 2.27 shows the stationary joint PDF $p(q_1, q_2)$ of the system without time-delay control force. Figure 2.28 shows the stationary PDFs $p(q_1)$ and $p(q_2)$ of the system with different delay time control forces. Figure 2.29 shows the mean-square displacements $E[Q_1^2]$ and $E[Q_2^2]$ of the system with different delay time control forces. Figure 2.30 shows the stationary joint PDF $p(q_1, q_2)$ of the system without time-delay control forces. Figure 2.31 shows the stationary PDFs $p(q_1)$ and $p(q_2)$ of the system with different delay time control forces. Figure 2.32 shows the mean-square displacements $E[Q_1^2]$ and $E[Q_2^2]$ of the system with different delay time control forces. All the results obtained using the stochastic averaging method are in excellent agreement with those from Monte Carlo simulations. The results indicate that delay time in control forces significantly affects the response. For longer delay time, such as $\tau = 2, 3$, the time-delay control force even increases the system response.

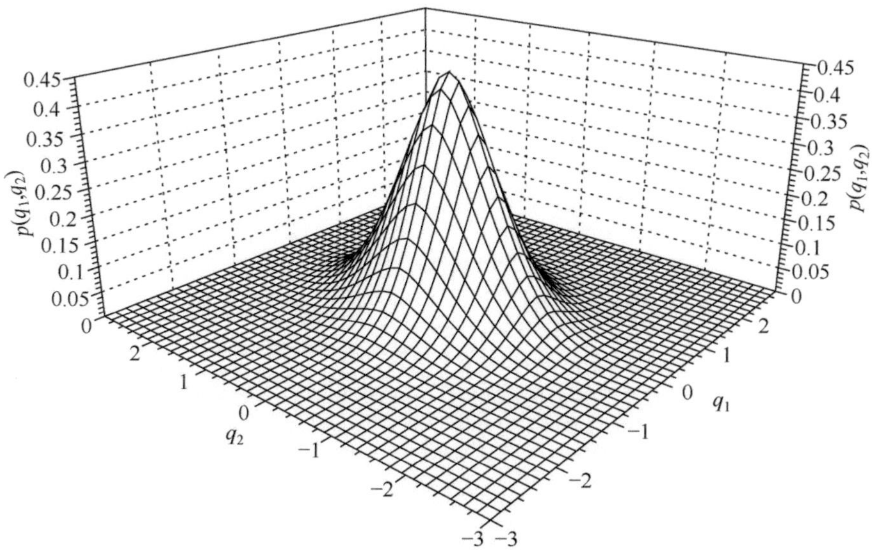

(a)

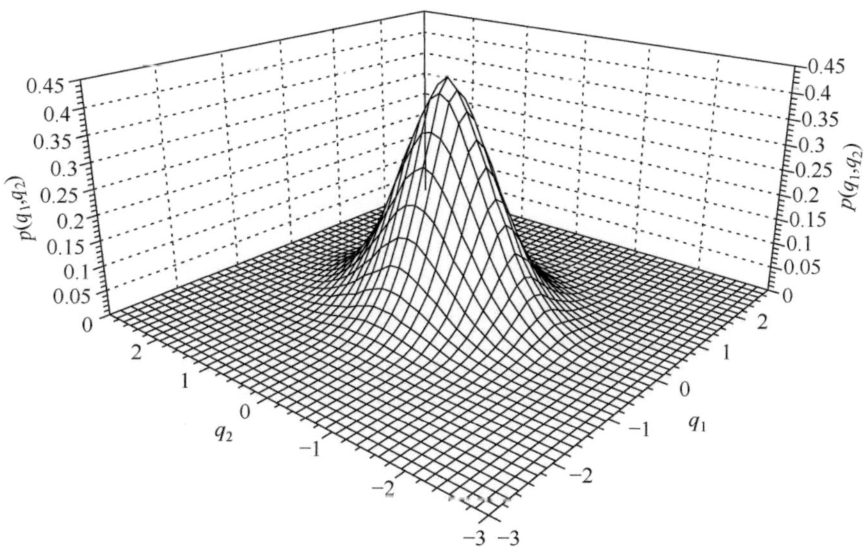

(b)

Fig. 2.27 Stationary joint PDF $p(q_1, q_2)$ of system (2.218) in non-resonant case and without time-delay control. **a** Result from the stochastic averaging method; **b** result from Monte Carlo simulation. System parameters are $\alpha_{11} = 0.01$, $\alpha_{12} = 0.01$, $\beta_1 = 0.01$, $\omega_1 = 1$, $D_{11} = 0.01$, $b_1 = 0.01$, $\alpha_{21} = 0.01$, $\alpha_{22} = 0.01$, $\beta_2 = 0.02$, $\omega_2 = 0.707$, $D_{22} = 0.01$, $b_2 = 0.01$, $\tau = 0$ (Zhu and Liu 2007)

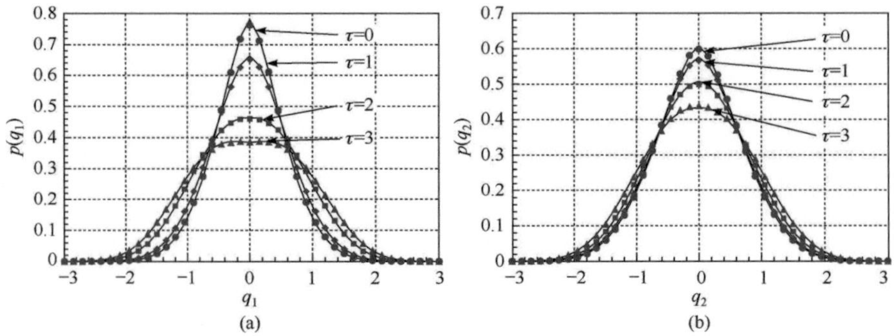

Fig. 2.28 Stationary PDF of system (2.218) in non-resonant case and with different delay time control forces. **a** PDF $p(q_1)$; **b** PDF $p(q_2)$. Lines —, results from the stochastic averaging method; symbols ●◆■▲, results from Monte Carlo simulation. System parameters are $\alpha_{11} = 0.01, \alpha_{12} = 0.01, \beta_1 = 0.01, \omega_1 = 1, D_{11} = 0.01, b_1 = 0.01, \alpha_{21} = 0.01, \alpha_{22} = 0.01, \beta_2 = 0.02, \omega_2 = 0.707, D_{22} = 0.01, b_2 = 0.01$ (Zhu and Liu 2007)

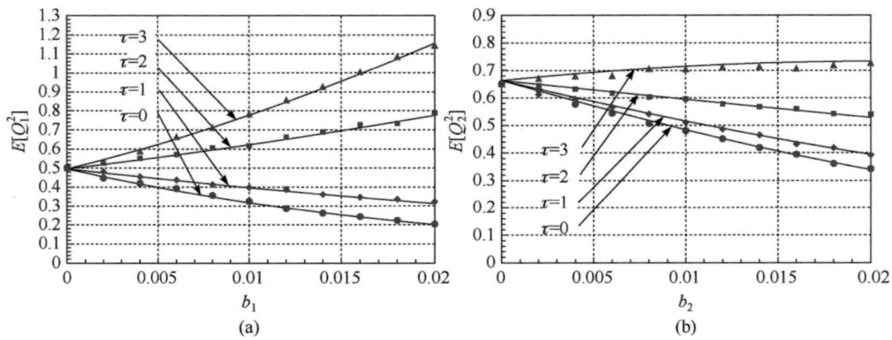

Fig. 2.29 Mean-square displacements of system (2.218) in non-resonant case and with different delay time control forces. **a** Mean-square $E[Q_1^2]$, **b** mean-square $E[Q_2^2]$. Lines —, results from the stochastic averaging method; symbols ●◆■▲, results from Monte Carlo simulation. System parameters are $\alpha_{11} = 0.01, \alpha_{12} = 0.01, \beta_1 = 0.01, \omega_1 = 1, D_{11} = 0.01, b_1 = 0.01, \alpha_{21} = 0.01, \alpha_{22} = 0.01, \beta_2 = 0.02, \omega_2 = 0.707, D_{22} = 0.01, b_2 = 0.01$ (Zhu and Liu 2007)

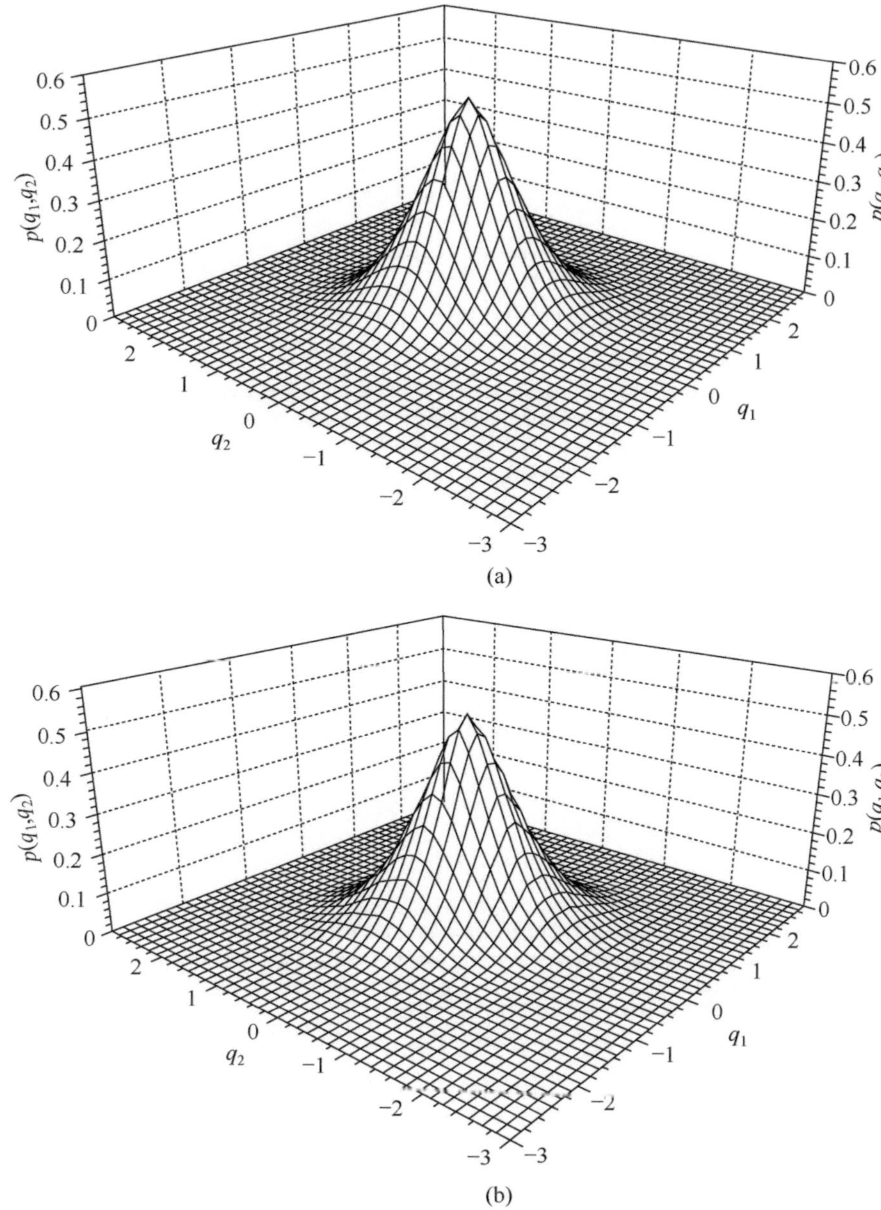

Fig. 2.30 Stationary joint PDF $p(q_1, q_2)$ of system (2.218) in resonant case and without time-delay control. **a** Result from the stochastic averaging method; **b** result from Monte Carlo simulation. System parameters are $\alpha_{11} = 0.01$, $\alpha_{12} = 0.01$, $\beta_1 = 0.01$, $\omega_1 = 1$, $D_{11} = 0.01$, $b_1 = 0.01$, $\alpha_{21} = 0.01$, $\alpha_{22} = 0.01$, $\beta_2 = 0.01$, $\omega_2 = 1$, $D_{22} = 0.01$, $b_2 = 0.01$, $\tau = 0$ (Zhu and Liu 2007)

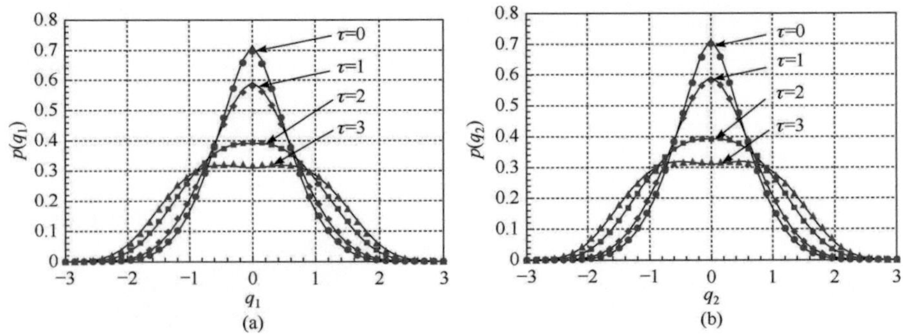

Fig. 2.31 Stationary PDF of system (2.218) in resonant case and with different delay time control forces. **a** PDF $p(q_1)$; **b** PDF $p(q_2)$. Lines —, results from the stochastic averaging method; symbols ●◆■▲, results from Monte Carlo simulation. System parameters are $\alpha_{11} = 0.01$, $\alpha_{12} = 0.01$, $\beta_1 = 0.01$, $\omega_1 = 1$, $D_{11} = 0.01$, $b_1 = 0.01$, $\alpha_{21} = 0.01$, $\alpha_{22} = 0.01$, $\beta_2 = 0.01$, $\omega_2 = 1$, $D_{22} = 0.01$, $b_2 = 0.01$ (Zhu and Liu 2007)

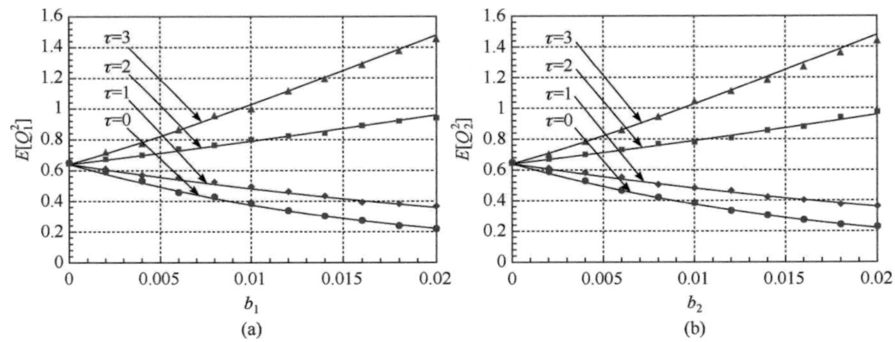

Fig. 2.32 Mean-square displacements of system (2.218) in resonant case and with different delay time control forces. **a** Mean-square $E[Q_1^2]$; **b** mean-value $E[Q_2^2]$. Lines —, results from the stochastic averaging method; symbols ●◆■▲, results from Monte Carlo simulation. System parameters are $\alpha_{11} = 0.01$, $\alpha_{12} = 0.01$, $\beta_1 = 0.01$, $\omega_1 = 1$, $D_{11} = 0.01$, $b_1 = 0.01$, $\alpha_{21} = 0.01$, $\alpha_{22} = 0.01$, $\beta_2 = 0.01$, $\omega_2 = 1$, $D_{22} = 0.01$, $b_2 = 0.01$ (Zhu and Liu 2007)

References

Chen L C, Li Z S, Zhuang Q Q, Zhu W Q. 2013a. First-passage failure of single-degree-of-freedom nonlinear oscillator with fractional derivative. Journal of Vibration and Control, 19(14): 2154–2163.

Chen L C, Wang W H, Li Z S, Zhu W Q. 2013b. Stationary response of Duffing oscillator with hardening stiffness and fractional derivative damping. International Journal of Non-Linear Mechanics, 48: 44–50.

Chen L C, Li H F, Mei Z, Zhu W Q. 2014. Reliability of van der Pol-Duffing oscillator with fractional derivative damping under wide-band noise excitations. Journal of Southwest Jiaotong University, 49(1): 45–51. (in Chinese)

Christensen R M. 1982. Theory of Viscoelasticity, An Introduction. New York: Academic Press.

Deng M L, Zhu W Q. 2004. Stationary motion of active Brownian particles. Physical Review E, 69(4): 046105.

Di Paola M, Pirrotta A. 2001. Time delay induced effects on control of linear systems under random excitation. Probabilistic Engineering Mechanics, 16: 43–51.

Drozdov A D. 1998. Viscoelastic Structures: Mechanics of Growth and Aging. San Diego, CA: Academic Press.

Grigoriu M. 1997. Control of time delay linear system with Gaussian white noise. Probabilistic Engineering Mechanics, 12(2): 89–96.

Huang Z L, Jin X L. 2009. Response and stability of SDOF strongly nonlinear stochastic system with light damping modeled by a fractional derivative. Journal of Sound and Vibration, 319: 1121–1135.

Khasminskii R Z. 1966. A limit theorem for the solution of differential equations with random right hand sides. Theory of Probability and Application, 11(3): 390–405.

Khasminskii R Z. 1968. On the averaging principle for Itô stochastic differential equations. Kibernetika, 3(4): 260–279. (in Russian)

Liu Z H. 2007. Study on the stochastic dynamics and compensation of time-delayed feedback controlled quasi-integrable Hamiltonian systems. Post-doctoral Work Report. Hangzhou: Zhejiang University. (in Chinese)

Spanos P D, Cacciola P, Muscolino G. 2004. Stochastic averaging of Preisach hysteretic systems. ASCE Journal of Engineering Mechanics, 130(11): 1257–1267.

Wang Y. 2008. Robustness of nonlinear stochastic optimal control and robust control for uncertain quasi Hamiltonian systems. Ph.D. Dissertation. Hangzhou: Zhejiang University. (in Chinese)

Wang Y, Ying Z G, Zhu W Q. 2009. Stochastic averaging of energy envelope of Preisach hysteretic systems. Journal of Sound and Vibration, 321: 976–993.

Zhang C Y. 1994. Viscoclastic Fracture Mechanics. Wuhan: Huazhong University of Technology Press. (in Chinese)

Zhu W Q, Huang Z L, Yang Y Q. 1997. Stochastic averaging of quasi integrable-Hamiltonian systems. ASME Journal of Applied Mechanics, 64(4): 975–984.

Zhu W Q, Lin Y K. 1991. Stochastic averaging of energy envelope. ASCE Journal of Engineering Mechanics, 117(8): 1890–1905.

Zhu W Q, Liu Z H. 2007. Response of quasi-integrable Hamiltonian systems with delayed feedback bang-bang control. Nonlinear Dynamics, 49:31–47.

Chapter 3
Stochastic Averaging Methods of Quasi-generalized Hamiltonian Systems Excited by Gaussian White Noises

Generalized Hamiltonian system is an extension of classical Hamiltonian system, The Hamiltonian system is even-dimensional system while the generalized Hamiltonian system may be even, odd or infinite-dimensional system. The main difference of generalized Hamiltonian system from Hamiltonian system is the presence of Casimir functions (Li et al. 1999). The generalized Poisson bracket of Casimir function with any continuous differentiable function is zero, as described in Sect. 3.3, vol. 1. In quasi-generalized Hamiltonian systems, Casimir functions are slowly-varying processes, and the averaged Itô stochastic differential equations will contain equations for Casimir functions, which is the main difference from the stochastic averaging methods of quasi-Hamiltonian systems. In this chapter, the stochastic averaging method of quasi-Hamiltonian systems excited by Gaussian white noises, as described in Chap. 5, is extended to quasi-generalized Hamiltonian systems (Huang 2005; Huang and Zhu 2009). These stochastic averaging methods have been used to study the reliability and first-passage problems of quasi-generalized Hamiltonian systems (Chen and Zhu 2010a, b, c).

Consider the following quasi-generalized Hamiltonian system (Huang and Zhu 2009),

$$\begin{aligned} \dot{X}_i &= [X_i, H'] + \varepsilon \sum_{j=1}^{m} d'_{ij}(\boldsymbol{X}) \frac{\partial H'}{\partial X_j} + \varepsilon^{1/2} \sum_{s=1}^{l} f_{is}(\boldsymbol{X}) W_{gs}(t), \\ i &= 1, \ldots, m, \end{aligned} \tag{3.1}$$

where $\boldsymbol{X} = [X_1, \ldots, X_m]^{\mathrm{T}}$, $H' = H'(\boldsymbol{X})$ represents the generalized Hamiltonian function; $[\cdot, \cdot]$ is the generalized Poisson bracket; the second term on the right-hand side of Eq. (3.1) is small perturbation; $\varepsilon^{1/2} f_{is}(\boldsymbol{X})$ are the amplitudes of weak random excitations; $W_{gs}(t) \quad (s = 1, 2, \ldots, l)$ are Gaussian white noises with correlation functions given by

W. Zhu et al., *Stochastic Averaging*, https://doi.org/10.1007/978-981-96-3833-8_3

$$E[W_{gs}(t)W_{gr}(t+\tau)] = 2\pi K_{sr}\delta(\tau), \quad s, r = 1, 2, \ldots, l. \tag{3.2}$$

The quasi-generalized Hamiltonian system (3.1) can be modeled as the following Stratonovich stochastic differential equation

$$\begin{aligned} \mathrm{d}X_i &= \left\{ [X_i, H'] + \varepsilon \sum_{j=1}^{m} d'_{ij} \frac{\partial H'}{\partial X_j} \right\} \mathrm{d}t + \varepsilon^{1/2} \sum_{s=1}^{l} \sigma_{is} \mathrm{d}^{\circ} B_s(t), \\ \sigma_{is} &= (\boldsymbol{fL})_{is}, \quad \boldsymbol{LL}^T = 2\pi \boldsymbol{K}, \quad i = 1, 2, \ldots, m. \end{aligned} \tag{3.3}$$

where $B_s(t)$ are independent unit Wiener processes, and $\boldsymbol{f} = [f_{is}]_{m\times l}$ and $2\pi\boldsymbol{K} = 2\pi[K_{sr}]_{l\times l}$ are matrices of excitation amplitudes and intensities of Gaussian white noises, respectively. Equation (3.3) can be further transformed into the following equivalent Itô stochastic differential equations

$$\begin{aligned} \mathrm{d}X_i = &\left\{ [X_i, H'] + \varepsilon \sum_{j=1}^{m} d'_{ij} \frac{\partial H'}{\partial X_j} + \frac{\varepsilon}{2} \sum_{j=1}^{m} \sum_{s=1}^{l} \sigma_{js} \frac{\partial \sigma_{is}}{\partial X_j} \right\} \mathrm{d}t \\ &+ \varepsilon^{1/2} \sum_{s=1}^{l} \sigma_{is} \mathrm{d}B_s(t), \quad i = 1, \ldots, m. \end{aligned} \tag{3.4}$$

where $\frac{\varepsilon}{2} \sum_{j=1}^{m} \sum_{s=1}^{l} \sigma_{js} \frac{\partial \sigma_{is}}{\partial X_j}$ is the Wong-Zakai correction term, which can be decomposed into two parts. One part can be combined with the original generalized Hamiltonian function to form a new generalized Hamiltonian function $H = H(\boldsymbol{X})$. The other part can be combined with $\varepsilon d'_{ij} \partial H' / \partial X_j$ to form $\varepsilon d_{ij}(\boldsymbol{X}) \partial H / \partial X_j$. After completing these steps, the Itô stochastic differential equation (3.4) can be rewritten as

$$\begin{aligned} \mathrm{d}X_i &= \left\{ [X_i, H] + \varepsilon \sum_{j=1}^{m} d_{ij} \frac{\partial H}{\partial X_j} \right\} \mathrm{d}t + \varepsilon^{1/2} \sum_{s=1}^{l} \sigma_{is} \mathrm{d}B_s(t), \\ i &= 1, \ldots, m. \end{aligned} \tag{3.5}$$

It is known from Sect. 3.3, in vol. 1 that, the generalized Hamiltonian system associated with Eq. (3.5) belongs to one of the following five classes: nonintegrable, completely integrable and non-resonant, completely integrable and resonant, partially integrable and non-resonant, and partially integrable and resonant. In the following sections, stochastic averaging method will be presented for each of these five cases of the quasi-generalized Hamiltonian system (3.5).

3.1 Quasi-nonintegrable Generalized Hamiltonian Systems

As described in Sect. 3.3, vol. 1, the nonintegrable generalized Hamiltonian system associated with Eq. (3.5) has $M+1$ $(M+1<m)$ first integrals including a generalized Hamiltonian function H and M Casimir functions $C_1, \ldots, C_M$. By using the Itô differential rule and the property of the generalized Poisson bracket, the following Itô stochastic differential equations governing $H(t), C_1(t), \ldots, C_M(t)$ can be derived from Eq. (3.5)

$$
\begin{aligned}
\mathrm{d}H &= \varepsilon\left(\sum_{i,j=1}^{m} d_{ij}\frac{\partial H}{\partial X_i}\frac{\partial H}{\partial X_j} + \frac{1}{2}\sum_{i,j=1}^{m}\sum_{s=1}^{l}\sigma_{is}\sigma_{js}\frac{\partial^2 H}{\partial X_i \partial X_j}\right)\mathrm{d}t \\
&\quad + \varepsilon^{1/2}\sum_{i=1}^{m}\sum_{s=1}^{l}\sigma_{is}\frac{\partial H}{\partial X_i}\mathrm{d}B_s(t), \\
\mathrm{d}C_v &= \varepsilon\left(\sum_{i,j=1}^{m} d_{ij}\frac{\partial C_v}{\partial X_i}\frac{\partial H}{\partial X_j} + \frac{1}{2}\sum_{i,j=1}^{m}\sum_{s=1}^{l}\sigma_{is}\sigma_{js}\frac{\partial^2 C_v}{\partial X_i \partial X_j}\right)\mathrm{d}t \\
&\quad + \varepsilon^{1/2}\sum_{i=1}^{m}\sum_{s=1}^{l}\sigma_{is}\frac{\partial C_v}{\partial X_i}\mathrm{d}B_s(t), \\
v &= 1, 2, \ldots, M.
\end{aligned} \tag{3.6}
$$

Replace $M+1$ variables $X_1, X_2, \ldots, X_{M+1}$ by $H, C_1, \ldots, C_M$. The remaining $m-M-1$ variables form a vector $\boldsymbol{X}' = [X_{M+2}, X_{M+3}, \ldots, X_m]^{\mathrm{T}}$. The new system is constituted by the $m-M-1$ equations from Eq. (3.5) and the $M+1$ equations in Eq. (3.6). In the new system, $\boldsymbol{X}'$ is fast-varying vector process while $[H, C_1, \ldots, C_M]^{\mathrm{T}}$ is slowly-varying vector process. According to Khasminskii's theorem (Khasminskii 1968), when $\varepsilon \to 0$, $[H, C_1, \ldots, C_M]^{\mathrm{T}}$ converges weakly to a $(M+1)$-dimensional vector Markov diffusion process. The averaged Itô stochastic differential equations govering this vector process is of the form

$$
\begin{aligned}
\mathrm{d}H &= m_H(H, \boldsymbol{C})\mathrm{d}t + \sum_{s=1}^{l}\sigma_{Hs}(H, \boldsymbol{C})\mathrm{d}B_s(t), \\
\mathrm{d}C_v &= m_v(H, \boldsymbol{C})\mathrm{d}t + \sum_{s=1}^{l}\sigma_{vs}(H, \boldsymbol{C})\mathrm{d}B_s(t), \\
v &= 1, 2, \ldots, M,
\end{aligned} \tag{3.7}
$$

where the drift coefficient and diffusion coefficient can be obtained according to Eqs. (4.29) and (4.30) in vol. 1. Since the nonintegrable generalized Hamiltonian system is ergodic on $(m-M-1)$-dimensional submanifold with H, C_v being constants. The time averaging can be replaced by spatial averaging, which results in

$$
\begin{aligned}
&m_H(H,\boldsymbol{C})=\frac{\varepsilon}{T}\int_{\Omega}\frac{1}{A}\left\{\sum_{i,j=1}^{m}\left(d_{ij}\frac{\partial H}{\partial x_i}\frac{\partial H}{\partial x_j}\right)+\frac{1}{2}\sum_{i,j=1}^{m}\sum_{s=1}^{l}\left(\sigma_{is}\sigma_{js}\frac{\partial^2 H}{\partial x_i\partial x_j}\right)\right\})\mathrm{d}\boldsymbol{x}',\\
&m_v(H,\boldsymbol{C})=\frac{\varepsilon}{T}\int_{\Omega}\frac{1}{A}\{\sum_{i,j=1}^{m}\left(d_{ij}\frac{\partial C_v}{\partial x_i}\frac{\partial H}{\partial x_j}\right)+\frac{1}{2}\sum_{i,j=1}^{m}\sum_{s=1}^{l}\left(\sigma_{is}\sigma_{js}\frac{\partial^2 C_v}{\partial x_i\partial x_j}\right)\})\mathrm{d}\boldsymbol{x}',\\
&\sum_{s=1}^{l}\sigma_{Hs}\sigma_{Hs}(H,\boldsymbol{C})=\frac{\varepsilon}{T}\int_{\Omega}\frac{1}{A}\sum_{i,j=1}^{m}\sum_{s=1}^{l}\sigma_{is}\sigma_{js}\frac{\partial H}{\partial x_i}\frac{\partial H}{\partial x_j}\mathrm{d}\boldsymbol{x}',\\
&\sum_{s=1}^{l}\sigma_{v_1s}\sigma_{v_2s}(H,\boldsymbol{C})=\frac{\varepsilon}{T}\int_{\Omega}\frac{1}{A}\sum_{i,j=1}^{m}\sum_{s=1}^{l}\sigma_{is}\sigma_{js}\frac{\partial C_{v_1}}{\partial x_i}\frac{\partial C_{v_2}}{\partial x_j}\mathrm{d}\boldsymbol{x}',\\
&\sum_{s=1}^{l}\sigma_{Hs}\sigma_{vs}(H,\boldsymbol{C})=\frac{\varepsilon}{T}\int_{\Omega}\frac{1}{A}\sum_{i,j=1}^{m}\sum_{s=1}^{l}\sigma_{is}\sigma_{js}\frac{\partial H}{\partial x_i}\frac{\partial C_v}{\partial x_j}\mathrm{d}\boldsymbol{x}',\\
&T=T(H,\boldsymbol{C})=\int_{\Omega}(1/A)\mathrm{d}\boldsymbol{x}',\\
&\Omega=\{\boldsymbol{x}'\big|H(\boldsymbol{x}')\le H;\ C_v(\boldsymbol{x}')\le C_v\},\\
&v,v_1,v_2=1,2,\ldots,M,
\end{aligned}
\tag{3.8}
$$

where $\boldsymbol{x}'=[x_{r_1},\ldots,x_{r_{m-M-1}}]^{\mathrm{T}}$. $A=\left|\partial(H,\boldsymbol{x}',\boldsymbol{c})/\partial\boldsymbol{x}\right|$ is the determinant of the Jacobian matrix for the transformation from $\boldsymbol{x}$ to $H,\boldsymbol{x}',\boldsymbol{c}$. The averaged FPK equation associated with Itô Eq. (3.7) is

$$
\begin{aligned}
\frac{\partial p}{\partial t}=&-\frac{\partial}{\partial h}(a_h p)-\sum_{v=1}^{M}\frac{\partial}{\partial c_v}(a_v p)+\frac{1}{2}\frac{\partial^2}{\partial h^2}(b_{hh}p)\\
&+\frac{1}{2}\sum_{v_1,v_2=1}^{M}\frac{\partial^2}{\partial c_{v_1}\partial c_{v_2}}(b_{v_1v_2}p)+\sum_{v=1}^{M}\frac{\partial^2}{\partial h\partial c_v}(b_{hv}p),
\end{aligned}
\tag{3.9}
$$

where

$$
\begin{aligned}
&a_h=a_h(h,\boldsymbol{c})=m_H(H,\boldsymbol{C})|_{\substack{H=h\\ \boldsymbol{C}=\boldsymbol{c}}},\\
&a_v=a_v(h,\boldsymbol{c})=m_v(H,\boldsymbol{C})|_{\substack{H=h\\ \boldsymbol{C}=\boldsymbol{c}}},\\
&b_{hh}=b_{hh}(h,\boldsymbol{c})=\left.\sum_{s=1}^{l}\sigma_{Hs}\sigma_{Hs}(H,\boldsymbol{C})\right|_{\substack{H=h\\ \boldsymbol{C}=\boldsymbol{c}}},\\
&b_{v_1v_2}=b_{v_1v_2}(h,\boldsymbol{c})=\left.\sum_{s=1}^{l}\sigma_{v_1s}\sigma_{v_2s}(H,\boldsymbol{C})\right|_{\substack{H=h\\ \boldsymbol{C}=\boldsymbol{c}}},
\end{aligned}
$$

$$b_{hv} = b_{hv}(h, \boldsymbol{c}) = \sum_{s=1}^{l} \sigma_{Hs}\sigma_{vs}(H, \boldsymbol{C})\Bigg|_{\substack{H=h \\ \boldsymbol{C}=\boldsymbol{c}}}. \tag{3.10}$$

In Eq. (3.9), $p = p(h, \boldsymbol{c}, t|h_0, \boldsymbol{c}_0)$ is the transition PDF of the vector Markov diffusion process $[H, C_1, \ldots, C_M]^{\mathrm{T}}$ with initial condition $p(h, \boldsymbol{c}, 0|h_0, \boldsymbol{c}_0) = \delta(h - h_0)\delta(\boldsymbol{c} - \boldsymbol{c}_0)$, or $p = p(h, \boldsymbol{c}, t)$ is the joint PDF for $H, \boldsymbol{C}$ with initial condition $p(h, \boldsymbol{c}, 0) = p(h_0, \boldsymbol{c}_0)$. Boundary conditions depend on the properties and constraint of the generalized Hamiltonian system associated with Eq. (3.7).

After the stationary PDF $p(h, \boldsymbol{c})$ of the FPK Eq. (3.9) is obtained, then the approximate stationary PDF of the original system (3.5) is given by

$$p(\boldsymbol{x}) = p(h, \boldsymbol{c}, \boldsymbol{x}')\left|\frac{\partial(h, \boldsymbol{x}', \boldsymbol{c})}{\partial \boldsymbol{x}}\right| = p(\boldsymbol{x}'|h, \boldsymbol{c})p(h, \boldsymbol{c})\left|\frac{\partial(h, \boldsymbol{x}', \boldsymbol{c})}{\partial \boldsymbol{x}}\right|. \tag{3.11}$$

For fixed values of $h, \boldsymbol{c}$, the conditional PDF $p(\boldsymbol{x}'|h, \boldsymbol{c})$ is inversely proportional to $|\partial(h, \boldsymbol{x}', \boldsymbol{c})/\partial \boldsymbol{x}|$, i.e.,

$$p(\boldsymbol{x}'|h, \boldsymbol{c}) = C'/A. \tag{3.12}$$

where C' is a constant. Normalize Eq. (3.12) with respect to $\boldsymbol{x}'$ yields $C' = T^{-1}$. Substituting it back into Eq. (3.12) and then substituting Eq. (3.12) into Eq. (3.11) result in the approximate stationary PDF of the original system (3.5)

$$p(\boldsymbol{x}) = \frac{p(h, \boldsymbol{c})}{T(h, \boldsymbol{c})}\bigg|_{h=h(\boldsymbol{x}),\ \boldsymbol{c}=\boldsymbol{c}(\boldsymbol{x})}. \tag{3.13}$$

where $T(h, \boldsymbol{c}) = T(H, \boldsymbol{C})|_{H=h,\ \boldsymbol{C}=\boldsymbol{c}}$, determined by Eq. (3.8).

Consider a special case that the variables associated with Eq. (3.5) are separable, i.e.,

$$\begin{aligned}
&\boldsymbol{x} = [\boldsymbol{x}_1^{\mathrm{T}}, \boldsymbol{x}_2^{\mathrm{T}}]^{\mathrm{T}}, \quad \boldsymbol{x}_1 = [x_1, x_2, \ldots, x_{2n}]^{\mathrm{T}} = [\boldsymbol{q}^{\mathrm{T}}, \boldsymbol{p}^{\mathrm{T}}]^{\mathrm{T}}, \\
&\boldsymbol{x}_2 = [x_{2n+1}, x_{2n+2}, \ldots, x_m]^{\mathrm{T}}, \\
&H = \overline{H}(\boldsymbol{x}_1), \quad C_1 = C_1(\boldsymbol{x}_2), \\
&C_2 = C_2(\boldsymbol{x}_2), \ldots, C_M = C_M(\boldsymbol{x}_2),
\end{aligned} \tag{3.14}$$

where $2n = m - M$. $\boldsymbol{q} = [q_1, q_2, \ldots, q_n]^{\mathrm{T}}$ and $\boldsymbol{p} = [p_1, p_2, \ldots, p_n]^{\mathrm{T}}$ are vectors of the generalized displacements and generalized momenta, respectively, of the nonintegrable Hamiltonian subsystem with Hamiltonian function $\overline{H}(\boldsymbol{q}, \boldsymbol{p}) = \overline{H}(\boldsymbol{x}_1)$. In this case, the governing Itô stochastic differential equations for $\overline{H}(t), C_1(t), \ldots, C_M(t)$, instead of Eq. (3.6), are given by

$$
\begin{aligned}
\mathrm{d}\overline{H} &= \varepsilon\left\{\sum_{j=1}^{n}\frac{\partial\overline{H}}{\partial P_j}\left[\sum_{k=1}^{n}d_{jk}\frac{\partial\overline{H}}{\partial P_k}+\sum_{i=2n+1}^{m}d_{ji}\frac{\partial}{\partial X_i}\left(\sum_{\beta=1}^{M}C_\beta\right)\right]\right.\\
&\quad\left.+\frac{1}{2}\sum_{j,k=1}^{n}\sum_{s=1}^{l}\sigma_{js}\sigma_{ks}\frac{\partial^2\overline{H}}{\partial P_j\partial P_k}\right\}\mathrm{d}t+\varepsilon^{1/2}\sum_{j=1}^{n}\sum_{s=1}^{l}\sigma_{js}\frac{\partial\overline{H}}{\partial P_j}\mathrm{d}B_s(t),\\
\mathrm{d}C_v &= \varepsilon\left\{\sum_{i_1=2n+1}^{m}\frac{\partial C_v}{\partial X_{i_1}}\left[\sum_{k=1}^{n}d_{i_1k}\frac{\partial\overline{H}}{\partial P_k}+\sum_{i=2n+1}^{m}d_{i_1i}\frac{\partial}{\partial X_i}\left(\sum_{\beta=1}^{M}C_\beta\right)\right]\right.\\
&\quad\left.+\frac{1}{2}\sum_{i,i_1=2n+1}^{m}\sum_{s=1}^{l}\sigma_{is}\sigma_{i_1s}\frac{\partial^2C_v}{\partial X_i\partial X_{i_1}}\right\}\mathrm{d}t+\varepsilon^{1/2}\sum_{i=2n+1}^{m}\sum_{s=1}^{l}\sigma_{is}\frac{\partial C_v}{\partial X_i}\mathrm{d}B_s(t),\\
v &= 1,2,\ldots,M.
\end{aligned}
\tag{3.15}
$$

Replacing Eq. (3.7), the averaged Itô stochastic differential equations for $\overline{H},\boldsymbol{C}$ are

$$
\begin{aligned}
\mathrm{d}\overline{H} &= \overline{m}_{\overline{H}}(\overline{H},\boldsymbol{C})\mathrm{d}t+\sum_{s=1}^{l}\overline{\sigma}_{\overline{H}s}(\overline{H},\boldsymbol{C})\mathrm{d}B_s(t),\\
\mathrm{d}C_v &= \overline{m}_v(\overline{H},\boldsymbol{C})\mathrm{d}t+\sum_{s=1}^{l}\overline{\sigma}_{vs}(\overline{H},\boldsymbol{C})\mathrm{d}B_s(t),\\
v &= 1,2,\ldots,M.
\end{aligned}
\tag{3.16}
$$

Instead of Eq. (3.8), the drift and diffusion coefficients are

$$
\begin{aligned}
\overline{m}_{\overline{H}}(\overline{H},\boldsymbol{C}) &= \frac{\varepsilon}{T'}\int_{\Omega'}\frac{1}{A'}\{\sum_{j=1}^{n}\frac{\partial\overline{H}}{\partial p_j}[\sum_{k=1}^{n}d_{jk}\frac{\partial\overline{H}}{\partial p_k}+\sum_{i=2n+1}^{m}d_{ji}\frac{\partial}{\partial x_i}\left(\sum_{\beta=1}^{M}C_\beta\right)]\\
&\quad+\frac{1}{2}\sum_{j,k=1}^{n}\sum_{s=1}^{l}\sigma_{js}\sigma_{ks}\frac{\partial^2\overline{H}}{\partial p_j\partial p_k}\}\mathrm{d}q_1\cdots\mathrm{d}q_n\mathrm{d}p_2\cdots\mathrm{d}p_n,\\
\overline{m}_v(\overline{H},\boldsymbol{C}) &= \frac{\varepsilon}{T'}\int_{\Omega'}\frac{1}{A'}\{\sum_{i_1=2n+1}^{m}\frac{\partial C_v}{\partial x_{i_1}}[\sum_{k=1}^{n}d_{i_1k}\frac{\partial\overline{H}}{\partial p_k}+\sum_{i=2n+1}^{m}d_{i_1i}\frac{\partial}{\partial x_i}(\sum_{\beta=1}^{M}C_\beta)]\\
&\quad+\frac{1}{2}\sum_{i,i_1=2n+1}^{m}\sum_{s=1}^{l}\sigma_{is}\sigma_{i_1s}\frac{\partial^2C_v}{\partial x_i\partial x_{i_1}}\}\mathrm{d}q_1\cdots\mathrm{d}q_n\mathrm{d}p_2\cdots\mathrm{d}p_n,
\end{aligned}
$$

$$
\sum_{s=1}^{l}\overline{\sigma}_{\overline{H}s}\overline{\sigma}_{\overline{H}s}(\overline{H},\boldsymbol{C})
$$

$$= \frac{\varepsilon}{T'}\int_{\Omega'}\frac{1}{A'}\sum_{j,k=1}^{n}\sum_{s=1}^{l}\overline{\sigma}_{js}\overline{\sigma}_{ks}\frac{\partial\overline{H}}{\partial p_j}\frac{\partial\overline{H}}{\partial p_k}\mathrm{d}q_1\cdots\mathrm{d}q_n\mathrm{d}p_2\cdots\mathrm{d}p_n,$$

$$\sum_{s=1}^{l}\overline{\sigma}_{\nu_1 s}\overline{\sigma}_{\nu_2 s}(\overline{H},\boldsymbol{C})$$

$$= \frac{\varepsilon}{T'}\int_{\Omega'}\frac{1}{A'}\sum_{i,i_1=2n+1}^{m}\sum_{s=1}^{l}\overline{\sigma}_{is}\overline{\sigma}_{i_1 s}\frac{\partial C_\nu}{\partial x_i}\frac{\partial C_\nu}{\partial x_{i_1}}\mathrm{d}q_1\cdots\mathrm{d}q_n\mathrm{d}p_2\cdots\mathrm{d}p_n,$$

$$\sum_{s=1}^{l}\overline{\sigma}_{\overline{H}s}\overline{\sigma}_{\nu_1 s}(\overline{H},\boldsymbol{C})$$

$$= \frac{\varepsilon}{T'}\int_{\Omega'}\frac{1}{A'}\sum_{j=1}^{n}\sum_{i=2n+1}^{m}\sum_{s=1}^{l}\overline{\sigma}_{js}\overline{\sigma}_{is}\frac{\partial\overline{H}}{\partial p_j}\frac{\partial C_\nu}{\partial x_i}\mathrm{d}q_1\cdots\mathrm{d}q_n\mathrm{d}p_2\cdots\mathrm{d}p_n, \tag{3.17}$$

$$A' = \left|\frac{\partial(\boldsymbol{q},\overline{H},p_2,\ldots,p_n,\boldsymbol{C})}{\partial(\boldsymbol{q},\boldsymbol{p},\boldsymbol{x}_2)}\right| = \left|\frac{\partial\overline{H}}{\partial p_1}\right|\left|\frac{\partial\boldsymbol{C}}{\partial\boldsymbol{x}_2}\right|,$$

$$\Omega' = \{(q_1,\ldots,q_n,p_2,\ldots,p_n)|\overline{H}(q_1,\ldots,q_n,p_2,\ldots,p_n)\le\overline{H}\},$$

$$T' = \int_{\Omega'}\frac{1}{A'}\mathrm{d}q_1\cdots\mathrm{d}q_n\mathrm{d}p_2\cdots\mathrm{d}p_n$$

$$= \int_{\Omega'}\left|\frac{\partial\overline{H}}{\partial p_1}\right|^{-1}\mathrm{d}q_1\cdots\mathrm{d}q_n\mathrm{d}p_2\cdots\mathrm{d}p_n.$$

where $|\partial\boldsymbol{C}/\partial\boldsymbol{x}_2|$ is the determinant of the Jacobian matrix for the transformation from $\boldsymbol{x}_2$ to $\boldsymbol{C} = [C_1, C_2, \cdots, C_M]^{\mathrm{T}}$.

The averaged FPK equation associated with Eq. (3.16) is

$$\frac{\partial p}{\partial t} = -\frac{\partial}{\partial\overline{h}}(\overline{a}_{\overline{h}}p) - \sum_{\nu=1}^{M}\frac{\partial}{\partial c_\nu}(\overline{a}_\nu p) + \frac{1}{2}\frac{\partial^2}{\partial\overline{h}^2}(\overline{b}_{\overline{hh}}p)$$
$$+\frac{1}{2}\sum_{\nu_1,\nu_2=1}^{M}\frac{\partial^2}{\partial c_{\nu_1}\partial c_{\nu_2}}(\overline{b}_{\nu_1\nu_2}p) + \sum_{\nu=1}^{M}\frac{\partial^2}{\partial\overline{h}\partial c_\nu}(\overline{b}_{\overline{h}\nu}p), \tag{3.18}$$

where

$$\overline{a}_{\overline{h}} = \overline{a}_{\overline{h}}(\overline{h},\boldsymbol{c}) = \overline{m}_{\overline{H}}(\overline{H},\boldsymbol{C})\Big|_{\substack{\overline{H}=\overline{h}\\ \boldsymbol{C}=\boldsymbol{c}}},$$

$$\overline{a}_\nu = \overline{a}_\nu(\overline{h},\boldsymbol{c}) = \overline{m}_\nu(\overline{H},\boldsymbol{C})\Big|_{\substack{\overline{H}=\overline{h}\\ \boldsymbol{C}=\boldsymbol{c}}},$$

$$\overline{b}_{\overline{hh}} = \overline{b}_{\overline{hh}}(\overline{h},\boldsymbol{c}) = \sum_{s=1}^{l}\overline{\sigma}_{\overline{H}s}\overline{\sigma}_{\overline{H}s}(\overline{H},\boldsymbol{C})\Bigg|_{\substack{\overline{H}=\overline{h}\\ \boldsymbol{C}=\boldsymbol{c}}},$$

$$\bar{b}_{v_1 v_v} = \bar{b}_{v_1 v_v}(\bar{h}, \boldsymbol{c}) = \sum_{s=1}^{l} \bar{\sigma}_{v_1 s} \bar{\sigma}_{v_2 s}(\overline{H}, \boldsymbol{C})\Big|_{\substack{\overline{H}=\bar{h} \\ \boldsymbol{C}=\boldsymbol{c}}},$$

$$\bar{b}_{\bar{h}v} = \bar{b}_{\bar{h}v}(\bar{h}, \boldsymbol{c}) = \sum_{s=1}^{l} \bar{\sigma}_{\overline{H}s} \bar{\sigma}_{v_1 s}(\overline{H}, \boldsymbol{C})\Big|_{\substack{\overline{H}=\bar{h} \\ \boldsymbol{C}=\boldsymbol{c}}}. \tag{3.19}$$

In Eq. (3.18), $p = p(\bar{h}, \boldsymbol{c}, t|\bar{h}_0, \boldsymbol{c}_0)$ is the transition PDF of $[\overline{H}^{\mathrm{T}}, \boldsymbol{C}^{\mathrm{T}}]^{\mathrm{T}}$ with initial condition $p(\bar{h}, \boldsymbol{c}, 0|\bar{h}_0, \boldsymbol{c}_0) = \delta(\bar{h} - \bar{h}_0)\delta(\boldsymbol{c} - \boldsymbol{c}_0)$, or $p = p(\bar{h}, \boldsymbol{c}, t)$ is the joint PDF of $[\overline{H}^{\mathrm{T}}, \boldsymbol{C}^{\mathrm{T}}]^{\mathrm{T}}$ with initial condition $p(\bar{h}, \boldsymbol{c}, 0) = p(\bar{h}_0, \boldsymbol{c}_0)$. Boundary conditions depend on the properties and constraint of the generalized Hamiltonian system associated with Eq. (3.15).

After the stationary solution $p(\bar{h}, \boldsymbol{c})$ of the FPK Eq. (3.18) is obtained, then similar to the derivation of Eqs. (3.11)–(3.13), the approximate stationary PDF of the original system (3.5) can be obtained as follows

$$p(\boldsymbol{x}) = \frac{p(\bar{h}, \boldsymbol{c})}{T(\bar{h})}\left|\frac{\partial \boldsymbol{c}}{\partial \boldsymbol{x}_2}\right|\Bigg|_{\bar{h}=\bar{h}(\boldsymbol{x}_1),\ \boldsymbol{c}=\boldsymbol{c}(\boldsymbol{x}_2)}. \tag{3.20}$$

where $T(\bar{h}) = T'(\overline{H})\big|_{\overline{H}=\bar{h}}$, determined by Eq. (3.17).

Example 3.1 Consider a single machine infinite power system with steam control governed by following equations (Chen and Zhu 2010a):

$$\begin{aligned}
\dot{\delta} &= \omega_0 \omega, \\
\dot{\omega} &= \frac{P_m}{M} - \frac{D\omega}{M} - P_0[1 + W_g(t)]\sin\delta, \\
\dot{P}_m &= -\frac{P_m - P_m^0}{T_s} + \frac{\delta - \delta_0}{T_s} + u.
\end{aligned} \tag{3.21}$$

where δ is the rotor angle; ω is the rotor speed; P_m is the mechanical power; $\omega_0 = 2\pi f_0$; M is the inertia coefficient of a generator; D is the damping coefficient; $P_0[1 + W_g(t)]$ is the load and its random perturbation; T_s is the time constant of steam volume; $W_g(t)$ is Gaussian white noise with intensity $D_1 = 0.25$; u is the steam control and here $u = 0$ is assumed.

Let $X_1 = \delta$, $X_2 = \omega$, $X_3 = P_m$, $x_1^0 = \delta_0$, $x_3^0 = P_m^0$, $a = 1/M$, $b = P_0$, $c = D/M\omega_0$, $d = 1/T_s$, $f = D/M$, $e = \omega_0$. Equation (3.21) can be rewritten as follows:

$$\begin{pmatrix} \dot{X}_1 \\ \dot{X}_2 \\ \dot{X}_3 \end{pmatrix} = \left(\begin{pmatrix} 0 & 1 & 0 \\ -1 & 0 & 0 \\ 0 & 0 & 0 \end{pmatrix} - \begin{pmatrix} 0 & 0 & 0 \\ 0 & c & 0 \\ 0 & 0 & d/a \end{pmatrix} \right) \begin{pmatrix} -aX_3 + ab\sin X_1 \\ eX_2 \\ a(X_3 - x_3^0 - X_1 + x_1^0) \end{pmatrix}$$

$$+\begin{pmatrix} 0 \\ -b\sin X_1 \\ 0 \end{pmatrix} W_g(t). \tag{3.22}$$

Then, Eq. (3.22) can be converted into the following Itô stochastic differential equations:

$$\begin{pmatrix} \mathrm{d}X_1 \\ \mathrm{d}X_2 \\ \mathrm{d}X_3 \end{pmatrix} = \left[J(\boldsymbol{X}) \frac{\partial H}{\partial \boldsymbol{X}} - R(\boldsymbol{X}) \frac{\partial H}{\partial \boldsymbol{X}} \right] \mathrm{d}t + \begin{pmatrix} 0 \\ -b\sin X_1 \\ 0 \end{pmatrix} \mathrm{d}B(t) \tag{3.23}$$

where

$$J(\boldsymbol{X}) = \begin{pmatrix} 0 & 1 & 0 \\ -1 & 0 & 0 \\ 0 & 0 & 0 \end{pmatrix}, \quad R(\boldsymbol{X}) = \begin{pmatrix} 0 & 0 & 0 \\ 0 & c & 0 \\ 0 & 0 & d/a \end{pmatrix},$$

$$\frac{\partial H}{\partial \boldsymbol{X}} = \begin{pmatrix} -aX_3 + ab\sin X_1 \\ eX_2 \\ a(X_3 - x_3^0 - X_1 + x_1^0) \end{pmatrix},$$

$$\begin{aligned} H(X_1, X_2, X_3) &= \frac{1}{2} e X_2^2 - a X_3 (X_1 - x_1^0) + \frac{1}{2} a (X_3 - x_3^0)^2 \\ &\quad - ab(\cos X_1 - \cos x_1^0). \end{aligned} \tag{3.24}$$

in which $x_e^s = [x_1^s, 0, x_3^s]^T$ is the stable equilibration point of system.

The generalized Hamiltonian system associated with system (3.23) has one Casimir function $C(\boldsymbol{X}) = (X_3 - x_3^0)^2 / 2$ and one integral $H(\boldsymbol{X})$, where $H(\boldsymbol{X})$ is the total energy of the system (3.23) and $C(\boldsymbol{X})$ is the energy of steam system associated with the system (3.23).

By using the stochastic averaging method of quasi-nonintegrable generalized Hamiltonian systems, the averaged Itô equations for C and H can be derived as follows:

$$\begin{aligned} \mathrm{d}C &= m_1(C, H)\mathrm{d}t, \\ \mathrm{d}H &= \overline{m}_H(C, H)\mathrm{d}t + \overline{\sigma}_H(C, H)\mathrm{d}B(t). \end{aligned} \tag{3.25}$$

where

$$
\begin{aligned}
\overline{m}_1 &= \frac{1}{T}\int_{x_{20}}^{x_{10}} \left[-d\left(x_3-x_3^0\right)\left(x_3-x_3^0-x_1+x_1^0\right)/x_2\right]\mathrm{d}x_1,\\
\overline{m}_H &= \frac{1}{T}\int_{x_{20}}^{x_{10}} \left\{\left[-fex_2^2-da\left(x_3-x_3^0-x_1+x_1^0\right)^2+D_1eb^2\sin^2 x_1\right]\Big/x_2\right\}\mathrm{d}x_1,\\
\overline{b}_{HH} &= \tilde{\sigma}_{Hs}\tilde{\sigma}_{Hs} = \frac{1}{T}\int_{x_{20}}^{x_{10}} \left[\left(2D_1e^2b^2x_2^2\sin^2 x_1\right)/x_2\right]\mathrm{d}x_1,\\
T &= \int_{x_{20}}^{x_{10}} \left[{}^1\!/_{x_2}\right]\mathrm{d}x_1.
\end{aligned}
\tag{3.26}
$$

In Eq. (3.26), $x_2=\sqrt{2\left(H-aC+ax_3\left(x_1-x_1^0\right)+ab\left(\cos x_1-\cos x_1^0\right)\right)/e}$ and $x_3=\sqrt{2C}+x_3^0$. The upper and lower limits of integrals, $x_{10}(H,C)$ and $x_{20}(H,C)$, are the two roots of equation $-ax_3\left(x_1-x_1^0\right)-ab\left(\cos x_1-\cos x_1^0\right)=H-aC$, respectively.

For conservative system associated with system (3.23), there are two equilibration points: stable equilibration point $\boldsymbol{x}_e^s=\left[x_1^s,0,x_3^s\right]^T$ and unstable equilibration point $\boldsymbol{x}_e^u=\left[x_1^u,0,x_3^u\right]^T$. From stable equilibration point we get the $h_{\min}=H\left(\boldsymbol{x}_e^s\right)=0$, while we can get the critical value $h_{\max}=H\left(\boldsymbol{x}_e^u\right)$ from unstable equilibration point. Besides, the energy of steam system, $C(x)$, has its normal limit, i.e., $c_{\min}\le c<c_{\max}$. Thus, the safety domain Ω_s of system (3.23) is the interior of a rectangle with boundaries consisting of Γ_1,Γ_2,Γ_3 and Γ_4, i.e.,

$$
\begin{aligned}
\Gamma_1:&\ \ c=c_{\max},\ \ h_{\min}\le h<h_{\max},\\
\Gamma_2:&\ \ c_{\min}\le c<c_{\max},\ \ h=h_{\max},\\
\Gamma_3:&\ \ c=c_{\min},\ \ h_{\min}<h<h_{\max},\\
\Gamma_4:&\ \ c_{\min}\le c<c_{\max},\ \ h=h_{\min}.
\end{aligned}
\tag{3.27}
$$

where the boundaries Γ_1 and Γ_2 are absorbing boundaries while the boundaries Γ_3 and Γ_4 are reflecting boundaries (see Fig. 3.1a). However, since the electrical power should not be larger than the mechanical power (Venikov 1980), i.e., $b\sin x_1\le x_3$, the reflecting boundary Γ_4 should be replaced by the reflecting boundary Γ_5 as follows:

$$
\Gamma_5:\ \ c_{\min}\le c<c_{\max},\ \ h_{\min}\le h<h_{\max}.
\tag{3.28}
$$

where $H=H(x_1,x_2,x_3)$ is defined in Eq. (3.24) with $x_1=\sin^{-1}\left(x_3/b\right)$, $x_2=0$, $x_3=\sqrt{2c}+x_3^0$ (see Fig. 3.1b).

The conditional reliability function is governed by following backward Kolmogorov equation:

$$
\frac{\partial R}{\partial t}=\overline{m}_1\frac{\partial R}{\partial c_0}+\overline{m}_H\frac{\partial R}{\partial h_0}+\frac{\overline{b}_{HH}}{2}\frac{\partial^2 R}{\partial h_0^2}
\tag{3.29}
$$

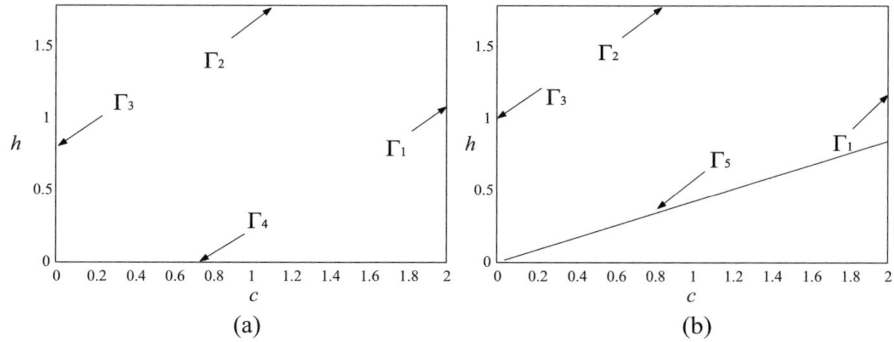

Fig. 3.1 The safety domain of system (3.22). Γ_1 and Γ_2 are the absorbing boundaries while Γ_4 and Γ_5 are the reflecting boundary. The parameters are taken as follows: $P_0 = 8, M = 2.0\text{s}, x_1^0 = \pi/6$, $x_3^0 = 4, T_s = 4\text{s}, a = 0.5, b = 8.0, c = 0.0001, d = 0.25, f = 0.0314, e = 100\pi, H_{\max} = 1.7889$, $H_{\min} = H_0 = 0.0, C_{\min} = C_0 = 0.0, C_{\max} = 2.0$ (Chen and Zhu 2010a).

where the coefficients are defined by Eq. (3.26) with $[c, h]^{\mathrm{T}}$ replaced by $[c_0, h_0]^{\mathrm{T}}$. The associated boundary conditions are

$$R(t|c_0, h_0) = 0 \quad \text{at } \Gamma_1 \text{ and } \Gamma_2. \tag{3.30}$$

$$R(t|c_0, h_0) = \text{finite} \quad \text{at } \Gamma_3 \text{ and } \Gamma_5. \tag{3.31}$$

The initial condition is defined by

$$R(0|c_0, h_0) = 1, \quad c_0, h_0 \in \Omega_1. \tag{3.32}$$

Note that the boundary conditions at Γ_3 and Γ_5 are qualitative and have to be replaced by quantitative ones for solving Eq. (3.29) numerically. They are

$$\frac{\partial R}{\partial t} = m_{20}\frac{\partial R}{\partial h_0} + \frac{b_{20}}{2}\frac{\partial^2 R}{\partial h_0^2} \tag{3.33}$$

where

$$\begin{aligned} m_{20} &= \frac{1}{T}\int_{y_{20}}^{y_{10}} \left\{\left[-fex_2^2 - da\left(-x_1 + x_1^0\right)^2 + D_1 eb^2 \sin^2 x_1\right] \Big/ x_2\right\} \mathrm{d}x_1, \\ b_{20} &= \frac{1}{T}\int_{y_{20}}^{y_{10}} \left[\left(2D_1 e^2 b^2 x_2^2 \sin^2 x_1\right)/x_2\right] \mathrm{d}x_1, \\ T &= \int_{y_{20}}^{y_{10}} \left[{}^{1}/_{x_2}\right] \mathrm{d}x_1. \end{aligned} \tag{3.34}$$

at Γ_3,

$$\frac{\partial R}{\partial t} = m_{10}\frac{\partial R}{\partial c_0} + m_{20}\frac{\partial R}{\partial h_0}. \tag{3.35}$$

where

$$\begin{aligned}
m_{10} &= -d\left(x_3 - x_3^0\right)\left(x_3 - x_3^0 - x_1^{\min} + x_1^0\right),\\
m_{20} &= -da\left(x_3 - x_3^0 - x_1^{\min} + x_1^0\right)^2 + D_1 eb^2 \sin^2 x_1^{\min},\\
x_{\min} &= \sin^{-1}\left(x_3/b\right); \quad x_3 = \sqrt{2c_0} + x_3^0.
\end{aligned} \tag{3.36}$$

Γ_5. Note that in Eq. (3.34), $x_2 = \sqrt{2\left[h_0 + ax_3^0(x_1 - x_1^0) + ab\left(\cos x_1 - \cos x_1^0\right)\right]/e}$, the upper and lower limits of integrals, $y_{10}(h_0)$ and $y_{20}(h_0)$, are the two roots of equation $-ax_3^0(x_1 - x_1^0) - ab\left(\cos x_1 - \cos x_1^0\right) = h_0$, respectively.

Similarly, the Pontryagin equation for the mean first passage time of system (3.25) is derived from Eq. (3.29) as follows:

$$\overline{m}_1\frac{\partial \mu}{\partial c_0} + \overline{m}_H\frac{\partial \mu}{\partial h_0} + \frac{\overline{b}_{HH}}{2}\frac{\partial^2 \mu}{\partial h_0^2} = -1. \tag{3.37}$$

together with the boundary conditions

$$\mu(c_0, h_0) = 0 \quad \text{at } \Gamma_1 \text{ and } \Gamma_2. \tag{3.38}$$

$$\mu(c_0, h_0) = \text{finite} \quad \text{at } \Gamma_3 \text{ and } \Gamma_5. \tag{3.39}$$

The qualitative conditions at Γ_3 and Γ_5 can be converted into the following quantitative ones:

$$m_{20}\frac{\partial \mu}{\partial h_0} + \frac{1}{2}b_{20}\frac{\partial^2 \mu}{\partial h_0^2} = -1. \tag{3.40}$$

at Γ_3,

$$-1 = m_{10}\frac{\partial \mu}{\partial c_0} + m_{20}\frac{\partial \mu}{\partial h_0}. \tag{3.41}$$

at Γ_5.

The backward Kolmogorov Eq. (3.29) can be solved with boundary conditions (3.30), (3.33), (3.35) and initial condition (3.32) numerically while the Pontryagin Eq. (3.37) can be solved with boundary conditions (3.38), (3.40) and (3.41) by using successive over-relaxation method.

Some numerical results for the conditional reliability function, the conditional probability density and mean first the results from Monte Carlo simulation of system in Figs. 3.2, 3.3, 3.4 and 3.5. To assess the validity and accuracy of the proposed procedure, the results from Monte Carlo simulation of system (3.22) are also obtained and shown in Figs. 3.2 and 3.3 using symbols ●, ■. It is seen that the analytical results agree quite well with simulation results and higher excitation intensity leads to lower reliability of system. Additional results of reliability function and mean first passage time are obtained by using the proposed method and shown in Figs. 3.2, 3.3, 3.4 and 3.5, respectively. Note that the reliability function and mean first passage time are monotonously decreasing function of initial energy.

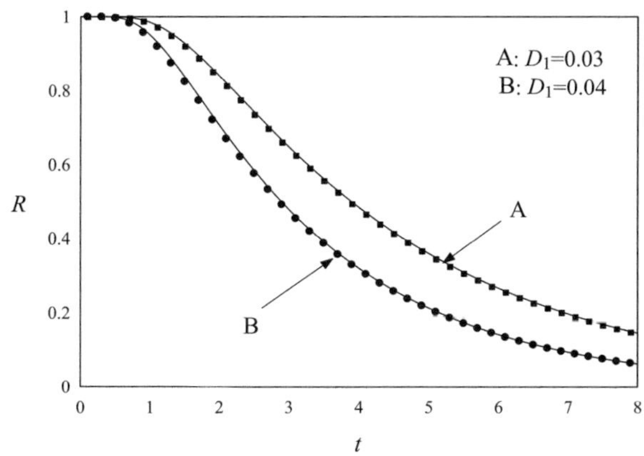

Fig. 3.2 The conditional reliability function $R(t|c_0, h_0)$ of system (3.22) for zero initial condition $c_0 = 0,\ h_0 = 0$. The parameters are the same as those in Fig. 3.1b. ——, analytical results; ●, ■, results from Monte Carlo simulation of original system (3.22) (Chen and Zhu 2010a)

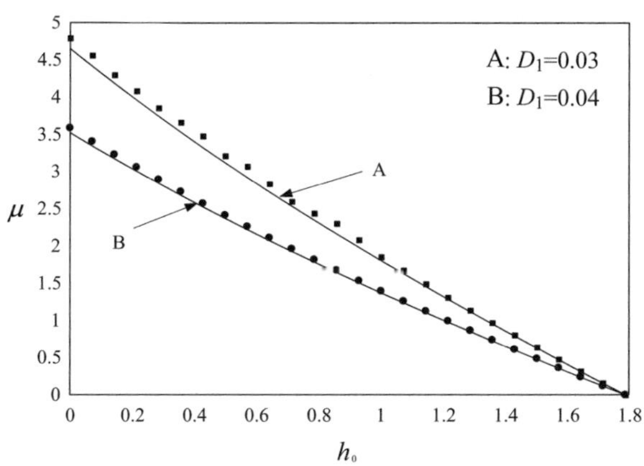

Fig. 3.3 The mean first passage time $\mu(c_0,\ h_0)$ of system (3.22) as function of initial energy h_0 for $c_0 = 0.0$. The parameters are the same as those in Fig. 3.1b. ——, analytical results; ●, ■, results from Monte Carlo simulation of original system (3.22) (Chen and Zhu 2010a)

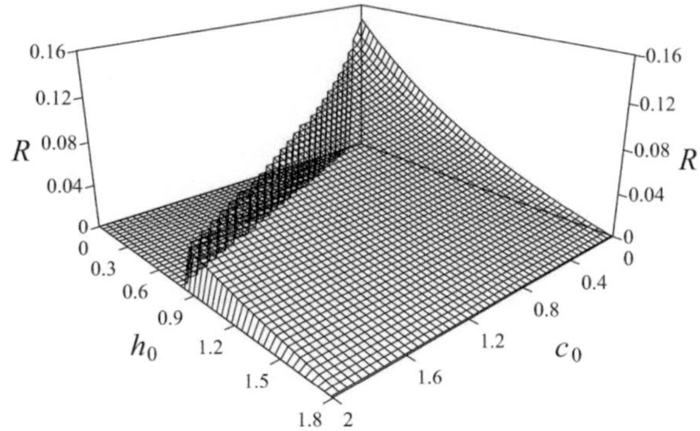

Fig. 3.4 The reliability function $R(t|c_0, h_0)$ of system (3.22) at $t = 8$ (second) as function of h_0 and c_0 with $D_1 = 0.03$. The other parameters are the same as those in Fig. 3.1b (Chen and Zhu 2010a)

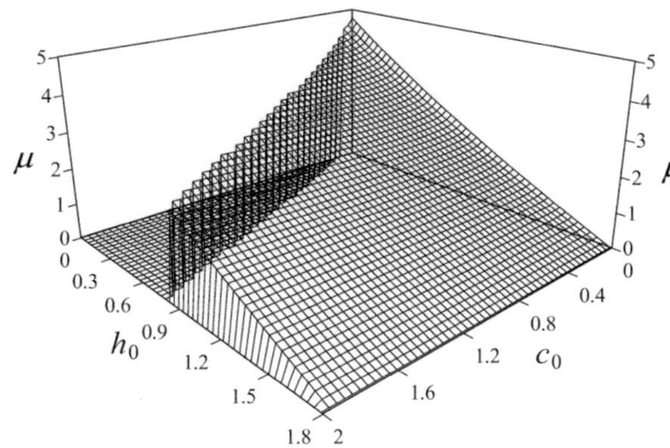

Fig. 3.5 The mean first passage time of system (3.22) as function of h_0 and c_0 with $D_1 = 0.03$. The other parameters are the same as those in Fig. 3.1b (Chen and Zhu 2010a)

3.2 Quasi-integrable Generalized Hamiltonian Systems

Assume that the Hamiltonian system associated with system (3.5) is completely integrable and has a vector $\boldsymbol{I} = [I_1(\boldsymbol{x}), \ldots, I_n(\boldsymbol{x})]^{\mathrm{T}}$ of n action variables, a vector $\boldsymbol{\Theta} = [\Theta_1(\boldsymbol{x}), \ldots, \Theta_n(\boldsymbol{x})]^T$ of n angle variables and a vector $\boldsymbol{C} = [C_1(\boldsymbol{x}), \ldots, C_M(\boldsymbol{x})]^{\mathrm{T}}$ of M Casimir functions, where $2n + M = m$. Using the Itô differential rule, the following Itô stochastic differential equations governing the stochastic process $[\boldsymbol{I}^T(t), \boldsymbol{\Theta}^T(t), \boldsymbol{C}^T(t)]^T$ can be derived from Eq. (3.5)

$$\mathrm{d}I_k = \varepsilon\left(\sum_{i,j=1}^{m} d_{ij}\frac{\partial I_k}{\partial X_i}\frac{\partial H}{\partial X_j} + \frac{1}{2}\sum_{i,j=1}^{m}\sum_{s=1}^{l}\sigma_{is}\sigma_{js}\frac{\partial^2 I_k}{\partial X_i \partial X_j}\right)\mathrm{d}t$$
$$+ \varepsilon^{1/2}\sum_{i=1}^{m}\sum_{s=1}^{l}\sigma_{is}\frac{\partial I_k}{\partial X_i}\mathrm{d}B_s(t),$$

$$
\begin{aligned}
\mathrm{d}\Theta_k &= \left[\omega_k + \varepsilon\left(\sum_{i,j=1}^{m} d_{ij}\frac{\partial\Theta_k}{\partial X_i}\frac{\partial H}{\partial X_j} + \frac{1}{2}\sum_{i,j=1}^{m}\sum_{s=1}^{l}\sigma_{is}\sigma_{js}\frac{\partial^2\Theta_k}{\partial X_i\partial X_j}\right)\right]\mathrm{d}t \\
&\quad + \varepsilon^{1/2}\sum_{i=1}^{m}\sum_{s=1}^{l}\sigma_{is}\frac{\partial\Theta_k}{\partial X_i}\mathrm{d}B_s(t), \\
\mathrm{d}C_v &= \varepsilon\left(\sum_{i,j=1}^{m} d_{ij}\frac{\partial C_v}{\partial X_i}\frac{\partial H}{\partial X_j} + \frac{1}{2}\sum_{i,j=1}^{m}\sum_{s=1}^{l}\sigma_{is}\sigma_{js}\frac{\partial^2 C_v}{\partial X_i\partial X_j}\right)\mathrm{d}t \\
&\quad + \varepsilon^{1/2}\sum_{i=1}^{m}\sum_{s=1}^{l}\sigma_{is}\frac{\partial C_v}{\partial X_i}\mathrm{d}B_s(t), \\
k &= 1,2,\ldots,n;\quad v = 1,2,\ldots,M.
\end{aligned}
\tag{3.42}
$$

In this case, the dimension and form of the averaged Itô stochastic differential equations depend on the resonances of the integrable generalized Hamiltonian subsystem with Hamiltonians $\boldsymbol{H}_1(\boldsymbol{I})$.

3.2.1 Non-internal Resonance

Assume that the n frequencies $\omega_r(\boldsymbol{I},\boldsymbol{C})$ of the integrable Hamiltonian subsystem with Hamiltonians $\boldsymbol{H}_1(\boldsymbol{I})$ do not satisfy the following internal resonance relations

$$
\sum_{r=1}^{n} k_r^u\omega_r = O_u(\varepsilon),\quad u = 1,2,\ldots,\alpha. \tag{3.43}
$$

where k_r^u are integers of not all zero; $O_u(\varepsilon)$ denote small detuning quantities of ε order. In this case, the generalized Hamiltonian system associated with system (3.5) is completely integrable and non-resonant. $I_k(t)$, $C_v(t)$ in Eq. (3.42) are slowly-varying processes, and $\Theta_k(t)$ are rapidly-varying processes. According to Khasminskii's theorem (Khasminskii 1968), as $\varepsilon\to 0$, $[I_1,\ldots,I_n,C_1,\ldots,C_M]^{\mathrm{T}}$ converges weakly to a $(n+M)$-dimensional vector Markov diffusion process. The averaged Itô stochastic differential equations for this process are

$$
\begin{aligned}
\mathrm{d}I_k &= m_k(\boldsymbol{I},\boldsymbol{C})\mathrm{d}t + \sum_{s=1}^{l}\sigma_{ks}(\boldsymbol{I},\boldsymbol{C})\mathrm{d}B_s(t), \\
\mathrm{d}C_v &= m_v(\boldsymbol{I},\boldsymbol{C})\mathrm{d}t + \sum_{s=1}^{l}\sigma_{vs}(\boldsymbol{I},\boldsymbol{C})\mathrm{d}B_s(t), \\
k &= 1,2,\ldots,n;\quad v = 1,2,\ldots,M,
\end{aligned}
\tag{3.44}
$$

where the drift and diffusion coefficients can be obtained by performing time averaging over the corresponding coefficients in system (3.42). Since the non-resonant generalized Hamiltonian system with constant I_k, C_v are ergodic on the n-dimensional torus, spatial averaging over $\boldsymbol{\theta}$ can be used instead of time averaging. Thus, the coefficients are

$$
\begin{aligned}
& m_k(\boldsymbol{I}, \boldsymbol{C}) = \frac{\varepsilon}{T_1}\int_0^{2\pi}\frac{1}{A_1}\left[\sum_{i,j=1}^{m}\left(d_{ij}\frac{\partial I_k}{\partial x_i}\frac{\partial H}{\partial x_j}\right) + \frac{1}{2}\sum_{i,j=1}^{m}\sum_{s=1}^{l}\left(\sigma_{is}\sigma_{js}\frac{\partial^2 I_k}{\partial x_i \partial x_j}\right)\right]\mathrm{d}\boldsymbol{\theta},\\
& m_v(\boldsymbol{I}, \boldsymbol{C}) = \frac{\varepsilon}{T_1}\int_0^{2\pi}\frac{1}{A_1}\left[\sum_{i,j=1}^{m}\left(d_{ij}\frac{\partial C_v}{\partial x_i}\frac{\partial H}{\partial x_j}\right) + \frac{1}{2}\sum_{i,j=1}^{m}\sum_{s=1}^{l}\left(\sigma_{is}\sigma_{js}\frac{\partial^2 C_v}{\partial x_i \partial x_j}\right)\right]\mathrm{d}\boldsymbol{\theta},\\
& \sum_{s=1}^{l}\sigma_{k_1 s}\sigma_{k_2 s}(\boldsymbol{I}, \boldsymbol{C}) = \frac{\varepsilon}{T_1}\int_0^{2\pi}\frac{1}{A_1}\sum_{i,j=1}^{m}\sum_{s=1}^{l}\left(\sigma_{is}\sigma_{js}\frac{\partial I_{k_1}}{\partial x_i}\frac{\partial I_{k_2}}{\partial x_j}\right)\mathrm{d}\boldsymbol{\theta},\\
& \sum_{s=1}^{l}\sigma_{v_1 s}\sigma_{v_2 s}(\boldsymbol{I}, \boldsymbol{C}) = \frac{\varepsilon}{T_1}\int_0^{2\pi}\frac{1}{A_1}\sum_{i,j=1}^{m}\sum_{s=1}^{l}\left(\sigma_{is}\sigma_{js}\frac{\partial C_{v_1}}{\partial x_i}\frac{\partial C_{v_2}}{\partial x_j}\right)\mathrm{d}\boldsymbol{\theta},\\
& \sum_{s=1}^{l}\sigma_{k_1 s}\sigma_{v_1 s}(\boldsymbol{I}, \boldsymbol{C}) = \frac{\varepsilon}{T_1}\int_0^{2\pi}\frac{1}{A_1}\sum_{i,j=1}^{m}\sum_{s=1}^{l}\left(\sigma_{is}\sigma_{js}\frac{\partial I_{k_1}}{\partial x_i}\frac{\partial C_{v_1}}{\partial x_j}\right)\mathrm{d}\boldsymbol{\theta},\\
& T_1 = T_1(\boldsymbol{I}, \boldsymbol{C}) = \int_0^{2\pi}(1/A_1)\mathrm{d}\boldsymbol{\theta},\\
& k, k_1, k_2 = 1, 2, \ldots, n; \quad v, v_1, v_2 = 1, 2, \ldots, M,
\end{aligned}
\tag{3.45}
$$

where $\int_0^{2\pi}(\cdot)d\theta$ is the n-fold integral with respective to $\theta_1, \theta_2, \ldots, \theta_n$, and $A_1 = \left|\partial(\boldsymbol{I}, \boldsymbol{\theta}, \boldsymbol{c})/\partial \boldsymbol{x}\right|$ denotes the determinant of the Jacobian matrix for the transformation from $\boldsymbol{x}$ to $\boldsymbol{I}, \boldsymbol{\theta}, \boldsymbol{c}$. The averaged FPK equation associated with averaged Itô Eq. (3.44) is given by

$$
\begin{aligned}
\frac{\partial p}{\partial t} = & -\sum_{k=1}^{n}\frac{\partial}{\partial I_k}(a_k p) - \sum_{v=1}^{M}\frac{\partial}{\partial c_v}(a_v p) + \frac{1}{2}\sum_{k_1,k_2=1}^{n}\frac{\partial^2}{\partial I_{k_1}\partial I_{k_2}}(b_{k_1k_2}p)\\
& + \frac{1}{2}\sum_{v_1,v_2=1}^{M}\frac{\partial^2}{\partial c_{v_1}\partial c_{v_2}}(b_{v_1v_2}p) + \sum_{k_1=1}^{n}\sum_{v_1=1}^{M}\frac{\partial^2}{\partial I_{k_1}\partial c_{v_1}}(b_{k_1v_1}p),
\end{aligned}
\tag{3.46}
$$

where

$$
\begin{aligned}
a_k &= a_k(\boldsymbol{I}, \boldsymbol{c}) = m_k(\boldsymbol{I}, \boldsymbol{C})|_{\boldsymbol{C}=\boldsymbol{c}},\\
a_v &= a_v(\boldsymbol{I}, \boldsymbol{c}) = m_v(\boldsymbol{I}, \boldsymbol{C})|_{\boldsymbol{C}=\boldsymbol{c}},
\end{aligned}
$$

$$
\begin{aligned}
b_{k_1k_2} &= b_{k_1k_2}(\boldsymbol{I}, \boldsymbol{c}) = \sum_{s=1}^{l} \sigma_{k_1s}\sigma_{k_2s}(\boldsymbol{I}, \boldsymbol{C})\bigg|_{\boldsymbol{C}=\boldsymbol{c}}, \\
b_{v_1v_2} &= b_{v_1v_2}(\boldsymbol{I}, \boldsymbol{c}) = \sum_{s=1}^{l} \sigma_{v_1s}\sigma_{v_2s}(\boldsymbol{I}, \boldsymbol{C})\bigg|_{\boldsymbol{C}=\boldsymbol{c}}, \\
b_{k_1v_1} &= b_{k_1v_1}(\boldsymbol{I}, \boldsymbol{c}) = \sum_{s=1}^{l} \sigma_{k_1s}\sigma_{v_1s}(\boldsymbol{I}, \boldsymbol{C})\bigg|_{\boldsymbol{C}=\boldsymbol{c}}.
\end{aligned}
\tag{3.47}
$$

In Eq. (3.46), $p = p(\boldsymbol{I}, \boldsymbol{c}, t|\boldsymbol{I}_0, \boldsymbol{c}_0)$ is the transition PDF of the vector Markov diffusion process $[\boldsymbol{I}^{\mathrm{T}}, \boldsymbol{C}^{\mathrm{T}}]^{\mathrm{T}}$, or $p = p(\boldsymbol{I}, \boldsymbol{c}, t)$ is the joint PDF for $[\boldsymbol{I}^{\mathrm{T}}, \boldsymbol{C}^{\mathrm{T}}]^{\mathrm{T}}$. The FPK Eq. (3.46) can be solved under certain initial condition and boundary condition.

Once the stationary PDF $p(\boldsymbol{I}, \boldsymbol{c})$ of the FPK Eq. (3.46) has been obtained, the approximate stationary PDF of the original system (3.5) is given by

$$
p(\boldsymbol{x}) = p(\boldsymbol{I}, \boldsymbol{\theta}, \boldsymbol{c})\left|\frac{\partial(\boldsymbol{I}, \boldsymbol{\theta}, \boldsymbol{c})}{\partial \boldsymbol{x}}\right| = p(\boldsymbol{\theta}|\boldsymbol{I}, \boldsymbol{c})p(\boldsymbol{I}, \boldsymbol{c})\left|\frac{\partial(\boldsymbol{I}, \boldsymbol{\theta}, \boldsymbol{c})}{\partial \boldsymbol{x}}\right|. \tag{3.48}
$$

For fixed values of $\boldsymbol{I}, \boldsymbol{c}$, the conditional PDF $p(\boldsymbol{\theta}|\boldsymbol{I}, \boldsymbol{c})$ is inversely proportional to $\left|\partial(\boldsymbol{I}, \boldsymbol{\theta}, \boldsymbol{c})/\partial \boldsymbol{x}\right|$, thus,

$$
p(\boldsymbol{\theta}|\boldsymbol{I}, \boldsymbol{c}) = C'/A_1, \tag{3.49}
$$

where C' is a constant. Because the non-resonant generalized Hamiltonian system with constant $\boldsymbol{I}, \boldsymbol{c}$ is ergodic in a n-dimensional torus, normalizing Eq. (3.49) results in $C' = T_1^{-1}$, which can be substituted back into Eq. (3.49) and then into Eq. (3.48) to obtain

$$
p(\boldsymbol{x}) = \left.\frac{p(\boldsymbol{I}, \boldsymbol{c})}{T_1(\boldsymbol{I}, \boldsymbol{c})}\right|_{\boldsymbol{I}=\boldsymbol{I}(\boldsymbol{x}), \boldsymbol{c}=\boldsymbol{c}(\boldsymbol{x})}, \tag{3.50}
$$

where $T_1(\boldsymbol{I}, \boldsymbol{c}) = T_1(\boldsymbol{I}, \boldsymbol{C})|_{\boldsymbol{C}=\boldsymbol{c}}$, determined by Eq. (3.45).

Consider the case that variables of the generalized Hamiltonian system associated with Eq. (3.42) are separable, i.e.,

$$
\begin{aligned}
&\boldsymbol{x} = [\boldsymbol{x}_1^{\mathrm{T}}, \boldsymbol{x}_2^{\mathrm{T}}]^{\mathrm{T}}, \quad \boldsymbol{x}_1 = [x_1, x_2, \ldots, x_{2n}]^{\mathrm{T}} = [\boldsymbol{q}^{\mathrm{T}}, \boldsymbol{p}^{\mathrm{T}}]^{\mathrm{T}}, \\
&\boldsymbol{q} = [q_1, q_2, \ldots, q_n]^{\mathrm{T}}, \quad \boldsymbol{p} = [p_1, p_2, \ldots, p_n]^{\mathrm{T}}, \\
&\boldsymbol{x}_2 = [x_{2n+1}, x_{2n+2}, \ldots, x_m]^{\mathrm{T}}.
\end{aligned}
\tag{3.51}
$$

The subsystem corresponding to $\boldsymbol{x}_1$ has Hamiltonian function $H_1(\boldsymbol{q}, \boldsymbol{p})$, a vector $\boldsymbol{I}(\boldsymbol{x}_1) = [I_1, I_2, \ldots, I_n]^{\mathrm{T}}$ of action variables and a vector $\boldsymbol{\theta}(\boldsymbol{x}_1) = [\theta_1, \theta_2, \ldots, \theta_n]^{\mathrm{T}}$ of angle variables. The subsystem corresponding to $\boldsymbol{x}_2$ has Casimir functions $C_v(\boldsymbol{x}_2),\ \ v = 1, 2, \ldots, M$, where $M = m - 2n$. In this case, Eq. (3.42) is replaced by

$$
\begin{aligned}
\mathrm{d}I_k &= \varepsilon\left\{\sum_{j=1}^{n}\frac{\partial I_k}{\partial P_j}[\sum_{r=1}^{n} d_{jr}\frac{\partial H_1}{\partial P_r}+\sum_{i=2n+1}^{m} d_{ji}\frac{\partial}{\partial X_i}(\sum_{\beta=1}^{M} C_\beta)]\right.\\
&\left.+\frac{1}{2}\sum_{j,r=1}^{n}\sum_{s=1}^{l}\sigma_{js}\sigma_{rs}\frac{\partial^2 I_k}{\partial P_j\partial P_r}\right\}\mathrm{d}t+\varepsilon^{1/2}\sum_{j=1}^{n}\sum_{s=1}^{l}\sigma_{js}\frac{\partial I_k}{\partial P_j}\mathrm{d}B_s(t),\\
\mathrm{d}\Theta_k &= \left\{\omega_k+\varepsilon\sum_{j=1}^{n}\frac{\partial \Theta_k}{\partial P_j}[\sum_{r=1}^{n} d_{jr}\frac{\partial H_1}{\partial P_r}+\sum_{i=2n+1}^{m} d_{ji}\frac{\partial}{\partial X_i}(\sum_{\beta=1}^{M} C_\beta)]\right.\\
&\left.+\frac{\varepsilon}{2}\sum_{j,r=1}^{n}\sum_{s=1}^{l}\sigma_{js}\sigma_{rs}\frac{\partial^2 \Theta_k}{\partial P_j\partial P_r}\right\}\mathrm{d}t+\varepsilon^{1/2}\sum_{j=1}^{n}\sum_{s=1}^{l}\sigma_{js}\frac{\partial \Theta_k}{\partial P_j}\mathrm{d}B_s(t),\\
\mathrm{d}C_v &= \varepsilon\{\sum_{i_1=2n+1}^{m}\frac{\partial C_v}{\partial X_{i_1}}[\sum_{r=1}^{n} d_{i_1 r}\frac{\partial H_1}{\partial P_r}+\sum_{i=2n+1}^{m} d_{i_1 i}\frac{\partial}{\partial X_i}(\sum_{\beta=1}^{M} C_\beta)]\\
&+\frac{1}{2}\sum_{i,i_1=2n+1}^{m}\sum_{s=1}^{l}\sigma_{is}\sigma_{i_1 s}\frac{\partial^2 C_v}{\partial X_i\partial X_{i_1}}\}\mathrm{d}t+\varepsilon^{1/2}\sum_{i=2n+1}^{m}\sum_{s=1}^{l}\sigma_{is}\frac{\partial C_v}{\partial X_i}\mathrm{d}B_s(t),\\
&k=1,2,\ldots,n,\quad v=1,2,\ldots,M.
\end{aligned}
\tag{3.52}
$$

In non-resonant case, the averaged Itô stochastic differential equations are still of the form of Eq. (3.44), but the coefficients Eq. (3.45) have to be replaced by

$$
\begin{aligned}
m'_k(\mathbf{I},\mathbf{C}) &= \frac{\varepsilon}{T'_1}\int_0^{2\pi}\frac{1}{A'_1}\{\sum_{j=1}^{n}\frac{\partial I_k}{\partial p_j}[\sum_{r=1}^{n} d_{jr}\frac{\partial H_1}{\partial p_r}\\
&+\sum_{i=2n+1}^{m} d_{ji}\frac{\partial}{\partial x_i}(\sum_{\beta=1}^{M} C_\beta)]+\frac{1}{2}\sum_{j,r=1}^{n}\sum_{s=1}^{l}\sigma_{js}\sigma_{rs}\frac{\partial^2 I_k}{\partial p_j\partial p_r}\}\mathrm{d}\boldsymbol{\theta},\\
m'_v(\mathbf{I},\mathbf{C}) &= \frac{\varepsilon}{T'_1}\int_0^{2\pi}\frac{1}{A'_1}\{\sum_{i_1=2n+1}^{m}\frac{\partial C_v}{\partial x_{i_1}}[\sum_{r=1}^{n} d_{i_1 r}\frac{\partial H_1}{\partial p_r}\\
&+\sum_{i=2n+1}^{m} d_{i_1 i}\frac{\partial}{\partial x_i}(\sum_{\beta=1}^{M} C_\beta)]+\frac{1}{2}\sum_{i,i_1=2n+1}^{m}\sum_{s=1}^{l}\sigma_{is}\sigma_{i_1 s}\frac{\partial^2 C_v}{\partial x_i\partial x_{i_1}}\}\mathrm{d}\boldsymbol{\theta},\\
\sum_{s=1}^{l}\sigma'_{k_1 s}\sigma'_{k_2 s}(\mathbf{I},\mathbf{C}) &= \frac{\varepsilon}{T'_1}\int_0^{2\pi}\frac{1}{A'_1}\sum_{j,r=1}^{n}\sum_{s=1}^{l}(\sigma_{js}\sigma_{rs}\frac{\partial I_{k_1}}{\partial p_j}\frac{\partial I_{k_2}}{\partial p_r})\mathrm{d}\boldsymbol{\theta},\\
\sum_{s=1}^{l}\sigma'_{v_1 s}\sigma'_{v_2 s}(\mathbf{I},\mathbf{C}) &= \frac{\varepsilon}{T'_1}\int_0^{2\pi}\frac{1}{A'_1}\sum_{i,i_1=2n+1}^{m}\sum_{s=1}^{l}(\sigma_{is}\sigma_{i_1 s}\frac{\partial C_{v_1}}{\partial x_i}\frac{\partial C_{v_2}}{\partial x_{i_1}})\mathrm{d}\boldsymbol{\theta},
\end{aligned}
$$

$$\sum_{s=1}^{l}\sigma'_{k_1s}\sigma'_{v_1s}(\mathbf{I},\mathbf{C})=\frac{\varepsilon}{T'_1}\int_0^{2\pi}\frac{1}{A'_1}\sum_{j=1}^{n}\sum_{i=2n+1}^{m}\sum_{s=1}^{l}(\sigma_{js}\sigma_{is}\frac{\partial I_{k_1}}{\partial p_j}\frac{\partial C_{v_1}}{\partial x_i})\mathrm{d}\boldsymbol{\theta},$$
$$A'_1=\left|\frac{\partial(\mathbf{I},\boldsymbol{\theta},\mathbf{c})}{\partial\mathbf{x}}\right|=\left|\frac{\partial(\mathbf{I},\boldsymbol{\theta},\mathbf{c})}{\partial(\mathbf{q},\mathbf{p},\mathbf{x}_2)}\right|=\left|\frac{\partial(\mathbf{I},\boldsymbol{\theta})}{\partial(\mathbf{q},\mathbf{p})}\right|\left|\frac{\partial\mathbf{c}}{\partial\mathbf{x}_2}\right|=\left|\frac{\partial\mathbf{c}}{\partial\mathbf{x}_2}\right|,$$
$$T'_1=\int_0^{2\pi}(1/A'_1)\mathrm{d}\boldsymbol{\theta}=(2\pi)^n\left|\frac{\partial\mathbf{c}}{\partial\mathbf{x}_2}\right|^{-1},$$
$$k,k_1,k_2=1,2,\ldots,n;\quad v,v_1,v_2=1,2,\ldots,M, \tag{3.53}$$

where $\int_0^{2\pi}(\cdot)\mathrm{d}\boldsymbol{\theta}$ is a n-fold integral with respective to $\theta_1,\theta_2,\ldots,\theta_n$. The averaged FPK equation associated with Eq. (3.52) has the form of Eq. (3.46), while the first and second derivative moments are

$$\begin{aligned}
a'_k&=a'_k(\boldsymbol{I},\boldsymbol{c})=m'_k(\boldsymbol{I},\boldsymbol{C})\big|_{\boldsymbol{C}=\boldsymbol{c}},\\
a'_v&=a'_v(\boldsymbol{I},\boldsymbol{c})=m'_v(\boldsymbol{I},\boldsymbol{C})\big|_{\boldsymbol{C}=\boldsymbol{c}},\\
b'_{k_1k_2}&=b'_{k_1k_2}(\boldsymbol{I},\boldsymbol{c})=\sum_{s=1}^{l}\sigma'_{k_1s}\sigma'_{k_2s}(\boldsymbol{I},\boldsymbol{C})\Big|_{\boldsymbol{C}=\boldsymbol{c}},\\
b'_{v_1v_2}&=b'_{v_1v_2}(\boldsymbol{I},\boldsymbol{c})=\sum_{s=1}^{l}\sigma'_{v_1s}\sigma'_{v_2s}(\boldsymbol{I},\boldsymbol{C})\Big|_{\boldsymbol{C}=\boldsymbol{c}},\\
b'_{k_1v_1}&=b'_{k_1v_1}(\boldsymbol{I},\boldsymbol{c})=\sum_{s=1}^{l}\sigma'_{k_1s}\sigma'_{v_1s}(\boldsymbol{I},\boldsymbol{C})\Big|_{\boldsymbol{C}=\boldsymbol{c}}.
\end{aligned} \tag{3.54}$$

Once the stationary solution $p(\boldsymbol{I},\boldsymbol{c})$ of the averaged FPK equation has been obtained, the approximate PDF of the original system (3.5) is given by

$$p(\boldsymbol{x})=\frac{p(\boldsymbol{I},\boldsymbol{c})}{T'_1(\boldsymbol{I},\boldsymbol{c})}=\frac{p(\boldsymbol{I},\boldsymbol{c})}{(2\pi)^n}\left|\frac{\partial\boldsymbol{c}}{\partial\boldsymbol{x}_2}\right|. \tag{3.55}$$

In the case that the integrable generalized Hamiltonian system associated with Eq. (3.42) is completely separable, i.e., $\boldsymbol{x}=[\boldsymbol{x}_1^{\mathrm{T}},\boldsymbol{x}_2^{\mathrm{T}}]^{\mathrm{T}}$, $\boldsymbol{x}_1=[x_1,x_2,\ldots,x_{2n}]^{\mathrm{T}}=[q_1,q_2,\ldots,q_n,p_1,p_2,\ldots,p_n]^{\mathrm{T}}$, $H_k=H_k(q_k,p_k)$ $(k=1,2,\ldots,n)$. The governing Itô stochastic differential equations for $H_k=H_k(q_k,p_k)$ and $C_v(\boldsymbol{x}_2)$ can be derived from Eq. (3.5) as follows

$$\begin{aligned}
\mathrm{d}H_k=\varepsilon\{&\frac{\partial H_k}{\partial P_k}[\sum_{r=1}^{n}d_{kr}\frac{\partial H_r}{\partial P_r}+\sum_{i=2n+1}^{m}d_{ki}\frac{\partial}{\partial X_i}(\sum_{\beta=1}^{M}C_\beta)]\\
&+\frac{1}{2}\sum_{s=1}^{l}\sigma_{ks}\sigma_{ks}\frac{\partial^2H_k}{\partial P_k^2}\}\mathrm{d}t+\varepsilon^{1/2}\sum_{s=1}^{l}\sigma_{js}\frac{\partial H_k}{\partial P_k}\mathrm{d}B_s(t),
\end{aligned}$$

$$
\begin{aligned}
\mathrm{d}C_v = {} & \varepsilon\{\sum_{i_1=2n+1}^{m} \frac{\partial C_v}{\partial X_{i_1}}[\sum_{r=1}^{n} d_{i_1 r}\frac{\partial H_r}{\partial P_r} + \sum_{i=2n+1}^{m} d_{i_1 i}\frac{\partial}{\partial X_i}(\sum_{\beta=1}^{M} C_\beta)] \\
& + \frac{1}{2}\sum_{i,i_1=2n+1}^{m}\sum_{s=1}^{l}\sigma_{is}\sigma_{i_1 s}\frac{\partial^2 C_v}{\partial X_i \partial X_{i_1}}\}\mathrm{d}t + \varepsilon^{1/2}\sum_{i=2n+1}^{m}\sum_{s=1}^{l}\sigma_{is}\frac{\partial C_v}{\partial X_i}\mathrm{d}B_s(t), \\
& k = 1, 2, \ldots, n, \quad v = 1, 2, \ldots, M.
\end{aligned} \tag{3.56}
$$

In this case, the new system is constituted of Eq. (3.56) with the equations for Q_k in Eq. (3.5). $Q_k(t)$ $(k = 1, 2, \ldots, n)$ are fast-varying processes while $H_k(t)$ and $C_v(t)$ are slowly-varying processes. According to Khasminskii theorem (Khasminskii 1968), as $\varepsilon \to 0$, $H_k(t)$, $C_v(t)$ converge weakly to a vector of $(n + M)$–dimensional Markov diffusion process. The averaged Itô stochastic differential equations are

$$
\begin{aligned}
\mathrm{d}H_k &= \overline{m}_k(\boldsymbol{H}, \boldsymbol{C})\mathrm{d}t + \sum_{s=1}^{l}\overline{\sigma}_{ks}(\boldsymbol{H}, \boldsymbol{C})\mathrm{d}B_s(t), \\
\mathrm{d}C_v &= \overline{m}_v(\boldsymbol{H}, \boldsymbol{C})\mathrm{d}t + \sum_{s=1}^{l}\overline{\sigma}_{vs}(\boldsymbol{H}, \boldsymbol{C})\mathrm{d}B_s(t), \\
k &= 1, 2, \ldots, n; \quad v = 1, 2, \ldots, M.
\end{aligned} \tag{3.57}
$$

According to Eqs. (4.29) and (4.30) in vol. 1, the drift and diffusion coefficients can be derived as follows

$$
\begin{aligned}
\overline{m}_k(\boldsymbol{H}, \boldsymbol{C}) = {} & \frac{\varepsilon}{T_2}\oint \frac{1}{A_2}\{\frac{\partial H_k}{\partial p_k}[\sum_{r=1}^{n} d_{kr}\frac{\partial H_r}{\partial p_r} + \sum_{i=2n+1}^{m} d_{ki}\frac{\partial}{\partial x_i}(\sum_{\beta=1}^{M} C_\beta)] \\
& + \frac{1}{2}\sum_{s=1}^{l}\sigma_{ks}\sigma_{ks}\frac{\partial^2 H_k}{\partial p_k^2}\}\prod_{r_1=1}^{n}\mathrm{d}q_{r_1}, \\
\overline{m}_v(\boldsymbol{H}, \boldsymbol{C}) = {} & \frac{\varepsilon}{T_2}\oint \frac{1}{A_2}\{\sum_{i_1=2n+1}^{m}\frac{\partial C_v}{\partial x_{i_1}}[\sum_{r=1}^{n} d_{i_1 r}\frac{\partial H_r}{\partial p_r} + \sum_{i=2n+1}^{m} d_{i_1 i}\frac{\partial}{\partial x_i}(\sum_{\beta=1}^{M} C_\beta)] \\
& + \frac{1}{2}\sum_{i,i_1=2n+1}^{m}\sum_{s=1}^{l}\sigma_{is}\sigma_{i_1 s}\frac{\partial^2 C_v}{\partial x_i \partial x_{i_1}}\}\prod_{r_1=1}^{n}\mathrm{d}q_{r_1}, \\
\sum_{s=1}^{l}\overline{\sigma}_{k_1 s}\overline{\sigma}_{k_2 s}(\boldsymbol{H}, \boldsymbol{C}) = {} & \frac{\varepsilon}{T_2}\oint \frac{1}{A_2}\sum_{s=1}^{l}(\sigma_{k_1 s}\sigma_{k_2 s}\frac{\partial H_{k_1}}{\partial p_{k_1}}\frac{\partial H_{k_2}}{\partial p_{k_2}})\prod_{r_1=1}^{n}\mathrm{d}q_{r_1}, \\
\sum_{s=1}^{l}\overline{\sigma}_{v_1 s}\overline{\sigma}_{v_2 s}(\boldsymbol{H}, \boldsymbol{C}) = {} & \frac{\varepsilon}{T_2}\oint \frac{1}{A_2}\sum_{i,i_1=2n+1}^{m}\sum_{s=1}^{l}(\sigma_{is}\sigma_{i_1 s}\frac{\partial C_{v_1}}{\partial x_i}\frac{\partial C_{v_2}}{\partial x_{i_1}})\prod_{r_1=1}^{n}\mathrm{d}q_{r_1},
\end{aligned}
$$

$$\sum_{s=1}^{l}\overline{\sigma}_{k_1 s}\overline{\sigma}_{v_1 s}(\boldsymbol{H},\boldsymbol{C})=\frac{\varepsilon}{T_2}\oint\frac{1}{A_2}\sum_{i=2n+1}^{m}\sum_{s=1}^{l}(\sigma_{k_1 s}\sigma_{is}\frac{\partial H_{k_1}}{\partial p_{k_1}}\frac{\partial C_{v_1}}{\partial x_i})\prod_{r_1=1}^{n}\mathrm{d}q_{r_1},$$

$$A_2=\left|\frac{\partial(q_1,\ldots,q_n,H_1,\ldots,H_n,\boldsymbol{c})}{\partial(q_1,\ldots,q_n,p_1,\ldots,p_n,\boldsymbol{x}_2)}\right|=\prod_{r_1=1}^{n}\frac{\partial H_{r_1}}{\partial p_{r_1}}\left|\frac{\partial\boldsymbol{c}}{\partial\boldsymbol{x}_2}\right|,$$

$$T_2=\oint(1/A_2)\prod_{r_1=1}^{n}\mathrm{d}q_{r_1}=[\oint\prod_{r_1=1}^{n}(\frac{\partial H_{r_1}}{\partial p_{r_1}})^{-1}\mathrm{d}q_{r_1}]\left|\frac{\partial\boldsymbol{c}}{\partial\boldsymbol{x}_2}\right|^{-1}=(\prod_{r_1=1}^{n}T_{r_1})\left|\frac{\partial\boldsymbol{c}}{\partial\boldsymbol{x}_2}\right|^{-1},$$

$$k,k_1,k_2=1,2,\ldots,n;\quad v,v_1,v_2=1,2,\ldots,M. \tag{3.58}$$

Note that in deriving Eq. (3.58) it is assumed each sub-Hamiltonian system associated with $H_k(t)$ has periodic solution with period T. The time averaging is replaced by spatial averaging with respect to $q_1,q_2,\ldots,q_n$. The averaged FPK equation associated with Itô Eq. (3.57) is

$$\begin{aligned}\frac{\partial p}{\partial t}=&-\sum_{k=1}^{n}\frac{\partial}{\partial h_k}(\overline{a}_k p)-\sum_{v=1}^{M}\frac{\partial}{\partial c_v}(\overline{a}_v p)+\frac{1}{2}\sum_{k_1,k_2=1}^{n}\frac{\partial^2}{\partial h_{k_1}\partial h_{k_2}}(\overline{b}_{k_1k_2}p)\\&+\frac{1}{2}\sum_{v_1,v_2=1}^{M}\frac{\partial^2}{\partial c_{v_1}\partial c_{v_2}}(\overline{b}_{v_1v_2}p)+\sum_{k_1=1}^{n}\sum_{v_1=1}^{M}\frac{\partial^2}{\partial h_{k_1}\partial c_{v_1}}(\overline{b}_{k_1v_1}p),\end{aligned} \tag{3.59}$$

where the first and second derivative moments are

$$\begin{aligned}\overline{a}_k&=\overline{a}_k(\boldsymbol{h},\boldsymbol{c})=\overline{m}_k(\boldsymbol{H},\boldsymbol{C})|_{\boldsymbol{H}=\boldsymbol{h},\ \boldsymbol{C}=\boldsymbol{c}},\\ \overline{a}_v&=\overline{a}_v(\boldsymbol{h},\boldsymbol{c})=\overline{m}_v(\boldsymbol{H},\boldsymbol{C})|_{\boldsymbol{H}=\boldsymbol{h},\ \boldsymbol{C}=\boldsymbol{c}},\\ \overline{b}_{k_1k_2}&=\overline{b}_{k_1k_2}(\boldsymbol{h},\boldsymbol{c})=\left.\sum_{s=1}^{l}\overline{\sigma}_{k_1s}\overline{\sigma}_{k_2s}(\boldsymbol{H},\boldsymbol{C})\right|_{\boldsymbol{H}=\boldsymbol{h},\ \boldsymbol{C}=\boldsymbol{c}},\\ \overline{b}_{v_1v_2}&=\overline{b}_{v_1v_2}(\boldsymbol{h},\boldsymbol{c})=\left.\sum_{s=1}^{l}\overline{\sigma}_{v_1s}\overline{\sigma}_{v_2s}(\boldsymbol{H},\boldsymbol{C})\right|_{\boldsymbol{H}=\boldsymbol{h},\ \boldsymbol{C}=\boldsymbol{c}},\\ \overline{b}_{k_1v_1}&=\overline{b}_{k_1v_1}(\boldsymbol{h},\boldsymbol{c})=\left.\sum_{s=1}^{l}\overline{\sigma}_{k_1s}\overline{\sigma}_{v_1s}(\boldsymbol{H},\boldsymbol{C})\right|_{\boldsymbol{H}=\boldsymbol{h},\ \boldsymbol{C}=\boldsymbol{c}}.\end{aligned} \tag{3.60}$$

Once the stationary solution $p(\boldsymbol{h},\boldsymbol{c})$ of the averaged FPK Eq. (3.59) has been obtained, then the approximate PDF of the original system (3.5) is given by

$$p(\boldsymbol{x})=\left.\frac{p(\boldsymbol{h},\boldsymbol{c})}{T_2}=\frac{p(\boldsymbol{h},\boldsymbol{c})}{\prod_{r_1=1}^{n}T_{r_1}}\left|\frac{\partial\boldsymbol{c}}{\partial\boldsymbol{x}_2}\right|\right|_{\boldsymbol{h}=\boldsymbol{h}(\boldsymbol{x}_1),\ \boldsymbol{c}=\boldsymbol{c}(\boldsymbol{x}_2)}. \tag{3.61}$$

3.2.2 Internal Resonant Case

Suppose that the generalized Hamiltonian system (3.5) is completely integrable and resonant. There exist α $(1 \le \alpha \le n-1)$ weak internal resonant relations in Eq. (3.43). Introducing α combinations of angular variables

$$\Psi_u = \sum_{r=1}^{n} k_r^u \Theta_r, \quad u = 1, 2, \ldots, \alpha. \tag{3.62}$$

and using the Itô stochastic differential rule, the Itô stochastic differential equations for Ψ_u can be obtained from Θ_k equations in Eq. (3.42) as follows

$$\begin{aligned} \mathrm{d}\Psi_u = &\{O_u(\varepsilon) + \varepsilon \sum_{i,j=1}^{m} (d_{ij} \frac{\partial \Psi_u}{\partial X_i} \frac{\partial H}{\partial X_j}) + \frac{\varepsilon}{2} \sum_{i,j=1}^{m} \sum_{s=1}^{l} (\sigma_{is}\sigma_{js} \frac{\partial^2 \Psi_u}{\partial X_i \partial X_j})\}\mathrm{d}t \\ &+ \varepsilon^{1/2} \sum_{i=1}^{m} \sum_{s=1}^{l} \sigma_{is} \frac{\partial \Psi_u}{\partial X_i} \mathrm{d}B_s(t), \quad u = 1, 2, \ldots, \alpha. \end{aligned} \tag{3.63}$$

In the new system, with Θ_k $(k = 1, 2, \ldots, \alpha)$ equations in Eq. (3.42) replaced by the α equations in Eq. (3.63). $\boldsymbol{I} = [I_1, \ldots, I_n]^{\mathrm{T}}$, $\boldsymbol{\Psi} = [\Psi_1, \ldots, \Psi_\alpha]^{\mathrm{T}}$ and $\boldsymbol{C} = [C_1, \ldots, C_M]^{\mathrm{T}}$ are slowly-varying processes while the vector $\Theta' = [\Theta_{r_1}, \ldots, \Theta_{r_{n-\alpha}}]^{\mathrm{T}}$ of remaining angles are rapidly-varying processes. According to the Khasminskii theorem (Khasminskii 1968), as $\varepsilon \to 0$, $[\boldsymbol{I}^{\mathrm{T}}, \boldsymbol{\Psi}^{\mathrm{T}}, \boldsymbol{C}^{\mathrm{T}}]^{\mathrm{T}}$ converges weakly to a $(n+\alpha+M)$-dimensional vector Markov diffusion process. The averaged Itô stochastic differential equations for this process are

$$\begin{aligned} \mathrm{d}I_k &= \overline{\overline{m}}_k(\boldsymbol{I}, \boldsymbol{\Psi}, \boldsymbol{C})\mathrm{d}t + \sum_{s=1}^{l} \overline{\overline{\sigma}}_{ks}(\boldsymbol{I}, \boldsymbol{\Psi}, \boldsymbol{C})\mathrm{d}B_s(t), \\ \mathrm{d}\Psi_u &= \overline{\overline{m}}_u(\boldsymbol{I}, \boldsymbol{\Psi}, \boldsymbol{C})\mathrm{d}t + \sum_{s=1}^{l} \overline{\overline{\sigma}}_{us}(\boldsymbol{I}, \boldsymbol{\Psi}, \boldsymbol{C})\mathrm{d}B_s(t), \\ \mathrm{d}C_v &= \overline{\overline{m}}_v(\boldsymbol{I}, \boldsymbol{\Psi}, \boldsymbol{C})\mathrm{d}t + \sum_{s=1}^{l} \overline{\overline{\sigma}}_{vs}(\boldsymbol{I}, \boldsymbol{\Psi}, \boldsymbol{C})\mathrm{d}B_s(t), \\ k &= 1, 2, \ldots, n; \quad u = 1, 2, \ldots, \alpha; \quad v = 1, 2, \ldots, M. \end{aligned} \tag{3.64}$$

According to Eqs. (4.29) and (4.30) in vol. 1, the drift and diffusion coefficients in Eq. (3.64) are obtained by performing time averaging over the corresponding coefficients in Eqs. (3.42) and (3.63). Since the generalized Hamiltonian system associated with original system (3.5) is complete integrable and resonant, it is ergodic on the $(n-\alpha)$-dimensional sub-torus with constant $\boldsymbol{I}$, $\boldsymbol{\Psi}$, $\boldsymbol{C}$. The time averaging can be replaced by phase averaging with respect to θ'. Therefore, the coefficients are

$$
\begin{aligned}
\overline{\overline{m}}_k(\boldsymbol{I}, \boldsymbol{\Psi}, \boldsymbol{C}) &= \frac{\varepsilon}{T_3}\int_0^{2\pi}\frac{1}{A_3}\Big(\sum_{i,j=1}^{m} d_{ij}\frac{\partial I_k}{\partial x_i}\frac{\partial H}{\partial x_j} \\
&\quad + \frac{1}{2}\sum_{i,j=1}^{m}\sum_{s=1}^{l}\sigma_{is}\sigma_{js}\frac{\partial^2 I_k}{\partial x_i \partial x_j}\Big)\mathrm{d}\theta', \\
\overline{\overline{m}}_u(\boldsymbol{I}, \boldsymbol{\Psi}, \boldsymbol{C}) &= \frac{\varepsilon}{T_3}\int_0^{2\pi}\frac{1}{A_3}\Big[\frac{O(\varepsilon)}{\varepsilon} + \sum_{i,j=1}^{m} d_{ij}\frac{\partial \Psi_u}{\partial x_i}\frac{\partial H}{\partial x_j} \\
&\quad + \frac{1}{2}\sum_{i,j=1}^{m}\sum_{s=1}^{l}\sigma_{is}\sigma_{js}\frac{\partial^2 \Psi_u}{\partial x_i \partial x_j}\Big]\mathrm{d}\theta', \\
\overline{\overline{m}}_v(\boldsymbol{I}, \boldsymbol{\Psi}, \boldsymbol{C}) &= \frac{\varepsilon}{T_3}\int_0^{2\pi}\frac{1}{A_3}\Big(\sum_{i,j=1}^{m} d_{ij}\frac{\partial C_v}{\partial x_i}\frac{\partial H}{\partial x_j} \\
&\quad + \frac{1}{2}\sum_{i,j=1}^{m}\sum_{s=1}^{l}\sigma_{is}\sigma_{js}\frac{\partial^2 C_v}{\partial x_i \partial x_j}\Big)\mathrm{d}\theta',
\end{aligned}
$$

$$
\begin{aligned}
&\sum_{s=1}^{l}\overline{\overline{\sigma}}_{k_1 s}\overline{\overline{\sigma}}_{k_2 s}(\boldsymbol{I}, \boldsymbol{\Psi}, \boldsymbol{C}) = \frac{\varepsilon}{T_3}\int_0^{2\pi}\frac{1}{A_3}\sum_{i,j=1}^{m}\sum_{s=1}^{l}\sigma_{is}\sigma_{js}\frac{\partial I_{k_1}}{\partial x_i}\frac{\partial I_{k_2}}{\partial x_j}\mathrm{d}\theta', \\
&\sum_{s=1}^{l}\overline{\overline{\sigma}}_{u_1 s}\overline{\overline{\sigma}}_{u_2 s}(\boldsymbol{I}, \boldsymbol{\Psi}, \boldsymbol{C}) = \frac{\varepsilon}{T_3}\int_0^{2\pi}\frac{1}{A_3}\sum_{i,j=1}^{m}\sum_{s=1}^{l}\sigma_{is}\sigma_{js}\frac{\partial \Psi_{u_1}}{\partial x_i}\frac{\partial \Psi_{u_2}}{\partial x_j}\mathrm{d}\theta', \\
&\sum_{s=1}^{l}\overline{\overline{\sigma}}_{v_1 s}\overline{\overline{\sigma}}_{v_2 s}(\boldsymbol{I}, \boldsymbol{\Psi}, \boldsymbol{C}) = \frac{\varepsilon}{T_3}\int_0^{2\pi}\frac{1}{A_3}\sum_{i,j=1}^{m}\sum_{s=1}^{l}\sigma_{is}\sigma_{js}\frac{\partial C_{v_1}}{\partial x_i}\frac{\partial C_{v_2}}{\partial x_j}\mathrm{d}\theta', \\
&\sum_{s=1}^{l}\overline{\overline{\sigma}}_{k_1 s}\overline{\overline{\sigma}}_{u_1 s}(\boldsymbol{I}, \boldsymbol{\Psi}, \boldsymbol{C}) = \frac{\varepsilon}{T_3}\int_0^{2\pi}\frac{1}{A_3}\sum_{i,j=1}^{m}\sum_{s=1}^{l}\sigma_{is}\sigma_{js}\frac{\partial I_{k_1}}{\partial x_i}\frac{\partial \Psi_{u_1}}{\partial x_j}\mathrm{d}\theta', \\
&\sum_{s=1}^{l}\overline{\overline{\sigma}}_{k_1 s}\overline{\overline{\sigma}}_{v_1 s}(\boldsymbol{I}, \boldsymbol{\Psi}, \boldsymbol{C}) = \frac{\varepsilon}{T_3}\int_0^{2\pi}\frac{1}{A_3}\sum_{i,j=1}^{m}\sum_{s=1}^{l}\sigma_{is}\sigma_{js}\frac{\partial I_{k_1}}{\partial x_i}\frac{\partial C_{v_1}}{\partial x_j}\mathrm{d}\theta', \\
&\sum_{s=1}^{l}\overline{\overline{\sigma}}_{v_1 s}\overline{\overline{\sigma}}_{u_1 s}(\boldsymbol{I}, \boldsymbol{\Psi}, \boldsymbol{C}) = \frac{\varepsilon}{T_3}\int_0^{2\pi}\frac{1}{A_3}\sum_{i,j=1}^{m}\sum_{s=1}^{l}\sigma_{is}\sigma_{js}\frac{\partial C_{v_1}}{\partial x_i}\frac{\partial \Psi_{u_1}}{\partial x_j}\mathrm{d}\theta', \\
&T_3 = \int_0^{2\pi}(1/A_3)\mathrm{d}\theta', \\
&k, k_1, k_2 = 1, 2, \ldots, n, \quad u, u_1, u_2 = 1, 2, \ldots, \alpha, \quad v, v_1, v_2 = 1, 2, \ldots, M,
\end{aligned}
\tag{3.65}
$$

where $\int_0^{2\pi}(\cdot)\mathrm{d}\theta'$ is a $(n-\alpha)$-fold integral with respect to $\theta_{r_1}, \ldots, \theta_{r_{n-\alpha}}$, and $A_3 = \left|\partial(\boldsymbol{I}, \boldsymbol{\psi}, \theta', \boldsymbol{c}) / \partial \boldsymbol{x}\right|$ is the Jacobian determinant of the transformation from $\boldsymbol{x}$ to $\boldsymbol{I}, \boldsymbol{\psi}, \theta', \boldsymbol{c}$.

The FPK equation associated with Itô Eq. (3.64) is

$$\begin{aligned}\frac{\partial p}{\partial t} &= -\sum_{k=1}^{n}\frac{\partial}{\partial I_k}(\bar{\bar{a}}_k p) - \sum_{u=1}^{\alpha}\frac{\partial}{\partial \psi_u}(\bar{\bar{a}}_u p) - \sum_{v=1}^{M}\frac{\partial}{\partial c_v}(\bar{\bar{a}}_v p) + \frac{1}{2}\sum_{k_1,k_2=1}^{n}\frac{\partial^2}{\partial I_{k_1}\partial I_{k_2}}(\bar{\bar{b}}_{k_1k_2}p)\\ &+\frac{1}{2}\sum_{u_1,u_2=1}^{\alpha}\frac{\partial^2}{\partial \psi_{u_1}\partial \psi_{u_2}}(\bar{\bar{b}}_{u_1u_2}p) + \frac{1}{2}\sum_{v_1,v_2=1}^{M}\frac{\partial^2}{\partial c_{v_1}\partial c_{v_2}}(\bar{\bar{b}}_{v_1v_2}p) + \sum_{k_1=1}^{n}\sum_{u_1=1}^{\alpha}\frac{\partial^2}{\partial I_{k_1}\partial \psi_{u_1}}(\bar{\bar{b}}_{k_1u_1}p)\\ &+\sum_{u_1=1}^{\alpha}\sum_{v_1=1}^{M}\frac{\partial^2}{\partial \psi_{u_1}\partial c_{v_1}}(\bar{\bar{b}}_{u_1v_1}p) + \sum_{k_1=1}^{n}\sum_{v_1=1}^{M}\frac{\partial^2}{\partial I_{k_1}\partial c_{v_1}}(\bar{\bar{b}}_{k_1v_1}p),\end{aligned} \tag{3.66}$$

where

$$\begin{aligned}\bar{\bar{a}}_k &= \bar{\bar{a}}_k(\boldsymbol{I}, \boldsymbol{\psi}, \boldsymbol{c}) = \bar{\bar{m}}_k(\boldsymbol{I}, \boldsymbol{\Psi}, \boldsymbol{C})\big|_{\boldsymbol{\Psi}=\boldsymbol{\psi},\, \boldsymbol{C}=\boldsymbol{c}},\\ \bar{\bar{a}}_u &= \bar{\bar{a}}_u(\boldsymbol{I}, \boldsymbol{\psi}, \boldsymbol{c}) = \bar{\bar{m}}_u(\boldsymbol{I}, \boldsymbol{\Psi}, \boldsymbol{C})\big|_{\boldsymbol{\Psi}=\boldsymbol{\psi},\, \boldsymbol{C}=\boldsymbol{c}},\\ \bar{\bar{a}}_v &= \bar{\bar{a}}_v(\boldsymbol{I}, \boldsymbol{\psi}, \boldsymbol{c}) = \bar{\bar{m}}_v(\boldsymbol{I}, \boldsymbol{\Psi}, \boldsymbol{C})\big|_{\boldsymbol{\Psi}=\boldsymbol{\psi},\, \boldsymbol{C}=\boldsymbol{c}},\\ \bar{\bar{b}}_{k_1k_2} &= \bar{\bar{b}}_{k_1k_2}(\boldsymbol{I}, \boldsymbol{\psi}, \boldsymbol{c}) = \left.\sum_{s=1}^{l}\bar{\bar{\sigma}}_{k_1s}\bar{\bar{\sigma}}_{k_2s}(\boldsymbol{I}, \boldsymbol{\Psi}, \boldsymbol{C})\right|_{\boldsymbol{\Psi}=\boldsymbol{\psi},\, \boldsymbol{C}=\boldsymbol{c}},\\ \bar{\bar{b}}_{u_1u_2} &= \bar{\bar{b}}_{u_1u_2}(\boldsymbol{I}, \boldsymbol{\psi}, \boldsymbol{c}) = \left.\sum_{s=1}^{l}\bar{\bar{\sigma}}_{u_1s}\bar{\bar{\sigma}}_{u_2s}(\boldsymbol{I}, \boldsymbol{\Psi}, \boldsymbol{C})\right|_{\boldsymbol{\Psi}=\boldsymbol{\psi},\, \boldsymbol{C}=\boldsymbol{c}},\\ \bar{\bar{b}}_{v_1v_2} &= \bar{\bar{b}}_{v_1v_2}(\boldsymbol{I}, \boldsymbol{\psi}, \boldsymbol{c}) = \left.\sum_{s=1}^{l}\bar{\bar{\sigma}}_{v_1s}\bar{\bar{\sigma}}_{v_2s}(\boldsymbol{I}, \boldsymbol{\Psi}, \boldsymbol{C})\right|_{\boldsymbol{\Psi}=\boldsymbol{\psi},\, \boldsymbol{C}=\boldsymbol{c}},\\ \bar{\bar{b}}_{k_1u_1} &= \bar{\bar{b}}_{k_1u_1}(\boldsymbol{I}, \boldsymbol{\psi}, \boldsymbol{c}) = \left.\sum_{s=1}^{l}\bar{\bar{\sigma}}_{k_1s}\bar{\bar{\sigma}}_{u_1s}(\boldsymbol{I}, \boldsymbol{\Psi}, \boldsymbol{C})\right|_{\boldsymbol{\Psi}=\boldsymbol{\psi},\, \boldsymbol{C}=\boldsymbol{c}},\\ \bar{\bar{b}}_{k_1v_1} &= \bar{\bar{b}}_{k_1v_1}(\boldsymbol{I}, \boldsymbol{\psi}, \boldsymbol{c}) = \left.\sum_{s=1}^{l}\bar{\bar{\sigma}}_{k_1s}\bar{\bar{\sigma}}_{v_1s}(\boldsymbol{I}, \boldsymbol{\Psi}, \boldsymbol{C})\right|_{\boldsymbol{\Psi}=\boldsymbol{\psi},\, \boldsymbol{C}=\boldsymbol{c}},\\ \bar{\bar{b}}_{u_1v_1} &= \bar{\bar{b}}_{u_1v_1}(\boldsymbol{I}, \boldsymbol{\psi}, \boldsymbol{c}) = \left.\sum_{s=1}^{l}\bar{\bar{\sigma}}_{u_1s}\bar{\bar{\sigma}}_{v_1s}(\boldsymbol{I}, \boldsymbol{\Psi}, \boldsymbol{C})\right|_{\boldsymbol{\Psi}=\boldsymbol{\psi},\, \boldsymbol{C}=\boldsymbol{c}}.\end{aligned} \tag{3.67}$$

In Eq. (3.66), p could be the transition PDF $p = p(\boldsymbol{I}, \boldsymbol{\psi}, \boldsymbol{c}, t|\boldsymbol{I}_0, \boldsymbol{\psi}_0, \boldsymbol{c}_0)$, or the joint PDF $p = p(\boldsymbol{I}, \boldsymbol{\psi}, \boldsymbol{c}, t)$. Equation (3.66) can be solved under certain initial condition and boundary conditions including periodic boundary conditions $p|_{\psi_u+2k\pi} = p|_{\psi_u}$ for ψ_u, where k is an integer.

After obtaining the stationary solution of Eq. (3.66), the approximate stationary PDF for the original system (3.5) can be calculated as follows

$$p(\boldsymbol{x}) = p(\boldsymbol{I}, \boldsymbol{\psi}, \theta', \boldsymbol{c})\left|\frac{\partial(\boldsymbol{I}, \boldsymbol{\psi}, \theta', \boldsymbol{c})}{\partial \boldsymbol{x}}\right| = p(\theta'|\boldsymbol{I}, \boldsymbol{\psi}, \boldsymbol{c})p(\boldsymbol{I}, \boldsymbol{\psi}, \boldsymbol{c})A_3. \tag{3.68}$$

For a fixed $\boldsymbol{I}, \boldsymbol{\psi}, \boldsymbol{c}$, the conditional PDF $p(\theta'|\boldsymbol{I}, \boldsymbol{\psi}, \boldsymbol{c})$ is inversely proportional to $A_3 = \left|\partial(\boldsymbol{I}, \boldsymbol{\psi}, \theta', \boldsymbol{c})/\partial \boldsymbol{x}\right|$, thus,

$$p(\theta'|\boldsymbol{I}, \boldsymbol{\psi}, \boldsymbol{c}) = C'/A_3. \tag{3.69}$$

where C' is a constant. Normalizing over Eq. (3.69) with respect to θ', then substituting it into Eq. (3.68) yield the following approximate stationary PDF for the original system (3.5)

$$p(\boldsymbol{x}) = \left.\frac{p(\boldsymbol{I}, \boldsymbol{\psi}, \boldsymbol{c})}{T_3(\boldsymbol{I}, \boldsymbol{\psi}, \boldsymbol{c})}\right|_{\boldsymbol{I}=\boldsymbol{I}(\boldsymbol{x}),\ \boldsymbol{\psi}=\boldsymbol{\psi}(\boldsymbol{x}),\ \boldsymbol{c}=\boldsymbol{c}(\boldsymbol{x})}. \tag{3.70}$$

Similar to Eq. (3.51), if the variables of generalized Hamiltonian system associated with Eqs. (3.42) and (3.63) are separable, the Hamiltonian subsystem with Hamiltonian function H_1 is completely integrable and resonant, the action angle vectors $\boldsymbol{I}(\boldsymbol{x}_1), \boldsymbol{\theta}(\boldsymbol{x}_1)$ have been known, the Itô stochastic differential equation (3.52) governing vector process $[\boldsymbol{I}^{\mathrm{T}}, \boldsymbol{\Theta}^{\mathrm{T}}, \boldsymbol{C}^{\mathrm{T}}]^{\mathrm{T}}$ is still valid, then using Eq. (3.62), the Itô stochastic differential equations for Ψ_u can be derived from the Θ_k equations in Eq. (3.52)

$$\begin{aligned}
\mathrm{d}\Psi_u = &\{O_u(\varepsilon) + \varepsilon\sum_{j=1}^{n}\frac{\partial\Psi_u}{\partial P_j}[\sum_{r=1}^{n}d_{jr}\frac{\partial H_1}{\partial P_r} + \sum_{i=2n+1}^{m}d_{ji}\frac{\partial}{\partial X_i}(\sum_{\beta=1}^{M}C_\beta)] \\
&+\frac{1}{2}\sum_{j,r=1}^{n}\sum_{s=1}^{l}\sigma_{js}\sigma_{rs}\frac{\partial^2\Psi_u}{\partial P_j\partial P_r}\}\mathrm{d}t + \varepsilon^{1/2}\sum_{j=1}^{n}\sum_{s=1}^{l}\sigma_{js}\frac{\partial\Psi_u}{\partial P_j}\mathrm{d}B_s(t), \\
&u = 1, 2, \ldots, \alpha.
\end{aligned} \tag{3.71}$$

The new system is Eq. (3.52) with α equations for Θ_k replaced by Eq. (3.71). In this system, $[\boldsymbol{I}^{\mathrm{T}}, \boldsymbol{\Psi}^{\mathrm{T}}, \boldsymbol{C}^{\mathrm{T}}]^{\mathrm{T}}$ is slowly-varying process. According to the Khasminskii theorem (Khasminskii 1968), it converges weakly to a $(n+\alpha+M)$-dimensional vector Markov diffusion process as $\varepsilon \to 0$. The averaged Itô stochastic differential equations for this $(n+\alpha+M)$-dimensional vector Markov diffusion process $[\boldsymbol{I}^{\mathrm{T}}, \boldsymbol{\Psi}^{\mathrm{T}}, \boldsymbol{C}^{\mathrm{T}}]^{\mathrm{T}}$ are still of the form as Eq. (3.64), but the drift and diffusion coefficients are modified as follows

$$\overline{\overline{m'}}_k(\mathbf{I},\boldsymbol{\Psi},\mathbf{C})=\frac{\varepsilon}{T_3'}\int_0^{2\pi}\frac{1}{A_3'}\{\sum_{j=1}^{n}\frac{\partial I_k}{\partial p_j}[\sum_{r=1}^{n}d_{jr}\frac{\partial H_1}{\partial p_r}+\sum_{i=2n+1}^{m}d_{ji}\frac{\partial}{\partial x_i}(\sum_{\beta=1}^{M}C_\beta)]$$
$$+\frac{1}{2}\sum_{j,r=1}^{n}\sum_{s=1}^{l}\sigma_{js}\sigma_{rs}\frac{\partial^2 I_k}{\partial p_j\partial p_r}\}\mathrm{d}\theta',$$
$$\overline{\overline{m'}}_u(\mathbf{I},\boldsymbol{\Psi},\mathbf{C})=\frac{\varepsilon}{T_3'}\int_0^{2\pi}\frac{1}{A_3'}\{\frac{O_u(\varepsilon)}{\varepsilon}+\sum_{j=1}^{n}\frac{\partial\Psi_u}{\partial p_j}[\sum_{r=1}^{n}d_{jr}\frac{\partial H_1}{\partial p_r}+\sum_{i=2n+1}^{m}d_{ji}\frac{\partial}{\partial x_i}(\sum_{\beta=1}^{M}C_\beta)]$$
$$+\frac{1}{2}\sum_{j,r=1}^{n}\sum_{s=1}^{l}\sigma_{js}\sigma_{rs}\frac{\partial^2\Psi_u}{\partial x_j\partial x_r}\}\mathrm{d}\theta',$$
$$\overline{\overline{m'}}_v(\mathbf{I},\boldsymbol{\Psi},\mathbf{C})=\frac{\varepsilon}{T_3'}\int_0^{2\pi}\frac{1}{A_3'}\{\sum_{i_1=2n+1}^{m}\frac{\partial C_v}{\partial x_{i_1}}[\sum_{r=1}^{n}d_{i_1 r}\frac{\partial H_1}{\partial p_r}+\sum_{i=2n+1}^{m}d_{i_1 i}\frac{\partial}{\partial x_i}(\sum_{\beta=1}^{M}C_\beta)]$$
$$+\frac{1}{2}\sum_{i,i_1=2n+1}^{m}\sum_{s=1}^{l}\sigma_{is}\sigma_{i_1 s}\frac{\partial^2 C_v}{\partial x_i\partial x_{i_1}}\}\mathrm{d}\theta',$$
$$\sum_{s=1}^{l}\overline{\overline{\sigma'}}_{k_1 s}\overline{\overline{\sigma'}}_{k_2 s}(\mathbf{I},\boldsymbol{\Psi},\mathbf{C})=\frac{\varepsilon}{T_3'}\int_0^{2\pi}\frac{1}{A_3'}\sum_{j,r=1}^{n}\sum_{s=1}^{l}(\sigma_{js}\sigma_{rs}\frac{\partial I_{k_1}}{\partial p_j}\frac{\partial I_{k_2}}{\partial p_r})\mathrm{d}\theta', \tag{3.72}$$
$$\sum_{s=1}^{l}\overline{\overline{\sigma'}}_{u_1 s}\overline{\overline{\sigma'}}_{u_2 s}(\mathbf{I},\boldsymbol{\Psi},\mathbf{C})=\frac{\varepsilon}{T_3'}\int_0^{2\pi}\frac{1}{A_3'}\sum_{j,r=1}^{n}\sum_{s=1}^{l}(\sigma_{js}\sigma_{rs}\frac{\partial\Psi_{u_1}}{\partial p_j}\frac{\partial\Psi_{u_2}}{\partial p_r})\mathrm{d}\theta',$$
$$\sum_{s=1}^{l}\overline{\overline{\sigma'}}_{v_1 s}\overline{\overline{\sigma'}}_{v_2 s}(\mathbf{I},\boldsymbol{\Psi},\mathbf{C})=\frac{\varepsilon}{T_3'}\int_0^{2\pi}\frac{1}{A_3'}\sum_{i,i_1=2n+1}^{m}\sum_{s=1}^{l}(\sigma_{is}\sigma_{i_1 s}\frac{\partial C_{v_1}}{\partial x_i}\frac{\partial C_{v_2}}{\partial x_{i_1}})\mathrm{d}\theta',$$
$$\sum_{s=1}^{l}\overline{\overline{\sigma'}}_{k_1 s}\overline{\overline{\sigma'}}_{u_1 s}(\mathbf{I},\boldsymbol{\Psi},\mathbf{C})=\frac{\varepsilon}{T_3'}\int_0^{2\pi}\frac{1}{A_3'}\sum_{j,r=1}^{n}\sum_{s=1}^{l}(\sigma_{js}\sigma_{rs}\frac{\partial I_{k_1}}{\partial p_j}\frac{\partial\Psi_{u_1}}{\partial p_r})\mathrm{d}\theta',$$
$$\sum_{s=1}^{l}\overline{\overline{\sigma'}}_{k_1 s}\overline{\overline{\sigma'}}_{v_1 s}(\mathbf{I},\boldsymbol{\Psi},\mathbf{C})=\frac{\varepsilon}{T_3'}\int_0^{2\pi}\frac{1}{A_3'}\sum_{j=1}^{n}\sum_{i=2n+1}^{m}\sum_{s=1}^{l}(\sigma_{js}\sigma_{is}\frac{\partial I_{k_1}}{\partial p_j}\frac{\partial C_{v_1}}{\partial x_i})\mathrm{d}\theta',$$
$$\sum_{s=1}^{l}\overline{\overline{\sigma'}}_{u_1 s}\overline{\overline{\sigma'}}_{v_1 s}(\mathbf{I},\boldsymbol{\Psi},\mathbf{C})=\frac{\varepsilon}{T_3'}\int_0^{2\pi}\frac{1}{A_3'}\sum_{j=1}^{n}\sum_{i=2n+1}^{m}\sum_{s=1}^{l}(\sigma_{js}\sigma_{is}\frac{\partial\Psi_{u_1}}{\partial p_j}\frac{\partial C_{v_1}}{\partial x_i})\mathrm{d}\theta',$$
$$T_3'=\int_0^{2\pi}(1/A_3')\mathrm{d}\theta',$$
$$k,k_1,k_2=1,2,\ldots,n,\quad u,u_1,u_2=1,2,\ldots,\alpha,\quad v,v_1,v_2=1,2,\ldots,M,$$

where $A_3'=\left|\frac{\partial(\boldsymbol{I},\boldsymbol{\psi},\theta',\boldsymbol{c})}{\partial(\boldsymbol{x}_1,\boldsymbol{x}_2)}\right|=\left|\frac{\partial(\boldsymbol{I},\boldsymbol{\psi},\theta')}{\partial(\boldsymbol{q},\boldsymbol{p})}\right|\left|\frac{\partial\boldsymbol{c}}{\partial\boldsymbol{x}_2}\right|$, $\left|\frac{\partial(\boldsymbol{I},\boldsymbol{\psi},\theta')}{\partial(\boldsymbol{q},\boldsymbol{p})}\right|$ is a linear combination of $\left|\frac{\partial(\boldsymbol{I},\boldsymbol{\theta})}{\partial(\boldsymbol{q},\boldsymbol{p})}\right|$, value of the latter is 1. Thus, $\left|\frac{\partial(\boldsymbol{I},\boldsymbol{\psi},\theta')}{\partial(\boldsymbol{q},\boldsymbol{p})}\right|$ should be a constant $\overline{\overline{C}}$, and $T_3'=(2\pi)^{n-\alpha}(\overline{\overline{C}}\left|\frac{\partial\boldsymbol{c}}{\partial\boldsymbol{x}_2}\right|)^{-1}$.

The averaged FPK equation associated with Itô Eq. (3.64) remains of the form of Eq. (3.66), but the first and second derivative moments are changed to

$$\begin{aligned}
\overline{\overline{a}}'_k &= \overline{\overline{a}}'_k(\boldsymbol{I},\boldsymbol{\psi},\boldsymbol{c}) = \overline{\overline{m}}'_k(\boldsymbol{I},\boldsymbol{\Psi},\boldsymbol{C})\Big|_{\boldsymbol{\Psi}=\boldsymbol{\psi},\ \boldsymbol{C}=\boldsymbol{c}},\\
\overline{\overline{a}}'_u &= \overline{\overline{a}}'_u(\boldsymbol{I},\boldsymbol{\psi},\boldsymbol{c}) = \overline{\overline{m}}'_u(\boldsymbol{I},\boldsymbol{\Psi},\boldsymbol{C})\Big|_{\boldsymbol{\Psi}=\boldsymbol{\psi},\ \boldsymbol{C}=\boldsymbol{c}},\\
\overline{\overline{a}}'_v &= \overline{\overline{a}}'_v(\boldsymbol{I},\boldsymbol{\psi},\boldsymbol{c}) = \overline{\overline{m}}'_v(\boldsymbol{I},\boldsymbol{\Psi},\boldsymbol{C})\Big|_{\boldsymbol{\Psi}=\boldsymbol{\psi},\ \boldsymbol{C}=\boldsymbol{c}},\\
\overline{\overline{b}}'_{k_1k_2} &= \overline{\overline{b}}'_{k_1k_2}(\boldsymbol{I},\boldsymbol{\psi},\boldsymbol{c}) = \sum_{s=1}^{l}\overline{\overline{\sigma}}'_{k_1s}\overline{\overline{\sigma}}'_{k_2s}(\boldsymbol{I},\boldsymbol{\Psi},\boldsymbol{C})\Bigg|_{\boldsymbol{\Psi}=\boldsymbol{\psi},\ \boldsymbol{C}=\boldsymbol{c}},\\
\overline{\overline{b}}'_{u_1u_2} &= \overline{\overline{b}}'_{u_1u_2}(\boldsymbol{I},\boldsymbol{\psi},\boldsymbol{c}) = \sum_{s=1}^{l}\overline{\overline{\sigma}}'_{u_1s}\overline{\overline{\sigma}}'_{u_2s}(\boldsymbol{I},\boldsymbol{\Psi},\boldsymbol{C})\Bigg|_{\boldsymbol{\Psi}=\boldsymbol{\psi},\ \boldsymbol{C}=\boldsymbol{c}},\\
\overline{\overline{b}}'_{v_1v_2} &= \overline{\overline{b}}'_{v_1v_2}(\boldsymbol{I},\boldsymbol{\psi},\boldsymbol{c}) = \sum_{s=1}^{l}\overline{\overline{\sigma}}'_{v_1s}\overline{\overline{\sigma}}'_{v_2s}(\boldsymbol{I},\boldsymbol{\Psi},\boldsymbol{C})\Bigg|_{\boldsymbol{\Psi}=\boldsymbol{\psi},\ \boldsymbol{C}=\boldsymbol{c}},\\
\overline{\overline{b}}'_{k_1u_1} &= \overline{\overline{b}}'_{k_1u_1}(\boldsymbol{I},\boldsymbol{\psi},\boldsymbol{c}) = \sum_{s=1}^{l}\overline{\overline{\sigma}}'_{k_1s}\overline{\overline{\sigma}}'_{u_1s}(\boldsymbol{I},\boldsymbol{\Psi},\boldsymbol{C})\Bigg|_{\boldsymbol{\Psi}=\boldsymbol{\psi},\ \boldsymbol{C}=\boldsymbol{c}},\\
\overline{\overline{b}}'_{k_1v_1} &= \overline{\overline{b}}'_{k_1v_1}(\boldsymbol{I},\boldsymbol{\psi},\boldsymbol{c}) = \sum_{s=1}^{l}\overline{\overline{\sigma}}'_{k_1s}\overline{\overline{\sigma}}'_{v_1s}(\boldsymbol{I},\boldsymbol{\Psi},\boldsymbol{C})\Bigg|_{\boldsymbol{\Psi}=\boldsymbol{\psi},\ \boldsymbol{C}=\boldsymbol{c}},\\
\overline{\overline{b}}'_{u_1v_1} &= \overline{\overline{b}}'_{u_1v_1}(\boldsymbol{I},\boldsymbol{\psi},\boldsymbol{c}) = \sum_{s=1}^{l}\overline{\overline{\sigma}}'_{u_1s}\overline{\overline{\sigma}}'_{v_1s}(\boldsymbol{I},\boldsymbol{\Psi},\boldsymbol{C})\Bigg|_{\boldsymbol{\Psi}=\boldsymbol{\psi},\ \boldsymbol{C}=\boldsymbol{c}}.
\end{aligned} \tag{3.73}$$

Once the stationary solution of the averaged FPK equation has been obtained, similar to the derivation from Eq. (3.68) to Eq. (3.70), the approximate stationary PDF for the original system (3.5) can be derived as

$$p(\boldsymbol{x}) = \frac{\overline{\overline{C}}}{(2\pi)^{n-\alpha}}p(\boldsymbol{I},\boldsymbol{\psi},\boldsymbol{c})\left|\frac{\partial \boldsymbol{c}}{\partial \boldsymbol{x}_2}\right|. \tag{3.74}$$

where $\overline{\overline{C}}$ is a normalization constant.

Example 3.2 Consider a 5-dimensional linear generalized Hamiltonian system subjected to non-linear dampings and Gaussian white noise excitations governed by (Chen and Zhu 2010b).

$$\begin{aligned}
\dot{X}_1 &= X_2,\\
\dot{X}_2 &= -(\alpha_{10}+\alpha_{11}X_2^2+\alpha_{12}X_4^2+\alpha_{13}X_5^2)X_2-\alpha_{14}X_4-\omega_1^2X_1+W_{g1}(t),\\
\dot{X}_3 &= X_4,\\
\dot{X}_4 &= -(\alpha_{20}+\alpha_{21}X_2^2+\alpha_{22}X_4^2+\alpha_{23}X_5^2)X_4-\alpha_{24}X_2-\omega_2^2X_4+W_{g2}(t),\\
\dot{X}_5 &= -(\alpha_{30}+\alpha_{31}X_2^2+\alpha_{32}X_4^2+\alpha_{33}X_5^2)X_5+W_{g3}(t).
\end{aligned} \tag{3.75}$$

where α_{ij} and ω_k $(i, k = 1, 2, 3;\ j = 0, \ldots, 4)$ are constants; $W_{gk}(t)$ $(k = 1, 2, 3)$ are independent Gaussian white noises with intensities $2D_k$.

Equation (3.75) can be rewritten as equations of the form of Eq. (3.1). The generalized Hamiltonian system associated with the system (3.75) has one Casimir function $C = X_5^2/2$ and two independent first integrals $H_1 = (X_2^2 + \omega_1^2 X_1^2)/2$ and $H_2 = (X_4^2 + \omega_2^2 X_3^2)/2$. They are involution. It is integrable and non-resonant if ω_1, ω_2 are not rationally related. Furthermore, H_1, H_2, C represent the energies of three sub-systems while $H = H_1 + H_2 + C$ represents the total energy of system (3.75). In this example, two pairs of action-angle variables I_k, Θ_k $(k = 1, 2)$ are available, i.e.,

$$\begin{aligned} I_1 &= \frac{1}{2\omega_1}(X_2^2 + \omega_1^2 X_1^2), \quad \Theta_1 = \tan^{-1}\left(\frac{X_2}{\omega_1 X_1}\right), \\ I_2 &= \frac{1}{2\omega_2}(X_4^2 + \omega_2^2 X_3^2), \quad \Theta_2 = \tan^{-1}\left(\frac{X_4}{\omega_2 X_3}\right). \end{aligned} \tag{3.76}$$

By using the stochastic averaging method of quasi-integrable generalized Hamiltonian systems, the averaged Itô equations for C, I_1, I_2 can be obtained as follows:

$$\begin{aligned} \mathrm{d}C &= \overline{m'}_1(I_1, I_2, C)\mathrm{d}t + \overline{\sigma'}_{11}(I_1, I_2, C)\mathrm{d}B_1, \\ \mathrm{d}I_1 &= \overline{m'}_2(I_1, I_2, C)\mathrm{d}t + \overline{\sigma'}_{22}(I_1, I_2, C)\mathrm{d}B_2, \\ \mathrm{d}I_2 &= \overline{m'}_3(I_1, I_2, C)\mathrm{d}t + \overline{\sigma'}_{33}(I_1, I_2, C)\mathrm{d}B_3. \end{aligned} \tag{3.77}$$

where

$$\begin{aligned} \overline{m'}_1 &= D_3 - 2C(\alpha_{30} + \alpha_{31}\omega_1 I_1 + \alpha_{32}\omega_2 I_2 + \alpha_{33}C), \\ \overline{m'}_2 &= \frac{D_1}{\omega_1} - I_1(\alpha_{10} + 1.5\alpha_{11}\omega_1 I_1 + \alpha_{12}\omega_2 I_2 + \alpha_{13}C), \\ \overline{m'}_3 &= \frac{D_2}{\omega_2} - I_2(\alpha_{20} + \alpha_{21}\omega_1 I_1 + 1.5\alpha_{22}\omega_2 I_2 + \alpha_{23}C), \\ \overline{b}'_{11} &= \overline{\sigma'}_{11}\overline{\sigma'}_{11} = 4D_3 C, \quad \overline{b}'_{22} = \overline{\sigma'}_{22}\overline{\sigma'}_{22} = 2D_1 I_1/\omega_1, \\ \overline{b}'_{33} &= \overline{\sigma'}_{33}\overline{\sigma'}_{33} = 2D_2 I_2/\omega_2, \\ \overline{b}'_{ij} &= 0, \quad i, j = 1, 2, 3 \text{ and } i \neq j. \end{aligned} \tag{3.78}$$

Since $H_k = \omega_k I_k$ $(k = 1, 2)$, the equations for I_k in Eq. (3.77) can be replaced by those for H_k. The later can be obtained from the former by using Itô differential rule. The resultant averaged Itô equations for C, H_1, H_2 are of the following form:

$$\begin{aligned} \mathrm{d}C &= \overline{m}_1(H_1, H_2, C)\mathrm{d}t + \overline{\sigma}_{11}(H_1, H_2, C)\mathrm{d}B_1, \\ \mathrm{d}H_1 &= \overline{m}_2(H_1, H_2, C)\mathrm{d}t + \overline{\sigma}_{22}(H_1, H_2, C)\mathrm{d}B_2, \\ \mathrm{d}H_2 &= \overline{m}_3(H_1, H_2, C)\mathrm{d}t + \overline{\sigma}_{33}(H_1, H_2, C)\mathrm{d}B_3. \end{aligned} \tag{3.79}$$

where

$$\begin{aligned}
\overline{m}_1 &= D_3 - 2C(\alpha_{30} + \alpha_{31}H_1 + \alpha_{32}H_2 + \alpha_{33}C),\\
\overline{m}_2 &= D_1 - H_1(\alpha_{10}+1.5\alpha_{11}H_1 + \alpha_{12}H_2 + \alpha_{13}C),\\
\overline{m}_3 &= D_2 - H_2(\alpha_{20} + \alpha_{21}H_1 + 1.5\alpha_{22}H_2 + \alpha_{23}C),\\
\overline{b}_{11} &= \overline{\sigma}_{11}^2 = 4D_3C, \quad \overline{b}_{22} = \overline{\sigma}_{22}^2 = 2D_1H_1,\\
\overline{b}_{33} &= \overline{\sigma}_{33}^2 = 2D_2H_2,\\
\overline{b}_{ij} &= 0, \quad i,j = 1,2,3 \text{ and } i \neq j.
\end{aligned} \tag{3.80}$$

Since the energies of three sub-systems vary from 0 to ∞, the state of the averaged system (3.79) varies randomly in the first quadrant of three-dimensional space C, H_1, H_2. Suppose that the limit state of the system is $H = H_1 + H_2 + C = H_c$, i.e.,

$$\Gamma_c: \quad H_1 + H_2 + C = H_c, \quad H_1, H_2, C \geq 0. \tag{3.81}$$

The safety domain of the system is inside of a tetrahedron with boundaries Γ_c and Γ_0, where Γ_0 consists of

$$\begin{aligned}
\Gamma_{10}: &\quad C = 0, \quad 0 \leq H_1 + H_2 < H_c,\\
\Gamma_{20}: &\quad H_1 = 0, \quad 0 \leq C + H_2 < H_c,\\
\Gamma_{30}: &\quad H_2 = 0, \quad 0 \leq H_1 + C < H_c.
\end{aligned} \tag{3.82}$$

(see Fig. 3.6).

The conditional reliability function $R(t|c_0, h_{10}, h_{20})$ of the system (3.79) is governed by following backward Kolmogorov equation:

$$\frac{\partial R}{\partial t} = \overline{m}_1 \frac{\partial R}{\partial c_0} + \overline{m}_2 \frac{\partial R}{\partial h_{10}} + \overline{m}_3 \frac{\partial R}{\partial h_{20}} + \frac{1}{2}\overline{b}_{11}\frac{\partial^2 R}{\partial c_0^2}$$

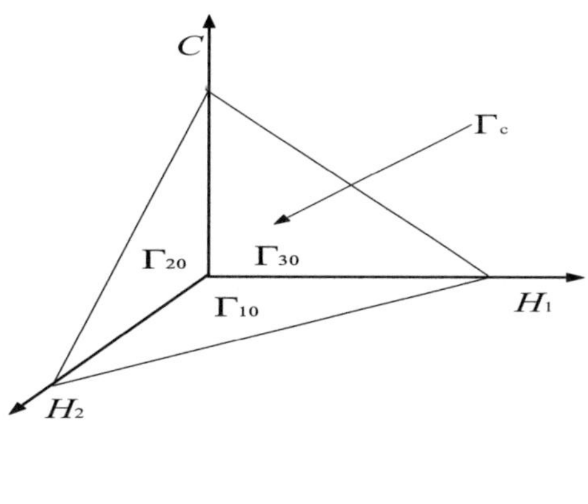

Fig. 3.6 Safety domain of system (3.79). The boundaries Γ_{10}, Γ_{20} and Γ_{30} are the reflecting boundaries. Γ_c is the absorbing boundary (Chen and Zhu 2010b)

$$+\frac{1}{2}\bar{b}_{22}\frac{\partial^2 R}{\partial h_{10}^2}+\frac{1}{2}\bar{b}_{33}\frac{\partial^2 R}{\partial h_{20}^2}. \tag{3.83}$$

where the coefficients are defined by Eq. (3.80) with c, h_1, h_2 replaced by c_0, h_{10}, h_{20}, respectively. The initial condition is defined by

$$R(0|c_0,h_{10},h_{20})=1,\quad (c_0,h_{10},h_{20})\in\Omega_1. \tag{3.84}$$

One boundary condition on Γ_c is defined by

$$R(t|c_0,h_{10},h_{20})=0\quad \text{at}\quad \Gamma_c \tag{3.85}$$

The other qualitative boundary condition on Γ_0 defined by $R(t|c_0,h_{10},h_{20})=$ finite can be converted into the following equations:

$$\begin{aligned}\frac{\partial R}{\partial t}&=[D_1-h_{10}(\alpha_{10}+1.5\alpha_{11}h_{10}+\alpha_{12}h_{20})]\frac{\partial R}{\partial h_{10}}\\&+[D_2-h_{20}(\alpha_{20}+\alpha_{21}h_{10}\ +1.5\alpha_{22}h_{20})]\frac{\partial R}{\partial h_{20}}+D_3\frac{\partial R}{\partial c_0}\\&+D_1h_{10}\frac{\partial^2 R}{\partial h_{10}^2}+D_2h_{20}\frac{\partial^2 R}{\partial h_{20}^2}.\end{aligned} \tag{3.86}$$

at Γ_{10},

$$\begin{aligned}\frac{\partial R}{\partial t}&=[D_3-2c_0(\alpha_{30}+\alpha_{32}h_{20}+\alpha_{33}c_0)]\frac{\partial R}{\partial c_0}\\&+[D_2-h_{20}(\alpha_{20}+1.5\alpha_{22}h_{20}\ +\alpha_{23}c_0)]\frac{\partial R}{\partial h_{20}}+D_1\frac{\partial R}{\partial h_{10}}\\&+D_3c_0\frac{\partial^2 R}{\partial c_0^2}+D_2h_{20}\frac{\partial^2 R}{\partial h_{20}^2}.\end{aligned} \tag{3.87}$$

at Γ_{20},

$$\begin{aligned}\frac{\partial R}{\partial t}&=[D_3-2c_0(\alpha_{30}+\alpha_{31}h_{10}+\alpha_{33}c_0)]\frac{\partial R}{\partial c_0}\\&+[D_1-h_{10}(\alpha_{10}+1.5\alpha_{11}h_{10}\ +\alpha_{13}c_0)]\frac{\partial R}{\partial h_{10}}\\&+D_2\frac{\partial R}{\partial h_{20}}+2D_3c_0\frac{\partial^2 R}{\partial c_0^2}+D_2h_{10}\frac{\partial^2 R}{\partial h_{10}^2}.\end{aligned} \tag{3.88}$$

at Γ_{30}.

Equation (3.83) is a three-dimensional parabolic partial differential equation and can be solved numerically together with the initial condition and boundary conditions by using the finite difference method to yield the conditional reliability function of system (3.79).

Similarly, the Pontryagin equation for the mean first passage time of system (3.79) can be derived from the averaged Itô stochastic differential equations as follows:

$$-1=\overline{m}_1\frac{\partial\mu}{\partial c_0}+\overline{m}_2\frac{\partial\mu}{\partial h_{10}}+\overline{m}_3\frac{\partial\mu}{\partial h_{20}}+\frac{1}{2}\overline{b}_{11}\frac{\partial^2\mu}{\partial c_0^2}+\frac{1}{2}\overline{b}_{22}\frac{\partial^2\mu}{\partial h_{10}^2}+\frac{1}{2}\overline{b}_{33}\frac{\partial^2\mu}{\partial h_{20}^2}. \tag{3.89}$$

in which the coefficients are the same as those in Eq. (3.80). The boundary condition at Γ_c is defined by $\mu(c_0, h_{10}, h_{20})=0$. The boundary condition on Γ_0 is defined by $\mu(c_0, h_{10}, h_{20})=$ finite which can be converted into quantitative condition by using Eq. (3.89) and the drift and diffusion coefficients at Γ_0. And they are

$$\begin{aligned}-1&=[D_1-h_{10}(\alpha_{10}+1.5\alpha_{11}h_{10}+\alpha_{12}h_{20})]\frac{\partial\mu}{\partial h_{10}}\\&+[D_2-h_{20}(\alpha_{20}+\alpha_{21}h_{10}+1.5\alpha_{22}h_{20})]\frac{\partial\mu}{\partial h_{20}}+D_3\frac{\partial\mu}{\partial c_0}\\&+D_1h_{10}\frac{\partial^2\mu}{\partial h_{10}^2}+D_2h_{20}\frac{\partial^2\mu}{\partial h_{20}^2}.\end{aligned} \tag{3.90}$$

at Γ_{10},

$$\begin{aligned}-1&=[D_3-2c_0(\alpha_{30}+\alpha_{32}h_{20}+\alpha_{33}c_0)]\frac{\partial\mu}{\partial c_0}\\&+[D_2-h_{20}(\alpha_{20}+1.5\alpha_{22}h_{20}+\alpha_{23}c_0)]\frac{\partial\mu}{\partial h_{20}}+D_1\frac{\partial\mu}{\partial h_{10}}\\&+D_3c_0\frac{\partial^2\mu}{\partial c_0^2}+D_2h_{20}\frac{\partial^2\mu}{\partial h_{20}^2}.\end{aligned} \tag{3.91}$$

at Γ_{20}, and

$$\begin{aligned}-1&=[D_3-2c_0(\alpha_{30}+\alpha_{31}h_{10}+\alpha_{33}c_0)]\frac{\partial\mu}{\partial c_0}\\&+[D_1-h_{10}(\alpha_{10}+1.5\alpha_{11}h_{10}+\alpha_{13}c_0)]\frac{\partial\mu}{\partial h_{10}}+D_2\frac{\partial\mu}{\partial h_{20}}\\&+2D_3c_0\frac{\partial^2\mu}{\partial c_0^2}+D_2h_{10}\frac{\partial^2\mu}{\partial h_{10}^2}.\end{aligned} \tag{3.92}$$

at Γ_{30}.

The mean first passage time of system (3.79) can be obtained by solving Eq. (3.89) and its boundary conditions numerically.

Some numerical results for the conditional reliability function and the mean first passage time of system (3.79) for two different excitation intensities are shown in Figs. 3.7 and 3.8. The results from the Monte Carlo simulation of system (3.75) are also shown for comparison. In the calculation, the parameters are as follows: $\alpha_{10} = -0.03$, $\alpha_{11} = 0.01$, $\alpha_{12} = 0.02$, $\alpha_{13} = 0.01$, $\alpha_{14} = 0.01$, $\alpha_{20} = -0.03$, $\alpha_{21} = 0.02$, $\alpha_{22} = 0.01$, $\alpha_{23} = 0.01$, $\alpha_{24} = 0.01$, $\alpha_{30} = -0.02$, $\alpha_{31} = 0.02$, $\alpha_{32} = 0.02$, $\alpha_{33} = 0.01$, $\omega_1 = 1.0$, $\omega_2 = 1.142$, $H_c = 0.1$. It is seen from the figures that the two results are in good agreement and higher excitation intensity leads to lower reliability of system.

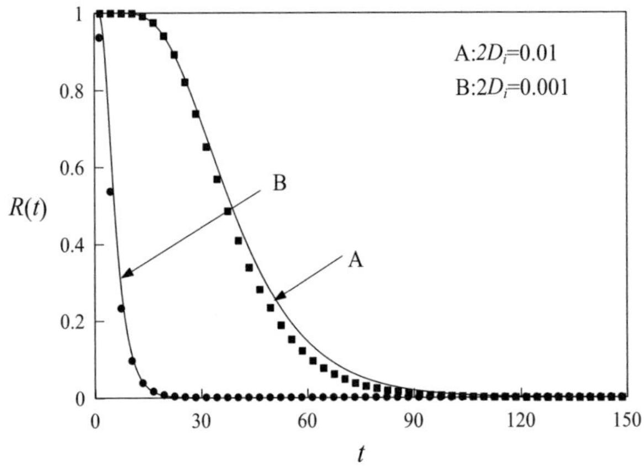

Fig. 3.7 The conditional reliability function $R(t|c_0, h_{10}, h_{20})$ of system (3.79) for zero initial condition $c_0 = 0,\ h_{10} = 0,\ h_{20} = 0$. ——, analytical results; ■, ●, results from Monte Carlo simulation of system (3.75) (Chen and Zhu 2010b)

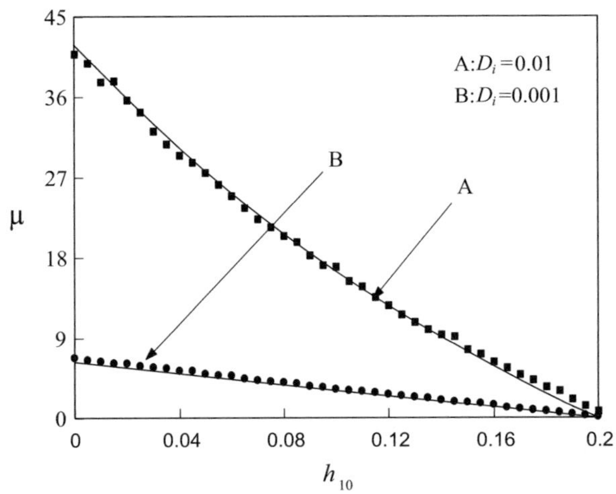

Fig. 3.8 The mean first passage time $\mu(c_{10},\ h_{10}, h_{20})$ of system (3.79) as function of h_{10} for given $h_{20} = 0$ and $c_0 = 0$. ——, analytical results; ■, ●, results from Monte Carlo simulation of system (3.75) (Chen and Zhu 2010b)

3.3 Quasi-partially Integrable Generalized Hamiltonian Systems

Assume that Eq. (3.5) is quasi-partially integrable generalized Hamiltonian system. In the generalized Hamiltonian system, associated with it, there are a $2n_1$-dimensional completely integrable Hamiltonian subsystem with Hamiltonian function H_1, which having action-angle vectors $\boldsymbol{I}_1 = [I_1, \ldots, I_{n_1}]^{\mathrm{T}}$ and $\boldsymbol{\Theta}_1 = [\Theta_1, \ldots, \Theta_{n_1}]^{\mathrm{T}}$, $2n_2$-dimensional completely non-integrable Hamiltonian subsystem with Hamiltonian function H_2, and a M-dimensional ($M = m - 2n_1 - 2n_2$) vector $\boldsymbol{C} = [C_1, \ldots, C_M]^{\mathrm{T}}$ of Casimir functions. Applying the Itô stochastic differential rule to the stochastic differential equation (3.5) yields the following Itô stochastic differential equations for the vector stochastic process $[\boldsymbol{I}_1^{\mathrm{T}}, \boldsymbol{\Theta}_1^{\mathrm{T}}, H_2, \boldsymbol{C}^{\mathrm{T}}]^{\mathrm{T}}$

$$
\begin{aligned}
\mathrm{d}I_k &= \varepsilon\Big(\sum_{i,j=1}^{m} d_{ij}\frac{\partial I_k}{\partial X_i}\frac{\partial H}{\partial X_j} + \frac{1}{2}\sum_{i,j=1}^{m}\sum_{s=1}^{l}\sigma_{is}\sigma_{js}\frac{\partial^2 I_k}{\partial X_i \partial X_j}\Big)\mathrm{d}t \\
&\quad + \varepsilon^{1/2}\sum_{i=1}^{m}\sum_{s=1}^{l}\sigma_{is}\frac{\partial I_k}{\partial X_i}\mathrm{d}B_s(t), \\
\mathrm{d}\Theta_k &= \Big[\omega_k + \varepsilon\Big(\sum_{i,j=1}^{m} d_{ij}\frac{\partial \Theta_k}{\partial X_i}\frac{\partial H}{\partial X_j} + \frac{1}{2}\sum_{i,j=1}^{m}\sum_{s=1}^{l}\sigma_{is}\sigma_{js}\frac{\partial^2 \Theta_k}{\partial X_i \partial X_j}\Big)\Big]\mathrm{d}t \\
&\quad + \varepsilon^{1/2}\sum_{i=1}^{m}\sum_{s=1}^{l}\sigma_{is}\frac{\partial \Theta_k}{\partial X_i}\mathrm{d}B_s(t), \\
\mathrm{d}H_2 &= \varepsilon\Big(\sum_{i,j=1}^{m} d_{ij}\frac{\partial H_2}{\partial X_i}\frac{\partial H}{\partial X_j} + \frac{1}{2}\sum_{i,j=1}^{m}\sum_{s=1}^{l}\sigma_{is}\sigma_{js}\frac{\partial^2 H_2}{\partial X_i \partial X_j}\Big)\mathrm{d}t \\
&\quad + \varepsilon^{1/2}\sum_{i=1}^{m}\sum_{s=1}^{l}\sigma_{is}\frac{\partial H_2}{\partial X_i}\mathrm{d}B_s(t), \\
\mathrm{d}C_v &= \varepsilon\Big(\sum_{i,j=1}^{m} d_{ij}\frac{\partial C_v}{\partial X_i}\frac{\partial H}{\partial X_j} + \frac{1}{2}\sum_{i,j=1}^{m}\sum_{s=1}^{l}\sigma_{is}\sigma_{js}\frac{\partial^2 C_v}{\partial X_i \partial X_j}\Big)\mathrm{d}t \\
&\quad + \varepsilon^{1/2}\sum_{s=1}^{l}\sum_{i=1}^{m}\sigma_{is}\frac{\partial C_v}{\partial X_i}\mathrm{d}B_s(t), \\
& k = 1, 2, \ldots, n_1; \quad v = 1, 2, \ldots, M.
\end{aligned}
\tag{3.93}
$$

In this case, the new system is governed by the $2n_1 + 1 + M$ equations in Eq. (3.93) and $2n_2 - 1$ equations for X_i in Eq. (3.5). The dimensions and forms of the averaged Itô stochastic differential equations depend on the resonant characteristics of the integrable Hamiltonian subsystem.

3.3.1 Non-resonant Case

In this case, $\boldsymbol{I}_1, H_2, \boldsymbol{C}$ are slowly-varying processes while $\boldsymbol{\theta}_1 = [\theta_1, \theta_2, \ldots, \theta_{n_1}]^{\mathrm{T}}$ and $\boldsymbol{X}' = [X_{2n_1+2}, \ldots, X_{2n_1+2n_2}]^{\mathrm{T}}$ are rapidly-varying processes. According to the Khasminskii's theorem (Khasminskii 1968), as $\varepsilon \to 0$, $[\boldsymbol{I}_1^{\mathrm{T}}, H_2, \boldsymbol{C}^{\mathrm{T}}]^{\mathrm{T}}$ converges weakly to a (n_1+1+M)-dimensional vector Markov diffusion process. The averaged Itô stochastic differential equations for $[\boldsymbol{I}_1^{\mathrm{T}}, H_2, \boldsymbol{C}^{\mathrm{T}}]^{\mathrm{T}}$ are

$$
\begin{aligned}
\mathrm{d}I_k &= m_k(\boldsymbol{I}_1, H_2, \boldsymbol{C})\mathrm{d}t + \sum_{s=1}^{l} \sigma_{ks}(\boldsymbol{I}_1, H_2, \boldsymbol{C})\mathrm{d}B_s(t), \\
\mathrm{d}H_2 &= m_{H_2}(\boldsymbol{I}_1, H_2, \boldsymbol{C})\mathrm{d}t + \sum_{s=1}^{l} \sigma_{H_2 s}(\boldsymbol{I}_1, H_2, \boldsymbol{C})\mathrm{d}B_s(t), \\
\mathrm{d}C_v &= m_v(\boldsymbol{I}_1, H_2, \boldsymbol{C})\mathrm{d}t + \sum_{s=1}^{l} \sigma_{vs}(\boldsymbol{I}_1, H_2, \boldsymbol{C})\mathrm{d}B_s(t), \\
k &= 1, 2, \ldots, n_1; \quad v = 1, 2, \ldots, M,
\end{aligned}
\tag{3.94}
$$

where the drift and diffusion coefficients are obtained from time averaging of the corresponding coefficients of Eq. (3.93) according to Eqs. (4.29) and (4.30) in vol. 1. Since the integrable Hamiltonian subsystem is ergodic on n-dimensional torus and the non-integrable Hamiltonian subsystem is ergodic on $(2n_2 - 1)$-dimensional equi-energy surface, the time averaging can be replaced by averaging with respect to $\boldsymbol{\theta}_1$ and $\boldsymbol{x}'$ with constant $\boldsymbol{I}_1, H_2, \boldsymbol{C}$ i.e.,

$$
\begin{aligned}
&m_k(\boldsymbol{I}_1, H_2, \boldsymbol{C}) \\
&= \frac{\varepsilon}{T_4} \int_0^{2\pi} \int_\Omega \frac{1}{A_4} \Big(\sum_{i,j=1}^{m} d_{ij} \frac{\partial I_k}{\partial x_i} \frac{\partial H}{\partial x_j} + \frac{1}{2} \sum_{i,j=1}^{m} \sum_{s=1}^{l} \sigma_{is}\sigma_{js} \frac{\partial^2 I_k}{\partial x_i \partial x_j}\Big)\mathrm{d}\boldsymbol{x}'\mathrm{d}\boldsymbol{\theta}_1, \\
&m_v(\boldsymbol{I}_1, H_2, \boldsymbol{C}) \\
&= \frac{\varepsilon}{T_4} \int_0^{2\pi} \int_\Omega \frac{1}{A_4} \Big(\sum_{i,j=1}^{m} d_{ij} \frac{\partial C_v}{\partial x_i} \frac{\partial H}{\partial x_j} + \frac{1}{2} \sum_{i,j=1}^{m} \sum_{s=1}^{l} \sigma_{is}\sigma_{js} \frac{\partial^2 C_v}{\partial x_i \partial x_j}\Big)\mathrm{d}\boldsymbol{x}'\mathrm{d}\boldsymbol{\theta}_1, \\
&m_{H_2}(\boldsymbol{I}_1, H_2, \boldsymbol{C}) \\
&= \frac{\varepsilon}{T_4} \int_0^{2\pi} \int_\Omega \frac{1}{A_4} \Big(\sum_{i,j=1}^{m} d_{ij} \frac{\partial H_2}{\partial x_i} \frac{\partial H}{\partial x_j} + \frac{1}{2} \sum_{i,j=1}^{m} \sum_{s=1}^{l} \sigma_{is}\sigma_{js} \frac{\partial^2 H_2}{\partial x_i \partial x_j}\Big)\mathrm{d}\boldsymbol{x}'\mathrm{d}\boldsymbol{\theta}_1,
\end{aligned}
$$

$$\begin{aligned}
&\sum_{s=1}^{l}\sigma_{k_1s}\sigma_{k_2s}(\boldsymbol{I}_1,H_2,\boldsymbol{C})=\frac{\varepsilon}{T_4}\int_0^{2\pi}\int_{\Omega}\frac{1}{A_4}\sum_{i,j=1}^{m}\sum_{s=1}^{l}\sigma_{is}\sigma_{js}\frac{\partial I_{k_1}}{\partial x_i}\frac{\partial I_{k_2}}{\partial x_j}\mathrm{d}\boldsymbol{x}'\mathrm{d}\boldsymbol{\theta}_1,\\
&\sum_{s=1}^{l}\sigma_{H_2s}\sigma_{H_2s}(\boldsymbol{I}_1,H_2,\boldsymbol{C})=\frac{\varepsilon}{T_4}\int_0^{2\pi}\int_{\Omega}\frac{1}{A_4}\sum_{i,j=1}^{m}\sum_{s=1}^{l}\sigma_{is}\sigma_{js}\frac{\partial H_2}{\partial x_i}\frac{\partial H_2}{\partial x_j}\mathrm{d}\boldsymbol{x}'\mathrm{d}\boldsymbol{\theta}_1,\\
&\sum_{s=1}^{l}\sigma_{v_1s}\sigma_{v_2s}(\boldsymbol{I}_1,H_2,\boldsymbol{C})=\frac{\varepsilon}{T_4}\int_0^{2\pi}\int_{\Omega}\frac{1}{A_4}\sum_{i,j=1}^{m}\sum_{s=1}^{l}\sigma_{is}\sigma_{js}\frac{\partial C_{v_1}}{\partial x_i}\frac{\partial C_{v_2}}{\partial x_j}\mathrm{d}\boldsymbol{x}'\mathrm{d}\boldsymbol{\theta}_1,\\
&\sum_{s=1}^{l}\sigma_{k_1s}\sigma_{H_2s}(\boldsymbol{I}_1,H_2,\boldsymbol{C})=\frac{\varepsilon}{T_4}\int_0^{2\pi}\int_{\Omega}\frac{1}{A_4}\sum_{i,j=1}^{m}\sum_{s=1}^{l}\sigma_{is}\sigma_{js}\frac{\partial I_{k_1}}{\partial x_i}\frac{\partial H_2}{\partial x_j}\mathrm{d}\boldsymbol{x}'\mathrm{d}\boldsymbol{\theta}_1,\\
&\sum_{s=1}^{l}\sigma_{k_1s}\sigma_{v_1s}(\boldsymbol{I}_1,H_2,\boldsymbol{C})=\frac{\varepsilon}{T_4}\int_0^{2\pi}\int_{\Omega}\frac{1}{A_4}\sum_{i,j=1}^{m}\sum_{s=1}^{l}\sigma_{is}\sigma_{js}\frac{\partial I_{k_1}}{\partial x_i}\frac{\partial C_{v_1}}{\partial x_j}\mathrm{d}\boldsymbol{x}'\mathrm{d}\boldsymbol{\theta}_1,\\
&\sum_{s=1}^{l}\sigma_{v_1s}\sigma_{H_2s}(\boldsymbol{I}_1,H_2,\boldsymbol{C})=\frac{\varepsilon}{T_4}\int_0^{2\pi}\int_{\Omega}\frac{1}{A_4}\sum_{i,j=1}^{m}\sum_{s=1}^{l}\sigma_{is}\sigma_{js}\frac{\partial C_{v_1}}{\partial x_i}\frac{\partial H_2}{\partial x_j}\mathrm{d}\boldsymbol{x}'\mathrm{d}\boldsymbol{\theta}_1,\\
&T_4=\int_0^{2\pi}\int_{\Omega}(1/A_4)\mathrm{d}\boldsymbol{x}'\mathrm{d}\boldsymbol{\theta}_1,\\
&\Omega=\{\boldsymbol{x}'\,|\,H_2(0,\boldsymbol{x}')\le H_2\},\\
&k,k_1,k_2=1,2,\ldots,n_1,\quad v,v_1,v_2=1,2,\ldots,M,
\end{aligned}\tag{3.95}$$

where $A_4=\left|\partial(\boldsymbol{I}_1,\boldsymbol{\theta}_1,H_2,\boldsymbol{x}',\boldsymbol{c})/\partial\boldsymbol{x}\right|$ is the Jacobian determinant of the transformation from $\boldsymbol{x}$ to $\boldsymbol{I}_1,\boldsymbol{\theta}_1,H_2,\boldsymbol{x}',\boldsymbol{c}$. The averaged FPK equation associated with Itô Eq. (3.94) is

$$\begin{aligned}
\frac{\partial p}{\partial t}=&-\sum_{k=1}^{n_1}\frac{\partial(a_kp)}{\partial I_k}-\frac{\partial(a_{h_2}p)}{\partial h_2}-\sum_{v=1}^{M}\frac{\partial(a_vp)}{\partial c_v}+\frac{1}{2}\sum_{k_1,k_2=1}^{n_1}\frac{\partial^2(b_{k_1k_2}p)}{\partial I_{k_1}\partial I_{k_2}}\\
&+\frac{1}{2}\frac{\partial^2(b_{h_2h_2}p)}{\partial h_2^2}+\frac{1}{2}\sum_{v_1,v_2=1}^{M}\frac{\partial^2(b_{v_1v_2}p)}{\partial c_{v_1}\partial c_{v_2}}+\sum_{k_1=1}^{n_1}\frac{\partial^2(b_{k_1h_2}p)}{\partial I_{k_1}\partial h_2}+\sum_{k_1=1}^{n_1}\sum_{v_1=1}^{M}\frac{\partial^2(b_{k_1v_1}p)}{\partial I_{k_1}\partial c_{v_1}}\\
&+\sum_{v_1=1}^{M}\frac{\partial^2(b_{v_1h_2}p)}{\partial c_{v_1}\partial h_2},
\end{aligned}\tag{3.96}$$

where

$$\begin{aligned}
&a_k=a_k(\boldsymbol{I}_1,h_2,\boldsymbol{c})=m_k(\boldsymbol{I}_1,H_2,\boldsymbol{C})|_{H_2=h_2,\ \boldsymbol{C}=\boldsymbol{c}},\\
&a_{h_2}=a_{h_2}(\boldsymbol{I}_1,h_2,\boldsymbol{c})=m_{h_2}(\boldsymbol{I}_1,H_2,\boldsymbol{C})\big|_{H_2=h_2,\ \boldsymbol{C}=\boldsymbol{c}},\\
&a_v=a_v(\boldsymbol{I}_1,h_2,\boldsymbol{c})=m_v(\boldsymbol{I}_1,H_2,\boldsymbol{C})|_{H_2=h_2,\ \boldsymbol{C}=\boldsymbol{c}},
\end{aligned}$$

$$
\begin{aligned}
b_{k_1k_2} &= b_{k_1k_2}(\boldsymbol{I}_1, h_2, \boldsymbol{c}) = \sum_{s=1}^{l} \sigma_{k_1s}\sigma_{k_2s}(\boldsymbol{I}_1, H_2, \boldsymbol{C})\bigg|_{H_2=h_2,\ \boldsymbol{C}=\boldsymbol{c}}, \\
b_{h_2h_2} &= b_{h_2h_2}(\boldsymbol{I}_1, h_2, \boldsymbol{c}) = \sum_{s=1}^{l} \sigma_{h_2s}\sigma_{h_2s}(\boldsymbol{I}_1, H_2, \boldsymbol{C})\bigg|_{H_2=h_2,\ \boldsymbol{C}=\boldsymbol{c}}, \\
b_{v_1v_2} &= b_{v_1v_2}(\boldsymbol{I}_1, h_2, \boldsymbol{c}) = \sum_{s=1}^{l} \sigma_{v_1s}\sigma_{v_2s}(\boldsymbol{I}_1, H_2, \boldsymbol{C})\bigg|_{H_2=h_2,\ \boldsymbol{C}=\boldsymbol{c}}, \\
b_{k_1h_2} &= b_{k_1h_2}(\boldsymbol{I}_1, h_2, \boldsymbol{c}) = \sum_{s=1}^{l} \sigma_{k_1s}\sigma_{h_2s}(\boldsymbol{I}_1, H_2, \boldsymbol{C})\bigg|_{H_2=h_2,\ \boldsymbol{C}=\boldsymbol{c}}, \\
b_{k_1v_1} &= b_{k_1v_1}(\boldsymbol{I}_1, h_2, \boldsymbol{c}) = \sum_{s=1}^{l} \sigma_{k_1s}\sigma_{v_1s}(\boldsymbol{I}_1, H_2, \boldsymbol{C})\bigg|_{H_2=h_2,\ \boldsymbol{C}=\boldsymbol{c}}, \\
b_{v_1h_2} &= b_{v_1h_2}(\boldsymbol{I}_1, h_2, \boldsymbol{c}) = \sum_{s=1}^{l} \sigma_{v_1s}\sigma_{h_2s}(\boldsymbol{I}_1, H_2, \boldsymbol{C})\bigg|_{H_2=h_2,\ \boldsymbol{C}=\boldsymbol{c}}.
\end{aligned}
\tag{3.97}
$$

In Eq. (3.96), $p = p(\boldsymbol{I}_1, h_2, \boldsymbol{c}, t|\boldsymbol{I}_{10}, h_{20}, \boldsymbol{c}_0)$ is the transition probability density of $[\boldsymbol{I}_1^{\mathrm{T}}, H_2, \boldsymbol{C}^{\mathrm{T}}]^{\mathrm{T}}$, or $p = p(\boldsymbol{I}_1, h_2, \boldsymbol{c}, t)$ is the joint stationary probability density of $[\boldsymbol{I}_1^{\mathrm{T}}, H_2, \boldsymbol{C}^{\mathrm{T}}]^{\mathrm{T}}$. Under certain initial and boundary conditions, Eq. (3.96) can be solved at least numerically. It can be proven that if the stationary solution $p = p(\boldsymbol{I}_1, h_2, \boldsymbol{c})$ has been obtained, then the approximate stationary PDF for the original system (3.5) can be given by

$$
p(\boldsymbol{x}) = \frac{p(\boldsymbol{I}_1, h_2, \boldsymbol{c})}{T_4(\boldsymbol{I}_1, h_2, \boldsymbol{c})}\bigg|_{\boldsymbol{I}_1=\boldsymbol{I}_1(\boldsymbol{x}),\ h_2=h_2(\boldsymbol{x}),\ \boldsymbol{c}=\boldsymbol{c}(\boldsymbol{x})}, \tag{3.98}
$$

where $T_4(\boldsymbol{I}_1, h_2, \boldsymbol{c}) = T_4(\boldsymbol{I}_1, H_2, \boldsymbol{C})|_{H_2=h_2, \boldsymbol{C}=\boldsymbol{c}}$, determined by Eq. (3.95).

Assume that the partially integrable and resonant generalized Hamiltonian system associated with system (3.93) is separable, i.e.,

$$
\begin{aligned}
&\boldsymbol{x} = [\boldsymbol{x}_1^{\mathrm{T}}, \boldsymbol{x}_2^{\mathrm{T}}, \boldsymbol{x}_3^{\mathrm{T}}]^{\mathrm{T}}, \quad \boldsymbol{x}_1 = [x_1, x_2, \ldots, x_{2n_1}]^{\mathrm{T}} = [\boldsymbol{q}_1^{\mathrm{T}}, \boldsymbol{p}_1^{\mathrm{T}}]^{\mathrm{T}}, \\
&\boldsymbol{q}_1 = [q_1, q_2, \ldots, q_{n_1}]^{\mathrm{T}}, \quad \boldsymbol{p}_1 = [p_1, p_2, \ldots, p_{n_1}]^{\mathrm{T}}, \\
&\boldsymbol{x}_2 = [x_{2n_1+1}, x_{2n_1+2}, \ldots, x_{2n_1+2n_2}]^{\mathrm{T}} = [\boldsymbol{q}_2^{\mathrm{T}}, \boldsymbol{p}_2^{\mathrm{T}}]^{\mathrm{T}}, \\
&\boldsymbol{q}_2 = [q_{n_1+1}, q_{n_1+2}, \ldots, q_{n_1+n_2}]^{\mathrm{T}}, \quad \boldsymbol{p}_2 = [p_{n_1+1}, p_{n_1+2}, \ldots, p_{n_1+n_2}]^{\mathrm{T}}, \\
&\boldsymbol{x}_3 = [x_{2n_1+2n_2+1}, x_{2n_1+2n_2+2}, \ldots, x_m]^{\mathrm{T}},
\end{aligned}
\tag{3.99}
$$

Suppose that the subsystem associated with $H_1(\boldsymbol{q}_1, \boldsymbol{p}_1)$ is completely integrable, the action vector $\boldsymbol{I}_1(\boldsymbol{x}_1) = [I_1, I_2, \ldots, I_{n_1}]^{\mathrm{T}}$ and angle vector $\boldsymbol{\theta}_1(\boldsymbol{x}_1) = [\theta_1, \theta_2, \ldots, \theta_{n_1}]^{\mathrm{T}}$ can be obtained. Hamiltonian subsystem with Hamiltonian function $H_2(\boldsymbol{x}_2) = H(\boldsymbol{q}_2, \boldsymbol{p}_2)$ is $2n_2$-dimensional completely non-integrable. The third subsystem has

M Casimir functions $\boldsymbol{C}(\boldsymbol{x}_3) = [C_1, C_2, \ldots, C_M]^{\mathrm{T}}$. In this case, instead of Eq. (3.93), the Itô stochastic differential equations governing vector process $[\boldsymbol{I}_1^{\mathrm{T}}, \boldsymbol{\theta}_1^{\mathrm{T}}, H_2, \boldsymbol{C}^{\mathrm{T}}]^{\mathrm{T}}$ are

$$
\begin{aligned}
\mathrm{d}I_k &= \varepsilon\{\sum_{j=1}^{n_1} \frac{\partial I_k}{\partial P_j}[\sum_{r=1}^{n_1} d_{jr}\frac{\partial H_1}{\partial P_r} + \sum_{r_1=n_1+1}^{n_1+n_2} d_{jr_1}\frac{\partial H_2}{\partial P_{r_1}} + \sum_{i=2n_1+2n_2+1}^{m} d_{ji}\frac{\partial}{\partial X_i}(\sum_{\beta=1}^{M} C_\beta)] \\
&\quad + \frac{1}{2}\sum_{j,r=1}^{n_1}\sum_{s=1}^{l}\sigma_{js}\sigma_{rs}\frac{\partial^2 I_k}{\partial P_j \partial P_r}\}\mathrm{d}t + \varepsilon^{1/2}\sum_{j=1}^{n_1}\sum_{s=1}^{l}\sigma_{js}\frac{\partial I_k}{\partial P_j}\mathrm{d}B_s(t), \\
\mathrm{d}\Theta_k &= \varepsilon\{\omega_k + \sum_{j=1}^{n_1}\frac{\partial \Theta_k}{\partial P_j}[\sum_{r=1}^{n_1} d_{jr}\frac{\partial H_1}{\partial P_r} + \sum_{r_1=n_1+1}^{n_1+n_2} d_{jr_1}\frac{\partial H_2}{\partial P_{r_1}} \\
&\quad + \sum_{i=2n_1+2n_2+1}^{m} d_{ji}\frac{\partial}{\partial X_i}(\sum_{\beta=1}^{M} C_\beta)] \\
&\quad + \frac{1}{2}\sum_{j,r=1}^{n_1}\sum_{s=1}^{l}\sigma_{js}\sigma_{rs}\frac{\partial^2 \Theta_k}{\partial P_j \partial P_r}\}\mathrm{d}t + \varepsilon^{1/2}\sum_{j=1}^{n_1}\sum_{s=1}^{l}\sigma_{js}\frac{\partial \Theta_k}{\partial P_j}\mathrm{d}B_s(t), \qquad (3.100) \\
\mathrm{d}H_2 &= \varepsilon\{\sum_{r_2=n_1+1}^{n_1+n_2}\frac{\partial H_2}{\partial P_{r_2}}[\sum_{r=1}^{n_1} d_{r_2 r}\frac{\partial H_1}{\partial P_r} \\
&\quad + \sum_{r_1=n_1+1}^{n_1+n_2} d_{r_2 r_1}\frac{\partial H_2}{\partial P_{r_1}} \\
&\quad + \sum_{i=2n_1+2n_2+1}^{m} d_{r_2 i}\frac{\partial}{\partial X_i}(\sum_{\beta=1}^{M} C_\beta)] \\
&\quad + \frac{1}{2}\sum_{r_2,r_1=n_1+1}^{n_1+n_2}\sum_{s=1}^{l}\sigma_{r_2 s}\sigma_{r_1 s}\frac{\partial^2 H_2}{\partial P_{r_2}\partial P_{r_1}}\}\mathrm{d}t + \varepsilon^{1/2}\sum_{r_1=n_1+1}^{n_1+n_2}\sum_{s=1}^{l}\sigma_{r_1 s}\frac{\partial H_2}{\partial P_{r_1}}\mathrm{d}B_s(t), \\
\mathrm{d}C_v &= \varepsilon\{\sum_{i_1=2n_1+2n_2+1}^{m}\frac{\partial C_v}{\partial X_{i_1}}[\sum_{r=1}^{n_1} d_{i_1 r}\frac{\partial H_1}{\partial P_r} + \sum_{r_1=n_1+1}^{n_1+n_2} d_{jr_1}\frac{\partial H_2}{\partial P_{r_1}} \\
&\quad + \sum_{i=2n_1+2n_2+1}^{m} d_{i_1 i}\frac{\partial}{\partial X_i}(\sum_{\beta=1}^{M} C_\beta)] \\
&\quad + \frac{1}{2}\sum_{i,i_1=2n_1+2n_2+1}^{m}\sum_{s=1}^{l}\sigma_{is}\sigma_{i_1 s}\frac{\partial^2 C_v}{\partial X_i \partial X_{i_1}}\}\mathrm{d}t + \varepsilon^{1/2}\sum_{i=2n_1+2n_2+1}^{m}\sum_{s=1}^{l}\sigma_{is}\frac{\partial C_v}{\partial X_i}\mathrm{d}B_s(t),
\end{aligned}
$$

$k = 1, 2, \ldots, n; \quad v = 1, 2, \ldots, M.$

In this case, system is described by the $2n_1 + 1 + M$ equations in Eq. (3.100) and $2n_2 - 1$ equations for $X_{2n_1+2}, \ldots, X_{2n_1+2n_2}$ in Eq. (3.5). According to the Khasminskii's theorem (Khasminskii 1968), as $\varepsilon \to 0$, $[\boldsymbol{I}_1^{\mathrm{T}}, H_2, \boldsymbol{C}^{\mathrm{T}}]^{\mathrm{T}}$ is slowly-varying vector process and converges weakly to a $(n_1 + 1 + M)$-dimensional vector Markov diffusion process. The averaged Itô stochastic differential equations are of the same as Eq. (3.94), while the drift and diffusion coefficient functions are changed to

$$\begin{aligned}
\overline{m}_k(\boldsymbol{I}_1, H_2, \boldsymbol{C}) = \frac{\varepsilon}{T_4'} \int_0^{2\pi} \int_\Omega \frac{1}{A_4'} \Bigg\{ \sum_{j=1}^{n_1} \frac{\partial I_k}{\partial p_j} \Bigg[\sum_{r=1}^{n_1} d_{jr} \frac{\partial H_1}{\partial p_r} \\
+ \sum_{r_1=n_1+1}^{n_1+n_2} d_{jr_1} \frac{\partial H_2}{\partial p_{r_1}} + \sum_{i=2n_1+2n_2+1}^{M} d_{ji} \frac{\partial}{\partial x_i} \left(\sum_{\beta=1}^{M} C_\beta \right) \Bigg]] \\
+ \frac{1}{2} \sum_{j,r=1}^{n_1} \sum_{s=1}^{l} \sigma_{js}\sigma_{rs} \frac{\partial^2 I_k}{\partial p_j \partial p_r} \Bigg\} \mathrm{d}\boldsymbol{q}_2 \mathrm{d}p_{n_1+2} \cdots \mathrm{d}p_{n_1+n_2} \mathrm{d}\boldsymbol{\theta}_1,
\end{aligned}$$

$$\begin{aligned}
\overline{m}_{H_2}(\boldsymbol{I}_1, H_2, \boldsymbol{C}) = \frac{\varepsilon}{T_4'} \int_0^{2\pi} \int_\Omega \frac{1}{A_4'} \Bigg\{ \sum_{r_2=n_1+1}^{n_1+n_2} \frac{\partial H_2}{\partial P_{r_2}} \Bigg[\sum_{r=1}^{n_1} d_{r_2 r} \frac{\partial H_1}{\partial P_r} \\
+ \sum_{r_1=n_1+1}^{n_1+n_2} d_{r_2 r_1} \frac{\partial H_2}{\partial P_{r_1}} + \sum_{i=2n_1+2n_2+1}^{m} d_{r_2 i} \frac{\partial}{\partial X_i} \left(\sum_{\beta=1}^{M} C_\beta \right) \Bigg] \\
+ \frac{1}{2} \sum_{r_1,r_2=n_1+1}^{n_1+n_2} \sum_{s=1}^{l} \sigma_{r_1 s}\sigma_{r_2 s} \frac{\partial^2 H_2}{\partial P_{r_1} \partial P_{r_2}} \Bigg\} \mathrm{d}\boldsymbol{q}_2 \mathrm{d}p_{n_1+2} \cdots \mathrm{d}p_{n_1+n_2} \mathrm{d}\boldsymbol{\theta}_1,
\end{aligned}$$

$$\begin{aligned}
\overline{m}_v(\boldsymbol{I}_1, H_2, \boldsymbol{C}) = \frac{\varepsilon}{T_4'} \int_0^{2\pi} \int_\Omega \frac{1}{A_4'} \Bigg\{ \sum_{i_1=2n_1+2n_2+1}^{m} \frac{\partial C_v}{\partial x_{i_1}} \Bigg[\sum_{r=1}^{n_1} d_{i_1 r} \frac{\partial H_1}{\partial p_r} \\
+ \sum_{r_1=n_1+1}^{n_1+n_2} d_{jr_1} \frac{\partial H_2}{\partial p_{r_1}} + \sum_{i=2n_1+2n_2+1}^{m} d_{i_1 i} \frac{\partial}{\partial x_i} \left(\sum_{\beta=1}^{M} C_\beta \right) \Bigg] \\
+ \frac{1}{2} \sum_{i,i_1=2n_1+2n_2+1}^{m} \sum_{s=1}^{l} \sigma_{is}\sigma_{i_1 s} \frac{\partial^2 C_v}{\partial x_i \partial x_{i_1}} \Bigg\} \mathrm{d}\boldsymbol{q}_2 \mathrm{d}p_{n_1+2} \cdots \mathrm{d}p_{n_1+n_2} \mathrm{d}\boldsymbol{\theta}_1,
\end{aligned}$$

$$\begin{aligned}
\sum_{s=1}^{l} \overline{\sigma}_{k_1 s} \overline{\sigma}_{k_2 s}(\boldsymbol{I}_1, H_2, \boldsymbol{C}) \\
= \frac{\varepsilon}{T_4'} \int_0^{2\pi} \int_\Omega \frac{1}{A_4'} \sum_{j,r=1}^{n_1} \sum_{s=1}^{l} \sigma_{js}\sigma_{rs} \frac{\partial I_{k_1}}{\partial p_j} \frac{\partial I_{k_2}}{\partial p_r} \mathrm{d}\boldsymbol{q}_2 \mathrm{d}p_{n_1+2} \cdots \mathrm{d}p_{n_1+n_2} \mathrm{d}\boldsymbol{\theta}_1,
\end{aligned}$$

$$\sum_{s=1}^{l} \overline{\sigma}_{H_2 s} \overline{\sigma}_{H_2 s}(\boldsymbol{I}_1, H_2, \boldsymbol{C})$$

$$= \frac{\varepsilon}{T_4'} \int_0^{2\pi} \int_\Omega \frac{1}{A_4'} \sum_{r_1,r_2=n_1+1}^{n_1+n_2} \sum_{s=1}^{l} \sigma_{r_1 s}\sigma_{r_2 s} \frac{\partial H_2}{\partial p_{r_1}} \frac{\partial H_2}{\partial p_{r_2}} \mathrm{d}\boldsymbol{q}_2 \mathrm{d}p_{n_1+2} \cdots \mathrm{d}p_{n_1+n_2} \mathrm{d}\boldsymbol{\theta}_1,$$

$$\sum_{s=1}^{l} \overline{\sigma}_{v_1 s}\overline{\sigma}_{v_2 s}(\boldsymbol{I}_1, H_2, \boldsymbol{C})$$

$$= \frac{\varepsilon}{T_4'} \int_0^{2\pi} \int_\Omega \frac{1}{A_4'} \sum_{i,i_1=2n_1+2n_2+1}^{m} \sum_{s=1}^{l} \sigma_{is}\sigma_{i_1 s} \frac{\partial C_{v_1}}{\partial x_i} \frac{\partial C_{v_2}}{\partial x_{i_1}} \mathrm{d}\boldsymbol{q}_2 \mathrm{d}p_{n_1+2} \cdots \mathrm{d}p_{n_1+n_2} \mathrm{d}\boldsymbol{\theta}_1,$$

$$\sum_{s=1}^{l} \overline{\sigma}_{k_1 s}\overline{\sigma}_{H_2 s}(\boldsymbol{I}_1, H_2, \boldsymbol{C})$$

$$= \frac{\varepsilon}{T_4'} \int_0^{2\pi} \int_\Omega \frac{1}{A_4'} \sum_{j=1}^{n_1} \sum_{r_1=n_1+1}^{n_1+n_2} \sum_{s=1}^{l} \sigma_{js}\sigma_{r_1 s} \frac{\partial I_{k_1}}{\partial p_j} \frac{\partial H_2}{\partial p_{r_1}} \mathrm{d}\boldsymbol{q}_2 \mathrm{d}p_{n_1+2} \cdots \mathrm{d}p_{n_1+n_2} \mathrm{d}\boldsymbol{\theta}_1,$$

$$\sum_{s=1}^{l} \overline{\sigma}_{k_1 s}\overline{\sigma}_{v_1 s}(\boldsymbol{I}_1, H_2, \boldsymbol{C})$$

$$= \frac{\varepsilon}{T_4'} \int_0^{2\pi} \int_\Omega \frac{1}{A_4'} \sum_{j=1}^{n_1} \sum_{i=2n_1+2n_2+1}^{m} \sum_{s=1}^{l} \sigma_{js}\sigma_{is} \frac{\partial I_{k_1}}{\partial p_j} \frac{\partial C_{v_1}}{\partial x_i} \mathrm{d}\boldsymbol{q}_2 \mathrm{d}p_{n_1+2} \cdots \mathrm{d}p_{n_1+n_2} \mathrm{d}\boldsymbol{\theta}_1,$$

$$\sum_{s=1}^{l} \overline{\sigma}_{v_1 s}\overline{\sigma}_{H_2 s}(\boldsymbol{I}_1, H_2, \boldsymbol{C})$$

$$= \frac{\varepsilon}{T_4'} \int_0^{2\pi} \int_\Omega \frac{1}{A_4'} \sum_{i=2n_1+2n_2+1}^{m} \sum_{r_1=n_1+1}^{n_1+n_2} \sum_{s=1}^{l} \sigma_{is}\sigma_{r_1 s} \frac{\partial C_{v_1}}{\partial X_i} \frac{\partial H_2}{\partial P_{r_1}} \mathrm{d}\boldsymbol{q}_2 \mathrm{d}p_{n_1+2} \cdots \mathrm{d}p_{n_1+n_2} \mathrm{d}\boldsymbol{\theta}_1,$$

$$T_4' = \int_0^{2\pi} \int_\Omega (1/A_4') \mathrm{d}\boldsymbol{q}_2 \mathrm{d}p_{n_1+2} \cdots \mathrm{d}p_{n_1+n_2} \mathrm{d}\boldsymbol{\theta}_1,$$
$$\Omega = \{ (\boldsymbol{q}_2, p_{n_1+2}, \ldots, p_{n_1+n_2}) \big| H_2(\boldsymbol{q}_2, 0, p_{n_1+2}, \ldots, p_{n_1+n_2}) \le H_2 \},$$
$$k, k_1, k_2 = 1, 2, \ldots, n_1, \quad v, v_1, v_2 = 1, 2, \ldots, M, \tag{3.101}$$

where

$$\begin{aligned} A_4' &= \left| \frac{\partial(\boldsymbol{I}_1, \boldsymbol{\theta}_1, \boldsymbol{q}_2, H_2, p_{n_1+2}, \ldots, p_{n_1+n_2}, \boldsymbol{c})}{\partial(\boldsymbol{q}_1, \boldsymbol{p}_1, \boldsymbol{q}_2, \boldsymbol{p}_2, \boldsymbol{x}_3)} \right| \\ &= \left| \frac{\partial(\boldsymbol{I}_1, \boldsymbol{\theta}_1)}{\partial(\boldsymbol{q}_1, \boldsymbol{p}_1)} \right| \left| \frac{\partial(\boldsymbol{q}_2, H_2, p_{n_1+2}, \ldots, p_{n_1+n_2})}{\partial(\boldsymbol{q}_2, \boldsymbol{p}_2)} \right| \left| \frac{\partial \boldsymbol{c}}{\partial \boldsymbol{x}_3} \right| \\ &= \left| \frac{\partial H_2}{\partial p_{n_1+1}} \right| \left| \frac{\partial \boldsymbol{c}}{\partial \boldsymbol{x}_3} \right|. \end{aligned} \tag{3.102}$$

Thus,

$$T_4' = (2\pi)^{n_1} \left| \frac{\partial \boldsymbol{c}}{\partial \boldsymbol{x}_3} \right|^{-1} \int_{\Omega} \left| \frac{\partial H_2}{\partial p_{n_1+1}} \right|^{-1} \mathrm{d}\boldsymbol{q}_2 \mathrm{d}p_{n_1+2} \cdots \mathrm{d}p_{n_1+n_2}. \tag{3.103}$$

The averaged FPK equation has the form as Eq. (3.96) while its first and second derivative moments are changed into

$$\begin{aligned} \overline{a}_k &= \overline{a}_k(\boldsymbol{I}_1, h_2, \boldsymbol{c}) = \overline{m}_k(\boldsymbol{I}_1, H_2, \boldsymbol{C})\big|_{H_2=h_2,\ \boldsymbol{C}=\boldsymbol{c}}, \\ \overline{a}_{h_2} &= \overline{a}_{h_2}(\boldsymbol{I}_1, h_2, \boldsymbol{c}) = \overline{m}_{H_2}(\boldsymbol{I}_1, H_2, \boldsymbol{C})\big|_{H_2=h_2,\ \boldsymbol{C}=\boldsymbol{c}}, \\ \overline{a}_v &= \overline{a}_v(\boldsymbol{I}_1, h_2, \boldsymbol{c}) = \overline{m}_v(\boldsymbol{I}_1, H_2, \boldsymbol{C})\big|_{H_2=h_2,\ \boldsymbol{C}=\boldsymbol{c}}, \\ \overline{b}_{k_1k_2} &= \overline{b}_{k_1k_2}(\boldsymbol{I}_1, h_2, \boldsymbol{c}) = \sum_{s=1}^{l} \overline{\sigma}_{k_1s}\overline{\sigma}_{k_2s}(\boldsymbol{I}_1, H_2, \boldsymbol{C})\Big|_{H_2=h_2,\ \boldsymbol{C}=\boldsymbol{c}}, \\ \overline{b}_{h_2h_2} &= \overline{b}_{h_2h_2}(\boldsymbol{I}_1, h_2, \boldsymbol{c}) = \sum_{s=1}^{l} \overline{\sigma}_{H_2s}\overline{\sigma}_{H_2s}(\boldsymbol{I}_1, H_2, \boldsymbol{C})\Big|_{H_2=h_2,\ \boldsymbol{C}=\boldsymbol{c}}, \\ \overline{b}_{v_1v_2} &= \overline{b}_{v_1v_2}(\boldsymbol{I}_1, h_2, \boldsymbol{c}) = \sum_{s=1}^{l} \overline{\sigma}_{v_1s}\overline{\sigma}_{v_2s}(\boldsymbol{I}_1, H_2, \boldsymbol{C})\Big|_{H_2=h_2,\ \boldsymbol{C}=\boldsymbol{c}}, \\ \overline{b}_{k_1h_2} &= \overline{b}_{k_1h_2}(\boldsymbol{I}_1, h_2, \boldsymbol{c}) = \sum_{s=1}^{l} \overline{\sigma}_{k_1s}\overline{\sigma}_{H_2s}(\boldsymbol{I}_1, H_2, \boldsymbol{C})\Big|_{H_2=h_2,\ \boldsymbol{C}=\boldsymbol{c}}, \\ \overline{b}_{k_1v_1} &= \overline{b}_{k_1v_1}(\boldsymbol{I}_1, h_2, \boldsymbol{c}) = \sum_{s=1}^{l} \overline{\sigma}_{k_1s}\overline{\sigma}_{v_1s}(\boldsymbol{I}_1, H_2, \boldsymbol{C})\Big|_{H_2=h_2,\ \boldsymbol{C}=\boldsymbol{c}}, \\ \overline{b}_{v_1h_2} &= \overline{b}_{v_1h_2}(\boldsymbol{I}_1, h_2, \boldsymbol{c}) = \sum_{s=1}^{l} \overline{\sigma}_{v_1s}\overline{\sigma}_{H_2s}(\boldsymbol{I}_1, H_2, \boldsymbol{C})\Big|_{H_2=h_2,\ \boldsymbol{C}=\boldsymbol{c}}. \end{aligned} \tag{3.104}$$

It has been proven that, if the stationary solution $p(\boldsymbol{I}_1, h_2, \boldsymbol{c})$ for averaged FPK equation has been obtained, then the approximated stationary PDF for the original system (3.5) is

$$p(\boldsymbol{x}) = \left.\frac{p(\boldsymbol{I}_1, h_2, \boldsymbol{c})}{T_4'}\right|_{\boldsymbol{I}_1=\boldsymbol{I}_1(\boldsymbol{x}_1),\ h_2=h_2(\boldsymbol{x}_2),\ \boldsymbol{c}=\boldsymbol{c}(\boldsymbol{x}_3)}. \tag{3.105}$$

where $T_4'(\boldsymbol{I}_1, h_2, \boldsymbol{c}) = T_4'(\boldsymbol{I}_1, H_2, \boldsymbol{C})\big|_{H_2=h_2, \boldsymbol{C}=\boldsymbol{c}}$ determined in Eq. (3.103).

3.3.2 Internal Resonant Case

Consider the case that the integrable Hamiltonian subsystem with action-angle vectors $\boldsymbol{I}_1, \boldsymbol{\theta}_1$ are internal resonant, and there exist α $(1 \le \alpha \le n_1 - 1)$ weak internal resonance relations between the frequencies ω_r $(r = 1, 2, \ldots, n_1)$, i.e.,

$$\sum_{r=1}^{n_1} k_r^u \omega_r = O_u(\varepsilon), \quad u = 1, 2, \ldots, \alpha. \tag{3.106}$$

Introducing α combinations of angle variables

$$\Psi_u = \sum_{r=1}^{n_1} k_r^u \Theta_r, \quad u = 1, 2, \ldots, \alpha, \tag{3.107}$$

and using the Itô stochastic differential rule, the Itô stochastic differential equations for $\boldsymbol{\Psi} = [\Psi_1, \ldots, \Psi_\alpha]^{\mathrm{T}}$ can be derived from Eq. (3.93) as follows

$$\begin{aligned} \mathrm{d}\Psi_u = {} & \{O_u(\varepsilon) + \varepsilon(\sum_{i,j=1}^{m} d_{ij} \frac{\partial \Psi_u}{\partial X_i}\frac{\partial H}{\partial X_j} + \frac{1}{2}\sum_{i,j=1}^{m}\sum_{s=1}^{l} \sigma_{is}\sigma_{js}\frac{\partial^2 \Psi_u}{\partial X_i \partial X_j})]\mathrm{d}t \\ & + \varepsilon^{1/2}\sum_{i=1}^{m}\sum_{s=1}^{l} \sigma_{is}\frac{\partial \Psi_u}{\partial X_i}\mathrm{d}B_s(t), \quad u = 1, 2, \ldots, \alpha. \end{aligned} \tag{3.108}$$

In this case, the quasi-partially integrable and resonant generalized Hamiltonian system is governed by the α equations in Eq. (3.108), $n_1 - \alpha$ equations of Θ_k and the equations for I_k, H_2, C_v in Eq. (3.93) and $2n_2 - 1$ equations for X_i in Eq. (3.5). $[\boldsymbol{I}_1^{\mathrm{T}}, \boldsymbol{\Psi}^{\mathrm{T}}, H_2, \boldsymbol{C}^{\mathrm{T}}]^{\mathrm{T}}$ is a slowly-varying vector process. According to the Khasminskii's theorem (Khasminskii, 1968), as $\varepsilon \to 0$, it converges weakly to a $(n_1 + \alpha + 1 + M)$-dimensional vector Markov diffusion process. The averaged Itô stochastic differential equations are of the form

$$
\begin{aligned}
&\mathrm{d}I_k = \overline{\overline{m}}_k(\boldsymbol{I}_1, \boldsymbol{\Psi}, H_2, \boldsymbol{C})\mathrm{d}t + \sum_{s=1}^{l} \overline{\overline{\sigma}}_{ks}(\boldsymbol{I}_1, \boldsymbol{\Psi}, H_2, \boldsymbol{C})\mathrm{d}B_s(t), \\
&\mathrm{d}\Psi_u = \overline{\overline{m}}_u(\boldsymbol{I}_1, \boldsymbol{\Psi}, H_2, \boldsymbol{C})\mathrm{d}t + \sum_{s=1}^{l} \overline{\overline{\sigma}}_{us}(\boldsymbol{I}_1, \boldsymbol{\Psi}, H_2, \boldsymbol{C})\mathrm{d}B_s(t), \\
&\mathrm{d}H_2 = \overline{\overline{m}}_{H_2}(\boldsymbol{I}_1, \boldsymbol{\Psi}, H_2, \boldsymbol{C})\mathrm{d}t + \sum_{s=1}^{l} \overline{\overline{\sigma}}_{H_2 s}(\boldsymbol{I}_1, \boldsymbol{\Psi}, H_2, \boldsymbol{C})\mathrm{d}B_s(t), \\
&\mathrm{d}C_v = \overline{\overline{m}}_v(\boldsymbol{I}_1, \boldsymbol{\Psi}, H_2, \boldsymbol{C})\mathrm{d}t + \sum_{s=1}^{l} \overline{\overline{\sigma}}_{vs}(\boldsymbol{I}_1, \boldsymbol{\Psi}, H_2, \boldsymbol{C})\mathrm{d}B_s(t), \\
&k = 1, 2, \ldots, n_1; \quad u = 1, 2, \ldots, \alpha; \quad v = 1, 2, \ldots, M.
\end{aligned}
\tag{3.109}
$$

According to Eqs. (4.29) and (4.30) in vol. 1 and replacing the time averaging by special averaging with respect to $\overline{\theta} = [\theta_{\alpha+1}, \ldots, \theta_{n_1}]^{\mathrm{T}}$ and $\boldsymbol{x'} = [x_{2n_1+M+2}, \ldots, x_m]^{\mathrm{T}}$, the drift and diffusion coefficients are as follows

$$
\begin{aligned}
&\overline{\overline{m}}_k(\boldsymbol{I}_1, \boldsymbol{\Psi}, H_2, \boldsymbol{C}) \\
&= \frac{\varepsilon}{T_5}\int_0^{2\pi}\int_\Omega \frac{1}{A_5}\left(\sum_{i,j=1}^{m} d_{ij}\frac{\partial I_k}{\partial x_i}\frac{\partial H}{\partial x_j} + \frac{1}{2}\sum_{i,j=1}^{m}\sum_{s=1}^{l}\sigma_{is}\sigma_{js}\frac{\partial^2 I_k}{\partial x_i \partial x_j}\right)\mathrm{d}\boldsymbol{x}'\mathrm{d}\overline{\theta}, \\
&\overline{\overline{m}}_u(\boldsymbol{I}_1, \boldsymbol{\Psi}, H_2, \boldsymbol{C}) \\
&= \frac{\varepsilon}{T_5}\int_0^{2\pi}\int_\Omega \frac{1}{A_5}\left(\frac{O(\varepsilon)}{\varepsilon} + \sum_{i,j=1}^{m} d_{ij}\frac{\partial \Psi_u}{\partial x_i}\frac{\partial H}{\partial x_j} + \frac{1}{2}\sum_{i,j=1}^{m}\sum_{s=1}^{l}\sigma_{is}\sigma_{js}\frac{\partial^2 \Psi_u}{\partial x_i \partial x_j}\right)\mathrm{d}\boldsymbol{x}'\mathrm{d}\overline{\theta}, \\
&\overline{\overline{m}}_{H_2}(\boldsymbol{I}_1, \boldsymbol{\Psi}, H_2, \boldsymbol{C}) \\
&= \frac{\varepsilon}{T_5}\int_0^{2\pi}\int_\Omega \frac{1}{A_5}\left(\sum_{i,j=1}^{m} d_{ij}\frac{\partial H_2}{\partial x_i}\frac{\partial H}{\partial x_j} + \frac{1}{2}\sum_{i,j=1}^{m}\sum_{s=1}^{l}\sigma_{is}\sigma_{js}\frac{\partial^2 H_2}{\partial x_i \partial x_j}\right)\mathrm{d}\boldsymbol{x}'\mathrm{d}\overline{\theta}, \\
&\overline{\overline{m}}_v(\boldsymbol{I}_1, \boldsymbol{\Psi}, H_2, \boldsymbol{C}) \\
&= \frac{\varepsilon}{T_5}\int_0^{2\pi}\int_\Omega \frac{1}{A_5}\left(\sum_{i,j=1}^{m} d_{ij}\frac{\partial C_v}{\partial x_i}\frac{\partial H}{\partial x_j} + \frac{1}{2}\sum_{i,j=1}^{m}\sum_{s=1}^{l}\sigma_{is}\sigma_{js}\frac{\partial^2 C_v}{\partial x_i \partial x_j}\right)\mathrm{d}\boldsymbol{x}'\mathrm{d}\overline{\theta},
\end{aligned}
$$

$$
\begin{aligned}
&\sum_{s=1}^{l}\overline{\overline{\sigma}}_{k_1 s}\overline{\overline{\sigma}}_{k_2 s}(\mathbf{I}_1, \boldsymbol{\Psi}, H_2, \mathbf{C}) = \frac{\varepsilon}{T_5}\int_0^{2\pi}\int_\Omega \frac{1}{A_5}(\sum_{i,j=1}^{m}\sum_{s=1}^{l}\sigma_{is}\sigma_{js}\frac{\partial I_{k_1}}{\partial x_i}\frac{\partial I_{k_2}}{\partial x_j})\mathrm{d}\mathbf{x}'\mathrm{d}\overline{\theta}, \\
&\sum_{s=1}^{l}\overline{\overline{\sigma}}_{u_1 s}\overline{\overline{\sigma}}_{u_2 s}(\mathbf{I}_1, \boldsymbol{\Psi}, H_2, \mathbf{C}) = \frac{\varepsilon}{T_5}\int_0^{2\pi}\int_\Omega \frac{1}{A_5}(\sum_{i,j=1}^{m}\sum_{s=1}^{l}\sigma_{is}\sigma_{js}\frac{\partial \Psi_{u_1}}{\partial x_i}\frac{\partial \Psi_{u_2}}{\partial x_j})\mathrm{d}\mathbf{x}'\mathrm{d}\overline{\theta},
\end{aligned}
$$

$$\sum_{s=1}^{l}\overline{\overline{\sigma}}_{H_2 s}\overline{\overline{\sigma}}_{H_2 s}(\mathbf{I}_1,\boldsymbol{\Psi},H_2,\mathbf{C})=\frac{\varepsilon}{T_5}\int_0^{2\pi}\int_{\Omega}\frac{1}{A_5}(\sum_{i,j=1}^{m}\sum_{s=1}^{l}\sigma_{is}\sigma_{js}\frac{\partial H_2}{\partial x_i}\frac{\partial H_2}{\partial x_j})\mathrm{d}\mathbf{x}'\mathrm{d}\overline{\theta},$$

$$\sum_{s=1}^{l}\overline{\overline{\sigma}}_{v_1 s}\overline{\overline{\sigma}}_{v_2 s}(\mathbf{I}_1,\boldsymbol{\Psi},H_2,\mathbf{C})=\frac{\varepsilon}{T_5}\int_0^{2\pi}\int_{\Omega}\frac{1}{A_5}(\sum_{i,j=1}^{m}\sum_{s=1}^{l}\sigma_{is}\sigma_{js}\frac{\partial C_{v_1}}{\partial x_i}\frac{\partial C_{v_2}}{\partial x_j})\mathrm{d}\mathbf{x}'\mathrm{d}\overline{\theta},$$

$$\sum_{s=1}^{l}\overline{\overline{\sigma}}_{k_1 s}\overline{\overline{\sigma}}_{u_1 s}(\mathbf{I}_1,\boldsymbol{\Psi},H_2,\mathbf{C})=\frac{\varepsilon}{T_5}\int_0^{2\pi}\int_{\Omega}\frac{1}{A_5}(\sum_{i,j=1}^{m}\sum_{s=1}^{l}\sigma_{is}\sigma_{js}\frac{\partial I_{k_1}}{\partial x_i}\frac{\partial \Psi_{u_1}}{\partial x_j})\mathrm{d}\mathbf{x}'\mathrm{d}\overline{\theta},$$

$$\sum_{s=1}^{l}\overline{\overline{\sigma}}_{k_1 s}\overline{\overline{\sigma}}_{H_2 s}(\mathbf{I}_1,\boldsymbol{\Psi},H_2,\mathbf{C})=\frac{\varepsilon}{T_5}\int_0^{2\pi}\int_{\Omega}\frac{1}{A_5}(\sum_{i,j=1}^{m}\sum_{s=1}^{l}\sigma_{is}\sigma_{js}\frac{\partial I_{k_1}}{\partial x_i}\frac{\partial H_2}{\partial x_j})\mathrm{d}\mathbf{x}'\mathrm{d}\overline{\theta},$$

$$\sum_{s=1}^{l}\overline{\overline{\sigma}}_{k_1 s}\overline{\overline{\sigma}}_{v_1 s}(\mathbf{I}_1,\boldsymbol{\Psi},H_2,\mathbf{C})=\frac{\varepsilon}{T_5}\int_0^{2\pi}\int_{\Omega}\frac{1}{A_5}(\sum_{i,j=1}^{m}\sum_{s=1}^{l}\sigma_{is}\sigma_{js}\frac{\partial I_{k_1}}{\partial x_i}\frac{\partial C_{v_1}}{\partial x_j})\mathrm{d}\mathbf{x}'\mathrm{d}\overline{\theta},$$

$$\sum_{s=1}^{l}\overline{\overline{\sigma}}_{u_1 s}\overline{\overline{\sigma}}_{H_2 s}(\mathbf{I}_1,\boldsymbol{\Psi},H_2,\mathbf{C})=\frac{\varepsilon}{T_5}\int_0^{2\pi}\int_{\Omega}\frac{1}{A_5}(\sum_{i,j=1}^{m}\sum_{s=1}^{l}\sigma_{is}\sigma_{js}\frac{\partial \Psi_{u_1}}{\partial x_i}\frac{\partial H_2}{\partial x_j})\mathrm{d}\mathbf{x}'\mathrm{d}\overline{\theta},$$

$$\sum_{s=1}^{l}\overline{\overline{\sigma}}_{u_1 s}\overline{\overline{\sigma}}_{v_1 s}(\mathbf{I}_1,\boldsymbol{\Psi},H_2,\mathbf{C})=\frac{\varepsilon}{T_5}\int_0^{2\pi}\int_{\Omega}\frac{1}{A_5}(\sum_{i,j=1}^{m}\sum_{s=1}^{l}\sigma_{is}\sigma_{js}\frac{\partial C_{v_1}}{\partial x_i}\frac{\partial \Psi_{u_1}}{\partial x_j})\mathrm{d}\mathbf{x}'\mathrm{d}\overline{\theta},$$

$$\sum_{s=1}^{l}\overline{\overline{\sigma}}_{v_1 s}\overline{\overline{\sigma}}_{H_2 s}(\mathbf{I}_1,\boldsymbol{\Psi},H_2,\mathbf{C})=\frac{\varepsilon}{T_5}\int_0^{2\pi}\int_{\Omega}\frac{1}{A_5}(\sum_{i,j=1}^{m}\sum_{s=1}^{l}\sigma_{is}\sigma_{js}\frac{\partial C_{v_1}}{\partial x_i}\frac{\partial H_2}{\partial x_j})\mathrm{d}\mathbf{x}'\mathrm{d}\overline{\theta},$$

$$T_5=\int_0^{2\pi}\int_{\Omega}(1/A_5)\mathrm{d}\mathbf{x}'\mathrm{d}\overline{\theta},$$

$$\Omega=\{\mathbf{x}'\big|H_2(0,\mathbf{x}')\le H_2\},$$

$$k,k_1,k_2=1,2,\ldots,n_1;\quad u,u_1,u_2=1,2,\ldots,\alpha;\quad v,v_1,v_2=1,2,\ldots,M, \tag{3.110}$$

where $\int_0^{2\pi}(\cdot)\mathrm{d}\overline{\theta}$ represents the $(n_1-\alpha)$-fold integral with respect to $\overline{\theta}$, $A_5=\left|\partial(\boldsymbol{I}_1,\boldsymbol{\psi},\overline{\theta},H_2,\boldsymbol{x}',\boldsymbol{c})/\partial\boldsymbol{x}\right|$ is the determinant of the Jacobian matrix for the transformation from $\boldsymbol{x}$ to $\boldsymbol{I}_1,\boldsymbol{\psi},\overline{\theta},H_2,\boldsymbol{x}',\boldsymbol{c}$.

The averaged FPK equation associated with Itô Eq. (3.109) is

$$\frac{\partial p}{\partial t}=-\sum_{k=1}^{n_1}\frac{\partial}{\partial I_k}(\overline{\overline{a}}_k p)-\sum_{u=1}^{\alpha}\frac{\partial}{\partial\psi_u}(\overline{\overline{a}}_u p)-\frac{\partial}{\partial h_2}(\overline{\overline{a}}_{h_2}p)$$
$$-\sum_{v=1}^{M}\frac{\partial}{\partial c_v}(\overline{\overline{a}}_v p)+\frac{1}{2}\sum_{k_1,k_2=1}^{n_1}\frac{\partial^2}{\partial I_{k_1}\partial I_{k_2}}(\overline{\overline{b}}_{k_1k_2}p)+\frac{1}{2}\sum_{u_1,u_2=1}^{\alpha}\frac{\partial^2}{\partial\psi_{u_1}\partial\psi_{u_2}}(\overline{\overline{b}}_{u_1u_2}p)$$

$$
\begin{aligned}
&+\frac{1}{2}\frac{\partial^2}{\partial h_2^2}(\overline{\overline{b}}_{h_2h_2}p)+\frac{1}{2}\sum_{v_1,v_2=1}^{M}\frac{\partial^2}{\partial c_{v_1}\partial c_{v_2}}(\overline{\overline{b}}_{v_1v_2}p)+\sum_{k_1=1}^{n_1}\sum_{u_1=1}^{\alpha}\frac{\partial^2}{\partial I_{k_1}\partial\psi_{u_1}}(\overline{\overline{b}}_{k_1u_1}p)\\
&+\sum_{k_1=1}^{n_1}\frac{\partial^2}{\partial I_{k_1}\partial h_2}(\overline{\overline{b}}_{k_1h_2}p)+\sum_{k_1=1}^{n_1}\sum_{v_1=1}^{M}\frac{\partial^2}{\partial I_{k_1}\partial c_{v_1}}(\overline{\overline{b}}_{k_1v_1}p)+\sum_{u_1=1}^{\alpha}\frac{\partial^2}{\partial\psi_{u_1}\partial h_2}(\overline{\overline{b}}_{u_1h_2}p)\\
&+\sum_{u_1=1}^{\alpha}\sum_{v_1=1}^{M}\frac{\partial^2}{\partial\psi_{u_1}\partial c_{v_1}}(\overline{\overline{b}}_{u_1v_1}p)+\sum_{v_1=1}^{M}\frac{\partial^2}{\partial c_{v_1}\partial h_2}(\overline{\overline{b}}_{v_1h_2}p).
\end{aligned}
\tag{3.111}
$$

where

$$
\begin{aligned}
&\overline{\overline{a}}_h=\overline{\overline{a}}_h(\boldsymbol{I}_1,\boldsymbol{\psi},h_2,\boldsymbol{c})=\overline{\overline{m}}_k(\boldsymbol{I}_1,\boldsymbol{\Psi},H_2,\boldsymbol{C})\big|_{\boldsymbol{\Psi}=\boldsymbol{\psi},\,H_2=h_2,\,\boldsymbol{C}=\boldsymbol{c}},\\
&\overline{\overline{a}}_u=\overline{\overline{a}}_u(\boldsymbol{I}_1,\boldsymbol{\psi},h_2,\boldsymbol{c})=\overline{\overline{m}}_u(\boldsymbol{I}_1,\boldsymbol{\Psi},H_2,\boldsymbol{C})\big|_{\boldsymbol{\Psi}=\boldsymbol{\psi},\,H_2=h_2,\,\boldsymbol{C}=\boldsymbol{c}},\\
&\overline{\overline{a}}_v=\overline{\overline{a}}_v(\boldsymbol{I}_1,\boldsymbol{\psi},h_2,\boldsymbol{c})=\overline{\overline{m}}_v(\boldsymbol{I}_1,\boldsymbol{\Psi},H_2,\boldsymbol{C})\big|_{\boldsymbol{\Psi}=\boldsymbol{\psi},\,H_2=h_2,\,\boldsymbol{C}=\boldsymbol{c}},\\
&\overline{\overline{a}}_{h_2}=\overline{\overline{a}}_{h_2}(\boldsymbol{I}_1,\boldsymbol{\psi},h_2,\boldsymbol{c})=\overline{\overline{m}}_{H_2}(\boldsymbol{I}_1,\boldsymbol{\Psi},H_2,\boldsymbol{C})\big|_{\boldsymbol{\Psi}=\boldsymbol{\psi},\,H_2=h_2,\,\boldsymbol{C}=\boldsymbol{c}},\\
&\overline{\overline{b}}_{k_1k_2}=\overline{\overline{b}}_{k_1k_2}(\boldsymbol{I}_1,\boldsymbol{\psi},h_2,\boldsymbol{c})=\sum_{s=1}^{l}\overline{\overline{\sigma}}_{k_1s}\overline{\overline{\sigma}}_{k_2s}(\boldsymbol{I}_1,\boldsymbol{\Psi},H_2,\boldsymbol{C})\Bigg|_{\boldsymbol{\Psi}=\boldsymbol{\psi},\,H_2=h_2,\,\boldsymbol{C}=\boldsymbol{c}},\\
&\overline{\overline{b}}_{u_1u_2}=\overline{\overline{b}}_{u_1u_2}(\boldsymbol{I}_1,\boldsymbol{\psi},h_2,\boldsymbol{c})=\sum_{s=1}^{l}\overline{\overline{\sigma}}_{u_1s}\overline{\overline{\sigma}}_{u_2s}(\boldsymbol{I}_1,\boldsymbol{\Psi},H_2,\boldsymbol{C})\Bigg|_{\boldsymbol{\Psi}=\boldsymbol{\psi},\,H_2=h_2,\,\boldsymbol{C}=\boldsymbol{c}},\\
&\overline{\overline{b}}_{h_2h_2}=\overline{\overline{b}}_{h_2h_2}(\boldsymbol{I}_1,\boldsymbol{\psi},h_2,\boldsymbol{c})=\sum_{s=1}^{l}\overline{\overline{\sigma}}_{H_2s}\overline{\overline{\sigma}}_{H_2s}(\boldsymbol{I}_1,\boldsymbol{\Psi},H_2,\boldsymbol{C})\Bigg|_{\boldsymbol{\Psi}=\boldsymbol{\psi},\,H_2=h_2,\,\boldsymbol{C}=\boldsymbol{c}},\\
&\overline{\overline{b}}_{v_1v_2}=\overline{\overline{b}}_{v_1v_2}(\boldsymbol{I}_1,\boldsymbol{\psi},h_2,\boldsymbol{c})=\sum_{s=1}^{l}\overline{\overline{\sigma}}_{v_1s}\overline{\overline{\sigma}}_{v_2s}(\boldsymbol{I}_1,\boldsymbol{\Psi},H_2,\boldsymbol{C})\Bigg|_{\boldsymbol{\Psi}=\boldsymbol{\psi},\,H_2=h_2,\,\boldsymbol{C}=\boldsymbol{c}},
\end{aligned}
\tag{3.112}
$$

$$\overline{\overline{b}}_{k_1u_1} = \overline{\overline{b}}_{k_1u_1}(\boldsymbol{I}_1, \boldsymbol{\psi}, h_2, \boldsymbol{c}) = \sum_{s=1}^{l} \overline{\overline{\sigma}}_{k_1s}\overline{\overline{\sigma}}_{u_1s}(\boldsymbol{I}_1, \boldsymbol{\Psi}, H_2, \boldsymbol{C})\Big|_{\boldsymbol{\Psi}=\boldsymbol{\psi},\, H_2=h_2,\, \boldsymbol{C}=\boldsymbol{c}},$$

$$\overline{\overline{b}}_{k_1h_2} = \overline{\overline{b}}_{k_1h_2}(\boldsymbol{I}_1, \boldsymbol{\psi}, h_2, \boldsymbol{c}) = \sum_{s=1}^{l} \overline{\overline{\sigma}}_{k_1s}\overline{\overline{\sigma}}_{H_2s}(\boldsymbol{I}_1, \boldsymbol{\Psi}, H_2, \boldsymbol{C})\Big|_{\boldsymbol{\Psi}=\boldsymbol{\psi},\, H_2=h_2,\, \boldsymbol{C}=\boldsymbol{c}},$$

$$\overline{\overline{b}}_{k_1v_1} = \overline{\overline{b}}_{k_1v_1}(\boldsymbol{I}_1, \boldsymbol{\psi}, h_2, \boldsymbol{c}) = \sum_{s=1}^{l} \overline{\overline{\sigma}}_{k_1s}\overline{\overline{\sigma}}_{v_1s}(\boldsymbol{I}_1, \boldsymbol{\Psi}, H_2, \boldsymbol{C})\Big|_{\boldsymbol{\Psi}=\boldsymbol{\psi},\, H_2=h_2,\, \boldsymbol{C}=\boldsymbol{c}},$$

$$\overline{\overline{b}}_{u_1h_2} = \overline{\overline{b}}_{u_1h_2}(\boldsymbol{I}_1, \boldsymbol{\psi}, h_2, \boldsymbol{c}) = \sum_{s=1}^{l} \overline{\overline{\sigma}}_{u_1s}\overline{\overline{\sigma}}_{H_2s}(\boldsymbol{I}_1, \boldsymbol{\Psi}, H_2, \boldsymbol{C})\Big|_{\boldsymbol{\Psi}=\boldsymbol{\psi},\, H_2=h_2,\, \boldsymbol{C}=\boldsymbol{c}},$$

$$\overline{\overline{b}}_{u_1v_1} = \overline{\overline{b}}_{u_1v_1}(\boldsymbol{I}_1, \boldsymbol{\psi}, h_2, \boldsymbol{c}) = \sum_{s=1}^{l} \overline{\overline{\sigma}}_{u_1s}\overline{\overline{\sigma}}_{v_1s}(\boldsymbol{I}_1, \boldsymbol{\Psi}, H_2, \boldsymbol{C})\Big|_{\boldsymbol{\Psi}=\boldsymbol{\psi},\, H_2=h_2,\, \boldsymbol{C}=\boldsymbol{c}},$$

$$\overline{\overline{b}}_{v_1h_2} = \overline{\overline{b}}_{v_1h_2}(\boldsymbol{I}_1, \boldsymbol{\psi}, h_2, \boldsymbol{c}) = \sum_{s=1}^{l} \overline{\overline{\sigma}}_{v_1s}\overline{\overline{\sigma}}_{H_2s}(\boldsymbol{I}_1, \boldsymbol{\Psi}, H_2, \boldsymbol{C})\Big|_{\boldsymbol{\Psi}=\boldsymbol{\psi},\, H_2=h_2,\, \boldsymbol{C}=\boldsymbol{c}}.$$

In Eq. (3.111), $p = p(\boldsymbol{I}_1, \boldsymbol{\psi}, h_2, \boldsymbol{c}, t|\boldsymbol{I}_{10}, \boldsymbol{\psi}_0, h_{20}, \boldsymbol{c}_0)$ is the transition probability density, or $p = p(\boldsymbol{I}_1, \boldsymbol{\psi}, h_2, \boldsymbol{c}, t)$ is the joint probability density. Equation (3.111) can be solved under certain boundary conditions and initial condition. It can be proven that, once the stationary solution $p(\boldsymbol{I}_1, \boldsymbol{\psi}, h_2, \boldsymbol{c})$ is obtained, the approximated stationary PDF of system (3.5) can then be obtained by using

$$p(\boldsymbol{x}) = \frac{p(\boldsymbol{I}_1, \boldsymbol{\psi}, h_2, \boldsymbol{c})}{T_5(\boldsymbol{I}_1, \boldsymbol{\psi}, h_2, \boldsymbol{c})}\Bigg|_{\boldsymbol{I}_1=\boldsymbol{I}_1(\boldsymbol{x}),\ \boldsymbol{\psi}=\boldsymbol{\psi}(\boldsymbol{x}),\ h_2=h_2(\boldsymbol{x}),\ \boldsymbol{c}=\boldsymbol{c}(\boldsymbol{x})}. \tag{3.113}$$

where $T_5(\boldsymbol{I}_1, \boldsymbol{\psi}, h_2, \boldsymbol{c}) = T_5(\boldsymbol{I}_1, \boldsymbol{\Psi}, H_2, \boldsymbol{C})|_{\boldsymbol{\Psi}=\boldsymbol{\psi}, H_2=h_2, \boldsymbol{C}=\boldsymbol{c}}$, which is given in Eq. (3.110).

Assume that the partially integrable and resonant generalized Hamiltonian system is separable as Eq. (3.99), and the Hamiltonian subsystem with $\boldsymbol{I}_1(\boldsymbol{x}_1) = [I_1, I_2, \ldots, I_{n_1}]^{\mathrm{T}}$ and $\boldsymbol{\theta}_1(\boldsymbol{x}_1) = [\theta_1, \theta_2, \ldots, \theta_{n_1}]^{\mathrm{T}}$ has α weak internal resonance relations (3.106). Then the following Itô stochastic differential equations for the combinations in Eq. (3.107) of angle variables can be derived from Θ_k equations in Eq. (3.100)

$$\mathrm{d}\Psi_u = \{O_u(\varepsilon) + \varepsilon \sum_{j=1}^{n_1} \frac{\partial \Psi_u}{\partial P_j}[\sum_{r=1}^{n_1} d_{jr}\frac{\partial H_1}{\partial P_r} + \sum_{r_1=n_1+1}^{n_1+n_2} d_{jr_1}\frac{\partial H_2}{\partial P_{r_1}}$$
$$+ \sum_{i=2n_1+2n_2+1}^{m} d_{ji}\frac{\partial}{\partial X_i}(\sum_{\beta=1}^{M} C_\beta)] + \frac{1}{2}\sum_{j,r=1}^{n_1}\sum_{s=1}^{l} \sigma_{js}\sigma_{rs}\frac{\partial^2 \Psi_u}{\partial P_j \partial P_r}\}\mathrm{d}t$$

$$+\varepsilon^{1/2}\sum_{j=1}^{n_1}\sum_{s=1}^{l}\sigma_{js}\frac{\partial\Psi_u}{\partial P_j}\mathrm{d}B_s(t),\quad u=1,2,\ldots,\alpha. \tag{3.114}$$

In this case, system is governed by α equations in Eq. (3.114), $n_1-\alpha$ equations of Θ_k $(k=\alpha+1,\ldots,n_1)$ and equations for I_k, H_2, C_v in Eq. (3.100), and $(2n_2-1)$ equations for $X_{2n_1+2},\ldots,X_{2n_1+2n_2}$ in Eq. (3.5). $[\boldsymbol{I}_1^{\mathrm{T}},\boldsymbol{\Psi}^{\mathrm{T}},H_2,\boldsymbol{C}^{\mathrm{T}}]^{\mathrm{T}}$ is slowly-varying vector process. According to the Khasminskii's theorem (Khasminskii 1968), as $\varepsilon\to 0$, $[\boldsymbol{I}_1^{\mathrm{T}},\boldsymbol{\Psi}^{\mathrm{T}},H_2,\boldsymbol{C}^{\mathrm{T}}]^{\mathrm{T}}$ converges weakly to a $(n_1+\alpha+1+M)$-dimensional vector Markov diffusion process and $[\overline{\boldsymbol{\Theta}}^{\mathrm{T}},\boldsymbol{X}'^{\mathrm{T}}]^{\mathrm{T}}$ is a fast-varying vector process. The averaged Itô stochastic differential equation is of the same as that in Eq. (3.109), while the drift and diffusion coefficients change to

$$\overline{\overline{m'}}_k(\boldsymbol{I}_1,\boldsymbol{\Psi},H_2,\boldsymbol{C})=\frac{\varepsilon}{T_5'}\int_0^{2\pi}\int_{\Omega}\frac{1}{A_5'}\{\sum_{j=1}^{n_1}\frac{\partial I_k}{\partial p_j}[\sum_{r=1}^{n_1}d_{jr}\frac{\partial H_1}{\partial p_r}+\sum_{r_1=n_1+1}^{n_1+n_2}d_{jr_1}\frac{\partial H_2}{\partial p_{r_1}}$$
$$+\sum_{i=2n_1+2n_2+1}^{m}d_{ji}\frac{\partial}{\partial x_i}(\sum_{\beta=1}^{M}C_\beta)]+\frac{1}{2}\sum_{j,r=1}^{n_1}\sum_{s=1}^{l}\sigma_{js}\sigma_{rs}\frac{\partial^2 I_k}{\partial p_j\partial p_r}\}\mathrm{d}\boldsymbol{q}_2\mathrm{d}p_{n_1+2}\cdots\mathrm{d}p_{n_1+n_2}\mathrm{d}\theta'_1,$$

$$\overline{\overline{m'}}_u(\boldsymbol{I}_1,\boldsymbol{\Psi},H_2,\boldsymbol{C})=\frac{\varepsilon}{T_5'}\int_0^{2\pi}\int_{\Omega}\frac{1}{A_5'}\{\frac{O(\varepsilon)}{\varepsilon}+\sum_{j=1}^{n_1}\frac{\partial\Psi_u}{\partial p_j}[\sum_{r=1}^{n_1}d_{jr}\frac{\partial H_1}{\partial p_r}$$
$$+\sum_{r_1=n_1+1}^{n_1+n_2}d_{jr_1}\frac{\partial H_2}{\partial p_{r_1}}\sum_{i=2n_1+2n_2+1}^{m}d_{ji}\frac{\partial}{\partial x_i}(\sum_{\beta=1}^{M}C_\beta)]$$
$$+\frac{1}{2}\sum_{j,r=1}^{n_1}\sum_{s=1}^{l}\sigma_{js}\sigma_{rs}\frac{\partial^2\Psi_k}{\partial p_j\partial p_r}\}\mathrm{d}\boldsymbol{q}_2\mathrm{d}p_{n_1+2}\cdots\mathrm{d}p_{n_1+n_2}\mathrm{d}\theta'_1, \tag{3.115}$$

$$\overline{\overline{m'}}_{H_2}(\boldsymbol{I}_1,\boldsymbol{\Psi},H_2,\boldsymbol{C})=\frac{\varepsilon}{T_5'}\int_0^{2\pi}\int_{\Omega}\frac{1}{A_5'}\{\sum_{r_2=n_1+1}^{n_1+n_2}\frac{\partial H_2}{\partial p_{r_2}}[\sum_{r=1}^{n_1}d_{r_2r}\frac{\partial H_1}{\partial p_r}+\sum_{r_1=n_1+1}^{n_1+n_2}d_{r_2r_1}\frac{\partial H_2}{\partial p_{r_1}}$$
$$+\sum_{i=2n_1+2n_2+1}^{m}d_{r_2i}\tfrac{\partial}{\partial x_i}(\sum_{\beta=1}^{M}C_\beta)]+\frac{1}{2}\sum_{r_2,r_1=n_1+1}^{n_1+n_2}\sum_{s=1}^{l}\sigma_{r_2s}\sigma_{r_1s}\frac{\partial^2 H_2}{\partial p_{r_2}\partial p_{r_1}}\}\mathrm{d}\boldsymbol{q}_2\mathrm{d}p_{n_1+2}$$
$$\cdots\mathrm{d}p_{n_1+n_2}\mathrm{d}\theta'_1,$$

$$\overline{\overline{m'}}_v(\boldsymbol{I}_1,\boldsymbol{\Psi},H_2,\boldsymbol{C})=\frac{\varepsilon}{T_5'}\int_0^{2\pi}\int_{\Omega}\frac{1}{A_5'}\{\sum_{i_1=2n_1+2n_2+1}^{m}\frac{\partial C_v}{\partial x_{i_1}}[\sum_{r=1}^{n_1}d_{i_1r}\frac{\partial H_1}{\partial p_r}+\sum_{r_1=n_1+1}^{n_1+n_2}d_{jr_1}\frac{\partial H_2}{\partial p_{r_1}}$$
$$+\sum_{i=2n_1+2n_2+1}^{m}d_{i_1i}\tfrac{\partial}{\partial x_i}(\sum_{\beta=1}^{M}C_\beta)]+\frac{1}{2}\sum_{i,i_1=2n_1+2n_2+1}^{m}\sum_{s=1}^{l}\sigma_{is}\sigma_{i_1s}\frac{\partial^2 C_v}{\partial x_i\partial x_{i_1}}\}\mathrm{d}\boldsymbol{q}_2\mathrm{d}p_{n_1+2}$$
$$\cdots\mathrm{d}p_{n_1+n_2}\mathrm{d}\theta'_1,$$

$$\sum_{s=1}^{l}\overline{\overline{\sigma}}'_{k_1 s}\overline{\overline{\sigma}}'_{k_2 s}(\boldsymbol{I}_1,\boldsymbol{\Psi},H_2,\boldsymbol{C})$$

$$=\frac{\varepsilon}{T_5'}\int_0^{2\pi}\int_\Omega \frac{1}{A_5'}\sum_{j,r=1}^{n_1}\sum_{s=1}^{l}\sigma_{js}\sigma_{rs}\frac{\partial I_{k_1}}{\partial p_j}\frac{\partial I_{k_2}}{\partial p_r}\mathrm{d}\boldsymbol{q}_2\mathrm{d}p_{n_1+2}\cdots\mathrm{d}p_{n_1+n_2}\mathrm{d}\theta_1',$$

$$\sum_{s=1}^{l}\overline{\overline{\sigma}}'_{u_1 s}\overline{\overline{\sigma}}'_{u_2 s}(\boldsymbol{I}_1,\boldsymbol{\Psi},H_2,\boldsymbol{C})$$

$$=\frac{\varepsilon}{T_5'}\int_0^{2\pi}\int_\Omega \frac{1}{A_5'}\sum_{j,r=1}^{n_1}\sum_{s=1}^{l}\sigma_{js}\sigma_{rs}\frac{\partial \Psi_{u_1}}{\partial p_j}\frac{\partial \Psi_{u_2}}{\partial p_r}\mathrm{d}\boldsymbol{q}_2\mathrm{d}p_{n_1+2}\cdots\mathrm{d}p_{n_1+n_2}\mathrm{d}\theta_1',$$

$$\sum_{s=1}^{l}\overline{\overline{\sigma}}'_{H_2 s}\overline{\overline{\sigma}}'_{H_2 s}(\boldsymbol{I}_1,\boldsymbol{\Psi},H_2,\boldsymbol{C})$$

$$=\frac{\varepsilon}{T_5'}\int_0^{2\pi}\int_\Omega \frac{1}{A_5'}\sum_{r_1,r_2=n_1+1}^{n_1+n_2}\sum_{s=1}^{l}\sigma_{r_1 s}\sigma_{r_2 s}\frac{\partial H_2}{\partial p_{r_1}}\frac{\partial H_2}{\partial p_{r_2}}\mathrm{d}\boldsymbol{q}_2\mathrm{d}p_{n_1+2}\cdots\mathrm{d}p_{n_1+n_2}\mathrm{d}\theta_1',$$

$$\sum_{s=1}^{l}\overline{\overline{\sigma}}'_{v_1 s}\overline{\overline{\sigma}}'_{v_2 s}(\boldsymbol{I}_1,\boldsymbol{\Psi},H_2,\boldsymbol{C})$$

$$=\frac{\varepsilon}{T_5'}\int_0^{2\pi}\int_\Omega \frac{1}{A_5'}\sum_{i,i_1=2n_1+2n_2+1}^{m}\sum_{s=1}^{l}\sigma_{is}\sigma_{i_1 s}\frac{\partial C_{v_1}}{\partial x_i}\frac{\partial C_{v_2}}{\partial x_{i_1}}\mathrm{d}\boldsymbol{q}_2\mathrm{d}p_{n_1+2}\cdots\mathrm{d}p_{n_1+n_2}\mathrm{d}\theta_1',$$

$$\sum_{s=1}^{l}\overline{\overline{\sigma}}'_{k_1 s}\overline{\overline{\sigma}}'_{u_1 s}(\boldsymbol{I}_1,\boldsymbol{\Psi},H_2,\boldsymbol{C})$$

$$=\frac{\varepsilon}{T_5'}\int_0^{2\pi}\int_\Omega \frac{1}{A_5'}\sum_{j,r=1}^{n_1}\sum_{s=1}^{l}\sigma_{js}\sigma_{rs}\frac{\partial I_{k_1}}{\partial p_j}\frac{\partial \Psi_{u_1}}{\partial p_r}\mathrm{d}\boldsymbol{q}_2\mathrm{d}p_{n_1+2}\cdots\mathrm{d}p_{n_1+n_2}\mathrm{d}\theta_1',$$

$$\sum_{s=1}^{l}\overline{\overline{\sigma}}'_{k_1 s}\overline{\overline{\sigma}}'_{H_2 s}(\boldsymbol{I}_1,\boldsymbol{\Psi},H_2,\boldsymbol{C})$$

$$=\frac{\varepsilon}{T_5'}\int_0^{2\pi}\int_\Omega \frac{1}{A_5'}\sum_{j=1}^{n_1}\sum_{r_1=n_1+1}^{n_1+n_2}\sum_{s=1}^{l}\sigma_{js}\sigma_{r_1 s}\frac{\partial I_{k_1}}{\partial p_j}\frac{\partial H_2}{\partial p_{r_1}}\mathrm{d}\boldsymbol{q}_2\mathrm{d}p_{n_1+2}\cdots\mathrm{d}p_{n_1+n_2}\mathrm{d}\theta_1',$$

$$\sum_{s=1}^{l}\overline{\overline{\sigma}}'_{k_1 s}\overline{\overline{\sigma}}'_{v_1 s}(\boldsymbol{I}_1,\boldsymbol{\Psi},H_2,\boldsymbol{C})$$

$$=\frac{\varepsilon}{T_5'}\int_0^{2\pi}\int_\Omega \frac{1}{A_5'}\sum_{j=1}^{n_1}\sum_{i=2n_1+2n_2+1}^{m}\sum_{s=1}^{l}\sigma_{js}\sigma_{is}\frac{\partial I_{k_1}}{\partial p_j}\frac{\partial C_{v_1}}{\partial x_i}\mathrm{d}\boldsymbol{q}_2\mathrm{d}p_{n_1+2}\cdots\mathrm{d}p_{n_1+n_2}\mathrm{d}\theta_1',$$

$$\sum_{s=1}^{l}\overline{\overline{\sigma}}'_{u_1 s}\overline{\overline{\sigma}}'_{H_2 s}(\boldsymbol{I}_1,\boldsymbol{\Psi},H_2,\boldsymbol{C})$$
$$=\frac{\varepsilon}{T_5'}\int_0^{2\pi}\int_{\Omega}\frac{1}{A_5'}\sum_{j=1}^{n_1}\sum_{r_1=n_1+1}^{n_1+n_2}\sum_{s=1}^{l}\sigma_{js}\sigma_{rs}\frac{\partial\Psi_{u_1}}{\partial p_j}\frac{\partial H_2}{\partial p_{r_1}}\mathrm{d}\boldsymbol{q}_2\mathrm{d}p_{n_1+2}\cdots\mathrm{d}p_{n_1+n_2}\mathrm{d}\theta_1',$$

$$\sum_{s=1}^{l}\overline{\overline{\sigma}}'_{u_1 s}\overline{\overline{\sigma}}'_{v_1 s}(\boldsymbol{I}_1,\boldsymbol{\Psi},H_2,\boldsymbol{C})$$
$$=\frac{\varepsilon}{T_5'}\int_0^{2\pi}\int_{\Omega}\frac{1}{A_5'}\sum_{j=1}^{n_1}\sum_{i=2n_1+2n_2+1}^{m}\sum_{s=1}^{l}\sigma_{js}\sigma_{is}\frac{\partial\Psi_{u_1}}{\partial p_j}\frac{\partial C_{v_1}}{\partial x_i}\mathrm{d}\boldsymbol{q}_2\mathrm{d}p_{n_1+2}\cdots\mathrm{d}p_{n_1+n_2}\mathrm{d}\theta_1',$$

$$\sum_{s=1}^{l}\overline{\overline{\sigma}}'_{v_1 s}\overline{\overline{\sigma}}'_{H_2 s}(\boldsymbol{I}_1,\boldsymbol{\Psi},H_2,\boldsymbol{C})$$
$$=\frac{\varepsilon}{T_5'}\int_0^{2\pi}\int_{\Omega}\frac{1}{A_5'}\sum_{i=2n_1+2n_2+1}^{m}\sum_{r_1=n_1+1}^{n_1+n_2}\sum_{s=1}^{l}\sigma_{is}\sigma_{r_1 s}\frac{\partial C_{v_1}}{\partial x_i}\frac{\partial H_2}{\partial p_{r_1}}\mathrm{d}\boldsymbol{q}_2\mathrm{d}p_{n_1+2}\cdots\mathrm{d}p_{n_1+n_2}\mathrm{d}\theta_1',$$

$$T_5'=\int_0^{2\pi}\int_{\Omega}(1/A_5')\mathrm{d}\boldsymbol{q}_2\mathrm{d}p_{n_1+2}\cdots\mathrm{d}p_{n_1+n_2}\mathrm{d}\theta_1',$$
$$\Omega=\{(\boldsymbol{q}_2,p_{n_1+2},\ldots,p_{n_1+n_2})\big|H_2(\boldsymbol{q}_2,0,p_{n_1+2},\ldots,p_{n_1+n_2})\le H_2\},$$
$$k,k_1,k_2=1,2,\ldots,n_1;\quad u,u_1,u_2=1,2,\ldots,\alpha;\quad v,v_1,v_2=1,2,\ldots,M,$$

where $\theta'_1=[\theta_{\alpha+1},\ldots,\theta_{n_1}]^{\mathrm{T}}$,

$$A_5'=\left|\frac{\partial(\boldsymbol{I}_1,\boldsymbol{\psi},\theta'_1,\boldsymbol{q}_2,H_2,p_{n_1+2},\ldots,p_{n_1+n_2},\boldsymbol{c})}{\partial(\boldsymbol{q}_1,\boldsymbol{p}_1,\boldsymbol{q}_2,\boldsymbol{p}_2,\boldsymbol{x}_3)}\right|$$
$$=\left|\frac{\partial(\boldsymbol{I}_1,\boldsymbol{\psi},\theta'_1)}{\partial(\boldsymbol{q}_1,\boldsymbol{p}_1)}\right|\left|\frac{\partial(\boldsymbol{q}_2,H_2,p_{n_1+2},\ldots,p_{n_1+n_2})}{\partial(\boldsymbol{q}_2,\boldsymbol{p}_2)}\right|\left|\frac{\partial\boldsymbol{c}}{\partial\boldsymbol{x}_3}\right|=\overline{\overline{C}}\left|\frac{\partial H_2}{\partial p_{n_1+1}}\right|\left|\frac{\partial\boldsymbol{c}}{\partial\boldsymbol{x}_3}\right|. \tag{3.116}$$

Since $\left|\partial(\boldsymbol{I}_1,\boldsymbol{\psi},\theta'_1)/\partial(\boldsymbol{q}_1,\boldsymbol{p}_1)\right|$ is a linear combination of $\left|\partial(\boldsymbol{I}_1,\boldsymbol{\theta}_1)/\partial(\boldsymbol{q}_1,\boldsymbol{p}_1)\right|$, and the latter is equal to 1, it is a constant. The result is

$$T_5'=(2\pi)^{n_1-\alpha}\left|\frac{\partial\boldsymbol{c}}{\partial\boldsymbol{x}_3}\right|\int_{\Omega}\left|\frac{\partial H_2}{\partial p_{n_1+1}}\right|^{-1}\mathrm{d}\boldsymbol{q}_2\mathrm{d}p_{n_1+2}\cdots\mathrm{d}p_{n_1+n_2}. \tag{3.117}$$

The averaged FPK equation associated with Itô equation (3.109) still takes the form of Eq. (3.111), but the first and second derivative moments are modified as follows

$$\bar{\bar{a}}_k'=\bar{\bar{a}}_k'(\mathbf{I}_1,\psi,h_2,\mathbf{c})=\bar{\bar{m}}_k'(\mathbf{I}_1,\Psi,H_2,\mathbf{C})\big|_{\Psi=\psi,\ H_2=h_2,\ \mathbf{C}=\mathbf{c}},$$

$$\bar{\bar{a}}'_u = \bar{\bar{a}}'_u(\mathbf{I}_1, \psi, h_2, \mathbf{c}) = \bar{\bar{m}}'_u(\mathbf{I}_1, \Psi, H_2, \mathbf{C})\big|_{\Psi=\psi,\ H_2=h_2,\ \mathbf{C}=\mathbf{c}},$$

$$\bar{\bar{a}}'_{H_2} = \bar{\bar{a}}'_{H_2}(\mathbf{I}_1, \psi, h_2, \mathbf{c}) = \bar{\bar{m}}'_{H_2}(\mathbf{I}_1, \Psi, H_2, \mathbf{C})\big|_{\Psi=\psi,\ H_2=h_2,\ \mathbf{C}=\mathbf{c}},$$

$$\bar{\bar{a}}'_v = \bar{\bar{a}}'_v(\mathbf{I}_1, \psi, h_2, \mathbf{c}) = \bar{\bar{m}}'_v(\mathbf{I}_1, \Psi, H_2, \mathbf{C})\big|_{\Psi=\psi,\ H_2=h_2,\ \mathbf{C}=\mathbf{c}},$$

$$\bar{\bar{b}}'_{k_1k_2} = \bar{\bar{b}}'_{k_1k_2}(\mathbf{I}_1, \psi, h_2, \mathbf{c}) = \sum_{s=1}^{l} \bar{\bar{\sigma}}'_{k_1s}\bar{\bar{\sigma}}'_{k_2s}(\mathbf{I}_1, \Psi, H_2, \mathbf{C})\Bigg|_{\Psi=\psi,\ H_2=h_2,\ \mathbf{C}=\mathbf{c}},$$

$$\bar{\bar{b}}'_{u_1u_2} = \bar{\bar{b}}'_{u_1u_2}(\mathbf{I}_1, \psi, h_2, \mathbf{c}) = \sum_{s=1}^{l} \bar{\bar{\sigma}}'_{u_1s}\bar{\bar{\sigma}}'_{u_2s}(\mathbf{I}_1, \Psi, H_2, \mathbf{C})\Bigg|_{\Psi=\psi,\ H_2=h_2,\ \mathbf{C}=\mathbf{c}},$$

$$\bar{\bar{b}}'_{H_2H_2} = \bar{\bar{b}}'_{H_2H_2}(\mathbf{I}_1, \psi, h_2, \mathbf{c}) = \sum_{s=1}^{l} \bar{\bar{\sigma}}'_{H_2s}\bar{\bar{\sigma}}'_{H_2s}(\mathbf{I}_1, \Psi, H_2, \mathbf{C})\Bigg|_{\Psi=\psi,\ H_2=h_2,\ \mathbf{C}=\mathbf{c}},$$

$$\bar{\bar{b}}'_{v_1v_2} = \bar{\bar{b}}'_{v_1v_2}(\mathbf{I}_1, \psi, h_2, \mathbf{c}) = \sum_{s=1}^{l} \bar{\bar{\sigma}}'_{v_1s}\bar{\bar{\sigma}}'_{v_2s}(\mathbf{I}_1, \Psi, H_2, \mathbf{C})\Bigg|_{\Psi=\psi,\ H_2=h_2,\ \mathbf{C}=\mathbf{c}},$$

$$\bar{\bar{b}}'_{k_1u_1} = \bar{\bar{b}}'_{k_1u_1}(\mathbf{I}_1, \psi, h_2, \mathbf{c}) = \sum_{s=1}^{l} \bar{\bar{\sigma}}'_{k_1s}\bar{\bar{\sigma}}'_{u_1s}(\mathbf{I}_1, \Psi, H_2, \mathbf{C})\Bigg|_{\Psi=\psi,\ H_2=h_2,\ \mathbf{C}=\mathbf{c}},$$

$$\bar{\bar{b}}'_{k_1H_2} = \bar{\bar{b}}'_{k_1H_2}(\mathbf{I}_1, \psi, h_2, \mathbf{c}) = \sum_{s=1}^{l} \bar{\bar{\sigma}}'_{k_1s}\bar{\bar{\sigma}}'_{H_2s}(\mathbf{I}_1, \Psi, H_2, \mathbf{C})\Bigg|_{\Psi=\psi,\ H_2=h_2,\ \mathbf{C}=\mathbf{c}},$$

$$\bar{\bar{b}}'_{k_1v_1} = \bar{\bar{b}}'_{k_1v_1}(\mathbf{I}_1, \psi, h_2, \mathbf{c}) = \sum_{s=1}^{l} \bar{\bar{\sigma}}'_{k_1s}\bar{\bar{\sigma}}'_{v_1s}(\mathbf{I}_1, \Psi, H_2, \mathbf{C})\Bigg|_{\Psi=\psi,\ H_2=h_2,\ \mathbf{C}=\mathbf{c}},$$

$$\bar{\bar{b}}'_{u_1H_2} = \bar{\bar{b}}'_{u_1H_2}(\mathbf{I}_1, \psi, h_2, \mathbf{c}) = \sum_{s=1}^{l} \bar{\bar{\sigma}}'_{u_1s}\bar{\bar{\sigma}}'_{H_2s}(\mathbf{I}_1, \Psi, H_2, \mathbf{C})\Bigg|_{\Psi=\psi,\ H_2=h_2,\ \mathbf{C}=\mathbf{c}},$$

$$\bar{\bar{b}}'_{u_1v_1} = \bar{\bar{b}}'_{u_1v_1}(\mathbf{I}_1, \psi, h_2, \mathbf{c}) = \sum_{s=1}^{l} \bar{\bar{\sigma}}'_{u_1s}\bar{\bar{\sigma}}'_{v_1s}(\mathbf{I}_1, \Psi, H_2, \mathbf{C})\Bigg|_{\Psi=\psi,\ H_2=h_2,\ \mathbf{C}=\mathbf{c}},$$

$$\bar{\bar{b}}'_{v_1H_2} = \bar{\bar{b}}'_{v_1H_2}(\mathbf{I}_1, \psi, h_2, \mathbf{c}) = \sum_{s=1}^{l} \bar{\bar{\sigma}}'_{v_1s}\bar{\bar{\sigma}}'_{H_2s}(\mathbf{I}_1, \Psi, H_2, \mathbf{C})\Bigg|_{\Psi=\psi,\ H_2=h_2,\ \mathbf{C}=\mathbf{c}}. \quad (3.118)$$

It can be proven that if the stationary solution $p(\boldsymbol{I}_1, \boldsymbol{\psi}, h_2, \boldsymbol{c})$ of the averaged FPK Eq. (3.111) has been obtained, then the approximate stationary PDF for the original system (3.5) is given by

$$p(\boldsymbol{x}) = \left.\frac{p(\boldsymbol{I}_1, \boldsymbol{\psi}, h_2, \boldsymbol{c})}{T_5'}\right|_{\boldsymbol{I}_1=\boldsymbol{I}_1(\boldsymbol{x}_1),\ \boldsymbol{\psi}=\boldsymbol{\psi}(\boldsymbol{x}_1),\ h_2=h_2(\boldsymbol{x}_2),\ \boldsymbol{c}=\boldsymbol{c}(\boldsymbol{x}_3)}, \tag{3.119}$$

where $T_5'(\boldsymbol{I}_1, \boldsymbol{\psi}, h_2, \boldsymbol{c}) = T_5'(\boldsymbol{I}_1, \boldsymbol{\Psi}, H_2, \boldsymbol{C})\big|_{\boldsymbol{\Psi}=\boldsymbol{\psi}, H_2=h_2, \boldsymbol{C}=\boldsymbol{c}}$, which is determined by Eq. (3.117).

Example 3.3 Consider a multi-dimensional nonlinear system subjected to Gaussian white noises (Huang and Zhu 2009). The motion equations are given by

$$\begin{aligned}
&\ddot{Y}_1 + d_{11}\dot{Y}_1 + d_{12}\dot{Y}_2 + \omega_1^2 Y_1 = g_{11}W_{g1}(t),\\
&\ddot{Y}_2 + d_{21}\dot{Y}_1 + d_{22}\dot{Y}_2 + \omega_2^2 Y_2 = g_{22}W_{g2}(t),\\
&\ddot{Y}_3 + d_{33}\dot{Y}_3 + \frac{\partial U}{\partial Y_3} = g_{33}W_{g3}(t),\\
&\ddot{Y}_4 + d_{44}\dot{Y}_4 + \frac{\partial U}{\partial Y_4} = g_{44}W_{g4}(t),\\
&\dot{Y}_5 + d_{55}Y_5 = g_{55}W_{g5}(t),
\end{aligned} \tag{3.120}$$

where $d_{ij} = d_{ij}(Y_1, \ldots, Y_5, \dot{Y}_1, \ldots, \dot{Y}_4)$, $i, j = 1, 2, 3, 4, 5$, are small coefficients of linear or nonlinear dampings of order ε, $g_{ii} = g_{ii}(Y_1, \ldots, Y_5, \dot{Y}_1, \ldots, \dot{Y}_4)$ are the magnitudes of stochastic excitations of order $\varepsilon^{1/2}$, $U = U(Y_3, Y_4) = b(\omega_3^2 Y_3^2 + \omega_4^2 Y_4^2)^\delta$ $(b, \delta > 0, \delta \neq 1)$ is the potential function of 3th and 4th DOF, $W_{gi}(t)$ are independent Gaussian white noises with intensities $2\pi K_i$. The fifth equation in Eq. (3.120) can be considered as a degenerate system of the following oscillator with $\omega_5 = 0$,

$$\ddot{Z} + d_{55}\dot{Z} + \omega_5^2 Z = g_{55}W_{g5}(t) \tag{3.121}$$

Letting $\dot{Z} = Y_5$ in Eq. (3.121) yields the fifth equation in Eq. (3.120).

System (3.120) can be written in the form of Stratonovich stochastic differential equation, then by introducing Wong-Zakai correction terms, it can be rewritten into the following equivalent Itô stochastic differential equations

$$
\begin{aligned}
\mathrm{d}X_1 &= \frac{\partial H}{\partial X_5}\mathrm{d}t,\\
\mathrm{d}X_2 &= \frac{\partial H}{\partial X_6}\mathrm{d}t,\\
\mathrm{d}X_3 &= \frac{\partial H}{\partial X_7}\mathrm{d}t,\\
\mathrm{d}X_4 &= \frac{\partial H}{\partial X_8}\mathrm{d}t,\\
\mathrm{d}X_5 &= [-\frac{\partial H}{\partial X_1} - d_{11}X_5 - d_{12}X_6 + \frac{1}{4}\frac{\partial b_{11}}{\partial X_5}]\mathrm{d}t + \sigma_{11}\mathrm{d}B_1(t),\\
\mathrm{d}X_6 &= [-\frac{\partial H}{\partial X_2} - d_{21}X_5 - d_{22}X_6 + \frac{1}{4}\frac{\partial b_{22}}{\partial X_6}]\mathrm{d}t + \sigma_{22}\mathrm{d}B_2(t),\\
\mathrm{d}X_7 &= [-\frac{\partial H}{\partial X_3} - d_{33}X_7 + \frac{1}{4}\frac{\partial b_{33}}{\partial X_7}]\mathrm{d}t + \sigma_{33}\mathrm{d}B_3(t),\\
\mathrm{d}X_8 &= [-\frac{\partial H}{\partial X_4} - d_{44}X_8 + \frac{1}{4}\frac{\partial b_{44}}{\partial X_8}]\mathrm{d}t + \sigma_{44}\mathrm{d}B_4(t),\\
\mathrm{d}X_9 &= [-d_{55}X_9 + \frac{1}{4}\frac{\partial b_{55}}{\partial X_9}]\mathrm{d}t + \sigma_{55}\mathrm{d}B_5(t),
\end{aligned}
\tag{3.122}
$$

where $B_i(t)$ are independent unit Wiener processes

$$
\begin{aligned}
&\boldsymbol{X} = [X_1, X_2, \ldots, X_9]^{\mathrm{T}} = [Y_1, Y_2, Y_3, Y_4, \dot{Y}_1, \dot{Y}_2, \dot{Y}_3, \dot{Y}_4, Y_5]^{\mathrm{T}},\\
&H = H_1 + H_2 + C_1, \quad H_1 = \overline{H}_1 + \overline{H}_2,\\
&\overline{H}_1 = \omega_1 I_1 = \frac{X_5^2 + \omega_1^2 X_1^2}{2},\\
&\overline{H}_2 = \omega_2 I_2 = \frac{X_6^2 + \omega_2^2 X_2^2}{2}, \quad H_2 = \frac{X_7^2 + X_8^2}{2} + U(X_3, X_4),\\
&C_1 = \frac{X_9^2}{2}, \quad b_{ii} = \sigma_{ii}^2 = 2\pi K_i g_{ii}^2 \ (i = 1, 2, \ldots, 5).
\end{aligned}
\tag{3.123}
$$

The generalized Hamiltonian system associated with system (3.122) is partially integrable, and Eq. (3.99) holds. Referring to Eq. (3.94) in vol. 1, the structure matrix is

$$
\boldsymbol{J} = \begin{bmatrix} \boldsymbol{0} & \boldsymbol{I}_4 & 0 \\ -\boldsymbol{I}_4 & \boldsymbol{0} & 0 \\ 0 & 0 & 0 \end{bmatrix}
\tag{3.124}
$$

where $\boldsymbol{I}_4$ is a 4×4 identity matrix; $\boldsymbol{0}$ is a 4×4 zero matrix; the rank of the structure matrix $\boldsymbol{J}$ is 8. Therefore, there exists a Casimir function C_1.

In system (3.122), the sub-Hamiltonian system associated with H_2 is nonintegrable while the sub-Hamiltonian system associated with H_1 is integrable. The action-angle variables for H_1 are $I_1 = (X_5^2 + \omega_1^2 X_1^2)/2\omega_1$, $I_2 = (X_6^2 + \omega_2^2 X_2^2)/2\omega_2$, $\theta_1 = \tan^{-1}(X_5/\omega_1 X_1)$ and $\theta_2 = \tan^{-1}(X_6/\omega_2 X_2)$. It is internal resonant if the following resonance relation exists

$$k_1^1\omega_1 + k_2^1\omega_2 = O_1(\varepsilon) \tag{3.125}$$

where k_1^1, k_2^1 are integers of not all zero. Equation (3.125) is not satisfied, this Hamiltonian subsystem is non-internal resonant.

i. Non-internal resonance case

Assuming that the generalized Hamiltonian system associated with system (3.122) is not resonant, the Itô stochastic differential equations for $I_1, I_2, \Theta_1, \Theta_2, H_2, C_1$ are of the form of Eq. (3.100), and the Itô stochastic differential equations for X_3, X_4, X_8 are of form of Eq. (3.122). In this case, I_1, I_2, H_2, C_1 are slowly-varying processes while $\Theta_1, \Theta_2, X_3, X_4, X_8$ are rapidly-varying processes. Using the stochastic averaging method for quasi-partially integrable and non-resonant generalized Hamiltonian systems in Sect. 3.3.1, $[I_1, I_2, H_2, C_1]^{\mathrm{T}}$ converges weakly to 4-dimensional vector Markov diffusion process. According to Eq. (3.94), the averaged Itô stochastic differential equations are

$$\begin{aligned}
\mathrm{d}I_1 &= \overline{m}_1(I_1, I_2, H_2, C_1)\mathrm{d}t + \overline{\sigma}_{11}(I_1, I_2, H_2, C_1)\mathrm{d}B_1(t),\\
\mathrm{d}I_2 &= \overline{m}_2(I_1, I_2, H_2, C_1)\mathrm{d}t + \overline{\sigma}_{22}(I_1, I_2, H_2, C_1)\mathrm{d}B_2(t),\\
\mathrm{d}H_2 &= \overline{m}_3(I_1, I_2, H_2, C_1)\mathrm{d}t + \overline{\sigma}_{33}(I_1, I_2, H_2, C_1)\mathrm{d}B_3(t)\\
&\quad + \overline{\sigma}_{44}(I_1, I_2, C_1, H_2)\mathrm{d}B_4(t),\\
\mathrm{d}C_1 &= \overline{m}_4(I_1, I_2, H_2, C_1)\mathrm{d}t + \overline{\sigma}_{55}(I_1, I_2, H_2, C_1)\mathrm{d}B_5(t).
\end{aligned} \tag{3.126}$$

According to Eq. (3.101), the drift coefficient and diffusion coefficients are

$$\begin{aligned}
\overline{m}_1(I_1, I_2, H_2, C_1) &= -d_{110}I_1 - \frac{3}{2}d_{111}\omega_1 I_1^2 - d_{112}\omega_2 I_1 I_2 - 2d_{113}C_1 I_1\\
&\quad - \frac{\delta}{1+\delta}(d_{114} + d_{115})I_1 H_2 + \frac{\pi K_1 g_{11}^2}{\omega_1},\\
\overline{m}_2(I_1, I_2, H_2, C_1) &= -d_{220}I_2 - d_{221}\omega_1 I_1 I_2 - \frac{3}{2}d_{222}\omega_2 I_2^2 - 2d_{223}C_1 I_2\\
&\quad - \frac{\delta}{1+\delta}(d_{224} + d_{225})I_2 H_2 + \frac{\pi K_2 g_{22}^2}{\omega_2},\\
\overline{m}_3(I_1, I_2, H_2, C_1) &= \frac{-\delta}{1+\delta}[d_{330}H_2 + d_{331}\omega_1 I_1 H_2 + d_{332}\omega_2 I_2 H_2 + 2d_{333}C_1 H_2\\
&\quad + d_{440}H_2 + d_{441}\omega_1 I_1 H_2 + d_{442}\omega_2 I_2 H_2 + 2d_{443}C_1 H_2]\\
&\quad + \frac{\delta^2 H_2}{(1+\delta)(1+2\delta)}(3d_{334} + 3d_{445} + d_{335} + d_{444})
\end{aligned}$$

$$
\begin{aligned}
&+\pi K_3 g_{33}^2+\pi K_4 g_{44}^2,\\
\overline{m}_4(I_1,I_2,H_2,C_1)&=-2d_{550}C_1-2d_{551}C_1\omega_1 I_1-2d_{552}C_1\omega_2 I_2-4d_{553}C_1^2\\
&\quad-\frac{2\delta}{1+\delta}(d_{554}+d_{555})C_1H_2+\pi K_5 g_{55}^2,\\
\overline{\sigma}_{11}^2(I_1,I_2,H_2,C_1)&=\frac{2\pi K_1 g_{11}^2 I_1}{\omega_1},\quad \overline{\sigma}_{22}^2(I_1,I_2,H_2,C_1)=\frac{2\pi K_2 g_{22}^2 I_2}{\omega_2},\\
(\overline{\sigma}_{33}^2+\overline{\sigma}_{44}^2)(I_1,I_2,H_2,C_1)&=\frac{2\delta H_2}{1+\delta}(\pi K_3 g_{33}^2+\pi K_4 g_{44}^2),\\
\overline{\sigma}_{55}^2(I_1,I_2,H_2,C_1)&=4\pi K_5 g_{55}^2 C_1,\quad \sum_{s=1}^{5}\overline{\sigma}_{is}\overline{\sigma}_{js}(I_1,I_2,H_2,C_1)=0,\\
&\quad i,j=1,2,3,4;\quad i\neq j.
\end{aligned}
\tag{3.127}
$$

where d_{iij} $(i,j=1,2,\ldots,5)$ are the coefficients of the following expressions

$$
d_{ii}=d_{ii0}+d_{ii1}X_5^2+d_{ii2}X_6^2+d_{ii3}X_9^2+d_{ii4}X_7^2+d_{ii5}X_8^2. \tag{3.128}
$$

When the damping coefficients and excitation intensities satisfy the following compatibility conditions

$$
\begin{aligned}
&\frac{d_{112}}{K_1g_{11}^2}=\frac{d_{221}}{K_2g_{22}^2},\quad \frac{2d_{113}}{K_1g_{11}^2}=\frac{d_{551}}{K_5g_{55}^2},\quad \frac{(d_{114}+d_{115})\delta}{K_1g_{11}^2(1+\delta)}=\frac{d_{331}+d_{441}}{K_3g_{33}^2+K_4g_{44}^2},\\
&\frac{2d_{223}}{K_2g_{22}^2}=\frac{d_{552}}{K_5g_{55}^2},\quad \frac{(d_{224}+d_{225})\delta}{K_2g_{22}^2(1+\delta)}=\frac{d_{332}+d_{442}}{K_3g_{33}^2+K_4g_{44}^2},\\
&\frac{(d_{554}+d_{555})\delta}{K_5g_{55}^2(1+\delta)}=\frac{2(d_{333}+d_{443})}{K_3g_{33}^2+K_4g_{44}^2},
\end{aligned}
\tag{3.129}
$$

the exact stationary solution for FPK equation of the form (3.96) associated with Itô stochastic differential equation (3.126) is

$$
\begin{aligned}
p(I_1,I_2,h_2,c_1)&=\frac{\overline{C}H_2^{1/\delta}}{\sqrt{c_1}}\exp\Bigg\{-\Bigg[\frac{d_{110}\omega_1 I_1}{K_1g_{11}^2}+\frac{d_{220}\omega_2 I_2}{K_2g_{22}^2}+\frac{d_{330}+d_{440}}{K_3g_{33}^2+K_4g_{44}^2}h_2\\
&+\frac{d_{550}c_1}{K_5g_{55}^2}+\frac{3d_{111}\omega_1^2I_1^2}{4K_1g_{11}^2}+\frac{3d_{222}\omega_2^2I_2^2}{4K_2g_{22}^2}+\frac{\delta(3d_{334}+3d_{445}+d_{335}+d_{444})}{2(K_3g_{33}^2+K_4g_{44}^2)(1+2\delta)}h_2\\
&+\frac{d_{553}c_1^2}{K_5g_{55}^2}+\frac{d_{112}\omega_1\omega_2I_1I_2}{K_1g_{11}^2}+\frac{(d_{114}+d_{115})\delta\omega_1}{K_1g_{11}^2(1+\delta)}I_1h_2+\frac{d_{113}\omega_1I_1c_1}{K_1g_{11}^2}\\
&+\frac{(d_{224}+d_{225})\delta\omega_2}{K_2g_{22}^2(1+\delta)}I_2h_2+\frac{2d_{223}\omega_2I_2c_1}{K_2g_{22}^2}+\frac{(d_{554}+d_{555})\delta}{K_5g_{55}^2(1+\delta)}c_1h_2\Bigg]\Bigg\}.
\end{aligned}
\tag{3.130}
$$

In this example, according to Eqs. (3.101)–(3.103), the Jacobian matrix determinant A'_4 and the period T'_4 are

$$A'_4 = \left| \frac{\partial(I_1, I_2, \theta_1, \theta_2, X_3, X_4, H_2, X_8, C_1)}{\partial \boldsymbol{X}} \right| = \left| \frac{\partial H_2}{\partial X_7} \right| \left| \frac{\partial C_1}{\partial X_9} \right| = |X_7 X_9|,$$

$$T'_4 = \int_0^{2\pi} \int_\Omega \left(\frac{1}{A'_4} \right) dX_3 dX_4 dX_8 d\theta_1 d\theta_2 = \frac{4\sqrt{2}\pi^4}{\omega_1 \omega_2 b^{1/\delta}} \frac{H_2^{1/\delta}}{\sqrt{C_1}},$$

$$\Omega = \{ (X_3, X_4, X_8) | H_2(X_3, X_4, 0, X_8) \le H_2 \}. \tag{3.131}$$

Using Eq. (3.105), the approximate stationary PDF of the original system (3.122) is

$$p(\boldsymbol{x}) = \frac{p(I_1, I_2, h_2, c_1)}{T'_4} \Bigg|_{I_1 = I_1(x_1, x_5),\ I_2 = I_2(x_2, x_6),\ h_2 = h_2(x_3, x_4, x_7, x_8),\ c_1 = c_1(x_9)} . \tag{3.132}$$

ii. Internal resonance case

Assume that the generalized Hamiltonian system (3.122) has an internal resonant relation (3.125), e.g., main internal resonance $\omega_1 - \omega_2 = 0$. Introduce a combination of angle variables

$$\Psi_1 = \Theta_1 - \Theta_2. \tag{3.133}$$

The Itô stochastic differential equation for Ψ_1 is of the form of Eq. (3.114). In this case, $I_1, I_2, \Psi_1, H_2, C_1$ are slowly-varying processes while Θ_2, X_3, X_4, X_8 are rapidly-varying processes. As the damping coefficients and excitation intensities approach to zero, $[I_1, I_2, \Psi_1, H_2, C_1]^{\mathrm{T}}$ converges weakly to a 5-dimensional vector Markov diffusion process. The averaged Itô stochastic differential equations for $I_1, I_2, \Psi_1, H_2, C_1$ are of the form in Eq. (3.109), i.e.,

$$\begin{aligned}
\mathrm{d}I_1 &= \overline{\overline{m}}'_1(I_1, I_2, \Psi_1, H_2, C_1)\mathrm{d}t + \overline{\overline{\sigma}}'_{11}(I_1, I_2, \Psi_1, H_2, C_1)\mathrm{d}B_1(t), \\
\mathrm{d}I_2 &= \overline{\overline{m}}'_2(I_1, I_2, \Psi_1, H_2, C_1)\mathrm{d}t + \overline{\overline{\sigma}}'_{22}(I_1, I_2, \Psi_1, H_2, C_1)\mathrm{d}B_2(t), \\
\mathrm{d}\Psi_1 &= \overline{\overline{m}}'_3(I_1, I_2, \Psi_1, H_2, C_1)\mathrm{d}t + \overline{\overline{\sigma}}'_{31}(I_1, I_2, \Psi_1, H_2, C_1)\mathrm{d}B_1(t) \\
&\quad + \overline{\overline{\sigma}}'_{32}(I_1, I_2, \Psi_1, H_2, C_1)\mathrm{d}B_2(t), \\
\mathrm{d}H_2 &= \overline{\overline{m}}'_4(I_1, I_2, \Psi_1, H_2, C_1)\mathrm{d}t + \overline{\overline{\sigma}}'_{53}(I_1, I_2, \Psi_1, H_2, C_1)\mathrm{d}B_3(t) \\
&\quad + \overline{\overline{\sigma}}'_{54}(I_1, I_2, \Psi_1, H_2, C_1)\mathrm{d}B_4(t), \\
\mathrm{d}C_1 &= \overline{\overline{m}}'_5(I_1, I_2, \Psi_1, H_2, C_1)\mathrm{d}t + \overline{\overline{\sigma}}'_{45}(I_1, I_2, \Psi_1, H_2, C_1)\mathrm{d}B_5(t),
\end{aligned} \tag{3.134}$$

where the drift coefficient and diffusion coefficients can be obtained using Eq. (3.115) as follows

$$
\begin{aligned}
&\overline{\overline{m}}'_1(I_1, I_2, \Psi_1, H_2, C_1) = \overline{m}_1(I_1, I_2, H_2, C_1) \\
&\quad - \frac{d_{112}\omega_2 I_1 I_2}{2}\cos(2\Psi_1) - d_{12}\sqrt{\omega_2 I_1 I_2/\omega_1}\cos\Psi_1, \\
&\overline{\overline{m}}'_2(I_1, I_2, \Psi_1, H_2, C_1) = \overline{m}_2(I_1, I_2, H_2, C_1) \\
&\quad - \frac{d_{221}\omega_1 I_1 I_2}{2}\cos(2\Psi_1) - d_{21}\sqrt{\omega_1 I_1 I_2/\omega_2}\cos\Psi_1, \\
&\overline{\overline{m}}'_3(I_1, I_2, \Psi_1, H_2, C_1) = \frac{d_{112}\omega_2 I_2}{4}\sin(2\Psi_1) \\
&\quad + \frac{d_{221}\omega_1 I_1}{4}\sin(2\Psi_1) + \frac{d_{21}}{4}\sqrt{\frac{\omega_2 I_2}{\omega_1 I_1}}\sin\Psi_1 + \frac{d_{12}}{4}\sqrt{\frac{\omega_1 I_1}{\omega_2 I_2}}\sin\Psi_1, \\
&\overline{\overline{m}}'_4(I_1, I_2, \Psi_1, H_2, C_1) = \overline{m}_3(I_1, I_2, H_2, C_1), \\
&\overline{\overline{m}}'_5(I_1, I_2, \Psi_1, H_2, C_1) = \overline{m}_4(I_1, I_2, H_2, C_1), \\
&(\overline{\overline{\sigma}}'_{31}\overline{\overline{\sigma}}'_{31} + \overline{\overline{\sigma}}'_{32}\overline{\overline{\sigma}}'_{32})(I_1, I_2, \Psi_1, H_2, C_1) = \frac{\pi K_1 g_{11}^2}{2\omega_1 I_1} + \frac{\pi K_2 g_{22}^2}{2\omega_2 I_2}, \\
&\overline{\overline{\sigma}}{}'^2_{11} = \overline{\sigma}^2_{11}, \overline{\overline{\sigma}}{}'^2_{22} = \overline{\sigma}^2_{22}, \overline{\overline{\sigma}}{}'^2_{44} = \overline{\sigma}^2_{33}, \overline{\overline{\sigma}}{}'^2_{53} + \overline{\overline{\sigma}}{}'^2_{54} = \overline{\sigma}^2_{43} + \overline{\sigma}^2_{44}, \\
&\sum_{s=1}^{5}\overline{\overline{\sigma}}'_{is}\overline{\overline{\sigma}}'_{js} = 0, \; (i, j = 1, 2, 3, 4, 5;\, i \neq j).
\end{aligned} \tag{3.135}
$$

When the damping coefficients and excitation intensities satisfy the compatibility condition (3.129) and

$$
\frac{d_{12}}{K_1 g_{11}^2} = \frac{d_{21}}{K_2 g_{22}^2}, \tag{3.136}
$$

the exact stationary solution for FPK equation of the form (3.111) associated with Itô stochastic differential equation (3.134) is

$$
\begin{aligned}
p(I_1, I_2, \psi_1, h_2, c_1) &= \frac{\overline{C} h_2^{1/\delta}}{\sqrt{c_1}}\exp\Bigg\{-\Bigg[\frac{d_{110}\omega_1 I_1}{K_1 g_{11}^2} + \frac{d_{220}\omega_2 I_2}{K_2 g_{22}^2} + \frac{d_{550} c_1}{K_5 g_{55}^2} \\
&+ \frac{d_{330} + d_{440}}{K_3 g_{33}^2 + K_4 g_{44}^2} h_2 + \frac{3 d_{111}\omega_1^2 I_1^2}{4 K_1 g_{11}^2} + \frac{3 d_{222}\omega_2^2 I_2^2}{4 K_2 g_{22}^2} + \frac{d_{553} c_1^2}{K_5 g_{55}^2} \\
&+ \frac{\delta(3d_{334} + 3d_{445} + d_{335} + d_{444})}{2(K_3 g_{33}^2 + K_4 g_{44}^2)(1 + 2\delta)} h_2 + \frac{d_{112}\omega_1\omega_2 I_1 I_2}{K_1 g_{11}^2} + \frac{d_{113}\omega_1 I_1 c_1}{K_1 g_{11}^2} \\
&+ \frac{(d_{114} + d_{115})\delta\omega_1}{K_1 g_{11}^2(1 + \delta)} I_1 h_2 + \frac{2 d_{223}\omega_2 I_2 c_1}{K_2 g_{22}^2} + \frac{(d_{224} + d_{225})\delta\omega_2}{K_2 g_{22}^2(1 + \delta)} I_2 h_2 \\
&+ \frac{(d_{554} + d_{555})\delta}{K_5 g_{55}^2(1 + \delta)} c_1 h_2 + \frac{d_{112}\omega_1\omega_2 I_1 I_2}{2 K_1 g_{11}^2}\cos(2\psi_1) + \frac{2 d_{12}\sqrt{\omega_1\omega_2 I_1 I_2}}{K_1 g_{11}^2}\cos\psi_1\Bigg]\Bigg\}.
\end{aligned} \tag{3.137}
$$

In this case, the Jacobian determinant A'_5 in Eq. (3.116) and T'_5 in Eq. (3.117) are

$$\begin{aligned}
A'_5 &= \left| \frac{\partial(\psi_1, \theta_2, X_3, X_4, I_1, I_2, H_2, X_8, C_1)}{\partial(X_1, X_2, X_3, X_4, X_5, X_6, X_7, X_8, X_9)} \right| \\
&= \left| \frac{\partial(\psi_1, \theta_2, I_1, I_2)}{\partial(X_1, X_2, X_5, X_6)} \right| \left| \frac{\partial(X_3, X_4, H_2, X_8)}{\partial(X_3, X_4, X_7, X_8)} \right| \left| \frac{\partial C_1}{\partial X_9} \right| = \overline{\overline{C}} X_7 X_9, \\
T'_5(I_1, I_2, \psi_1, H_2, C_1) &= 2\pi \int_\Omega \left(\frac{1}{A'_5} \right) dX_3 dX_4 dX_8 = \frac{2\sqrt{2}\pi^3}{\omega_1 \omega_2 b^{1/\delta}} \frac{H_2^{1/\delta}}{\sqrt{C_1}}, \\
\Omega &= \{ (X_3, X_4, X_8) | H_2(X_3, X_4, 0, X_8) \le H_2 \}.
\end{aligned} \tag{3.138}$$

Using Eq. (3.119), the approximate stationary PDF of the original system (3.122) is

$$p(\boldsymbol{x}) = \left. \frac{p(I_1, I_2, \psi_1, h_2, c_1)}{T'_5} \right|_{I_1 = I_1(x_1, x_5), I_2 = I_2(x_2, x_6), \psi_1 = \psi_1(x_1, x_2, x_5, x_6), h_2 = h_2(x_3, x_4, x_7, x_8), c_1 = c_1(x_9)}. \tag{3.139}$$

Numerical calculations were carried out for the Example 3.3 using both the stochastic averaging method and Monte Carlo simulations of system (3.120). Figure 3.9a–c shows the stationary marginal PDFs of action variables I_1, I_2 and X_9, respectively, for the case of non-internal resonance. Figure 3.10a–d shows the joint stationary PDFs $p(x_1, x_5)$ and $p(x_3, x_7)$ for the case of non-internal resonance. Figure 3.11a–c illustrates the stationary marginal PDFs of action variable I_1, phase angle difference Ψ_1, and X_9, respectively for the case of internal resonance. Figure 3.12a and bnshows the joint stationary PDF $p(x_1, x_5)$ for the case of internal resonance. It can be observed from Figs. 3.9, 3.10, 3.11, and 3.12 that the results obtained by using the stochastic averaging method for quasi-generalized Hamiltonian systems are in good agreement with the results obtained from the Monte Carlo simulations of the original system (3.120) both in non-resonant and resonant cases.

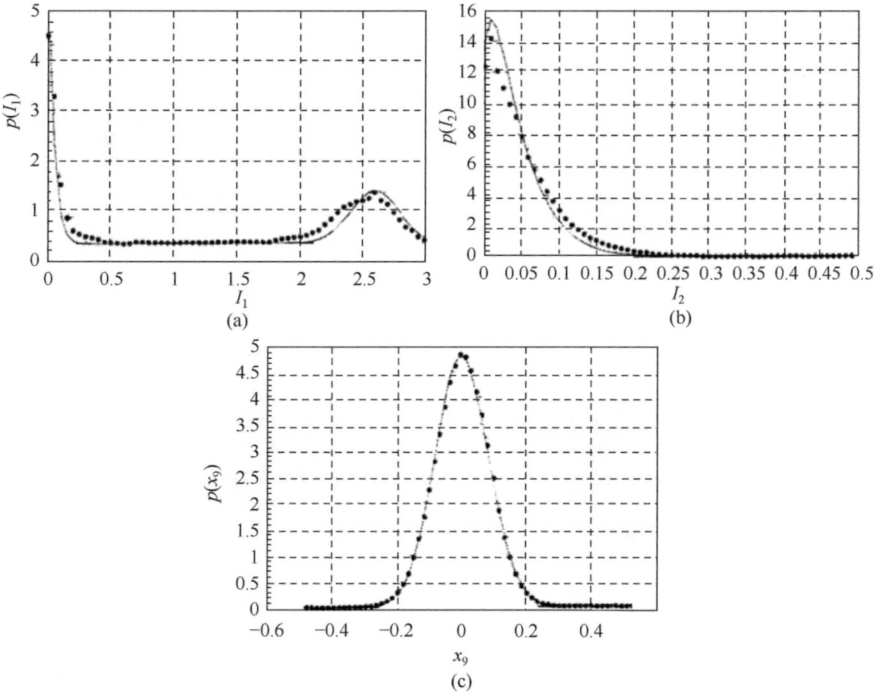

Fig. 3.9 Stationary marginal PDFs of action variables I_1, I_2 and X_9 in the non-internal resonant case. ——, results from the stochastic averaging method; ●, results from Monte Carlo simulation of system (3.120). System parameters are $\omega_1 = 1, \omega_2 = 1.414, \omega_3 = 1, \omega_4 = 1.5, g_{ii} = 1,\ \pi K_{ii} = 0.0005\ (i = 1, \ldots, 5),\ b = 1,\ \delta = 2,\ d_{110} = d_{220} = -0.04,\ d_{330} = d_{440} = d_{550} = -0.02,\ d_{12} = d_{21} = 0.02,\ d_{111} = d_{113} = d_{114} = d_{115} = d_{222} = d_{223} = d_{224} = d_{225} = d_{334} = d_{335} = d_{444} = d_{445} = d_{553} = d_{554} = d_{555} = 0.01,\ d_{112} = d_{221} = d_{551} = d_{552} = 0.02,\ d_{331} = d_{332} = d_{441} = d_{442} = 0.0133,\ d_{333} = d_{443} = 0.00667$ (Huang and Zhu 2009)

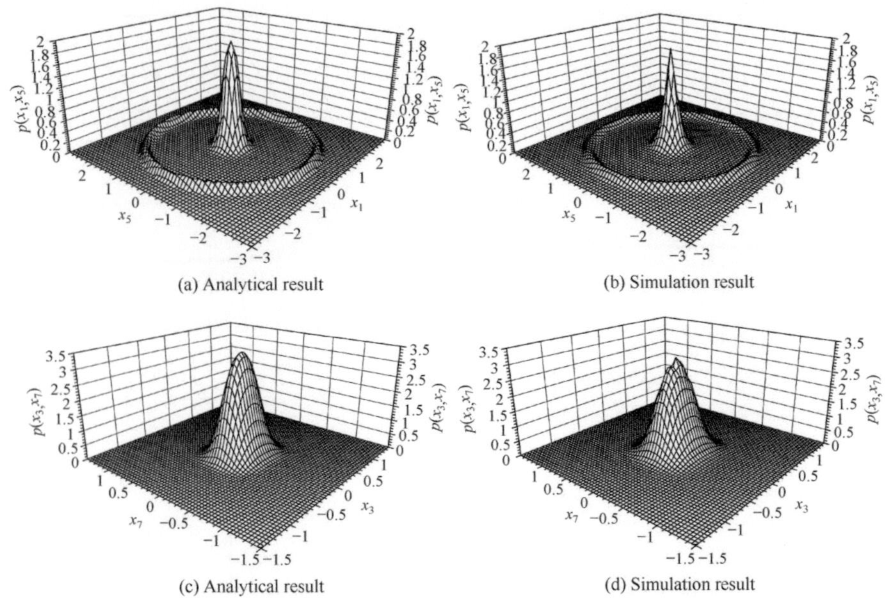

Fig. 3.10 Joint PDFs of displacement and velocity in the non-internal resonant case. **a** for the first oscillator using the stochastic averaging method; **b** for the first oscillator obtained from Monte Carlo simulation of system (3.120); **c** for the third oscillator using the stochastic averaging method; **d** for the third oscillator obtained from Monte Carlo simulation of system (3.120). Parameters are the same as those in Fig. 3.9 (Huang and Zhu 2009)

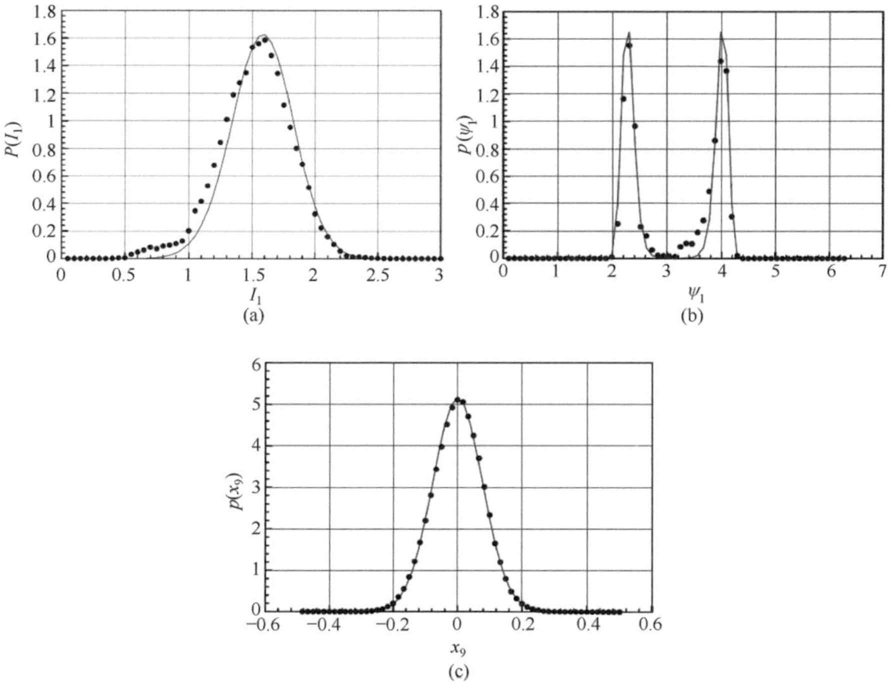

Fig. 3.11 Stationary marginal PDF of action variable I_1, phase angle difference Ψ_1 and X_9 in the internal resonant case. ——, results from the stochastic averaging method; ●, results from Monte Carlo simulation of system (3.120). The parameters are the same as those in Fig. 3.9, except for $\omega_2 = 1$ (Huang and Zhu 2009)

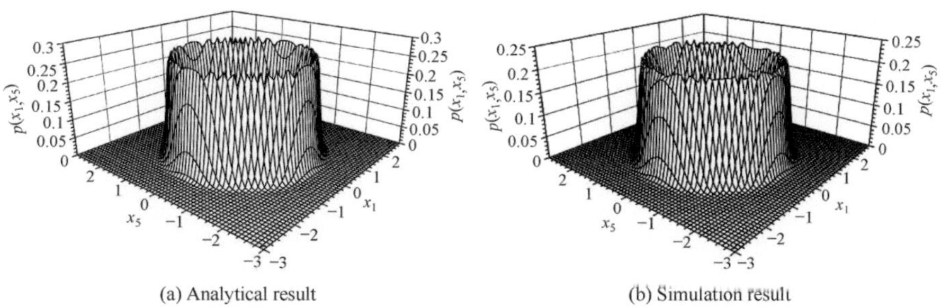

Fig. 3.12 Joint PDF of displacement and velocity for the first oscillator in the internal resonant case. **a** Results obtained from the stochastic averaging method; **b** results obtained from Monte Carlo simulation of system (3.120). The parameters are the same as those in Fig. 3.9 except for $\omega_2 = 1$ (Huang and Zhu 2009)

References

Chen L C, Zhu W Q. 2010a. First passage failure of quasi non-integrable generalized Hamiltonian systems. Archive of Applied Mechanics, 80: 883–893.
Chen L C, Zhu W Q. 2010b. Reliability of quasi integrable generalized Hamiltonian systems. Probabilistic Engineering Mechanics, 25(1): 61–66.
Chen L C, Zhu W Q. 2010c. First passage failure of quasi-partial integrable generalized Hamiltonian systems. International Journal of Non-Linear Mechanics, 45(1): 56–62.
Huang Z L. 2005. Dynamics and control of some classes of non-linear stochastic systems. Ph.D. Dissertation. Hangzhou: Zhejiang University. (in Chinese)
Huang Z L, Zhu W Q. 2009. Stochastic averaging of quasi-generalized Hamiltonian systems. International Journal of Non-Linear Mechanics, 44: 71–80.
Khasminskii R Z. 1968. On the averaging principle for Itô stochastic differential equations. Kibernetika, 3(4): 260–279. (in Russian)
Li J B, Zhao X H, Liu Z R. 1999. Theory and Application of Generalized Hamiltonian System. Beijing: Science Press.
Venikov V A. 1980. Transient Processes in Electrical Power System. Moscow: Mir Publisher.

Chapter 4
Stochastic Averaging Method of Predator–Prey Ecosystems

Evolution of species, either ecological or biological, is of great importance in various areas. Qualitatively, populations of species have been investigated for different situations (May 1973, 1981; Murray 1993; Bazykin 1998). One typical case is predator–prey type, in which the prey species (or host) and predator species (or parasite) are interacted with each other, and are influenced by environments. Since uncertainties are always present in the environment changes, the systems are essentially stochastic. It is found that the stochastic averaging method is an effective tool to investigate the dynamics of predator–prey systems.

In this chapter, the classical deterministic model for the predator–prey type ecosystems, known as Lotka-Volterra model, and its improved model are described first. Based on the deterministic model, the corresponding stochastic model is formed by introducing random disturbances in the prey birth (growth) rate and predator death rate. The stochastic averaging method is then applied to investigate the dynamical behaviors of the system. Several more advantage topics, including situations of predator-saturation and predator-competition, colored noise disturbances, time-delayed system, and habitat complexity, are treated also by the stochastic averaging method.

4.1 Classical Lotka-Volterra Predator–Prey Ecosystem

4.1.1 Deterministic Models

The earliest mathematical model describing the dynamics of predator–prey type ecosystems is the Lotka-Volterra model (Lotka 1925; Volterra 1926), governed by the following differential equations

W. Zhu et al., *Stochastic Averaging*, https://doi.org/10.1007/978-981-96-3833-8_4

$$\begin{aligned} \dot{x}_1 &= x_1(a - bx_2) \\ \dot{x}_2 &= x_2(-c + fx_1) \end{aligned} \tag{4.1}$$

where x_1 and x_2 are the population densities of preys and predators, respectively, and a, b, c, and f are positive constants. The term ax_1 on the right-hand side in the first equation indicates that the prey population would grow exponentially without the predators, while the term $-cx_2$ in the second equation indicates that the population of the predators would decrease exponentially without the preys. The parameters a and c are known as the prey birth (or growth) rate and predator death rate, respectively. The interactive terms x_1x_2 in the two equations provide a balance between the two populations.

System (4.1) has an unstable equilibrium state (0, 0), and a stable, but non-asymptotic stable equilibrium state (center)

$$x_{10} = \frac{c}{f}, \quad x_{20} = \frac{a}{b} \tag{4.2}$$

Moreover, the system has a first integral

$$r(x_1, x_2) = fx_1 - c - c\ln\frac{fx_1}{c} + bx_2 - a - a\ln\frac{bx_2}{a} \tag{4.3}$$

It can be shown that $r(x_{10}, x_{20}) = 0$, and $r(x_1, x_2) \geq 0$ for any positive x_1 and x_2. For a positive constant R, $r(x_1, x_2) = R$ represents a periodic trajectory. Its period can be determined from

$$T(R) = \oint \mathrm{d}t = \oint \frac{\mathrm{d}x_2}{x_2(fx_1 - c)} = \oint \frac{\mathrm{d}x_1}{x_1(a - bx_2)} \tag{4.4}$$

where x_1 and x_2 are related by $r(x_1, x_2) = R$.

Figure 4.1 depicts the equilibrium point O corresponding to $R = 0$, and three periodic trajectories corresponding to three R values, determined with system parameters $a = 0.9$, $b = 1$, $c = 0.5$, and $f = 0.5$. It shows that the prey and predator populations are changing periodically with time, along a path in the phase plane (x_1, x_2), which depends only on the initial states of x_1 and x_2. It also shows that an initial high level of prey density and/or predator density can lead to very low levels for both, even in an invariant environment.

System (4.1) shows that the prey population will grow without limit in the absence of the predators, contrary to what is expected of a realistic predator–prey ecosystem. To improve the classical Lotka-Volterra model, a self-competition term $-sx_1^2$ was added to the prey equation in (4.1) (Volterra 1931; May and Verga 1973); namely,

$$\begin{aligned} \dot{x}_1 &= x_1(a_1 - sx_1 - bx_2), \\ \dot{x}_2 &= x_2(-c + fx_1) \end{aligned} \tag{4.5}$$

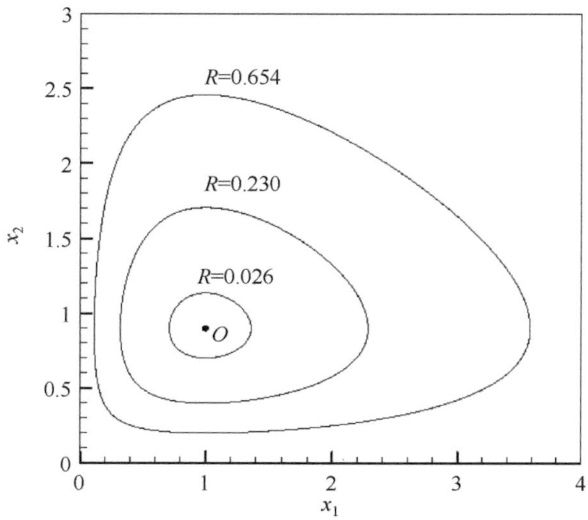

Fig. 4.1 Equilibrium point and periodic solutions of system (4.1)

System (4.5) has an asymptotic stable equilibrium at $x_1 = c/f$ and $x_2 = (a_1 - sc/f)/b$. Note that systems (4.1) and (4.5) have the same equilibrium state $(c/f, a/b)$ if

$$a = a_1 - \frac{sc}{f} \tag{4.6}$$

Therefore, equation set (4.5) may be rewritten as

$$\begin{aligned} \dot{x}_1 &= x_1\left[a - bx_2 - \frac{s}{f}(-c + fx_1)\right], \\ \dot{x}_2 &= x_2(-c + fx_1). \end{aligned} \tag{4.7}$$

In the absence of the predators, the prey density reaches its equilibrium state of a_1/s, which is inversely proportional to s, as expected. When the predators are present, however, the interaction between the prey and the predator populations is the more important factor, whereas the value of s affects only the prey density at the equilibrium state.

Figure 4.2 depicts two trajectories of system (4.5), corresponding to two different values of $s = 0.1$ and 0.02, respectively, and with the same $a_1 = 1$, $b = 1$, $c = 0.5$, $f = 0.5$. The motion of the system begins from point (3.5, 0.5). The system moves around the stable equilibrium state with a decreasing amplitude, and finally reaches the equilibrium state. The term $-sx_1^2$ models the effect of inner-species competition. With the larger $s = 0.1$, the system reaches its equilibrium faster, while with the smaller $s = 0.02$, the system moves slowly around the stable equilibrium. By comparing Figs. 4.1 and 4.2, it can be seen that the prey self-competition plays

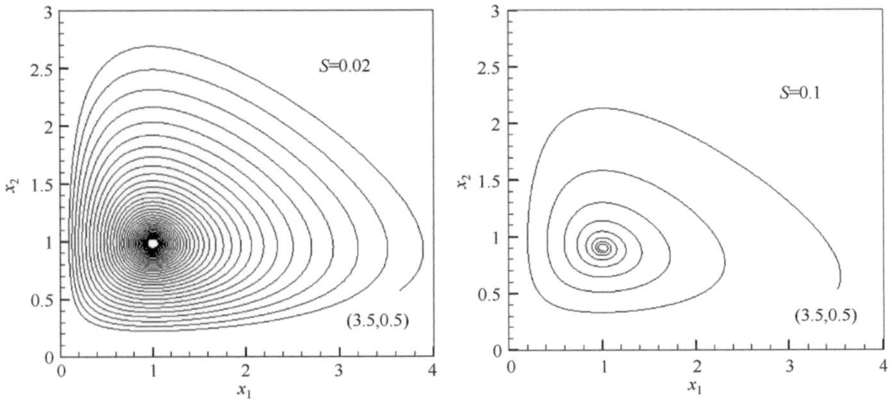

Fig. 4.2 Trajectories of system (4.5) for two different values of s

a dissipation role to the original system (4.1), and it changes a periodic trajectory to the final state of a single stable focus.

4.1.2 Stochastic Model

The ecosystem model described by equation set (4.7) fails to describe a basic phenomenon of a natural ecosystem, namely, the changing environment may cause random variations in the prey growth rate and the predator death rate. Based on the deterministic model (4.7), a stochastic model is proposed as follows (Cai and Lin 2004)

$$\begin{aligned} \dot{X}_1 &= X_1\left[a - bX_2 - \frac{s}{f}(-c + fX_1) + W_{g1}(t)\right], \\ \dot{X}_2 &= X_2\left[-c + fX_1 + W_{g2}(t)\right], \end{aligned} \tag{4.8}$$

where $X_1(t)$ and $X_2(t)$ are two stochastic processes, representing the prey and the predator population densities, respectively, and $W_{g1}(t)$ and $W_{g2}(t)$ are two independent Gaussian white noises with spectral densities K_1 and K_2, respectively. The noises $W_{g1}(t)$ and $W_{g2}(t)$ are introduced to model the random variations in the prey birth rate and the predator death rate, respectively.

Equation set (4.8) can be re-written in the form of Itô stochastic differential equations as follows

$$
\begin{aligned}
\mathrm{d}X_1 &= X_1\left[a - bX_2 - \frac{s}{f}(-c + fX_1) + \pi K_1\right]\mathrm{d}t + \sqrt{2\pi K_1}X_1\mathrm{d}B_1(t),\\
\mathrm{d}X_2 &= X_2\left[-c + fX_1 + \pi K_2\right]\mathrm{d}t + \sqrt{2\pi K_2}X_2\mathrm{d}B_2(t)
\end{aligned}
\tag{4.9}
$$

where $B_1(t)$ and $B_2(t)$ are two independent unit Wiener processes. Compared with (4.8), the additional terms $\pi K_1 X_1 dt$ and $\pi K_2 X_2 dt$ are the Wong-Zakai correction terms (Wong and Zakai 1965), which must be incorporated when $W_{g1}(t)$ and $W_{g2}(t)$ in (4.8) are interpreted as white noises in the physical Stratonovich sense (Stratonovich 1963).

Consider the following stochastic process

$$
R(X_1, X_2) = fX_1 - c - c\ln\frac{fX_1}{c} + bX_2 - a - a\ln\frac{bX_2}{a} \tag{4.10}
$$

which is the random counterpart of the first integral (4.3). The Itô equation for $R(X_1, X_2)$ can be obtained as follows upon applying the Itô differential rule,

$$
\begin{aligned}
\mathrm{d}R = &\left[-\frac{s}{f}(fX_1 - c)^2 + \pi fK_1X_1 + \pi bK_2X_2\right]\mathrm{d}t\\
&+ \sqrt{2\pi K_1}(fX_1 - c)\mathrm{d}B_1(t) + \sqrt{2\pi K_2}(bX_2 - a)\mathrm{d}B_2(t)
\end{aligned}
\tag{4.11}
$$

4.1.3 Stochastic Averaging

Assume that the coefficient s of the self-competition term is small, indicating that the term has a small influence when the prey density is small. Assume also that K_1 and K_2 are small, namely, the random disturbances are small. Under these assumptions, the right-hand side of (4.11) is small, and $R(t)$ is a slowly varying stochastic process. In this case, the stochastic averaging method can be applied to obtain an averaged Itô stochastic differential equation for R

$$
\mathrm{d}R = m(R)\mathrm{d}t + \sigma(R)\mathrm{d}B(t) \tag{4.12}
$$

where the drift coefficient m and the diffusion coefficient σ are obtained as follows

$$
m(R) = \pi fK_1\langle X_1\rangle_t + \pi bK_2\langle X_2\rangle_t - \frac{s}{f}\langle (fX_1 - c)^2\rangle_t \tag{4.13}
$$

$$
\sigma^2(R) = 2\pi K_1\langle (fX_1 - c)^2\rangle_t + 2\pi K_2\langle (bX_2 - a_1)^2\rangle_t \tag{4.14}
$$

and where $\langle[\cdot]\rangle_t$ denotes the time average in one quasi-period, defined as

$$[\cdot]_t = \frac{1}{T}\oint[\cdot]\mathrm{d}t = \frac{1}{T}\oint\frac{[\cdot]\mathrm{d}X_2}{X_2(fX_1-c)} = \frac{1}{T}\oint\frac{[\cdot]\mathrm{d}X_1}{X_1(a-bX_2)} \tag{4.15}$$

The quasi-period T in (4.15) is given in (4.4), with x_1 and x_2 replaced by their random counterparts X_1 and X_2, respectively. The result obtained from each time average is a function of R. Note that each integral in (4.15) is along the closed loop determined by (4.10) according to system (4.1). Change (4.1) to the following form

$$\begin{aligned}\frac{\mathrm{d}}{\mathrm{d}t}\ln x_1 &= a-bx_2\\ \frac{\mathrm{d}}{\mathrm{d}t}\ln x_2 &= -c+fx_1.\end{aligned} \tag{4.16}$$

Carry out the time averaging defined in (4.15) for the two equations in (4.16) to yield

$$\langle X_1\rangle_t = \frac{c}{f}, \quad \langle X_2\rangle_t = \frac{a}{b} \tag{4.17}$$

The same time averaging is performed to the first equation of (4.1), resulting in

$$\langle X_1X_2\rangle_t = \frac{ac}{bf} \tag{4.18}$$

From (4.1), we obtain

$$\begin{aligned}\frac{\mathrm{d}x_1^2}{\mathrm{d}t} &= 2x_1^2(a-bx_2)\\ \frac{\mathrm{d}(x_1x_2)}{\mathrm{d}t} &= (a-c)x_1x_2+fx_1^2x_2-bx_1x_2^2\\ \frac{\mathrm{d}x_2^2}{\mathrm{d}t} &= 2x_2^2(-c+fx_1).\end{aligned} \tag{4.19}$$

Taking the time averaging and using (4.18), we have

$$\begin{aligned}&\langle X_1^2\rangle_t = \frac{b}{a}\langle X_1^2X_2\rangle_t, \quad \langle X_2^2\rangle_t = \frac{f}{c}\langle X_1X_2^2\rangle_t,\\ &bf^2\langle X_1^2X_2\rangle_t - fb^2\langle X_1X_2^2\rangle_t = ac(c-a),\end{aligned} \tag{4.20}$$

which leads to

$$a\langle(fX_1-c)^2\rangle_t = c\langle(bX_2-a)^2\rangle_t \tag{4.21}$$

Defining

$$g(R) = a \oint \frac{fX_1 - c}{X_2} \mathrm{d}X_2 \tag{4.22}$$

we have from (4.15)

$$\langle (fX_1 - c)^2 \rangle_t = \frac{g(R)}{aT(R)}, \quad \langle (bX_2 - a)^2 \rangle_t = \frac{g(R)}{cT(R)} \tag{4.23}$$

Using Eqs. (4.17) and (4.23), the drift and diffusion coefficients in Eqs. (4.13) and (4.14) can be written as

$$m(R) = \pi c K_1 + \pi a K_2 - \frac{s}{af} \frac{g(R)}{T(R)} \tag{4.24}$$

$$\sigma^2(R) = \frac{2\pi}{ac}(cK_1 + aK_2)\frac{g(R)}{T(R)} \tag{4.25}$$

Equations (4.12), (4.24) and (4.25) constitute the governing law for the one-dimensional Markov diffusion process $R(t)$. This stochastic process $R(t)$, as defined in (4.10), can be considered as a representation of the system state. It is a function of two stochastic processes; namely, the prey density X_1 and the predator density X_2.

4.1.4 Stationary Probability Density

The reduced FPK equation for process $R(t)$ is given by

$$\frac{\mathrm{d}}{\mathrm{d}r}\left[m(r)p(r)\right] - \frac{1}{2}\frac{\mathrm{d}^2}{\mathrm{d}r^2}\left[\sigma^2(r)p(r)\right] = 0 \tag{4.26}$$

which is solved to obtain

$$\begin{aligned} p(r) &= \frac{C_1}{\sigma^2(r)} \exp \int \frac{2m(r)}{\sigma^2(r)} \mathrm{d}r \\ &= C\frac{T(r)}{g(r)} \exp \int \frac{\pi ac(cK_1 + aK_2)T(r) - \frac{sc}{f} g(r)}{\pi (cK_1 + aK_2) g(r)} \mathrm{d}r \end{aligned} \tag{4.27}$$

where C and C_1 are two normalization constants. By noticing that

$$\frac{\mathrm{d}g(r)}{\mathrm{d}r} = a\oint \frac{f}{x_2}\frac{\partial x_1}{\partial r}\mathrm{d}x_2 = a\oint \frac{fx_1\mathrm{d}x_2}{x_2(fx_1 - c)} = afT(r)x_{1t} = acT(r) \tag{4.28}$$

we have

$$\int \frac{acT(r)}{g(r)}\mathrm{d}r = \int \frac{\mathrm{d}g(r)}{g(r)} = \ln g(r) \tag{4.29}$$

and Eq. (4.27) is simplified to

$$p(r) = CT(r)\exp(-\beta r) \tag{4.30}$$

where β is a constant given by

$$\beta = \frac{sc}{\pi f(cK_1 + aK_2)} \tag{4.31}$$

The joint PDF of $R(t)$ and $X_1(t)$ can be written as

$$p(r, x_1) = p(r)p(x_1|r) \tag{4.32}$$

where $p(x_1|r)$ is the conditional PDF of $X_1(t)$ given $R(t) = r$. It can be obtained as follows

$$p(x_1|r)\mathrm{d}x_1 = \frac{\mathrm{d}t}{T(r)} = \frac{\mathrm{d}x_1}{|\dot{x}_1|T(r)} = \frac{\mathrm{d}x_1}{|x_1(a - bx_2)|T(r)} \tag{4.33}$$

Substituting (4.33) into (4.32), we have

$$p(r, x_1) = \frac{p(r)}{|x_1(a - bx_2)|T(r)} \tag{4.34}$$

in which x_2 is treated as a function of x_1 and r. Thus, the joint PDF $p(x_1, x_2)$ follows as

$$p(x_1, x_2) = p(r, x_1)\left|\frac{\partial(r, x_1)}{\partial(x_1, x_2)}\right| = \frac{p(r)}{x_1x_2T(r)} = \frac{C}{x_1x_2}\exp[-\beta r(x_1, x_2)] \tag{4.35}$$

where $\left|\frac{\partial(r,x_1)}{\partial(x_1,x_2)}\right|$ is the Jacobian determinant. Upon substituting (4.3) into (4.35), we obtain

$$p(x_1, x_2) = p(x_1)p(x_2) \tag{4.36}$$

where

$$p(x_1) = \frac{(\beta f)^{\beta c}}{\Gamma(\beta c)} x_1^{\beta c - 1} \exp(-\beta f x_1) \tag{4.37}$$

$$p(x_2) = \frac{(\beta b)^{\beta a}}{\Gamma(\beta \alpha)} x_2^{\beta a - 1} \exp(-\beta b x_2) \tag{4.38}$$

and $\Gamma(\cdot)$ is the gamma function. Equation (4.36) implies that $X_1(t)$ and $X_2(t)$ are independent when they reach the stationary state, which is an unexpected result for system (4.8) with nonlinear coupling between X_1 and X_2. Equations (4.37) and (4.38) show that non-trivial $p(x_1)$ and $p(x_2)$ exist only if $\beta > 0$ and $a > 0$. These conditions lead to

$$0 < s < \frac{fa_1}{c} \tag{4.39}$$

The system will diverge without the self-competition term, namely, if $s = 0$, since no restriction will be imposed on the growth of the prey population. On the other hand, if $s > \frac{fa_1}{c}$, the growth of the prey population will be over restricted, leading to extinction of the predators.

The existence of the PDFs of the prey and predator densities under condition (4.39) indicates that (i) the ecosystem is dynamic, it will not approach a single state of fixed predator and prey populations; (ii) neither the prey nor the predator will be extinct, except some unexpected events occur, which is not considered in the model; and (iii) it is not possible to predict the exact prey and predator populations, only their PDFs can be estimated.

Numerical calculations were carried out in (Cai and Lin 2004). Shown in Fig. 4.3 is the PDF $p(x_1)$ of the prey population X_1 for the stochastic system (4.8), calculated from (4.37), with $a_1 = 1$, $b = 1$, $c = 0.5$, $f = 0.5$, $\pi K_1 = \pi K_2 = 0.01$ and for two different s values of 0.1 and 0.02. Also depicted in Fig. 4.3 are results obtained from the Monte Carlo simulation. The theoretical and simulation results agree very well in both cases. With a large self-competition coefficient $s = 0.1$, the prey density is nearly centered around $x_1 = c/f = 1$, which is the equilibrium point for the deterministic counterpart without random variations in the prey birth rate and the predator death rate. With a small $s = 0.02$, the peak of the prey PDF is shifted to a value smaller than the equilibrium point $x_1 = 1$, and the probability for a high prey population becomes higher, indicating that the system is less stable.

The system asymptotic behaviors and the transition time to near extinction were investigated in (Cai and Lin 2004).

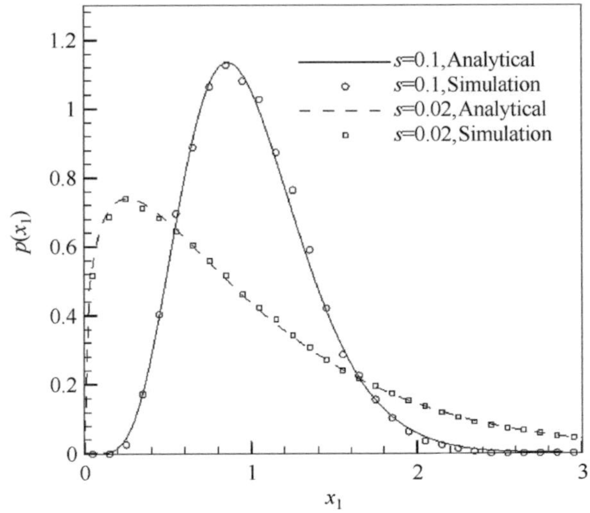

Fig. 4.3 Stationary PDFs of the prey population $X_1(t)$ (Cai and Lin 2004)

4.2 Ecosystem with Predator-Saturation and Predator-Competition

4.2.1 Deterministic Model

The coupling terms in system (4.5) indicate that the changing rates for the preys and the predators depend on both the prey and the predator populations. Consider a case where the preys are abundantly available. Then the consumption of preys depends only on the predator population and is no longer related to the prey population. In this case, the prey growth rate and the predator death rate are no longer dependent on the prey supply, and the interaction terms on the right-hand sides of the two equations in (4.5) are proportional only to x_2, not x_1. To take this into consideration, equation set (4.5) may be modified to

$$
\begin{aligned}
\dot{x}_1 &= a_1 x_1 - s x_1^2 - \frac{b_1 x_1 x_2}{1 + A x_1} \\
\dot{x}_2 &= -c x_2 + \frac{f_1 x_1 x_2}{1 + A x_1}.
\end{aligned}
\tag{4.40}
$$

When the prey supply is very large, i.e., $Ax_1 > > 1$, the first equation in (4.40) is degenerated to that in (4.5), and the predator population does not depends on the prey population as shown in the second equation of (4.40). The asymptotic stable equilibrium point of system (4.40) is found to be

$$x_{10} = \frac{c}{f_1 - cA}, \quad x_{20} = \frac{f_1}{b_1(f_1 - cA)}\left(a_1 - \frac{sc}{f_1 - cA}\right) \tag{4.41}$$

Denote

$$f = f_1 - cA, \quad a = a_1 - \frac{sc}{f} = a_1 - \frac{sc}{f_1 - cA}, \quad b = \frac{fb_1}{f + cA} = \frac{b_1}{f_1}(f_1 - cA) \tag{4.42}$$

The equilibrium point in (4.41) is simplified to

$$x_{10} = \frac{c}{f}, \qquad x_{20} = \frac{a}{b} \tag{4.43}$$

indicating that system (4.40) has the same equilibrium point as system (4.1) and (4.5) if the parameters have the relationships of (4.42). In fact, (4.40) can be rewritten as

$$\begin{aligned} \dot{x}_1 &= x_1\left[a - bx_2 - \frac{s}{f}(-c + fx_1) + g_1(x_1, x_2)\right] \\ \dot{x}_2 &= x_2\left[-c + fx_1 + g_2(x_1, x_2)\right] \end{aligned} \tag{4.44}$$

where

$$g_1(x_1, x_2) = \frac{b}{f}Ax_2\frac{fx_1 - c}{1 + Ax_1}, \quad g_2(x_1, x_2) = -Ax_1\frac{fx_1 - c}{1 + Ax_1} \tag{4.45}$$

Consider another situation in which the predator population is very large. Then the consumption of preys depends only on the prey supply, and the predators compete for the limited prey. The prey birth rate and the predator death rate are dependent only on the prey supply. In this case, the system equations can be modified as

$$\begin{aligned} \dot{x}_1 &= a_1x_1 - sx_1^2 - \frac{b_1x_1x_2}{1 + Bx_2} \\ \dot{x}_2 &= -cx_2 + \frac{f_1x_1x_2}{1 + Bx_2} \end{aligned} \tag{4.46}$$

Systems (4.46) and (4.5) have the same equilibrium point $x_{10} = c/f, x_{20} = a/b$ if we denote

$$a = a_1 - \frac{sc}{f}, \quad b = b_1 - aB, \quad f = \frac{bf_1}{b + aB} \tag{4.47}$$

In terms of the new coefficients *a, b,* and f, (4.46) can also be rewritten in the form (4.44) with another set of $g_1(x_1, x_2)$ and $g_2(x_1, x_2)$ given by

$$g_1(x_1, x_2) = -Bx_2 \frac{a - bx_2}{1 + Bx_2}, \quad g_2(x_1, x_2) = \frac{f}{b} Bx_1 \frac{a - bx_2}{1 + Bx_2} \tag{4.48}$$

In summary, the improved models (4.40) and (4.46) can all be described by a unified set of Eqs. (4.44), with different set of coefficients a, b, f and different functions $g_1(x_1, x_2)$ and $g_2(x_1, x_2)$. They have the same stable equilibrium point $x_{10} = c/f, x_{20} = a/b$.

4.2.2 Stochastic Model

The stochastic model corresponding to (4.44) is proposed as follows (Cai and Lin 2007a)

$$\begin{aligned} \dot{X}_1 &= X_1\left[a - bX_2 - \frac{s}{f}(-c + fX_1) + g_1(X_1, X_2) + W_{g1}(t)\right], \\ \dot{X}_2 &= X_2\left[-c + fX_1 + g_2(X_1, X_2) + W_{g2}(t)\right] \end{aligned} \tag{4.49}$$

where $W_{g1}(t)$ and $W_{g2}(t)$ are two independent Gaussian white noises with spectral densities K_1 and K_2, respectively. Equation set (4.44) can be re-written in the form of Itô stochastic differential equations as follows

$$\begin{aligned} \mathrm{d}X_1 &= X_1\left[a - bX_2 - \frac{s}{f}(-c + fX_1) + g_1(X_1, X_2) + \pi K_1\right]\mathrm{d}t + \sqrt{2\pi K_1}X_1\mathrm{d}B_1(t), \\ \mathrm{d}X_2 &= X_2\left[-c + fX_1 + g_2(X_1, X_2) + \pi K_2\right]\mathrm{d}t + \sqrt{2\pi K_2}X_2\mathrm{d}B_2(t) \end{aligned} \tag{4.50}$$

where $B_1(t)$ and $B_2(t)$ are two independent unit Wiener processes. Consider the stochastic process defined in (4.49), i.e.,

$$R(t) = fX_1 - c - c\ln\frac{fX_1}{c} + bX_2 - a - a\ln\frac{bX_2}{a} \tag{4.51}$$

By applying the Itô differential rule, the Itô equation for $R(X_1, X_2)$ can be obtained as follows

$$\begin{aligned} \mathrm{d}R = &\left[-\frac{s}{f}(fX_1 - c)^2 + (fX_1 - c)g_1(X_1, X_2) + (bX_2 - a)g_2(X_1, X_2) - \frac{s}{f}(fX_1 - c)^2\right. \\ &\left.+(fX_1 - c)g_1(X_1, X_2) + (bX_2 - a)g_2(X_1, X_2) + f\pi K_1X_1 + \frac{1}{2}b\pi K_2X_2\right]\mathrm{d}t \\ &+\sqrt{2\pi K_1}(fX_1 - c)\mathrm{d}B_1(t) + \sqrt{2\pi K_2}(bX_2 - a)\mathrm{d}B_2(t) \end{aligned} \tag{4.52}$$

4.2.3 Stochastic Averaging

Similar to the situation in Sect. 4.1, we assume that the coefficient s of the self-competition term is small and the noise intensities represented by K_1 and K_2 are small. Furthermore, we assume that the coefficient A in (4.40) is small since the predator saturation occurs only when the prey population is large, and the coefficient B in (4.46) is also small so that only a large predator population can cause their own competition. Thus, functions g_1 and g_2 are small. Under these assumptions, the right-hand side of (4.52) is small, and $R(t)$ is a slowly varying stochastic process. In this case, the stochastic averaging method can be applied to obtain an averaged Itô stochastic differential equation for R in then following form

$$\mathrm{d}R = m(R)\mathrm{d}t + \sigma(R)\mathrm{d}B(t) \tag{4.53}$$

where the drift coefficient $m(R)$ and the diffusion coefficient $\sigma^2(R)$ are obtained as follows

$$\begin{aligned} m(R) =& f\pi K_1\langle X_1\rangle_t + \frac{1}{2}b\pi K_2\langle X_2\rangle_t - \frac{s}{f}\langle (fX_1 - c)^2\rangle_t \\ &+\langle (fX_1 - c)g_1(X_1, X_2) + (bX_2 - a)g_2(X_1, X_2)\rangle_t, \end{aligned} \tag{4.54}$$

$$\sigma^2(R) = 2\pi K_1\langle (fX_1 - c)^2\rangle_t + 2\pi K_2\langle (bX_2 - a_1)^2\rangle_t, \tag{4.55}$$

and where the time averaging $\langle[\cdot]\rangle_t$ is given in (4.15). The result obtained from each time average is a function of R. Equations (4.53), (4.54), and (4.55) constitute the governing law for the one-dimensional Markov process R(t).

Denote

$$\langle (fX_1 - c)g_1(X_1, X_2)\rangle_t + \langle (bX_2 - a)g_2(X_1, X_2)\rangle_t = \frac{G(R)}{T(R)} \tag{4.56}$$

where $G(R)$ can be derived from the functions g_1 and g_2 defined in (4.45) for the predator-saturation or in (4.48) for the predator-competition as follows

$$G(R) = \begin{cases} \frac{bA}{f}\oint \frac{fX_1 - c}{1+AX_1}\mathrm{d}X_2, & \text{for predator} - \text{saturation} \\ \frac{fB}{b}\oint \frac{bX_2 - a}{1+BX_2}\mathrm{d}X_1, & \text{for predator} - \text{competetion} \end{cases} \tag{4.57}$$

Using (4.23) and (4.56), we have from Eqs. (4.54) and (4.55)

$$m(R) = c\pi K_1 + a\pi K_2 - \frac{s}{af}\frac{g(R)}{T(R)} + \frac{G(R)}{T(R)} \tag{4.58}$$

$$\sigma^2(R) = \frac{2\pi}{ac}(cK_1 + aK_2)\frac{g(R)}{T(R)} \tag{4.59}$$

where $g(R)$ is given in (4.22).

The stationary PDF of $R(t)$, denoted by $p(r)$ can be obtained as

$$\begin{aligned} p(r) &= \frac{C_1}{\sigma^2(r)} \exp \int \frac{2m(r)}{\sigma^2(r)} \mathrm{d}r \\ &= C \frac{T(r)}{g(r)} \exp \int \frac{2\pi ac(cK_1 + aK_2)T(r) - \frac{2sc}{f} g(r) + 2acG(r)}{2\pi (cK_1 + aK_2) g(r)} \mathrm{d}r \end{aligned} \tag{4.60}$$

where C and C_1 are two normalization constants. By noticing (4.28) and introducing constant □ defined in (4.31), Eq. (4.60) is simplified as

$$p(r) = CT(r) \exp\left\{\beta\left[-r + \frac{af}{s} \int \frac{G(r)}{g(r)} \mathrm{d}r\right]\right\} \tag{4.61}$$

The joint PDF of $X_1(t)$ and $X_2(t)$, and the marginal PDFs of X_1 and X_2 can be obtained and calculated following the procedure described in Sect. 4.1.4.

Numerical calculations were carried out in (Cai and Lin 2007a). Shown in Fig. 4.4 are the PDF $p(x_1)$ of the prey population X_1 and $p(x_2)$ of the predator population X_2, respectively, for the stochastic predator-saturation model with $a_1 = 1$, $b_1 = 1$, $c = 0.5$, $f_1 = 0.5$, $s = 0.1$, $2\pi K_1 = 2\pi K_2 = 0.01$, and for two different A values of 0 and 0.05. Also depicted in Fig. 4.4 are results obtained from the Monte Carlo simulation. The theoretical and the simulation results agree well in both cases. It is seen that the system behavior changes significantly when the predator saturation occurs. It becomes more unstable even for a small value of parameter $A = 0.05$, namely, the peak heights are lower and the probabilities increase for either small or large populations. Therefore, it is important to include the predator-saturation term in the model.

Similar to Fig. 4.4, the PDFs $p(x_1)$ and $p(x_2)$ are depicted in Fig. 4.5, respectively, for the stochastic predator-competition model. As seen in Fig. 4.5, the effect of the predator competition term is insignificant.

4.3 Ecosystem Under Colored Noise Excitations

In Sects. 4.1 and 4.2, the random disturbances are assumed to be Gaussian white noises with constant spectral densities over the entire frequency range $[0, \infty)$. Although non-Gaussian Poisson white noises are used to model the stochastic noises (Wu and Zhu 2008), their spectral densities are still constant. The reasons to use the white noise model lie in that (i) it can indeed model some environmental noises approximately, and (ii) the mathematical treatment is possible and more convenient. However, many environmental noises are far from white, i.e., they do not have constant spectra, and are known as colored noises. Representative models for cored noises are required. For this purpose, a review paper (Vaseur and Yodzis 2004)

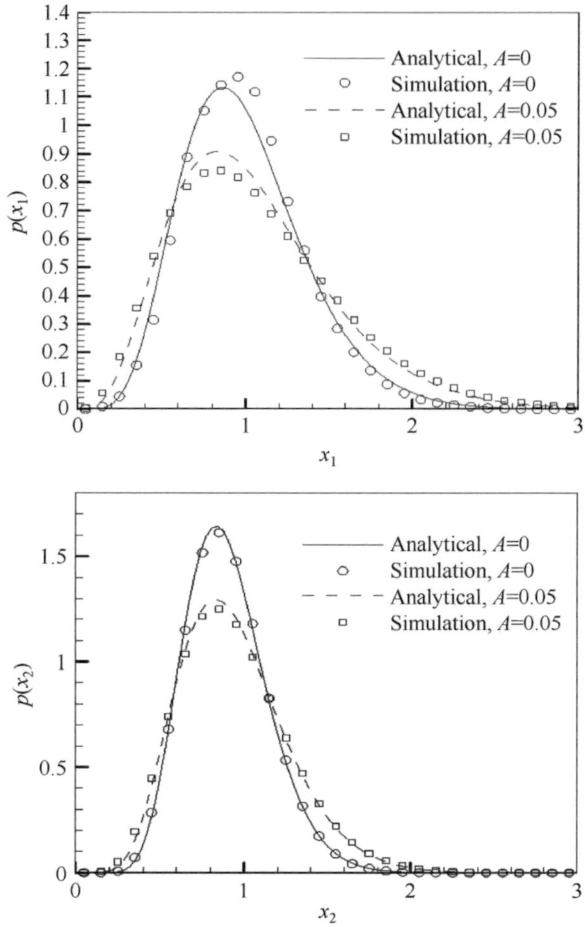

Fig. 4.4 PDFs of the prey population X_1 and the predator population X_2 for the predator-saturation model with $s = 0.1$ and two different A values (Cai and Lin 2007a)

was presented to discuss the color of environmental noises. It pointed out that the spectrum of a noise, representing the noise color, is an important characteristic, and different environmental noises have different spectra. It was concluded that the spectrum of exponential form, $1/f^{\beta}$, where f denotes the frequency, might be a good model. The terrestrial noise tends to be more white ($\beta < 0.5$), while the marine noise is red ($\beta \approx 1$) or brown ($\beta \approx 2$). To access the effects of noise color on the predator–prey systems, two different models were used in (Naess et al. 2008). One is the combination of a white noise process and a harmonic process, and another is the randomized harmonic process. The numerical method of path integration was applied to obtain the PDFs of the prey and predator populations.

In this section, colored noises of broad band are introduced in the stochastic models of the predator–prey ecosystems, taking into account of the two cases discussed in Sect. 4.2, i.e., predator–saturation (abundant prey supply) and predator–competition (large predator population). The stochastic model is a modification of that of

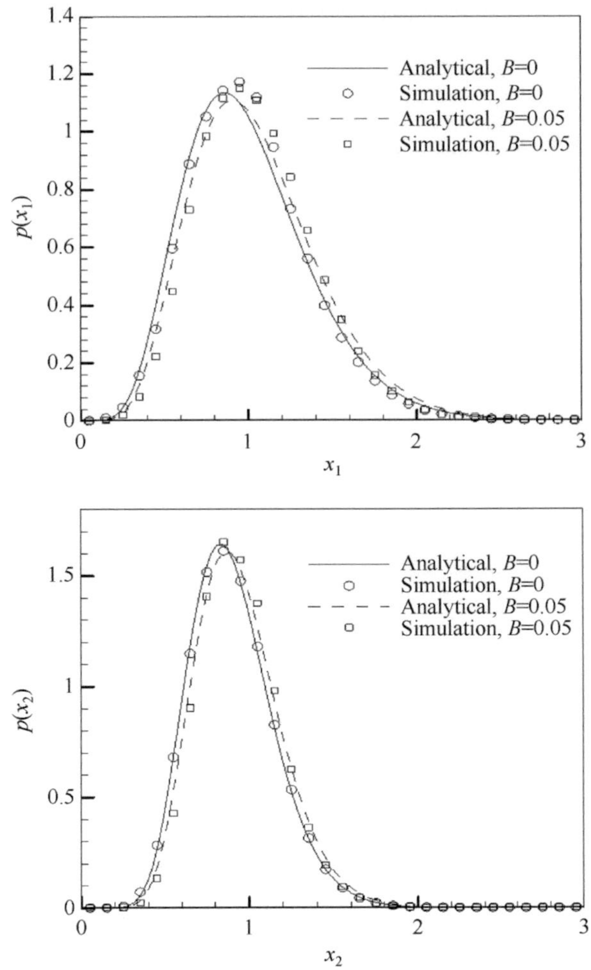

Fig. 4.5 PDFs of the prey population X_1 and the predator population X_2 for the predator-competition model with $s = 0.1$ and two different B values (Cai and Lin 2007a)

Eq. (4.49) as follows (Qi and Cai 2013)

$$\begin{aligned}\dot{X}_1 &= X_1\left[a - bX_2 - \frac{s}{f}(fX_1 - c) + g_1(X_1, X_2) + \xi_1(t)\right],\\ \dot{X}_2 &= X_2\big[fX_1 - c + g_2(X_1, X_2) + \xi_2(t)\big]\end{aligned} \tag{4.62}$$

where $\xi_1(t)$ and $\xi_2(t)$ are assumed to be two independent stationary stochastic processes with non-constant spectral densities, namely, colored noises. The level of the noise color is related to the shape of the spectral density. The flatter the spectrum shape appears, the broader the noise band is, and the lower level of the noise color is.

Consider the stochastic process $R(t)$ defined in (4.10). Using $R(t)$, instead $X_2(t)$, as a state variable of the system, system equations of motion in (4.62) can be transformed to another form as follows

$$\begin{aligned}\frac{\mathrm{d}}{\mathrm{d}t}R(t) &= -\frac{s}{f}(fX_1-c)^2+(fX_1-c)g_1(X_1,X_2)+(bX_2-a)g_2(X_1,X_2)\\&\quad+(fX_1-c)\xi_1(t)+(bX_2-a)\xi_2(t),\\\frac{\mathrm{d}}{\mathrm{d}t}X_1(t) &= X_1\left[a-bX_2-\frac{s}{f}(fX_1-c)+g_1(X_1,X_2)\right]+X_1\xi_1(t).\end{aligned} \tag{4.63}$$

Besides $R(t)$, $X_1(t)$ is chosen as the second state variable. In (4.63), $X_2(t)$ is treated as a function of $R(t)$ and $X_1(t)$, defined by (4.15) implicitly. Alternatively, $X_2(t)$ may be used for the second variable in the equations of motion. Assume that the noise intensities of $\xi_1(t)$ and $\xi_2(t)$ are small, namely, the random variations in the system dynamics are small. Under these assumptions, the right-hand side of the first equation of (4.63) is small, and $R(t)$ is a slowly varying stochastic process. In this case, the stochastic averaging method can be applied to obtain an averaged Itô stochastic differential equation for $R(t)$

$$\mathrm{d}R = m(R)\mathrm{d}t+\sigma(R)\mathrm{d}B(t) \tag{4.64}$$

where $m(R)$ and $\sigma^2(R)$ can be obtained as follows according to Eqs. (4.27) and (4.28) in vol. 1

$$\begin{aligned}m(R) &= -\frac{s}{f}\langle(fX_1-c)^2\rangle_t+\langle(fX_1-c)g_1(X_1,X_2)\rangle_t+\langle(bX_2-a)g_2(X_1,X_2)\rangle_t\\&\quad+\int_{-\infty}^{0}\langle h_{21}(t+\tau)\frac{\partial}{\partial X_1}h_{11}(t)\rangle_t R_{11}(\tau)\mathrm{d}\tau+\int_{-\infty}^{0}\langle h_{12}(t+\tau)\frac{\partial}{\partial R}h_{12}(t)\rangle_t R_{22}(\tau)\mathrm{d}\tau\end{aligned} \tag{4.65}$$

$$\sigma^2(R) = \int_{-\infty}^{\infty}\langle h_{11}(t+\tau)h_{11}(t)\rangle_t R_{11}(\tau)\mathrm{d}\tau+\int_{-\infty}^{\infty}\langle h_{12}(t+\tau)h_{12}(t)\rangle_t R_{22}(\tau)\mathrm{d}\tau \tag{4.66}$$

where functions h_{ij} are the coefficients before the excitations in the two equations of (4.63), i.e.,

$$h_{11}(t) = fX_1(t)-c,\quad h_{12}(t)=bX_2(t)-a,\quad h_{21}(t)=X_1(t),\quad h_{21}(t)=0 \tag{4.67}$$

In (4.65) and (4.66), $\langle[\cdot]\rangle_t$ denotes the time average over one quasi-period defined in (4.15), and $R_{11}(\tau)$ and $R_{22}(\tau)$ are the autocorrelation functions of $\xi_1(t)$ and $\xi_2(t)$, respectively, i.e.,

$$R_{11}(\tau)=E[\xi_1(t)\xi_1(t+\tau)],\quad R_{22}(\tau)=E[\xi_2(t)\xi_2(t+\tau)] \tag{4.68}$$

Equations (4.64) through (4.68) constitute the governing law for the one-dimensional Markov diffusion process R(t).

The first three terms on the right-hand-side of (4.65) are calculated in Sect. 4.2.3. The terms involving $R_{11}(\tau)$ and $R_{22}(\tau)$ in (4.65) and (4.66) are calculated as follows

$$\int_{-\infty}^{0} \langle h_{21}(t+\tau)\frac{\partial}{\partial X_1}h_{11}(t)\rangle_t R_{11}(\tau)\mathrm{d}\tau = \frac{1}{2}cD_1 \tag{4.69}$$

$$\int_{-\infty}^{0} \langle h_{12}(t+\tau)\frac{\partial}{\partial R}h_{12}(t)\rangle_t R_{22}(\tau)\mathrm{d}\tau = \frac{1}{2}aD_2 \tag{4.70}$$

$$\int_{-\infty}^{\infty} \langle h_{11}(t+\tau)h_{11}(t)\rangle_t R_{11}(\tau)\mathrm{d}\tau = f^2 \int_{-\infty}^{\infty} K_1(R,\tau)R_{11}(\tau)\mathrm{d}\tau - c^2D_1 \tag{4.71}$$

$$\int_{-\infty}^{\infty} \langle h_{12}(t+\tau)h_{12}(t)\rangle_t R_{22}(\tau)\mathrm{d}\tau = b^2 \int_{-\infty}^{\infty} K_2(R,\tau)R_{22}(\tau)\mathrm{d}\tau - a^2D_2 \tag{4.72}$$

where

$$D_1 = \int_{-\infty}^{\infty} R_{11}(\tau)\mathrm{d}\tau, \quad D_2 = \int_{-\infty}^{\infty} R_{22}(\tau)\mathrm{d}\tau \tag{4.73}$$

$$K_1(R,\tau) = \langle X_1(t+\tau)X_1(t)\rangle_t = \frac{1}{T(R)}\oint X_1(t+\tau)X_1(t)\mathrm{d}t \tag{4.74}$$

$$K_2(R,\tau) = \langle X_2(t+\tau)X_2(t)\rangle_t = \frac{1}{T(R)}\oint X_2(t+\tau)X_2(t)\mathrm{d}t \tag{4.75}$$

Substituting (4.23), (4.56), (4.69) and (4.67) into (4.65), and (4.71) and (4.72) into (4.66), we obtain

$$m(R) = -\frac{s}{af}\frac{g(R)}{T(R)} + \frac{G(R)}{T(R)} + \frac{1}{2}cD_1 + \frac{1}{2}aD_2 \tag{4.76}$$

$$\sigma^2(R) = f^2 \int_{-\infty}^{\infty} K_1(R,\tau)R_{11}(\tau)\mathrm{d}\tau + b^2 \int_{-\infty}^{\infty} K_2(R,\tau)R_{22}(\tau)\mathrm{d}\tau - \left(c^2D_1 + a^2D_2\right) \tag{4.77}$$

In (4.76) and (4.77), $T(R)$, $g(R)$, and $G(R)$ are given in (4.15), (4.22), and (4.57) respectively, D_1 and D_2 in (4.73), and $K_1(R,\tau)$ and $K_2(R,\tau)$ in (4.74) and (4.75). For a given R value, the drift coefficient $m(R)$ and diffusion coefficient $\sigma(R)$ can be calculated numerically.

In the special case of disturbances $\xi_1(t)$ and $\xi_2(t)$ being Gaussian white noises, we have

$$R_{11}(\tau) = 2\pi K_{11}\delta(\tau), \quad R_{22}(\tau) = 2\pi K_{22}\delta(\tau) \tag{4.78}$$

Then (4.76) remains the same and (4.77) reduces to

$$\sigma^2(R) = \left(\frac{1}{a}D_1 + \frac{1}{c}D_2\right)\frac{g(R)}{T(R)} \tag{4.79}$$

with $D_1 = 2\pi K_{11}$ and $D_2 = 2\pi K_{22}$. Equations (4.76) and (4.77) are the same as (4.58) and (4.59), as expected.

The stationary PDF $p(r)$ of $R(t)$, the joint PDF $p(x_1, x_2)$ of $X_1(t)$ and $X_2(t)$, and the marginal PDFs $p(x_1)$ of $X_1(t)$ and $p(x_2)$ of $X_2(t)$ can be obtained and calculated following the procedure described in Sect. 4.1.4.

Two different types of colored noises will be considered in the following numerical examples. One is the low-pass filtered process with a spectrum peak at zero frequency, and another is the randomized harmonic process with a spectrum peak at a non-zero frequency.

4.3.1 Low-Pass Filtered Stochastic Excitation

Consider low-pass filtered stochastic processes for the colored noises $\xi_1(t)$ and $\xi_2(t)$, generated from the first-order linear differential equations

$$\dot{\xi}_i + \alpha_i \xi_i = W_{gi}(t), \quad i = 1, 2 \tag{4.80}$$

where each $W_{gi}(t)$ is a Gaussian white noise with a spectral density K_i, and $W_{g1}(t)$ and $W_{g2}(t)$ are assumed to be independent. The correlation function and the power spectral density of random excitation $\xi_i(t)$ are, respectively,

$$R_{ii}(\tau) = H_i e^{-\alpha_i|\tau|}, \quad S_{ii}(\omega) = \frac{H_i \alpha_i}{\pi\left(\omega^2 + \alpha_i^2\right)} \tag{4.81}$$

where

$$H_i = \frac{\pi K_i}{\alpha_l} = R_{ii}(0) = E\left[\xi_i^2(t)\right] \tag{4.82}$$

From Eq. (4.73), we have

$$D_i = \frac{2H_i}{\alpha_i} = \frac{2\pi K_i}{\alpha_i^2} \tag{4.83}$$

It is clear that α_i and H_i are the band width and intensity parameters, respectively. A higher H_i corresponds to a stronger excitation, while a larger α_i results in a broader

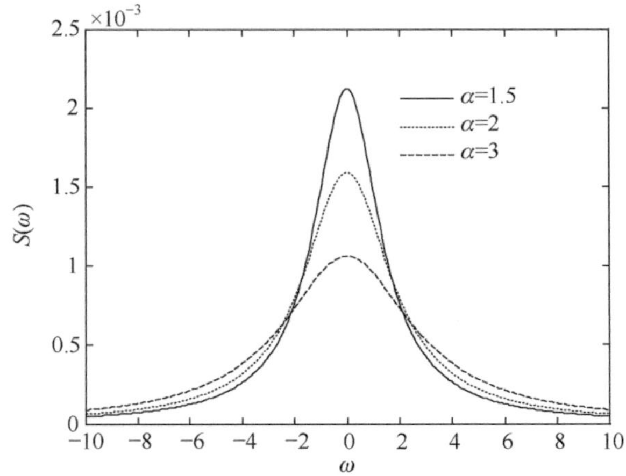

Fig. 4.6 Power spectral densities of the low-pass filtered processes (4.80) for different values of band width parameter α (Qi and Cai 2013)

band width. The level of the noise color directly corresponds to the band width parameter α_i. A larger α_i indicates a lower level of the noise color.

Figure 4.6 shows the spectral densities of the low-pass filtered noises with $H_1 = H_2 = 0.01$, and three different values of band width parameter α ($= \alpha_1 = \alpha_2$). As shown in Eq. (4.82), the same H_i value indicates the identical mean-square value of the noises. The highest points are always at zero frequency, as expected from Eq. (4.81) for the low-pass filter. For the case of $\alpha = 3$, the noise is of broad-band nature, while the band width is much narrower in the case of $\alpha = 1.5$. Compared with the spectrum of exponential form $1/f^{\beta}$ ($\omega = 2\pi f$) suggested in (Vaseur and Yodzis 2004), both have the similar shapes of decreasing spectral value with increasing frequency. However, the spectrum value at zero frequency is not infinity for the filtered noise (4.80), which is more realistic than the exponential form $1/f^{\beta}$.

Using the above low-pass filtered noises, the PDFs of X_1 and X_2 are calculated respectively for model (4.60) with possible abundant prey supply. The results are depicted in Fig. 4.7 with system parameters $a_1 = 1, b_1 = 1, c = 1.5, f_1 = 0.5, s = 0.1$, and $A = 0.05$. When the noises are more broadly banded ($\alpha_1 = \alpha_2 = 3$), i.e., less colored, the PDFs of both X_1 and X_2 are more centered around the equilibrium point (3.53, 0.76) of the deterministic system without the random noises. For the case of more colored noises ($\alpha_1 = \alpha_2 = 1.5$), the peaks are slightly shifted to the left, and the deviations are larger, indicating that the system is less stable.

Figure 4.8 shows the PDFs of X_1 and X_2 for the model of predator-competition, respectively, with system parameters $a_1 = 1.5, b_1 = 1, c = 0.5, f_1 = 0.5, s = 0.1$, and $B = 0.05$. Similar phenomena are observed as from Fig. 4.6 The equilibrium point of the corresponding deterministic system without the random noises is now (1.07, 1.50).

Monte Carlo simulations are carried out using the random noises generated from the filters (4.80), and the results are drawn in Figs. 4.7 and 4.8 for comparison. The analytical results are in agreement with the simulation ones within reasonable error ranges. For the case of $\alpha_1 = \alpha_2 = 1.5$, the deviation of the analytical results from the

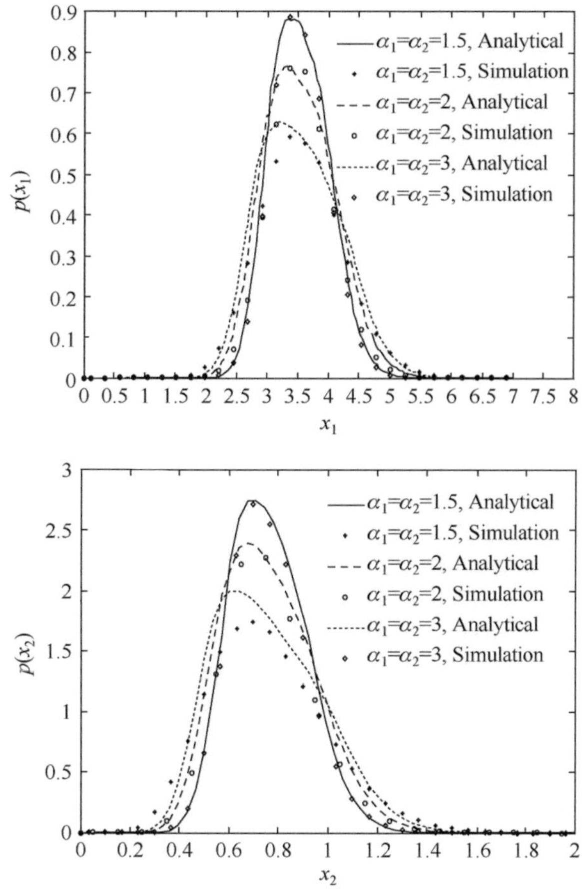

Fig. 4.7 PDFs of the prey and predator populations for the model of predator–saturation subjected to low-pass filtered noises (Qi and Cai 2013)

simulation ones is relatively large. This is because the stochastic averaging method requires the noises with broad-band spectra with respect to the system. The noise spectrum corresponding to $\alpha = \alpha_1 = \alpha_2 = 1.5$ is quite narrower as shown in Fig. 4.6

4.3.2 Excitation of Randomized Harmonic Process

Another type of colored stochastic processes considered here is the randomized harmonic process, modeled in Sect. 2.6.3, i.e.,

$$\xi_i(t) = A_i \sin[\omega_i t + \sigma_i B_i(t) + U_i], \quad i = 1, 2 \tag{4.84}$$

where each A_i is a positive constant specifying the intensity of the process, ω_i and σ_i are also positive constants representing the mean frequency and the randomness level

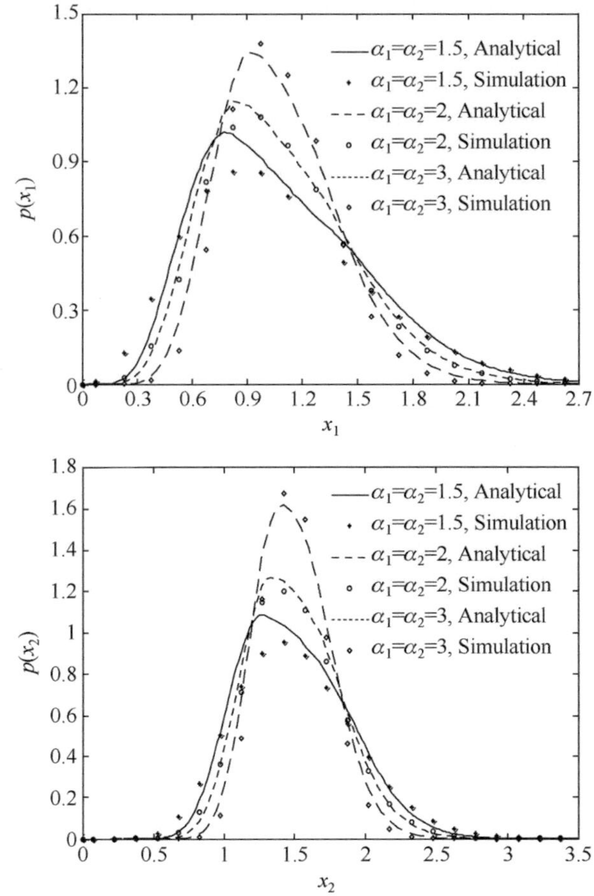

Fig. 4.8 PDFs of the prey and predator populations for the model of predator-competition subjected to low-pass filtered noises (Qi and Cai 2013)

in the phase, respectively, $B_i(t)$ is a unit Wiener process, and U_i is a random variable uniformly distributed in $[0, 2\pi)$. It is assumed that U_1 and U_2 are independent, and both are independent of $B_i(t)$. The incursion of the random variable U_i in (4.84) indicates that the initial phase is random, and also renders process $\xi_i(t)$ weakly stationary. It is known that parameter σ_i controls the spectrum band width. A larger σ_i corresponds to a broader band.

The correlation function of $\xi_i(t)$ is

$$R_{ii}(\tau) = \frac{1}{2}A_i^2\cos(\omega_i\tau)\exp\left(-\frac{1}{2}\sigma_i^2|\tau|\right) \tag{4.85}$$

and the power spectral density is

$$S_{ii}(\omega) = \frac{A_i^2\sigma_i^2\left(\omega^2+\omega_i^2+\sigma_i^4/4\right)}{4\pi\left[\left(\omega^2-\omega_i^2-\sigma_i^4/4\right)^2+\sigma_i^4\omega^2\right]} \tag{4.86}$$

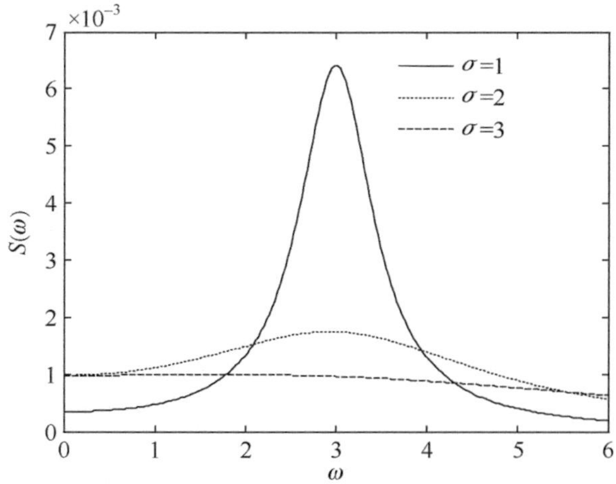

Fig. 4.9 Power spectral densities of randomized harmonic processes (4.84) for different values of band width parameter σ (Qi and Cai 2013)

Parameter D_i in Eq. (4.73) is then

$$D_i = \frac{2A_i^2\sigma_i^2}{4\omega_i^2 + \sigma_i^4} \tag{4.87}$$

In the numerical calculations, $A_1 = A_2 = 0.2$, so that all noises have the identical mean-square value. Also we use the same mean frequency $\omega_1 = \omega_2 = 3$ for the two excitations.

Figure 4.9 depicts the spectral densities of the randomized harmonic noises of three different values of σ ($=\sigma_1 = \sigma_2$). The spectral densities have peak values near $\omega = 3$. A smaller σ value corresponds to a noise of a narrower band width and a higher color level.

Figure 4.10 draws the PDFs of X_1 and X_2 respectively for the model with possible abundant prey. The system parameters are chosen as $a_1 = 1$, $b_1 = 1$, $c = 1$, $f_1 = 0.5$, $s = 0.1$, and $A = 0.05$. The results for the model with possible large predator population are shown in Fig. 4.10 with system parameters $a_1 = 2$, $b_1 = 1$, $c = 0.5$, $f_1 = 0.5$, $s = 0.1$, and $B = 0.05$. Similar observations can be drawn: (i) the peaks of the PDFs located near the equilibrium points of the deterministic system without the random noises, (ii) the narrower the noise bands are, the farther the PDFs deviate from the peaks, (iii) the peaks of the PDFs move away farther when the band width is narrower, (iv) the analytical results are accurate with admissible errors compared with the results from simulations, and (v) larger errors are observed for the narrower-band case of $\sigma_1 = \sigma_2 = 1$.

In summary, the term noise color is used to distinguish stochastic processes with different spectral densities, which describe the energy distribution over the entire frequency band. Two important characteristics are identified: one is the band width and another is the location of the spectrum peak. A narrow band width indicates a more colorful noise than the so called white noise, while the location of the spectrum

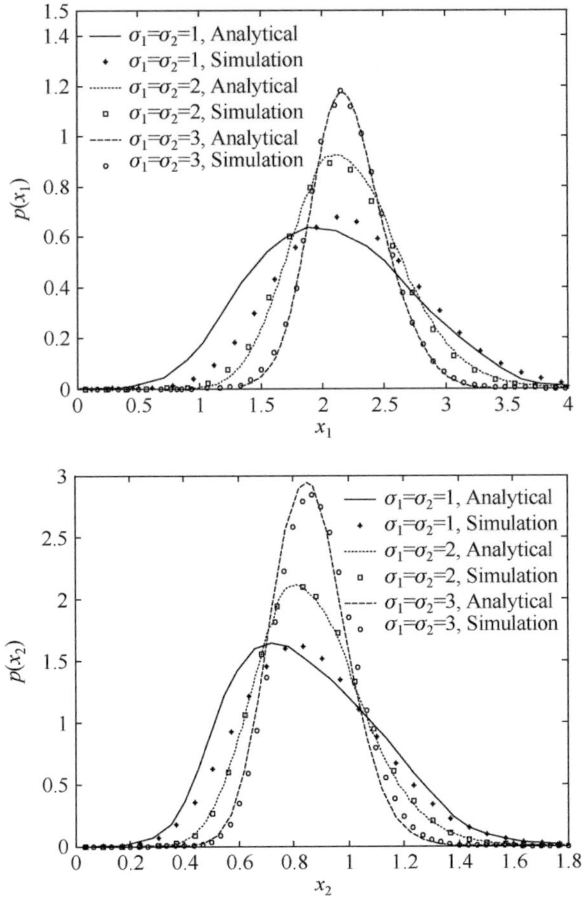

Fig. 4.10 PDFs of the prey and predator populations for the model of predator-saturation subjected to randomized harmonic noises (Qi and Cai 2013)

peak specifies what color the noise has. The two colored noises adopted in this section with peak locations at zero and nonzero frequencies have adjustable band widths.

For the two models of predator–prey systems investigated, the location of spectrum peak does not have large influence on the stationary state of the systems. However, the band width does have strong effects. The stochastic noises with narrower band widths make the system less stable in the sense that PDFs of the prey and predator populations are farther deviated from the peaks. Another minor effect of the band width is that the peak locations of the PDFs are shifted farther if the band width is narrower.

It should be pointed out that the colored noises used in the present investigation are of broad-band nature with respect to the systems as required by the stochastic averaging method. The probability distributions are generally unimodel with a single peak. It was found in (Naess et al. 2008) that if the band width is narrow even with respect to the system, bimodal probability distributions might occur.

4.4 Time-Delayed Ecosystem

In Sects. 4.1–4.3, the interaction between predators and preys are instantaneous, namely, an increase or decrease in prey population will affect the growth rate of the predator population immediately. In the real world, however, it is expected that the change of prey population will affect the growth rate of the predator population only after a time lag. Different types of models have been introduced (May 1973, 1981; Hutchinson 1948; MacDonald 1976, 1977; Nisbet and Gurney 1982; Korpimaki and Norrdahl 1991; Hornfeldt 1994; Turchin 2003) to account for the time-delay effect, and it has been found that the effect may indeed change the system behaviors substantially. However, all the above models were deterministic, namely, the coefficients in the governing equations were assumed to be known constants, and the inputs to the system were assumed to be precisely known. Such assumptions are idealistic since changes in the environment are always present, and in most cases, they cannot be predicted in advance. Although in Sects. 4.1–4.3, random variabilities of system properties have been considered, but only for the non-delayed type predator–prey models. To accounting the effects of time delay, stochastic modeling and analysis are extended in (Cai and Lin 2007b).

4.4.1 Deterministic Model

For the deterministic model (4.5), the term fx_1x_2 in the second equation implies that the change in the prey density x_1 will affect the change rate of the predator population instantaneously. On the other hand, a growth in predator population will consume more preys and lead to an immediate decrease of the prey population. Thus, the delay effect occurs essentially in the predator population. To account for such a time-delay effect, it has been suggested that the term fx_1x_2 be changed to (May 1973, 1981; Macdonald 1976, 1977)

$$fx_2 \int_{-\infty}^{t} F(t-\tau)x_1(\tau)\mathrm{d}\tau \tag{4.88}$$

where $F(t)$ is called the delay function, and is suitably modeled for a specific predator–prey system. Equation (4.88) implies that this delayed effect depends not on the population at a particular instant in the past; rather it depends on the average of the past populations. Thus, Eq. (4.5) is modified to

$$\begin{aligned} \dot{x}_1 &= a_1 x_1 - s x_1^2 - b x_1 x_2, \\ \dot{x}_2 &= -c x_2 + f x_2 \int_{-\infty}^{t} F(t-\tau) x_1(\tau) \mathrm{d}\tau \end{aligned} \tag{4.89}$$

It is convenient to write (4.90) in an alternative form

$$\begin{aligned} \dot{x}_1 &= a_1 x_1 - s x_1^2 - b x_1 x_2, \\ \dot{x}_2 &= -c x_2 + f x_2 \int_{0}^{\infty} F(\tau) x_1(t-\tau) \mathrm{d}\tau \end{aligned} \tag{4.90}$$

and normalize function $F(t)$ as follows

$$\int_0^\infty F(\tau) \mathrm{d}\tau = 1 \tag{4.91}$$

Define

$$\gamma = \int_0^\infty \tau F(\tau) \mathrm{d}\tau \tag{4.92}$$

which is a measure of the average delay time. Two reasonable choices for $F(t)$ are

$$F(t) = \frac{1}{\gamma} e^{-t/\gamma} \tag{4.93}$$

$$F(t) = \frac{4}{\gamma^2} t e^{-2t/\gamma} \tag{4.94}$$

Model (4.94) indicates that the effect of the prey population on the predator population diminishes gradually with an increasing time lag. In contrast, this effect reaches a maximum at a certain time lag in the delay model (4.94).

Expanding $x_1(t\text{-}\tau)$ in a Taylor series about $\tau = 0$, taking the first-order approximation, substituting into the second equation of (4.91), and applying Eq. (4.93), we obtain

$$\dot{x}_2 = -c x_2 + f x_1 x_2 - f \gamma x_2 \dot{x}_1 \tag{4.95}$$

Combination of the first equation of (4.90) and (4.95) results in

$$\begin{aligned} \dot{x}_1 &= a_1 x_1 - s x_1^2 - b x_1 x_2 \\ \dot{x}_2 &= -c x_2 + f(1 - \gamma a_1) x_1 x_2 + s f \gamma x_1^2 x_2 + b f \gamma x_1 x_2^2 \end{aligned} \tag{4.96}$$

It is noted that systems (4.7) and (4.96) have the same equilibrium point

$$x_{10} = \frac{c}{f}, \quad x_{20} = \frac{a}{b} \tag{4.97}$$

if we denote

$$a = a_1 - \frac{sc}{f} \tag{4.98}$$

System (4.97) will be used henceforth to investigate the effects of the time delay. It is of interest to note that only the mean value γ of the delay function $F(t)$ is required for the analysis, not the specific form of $F(t)$.

It can be shown that the equilibrium point (x_{10}, x_{20}) given in (4.98) is asymptotically stable if $\gamma = 0$ and the other parameters are not zero, and that it is unstable if $s = 0$ and the other parameters are not zero. Therefore, while the prey self-competition term stabilizes the system, the time delay term plays a role of destabilizing the system. Combining the two effects, the entire region of positive s and γ is divided into the stable and unstable sub-regions if other parameter values are fixed. Figure 4.12 shows the two sub-regions for system (4.97) with $a_1 = 0.9$, $b = 1$, $c = 0.5$ and $f = 0.5$. For a physically meaningful and stable ecological system, the pair of s and γ values must fall in the stable region.

Two trajectories of system (4.97) is depicted in Fig. 4.13, beginning from the same initial point (3.0, 1.0), and corresponding to the same γ value and two different s values in the stable region. One is $s = 0.07$ close to the stability boundary, while another is $s = 0.15$ far from the boundary. The other parameter values are the same as those in Fig. 4.12, i.e. $a_1 = 0.9$, $b = 1$, $c = 0.5$ and $f = 0.5$. It shows that the system reaches its equilibrium slowly with many cycles of decreasing amplitude for $s = 0.07$. On the other hand, the system approaches its equilibrium faster and with fewer cycles for case of $s = 0.15$.

Consider the conservative system corresponding to (4.97),

$$\begin{aligned} \dot{x}_1 &= x_1(a - bx_2), \\ \dot{x}_2 &= x_2(-c + fx_1)(1 + s\gamma x_1) \end{aligned} \tag{4.99}$$

System (4.100) has the same equilibrium point as that of systems (4.97). However, it is stable and non-asymptotic. System (4.100) possesses a first integral

$$r(x_1, x_2) = fx_1 - c - c\ln\frac{fx_1}{c} + bx_2 - a - a\ln\frac{bx_2}{a} - cs\gamma x_1 + \frac{1}{2}fs\gamma x_1^2 + \frac{c^2 s\gamma}{2f} \tag{4.100}$$

It can be shown that $r(x_1, x_2) = 0$ at the equilibrium point $(c/f,\ a/b)$, and that $r(x_1, x_2) > 0$ for any.

trajectory. Its period can be determined from

$$T(R) = \oint [\cdot]\mathrm{d}t = \oint \frac{\mathrm{d}x_2}{x_2(fx_1 - c)(1 + s\gamma x_1)} = \oint \frac{\mathrm{d}x_1}{x_1(a - bx_2)} \tag{4.101}$$

where x_1 and x_2 are related by $r(x_1, x_2) = R$. System (4.100) will serve as a base for applying the stochastic averaging technique.

Figure 4.14 depicts the equilibrium point O and three periodic trajectories for system (4.100) with parameters $a = 0.9$, $b = 1$, $c = 0.5, f = 0.5$, $s = 0.1$, and $\gamma = 0.2$. The equilibrium point O corresponds to $R = 0$, and the three different periodic trajectories correspond to three different R values, respectively.

4.4.2 Stochastic Model

Similar to the previous sections, the stochastic model extended from (4.97) is

$$\begin{aligned} \dot{X}_1 &= X_1\left[a_1 - sX_1 - bX_1X_2 + W_{g1}(t)\right], \\ \dot{X}_2 &= X_2\left[-c + f(1 - \gamma a_1)X_1 + sf\gamma X_1^2 + bf\gamma X_1X_2 + W_{g2}(t)\right] \end{aligned} \tag{4.102}$$

where $W_{g1}(t)$ and $W_{g2}(t)$ are two independent Gaussian white noises with spectral densities K_1 and K_2, respectively. Using Eq. (4.98), (4.102) can be rewritten as

$$\begin{aligned} \dot{X}_1 &= X_1\left[a - bX_2 - \frac{s}{f}(fX_1 - c) + W_{g1}(t)\right], \\ \dot{X}_2 &= X_2\left[(-c + fX_1)(1 + s\gamma X_1) + f\gamma X_1(bX_2 - a) + W_{g2}(t)\right] \end{aligned} \tag{4.103}$$

Due to the random variabilities in the prey growth rate and predator death rate, the behavior of the stochastic system (4.102) is substantially different from that of its deterministic counterpart. One significant change is that an equilibrium state of the deterministic system no longer exists; it is replaced by a distribution of states described by probability or statistical properties. Figure 4.15 illustrates such distributions for the stochastic system (4.102) with $a_1 = 0.9$, $b = 1$, $c = 0.5, f = 0.5$, $2\pi K_1 = 2\pi K_2 = 0.01$ and with two different sets of s and γ values. The case of $s = 0.07$ and $\gamma = 0.1$ corresponds to a state closer to the stability boundary shown in Fig. 4.12, while the case of $s = 0.2$ and $\gamma = 0.2$ corresponds to a state farther from the boundary. As expected, the first case has a larger area of distribution of the stationary states, indicating a less stable system.

The above observation shows that probabilistic measures are needed to describe the system random behaviors, which will be investigated here.

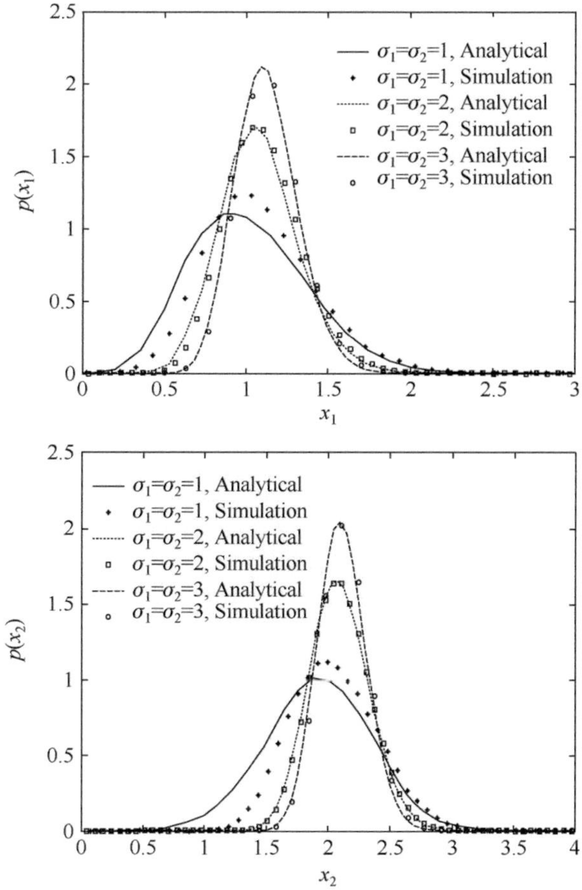

Fig. 4.11 PDFs of the prey and predator populations for the model of predator-competition subjected to randomized harmonic noises (Qi and Cai 2013)

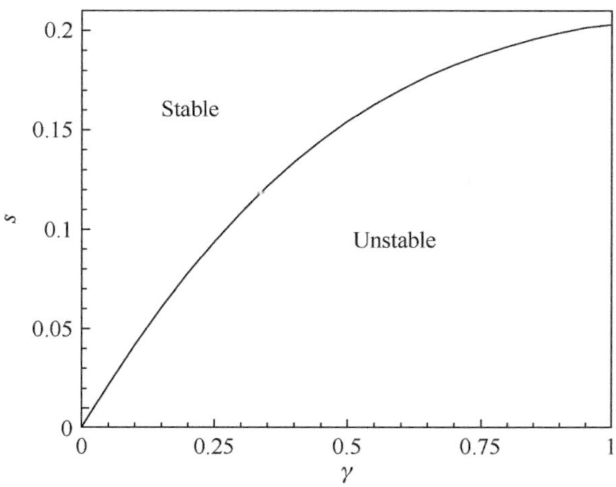

Fig. 4.12 Stable and unstable regions for the deterministic system (4.96) with $a_1 = 0.9$, $b = 1$, $c = 0.5$ and $f = 0.5$ (Cai and Lin 2007b)

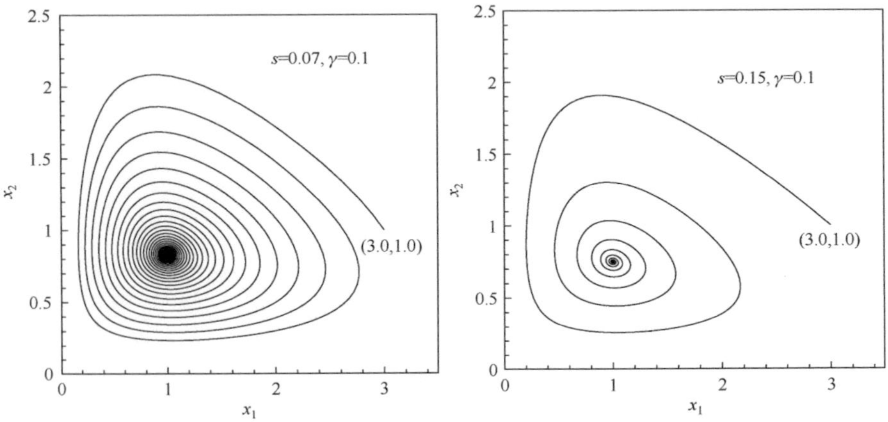

Fig. 4.13 Trajectories of the deterministic system (4.97) for two cases: $s = 0.07$ close to the stability boundary and $s = 0.015$ far from the stability boundary (Cai and Lin 2007b)

Fig. 4.14 Equilibrium and periodic trajectories of the deterministic system (4.100) (Cai and Lin 2007b)

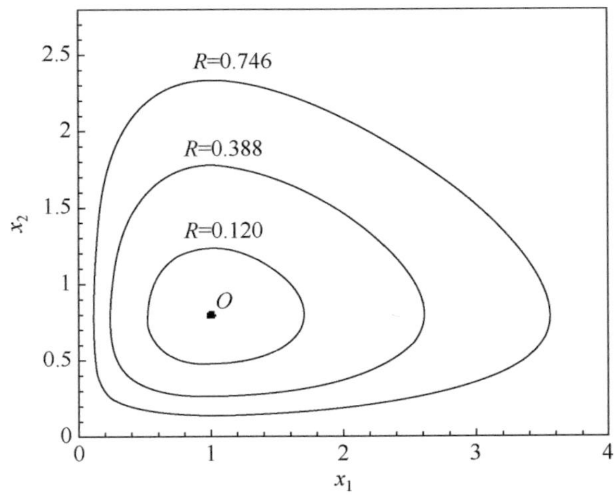

4.4.3 Stochastic Averaging

The Itô stochastic differential equations corresponding to (4.104) are

$$\begin{aligned} \mathrm{d}X_1 &= X_1\left[a - bX_2 + \frac{s}{f}(fX_1 - c) + \pi K_1\right]\mathrm{d}t + \sqrt{2\pi K_1}X_1\mathrm{d}B_1(t) \\ \mathrm{d}X_2 &= X_2\big[(-c + fX_1)(1 + s\gamma X_1) + f\gamma X_1(bX_2 - a) + \pi K_2\big]\mathrm{d}t \\ &+\sqrt{2\pi K_2}X_2\mathrm{d}B_2(t) \end{aligned} \tag{4.104}$$

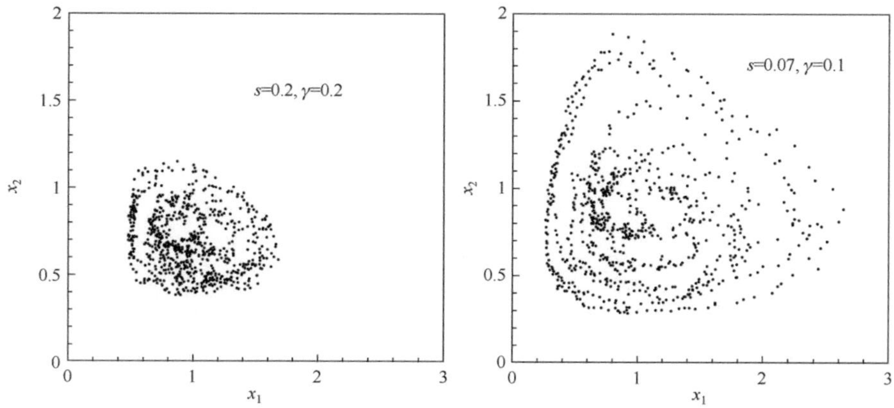

Fig. 4.15 Sample points of stochastic system (4.102) in the state of statistical stationarity for two different sets of parameter pair (s, γ). The set of $s = 0.2$ and $\gamma = 0.2$ is farther from the stability boundary; the set of $s = 0.07$ and $\gamma = 0.1$ is closer to the stability boundary (Cai and Lin 2007b)

The stochastic counterpart of (4.101) is

$$\begin{aligned} R(X_1, X_2) &= fX_1 - c - c\ln\frac{fX_1}{c} + bX_2 - a - a\ln\frac{bX_2}{a} \\ &-cs\gamma X_1 + \frac{1}{2}fs\gamma X_1^2 + \frac{c^2 s\gamma}{2f} \end{aligned} \tag{4.105}$$

which is a function of the stochastic processes X_1 and X_2; hence, it is also a stochastic process. The Itô equation for $R(X_1, X_2)$ can be obtained by applying the Itô differential rule, resulting in

$$\begin{aligned} \mathrm{d}R = &\Big[-\frac{s}{f}(fX_1 - c)^2(1 + s\gamma X_1) + f\gamma X_1(bX_2 - a)^2 \\ &+\pi K_1 X_1(f - sc\gamma + 2fs\gamma X_1) + b\pi K_2 X_2\Big]\mathrm{d}t \\ &+\sqrt{2\pi K_1}(fX_1 - c)(1 + s\gamma X_1)\mathrm{d}B_1(t) + \sqrt{2\pi K_2}(bX_2 - a)\mathrm{d}B_2(t) \end{aligned} \tag{4.106}$$

Assume that the coefficients s of the self-competition term and parameter γ of the time delay are small, and that the intensities of the white noises $W_{g1}(t)$ and $W_{g2}(t)$ are also small. These assumptions are generally valid for real ecosystems. Then the right-hand-side of (4.107) is small, indicating that $R(t)$ is a slowly varying random process. In this case, the stochastic averaging method is applicable. The averaged version of (4.107) can be cast in the standard form of an Itô equation as follows

$$\mathrm{d}R = m(R)\mathrm{d}t + \sigma(R)\mathrm{d}B(t) \tag{4.107}$$

in which $m(R)$ and $\sigma(R)$ are obtained as follows

$$m(R) = -\frac{s}{f}\langle (fX_1 - c)^2(1 + s\gamma X_1)\rangle_t + f\gamma\langle X_1(bX_2 - a)^2\rangle_t$$
$$+\pi K_1\langle X_1(f - sc\gamma + 2fs\gamma X_1)\rangle_t + a\pi K_2 \tag{4.108}$$

$$\sigma^2(R) = 2\pi K_1\langle (fX_1 - c)^2(1 + s\gamma X_1)^2\rangle_t + 2\pi K_2\langle (bX_2 - a)^2\rangle_t \tag{4.109}$$

where $\langle [\cdot]\rangle_t$ denotes the time average in one quasi-period, defined as

$$[\cdot]_t = \frac{1}{T}\oint [\cdot]\mathrm{d}t = \frac{1}{T}\oint \frac{[\cdot]\mathrm{d}X_2}{X_2(fX_1 - c)(1 + s\gamma X_1)} = \frac{1}{T}\oint \frac{[\cdot]\mathrm{d}X_1}{X_1(a - bX_2)} \tag{4.110}$$

For a given R, X_1 and X_2 are related according to (4.105). The time-average terms in (4.108) and (4.109) can be calculated numerically for given parameters a, b, c, f, s, and γ.

The stationary PDF $p(x_1)$ of $X_1(t)$ and $p(x_2)$ of $X_2(t)$ can be obtained and calculated following the procedure described in Sect. 4.1.4. Numerical calculations were carried out to obtain the results for the delayed stochastic system (4.102) with $2\pi K_1 = 2\pi K_2 = 0.01$ and with other parameters the same as those used when obtaining Fig. 4.12, i.e., $a_1 = 0.9$, $b = 1$, $c = 0.5$ and $f = 0.5$. Figure 4.15 shows the stationary PDFs of the prey and predator populations for several different combinations of s and γ values within the stable region shown in Fig. 4.12. For a given s, a larger γ value results in higher probabilities in both lower and higher prey populations, indicating a less stable system. On the other hand, a system with a larger s is more stable. Also depicted in the figure are results obtained using the Monte Carlo simulation technique. The theoretical and simulation results agree quite well Fig. 4.16.

Shown in Fig. 4.17 are the stationary PDFs of the prey and predator populations for two different sets of s and γ values, corresponding to two different points in the stable region. The point corresponding to $s = 0.2$ and $\gamma = 0.2$ is farther from the stability boundary, while the one corresponding to $s = 0.07$ and $\gamma = 0.1$ is closer to the stability boundary. In the latter case, the PDFs have lower peaks, which are compensated with higher values in the lower and higher population ranges. This is expected since the system state is closer to the stability boundary, and is less stable.

4.5 Ecosystem with Habitat Complexity

In Sects. 4.1—4.4, the habitat complexity was neglected, and the two species of the prey and predator were assumed to be fully coupled. In reality, however, the habitat complexity plays an important role on the coupling extent of the two species, as reported in literatures (e.g. Luckinbill 1973; Savino and Stein 1982; Manatunge et al. 2000; Alstad 2001; Grabowski 2004). To taking into account the habitat complexity, Bairagi and Jana (2011) investigated the stability and bifurcation of a deterministic predator–prey system with habitat complexity. In this section, the Stochastic predator–prey model with habitat complexity is investigated, following the paper by

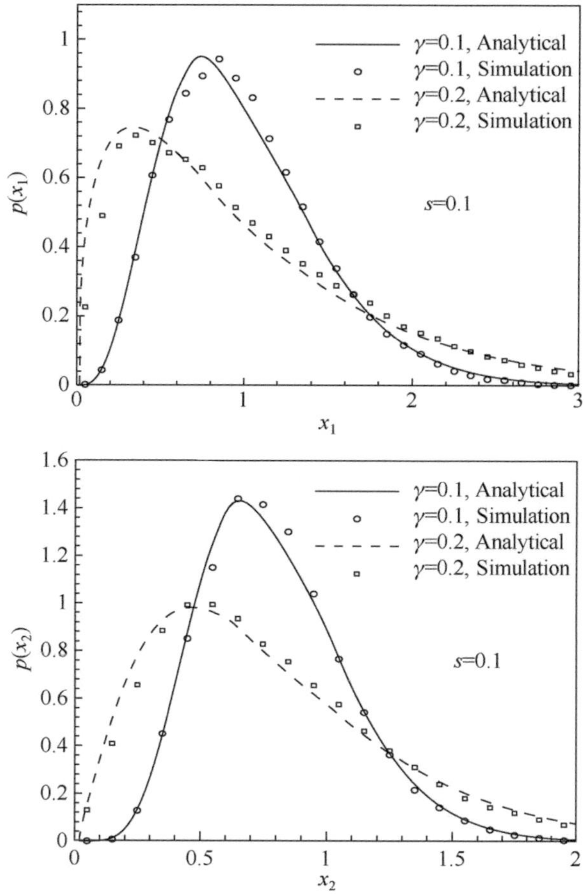

Fig. 4.16 PDFs of prey and predator populations for different values of parameter γ (Cai and Lin 2007b)

Qi and Cai (2013). The stochastic averaging method and the Monte Carlo simulation are applied to find the probability distributions of the predator and prey populations. Three cases corresponding to weak, moderate, and strong habitat complexities, are considered. Effects of the habitat complexity on the system dynamics are explored.

4.5.1 Deterministic Model

The mathematical model describing the dynamical behaviors of the predator–prey ecosystem with habitat complexity is (Bairagi and Jana 2011)

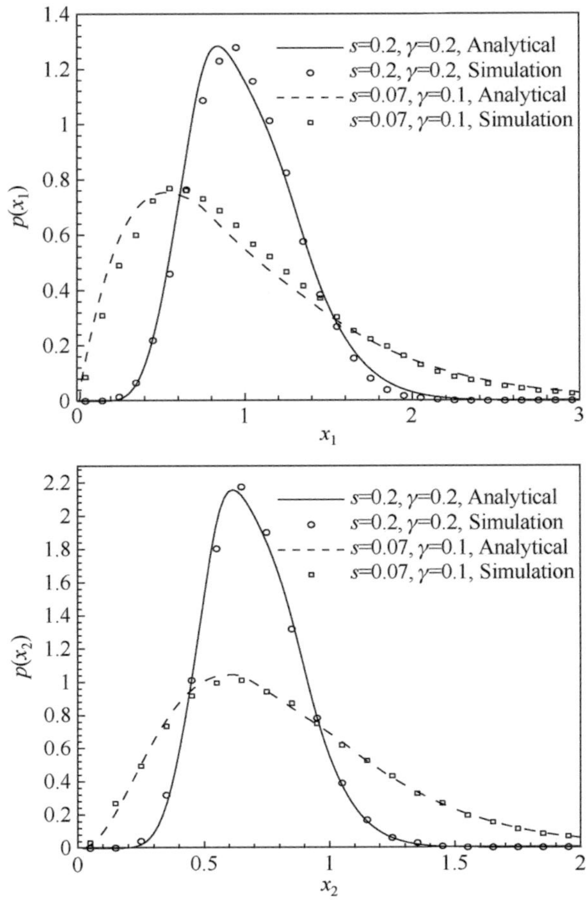

Fig. 4.17 PDFs of predator and prey populations for two different sets of parameter pair (s, γ). The set of $s = 0.2$ and $\gamma = 0.2$ is farther from the stability boundary; the set of $s = 0.07$ and $\gamma = 0.1$ is closer to the stability boundary (Cai and Lin 2007b)

$$\begin{aligned} \dot{x}_1 &= qx_1\left(1 - \frac{x_1}{k}\right) - \frac{\alpha(1-c)x_1x_2}{1+\alpha(1-c)hx_1}, \\ \dot{x}_2 &= -dx_2 + \frac{\theta\alpha(1-c)x_1x_2}{1+\alpha(1-c)hx_1} \end{aligned} \tag{4.111}$$

where x_1 and x_2 are the population densities of the prey and predator, respectively, and q, k, α, c, h, d, and θ are positive constants. Parameters q and k are known as the prey growth rate and prey carrying capacity, respectively. Without the presence of the predator, the prey population remains the same as soon as it reaches its carrying capacity k. Parameter c ($0 < c < 1$) reflects the strength of the habitat complexity. A larger c corresponds to a stronger habitat complexity, indicating a weaker interaction and stronger independence of each species. The limiting case of $c \to 0$ describes a negligible habitat complexity, and the two species interact fully, known as the Holling Type II model (Holling 1959). Parameters α, h, and θ are known as the attack rate, handling time, and conversion efficiency, respectively. The model (4.111) shows that (i) a growth in the predator population will consume more preys and lead to a decrease of the prey population and (ii) the effect of the prey population increase on the predator population is positive.

In this section, focus is on the effect of the habitat complexity; therefore, the other system parameters will be fixed to meaningful values according to a specific system.

4.5.2 Equilibriums and Stability

By letting the right-hand-sides of (4.111) be zero, three equilibriums can be found for the system:

(i) the trivial equilibrium $E_{0,}$ $x_1 = 0$ and $x_2 = 0$,
(ii) the predator-free equilibrium E_1, $x_1 = k$ and $x_2 = 0$, and
(iii) the coexistence equilibrium E^*,

$$x_1^* = \frac{d}{\alpha(1-c)(\theta - hd)}, \quad x_2^* = \frac{q\left(k - x_1^*\right)\left[1 + \alpha h(1-c)x_1^*\right]}{\alpha k(1-c)} \tag{4.112}$$

Equation (4.113) shows that the existence of E^* requires (i) $\theta > hd$, and (ii)$k > x_1^*$ since both x_1 and x_2 must be non-negative. The first condition $\theta > hd$ is satisfied for practical systems due to very small h. The second condition requires

$$k > \frac{d}{\alpha(1-c)(\theta - hd)} \quad \text{or} \quad c < c_2 = 1 - \frac{d}{\alpha k(\theta - hd)} \tag{4.113}$$

To investigate the local stability of the equilibriums, denote $\tilde{x}_1$ and $\tilde{x}_2$ as the two coordinates of an equilibrium point, and let

$$u = x_1 - \tilde{x}_1, \quad v = x_2 - \tilde{x}_2 \tag{4.114}$$

With the transformation (4.115), we obtain equations for u and v from Eq. (4.111), and linearize the equations to obtain

$$\begin{aligned} \dot{u} &= Au + Bv \\ \dot{v} &= Cu + Dv \end{aligned} \tag{4.115}$$

where

$$\begin{aligned} A &= q - \frac{2q}{k}\tilde{x}_1 - \frac{\alpha(1-c)\tilde{x}_2}{[1 + \alpha(1-c)h\tilde{x}_1]^2}, \quad B = -\frac{\alpha(1-c)\tilde{x}_1}{1 + \alpha h(1-c)\tilde{x}_1}, \\ C &= \frac{\theta\alpha(1-c)\tilde{x}_2}{[1 + \alpha(1-c)h\tilde{x}_1]^2}, \quad \mathrm{D} = \frac{\theta\alpha(1-c)\tilde{x}_1}{1 + \alpha h(1-c)\tilde{x}_1} - d \end{aligned} \tag{4.116}$$

The characteristic equation associated with system (4.116) is

$$\begin{vmatrix} \eta - A & -B \\ -C & \eta - D \end{vmatrix} = 0 \tag{4.117}$$

The two roots of Eq. (4.118) are obtained as

$$\eta_{1,2} = \frac{A+D}{2} \pm \frac{\sqrt{(A-D)^2 + 4BC}}{2} \tag{4.118}$$

The local stabilities of each equilibrium can be determined according to the signs of the real parts of η_1 and η_2.

First consider the trivial equilibrium E_0. Substitution of $\tilde{x}_1 = 0$ and $\tilde{x}_2 = 0$ into (4.117) and (4.119) results in $\eta_1 = q > 0$ and $\eta_2 = -d < 0$. It is clear that E_0 (0, 0) is a saddle point, and it is unstable.

For the predator-free equilibrium $E_1, \tilde{x}_1 = k$ and $\tilde{x}_2 = 0$, we have from Eqs. (4.116) and (4.118)

$$A = -q, \quad B = -\frac{\alpha k(1-c)}{1+\alpha hk(1-c)}, \quad C = 0, \quad D = \frac{\theta\alpha k(1-c)}{1+\alpha hk(1-c)} - d \tag{4.119}$$

$$\eta_1 = D = \frac{\theta\alpha k(1-c)}{1+\alpha hk(1-c)} - d, \quad \eta_2 = A = -q \tag{4.120}$$

Then the local stability of E_1 requires that $D < 0$, resulting in

$$c > c_2 = 1 - \frac{d}{\alpha k(\theta - hd)} \tag{4.121}$$

For the coexistence equilibrium E^*, Eqs. (4.116) and (4.118) lead to

$$A = \frac{q}{d}\left[hd - \frac{x_1^*(\theta + hd)}{k}\right], \quad B = -\frac{d}{k}, \quad C = \frac{q\left(k - x_1^*\right)(\theta - hd)}{k}, \quad D = 0, \tag{4.122}$$

$$\eta_{1,2} = \frac{A}{2} \pm \frac{\sqrt{A^2 + 4BC}}{2} \tag{4.123}$$

where x_1^* is given in Eq. (4.112). We consider the case in which $A^2 + 4BC < 0$ which is true for certain practical cases. Then, the stability of E^* requires both η_1 and η_2 to have negative real parts, i.e., $A < 0$, leading to

$$c > c_1 = 1 - \frac{\theta + hd}{\alpha kh(\theta - hd)} \tag{4.124}$$

Since it can be shown that $c_1 < c_2$, the stability of the three equilibriums is summarized as.

(i) For $0 < c < c_1$, all three equilibriums are unstable.
(ii) For $c_1 < c < c_2$, $E_0(0{,}0)$ and $E_1(k{,}0)$ are unstable, and $E^*(x_1^*, x_2^*)$ is locally asymptotically stable, indicating a coexisting state of the prey and predator.
(iii) For $c_2 < c < 1$, $E_0(0{,}0)$ is unstable, $E^*(x_1^*, x_2^*)$ does not exist, and $E_1(k{,}0)$ is asymptotically stable.

To illustrate the dynamical behaviors of system (4.111) more intuitively, consider a specific predator–prey system of paramecium aurelia and didinium nasutum. The following experimental data were available (Luckinbill 1973; Bairagi and Jana 2011): $q = 2.65$, $k = 898$, $\alpha = 0.045$, $h = 0.0437$, $d = 1.06$, and $\theta = 0.215$. These parameter values will be used for all numerical calculations in this section. The two threshold values c_1 and c_2 are then calculated as

$$c_1 = 0.1227, \quad c_2 = 0.8445 \tag{4.125}$$

Numerical calculation shows that a limit cycle takes place if $0 < c < c_1$. Figure 4.18a show the limit cycle for $c = 0.05$. The system starts from (150, 40) and (300, 30) respectively, and eventually moves along a trajectory of a limit cycle. Figure 4.18b depicts three limit cycles corresponding to three different c values. It is shown that an increasing c will result in a contraction of the limit cycle, indicating that the variation ranges of both population decrease as c approaches the limiting c_1 value.

When $c_1 < c < c_2$, only the coexistence equilibrium E^* is stable, and the system will converge to it, beginning from any permissible state. Figure 4.19a, b show the two trajectories of the system motion corresponding to two values of the habitat complex parameter c, respectively. The system moves around the stable equilibrium

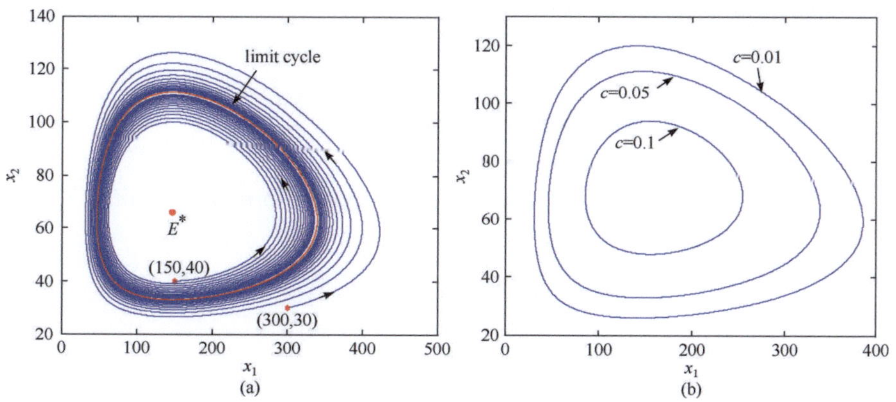

Fig. 4.18 limit cycles of the system (4.111) when $0 < c < c_1$, **a** $c = 0.05$, **b** three different c values (Cai and Qi 2016)

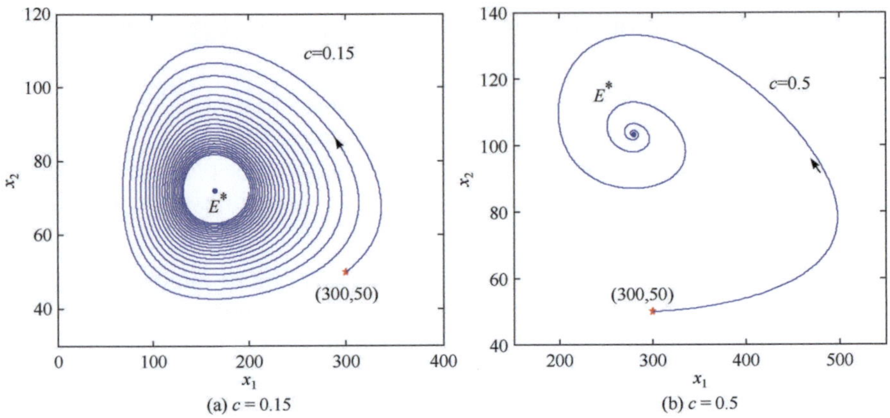

Fig. 4.19 Trajectories of system (4.116) when $c_1 < c < c_2$ (Cai and Qi, 2016)

point with a decreasing amplitude, and finally reaches the equilibrium state E^*. With a larger c value, it reaches the equilibrium faster. For a smaller c value, it reaches the equilibrium slower with more cycles.

The bifurcation diagrams of both populations, with the varying parameter c, are shown in Fig. 4.20 In the range of $0 < c < c_1$, the habitat complexity is weak, and the two species interact strongly, leading to their co-existence with varying populations. The variation ranges reduce with increasing c. In the range of $c_1 < c < c_2$, i.e., moderate habitat complexity, both species converge to fixed stable states, indicating the influences between the two species reduce. In both cases of weak and moderate habitat complexities, the two species coexist with different dynamical behaviors. When the habitat complex parameter c exceeds c_2, the interaction of the two species is so weak that they follow their own dynamics. After the transient stage, the prey population will reach its carrying capacity k, and the predator will extinct since the prey supply cannot maintain its survival.

4.5.3 Modified Lotka-Volterra Model

It turns out that system (4.111) can be written in another form of a modified Lotka-Volterra model as follows

$$\begin{aligned} \dot{x}_1 &= x_1[a - bx_2 + G_1(x_1, x_2)] \\ \dot{x}_2 &= x_2\big[fx_1 - d + G_2(x_1, x_2)\big] \end{aligned} \tag{4.126}$$

where

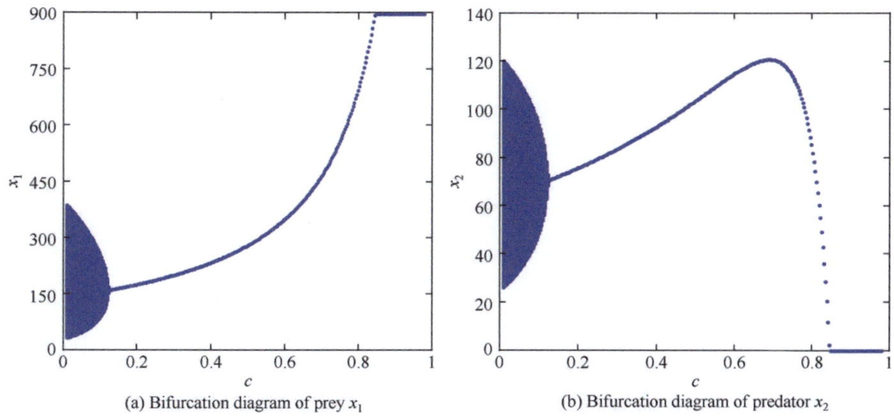

Fig. 4.20 Bifurcation diagrams for system (4.111) (Cai and Qi, 2016)

$$
\begin{aligned}
G_1(x_1, x_2) &= -\frac{q}{kf}(fx_1 - d) + Ex_2\frac{b(fx_1 - d)}{f(1 + Ex_1)} \\
G_2(x_1, x_2) &= -Ex_1\frac{fx_1 - d}{1 + Ex_1}
\end{aligned} \tag{4.127}
$$

and parameters a, b, f, and E in (4.127) and (4.128) are found as follows, in terms of the parameters in the original system (4.111),

$$
a = q - \frac{qd}{kf}, \quad b = \frac{f}{\theta}, \quad f = \alpha(1 - c)(\theta - hd), \quad E = \alpha h(1 - c) \tag{4.128}
$$

Since model (4.126) is the same as model (4.111), the system dynamic behaviors of (4.126) depend on the value of habitat parameter c. Consider the case of $c_1 < c < c_2$. It can be shown that systems (4.126) and the classical Lotka-Volterra model (4.1) have the same unstable equilibrium (0, 0) and stable equilibrium (d/f, a/b). The difference is that the stable equilibrium is asymptotic for system (4.126), but non-asymptotic for system (4.1).

For the classical Lotka-Volterra model (4.1), i.e., without G_1 and G_2 functions in (4.126), there are an equilibrium point O corresponding to $R = 0$, and periodic trajectories corresponding to different R values, i.e., different initial states, as shown in Fig. 4.1. From Figs. 4.18 and 4.19 for the modified model (4.126), it is shown that the G_1 and G_2 functions play the role of energy-dissipation mechanisms, resulting in smaller and smaller trajectories towards the equilibrium point.

Figure 4.21a, b plot the G_1 and G_2 functions for $c = 0.15$ and $c = 0.5$, respectively, starting from the same initial point (300, 100). As expected, both are vanishing with time, but with different paces. For the case of $c = 0.15$, they are oscillating many cycles, similar to underdamped case of an oscillator. On the other hand, the case of $c = 0.5$ corresponds to an overdamped oscillator. If the value of c is smaller

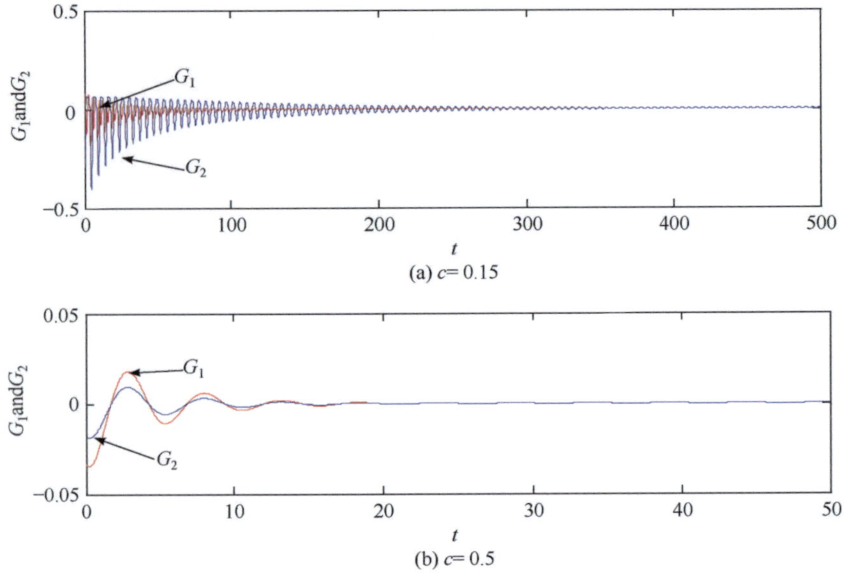

Fig. 4.21 Varying of functions G_1 and G_2 with time for two different values of c (Cai and Qi 2016)

indicating the system damping is weak, it takes longer time and more cycles to reach the equilibrium state.

The modified Lotka-Volterra model (4.126) will be used in the stochastic analyses in the following section.

4.5.4 Stochastic Model and Stochastic Averaging

As the same as in the previous sections, we introduce two disturbances into model (4.126), resulting in the following stochastic model

$$\begin{aligned} \dot{X}_1 &= X_1\big[a - bX_2 + G_1(X_1, X_2) + W_{g1}(t)\big] \\ \dot{X}_2 &= X_2\big[fX_1 - d + G_2(X_1, X_2) + W_{g2}(t)\big], \end{aligned} \tag{4.129}$$

where $W_{g1}(t)$ and $W_{g2}(t)$ are two independent Gaussian white noises with zero means and spectral densities K_1 and K_2, respectively. Since the dynamic behaviors of the deterministic system (4.126) depend on the habitat parameter c, analyses of the stochastic system (4.129) will also be investigated separately for the three cases: (i) weak habitat complexity $0 < c < c_1$, (ii) moderate habitat complexity $c_1 < c < c_2$, and (iii) strong habitat complexity $c_2 < c < 1$.

(i) Weak habitat complexity, $0 < c < c_1$

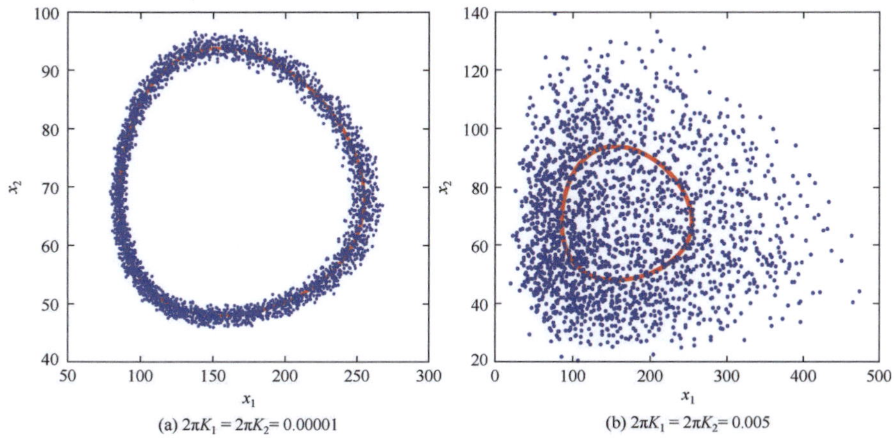

Fig. 4.22 Sample points of the stochastic model (4.129) with $c = 0.1$ and two different noise levels (Cai and Qi 2016)

In this case, no analytical method is found to be able to treat Eq. (4.129), and Monte Carlo simulation was carried out for investigation. Figure 4.22 depicts a comparison of the limit cycle of deterministic model (4.126) and the sample point sets of the stochastic model (4.129) with $c = 0.1$. Two different noise levels are selected: $2\pi K_1 = 2\pi K_2 = 0.00001$ and $2\pi K_1 = 2\pi K_2 = 0.005$. Obviously, the limit cycle is disrupted due to the noises, and the system state is diffused around the limit cycle of the deterministic system. The higher the noise level is, the more spread of the system state is. Corresponding to the two cases in Fig. 4.22, the PDFs of the prey and predator are calculated, and depicted in Fig. 4.23. The effect of the noise level on the distributions of the populations of both prey and predator are clearly observed.

(ii) Moderate habitat complexity, $c_1 < c < c_2$

As discussed in the deterministic analysis in Sect. 4.5.3, both prey and predator populations will converge to an equilibrium in the case of $c_1 < c < c_2$. Since the random disturbances are present in the stochastic model (4.129), the system will oscillate around this equilibrium, known as the dynamic balance. To carry out the analysis, we transform the system Eqs. (4.129) to the Itô differential equations,

$$\begin{aligned} dX_1 &= X_1[a - bX_2 + G_1(X_1, X_2) + \pi K_1]dt + \sqrt{2\pi K_1}X_1 dB_1(t), \\ dX_2 &= X_2\big[fX_1 - d + G_2(X_1, X_2) + \pi K_2\big]dt + \sqrt{2\pi K_2}X_2 dB_2(t) \end{aligned} \tag{4.130}$$

where $B_1(t)$ and $B_2(t)$ are independent unit Wiener processes. The two additional terms πK_1 and πK_2 are known as the Wong-Zakai correction terms due to the presence of the parametric excitations. Now consider function $R(X_1, X_2)$ defined in (4.10). Applying the Itô differential rule, we obtain from (4.130)

Fig. 4.23 PDFs of the prey and predator calculated from model (4.129) for weak habitat complexity, $c = 0.1$ (Cai and Qi 2016)

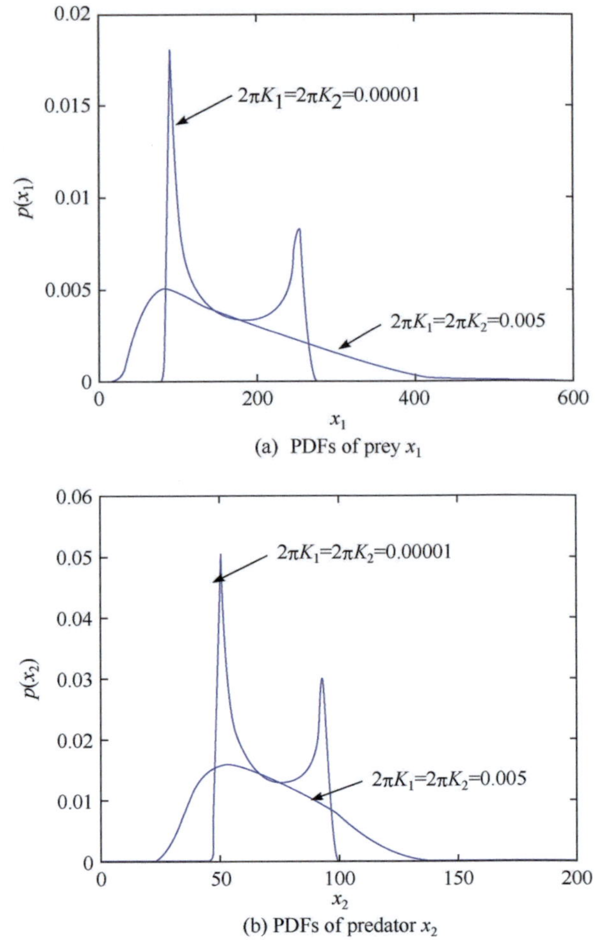

$$
\begin{aligned}
\mathrm{d}R(t) = &\left[(fX_1 - d)G_1(X_1, X_2) + (bX_2 - a)G_2(X_1, X_2) + \pi K_1 fX_1 + \pi K_2 bX_2\right]\mathrm{d}t \\
&+\sqrt{2\pi K_1}(fX_1 - d)\mathrm{d}B_1(t) + \sqrt{2\pi K_2}(bX_2 - a)\mathrm{d}B_2(t)
\end{aligned}
\tag{4.131}
$$

According to the analysis in Sect. 4.5.3 for the deterministic situation, functions G_1 and G_2 are converging to zero, as shown in Fig. 4.21. We also assume that the stochastic disturbances are weak; thus, the right hand of Eq. (4.131) is small and process $R(t)$ is a slowly varying stochastic process. In this case, the stochastic averaging method can be applied, and the averaged $R(t)$ process is approximated as a Markov diffusion process governed by an Itô stochastic differential equation

$$
\mathrm{d}R = m(R)\mathrm{d}t + \sigma(R)\mathrm{d}B(t) \tag{4.132}
$$

where $m(R)$ and $\sigma(R)$ are given by

$$m(R) = \langle (fX_1 - d)G_1(X_1, X_2) + (bX_2 - a)G_2(X_1, X_2)\rangle_t + \pi K_1 f\langle X_1\rangle_t + \pi K_2 b\langle X_2\rangle_t \tag{4.133}$$

$$\sigma^2(R) = 2\pi K_1\langle (fX_1 - d)^2\rangle_t + 2\pi K_2\langle (bX_2 - a)^2\rangle_t \tag{4.134}$$

where $[\cdot]_t$ denotes the time average in one quasi-period, defined in (4.15).

After carrying the time averages, Eqs. (4.133) and (4.134) are specified as

$$m(R) = \pi K_1 d + \pi K_2 a - \frac{q}{af}\frac{\varphi(R)}{T(R)} + \frac{\psi(R)}{T(R)} \tag{4.135}$$

$$\sigma^2(R) = \frac{(2\pi K_1 d + 2\pi K_2 a)}{ad}\frac{\varphi(R)}{T(R)} \tag{4.136}$$

where

$$\varphi(R) = a\oint \frac{fX_1 - d}{X_2}\mathrm{d}X_2, \quad \psi(R) = \frac{bE}{f}\oint \frac{fX_1 - d}{1 + EX_1}\mathrm{d}X_2 \tag{4.137}$$

The stationary PDF of the one-dimensional Markov process $R(t)$, the joint PDF $p(x_1, x_2)$, and the marginal PDFs of prey and predator can be obtained using the procedure similar as that in Sect. 4.1.4.

Numerical calculations are carried out to obtain the stationary PDFs of prey and predator from the above analytical procedure. The results are shown in Fig. 4.24 for noise level $2\pi K_1 = 2\pi K_2 = 0.005$ and two different levels of habitat parameter, $c = 0.15$ and 0.5. It is calculated that the equilibrium points for deterministic system without noises are (164, 72) and (279, 103) for $c = 0.15$ and 0.5, respectively. Figure 4.24 shows that the peaks are close to the equilibrium points, as expected. When the habitat complexity parameter c increases, the peaks of PDFs moves to the right, indicating the populations of both prey and predator increase. At the same time, the peaks become higher and the deviations are smaller. This is because the interaction between the prey and predator becomes weaker for a larger habitat parameter, and both species are more independent. Monte Carlo simulation results are also depicted in the figures for comparison. It is seen that the analytical procedure yields quite accurate results.

(iii) Strong habitat complexity, $c_2 < c < 1$

The interactions between prey and predator are weak at strong habitat complexity. Figure 4.25 simulate a sample function for each of the predator and prey for the case of $c = 0.9$, noise level $2\pi K_1 = 2\pi K_2 = 0.005$, and other parameters remain the same. It shows that the predator population will eventually vanish since (1) the prey supply is too small, and (2) the noise is not strong enough to overcome the effect from its death rate. On the other hand, the prey population fluctuated around its carrying

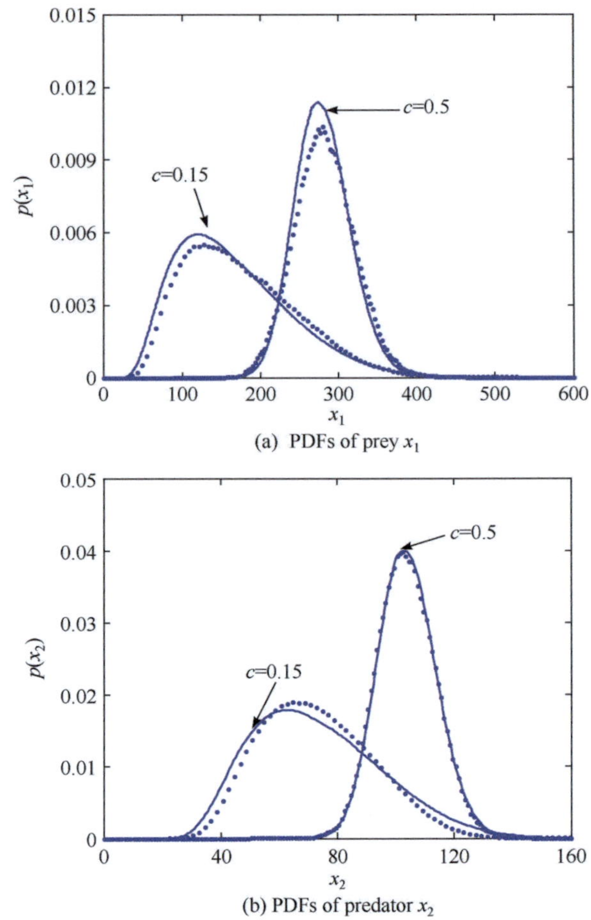

Fig. 4.24 PDFs of the prey and predator calculated from model (4.129) with noise level $2\pi K_1 = 2\pi K_2 = 0.005$ for moderate habitat complexity, $c = 0.15$ and 0.5. Lines: analytical results, symbols (dots): simulation results (Cai and Qi 2016)

capacity k. Letting $X_2 = 0$, the governing equation for the prey population density X_1 is simplified to

$$\dot{X}_1 = q\left(1 - \frac{X_1}{k}\right)X_1 + X_1 W_{g1}(t) \tag{4.138}$$

The Itô equation corresponding (4.119) is

$$\mathrm{d}X_1 = \left[(q + \pi K_1)X_1 - \frac{qX_1^2}{k}\right]\mathrm{d}t + \sqrt{2\pi K_1}X_1\mathrm{d}B_1(t) \tag{4.139}$$

The stationary PDF of X_1 is then

Fig. 4.25 Sample functions of $X_1(t)$ and $X_2(t)$ of the stochastic system (4.129) for strong habitat complexity of $c = 0.9$, and with noise level $2\pi K_1 = 2\pi K_2 = 0.005$

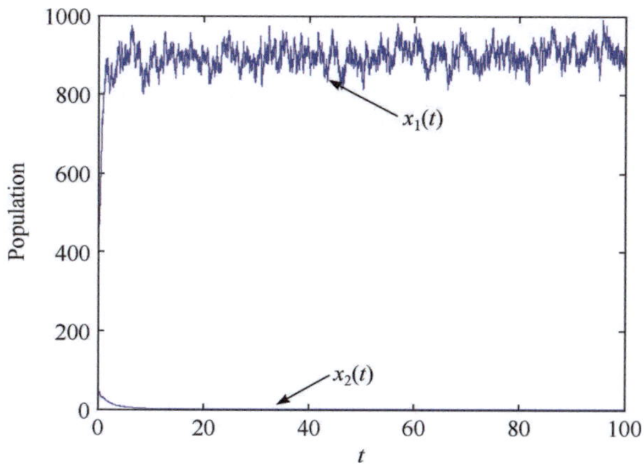

Fig. 4.26 Stationary PDFs of prey population $X_1(t)$ for strong habitat complexity of $c = 0.9$. Lines: analytical results, symbols (dots): simulation results (Cai and Qi 2016)

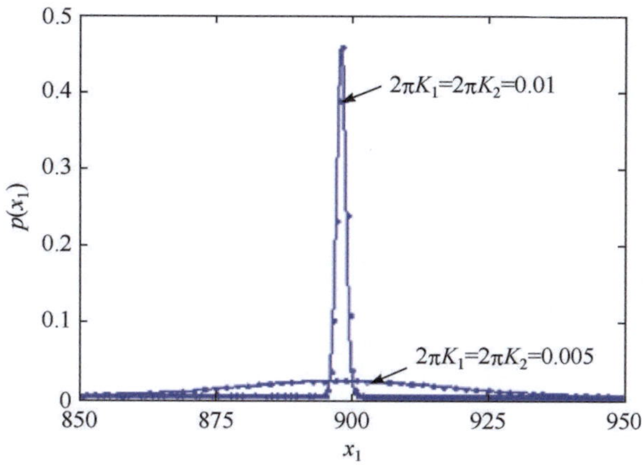

$$p(x_1) = C x_1^{q/(\pi K_1)-1} \exp\left(-\frac{q}{k\pi K_1} x_1\right) \tag{4.140}$$

where C is a normalization constant, given as

$$C = \left[\left(\frac{k\pi K_1}{q}\right)^{q/\pi K_1} \Gamma\left(\frac{q}{\pi K_1}\right)\right]^{-1} \tag{4.141}$$

and $\Gamma(\cdot)$ is the Gamma function.

Figure 4.26 shows that stationary PDFs of the prey population density X_1 for $c = 0.9$ and two different disturbance levels. The results calculated from Eq. (4.140) are depicted in lines, while those from simulations are represented by dots. Both agree very well.

References

Alstad D. 2001. Basic Populas Models of Ecology. NJ: Prentice Hall.
Bairagi N, Jana D. 2011. On the stability and Hopf bifurcation of a delay-induced predator-prey system with habitat complexity. Applied Mathematical Modeling, 35: 3255–3267.
Bazykin A D. 1998. Nonlinear dynamics of interacting populations. World Scientific Series on nonlinear science. Series A: monographs and treatises, 11. River Edge: World Scientific Publishing Co., Inc.
Cai G Q, Lin Y K. 2004. Stochastic analysis of the Lotka-Volterra model for ecosystems. Physical Review E, 70: 041910.
Cai G Q, Lin Y K. 2007a. Stochastic analysis of predator-prey type ecosystems. Ecological Complexity, 4: 241–249.
Cai G Q, Lin Y K. 2007b. Stochastic analysis of time-delayed ecosystems. Physical Review E, 76: 041913.
Cai G Q, Qi L. 2016. Effects of habitat complexity on stochastic nonlinear ecosystems. International Journal of Dynamics and Control, 4(3): 275–283.
Grabowski J G. 2004. Habitat complexity disrupts predator-prey interactions but not the trophic cascade on oyster reefs. Ecology, 85: 995–1004.
Holling C S. 1959. Some characteristics of some types of predation and parasitism. Canadian Entomologist, 91: 385–398.
Horndeldt B. 1994. Delayed density dependence as a determinant of vole cycles. Ecology, 75: 791–806.
Hutchinson G E. 1948. Circular causal systems in ecology. Annals of the New York Academy of Sciences, 50: 221–246.
Korpimaki E, Norrdahl K. 1991. Do breeding nomadic avian predators dampen population fluctuations of small mammals? Oikos, 62: 195–208.
Lotka A J. 1925. Elements of Physical Biology. Baltimore, NJ: William and Wilkins.
Luckinbill L. 1973. Coexistence in laboratory populations of paramecium aurelia and its predator didinium nasutum. Ecology, 54: 1320–1327.
Macdonald N. 1976. Time delay in prey-predator models. Mathematical Biosiences, 28: 321–330.
Macdonald N. 1977. Time delay in prey-predator models – II. Bifurcation Theory. Mathematical Biosiences, 33: 227–234.
Manatunge J, Asaeda T, Priyadarshana T. 2000. The influence of structural complexity on fish-zooplankton interactions: a study using artificial submerged microphytes. Environmental Biologo of Fishes, 59: 425–438.
May R M. 1981. Theoretical Ecology, Principles and Applications, Sunderland, MA: Sinauer Associates.
May R M, Verga A D. 1973. Stability and Complexity in Model Ecosystems. Princeton, NJ: Princeton University Press.
Murray J D. 1993. Mathematical Biology, 3rd Ed, Vol. I: An Introduction, Vol. II: Spatial Models and Biomedical Applications, Springer.
Naess A, Dimenberg M F, Gaidai O. 2008. Lotka-Volterra systems in environments with randomly disordered temporal periodicity. Physical Review E, 78: 021126.
Nisbet R, Gurney W. 1982. Modeling Fluctuating Populations. New York: Wiley.
Qi L, Cai G Q. 2013. Dynamics of nonlinear ecosystems under colored noise disturbances. Nonlinear Dynamics, 71: 463–474.
Savino J F, Stein R A. 1982. Predator-prey interaction between largemouth bass and bluegills as influenced by simulated, submersed vegetation. Trans American Fisheries Society, Vol. 111, 255–266.
Stratonovich R L. 1963. Topics in the Theory of Random Noise, Vol. 1, New York: Gordon and Breach.
Turchin P. 2003. Complex Population Dynamics. Princeton, NJ: Princeton University Press.
Vaseur D A, Yodzis P. 2004. The color of environmental noise. Ecology, 85: 1146–1152.

Volterra V. 1926. Variazioni e fluttuazioni del numerod'individui in specie d'animaniconviventi, Memorie della Accademia Nazionale dei Lincei 2: 31–113.

Volterra V. 1931. Leconssur la theoriemathematique de la lutte pour la vie. Paris: Gauthiers-Vilars.

Wong E, Zakai M. 1965. On the relation between ordinary and stochastic equations. International Journal of Engineering Sciences, 47(1): 150–154.

Wu Y, Zhu W Q. 2008. Stochastic analysis of a pulse-type prey-predator model. Physical Review E, 77: 041911.

Chapter 5
Several Applications of the Stochastic Averaging Methods in Natural Sciences

Many systems in physics, chemistry, and biology are nonlinear and influenced by various stochastic factors, such as thermal disturbances in biochemical reactions and climate effects on biological systems. To describe and study these systems more accurately, it is necessary to establish models as nonlinear stochastic dynamical systems. The stochastic averaging methods, especially the stochastic averaging method of quasi-Hamiltonian systems, have unique advantages to treat this type of systems. In this chapter, several applications of the stochastic averaging methods in physics chemistry, and life science (Deng and Zhu 2009a, 2009b) are described in detail.

5.1 Motion of Active Brownian Particles

Active motion, also known as self-driven motion, is prevalent in various species in the biological world, ranging from cells and microorganisms to birds and fish (Alt 1980; Dickinson and Tranquillo 1993; Schienbein and Gruler 1993). The active Brownian particle is an important model for describing such motion (Romanczuk et al. 2012). Active Brownian particles can absorb energy from the external environment and store energy in their internal energy reservoirs, and then convert it into mechanical energy for movement when needed. By extending the behavior of single active Brownian particle to swarm level and establishing coupling mechanisms between particles, a model for a swarm of active Brownian particles is obtained. In this section, the stochastic averaging method is used to study the stationary motion of both single active Brownian particle and a swarm of active Brownian particles.

W. Zhu et al., *Stochastic Averaging*, https://doi.org/10.1007/978-981-96-3833-8_5

5.1.1 Deterministic Motion of Active Brownian Particle

The motion of Brownian particles is affected by friction. The particles with positive friction consuming energy into the environment during their motion are called passive Brownian particles. In contrast, active Brownian particles can absorb energy from the environment to compensate for the energy consumed during their motion. The motion energy of active Brownian particles comes not only from noise but also from an energy reservoir, which can be converted into mechanical energy for motion when necessary.

Consider an active Brownian particle moving on a two-dimensional plane. Using the following parabolic harmonic potential (Schienbein and Gruler, 1993)

$$U(x_1, x_2) = \frac{1}{2}\omega^2(x_1^2 + x_2^2), \tag{5.1}$$

and introducing the Rayleigh damping coefficient

$$\alpha(x, \dot{x}) = -\gamma_1 + \gamma_2\dot{x}^2, \tag{5.2}$$

where γ_1 and γ_2 are positive constants. The motion equation of the particle can be established as

$$\begin{aligned} \dot{x}_1 &= v_1, \quad \dot{v}_1 + [-\gamma_1 + \gamma_2(v_1^2 + v_2^2)]v_1 + \omega^2 x_1 = 0, \\ \dot{x}_2 &= v_2, \quad \dot{v}_2 + [-\gamma_1 + \gamma_2(v_1^2 + v_2^2)]v_2 + \omega^2 x_2 = 0. \end{aligned} \tag{5.3}$$

In the case of a one-dimensional potential, Eq. (5.3) is reduced to a self-excited Rayleigh oscillator having a limit cycle in the phase plane. For the two-dimensional potential system described by Eq. (5.3), two limit cycles will appear in a four-dimensional phase space (x_1, x_2, v_1, v_2) (Ebeling et al. 1999). When these two limit cycles are projected onto the (v_1, v_2) plane, a circle is obtained as follows

$$v_1^2 + v_2^2 = v_0^2, \tag{5.4}$$

where v_0 is a constant, and in the case of Rayleigh damping, $v_0 = \sqrt{\gamma_1/\gamma_2}$. The projection of these two limit cycles on the (x_1, x_2) plane yields another circle

$$x_1^2 + x_2^2 = r_0^2, \tag{5.5}$$

where r_0 is also a constant. In the present deterministic case, the particle moves in circular path. Based on the balance between centrifugal force v_0^2/r_0 and centripetal forces $\mathrm{d}U(r)/\mathrm{d}r\big|_{r=r_0}$, the following constant can be derived

$$r_0 = v_0/\omega. \tag{5.6}$$

To have a clear understanding for the deterministic behavior of active Brownian particle in Eq. (5.3), a solution corresponding to one of the two limit cycles in the 4-dimensional phase space is provided as follows (Erdmann et al. 2000)

$$\begin{aligned} x_1 &= r_0 \cos(\omega_0 t + \varphi_0),\ v_1 = -r_0\omega \sin(\omega_0 t + \varphi_0), \\ x_2 &= r_0 \sin(\omega_0 t + \varphi_0),\ v_2 = r_0\omega \cos(\omega_0 t + \varphi_0), \end{aligned} \tag{5.7}$$

where ω_0 is the angular frequency at which the particle rotates with velocity v_0 and a radius r_0. Thus

$$\omega_0 = v_0 \big/ r_0 = \omega. \tag{5.8}$$

Eq. (5.8) implies that in the deterministic case, the motion frequency of active Brownian particle is determined by the natural frequency ω of system (5.3). In the 3-dimensional phase space (x_1, x_2, v_2), the trajectory of the solution (5.7) is like a hula hoop. Furthermore, through the following transformation

$$t \to -t, \quad v_1 \to -v_1, \quad v_2 \to -v_2. \tag{5.9}$$

another limit cycle can be obtained from (5.7)

$$\begin{aligned} x_1 &= r_0 \cos(-\omega_0 t + \varphi_0), \quad v_1 = r_0\omega \sin(-\omega_0 t + \varphi_0), \\ x_2 &= r_0 \sin(-\omega_0 t + \varphi_0), \quad v_2 = -r_0\omega \cos(-\omega_0 t + \varphi_0). \end{aligned} \tag{5.10}$$

In the 3-dimensional subspace (x_1, x_2, v_2), the trajectory of the solution (5.10) still likes a hula hoop. The difference is that the direction of rotation of limit cycle in Eq. (5.10) is exactly opposite to that of the previous limit cycle in Eq. (5.7). To get a more intuitive understanding of the two solutions, Figure 5.1a–c provide the trajectory of the two limit cycles in three or two dimensional subspaces. Figure 5.1a shows the projection of solutions (5.7) and (5.10) onto the (x_1, x_2, v_2) subspace, displaying two intersecting limit cycles. Figure 5.1b shows the projection onto the (v_1, v_2) plane, which is similar to the projection onto the (x_1, x_2), displaying circular path. Figure 5.1c shows the projection onto the (x_1, v_2) plane, resulting in two intersecting straight lines. The projection onto the (x_2, v_1) plane also show up as two intersecting straight lines.

The particle moving on the two limit cycles maintains a constant total energy

$$E_0 = \frac{1}{2}(v_1^2 + v_2^2) + \frac{1}{2}\omega^2(x_1^2 + x_2^2) = \frac{1}{2}v_0^2 + \frac{1}{2}\omega^2 r_0^2. \tag{5.11}$$

For any initial total energy H other than E_0, the total energy of a particle will eventually converge to the invariant E_0 (Ebeling et al. 1999).

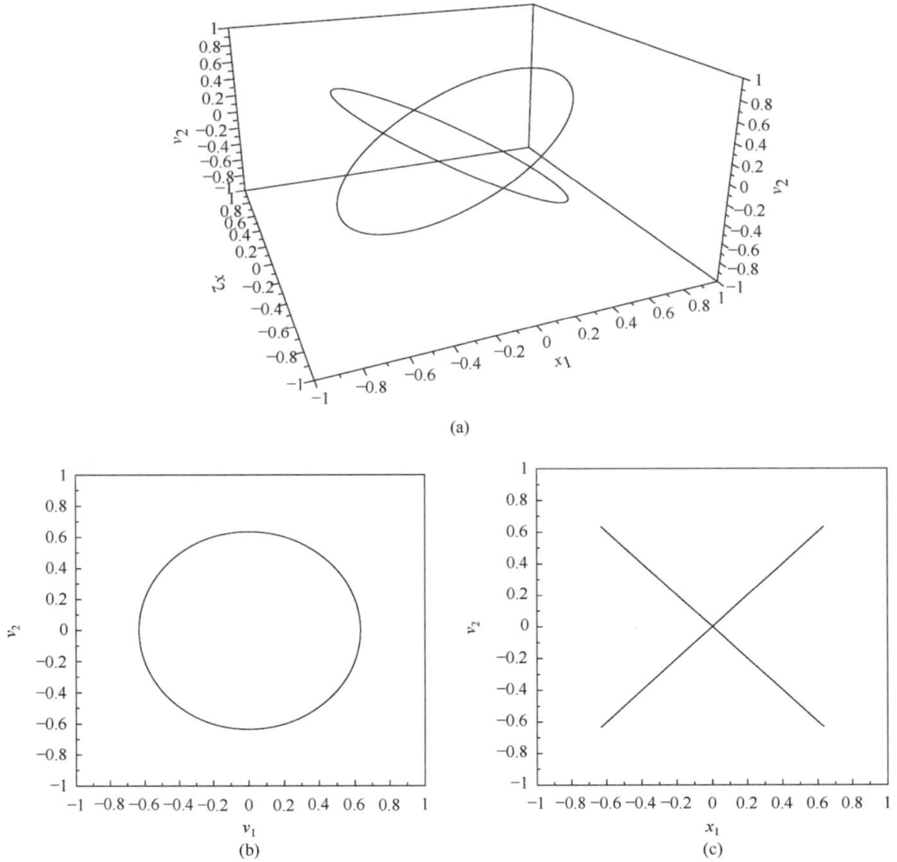

Fig. 5.1 Limit cycles of Eqs. (5.7) and (5.13). **a** Projection in subspace (x_1, x_2, v_2); **b** projection in plane (v_1, v_2); **c** projection in plane (x_1, v_2).

$$H \to E_0 = v_0^2, \quad t \to \infty. \tag{5.12}$$

Finally, concerning the deterministic motion equation of active Brownian particle in Eq. (5.3), the boundary line between the two limit cycles is

$$x_1 x_2 v_1 v_2 = 0. \tag{5.13}$$

When analyzing the stochastic dynamical characteristics of active Brownian particle, the boundary line plays a significant role. It separates two different motion states, and under the influence of random excitation, transitions between these states must pass this boundary line.

5.1.2 Stochastic Motion of Active Brownian Particle

(i) The case of parabolic harmonic potential

Real biological motion is inevitably influenced by random factors such as food distribution, environmental temperature, and climate. To describe the motion of biological entities perturbed by stochastic factors, it is necessary to introduce random factors into the motion of active Brownian particles. The random factors imposed on the motion of active Brownian particle can be various. They may affect the mechanism governing motion energy, the potential energy, or other system parameters. Currently, an external random excitation, particularly Gaussian white noise, is considered more frequently, as environmental noise. In this subsection, external excitations of Gaussian white noises are introduced to transform the original deterministic motion equation (5.3) into the following stochastic motion equation of active Brownian particle (Deng and Zhu 2004)

$$\begin{aligned}&\dot{X}_1 = V_1, \quad \dot{V}_1 + [-\gamma_1 + \gamma_2(V_1^2 + V_2^2)]V_1 + \omega^2 X_1 = \sqrt{2D}W_{g1}(t),\\&\dot{X}_2 = V_2, \quad \dot{V}_2 + [-\gamma_1 + \gamma_2(V_1^2 + V_2^2)]V_2 + \omega^2 X_2 = \sqrt{2D}W_{g2}(t),\end{aligned} \tag{5.14}$$

where ω is the natural frequency of the oscillator, $W_{g1}(t)$ and $W_{g2}(t)$ are independent unit Gaussian white noises, $2D$ is the excitation intensity, and γ_1, γ_2, D are small quantities of same order.

Let $Q_1 = X_1,\ Q_2 = X_2, P_1 = \dot{X}_1,\ P_2 = \dot{X}_2$, the original system (5.14) can be transformed into the following Itô stochastic differential equation

$$\begin{aligned}&\mathrm{d}Q_1 = P_1\mathrm{d}t, \quad \mathrm{d}P_1 = -\{\omega^2 Q_1 + [-\gamma_1 + \gamma_2(P_1^2 + P_2^2)]P_1\}\mathrm{d}t + \sqrt{2D}\mathrm{d}B_1(t),\\&\mathrm{d}Q_2 = P_2\mathrm{d}t, \quad \mathrm{d}P_2 = -\{\omega^2 Q_2 + [-\gamma_1 + \gamma_2(P_1^2 + P_2^2)]P_2\}\mathrm{d}t + \sqrt{2D}\mathrm{d}B_2(t),\end{aligned} \tag{5.15}$$

where $B_1(t)$ and $B_2(t)$ are independent unit Wiener processes. The Hamiltonian system associated with the original system (5.14) is integrable and has two independent and involutive motion integrals

$$H_1 = \frac{1}{2}P_1^2 + \frac{1}{2}\omega^2 Q_1^2, \quad H_2 = \frac{1}{2}P_2^2 + \frac{1}{2}\omega^2 Q_2^2. \tag{5.16}$$

Introduce the angle variables θ_1 and θ_2

$$\theta_1 = \tan^{-1}\left(\frac{P_1}{\omega Q_1}\right), \quad \theta_2 = \tan^{-1}\left(\frac{P_2}{\omega Q_2}\right). \tag{5.17}$$

Based on Eqs. (5.74) and (5.89) in vol. 1, the following Itô stochastic differential equation for $H_1, H_2, \theta_1, \theta_2$ can be established

$$
\begin{aligned}
dH_1 &= \{-[-\gamma_1 + \gamma_2(P_1^2 + P_2^2)]P_1^2 + D\}dt + \sqrt{2D}P_1 dB_1(t), \\
dH_2 &= \{-[-\gamma_1 + \gamma_2(P_1^2 + P_2^2)]P_2^2 + D\}dt + \sqrt{2D}P_2 dB_2(t), \\
d\theta_1 &= \{\omega - \frac{\omega[-\gamma_1 + \gamma_2(P_1^2 + P_2^2)]Q_1P_1}{P_1^2 + \omega^2 Q_1^2} + \frac{2\omega D Q_1 P_1}{(P_1^2 + \omega^2 Q_1^2)^2}\}dt + \frac{\omega Q_1}{P_1^2 + \omega^2 Q_1^2} dB_1(t), \\
d\theta_2 &= \{\omega - \frac{\omega[-\gamma_1 + \gamma_2(P_1^2 + P_2^2)]Q_2P_2}{P_2^2 + \omega^2 Q_2^2} + \frac{2\omega D Q_2 P_2}{(P_2^2 + \omega^2 Q_2^2)^2}\}dt + \frac{\omega Q_2}{P_2^2 + \omega^2 Q_2^2} dB_2(t).
\end{aligned} \tag{5.18}
$$

Since both natural frequencies in the system (5.14) are ω, an internal resonance relationship in Eq. (5.75) of vol. 1 exists, where $k_1 = -1$, $k_2 = 1$. The stochastic averaging method for quasi-integrable and resonant Hamiltonian systems described in Sect. 5.2.2 is applicable. Let the phase angle difference be $\Psi = \theta_2 - \theta_1$. The following Itô stochastic differential equation for $\Psi(t)$ can be derived from (5.18)

$$
\begin{aligned}
d\Psi = \{&\frac{\omega[-\gamma_1 + \gamma_2(P_1^2 + P_2^2)]Q_1P_1}{P_1^2 + \omega^2 Q_1^2} - \frac{\omega[-\gamma_1 + \gamma_2(P_1^2 + P_2^2)]Q_2P_2}{P_2^2 + \omega^2 Q_2^2} + \frac{2\omega D Q_2 P_2}{(P_2^2 + \omega^2 Q_2^2)^2} \\
&- \frac{2\omega D Q_1 P_1}{(P_1^2 + \omega^2 Q_1^2)^2}\}dt + \frac{\omega Q_2}{P_2^2 + \omega^2 Q_2^2} dB_2(t) - \frac{\omega Q_1}{P_1^2 + \omega^2 Q_1^2} dB_1(t).
\end{aligned} \tag{5.19}
$$

Following the stochastic averaging method for quasi-integrable and resonant Hamiltonian systems described in Sect. 5.2.2, as $\gamma_1, \gamma_2, D \to 0$, the slowly-varying vector process $[H_1, H_2, \Psi]^{\mathrm{T}}$ converges to a 3-dimensional Markov diffusion process, and its stationary PDF $p(h_1, h_2, \psi)$ satisfies the following reduced averaged FPK equation

$$
\begin{aligned}
0 = &-\frac{\partial}{\partial h_1}(a_1 p) - \frac{\partial}{\partial h_2}(a_2 p) - \frac{\partial}{\partial \psi}(a_\psi p) + \frac{1}{2}\frac{\partial^2}{\partial h_1^2}(b_{11}p) + \frac{1}{2}\frac{\partial^2}{\partial h_2^2}(b_{22}p) \\
&+ \frac{1}{2}\frac{\partial^2}{\partial \psi^2}(b_{\psi\psi}p) + \frac{1}{2}\frac{\partial^2}{\partial h_1 \partial h_2}(b_{12}p) + \frac{1}{2}\frac{\partial^2}{\partial h_1 \partial \psi}(b_{1\psi}p) + \frac{1}{2}\frac{\partial^2}{\partial h_2 \partial h_1}(b_{21}p) \\
&+ \frac{1}{2}\frac{\partial^2}{\partial h_2 \partial \psi}(b_{2\psi}p) + \frac{1}{2}\frac{\partial^2}{\partial \psi \partial h_1}(b_{\psi 1}p) + \frac{1}{2}\frac{\partial^2}{\partial \psi \partial h_2}(b_{\psi 2}p),
\end{aligned} \tag{5.20}
$$

where the first and second derivative moments are

$$
\begin{aligned}
a_1 &= \left\langle -[-\gamma_1 + \gamma_2(p_1^2 + p_2^2)]p_1^2 + D \right\rangle_t, \\
a_2 &= \left\langle -[-\gamma_1 + \gamma_2(p_1^2 + p_2^2)]p_2^2 + D \right\rangle_t, \\
a_\psi &= \left\langle -\frac{\omega}{2}[-\gamma_1 + \gamma_2(p_1^2 + p_2^2)](\frac{q_2p_2}{h_2} - \frac{q_1p_1}{h_1}) + \frac{\omega D}{2}(\frac{q_1p_1}{h_1^2} - \frac{q_2p_2}{h_2^2}) \right\rangle_t, \\
b_{11} &= \left\langle 2Dp_1^2 \right\rangle_t, \quad b_{12} = b_{1\psi} = 0, \\
b_{22} &= \left\langle 2Dp_2^2 \right\rangle_t, \quad b_{21} = b_{2\psi} = 0, \\
b_{\psi\psi} &= \left\langle \frac{\omega^2 D}{2}(\frac{q_1^2}{h_1^2} + \frac{q_2^2}{h_2^2}) \right\rangle_t, \quad b_{\psi 1} = b_{\psi 2} = 0.
\end{aligned} \tag{5.21}
$$

In the above equations, p_1, p_2 should be replaced by h_1, h_2, q_1, q_2 according to Eq. (5.16), $\langle[\cdot]\rangle_t$ denotes time averaging operations and can be replaced by the following spatial averaging with respect to θ_1

$$\langle[\cdot]\rangle_t = \frac{1}{2\pi}\int_0^{2\pi}[\cdot]\Big|_{q_1=\frac{\sqrt{2h_1}}{\omega}\cos\theta_1,\ q_2=\frac{\sqrt{2h_2}}{\omega}\cos(\psi+\theta_1)}\mathrm{d}\theta_1. \tag{5.22}$$

Finishing the averaging in Eq. (5.21), the first and second derivative moments are

$$\begin{aligned}
a_1 &= \gamma_1 h_1 - \frac{3}{2}\gamma_2 h_1^2 - \gamma_2(1+\frac{1}{2}\cos 2\psi)h_1 h_2 + D, \quad b_{11} = 2Dh_1,\\
a_2 &= \gamma_1 h_2 - \frac{3}{2}\gamma_2 h_2^2 - \gamma_2(1+\frac{1}{2}\cos 2\psi)h_2 h_1 + D, \quad b_{22} = 2Dh_2,\\
a_\psi &= \frac{1}{4}\gamma_2\sin 2\psi(h_1+h_2), \quad b_{\psi\psi} = \frac{1}{2}D(\frac{1}{h_1}+\frac{1}{h_2}).
\end{aligned} \tag{5.23}$$

Solving the reduced averaged FPK equation (5.20) yields the stationary PDF $p(h_1, h_2, \psi)$. Obviously, the solution of the following equations is also the solution of Eq. (5.20)

$$\begin{aligned}
&-a_1 p + \frac{1}{2}\frac{\partial}{\partial h_1}(b_{11}p) + \frac{1}{2}\frac{\partial}{\partial h_2}(b_{12}p) + \frac{1}{2}\frac{\partial}{\partial \psi}(b_{1\psi}p) = 0,\\
&-a_2 p + \frac{1}{2}\frac{\partial}{\partial h_2}(b_{22}p) + \frac{1}{2}\frac{\partial}{\partial \psi}(b_{2\psi}p) + \frac{1}{2}\frac{\partial}{\partial h_1}(b_{21}p) = 0,\\
&-a_\psi p + \frac{1}{2}\frac{\partial}{\partial \psi}(b_{\psi\psi}p) + \frac{1}{2}\frac{\partial}{\partial h_1}(b_{\psi 1}p) + \frac{1}{2}\frac{\partial}{\partial h_2}(b_{\psi 2}p) = 0.
\end{aligned} \tag{5.24}$$

Assume that the solution of Eq. (5.20) is of the form

$$p(h_1, h_2, \psi) = C\exp[-\lambda(h_1, h_2, \psi)], \tag{5.25}$$

where $\lambda(h_1, h_2, \psi)$ is the probability potential function, C is a normalization constant. Substituting solution (5.25) into Eq. (5.24) leads to the following linear partial differential equations for $\lambda(h_1, h_2, \psi)$

$$b_{11}\frac{\partial\lambda}{\partial h_1} = 2D - 2a_1, \quad b_{22}\frac{\partial\lambda}{\partial h_2} = 2D - 2a_2, \quad b_{\psi\psi}\frac{\partial\lambda}{\partial\psi} = -2a_\psi. \tag{5.26}$$

Substituting Eq. (5.23) into Eq. (5.26) yields

$$\frac{\partial\lambda}{\partial h_1} = \frac{1}{D}[-\gamma_1 + \frac{3}{2}\gamma_2 h_1 + \gamma_2(1+\frac{1}{2}\cos 2\psi)h_2],$$

$$\frac{\partial \lambda}{\partial h_2} = \frac{1}{D}[-\gamma_1 + \frac{3}{2}\gamma_2 h_2 + \gamma_2(1 + \frac{1}{2}\cos 2\psi)h_1],$$
$$\frac{\partial \lambda}{\partial \psi} = -\frac{1}{D}\gamma_2 h_1 h_2 \sin 2\psi. \tag{5.27}$$

Under the following compatibility conditions

$$\frac{\partial^2 \lambda}{\partial h_1 \partial h_2} = \frac{\partial^2 \lambda}{\partial h_2 \partial h_1}, \quad \frac{\partial^2 \lambda}{\partial h_1 \partial \psi} = \frac{\partial^2 \lambda}{\partial \psi \partial h_1}, \quad \frac{\partial^2 \lambda}{\partial \psi \partial h_2} = \frac{\partial^2 \lambda}{\partial h_2 \partial \psi}, \tag{5.28}$$

the probability potential $\lambda(h_1, h_2, \psi)$ can be obtained from Eq. (5.27) exactly. Substituting $\lambda(h_1, h_2, \psi)$ into Eq. (5.25), the exact stationary solution of FPK equation (5.20) is obtained as follows

$$p(h_1, h_2, \psi) = C\exp\{\frac{\gamma_1}{D}(h_1 + h_2) - \frac{\gamma_2}{D}[\frac{3}{4}(h_1^2 + h_2^2) + (1 + \frac{1}{2}\cos 2\psi)h_1 h_2]\}. \tag{5.29}$$

According to Eq. (5.127) of vol. 1, the joint stationary PDF $p(x_1, x_2, v_1, v_2)$ for the displacements and velocities of active Brownian particle can then be obtained from the stationary solution (5.29)

$$\begin{aligned} p(x_1, x_2, v_1, v_2) = C\exp\{&\frac{\gamma_1}{2D}[v_1^2 + v_2^2 + \omega^2(x_1^2 + x_2^2)] - \frac{\gamma_2}{16D}[3\omega^4(x_1^2 + x_2^2)^2 \\ &+ 3(v_1^2 + v_2^2)^2 + 6\omega^2(x_1 v_1 + x_2 v_2)^2 + 2\omega^2(x_1 v_2 - x_2 v_1)^2]\}. \end{aligned} \tag{5.30}$$

The stationary solution in Eq. (5.30) is the joint stationary PDF of displacements and velocities when active Brownian particle undergoes stationary motion. Stationary marginal PDFs and stationary statistics can then be obtained by integrating $p(x_1, x_2, v_1, v_2)$ with respect to x_i and/or v_i $(i = 1, 2)$.

Analytical solution (5.30) and the results from Monte Carlo simulation of original system (5.14) are shown in Figs. 5.2, 5.3, and 5.4. $D = 1, \omega = 6$. γ_1, γ_2 are the same as those in the reference (Erdmann et al., 2002), including near-equilibrium parameters $(\gamma_2 = 1, \gamma_1/\gamma_2 = -1)$ and far-from-equilibrium parameters $(\gamma_2 = 1, \gamma_1/\gamma_2 = 4)$, where near-equilibrium means that γ_1 is negative and the origin is stable while far-from-equilibrium means that γ_1 is positive and the origin is unstable when $D = 0$. The figures show that the theoretical results obtained using the stochastic averaging method are in good agreement with the simulation results.

In addition to the Rayleigh damping coefficient, the stochastic averaging method for quasi-integrable and resonant Hamiltonian systems is also applicable to other types of damping coefficients. By replacing the Rayleigh type damping coefficient in the system (5.3) with the Erdmann type damping coefficient $\alpha(x, \dot{x}) =$

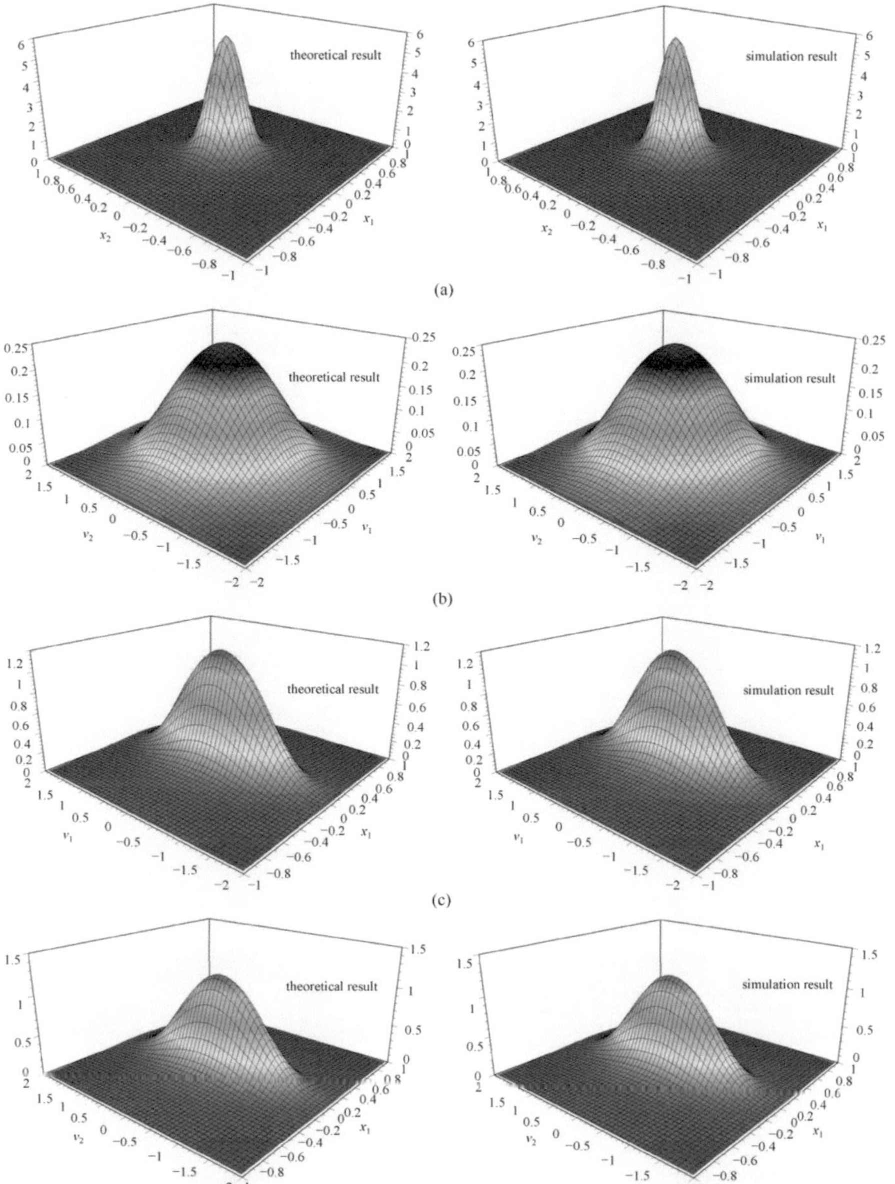

Fig. 5.2 Stationary PDF of system (5.14) with $\gamma_2 = 1$, $\gamma_1/\gamma_2 = -1$. **a** Displacements X_1, X_2; **b** velocities V_1, V_2; **c** displacement X_1 and velocity V_1; **d** displacement X_1 and velocity V_2 (Deng and Zhu 2004).

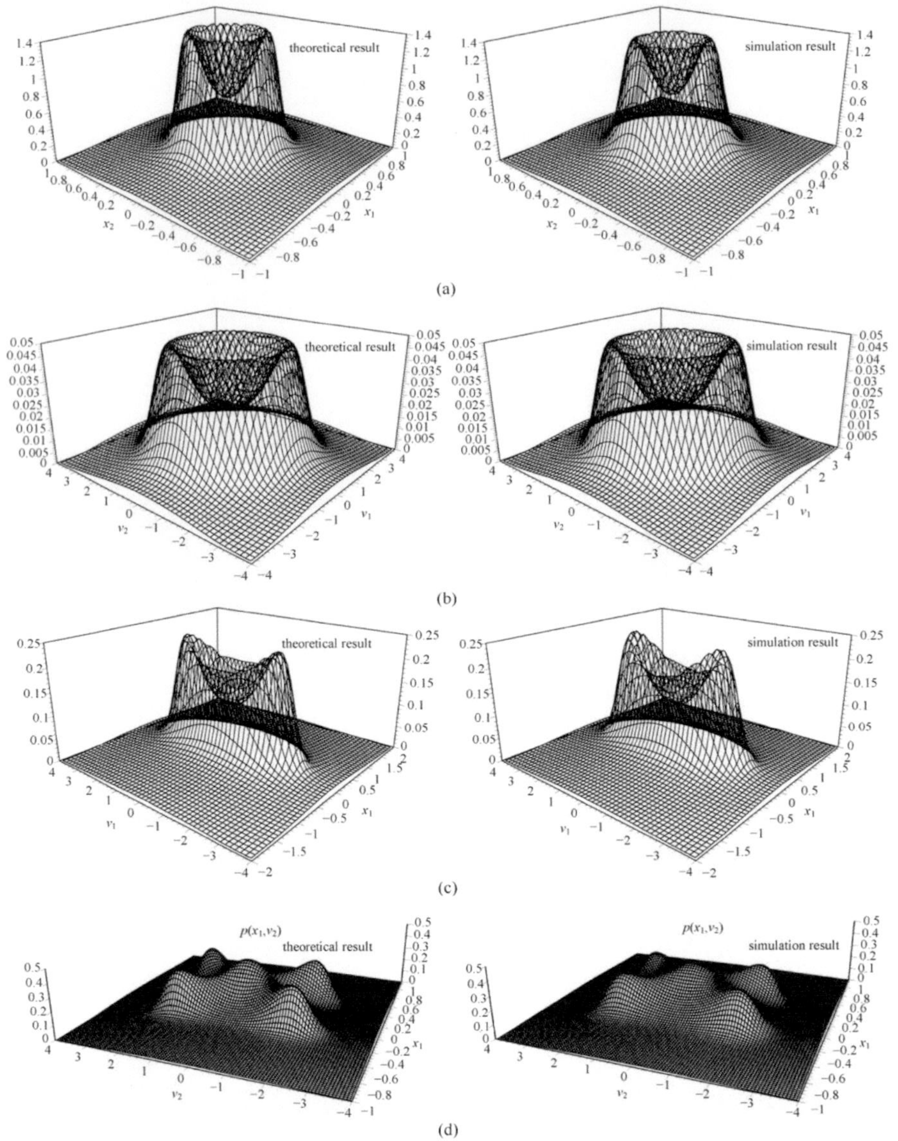

Fig. 5.3 Stationary PDF of system (5.14), $\gamma_2 = 1, \gamma_1 / \gamma_2 = 4$. **a** Displacements X_1, X_2; **b** velocities V_1, V_2; **c** displacement X_1 and velocity V_1; **d** displacement X_1 and velocity V_2 (Deng and Zhu 2004).

$\gamma_0 - d_2 q_0 / (c + d_2 \dot{x}^2)$ (Erdmann et al., 2000), a different energy management mechanism for active Brownian particle motion systems is obtained. Applying the same stochastic averaging method, the exact stationary response PDF for this system can be obtained (Deng and Zhu 2004)

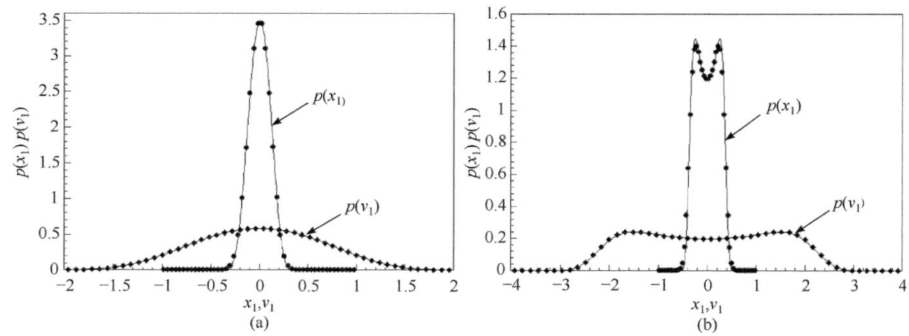

Fig. 5.4 Stationary PDF of displacement X_1 and velocity V_1 of system (5.14). Solid line —— represents results from the stochastic averaging method, symbols ●, ◆ represent simulation results. **a** $\gamma_2 = 1, \gamma_1 / \gamma_2 = -1$, **b** $\gamma_2 = 1, \gamma_1 / \gamma_2 = 4$ (Deng and Zhu 2004).

$$p(h_1, h_2, \psi) = C \exp\Big\{-\frac{\gamma_0}{D}(h_1 + h_2) + \frac{q_0}{\pi D}\int_0^{2\pi} \ln[\frac{c}{d_2} + 2h_1 \sin^2\theta + 2h_2 \sin^2(\psi + \phi)]\mathrm{d}\phi\Big\}. \tag{5.31}$$

Similarly, by replacing the damping coefficient in system (5.3) with the Schienbein-Gruler-type damping coefficient $\alpha(x, \dot{x}) = \gamma_0(1 - v_0/\dot{x})$ (Schienbein and Gruler, 1993), the same stochastic averaging method can be applied to obtain the following exact stationary PDF (Deng and Zhu, 2004)

$$p(h_1, h_2, \psi) = C \exp\left[-\frac{\gamma_0}{D}(h_1 + h_2) + \frac{\sqrt{2}v_0\gamma_0}{\pi D}\int_0^{2\pi}\sqrt{h_1 \sin^2\phi + h_2 \sin^2(\psi + \phi)\mathrm{d}\phi}\right]. \tag{5.32}$$

Using Eq. (5.127) of vol. 1, the joint stationary PDF $p(x_1, x_2, v_1, v_2)$ of the displacement and velocity of active Brownian particle and other stationary statistics can be then obtained from Eq. (5.32). Figure 5.5 shows the PDFs of cell movement velocity obtained from experimental data and solution (5.32), where experimental data comes from the results of Franke and Gruler on granulocyte motion (denoted by the symbol ●) (Franke and Gruler 1990), and the theoretical results (denoted by the solid line ——) comes from (5.32), i.e.,

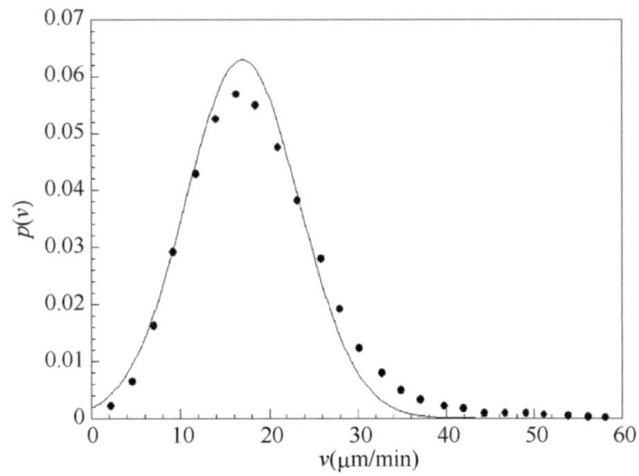

Fig. 5.5 Stationary PDF of particle velocity for system (5.14) with Schienbein-Gruler-type damping coefficients, $\gamma_0/2D = 81\mu m/\min$, and $v_0 = 17\mu m/\min$. Solid line —— represents results from the stochastic averaging method, and the symbol ● represents the experimental results (Deng and Zhu 2004)

$$p(v) = \frac{\mathrm{d}}{\mathrm{d}v}\left\{ \iint\limits_{\sqrt{v_1^2+v_2^2}<v} \left[\iint\limits_{-\infty}^{\infty} p(h_1, h_2, \psi)\big|_{\substack{h_1=(v_1^2+\omega^2x_1^2)/2 \\ h_2=(v_2^2+\omega^2x_2^2)/2 \\ \psi=\tan^{-1}(v_2^2/\omega x_2)-\tan^{-1}(v_1^2/\omega x_1)}} \mathrm{d}x_1\mathrm{d}x_2 \right] \mathrm{d}v_1\mathrm{d}v_2 \right\}. \tag{5.33}$$

The experimental data verified the applicability of the analytical solution (5.32).

(ii) **The case of elliptical harmonic potential**

Consider the plane motion of a Brownian particle with elliptical harmonic potential $U(x_1, x_2) = (\omega_1^2 x_1^2 + \omega_2^2 x_2^2)/2$, damping coefficient (5.2), and under Gaussian white noise excitations. The motion equations of the active Brownian particle is (Deng and Zhu 2004)

$$\begin{aligned} &\dot{X}_1 = V_1, \quad \dot{V}_1 + [-\gamma_1 + \gamma_2(V_1^2 + V_2^2)]V_1 + \omega_1^2 X_1 = \sqrt{2D}W_{g1}(t), \\ &\dot{X}_2 = V_2, \quad \dot{V}_2 + [-\gamma_1 + \gamma_2(V_1^2 + V_2^2)]V_2 + \omega_2^2 X_2 = \sqrt{2D}W_{g2}(t), \end{aligned} \tag{5.34}$$

where ω_1 and ω_2 do not satisfy the internal resonance relation in Eq. (5.75) of vol. 1; $W_{g1}(t)$ and $W_{g2}(t)$ are independent Gaussian white noises; $2D$ is the excitation intensity; γ_1, γ_2, D are small quantities of same order.

Let $Q_1 = X_1,\ Q_2 = X_2, P_1 = \dot{X}_1,\ P_2 = \dot{X}_2$. Eq. (5.34) can be transformed into the following Itô stochastic differential equation

$$\begin{aligned} &\mathrm{d}Q_1 = P_1\mathrm{d}t, \quad \mathrm{d}P_1 = -\{\omega_1^2 Q_1 + [-\gamma_1 + \gamma_2(P_1^2 + P_2^2)]P_1\}\mathrm{d}t + \sqrt{2D}\mathrm{d}B_1(t), \\ &\mathrm{d}Q_2 = P_2\mathrm{d}t, \quad \mathrm{d}P_2 = -\{\omega_2^2 Q_2 + [-\gamma_1 + \gamma_2(P_1^2 + P_2^2)]P_2\}\mathrm{d}t + \sqrt{2D}\mathrm{d}B_2(t), \end{aligned} \tag{5.35}$$

where $B_1(t)$ and $B_2(t)$ are independent unit Wiener processes. The Hamiltonian system associated with system (5.34) has the two independent and involutive motion integrals

$$H_1 = \frac{1}{2}P_1^2 + \frac{1}{2}\omega_1^2 Q_1^2, \quad H_2 = \frac{1}{2}P_2^2 + \frac{1}{2}\omega_2^2 Q_2^2. \tag{5.36}$$

Using the Itô differential rule, the following Itô stochastic differential equation for H_1, H_2 can be obtained from Eqs. (5.35) and (5.36)

$$\begin{aligned} \mathrm{d}H_1 &= \{-[-\gamma_1 + \gamma_2(P_1^2 + P_2^2)]P_1^2 + D\}\mathrm{d}t + \sqrt{2D}P_1\mathrm{d}B_1(t), \\ \mathrm{d}H_2 &= \{-[-\gamma_1 + \gamma_2(P_1^2 + P_2^2)]P_2^2 + D\}\mathrm{d}t + \sqrt{2D}P_2\mathrm{d}B_2(t). \end{aligned} \tag{5.37}$$

By applying the stochastic averaging method for quasi-integrable and non-resonant Hamiltonian systems presented in Sect. 5.2.1, the following reduced averaged FPK equation can be obtained

$$\begin{aligned} 0 = &-\frac{\partial}{\partial h_1}(a_1 p) - \frac{\partial}{\partial h_2}(a_2 p) + \frac{1}{2}\frac{\partial^2}{\partial h_1^2}(b_{11}p) + \frac{1}{2}\frac{\partial^2}{\partial h_2^2}(b_{22}p) \\ &+ \frac{1}{2}\frac{\partial^2}{\partial h_1 \partial h_2}(b_{12}p) + \frac{1}{2}\frac{\partial^2}{\partial h_2 \partial h_1}(b_{21}p), \end{aligned} \tag{5.38}$$

where the first and second derivative moments are given

$$\begin{aligned} a_1 &= \left\langle -[-\gamma_1 + \gamma_2(p_1^2 + p_2^2)]p_1^2 + D \right\rangle_t, \quad b_{11} = \left\langle 2Dp_1^2 \right\rangle_t, \quad b_{12} = 0, \\ a_2 &= \left\langle -[-\gamma_1 + \gamma_2(p_1^2 + p_2^2)]p_2^2 + D \right\rangle_t, \quad b_{22} = \left\langle 2Dp_2^2 \right\rangle_t, \quad b_{21} = 0. \end{aligned} \tag{5.39}$$

In Eq. (5.39), p_1, p_2 need to be replaced by $h_1, h_2, \theta_1, \theta_2$ according to Eq. (5.17), the time averaging $\langle[\cdot]\rangle_t$ can be replaced by space averaging with respective to θ_1 and θ_2

$$\langle[\cdot]\rangle_t = \frac{1}{4\pi^2}\int_0^{2\pi}\int_0^{2\pi}[\cdot]\Big|_{q_1=\frac{\sqrt{2h_1}}{\omega_1}\cos\theta_1,\ q_2=\frac{\sqrt{2h_2}}{\omega_2}\cos\theta_2}\mathrm{d}\theta_1\mathrm{d}\theta_2. \tag{5.40}$$

Completing the averaging in Eq. (5.39) yields the first and second derivative moments

$$\begin{aligned} a_1 &= \gamma_1 h_1 - \frac{3}{2}\gamma_2 h_1^2 - \gamma_2 h_1 h_2 + D, \quad b_{11} = 2Dh_1, \\ a_2 &= \gamma_1 h_2 - \frac{3}{2}\gamma_2 h_2^2 - \gamma_2 h_2 h_1 + D, \quad b_{22} = 2Dh_2. \end{aligned} \tag{5.41}$$

Similar to the derivation from Eq. (5.24) to Eq. (5.29), the exact stationary solution for the reduced averaged FPK equation (5.38) can be obtained as follows

$$p(h_1, h_2) = C\exp\left\{\frac{\gamma_1}{D}(h_1 + h_2) - \frac{\gamma_2}{D}\left[\frac{3}{4}(h_1^2 + h_2^2) + h_1 h_2\right]\right\}. \tag{5.42}$$

According to Eq. (5.106) in vol. 1, the joint stationary PDF of displacements and velocities of the Brownian particle can be derived from Eq. (5.42) as follows

$$\begin{aligned} p(x_1, x_2, v_1, v_2) = C\exp\Big\{&\frac{\gamma_1}{2D}(v_1^2 + v_2^2 + \omega_1^2 x_1^2 + \omega_2^2 x_2^2) - \frac{\gamma_2}{16D}\big[3(\omega_1^2 x_1^2 + \omega_2^2 x_2^2)^2 \\ &+ 3(v_1^2 + v_2^2)^2 + 6(\omega_1 x_1 v_1 + \omega_2 x_2 v_2)^2 \\ &+ 4(\omega_1 x_1 v_2 - \omega_2 x_2 v_1)^2 - 2(\omega_1\omega_2 x_1 x_2 + v_1 v_2)^2\big]\Big\}. \end{aligned} \tag{5.43}$$

Stationary marginal PDFs and statistics can then be obtained from $p(x_1, x_2, v_1, v_2)$.

Figures 5.6, 5.7 show a comparison between the analytical solution (5.43) and the results from Monte Carlo simulation of the original system (5.34) with parameters $\gamma_1 = 0.2$, $\gamma_2 = 0.1$, $\omega_1 = 1.414$, $\gamma_2 = 2.236$, $D = 0.02$. It can be observed that the analytical solution agrees well with the simulation results, demonstrating validity of the stochastic averaging method.

(iii) **The case of non-harmonic potential**

Erdmann et al. (2000) introduced the following centrally symmetric quartic potential into the motion of active Brownian particle

$$U(r) = \frac{1}{4}kr^4, \quad r = \sqrt{x_1^2 + x_2^2}. \tag{5.44}$$

If use Eq. (5.44) to replace Eq. (5.1), the solution form of Eq. (5.3) is the same as the solutions (5.7) and (5.10). However, here the radius r_0 of the limit cycles and the angular frequency ω_0 of particle rotation are given by

$$r_0 = v_0^{1/2}\Big/k^{1/4}, \quad \omega_0 = v_0^{1/2}k^{1/4}. \tag{5.45}$$

Consider an active Brownian particle with Rayleigh damping coefficients in Eq. (5.2) and under weak Gaussian white noise excitations moving in a quartic potential landscape of Eq. (5.44). The motion is governed by the following nonlinear stochastic dynamical equations (Deng 2005)

$$\begin{aligned} &\dot{X}_1 = V_1, \quad \dot{V}_1 + [-\gamma_1 + \gamma_2(V_1^2 + V_2^2)]V_1 + k\big(X_1^2 + X_2^2\big)X_1 = \sqrt{2D}W_{g1}(t), \\ &\dot{X}_2 = V_2, \quad \dot{V}_2 + [-\gamma_1 + \gamma_2(V_1^2 + V_2^2)]V_2 + k\big(X_1^2 + X_2^2\big)X_2 = \sqrt{2D}W_{g2}(t), \end{aligned} \tag{5.46}$$

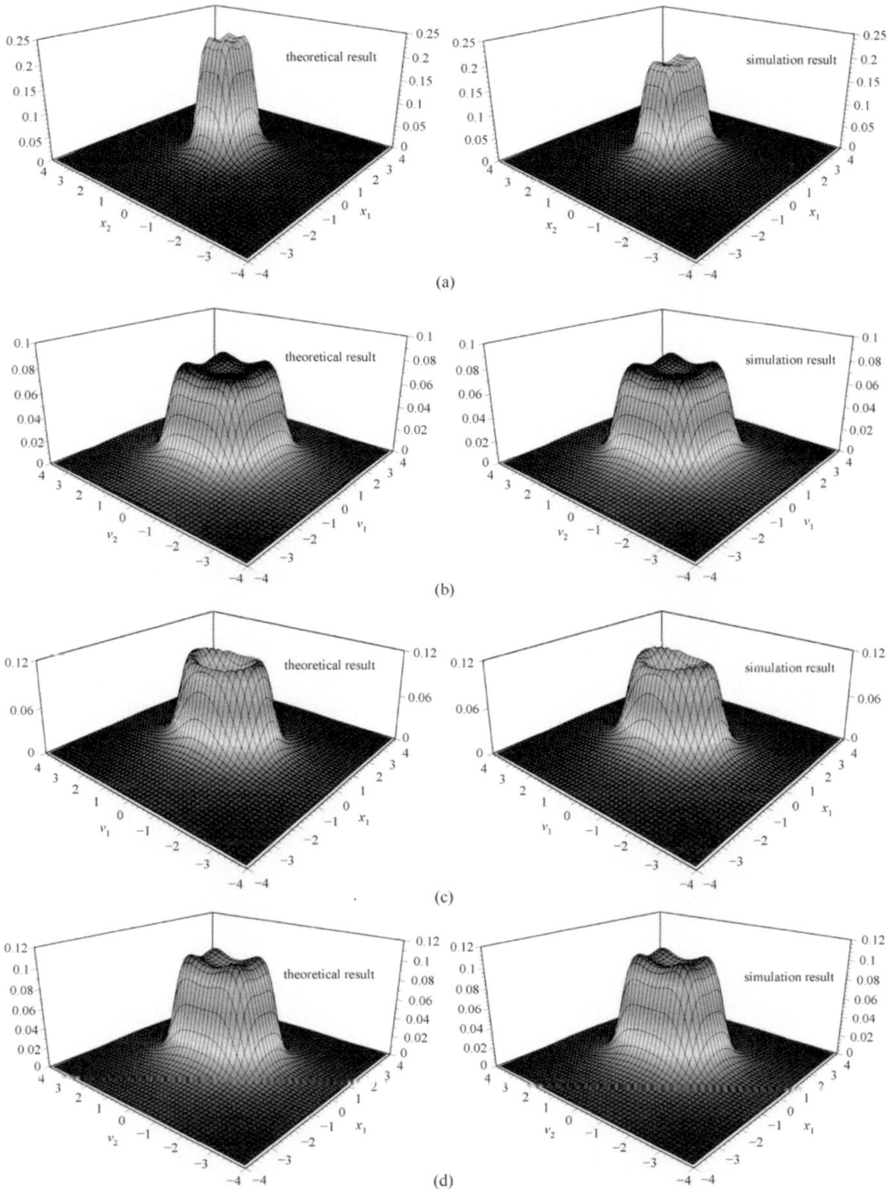

Fig. 5.6 Stationary joint PDF of system (5.34). **a** Displacements X_1, X_2; **b** velocities V_1, V_2; **c** displacement X_1 and velocity V_1; **d** displacement X_1 and velocity V_2 (Deng and Zhu 2004).

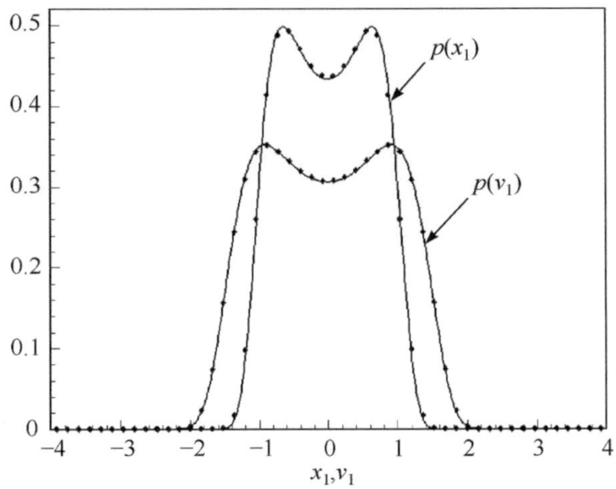

Fig. 5.7 Stationary PDF of displacement X_1 and velocity V_1 of system (5.34). Solid line —— represents analytical results, and symbols ●, ◆ represent simulation results (Deng and Zhu 2004).

where $W_{g1}(t)$ and $W_{g2}(t)$ are independent unit Gaussian white noises, $2D$ are the excitation intensity, and γ_1, γ_2, D are small quantities of same order. By introducing the transformation $Q_1 = X_1,\ Q_2 = X_2, P_1 = \dot{X}_1,\ P_2 = \dot{X}_2$, Eq. (5.46) can be converted into the following Itô stochastic differential equation

$$\begin{aligned} \mathrm{d}Q_1 &= P_1\mathrm{d}t,\quad \mathrm{d}P_1 = -\{k(Q_1^2+Q_2^2)Q_1 + [-\gamma_1+\gamma_2(P_1^2+P_2^2)]P_1\}\mathrm{d}t + \sqrt{2D}\mathrm{d}B_1(t), \\ \mathrm{d}Q_2 &= P_2\mathrm{d}t,\quad \mathrm{d}P_2 = -\{k(Q_1^2+Q_2^2)Q_2 + [-\gamma_1+\gamma_2(P_1^2+P_2^2)]P_2\}\mathrm{d}t + \sqrt{2D}\mathrm{d}B_2(t), \end{aligned} \tag{5.47}$$

where $B_1(t)$ and $B_2(t)$ are independent unit Wiener processes. The Hamiltonian system associated with Eq. (5.47) has only one independent first integral

$$H = \frac{1}{2}(P_1^2+P_2^2) + \frac{1}{4}k(Q_1^2+Q_2^2)^2, \tag{5.48}$$

Eq. (5.47) is a quasi-nonintegrable Hamiltonian system. By applying the stochastic averaging method for quasi-nonintegrable Hamiltonian systems described in Sect. 5.1, $H(t)$ converges weakly to a one-dimensional Markov diffusion process. The averaged Itô stochastic differential equation is

$$\mathrm{d}H = m(H)\mathrm{d}t + \sigma(H)\mathrm{d}B(t). \tag{5.49}$$

The reduced averaged FPK equation is

$$0 = -\frac{\partial}{\partial h}[m(h)p(h)] + \frac{1}{2}\frac{\partial}{\partial h^2}[\sigma^2(h)p(h)]. \tag{5.50}$$

According to Eqs. (5.10) to (5.13) of vol. 1, the first and second derivative moments can be obtained as

$$
\begin{aligned}
m(h) &= \frac{1}{T(h)} \int_{\Omega} \left(\{ -[-\gamma_1 + \gamma_2(p_1^2 + p_2^2)](p_1^2 + p_2^2) + 2D \} / |p_1| \right) \mathrm{d}q_1 \mathrm{d}q_2 \mathrm{d}p_2, \\
\sigma^2(h) &= \frac{1}{T(h)} \int_{\Omega} \left(2D(p_1^2 + p_2^2) / |p_1| \right) \mathrm{d}q_1 \mathrm{d}q_2 \mathrm{d}p_2, \\
T(h) &= \frac{1}{T(h)} \int_{\Omega} \left(1 / |p_1| \right) \mathrm{d}q_1 \mathrm{d}q_2 \mathrm{d}p_2, \quad \Omega = \{ (q_1, q_2, p_2) | p_2^2 / 2 + U(q_1, q_2) \leq h \},
\end{aligned}
\tag{5.51}
$$

where $p_1 = (2h - 2U - p_2^2)^{1/2}$. Transforming to polar coordinate and completing the integral in Eq. (5.51) yield

$$
m(h) = 2D + \frac{4}{3}\gamma_1 h - \frac{32}{15}\gamma_2 h^2, \quad \sigma^2(h) = \frac{8}{3} D h. \tag{5.52}
$$

The exact solution of the reduced averaged FPK equation (5.50) can be obtained as follows

$$
p(h) = C \exp\left(\frac{1}{2} \ln h + \frac{\gamma_1}{D} h - \frac{4\gamma_2}{5D} h^2 \right). \tag{5.53}
$$

Figure 5.8 shows the stationary probability functions of system energy for different excitation intensities. The theoretical results are from Eq. (5.53) and the Monte Carlo simulation results are for the original system (5.46). It can be observed that they match well, confirming validity of the stochastic averaging method.

(iv) **The case of damping coefficient depending on both of displacement and velocity**

During the movement of biological organisms, the opportunity to absorb energy within the motion area is not uniform. In some regions, an abundance of food can lead to a higher rate of energy absorption, while in other areas, a lack of food may result in less or no energy absorption. To describe this situation, a damping coefficient that depends on both displacement and velocity is introduced as follows

$$
\alpha(x, \dot{x}) = -\gamma_1 + \gamma_2 \dot{x}^2 + \gamma_3 x^2. \tag{5.54}
$$

Consider an active Brownian particle with quartic potential (5.44) and damping coefficients given by Eq. (5.54) subjected to weak Gaussian white noise excitation. The motion of the active Brownian particle is governed by the following nonlinear stochastic dynamical equations (Deng 2005)

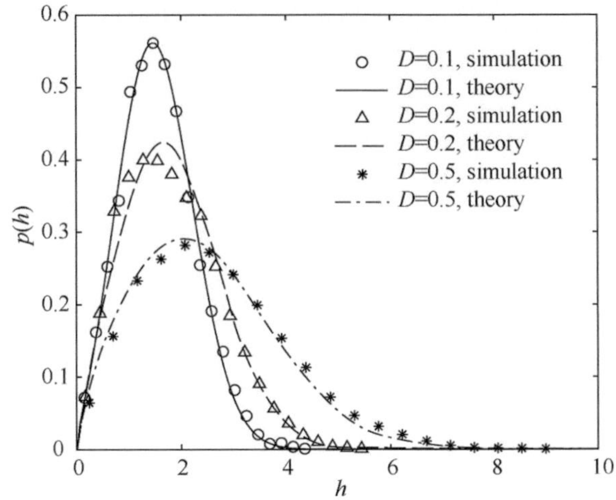

Fig. 5.8 Stationary PDF of energy of system (5.46) with parameters $\gamma_1 = 0.2$, $\gamma_2 = 0.1, k = 5$, where theoretical results are obtained from analytical solution (5.53).

$$
\begin{aligned}
\dot{X}_1 &= V_1, \dot{V}_1 + [-\gamma_1 + \gamma_2(V_1^2 + V_2^2) + \gamma_3(X_1^2 + X_2^2)]V_1 + k\left(X_1^2 + X_2^2\right)X_1 \\
&= \sqrt{2D}W_{g1}(t), \\
\dot{X}_2 &= V_2, \dot{V}_2 + [-\gamma_1 + \gamma_2(V_1^2 + V_2^2) + \gamma_3(X_1^2 + X_2^2)]V_2 + k\left(X_1^2 + X_2^2\right)X_2 \\
&= \sqrt{2D}W_{g2}(t).
\end{aligned} \tag{5.55}
$$

where $W_{g1}(t)$ and $W_{g2}(t)$ are independent unit Gaussian white noises; $2D$ is the excitation intensity; $\gamma_1, \gamma_2, \gamma_3, D$ are small quantities of same order.

Let $Q_1 = X_1,\ Q_2 = X_2, P_1 = \dot{X}_1,\ P_2 = \dot{X}_2$. Eq. (5.55) is converted into the following Itô stochastic differential equations

$$
\begin{aligned}
\mathrm{d}Q_1 &= P_1\mathrm{d}t, \\
\mathrm{d}P_1 &= -\{k(Q_1^2 + Q_2^2)Q_1 + [-\gamma_1 + \gamma_2(P_1^2 + P_2^2) + \gamma_3(Q_1^2 + Q_2^2)]P_1\}\mathrm{d}t + \sqrt{2D}\mathrm{d}B_1(t), \\
\mathrm{d}Q_2 &= P_2\mathrm{d}t, \\
\mathrm{d}P_2 &= -\{k(Q_1^2 + Q_2^2)Q_2 + [-\gamma_1 + \gamma_2(P_1^2 + P_2^2) + \gamma_3(Q_1^2 + Q_2^2)]P_2\}\mathrm{d}t + \sqrt{2D}\mathrm{d}B_2(t).
\end{aligned} \tag{5.56}
$$

where $B_1(t)$ and $B_2(t)$ are independent unit Wiener processes. The Hamiltonian system associated with Eq. (5.56) is non-integrable and has Hamiltonian function H in Eq. (5.48). Applying the stochastic averaging method for quasi-non-integrable Hamiltonian systems described in Sect. 5.1 leads to the following reduced averaged FPK equation

$$
0 = -\frac{\partial}{\partial h}\left[m(h)p\right] + \frac{1}{2}\frac{\partial}{\partial h^2}\left[\sigma^2(h)p\right]. \tag{5.57}
$$

where $p = p(h)$ is the stationary probability density. According to Eqs. (5.10)–(5.13) of vol. 1, the first and second derivative moments are

$$m(h) = \frac{1}{T(h)} \int_{\Omega} \left(\{-[-\gamma_1 + \gamma_2(p_1^2 + p_2^2) + \gamma_3(q_1^2 + q_2^2)](p_1^2 + p_2^2) + 2D\} / |p_1| \right) \mathrm{d}q_1 \mathrm{d}q_2 \mathrm{d}p_2,$$

$$\sigma^2(h) = \frac{1}{T(h)} \int_{\Omega} \left(2D(p_1^2 + p_2^2) / |p_1| \right) \mathrm{d}q \mathrm{d}q_2 \mathrm{d}p_2,$$

$$T(h) = \frac{1}{T(h)} \int_{\Omega} \left(1 / |p_1| \right) \mathrm{d}q_1 \mathrm{d}q_2 \mathrm{d}p_2, \quad \Omega = \{(q_1, q_2, p_2) | p_2^2/2 + U(q_1, q_2) \le h\}. \tag{5.58}$$

where $p_1 = (2h - 2U - p_2^2)^{1/2}$. After transforming into polar coordinate and completing the integral in Eq. (5.58), the following first and second derivative moments are obtained

$$m(h) = 2D + \frac{4}{3}\gamma_1 h - \frac{32}{15}\gamma_2 h^2 - \gamma_3 h \sqrt{\frac{h}{k}}, \quad \sigma^2(h) = \frac{8}{3} Dh. \tag{5.59}$$

Then the solution of Eq. (5.57) is

$$p(h) = C \exp\left(\frac{1}{2} \ln h + \frac{\gamma_1}{D} h - \frac{4\gamma_2}{5D} h^2 - \frac{\gamma_3}{2D} h \sqrt{\frac{h}{k}} \right). \tag{5.60}$$

Figure 5.9 show the analytical results from Eq. (5.60) and Monte Carlo simulation results from original system (5.55). The good agreement of the two results verified the stochastic averaging method.

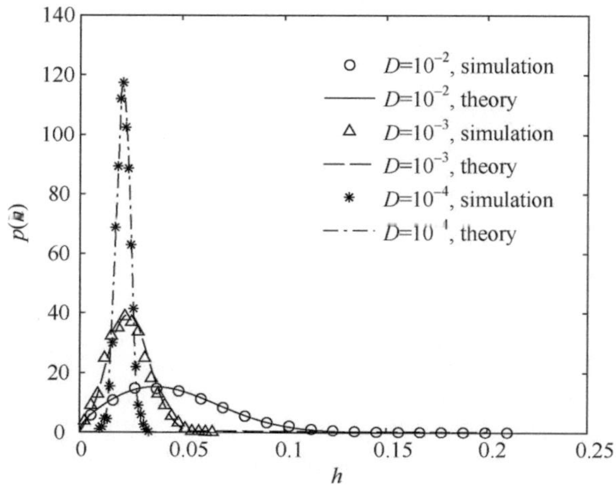

Fig. 5.9 Stationary PDF of energy H of system (5.55) with parameters $\gamma_1 = 0.2$, $\gamma_2 = 4.7$, $\gamma_3 = 0.4$, $k = 1$, where theoretical results are obtained from the analytical solution (5.60).

5.1.3 Random Swarm Motion of Active Brownian Particles

Understanding and studying the swarm motion of biological population are significant since the movement of lower single-celled organisms and animals like birds and insects often manifests as swarm motion. Extending the model of single active Brownian particle to a swarm model is a common approach in studying swarm motion (Czirok and Vicsek 2000; Mikhailov and Zanette 1999). Individuals in a swarm of Brownian particles are coupled through local or global connections, which can simultaneously affect individual and swarm motion.

By grouping many active Brownian particles together and introducing coupling mechanisms between particles, a model for studying swarm motion of Brownian particle swarm can be obtained. In this section, a coupling mechanism with respect to the mass center of swarm is introduced. For a swarm of n particles moving in a two-dimensional plane, the particle displacements are (x_{1i}, x_{2i}) $i = 1, 2, \ldots, n$ and the mass center of swarm is (C_1, C_2), where

$$C_1 = \frac{1}{n}\sum_{i=1}^{n} x_{1i}, \quad C_2 = \frac{1}{n}\sum_{i=1}^{n} x_{2i}. \tag{5.61}$$

Assume that each particle having Rayleigh type energy management mechanism in Eq. (5.2) and parabolic harmonic potential in Eq. (5.1) is subjected to independent Gaussian white noise excitations The coupling between particles is achieved through the mass center of swarm. The stochastic motion of the Brownian particle swarm can be described by the following equations (Zhu and Deng 2005)

$$\begin{aligned}
&\dot{X}_{1i} = V_{1i},\\
&\dot{V}_{1i} + [-\gamma_1 + \gamma_2(V_{1i}^2 + V_{2i}^2)]V_{1i} + \omega^2\left(X_{1i} - \frac{1}{n}\sum_{j=1}^{n} X_{1j}\right) = \sqrt{2D}W_{g1i}(t),\\
&\dot{X}_{2i} = V_{2i},\\
&\dot{V}_{2i} + [-\gamma_1 + \gamma_2(V_{1i}^2 + V_{2i}^2)]V_{2i} + \omega^2\left(X_{2i} - \frac{1}{n}\sum_{j=1}^{n} X_{2j}\right) = \sqrt{2D}W_{g2i}(t),\\
&i = 1, 2, \cdots, n.
\end{aligned} \tag{5.62}$$

The swarm motion of particles is generally divided into two kinds of motions, i.e., the motion of the mass center and the motion of individual particles relative to the mass center. Several quantities, including the displacement amplitude square $R^2(t)$ of mass center, the velocity square $V^2(t)$ of mass center, and the average $v^2(t)$ of particle velocity square, have been introduced as follows

$$R^2(t) = (\frac{1}{n}\sum_{j=1}^{n} X_{1j})^2 + (\frac{1}{n}\sum_{j=1}^{n} X_{2j})^2, \quad V^2(t) = (\frac{1}{n}\sum_{j=1}^{n} V_{1j})^2 + (\frac{1}{n}\sum_{j=1}^{n} V_{2j})^2,$$

$$v^2(t) = \frac{1}{n}\sum_{j=1}^{n} (V_{1j}^2 + V_{2j}^2). \tag{5.63}$$

Monte Carlo simulation were conducted for the motion of 20,000 active Brownian particles with parameters $\gamma_1 = 0.2$, $\gamma_2 = 0.1$, $\omega = 2$, $D = 1 \times 10^{-4}$, starting from (x_1, x_2, v_1, v_2)=(0, 0, 0, 0). Figure 5.10 illustrates that after a sufficient long time, the displacement amplitude square $R^2(t)$ of mass center for both cases reaches a constant value. Figure 5.11 shows that the velocity square $V^2(t)$ of mass center approaches zero gradually, and the velocity mean-square $v^2(t)$ of particles tend toward $v_0^2 = 2$. According to $R^2(t) \approx$ const and $V^2(t) \approx 0$, it is known that the limit motion of mass center is a slowly and randomly motion in a fixed distance from its initial position. When studying the stationary swarm motion of Brownian particles, the location of mass center can be regarded as the original point. Therefore, the swarm motion of particles relative to the mass center can be studied by using the dynamics of single Brownian particle, and many results obtained from the dynamics of single particle can be applied to swarm dynamics.

Figure 5.12a, b represent the projection of swarm motion states at four time moments onto subspace (x_1, x_2, v_2) and the plane (x_1, x_2) in coordinates relative to mass center. It is seen that after transient motion, the particles ultimately form two limit cycles similar to the crossed limit cycles formed by a single Brownian particle in 3-dimensional phase space (see Fig. 5.1a). Figure 5.13 indicates that with increasing excitation intensity, the limit cycles gradually become diffused limit cycles. In 2-dimensional plane (x_1, v_2), , two cross line segments (see Fig. 5.1c) become two cross diffused bars (see Fig. 5.14). Using the analytical solution in Eq. (5.30) obtained in the case of single Brownian particle, the PDF of the diffused limit

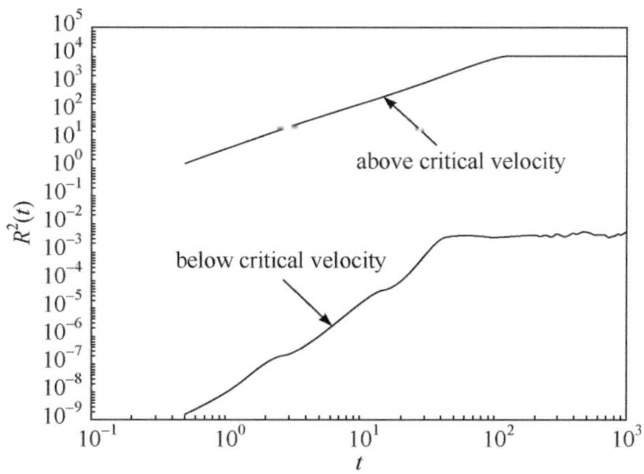

Fig. 5.10 Square amplitude $R^2(t)$ of mass center displacement as function of time t for system (5.62) (Zhu and Deng 2005).

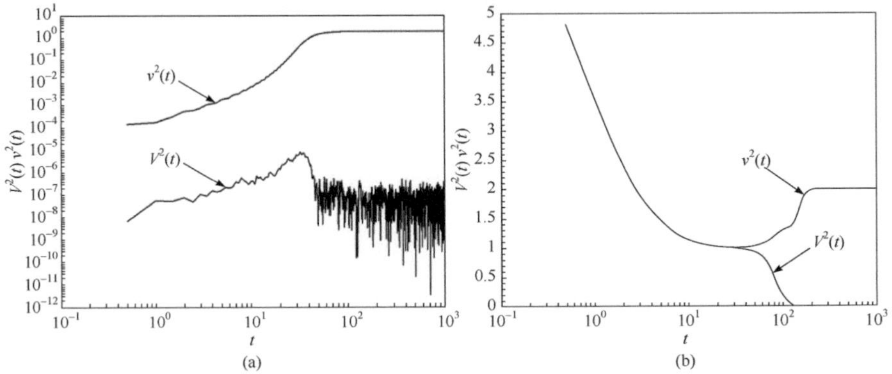

Fig. 5.11 Velocity square $V^2(t)$ of mass center and average of particle velocity square as functions of time t for swarm system (5.62). **a** Initial velocities are below the critical velocity v_0, **b** Initial velocities are above the critical velocity v_0 (Zhu and Deng 2005)

cycle in Figs. 5.13 and 5.14 can be obtained. As the excitation intensity increases, the probability for particles transition between these two limit cycles increases, leading to the gradual blurring boundary between the two limit cycles.

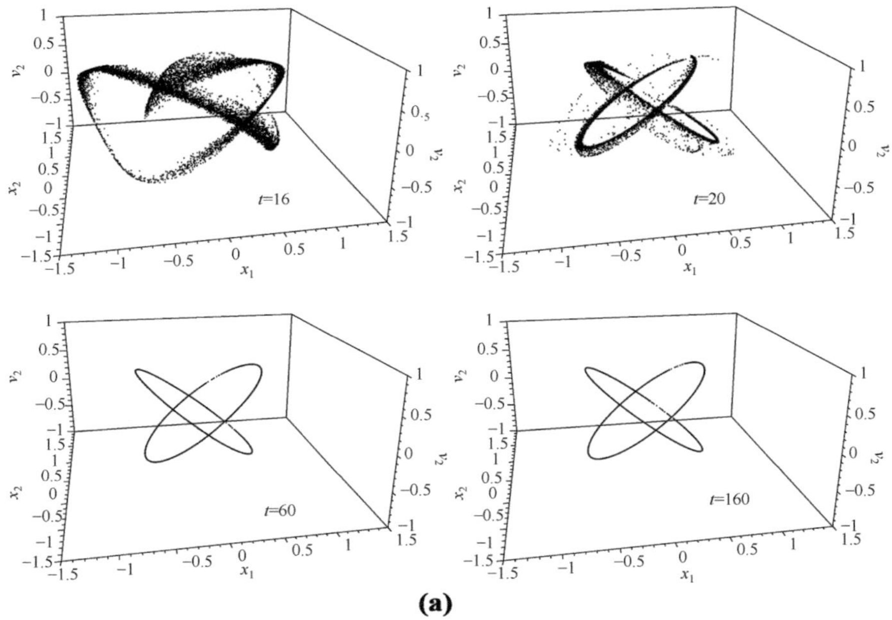

Fig. 5.12 **a** Projection of swarm motion trajectories in the subspace $\{x_1, x_2, v_2\}$ for a swarm of 10,000 Brownian particles (Zhu and Deng 2005). **b** Projection of swarm motion trajectories in the plane $\{x_1, x_2\}$ for a swarm of 10,000 Brownian particles (Zhu and Deng 2005).

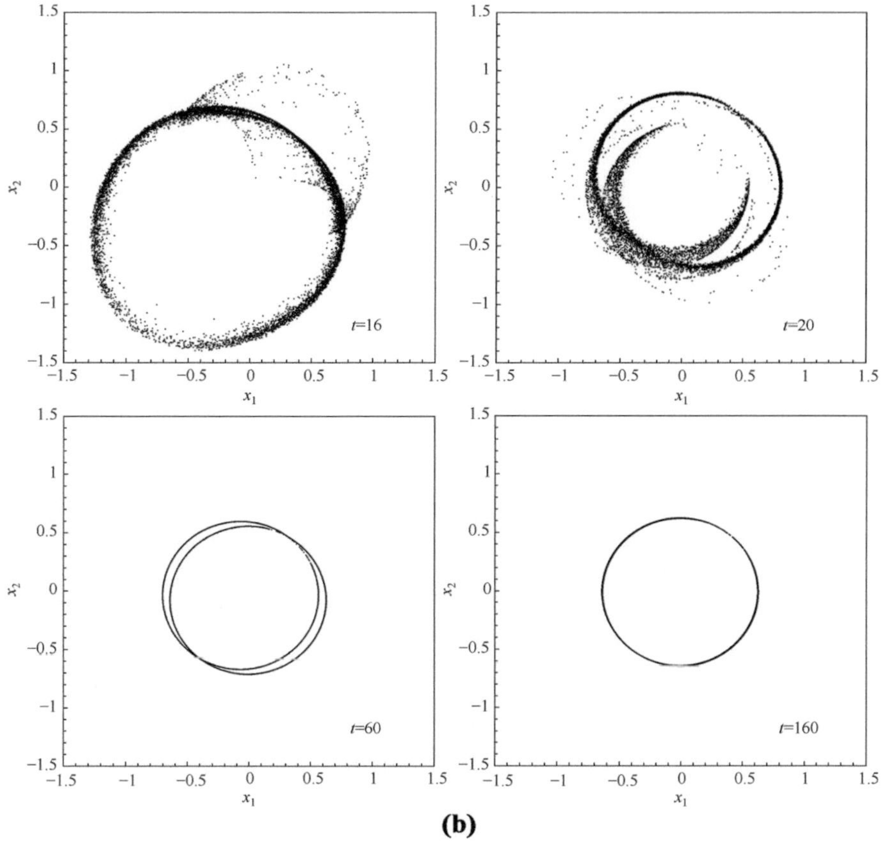

Fig. 5.12 (continued)

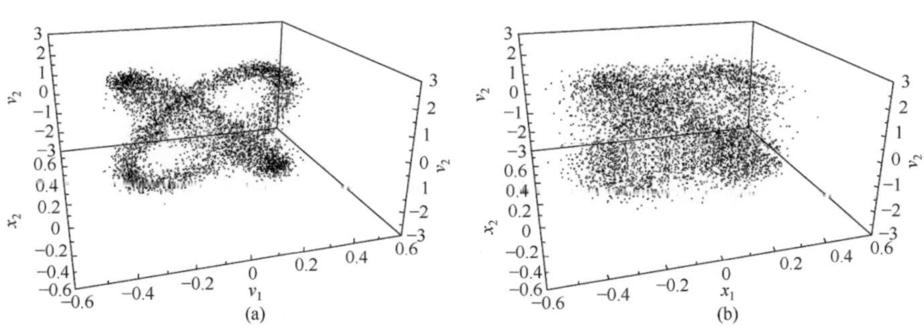

Fig. 5.13 Projection of stationary swarm motion trajectories in the subspace $\{x_1, x_2, v_2\}$ for a swarm of 5000 Brownian particles, $\gamma_1 = 4$, $\gamma_2 = 1$, $\omega = 6$, **a** $D = 0.2$; **b** $D = 0.8$ (Zhu and Deng 2005).

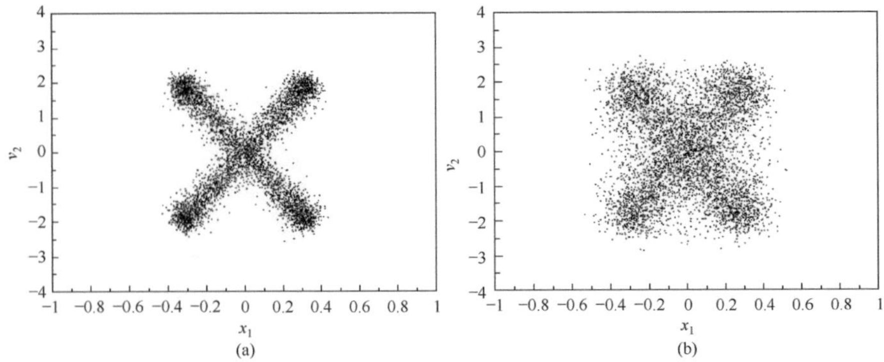

Fig. 5.14 Projection of stationary swarm motion trajectories in the plane $\{x_1, v_2\}$ for a swarm of 5000 Brownian particles, $\gamma_1 = 4$, $\gamma_2 = 1$, $\omega = 6$, **a** $D = 0.2$; **b** $D = 0.8$ (Zhu and Deng 2005).

Generally, research focus on the stationary response of total energy E, angular momentum L and angular position $\psi = \tan^{-1}(x_2/x_1)$ of the particle swarm (Ebeling and Röpke 2004). The stationary PDF of E, L, ψ can be derived from the exact solution in Eq. (5.29) for a single Brownian particle. For example, the stationary PDF of total energy of a Brownian particle can be obtained as

$$p(E^{(1)}) = \frac{\mathrm{d}}{\mathrm{d}E}\left\{ \iint\limits_{h_1+h_2\le E^{(1)}} \left[\int_0^{2\pi} p(h_1, h_2, \psi)\mathrm{d}\psi \right] \mathrm{d}h_1 \mathrm{d}h_2 \right\}. \tag{5.64}$$

For a swarm of n Brownian particles, in coordinates relative to the mass center, the energies of all particles are independently and identically distributed. The joint PDF for the energies of n Brownian particles can be obtained as follows

$$p(E^{(1)}, E^{(2)}, \ldots, E^{(n)}) = \prod_{i=1}^{n} p(E^{(i)}) \tag{5.65}$$

where $p(E^{(i)})$ is the PDF of energy of the i-th particle which is similar to Eq. (5.64). The PDF of total energy $E = \Sigma_{i=1}^{n} E^{(i)}$ of particle swarm can be obtained as follows

$$p(E) = \frac{\mathrm{d}}{\mathrm{d}E}\left\{ \overbrace{\int \cdots \int}^{n}_{\sum_{i=1}^{n} E^{(i)} \le E} \left[\prod_{i=1}^{n} p(E^{(i)}) \right] \mathrm{d}E^{(1)} \mathrm{d}E^{(2)} \cdots \mathrm{d}E^{(n)} \right\}. \tag{5.66}$$

For larger swarm of particle, it is difficult to complete the multiple integrals in Eq. (5.66). In such cases, the central limit theorem can be applied. When particle number $n \to \infty$, the total energy E follows a normal distribution $N(n\mu, n\sigma^2)$, i.e.,

$$p(E) = \frac{1}{\sqrt{2\pi n \sigma}} \exp[-\frac{(E - n\mu)^2}{2n\sigma^2}]. \tag{5.67}$$

where the mean value μ and variance σ^2 can be obtained from the PDF in Eq. (5.64).

For given parameters $\gamma_1 = 0.2$, $\omega = 2$ and $D = 0.05$, Figure 5.15 shows the PDFs $p(E)$ of the total energies of 1, 2 and 10 active Brownian particle swarms, respectively, where solid lines —— representing results from analytical solutions (5.64), (5.66), and (5.67), and symbols ● ◆ ▲ representing results from Monte Carlo simulation of system (5.62). It can be observed that the theoretical method can effectively predict the stationary PDF of total energy of active Brownian particle swarm.

The angular momentum L of a Brownian particle can be expressed as independent integrals of H_1, H_2 and ψ

$$L = \dot{x}_1 x_2 - x_1 \dot{x}_2 = \frac{2\sin\psi}{\omega}(H_1 H_2)^{1/2}. \tag{5.68}$$

The stationary PDF $p(L)$ of angular momentum can be obtained from the following expression

$$p(L) = \frac{\mathrm{d}}{\mathrm{d}L}[\iiint\limits_{\frac{2\sin\psi}{\omega}(h_1 h_2)^{1/2} \le L} p(h_1, h_2, \psi)\mathrm{d}h_1 \mathrm{d}h_2 \mathrm{d}\psi]. \tag{5.69}$$

Similarly, the stationary PDF $p(\psi)$ of the angular position of a Brownian particle can be obtained as

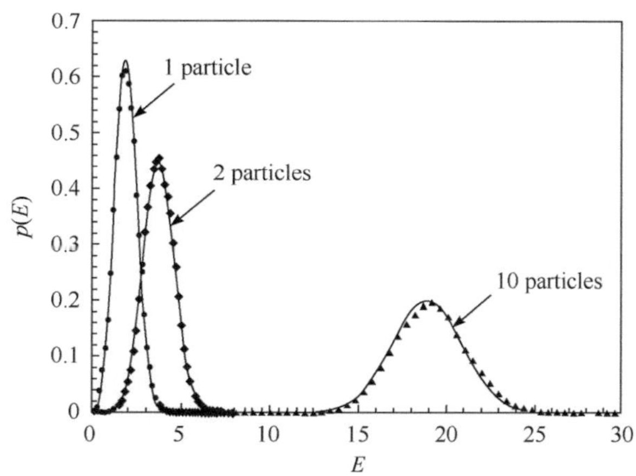

Fig. 5.15 Stationary PDF $p(E)$ of total energy for a swarm of Brownian particles (Zhu and Deng 2005).

$$p(\psi) = \frac{\mathrm{d}}{\mathrm{d}\psi}\left[\iint\limits_{\tan^{-1}(x_2/x_1)\le\psi} p(x_1, x_2)\mathrm{d}x_1\mathrm{d}x_2\right]. \tag{5.70}$$

where $p(x_1, x_2)$ is the analytical solution in Eq. (5.30). Given the parameters $\gamma_1 = 4$, $\gamma_2 = 1$, $\omega = 6$ and $D = 1$, Fig. 5.16 presents a comparison between the results from Eq. (5.70) and from Monte Carlo simulation of 10,000 particles, demonstrating a good agreement between the two results.

The following stationary PDF $p(r)$ of the displacement amplitude $r = \sqrt{x_1^2 + x_2^2}$ of a Brownian particle relative to the mass center can be obtained from Eq. (5.30)

$$\begin{aligned} F(r) &= \iint\limits_{(x_1^2+x_2^2)^{1/2}\le r} \int_{-\infty}^{\infty}\int_{-\infty}^{\infty} p(x_1, x_2, v_1, v_2)\mathrm{d}v_1\mathrm{d}v_2\mathrm{d}x_1\mathrm{d}x_2 \\ &= \int_0^{2\pi}\int_0^{r}\int_{-\infty}^{\infty}\int_{-\infty}^{\infty} ap(a\cos\theta, a\sin\theta, v_1, v_2)\mathrm{d}v_1\mathrm{d}v_2\mathrm{d}a\mathrm{d}\theta. \end{aligned} \tag{5.71}$$

$$p(r) = \frac{\mathrm{d}F(r)}{\mathrm{d}r} = \int_0^{2\pi}\int_{-\infty}^{\infty}\int_{-\infty}^{\infty} rp(r\cos\theta, r\sin\theta, v_1, v_2)\mathrm{d}v_1\mathrm{d}v_2.$$

Erdmann obtained the following expression for $p(r)$ using the standard stochastic averaging method (Erdmann et al. 2002)

$$p(r) = C\exp[\frac{\gamma_1\omega^2}{D}r^2(1 - \frac{r^2}{2r_0^2})]. \tag{5.72}$$

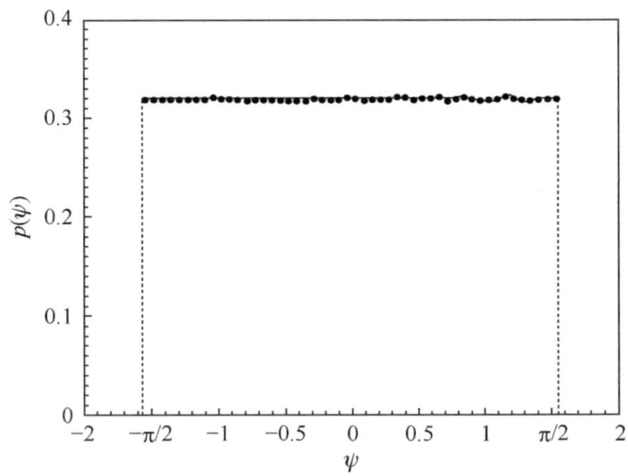

Fig. 5.16 Stationary PDF $p(\psi)$ of angular positions for a swarm of Brownian particles (Zhu and Deng 2005).

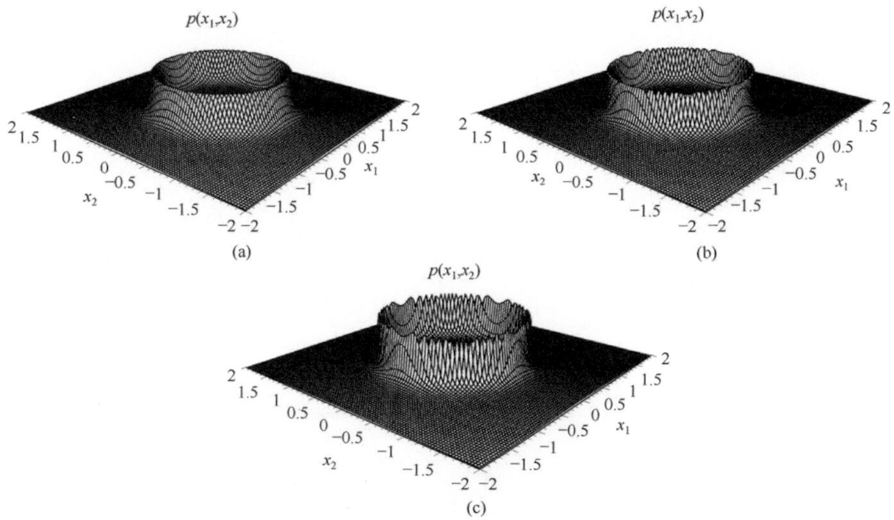

Fig. 5.17 Stationary PDF $p(x_1, x_2)$ of displacement X_1, X_2 for a swarm of 10,000 Brownian particles. **a** Simulation results, **b** results calculated from Eq. (5.71), **c** results calculated from Eq. (5.72). The parameters are $\gamma_1 = 4$, $\gamma_2 = 1$, $\omega = 2$, $D = 0.03$, $r_0 = \gamma_1 / (\gamma_2 \omega^2)$ (Zhu and Deng 2005).

Figure 5.17 displays the comparison of $p(x_1, x_2)$ from simulation, Eq. (5.71) and Eq. (5.72), respectively. Note that to get $p(r)$ from $p(x_1, x_2)$, the transformation $x_1 = r\cos\theta, \quad x_2 = r\sin\theta$ has been used. The analytical solution (5.71) is more complex than Eq. (5.72) but offers higher computational accuracy.

For the PDF of motion velocity of a Brownian particle, it can be obtained from Eq. (5.30) as follows

$$p(v) = \int_0^{2\pi} \int_{-\infty}^{\infty} \int_{-\infty}^{\infty} v p(x_1, x_2, v\cos\theta, v\sin\theta) \mathrm{d}x_1 \mathrm{d}x_2. \tag{5.73}$$

Figure 5.18 shows a comparison between the analytical results of $p(r)$ and $p(v)$ from Eqs. (5.71) and (5.73) and the results from Monte Carlo simulation of 10,000 particles, which further verifies the applicability of analytical solutions obtained from single particle to the swarm motion of particles.

5.2 Reaction Rate Theory

The classical Kramers reaction rate theory (Kramers 1940) views the reaction process as the event that a particle crosses a potential barrier under random perturbation. Depending on the magnitude of the damping force acting on the reacting particle,

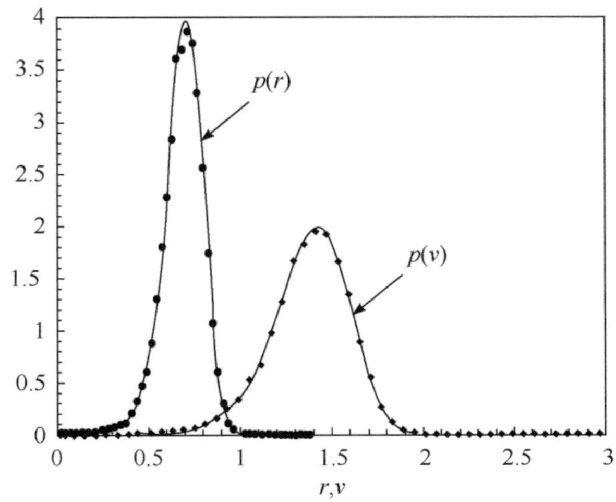

Fig. 5.18 Stationary PDF $p(r)$ and $p(v)$ of displacement r and velocity V, respectively. Parameters $\gamma_1 = 0.2$, $\gamma_2 = 0.1$, $\omega = 2$, $D = 0.01$. Line—denotes results calculated from Eqs. (5.71) and (5.73), the symbols ● ◆ denote Monte Carlo simulation results of 10,000 particles (Zhu and Deng 2005).

Kramers obtained the expressions for reaction rates in the cases of displacement diffusion and energy diffusion. Kramers reaction rate theory, after modifications and improvements, has been extensively applied in physics and life sciences. It has been demonstrated that the reciprocal of the mean first-passage time of the reacting particle is equivalent to the reaction rate (Reimann et al. 1999). Therefore, studying the first-passage problem of stochastic dynamical systems can contribute to the development of reaction rate theory. In this section, the stochastic averaging method for quasi-Hamiltonian systems is applied to study the first-passage problem of reaction dynamics models, leading to an analytical expression for the reaction rate (Deng and Zhu 2007a).

5.2.1 Kramers Reaction Rate Theory

The classic Kramers reaction rate theory can be described using the double-well system model depicted in Fig. 5.19. In this model, a unit-mass Brownian particle (reacting particle) moves from one well to anther well with a potential function $U(x)$ under thermal disturbance. The particle crossing the potential barrier represents the completion of the reaction. There are two kinds of barrier crossings, one is that the particle displacement crosses the barrier position, i.e., $x > x_b$, and the other is that the particle energy exceeds the barrier energy, i.e., $E > \Delta U$. The former corresponds to a reaction rate dominated by displacement diffusion, while the latter corresponds to a reaction rate dominated by energy diffusion.

The displacement $X(t)$ of the reacting particle is governed by the following Langevin equation

Fig. 5.19 Motion of reacting particle in the double potential well $U(x) = -x^2/2 + x^4/4$ under random excitation.

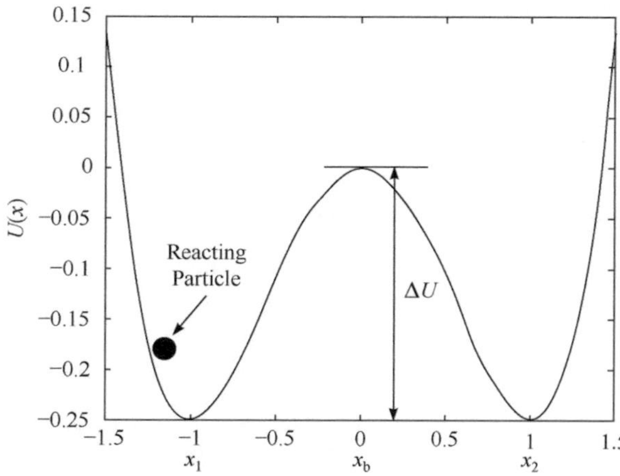

$$\frac{d^2X}{dt^2} + \gamma\frac{dX}{dt} + \frac{dU(X)}{dX} = \sqrt{2D}W_g(t). \tag{5.74}$$

where $W_g(t)$ is a unit Gaussian white noise. According to the fluctuation-dissipation theorem (Einstein relation), the following relationship can be established among the excitation intensity $2D$, the linear damping coefficient γ and the thermal disturbance temperature T

$$D = \gamma k_B T. \tag{5.75}$$

where k_B is Boltzmann constant. In a special case of Kramers reaction rate theory, the following symmetric double-well potential energy function is considered

$$U(x) = -\frac{a}{2}x^2 + \frac{b}{4}x^4, \quad a, b > 0. \tag{5.76}$$

It can be determined from Eq. (5.76) that $x_{1,2} = \mp\sqrt{a/b}$ is the positions of the left and right well bottoms and $x_b = 0$ is the barrier position. The linearized frequency at the bottom of the potential well is $\omega_0^2 = U''(x_1) = 2a$ and that at the top of the barrier is $\omega_b^2 = |U''(x_b)| = a$. The barrier height being $\Delta U = a^2/4b$.

The left well in Figure 5.19 can be considered as a "source" continually providing reacting particles, while the right well acts as a "trap" absorbing the reacting particles. When a reacting particle starts from the "source" under thermal disturbances, once it crosses the barrier position, i.e., $x > x_b$, it will be absorbed by the "trap" and will not return. When the reacting particle is in the "source" it represents reactant, and when it is in the "trap" it represents product. The Kramers reaction rate theory provides the crossing barrier rate of reacting particle, depending on the magnitude of the damping coefficient γ. When γ is large, reacting particle dissipate more energy, and according to Eq. (5.75), the thermal disturbance intensity $2D$ is also large. Kramers provided the crossing barrier rate of reacting particle as follows (Kramers 1940)

$$k_{\mathrm{M}} = \frac{\omega_0 \omega_{\mathrm{b}}}{2\pi\gamma} \exp\left(-\frac{\Delta U}{k_{\mathrm{B}} T}\right). \tag{5.77}$$

Notice that k_{M} is a reaction rate dominated by displacement diffusion.

If γ is small, both the energy dissipated by the reacting particles and the thermal disturbance energy are small, and the two energies approach a statistical balance. In this case, the crossing rate is dominated by energy diffusion. Based on this, Kramers derived the following equation governing the PDF $p(E, t)$ of particle energy

$$\frac{\partial p(E,t)}{\partial t} = \frac{\gamma}{2\pi} \frac{\partial}{\partial E}[I(E)(1 + k_{\mathrm{B}} T \frac{\partial}{\partial E})\omega(E) p(E,t)]. \tag{5.78}$$

where E is system energy;

$$I(E) = \oint \sqrt{2(E - U(x))} dx. \tag{5.79}$$

is the action variable and $\omega(E) = 2\pi \partial E / \partial I$ represents the angular frequency associated with conservative system, respectively. In the case of small damping, once the reacting particle gains energy slightly greater than the barrier energy, it will cross the barrier. This leads to an approximate expression for the reaction rate. Under high barrier assumption, i.e., $\Delta U / k_{\mathrm{B}} T \gg 1$, Eq. (5.78) can be used to derive the following reaction rate k_{W}

$$k_{\mathrm{W}} = \frac{\gamma \omega_0 I(\Delta U)}{2\pi k_{\mathrm{B}} T} \exp\left(-\frac{\Delta U}{k_{\mathrm{B}} T}\right). \tag{5.80}$$

After linearizing the potential energy at well bottom, i.e., setting $U(x) \approx \omega_0^2 x^2 / 2$, Eq. (5.80) can be further approximated as

$$k_{\mathrm{W}} = \frac{\gamma \Delta U}{k_{\mathrm{B}} T} \exp\left(\frac{-\Delta U}{k_{\mathrm{B}} T}\right). \tag{5.81}$$

Notice that k_{W} is a reaction rate dominated by energy diffusion.

When applying Kramers reaction rate theory in practice, one can combine Eqs. (5.77) and (5.81) to obtain the following expression for the reaction rate

$$k_{\mathrm{kramers}} = (k_{\mathrm{M}}^{-1} + k_{\mathrm{W}}^{-1})^{-1}. \tag{5.82}$$

Note that Kramers reaction rate k_{kramers}, combing the rates dominated by displacement diffusion and by energy diffusion, is applicable to the cases of small, medium, and large damping.

5.2.2 *Reaction Rate Dominated by Energy Diffusion*

In this section, the stochastic averaging method for quasi-Hamiltonian systems described in this chapter is applied to obtain a more general expression for the reaction rate dominated by energy diffusion. Under certain conditions, the results obtained in this subsection can be degenerated into the expression of Kramers reaction rate.

Consider system (5.74) as a stochastically excited and dissipated Hamiltonian system. It can be converted into the following Itô stochastic differential equations

$$\begin{aligned} \mathrm{d}Q &= P\mathrm{d}t, \\ \mathrm{d}P &= -\left(\frac{\mathrm{d}U(Q)}{\mathrm{d}Q} + \gamma P\right)\mathrm{d}t + \sqrt{2\gamma k_{\mathrm{B}}T}\mathrm{d}B(t), \end{aligned} \tag{5.83}$$

where $Q = X$ is the generalized displacement, $P = \mathrm{d}X/\mathrm{d}t$ is the generalized momentum, $B(t)$ is a unit Wiener process. The Hamiltonian function $H(t)$, , i.e., the system total energy, associated with system (5.83) is

$$H = P^2/2 + U(Q). \tag{5.84}$$

By applying Itô stochastic differential rule (Itô 1951a) and function relation (5.84), the Itô stochastic differential equation for H can be derived from Eq. (5.83) as follows

$$\mathrm{d}H = (-\gamma P^2 + \gamma k_{\mathrm{B}}T)\mathrm{d}t + \sqrt{2\gamma k_{\mathrm{B}}T}P\mathrm{d}B(t). \tag{5.85}$$

Now the system is governed by the first equation in Eqs. (5.83) and (5.85), with P replaced by $\pm\sqrt{2H - 2U(Q)}$ according to Eq. (5.84). As $\gamma \to 0$, the generalized displacement $Q(t)$ is fast-varying process while Hamiltonian function $H(t)$ is slowly-varying process and converges weakly to a Markov diffusion process. Applying the stochastic averaging method for quasi-non-integrable Hamiltonian systems in Sect. 5.1, the following averaged Itô equation can be obtained

$$\mathrm{d}H = m(H)\mathrm{d}t + \sigma(H)\mathrm{d}B(t). \tag{5.86}$$

According to Eqs. (5.10)–(5.13) of vol. 1 and using spatial averaging to replace time averaging, the drift coefficient $m(H)$ and diffusion coefficient $\sigma(H)$ are

$$m(H) = \frac{1}{T(H)} \int_{\Omega} (-\gamma p^2 + \gamma k_B T) \mathrm{d}q, \quad \sigma^2(H) = \frac{1}{T(H)} \int_{\Omega} (2\gamma k_B T p^2) \mathrm{d}q,$$
$$T(H) = \int_{\Omega} (1/p) \mathrm{d}q, \quad \Omega = \{q | U(q) \leq H\}. \tag{5.87}$$

After completing the integrals in Eq. (5.87) with respect to q, the expressions for $m(H)$ and $\sigma^2(H)$ are obtained as follows

$$m(H) = \gamma k_B T - \gamma S(H), \quad \sigma^2(H) = 2\gamma k_B T S(H),$$
$$S(H) = \frac{1}{3b}\left(a + \sqrt{a^2 - 4bH}\right)\left(a\frac{E(e)}{K(e)} - \sqrt{a^2 - 4bH}\right), \quad e = \frac{a - \sqrt{a^2 - 4bH}}{a + \sqrt{a^2 - 4bH}}, \tag{5.88}$$

where $K(\cdot)$ and $E(\cdot)$ are the first and second kinds of complete elliptic integrals, respectively. Eq. (5.88) can be equivalently written as

$$m(H) = \gamma k_B T - \frac{\gamma}{2\pi} I(H)\omega(H), \quad \sigma^2(H) = \frac{1}{\pi}\gamma k_B T I(H)\omega(H). \tag{5.89}$$

where $I(H)$ is the action variable defined in Eq. (5.79), and $\omega(H)$ is the angular frequency. The averaged FPK equation associated with averaged Itô equation (5.86) is

$$\frac{\partial p(h,t)}{\partial t} = \frac{-\partial}{\partial h}\left[m(h)p(h,t)\right] + \frac{1}{2}\frac{\partial^2}{\partial h^2}\left[\sigma^2(h)p(h,t)\right]. \tag{5.90}$$

By substituting the coefficients in Eq. (5.89) into the FPK equation (5.90) and expanding it, it can be proved that the FPK equation (5.90) obtained by applying the stochastic averaging method is the same as the Kramers equation (5.78) for the PDF of energy H in the case of small damping.

For a given threshold $h_C = \Delta U$, the first-passage time is defined as the time when the energy $H(t)$ reaches h_C for the first time under the condition that the initial energy $h_0 = H(0)$ is less than h_C. The mean first-passage time $\mu(h_0)$ is governed by the Pontryagin equation (Pontryagin et al. 1933)

$$\frac{1}{2}\sigma^2(h_0)\frac{\mathrm{d}^2\mu}{\mathrm{d}h_0^2} + m(h_0)\frac{\mathrm{d}\mu}{\mathrm{d}h_0} = -1. \tag{5.91}$$

The definition domain for Eq. (5.91) is $0 \leq h_0 < h_C$. According to Eq. (5.79), at the boundary $h_0 = 0$ (i.e., $E = 0$), $I(0) = 0$, $\sigma^2(0) = 0$, one boundary condition of Eq. (5.91) is

$$\left.\frac{\mathrm{d}\mu}{\mathrm{d}h_0}\right|_{h_0=0} = \frac{-1}{\gamma k_\mathrm{B} T}. \tag{5.92}$$

The other boundary condition at boundary $h_0 = h_\mathrm{C}$ is

$$\mu(h_0 = h_\mathrm{C}) = 0. \tag{5.93}$$

Under boundary conditions (5.92) and (5.93), the following analytical solution of Eq. (5.91) can be obtained

$$\mu(h_0) = 2\int_{h_0}^{h_\mathrm{C}} \mathrm{d}u \int_0^u \frac{1}{\sigma^2(v)} \exp\left[-2\int_v^u \frac{m(w)}{\sigma^2(w)}\mathrm{d}w\right]\mathrm{d}v. \tag{5.94}$$

The reaction rate is equivalent to the reciprocal of the mean first-passage time, i.e.,

$$k_\mathrm{E} = 1/\mu(0)\big|_{h_C=\Delta U}. \tag{5.95}$$

Figure 5.20 shows numerical results of reaction rate k_E in Eq. (5.95), k_W in Eq. (5.81) and k_Kramers in Eq. (5.82) and from Monte Carlo simulation of Eq. (5.74) for reaction rates dominated by displacement diffusion and by energy diffusion, respectively. It can be seen from simulation results that in the case of small damping, the two reaction rates dominated by displacement diffusion and by energy diffusion are consistent. k_Kramers is a good prediction formula for reaction rate dominated by displacement diffusion. Both k_E and k_W are reaction rates dominated by energy diffusion. k_E is more accurate than k_W while the cost is that k_E is more complicated to obtain than k_W.

Actually, using linearization approximations and high barrier assumptions, the theoretical formula k_E in Eq. (5.95) can be degenerated into Kramers prediction formula k_W in Eq. (5.81). First, consider the linearization approximation $U(x) \approx \omega_0^2 x^2/2$. In this case, the action $I(H)$ and angular frequency $\omega(H)$ are degenerated into $2\pi H/\omega_0$ and ω_0, respectively. The drift coefficient $m(H)$ and diffusion coefficient $\sigma(H)$ become

$$m(H) = \gamma k_\mathrm{B} T - \gamma H, \quad \sigma^2(H) = 2\gamma k_\mathrm{B} T H. \tag{5.96}$$

By substituting the coefficients in Eq. (5.96) into Eqs. (5.94) and (5.95), and after linearization approximation, the reaction rate is

$$k_\mathrm{L} = \gamma\left(\int_0^{\frac{\Delta U}{k_\mathrm{B} T}} \frac{e^t - 1}{t}\mathrm{d}t\right)^{-1}. \tag{5.97}$$

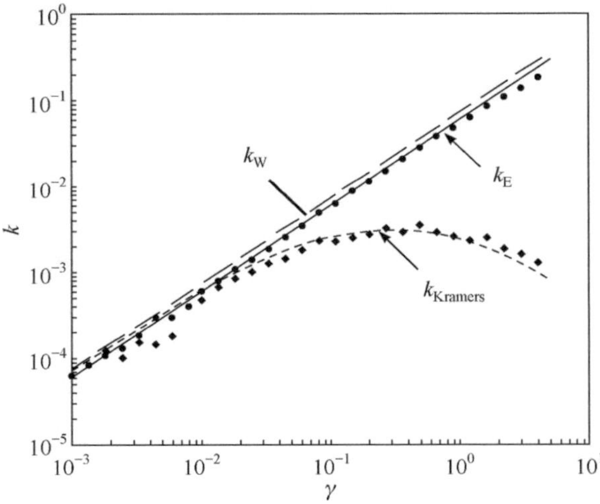

Fig. 5.20 Reaction rate of system (5.74) with parameters $a = 1, b = 1$, and $\Delta U / k_B T = 4$. Solid line —— is k_E from Eq. (5.95); short dashed line ----- is k_{Kramers} from Eq. (5.82); long dashed line — — is k_W from Eq. (5.81); symbol ◆ is simulation result under displacement diffusion; symbol ● is simulation result under energy diffusion (Deng and Zhu 2007a).

Next, consider the high barrier assumption $\Delta U / k_B T \gg 1$. Note that the following limit exists

$$\underset{x\to\infty}{\text{Limit}}\left[\left(\int_0^x \frac{e^t - 1}{t}\mathrm{d}t\right) \Big/ \left(\frac{1}{x}e^x\right)\right] = 1. \tag{5.98}$$

The reaction rate k_L in Eq. (5.97) is further degenerated into the expression for k_W in Eq. (5.81).

5.2.3 *Reaction Rate on Multi-dimensional Potential Energy Landscape*

In addition to one-dimensional reaction rate model, there are more complex reaction models describing the motion of reacting particles on multi-dimensional potential landscape. Langer obtained the reaction rate on multi-dimensional potential energy landscape for large damping case (Langer 1969; Hänggi et al. 1990). However, there are currently no well-established theoretical predictions for reaction rates on multi-dimensional potential energy landscape in small damping case. One advantage of the stochastic averaging methods of quasi-Hamiltonian systems is their ability of predicting the reaction rates of these complex models. Here, the stochastic averaging methods of quasi-Hamiltonian systems are used to predict the reaction rate on multi-dimensional potential energy landscape in small damping case.

For the sake of exposition, consider the motion of a reacting particle on a two-dimensional potential energy landscape. The equations governing this reaction process are (Deng 2007)

$$\frac{\mathrm{d}^2 X_1}{\mathrm{d}t^2} + \gamma \frac{\mathrm{d}X_1}{\mathrm{d}t} + \frac{\partial U(X_1, X_2)}{\partial X_1} = \sqrt{2\gamma k_B T} W_{g1}(t),$$
$$\frac{\mathrm{d}^2 X_2}{\mathrm{d}t^2} + \gamma \frac{\mathrm{d}X_2}{\mathrm{d}t} + \frac{\partial U(X_1, X_2)}{\partial X_2} = \sqrt{2\gamma k_B T} W_{g2}(t). \tag{5.99}$$

In these equations, the Einstein relation (5.75) is used to specify the intensities of random excitations. $W_{g1}(t)$ and $W_{g2}(t)$ in Eq. (5.101) are independent unit Gaussian white noises. The potential energy function is defined as

$$U(x_1, x_2) = -\frac{a x_1^2}{2} + \frac{b x_1^4}{4} + \frac{c x_2^2}{2} - \varepsilon x_1 x_2, \quad a, b, c, \varepsilon > 0. \tag{5.100}$$

As shown in Figure 5.21a, the potential energy $U(x_1, x_2)$ has two potential wells, and the reaction process can be described as a reacting particle in one potential well crossing the energy barrier into another potential well. The height of this barrier ΔU is given by

$$\Delta U = \left(ac + \varepsilon^2\right)^2 \Big/ 4bc^2 \tag{5.101}$$

By setting $Q_1 = X_1,\ Q_2 = X_2,\ P_1 = \mathrm{d}X_1 / \mathrm{d}t,\ P_2 = \mathrm{d}X_2 / \mathrm{d}t$, system (5.99) is transformed into the following Itô stochastic differential equations

$$\mathrm{d}Q_1 = P_1 \mathrm{d}t,$$

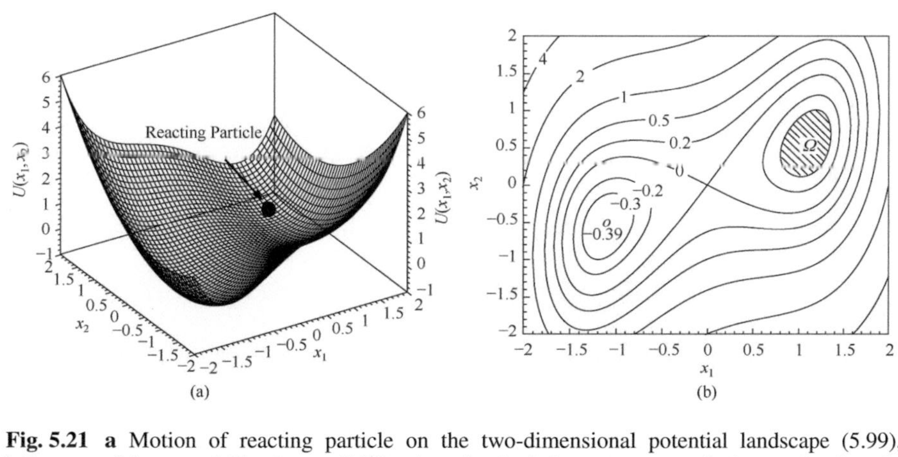

Fig. 5.21 **a** Motion of reacting particle on the two-dimensional potential landscape (5.99), **b** Contour of the potential landscape (5.99), where the shaded area represents the integration domain Ω in Eq. (5.106). Parameters $a = 1, b = 1, c = 1, \varepsilon = 0.5$ (Deng and Zhu 2007a).

$$
\begin{aligned}
\mathrm{d}Q_2 &= P_2\mathrm{d}t,\\
\mathrm{d}P_1 &= \left(-\partial U/\partial X_1 - \gamma P_1\right)\mathrm{d}t + \sqrt{2\gamma k_\mathrm{B}T}\mathrm{d}B_1(t),\\
\mathrm{d}P_2 &= \left(-\partial U/\partial X_2 - \gamma P_2\right)\mathrm{d}t + \sqrt{2\gamma k_\mathrm{B}T}\mathrm{d}B_2(t),
\end{aligned}
\tag{5.102}
$$

where $B_1(t)$ and $B_2(t)$ are independent unit Wiener processes. The Hamiltonian function associated with system (5.102) is

$$
H = (P_1^2 + P_2^2)/2 + U(X_1, X_2). \tag{5.103}
$$

Due to the inseparability of the two-dimensional potential (5.99), Eq. (5.103) represents the Hamiltonian function of a non-integrable Hamiltonian system. Applying Itô differential rules (Itô 1951a), the following Itô stochastic differential equation for $H(t)$ can be derived from Eq. (5.102),

$$
\mathrm{d}H = (-\gamma P_1^2 - \gamma P_2^2 + 2\gamma k_\mathrm{B}T)\mathrm{d}t + P_1\sqrt{2\gamma k_\mathrm{B}T}\mathrm{d}B_1(t) + P_2\sqrt{2\gamma k_\mathrm{B}T}\mathrm{d}B_2(t). \tag{5.104}
$$

In the case of weak damping and random excitations, Eq. (5.102) represents a quasi-non-integrable Hamiltonian system. Applying the stochastic averaging method of quasi-non-integrable Hamiltonian systems described in Sect. 5.1, an averaged Itô equation of the form of Eq. (5.86) can be obtained. According to Eqs. (5.10)–(5.13) of vol. 1, the drift and diffusion coefficients are

$$
\begin{aligned}
m(H) &= \frac{1}{T(H)}\int_\Omega (-\gamma p_1 - \gamma p_2 + 2\gamma k_\mathrm{B}T)\mathrm{d}q_1\mathrm{d}q_2\mathrm{d}p_2,\\
\sigma^2(H) &= \frac{1}{T(H)}\int_\Omega \left(2\gamma k_\mathrm{B}Tp_1^2 + 2\gamma k_\mathrm{B}Tp_2^2\right)\mathrm{d}q_1\mathrm{d}q_2\mathrm{d}p_2,\\
T(H) &= \int_\Omega (1/p_1)\mathrm{d}q_1\mathrm{d}q_2\mathrm{d}p_2, \quad \Omega = \left\{(q_1, q_2, p_2)|p_2^2/2 + U(q_1, q_2) \le H\right\}.
\end{aligned}
\tag{5.105}
$$

After completing integral with respect to P_2, the drift and diffusion coefficients are

$$
\begin{aligned}
&m(H) = 2\gamma k_\mathrm{B}T - 2\gamma H + 2\gamma G(H), \quad \sigma^2(H) = 4\gamma k_\mathrm{B}TH - 4\gamma k_\mathrm{B}TG(H),\\
&G(H) = \int_\Omega U(q_1, q_2)\mathrm{d}q_1\mathrm{d}q_2 \Big/ \int_\Omega \mathrm{d}q_1\mathrm{d}q_2, \quad \Omega = \{(q_1, q_2)|U(q_1, q_2) \le H\}.
\end{aligned}
\tag{5.106}
$$

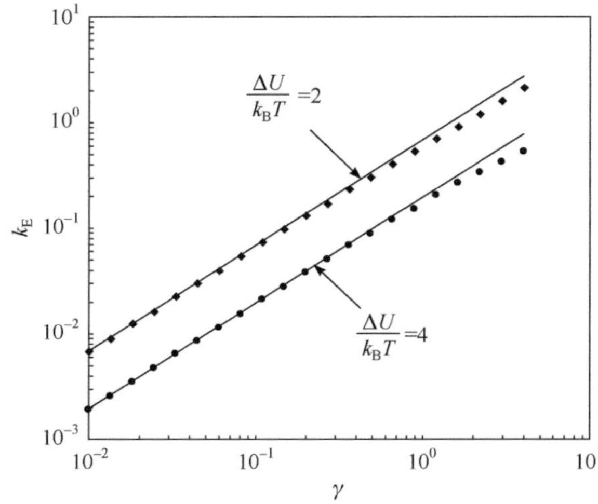

Fig. 5.22 Reaction rate dominated by energy diffusion of system (5.101). Solid line —— represents theoretical results from Eq. (5.95), symbols ◆, ● represent results from Monte Carlo simulation. Parameters are the same as those in Fig. 5.21 (Deng and Zhu, 2007a).

where the integration domain Ω is depicted in Figure 5.21b and $G(H)$ is the mean potential energy on the integration domain Ω. By evaluating the mean first-passage time using Eq. (5.94) and then using Eq. (5.95), the reaction rate k_E dominated by energy diffusion on the two-dimensional potential energy landscape can be determined. Figure 5.22 shows a comparison between theoretical results and results from Monte Carlo simulation of Eq. (5.99). It is seen that the two results match closely.

5.2.4 Reaction Rate Under Colored Noise Excitation

In the Kramers reaction rate theory, thermal perturbations or thermal noise are considered to be Gaussian white noise. However, real thermal perturbations often are colored noises. Thus, studying reaction rate under colored noise excitation has practical significance. In this case, the relationship between excitation intensity and damping is no longer simple (Hänggi et al. 1990). For simplicity, a type of colored noise known as low-pass noise (Cai and Zhu 2016), will be used. It can be generated from a Gaussian white noise passing through a first-order linear filter. Denote the low-pass noise as $\xi(t)$, the correlation function and power spectral density are given by

$$E\left[\xi(t)\xi(t')\right] = \frac{D}{\tau}\exp(-\frac{|t'-t|}{\tau}), \quad S(\omega) = \frac{D}{\pi(\tau^2\omega^2+1)}. \tag{5.107}$$

where $2D$ is excitation intensity; τ represents correlation time. Figure 5.23a displays the power spectral density of the low-pass noises in Eq. (5.107). It can be observed that when $\tau = 0$, the power spectral density becomes constant, and the noise becomes white noise with an intensity of $2D$. Currently, there is no unified expression for

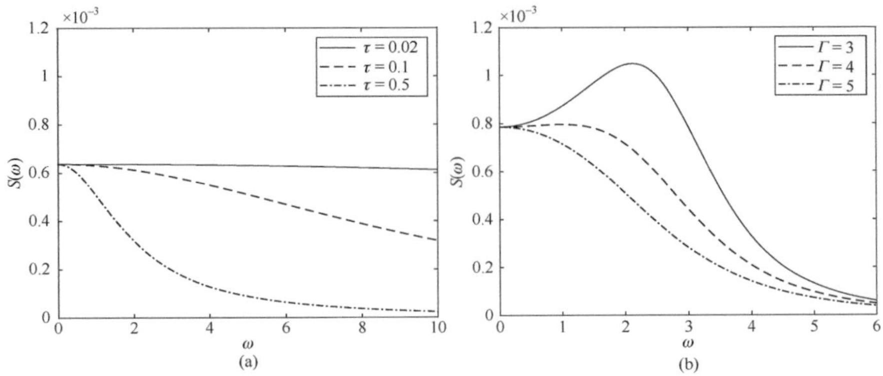

Fig. 5.23 **a** Power spectral densities of low-pass noises. **a** Eq. (5.107) with $D = 0.002$; **b** Eq. (5.140) with $D = 0.2$ and $\Omega = 3$.

reaction rate over the entire range $\tau \in [0, \infty)$. There are only expressions for two extreme cases, i.e., $\tau \to 0$ and $\tau \to \infty$.

Hänggi et al. (1984) studied the reaction rate of a particle excited by low-pass noise in a symmetric double-well potential $U(x) = -x^2/2 + x^4/4$ in the case of large damping (usually denoted as damping coefficient $\gamma = 1$) and $\tau = 0$. In this case, the low-pass noise becomes Gaussian white noise, the reaction rate in Eq. (5.77) can be applied (with $\omega_0 = \sqrt{2}$, $\omega_b = 1$, $\Delta U = 1/4$, $\gamma = 1$, $\gamma k_B T = D$). It can be expressed as

$$k_s = \frac{1}{\sqrt{2}\pi} \exp\left(-\frac{1}{4D}\right). \tag{5.108}$$

For the case of large damping and $\tau \neq 0$, Hänggi (Hänggi et al. 1984) combined the following two equations to describe the dynamics of a reacting particle

$$\frac{dX}{dt} - X + X^3 = \xi, \quad \tau\frac{d\xi}{dt} + \xi = \sqrt{2D}W(t). \tag{5.109}$$

The first equation represents the motion of reacting particle in the symmetric double-well potential $U(x) = -x^2/2 + x^4/4$. Note that under condition of large damping, the inertial force dX^2/dt^2 is neglected in the first equation of (5.109). The second equation generates a low-pass noise $\xi(t)$ by passing a Gaussian white noise $\sqrt{2D}W(t)$ through a first-order linear filter. By studying system (5.109), the final expression $k(\tau)$ for the reaction rate in the case of large damping and $\tau \neq 0$ can be obtained. It is a modified version for k_s (Jung and Hänggi 1988; Luciani and Verga 1988):

$$k(\tau) = k_s(1 - \frac{3}{2}\tau)\exp[-\frac{\tau^2}{8D} + \frac{3\tau^4}{10D} + O(\frac{\tau^6}{D})], \quad \text{when } \tau \to 0,$$

$$k(\tau) \propto \exp(-\frac{2\tau}{27D}), \quad \text{when } \tau \to \infty. \tag{5.110}$$

For a small damping γ and a low-pass noise $\xi(t)$, the Langevin equation governing the particle displacement $X(t)$ in the symmetric double-well potential (5.76) is given by

$$\frac{d^2X}{dt^2} + \gamma\frac{dX}{dt} - aX + bX^3 = \xi(t). \tag{5.111}$$

Marchesoni and others conducted experiments using simulated circuits and found the following empirical formula for the reaction rate (Marchesoni et al. 1988a, b)

$$k(\tau) = k(0)\exp(-\kappa_0\tau^2). \tag{5.112}$$

where (i) $k(0)$ represents the reaction rate when $\tau = 0$, which can be determined using the reaction rate $k(0) = k_W$ in Eq. (5.81) for small damping and Gaussian white noise; ((ii) The value of κ_0 does not depend on the damping coefficient γ; (iii) κ_0 is proportional to $\Delta U / k_B T$; (iv) Given $\Delta U / k_B T$, κ_0 is proportional to the potential parameter a, leading to the conclusion that $\kappa_0 \propto (\Delta U / k_B T)\omega_0^2$, where $\omega_0 = \sqrt{2a}$ is the frequency at the bottom of the potential well.

While the Einstein relation $D = \gamma k_B T$ is applicable only to model environmental thermal perturbations using Gaussian white noise, considering that exponential correlated noise can degenerate into white noise with power spectrum of D/π, a reasonable assumption is that the Einstein relation $D = \gamma k_B T$ can also be applied to the exponential correlated noise with power spectral density (5.107). Without violating the experimental results mentioned above, Marchesoni suggested the following substitution relationship between the reaction rate under white noise and that under exponential correlated noise

$$k_B T \leftrightarrow k_B T / (\tau^2\omega^2 + 1). \tag{5.113}$$

Introducing Eq. (5.113) into the expression of k_W in Eq. (5.81) and then to Eq. (5.112), the reaction rate under exponential correlated noise can then be obtained as follows (Marchesoni et al. 1988a, b):

$$k(\tau) = k(0)(1 + \omega_0^2\tau^2)\exp\left(\frac{-\Delta U}{k_B T}\omega_0^2\tau^2\right). \tag{5.114}$$

where $k(0) = k_W$ in Eq. (5.81). Equation (5.114) is an effective approximation for small damping γ and $\Delta U \gg k_B T$. In the following section, it will be found that the reaction rate obtained using the stochastic averaging method is consistent with Eq. (5.114).

5.2.5 Prediction of Reaction Rate Under Colored Noise Excitation Using the Stochastic Averaging Method in Sect. 1.1

In this section, the stochastic averaging method of quasi-integrable Hamiltonian system excited by wideband noise is used to study the rate of a reacting particle crossing a potential barrier in a one-dimensional potential landscape under colored noise excitation. The equation governing displacement $X(t)$ of a reacting particle is as follows (Deng and Zhu 2007b)

$$\frac{d^2X}{dt^2} + \gamma \frac{dX}{dt} + g(X) = \xi(t), \tag{5.115}$$

where the restoring force $g(X)$ and the corresponding potential energy $U(X)$ are

$$g(X) = \omega_0^2 X + \beta X^2 + \alpha X^3, \quad U(X) = \frac{1}{2}\omega_0^2 X^2 + \frac{1}{3}\beta X^3 + \frac{1}{4}\alpha X^4, \tag{5.116}$$

$\xi(t)$ is stationary wideband random excitation with power spectral density $S(\omega)$, and both of the damping coefficient γ and the excitation intensity are assumed to be weak. The potential in Eq. (5.116) can have a single potential well, a symmetric double-wells, or an asymmetric potential well by adjusting parameters ω_0, β and α.

Assuming that the restoring force function $g(X)$ and the potential function $U(X)$ in Eq. (5.116) satisfy the 4 conditions described below Eq. (1.5), the following transformation can be used

$$X(t) = A\cos\Phi(t) + B, \quad \dot{X}(t) = -Av(A, \Phi)\sin\Phi(t), \quad \Phi(t) = \Psi(t) + \Theta(t), \tag{5.117}$$

where both A, Φ and B are random processes. The instantaneous frequency is

$$\nu(A, \Phi) = \frac{1}{2}\nu_0(A) + \sum_{r=1}^{\infty} \nu_r(A)\cos r\Phi, \tag{5.118}$$

The mean frequency of system (5.115) is

$$\overline{\omega}_0(A) = \frac{1}{2\pi}\int_0^{2\pi} \nu(A, \Phi)d\Phi = \frac{1}{2}\nu_0(A). \tag{5.119}$$

Using transformation in Eq. (5.117) from $X, \dot{X}$ to A, Φ, the original system (5.115) can be transformed into the following first-order differential equation in terms of amplitude A and phase angle Φ

$$\frac{dA}{dt} = m_1(A, \Phi) + \sigma_1(A, \Phi)\xi(t), \quad \frac{d\Phi}{dt} = \nu(A, \Phi) + m_2(A, \Phi) + \sigma_2(A, \Phi)\xi(t), \tag{5.120}$$

where

$$\begin{aligned} m_1 &= -A^2 G\gamma\nu^2 \sin^2\Phi, \quad \sigma_1 = -AG\nu\sin\Phi, \\ m_2 &= -AG\gamma\nu^2(\cos\Phi + d)\sin\Phi, \quad \sigma_2 = -G\nu(\cos\Phi + d). \end{aligned} \tag{5.121}$$

and where $G = 1/[g(A+B)(1+d)]$, d is calculated according to Eq. (1.15). Applying the stochastic averaging method of quasi-integrable Hamiltonian system under wideband random excitation, the following averaged Itô equation governing $A(t)$ is obtained

$$dA = m(A)dt + \sigma(A)dB(t). \tag{5.122}$$

where $B(t)$ is a unit Wiener process, and according to Eq. (1.20) the drift coefficient and diffusion coefficient are obtained as follows

$$\begin{aligned} m(A) &= \left\langle m_1 + \int_{-\infty}^{0} \left((\partial\sigma_1/\partial A)\big|_t \sigma_1|_{t+\tau} + (\partial\sigma_1/\partial\Phi)\big|_t \sigma_2|_{t+\tau} \right) R(\tau)d\tau \right\rangle_t, \\ \sigma^2(A) &= \left\langle \int_{-\infty}^{\infty} \left(\sigma_1|_t \sigma_1|_{t+\tau} \right) R(\tau)d\tau \right\rangle_t, \end{aligned} \tag{5.123}$$

where $R(\tau)$ is the correlation function of colored noise $\xi(t)$ and $\langle[\cdot]\rangle_t$ is time averaging given by

$$\langle[\cdot]\rangle_t = \lim_{T\to\infty} \frac{1}{T} \int_0^T [\cdot]dt. \tag{5.124}$$

Expanding each term in Eq. (5.123) into Fourier series of Φ, integrating with respect to τ, and performing averaging with respect to Φ, using the approximation relationship $\Phi = \overline{\omega}_0(A)t + \Theta$, the following series expressions for $m(A)$ and $\sigma^2(A)$ are obtained

$$\begin{aligned} m(A) = &\frac{-\gamma G A^2}{16}(8\omega_0^2 + 16B\beta + 5A^2\alpha + 24B^2\alpha) + \frac{\pi G A}{8} \\ &\times \sum_{i=0}^{\infty} \left\{ (\nu_i - \nu_{i+2}) \left[\frac{d[GA(\nu_i - \nu_{i+2})]}{dA} + (i+1)G(\nu_i + 2h\nu_{i+1} + \nu_{i+2}) \right] S((i+1)\overline{\omega}_0) \right\}, \end{aligned}$$

$$\sigma^2(A) = \frac{\pi G^2 A^2}{4} \sum_{i=0}^{\infty} \left[(v_i - v_{i+2})^2 S((i+1)\overline{\omega}_0) \right]. \tag{5.125}$$

Using the relationship between the amplitude A and the system total energy E

$$E = U(A). \tag{5.126}$$

and the Itô differential rule, the Itô stochastic differential equation governing energy E can be obtained as follows

$$\mathrm{d}E = \overline{m}(E)\mathrm{d}t + \overline{\sigma}^2(E)\mathrm{d}B(t), \tag{5.127}$$

where the drift coefficient $\overline{m}(E)$ and the diffusion coefficient $\overline{\sigma}(E)$ are given by

$$\begin{aligned} \overline{m}(E) &= \left\{ \left(\frac{\mathrm{d}E}{\mathrm{d}A} \right) m(A) + \frac{1}{2} \frac{\mathrm{d}^2 E}{\mathrm{d}A^2} \sigma^2(A) \right\} \bigg|_{A=U^{-1}(E)}, \\ \overline{\sigma}^2(E) &= \left\{ \left(\frac{\mathrm{d}E}{\mathrm{d}A} \right)^2 \sigma^2(A) \right\} \bigg|_{A=U^{-1}(E)}. \end{aligned} \tag{5.128}$$

The mean first-passage time of system (5.127) is the solution of following Pontryagin equation (Andrnov et al. 1933)

$$\frac{1}{2}\overline{\sigma}^2(e_0)\frac{\partial^2 \mu}{\partial e_0^2} + \overline{m}(e_0)\frac{\partial \mu}{\partial e_0} = -1, \tag{5.129}$$

where the mean first-passage time $\mu(e_0)$ is a function of initial energy e_0. The boundary conditions for Eq. (5.129) are

$$\mu(e_0 = 0) = \text{finite}, \quad \mu(e_0 = e_C) = 0. \tag{5.130}$$

Equations (5.129) is a Bernoulli-type differential equation. Solving this equation yields the mean first-passage time

$$\mu(e_0) = 2 \int_{e_0}^{e_C} \mathrm{d}u \int_0^u \frac{1}{\overline{\sigma}^2(v)} \exp\left[-2 \int_v^u \frac{\overline{m}(w)}{\overline{\sigma}^2(w)} \mathrm{d}w \right] \mathrm{d}v. \tag{5.131}$$

Let

$$U(x) = \varepsilon_1 x^2 - \sqrt{\varepsilon_1 \varepsilon_2} x^3 + \frac{\varepsilon_2}{4} x^4, \quad \varepsilon_1, \varepsilon_2 > 0. \tag{5.132}$$

Figure 5.24 illustrates the schematic diagram of a reacting particle motion in the symmetric double-well potential (5.132). Comparing (5.132) and (5.116) yields

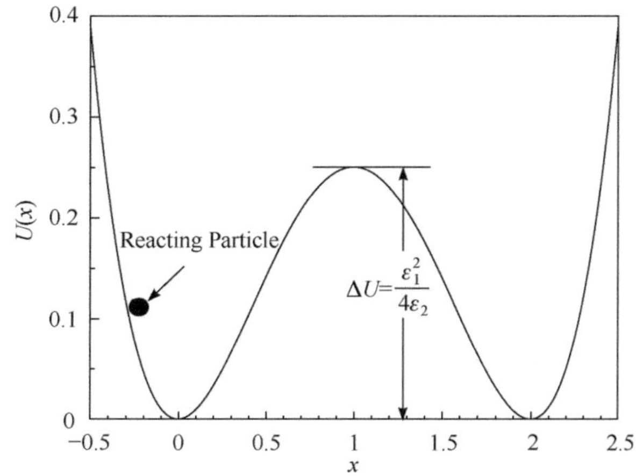

Fig. 5.24 A reacting particle in the double potential well potential (5.132) under colored noise excitation

$\omega_0 = \sqrt{2\varepsilon_1}$, $\beta = -3\sqrt{\varepsilon_1\varepsilon_2}$, $\alpha = \varepsilon_2$. Let $e_0 = 0$, and $e_C = \Delta U = \varepsilon_1^2/4\varepsilon_2$, representing the barrier height. By substituting the above parameters and the power spectral density $S(\omega)$ of colored noise into Eq. (5.131), the mean first-passage time $\mu(0)$ and the reaction rate k_{E} under colored noise in the double-well potential (5.132) are obtained (Deng and Zhu 2007b)

$$\mu(0) = 2\int_0^{\Delta U} \mathrm{d}u \int_0^u \frac{1}{\overline{\sigma}^2(v)} \exp\left[-2\int_v^u \frac{\overline{m}(w)}{\overline{\sigma}^2(w)}\mathrm{d}w\right]\mathrm{d}v, \quad k_{\mathrm{E}} = \frac{1}{\mu(0)}. \tag{5.133}$$

The reaction rate k_{E} calculated from (5.133) is general and can be reduced to Kramers formula (5.81) after the following three simplifications,

First is linear approximation. Letting $\varepsilon_2 = 0$ leads to $U(X) = \omega_0^2 X^2/2$. Then, using Eqs. (5.119), (1.8), (1.9) and (1.15) yields

$$\overline{\omega}_0(A) = \omega_0, \quad B = 0, \quad d = 0, \quad v_i(A) = 0, \quad (i = 1, 2, \cdots). \tag{5.134}$$

Substituting Eq. (5.134) into Eqs. (5.125) and (5.128) results in the following simplified coefficients

$$\begin{aligned} m(A) &= -\frac{1}{2}\gamma A + \frac{\pi}{2\omega_0^2 A} S(\omega_0), \quad \sigma^2(A) = \frac{\pi}{\omega_0^2} S(\omega_0), \\ \overline{m}(E) &= -\gamma E + \pi S(\omega_0), \quad \overline{\sigma}^2(E) = 2\pi E S(\omega_0). \end{aligned} \tag{5.135}$$

Substituting the coefficients $\overline{m}(E)$ and $\overline{\sigma}^2(E)$ in Eq. (5.135) into Eq. (5.133), the following linearized approximation for reaction rate is obtained

$$k_E = \left(\frac{1}{\gamma}\int_0^x \frac{e^t - 1}{t}\mathrm{d}t\right)^{-1}, \quad x = \frac{\gamma\Delta U}{\pi S(\omega_0)}. \tag{5.136}$$

Second is high barrier assumption. Based on the linear approximation, assume that the barrier height ΔU is sufficiently high (i.e., $\gamma\Delta U/\pi S(\omega_0) \gg 1$). By using Eq. (5.98), Eq. (5.136) is reduced to

$$k_E = \frac{\gamma^2\Delta U}{\pi S(\omega_0)}\exp\left(\frac{-\gamma\Delta U}{\pi S(\omega_0)}\right). \tag{5.137}$$

Third is white noise assumption. Substituting the power spectral density $S(\omega) = D/\pi$ of white noise into Eq. (5.137) and applying the Einstein relation $D = \gamma k_B T$ result in the reaction rate that is exactly the same as Kramers reaction rate in Eq. (5.81).

The reaction rate in Eq. (5.133) with Eq. (5.128) is applicable to any wideband noise including the low-pass noise. In the case of the low-pass noise, the power spectral density functions $S(\omega)$ in Eq. (5.107) and $D = \gamma k_B T$ are substituted into Eq. (5.137) to yield

$$k_E = \frac{\gamma\Delta U}{k_B T}(1 + \omega_0^2\tau^2)\exp\left[\frac{-\Delta U}{k_B T}(1 + \omega_0^2\tau^2)\right]. \tag{5.138}$$

Based on the explanations below Eq. (5.112), it is known that Eq. (5.138) is consistent with the result obtained by Marchesoni in Eq. (5.114).

Figures 5.25a, b provide some numerical results for the case of low-pass noise. In these figures, the solid line —— represents the theoretical result from Eq. (5.133), while the symbols ●, ◆, ■ represent the Monte Carlo simulation results of k_E dominated by energy diffusion. The symbols ○, ◇, □ represent the simulation results of k_d dominated by displacement diffusion. The error increases with increasing damping coefficient γ, indicating that reaction rate in the case of small damping is dominated by energy diffusion and the stochastic averaging method can provide good prediction. As the correlation time τ varies from 0 to 1.2, the power spectral density in Eq. (5.107) suggests that the low-pass noise gradually transitions from white noise to wideband noise. Figure 5.25b shows a comparison of the results obtained from the prediction formula (5.133) and the results from Eq. (5.114) (dashed line ------), as well as the results from Monte Carlo simulation. It is seen that the results in Eq. (5.133) obtained by using the stochastic averaging method is better than Eq. (5.114) for predicting the reaction rate dominated by energy diffusion. While the cost is that Eq. (5.133) is more complicated that Eq. (5.114)

Harmonic noise $\xi(t)$ is a colored noise produced by Gaussian white noise passing through a second-order linear filter, i.e.,

$$\frac{\mathrm{d}^2\xi}{\mathrm{d}t^2} + \Gamma\frac{\mathrm{d}\xi}{\mathrm{d}t} + \Omega^2\xi = \sqrt{2D}W_g(t). \tag{5.139}$$

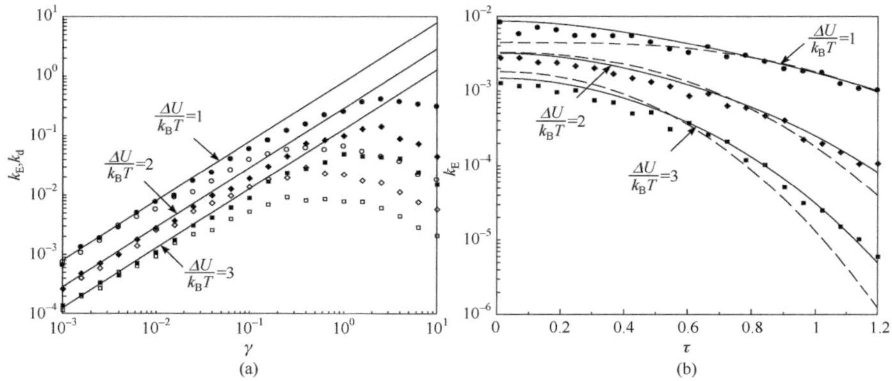

Fig. 5.25 Reaction rate dominated by energy diffusion of system (5.115) under low-pass noise excitation. **a** k_E and k_d versus γ, $\tau = 0.2$, **b** k_E versus τ, $\gamma = 0.01$ (Deng and Zhu 2007b)

where Γ and Ω are the damping coefficient and natural frequency, respectively, $W_g(t)$ is a unit Gaussian white noise, and $2D$ specifies the intensity of the excitation. The power spectral density function of $\xi(t)$ is

$$S(\omega) = \frac{D}{\pi\left(\omega^2\Gamma^2 + (\omega^2 - \Omega^2)^2\right)}. \tag{5.140}$$

By adjusting parameters D, Γ and Ω, (5.140) can be wideband or narrowband. Figure 5.23b shows the power spectral densities for three different Γ values.

Replace the noise in system (5.115) with the harmonic noise, and perform a similar treatment using the stochastic averaging method. Figure 5.26 provides numerical results for the reaction rate in the case of harmonic noise. The solid line —— represent the theoretical result in Eq. (5.133), the symbols ●, ◆, ■ represent simulation results for the reaction rate k_E dominated by energy diffusion, and ○, ◇, □ represent simulation results for the reaction rate k_d determined by the first-passage time of displacement. In the case of weak damping, i.e., small γ, k_E and k_d are quite close, and the Monte Carlo simulation results match the theoretical results quite well even for a large range of filter Γ values. Further research reveals that for predicting the reaction rate under narrowband noise, especially when the power spectral density curve has peak near the linearization frequency ω_0, the applicability of Eq. (5.133) is limited. In fact, under such conditions, the stochastic averaging method of quasi-integrable Hamiltonian systems under wideband stochastic excitation is no longer applicable. For large system damping, i.e., large γ values, both theoretical results k_E and k_d is not accurate, indicating that the stochastic averaging method is no longer applicable.

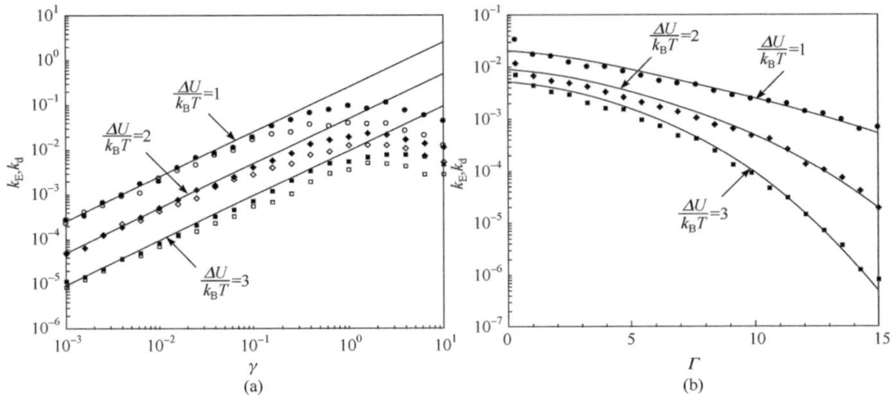

Fig. 5.26 Reaction rate dominated by energy diffusion of system (5.115) under harmonic noise excitation. **a** k_E and k_d versus γ, $\Omega = 3$, $\Gamma = 10$, **b** k_E and k_d versus Γ, $\Omega = 3$, $\gamma = 0.01$ (Deng and Zhu 2007b).

5.3 Fermi Resonance

Fermi resonance originates from the study of the Raman effect by the Italian-American physicist Fermi (Fermi 1931). According to Fermi's description, three atoms in CO_2 construct three oscillators, among which the natural frequency ratio of two oscillators is 1:2. Due to the coupling between oscillators, CO_2 exhibits a unique resonance phenomenon, known as Fermi resonance. Fermi resonance also exists in protein molecules (Volkenstein 1947; Shidlovskaya et al. 2000), which can explain why resonance can enhance the reaction rate of protein molecules at its active site. Theoretical studies on Fermi resonance can be conducted by establishing a vibrational system consisting of two coupled oscillators (Fermi 1931; Fermi and Rasetti 1931).

5.3.1 *Pippard Model of Fermi Resonance*

Usually, the reaction activities of biomolecules involve thousands of degrees of freedom. To explore the fundamental reaction mechanism, relatively simple models are often used. A popular model is the motion of a reacting particle with mass m in a two-dimensional potential landscape $U(x_1, x_2)$. The particle motion equation is given by

$$m\ddot{x}_1 + \frac{\partial U(x_1, x_2)}{\partial x_1} = 0, \quad m\ddot{x}_2 + \frac{\partial U(x_1, x_2)}{\partial x_2} = 0. \tag{5.141}$$

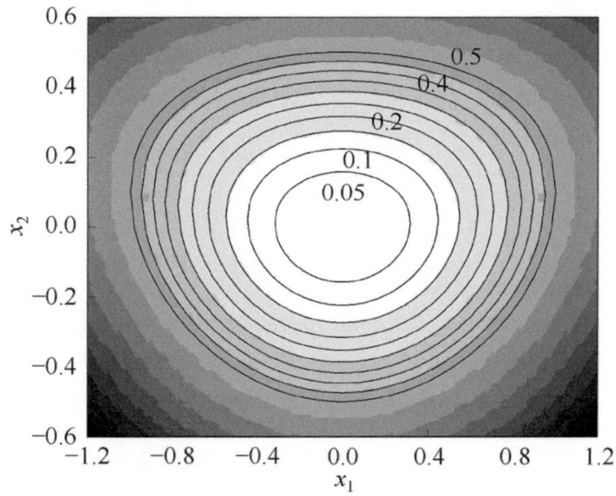

Fig. 5.27 Contour of Pippard potential energy in Eq. (5.142).

The Fermi resonance of the peptide bond in enzyme-catalyzed reactions is described by model (5.141). To describe Fermi resonance, various potential forms can be considered (Ebeling et al. 2003; Netrebko 1994; Shidlovskaya et al. 2000). Here, the following commonly used Pippard potential (Pippard et al. 1983) is taken

$$U(x_1, x_2) = \omega_1^2 x_1^2/2 + \omega_2^2 (x_2 - c x_1^2)^2/2. \tag{5.142}$$

Figure 5.27 shows the contour of Pippard potential (5.142), which is symmetric with respect to x_1 and asymmetric with respect to x_2. The Pippard model for studying the Fermi resonance of peptide bond in enzyme-catalyzed reaction is as follows

$$\ddot{x}_1 + \omega_1^2 x_1 - 2c\omega_2^2 x_1 x_2 + 2c^2\omega_2^2 x_1^3 = 0, \quad \ddot{x}_2 + \omega_2^2 x_2 - c\omega_2^2 x_1^2 = 0. \tag{5.143}$$

In system (5.143), the first oscillator is named as the reacting oscillator which represents the stretching vibrations and bond breaking behavior of peptide bond. The second oscillator is named as the exciting oscillator which represents vibrations occurring on the atomic clusters nearing to peptide bond and can influence the behavior of reacting oscillator through resonance relation. The coupling between the reacting oscillator and the exciting oscillator is characterized by the parameter c known as the coupling coefficient, which reflects the coupling strength between the two oscillators.

In the Pippard model, the linearized natural frequencies of the two oscillators are ω_1 and ω_2. When c is small, the high-order small terms $2c^2\omega_2^2 x_1^3$ are ignored, system (5.143) becomes

$$\ddot{x}_1 + \omega_1^2 x_1 - 2c\omega_2^2 x_1 x_2 = 0, \quad \ddot{x}_2 + \omega_2^2 x_2 - c\omega_2^2 x_1^2 = 0. \tag{5.144}$$

The deterministic averaging method of quasi-linear systems (Bogolyubov and Mitroplskii 1974) can be applied to obtain the change rate of energy e_1 of the reacting oscillator

$$\frac{\mathrm{d}e_1}{\mathrm{d}t} = \frac{-ce_1\sqrt{2e_2}m^2\left[4n\sin(m\pi)\sin(m\pi - \psi/n) + m\sin(4n\pi)\sin(\psi/n)\right]}{2\pi n^2(4n^2 - m^2)}. \tag{5.145}$$

where frequency ratio $\omega_1 : \omega_2 = n : m$; ψ represents the phase angle difference between the two oscillators. It can be observed through analyzing Eq. (5.145) that the influence of the ratio $n : m$ on the energy change rate $\mathrm{d}e_1/\mathrm{d}t$ is

$$\frac{\mathrm{d}e_1}{\mathrm{d}t} = \begin{cases} -2ce_1\sqrt{2e_2}\sin\psi, & n : m = 1 : 2, \\ 0, & n : m \neq 1 : 2. \end{cases} \tag{5.146}$$

Similarly, the following expressions can be obtained

$$\frac{\mathrm{d}e_2}{\mathrm{d}t} = 2ce_1\sqrt{2e_2}\sin\psi, \quad \frac{\mathrm{d}\psi}{\mathrm{d}t} = \frac{\sqrt{2}c(e_1 - 2e_2)\cos\psi}{\sqrt{e_2}}, \quad n : m = 1 : 2. \tag{5.147}$$

Equations (5.146) and (5.147) provide the energy exchange rate between the two oscillators. It can be seen that the energy change rate of the reacting oscillator is equal to that of the exciting oscillator with opposite sign, indicating conservation of the system total energy. This energy flow only exists when the frequency ratio $\omega_1 : \omega_2 = 1 : 2$ satisfies the internal resonance condition. Only under this frequency ratio, system (5.144) exhibits internal resonance, which is necessary for Fermi resonance. Furthermore, it can be seen that when the phase angle difference $\psi = 0$ or π, energy flow between the two oscillators does not occurs even the frequency ratio is satisfied.

5.3.2 *First-Passage Time of Pippard System Under Stochastic Excitation*

Temperature is an important factor in enzyme-catalyzed reactions. Therefore, it is more meaningful to study Fermi resonance in a thermal environment. By introducing thermal perturbations based on Einstein's relations $D = \gamma k_\mathrm{B} T$, where k_B is the Boltzmann constant and T is the thermal bath temperature, the original deterministic Pippard system (5.144) is transformed into the following stochastically excited Pippard system (Deng and Zhu 2008a, 2010, 2012)

$$\begin{aligned} &\ddot{X}_1 + \gamma\dot{X}_1 + \omega_1^2 X_1 - 2c\omega_2^2 X_1 X_2 = \sqrt{2\gamma T}W_{g1}(t), \\ &\ddot{X}_2 + \gamma\dot{X}_2 + \omega_2^2 X_2 - c\omega_2^2 X_1^2 = \sqrt{2\gamma T}W_{g2}(t), \end{aligned} \tag{5.148}$$

where γ is the damping coefficient; $W_{g1}(t)$ and $W_{g2}(t)$ are independent unit Gaussian white noises. By performing Monte Carlo simulation on system (5.148), its dynamic behavior can be observed. Figure 5.28 shows the energy $E_1(t)$, $E_2(t)$ of the stochastically excited oscillators under six different frequency ratios. It can be observed that energy exchange between the two oscillators is more frequent in the case of $\omega_1 : \omega_2 = 1 : 2$. Figure 5.29 shows the displacement $x_1(t)$ of the reacting oscillator. It can be seen that the reacting oscillator displacement changes more frequently in the case of $\omega_1 : \omega_2 = 1 : 2$. A similar report can be found in the reference (Ebeling et al. 2004), which indicates that this frequent changes of energy and displacement in reacting oscillators directly lead to a reduction in the first-passage time and an increase in the reaction rate. Since the energy change plays a crucial role in the first-passage of the reacting oscillators, it can be a suitable index for reaction rate study, and the stochastic averaging method of quasi-integrable Hamiltonian systems under wideband stochastic excitation presented in Sect. 1.1 can be applied to predict the mean first-passage time and then the reaction rate.

Combining the damping force and the nonlinear force in system (5.148) yields

$$F_1(\boldsymbol{X}, \dot{\boldsymbol{X}}) = \gamma \dot{X}_1 - 2c\omega_2^2 X_1 X_2, \quad F_2(\boldsymbol{X}, \dot{\boldsymbol{X}}) = \gamma \dot{X}_2 - c\omega_2^2 X_1^2. \tag{5.149}$$

where $\boldsymbol{X} = [X_1, X_2]^{\mathrm{T}}$. When γ and excitation intensity are weak, the following transformation can be used

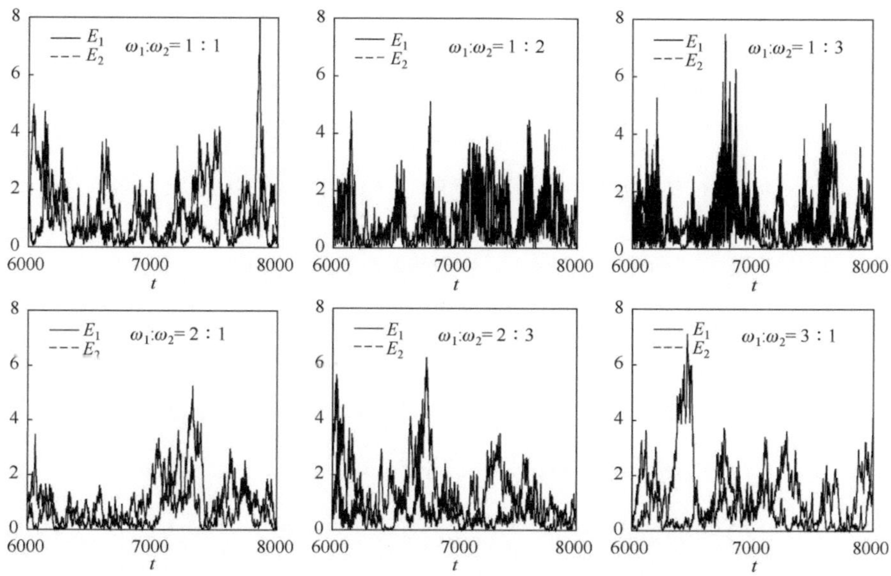

Fig. 5.28 The energy exchange processes of two oscillators in Pippard system (5.148) with different frequency ratios, $c = 0.1$, $\gamma = 0.01$, $\gamma T = 0.01$, $\omega_1 = 1$ (Deng and Zhu 2008a).

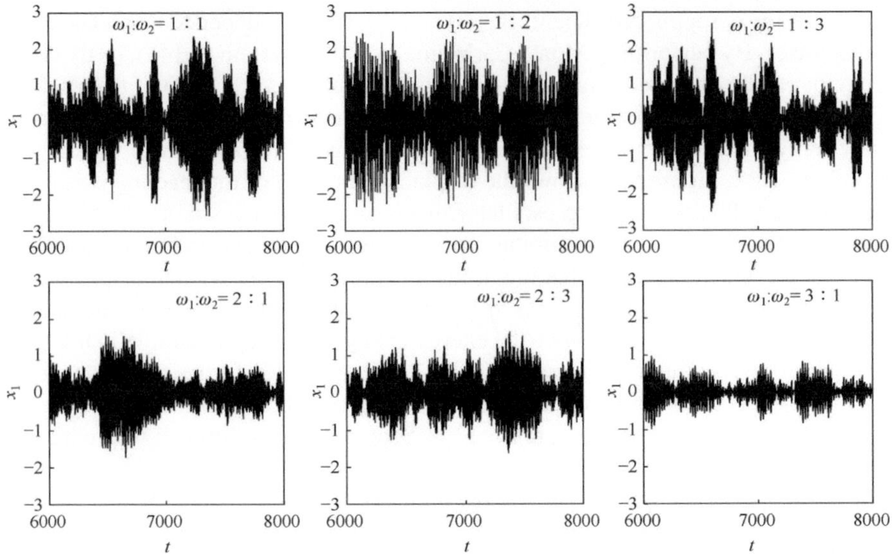

Fig. 5.29 Samples of displacement process of reacting oscillator in Pippard system (5.148) with different frequency ratios. Parameters are the same as those in Fig. 5.28 (Deng and Zhu 2008a).

$$X_i = A_i(t)\cos\Phi_i(t),\quad \dot{X}_i = -\omega_i A_i(t)\sin\Phi_i(t),\quad \Phi_i(t) = \omega_i t + \Theta_i(t),\quad i = 1, 2. \tag{5.150}$$

where A_i, Φ_i, Θ_i are stochastic processes. Eq. (5.150) is used to transform system (5.148) to

$$\begin{aligned} \dot{A}_1 &= F_1^A + G_{11}^A W_{g1}(t),\quad \dot{\Phi}_1 = \omega_1 + F_1^\Phi + G_{11}^\Phi W_{g1}(t), \\ \dot{A}_2 &= F_2^A + G_{22}^A W_{g2}(t),\quad \dot{\Phi}_2 = \omega_2 + F_2^\Phi + G_{22}^\Phi W_{g2}(t). \end{aligned} \tag{5.151}$$

where

$$\begin{aligned} F_1^A &= F_1\sin\Phi_1/\omega_1,\quad G_{11}^A = -\sqrt{2\gamma T}\sin\Phi_1\Big/\omega_1, \\ F_1^\Phi &= F_1\cos\Phi_1/(A_1\omega_1),\quad G_{11}^\Phi = -\sqrt{2\gamma T}\cos\Phi_1\Big/(A_1\omega_1), \\ F_2^A &= F_2\sin\Phi_2/\omega_2,\quad G_{22}^A = -\sqrt{2\gamma T}\sin\Phi_2\Big/\omega_2, \\ F_2^\Phi &= F_2\cos\Phi_2/(A_2\omega_2),\quad G_{22}^\Phi = -\sqrt{2\gamma T}\cos\Phi_2\Big/(A_2\omega_2). \end{aligned} \tag{5.152}$$

In the stochastic system (5.151), the phase angles Φ_1 and Φ_2 are rapidly-varying processes, while the amplitudes A_1 and A_2 are slowly-varying processes. The dimension of the averaged Itô stochastic differential equation depends on whether the system (5.148) has internal resonance.

In the non-resonant case, the frequency ratio $\omega_1 : \omega_2$ of the two oscillators is far from 1:2, and as $\gamma \to 0$, the amplitudes A_1 and A_2 of the system (5.151) weakly converge to a two-dimensional Markov diffusion process. Using the stochastic averaging method described in Sect. 8.1.2, the averaged Itô stochastic differential equations can be established. The corresponding averaged FPK equation governing the transition PDF $p = p(\boldsymbol{a}, t|\boldsymbol{a}_0, t_0)$ is (Deng and Zhu 2008a)

$$\frac{\partial p}{\partial t} = -\sum_{i=1}^{2} \frac{\partial}{\partial a_i}(a_i') + \frac{1}{2}\sum_{i,j=1}^{2} \frac{\partial^2}{\partial a_i \partial a_j}(b_{ij}p), \quad i, j = 1, 2, \tag{5.153}$$

where $\boldsymbol{a} = [a_1, a_2]^{\mathrm{T}}$, the first and second derivative moments are given by

$$\begin{aligned} a_i'(\boldsymbol{a}) &= \left\langle F_i^A + \gamma T \sum_{n,k=1}^{2} \frac{\partial G_{ik}^A}{\partial a_n} G_{nk}^A + \gamma T \sum_{n,k=1}^{2} \frac{\partial G_{ik}^A}{\partial \phi_n} G_{nk}^{\Phi} \right\rangle_t, \\ b_{ij}(\boldsymbol{a}) &= \left\langle 2\gamma T \sum_{k=1}^{2} G_{ik}^A G_{jk}^A \right\rangle_t, \quad i, j = 1, 2. \end{aligned} \tag{5.154}$$

In (5.154), the time averaging can be replaced by the averaging with respect to ϕ_1 and ϕ_2 as follows

$$\langle [\cdot] \rangle_t = \frac{1}{4\pi^2} \int_0^{2\pi}\int_0^{2\pi} [\cdot] \mathrm{d}\phi_1 \mathrm{d}\phi_2. \tag{5.155}$$

Substituting Eq. (5.152) into Eq. (5.154) and completing the operation of averaging result in

$$\begin{aligned} a_1' &= -\frac{\gamma a_1}{2} + \frac{\gamma T}{2a_1\omega_1^2}, \quad a_2' = -\frac{\gamma a_2}{2} + \frac{\gamma T}{8a_2\omega_1^2}, \\ b_{11} &= \frac{\gamma T}{\omega_1^2}, \quad b_{22} = \frac{\gamma T}{4\omega_1^2}, \quad b_{12} = b_{21} = 0. \end{aligned} \tag{5.156}$$

According to Itô differential rule and the following relationship between the amplitudes A_1, A_2 and the energy E_1, E_2

$$E_1 = \omega_1^2 A_1^2 / 2, \quad E_2 = \omega_2^2 A_2^2 / 2. \tag{5.157}$$

the averaged Itô stochastic differential equations governing the energy diffusion process $\mathbf{E} = [E_1, E_2]^{\mathrm{T}}$ can be established. The corresponding averaged FPK equation is

$$\frac{\partial p}{\partial t} = -\frac{\partial}{\partial e_1}(\bar{a}_1 p) - \frac{\partial}{\partial e_2}(\bar{a}_2 p) + \frac{1}{2}\frac{\partial^2}{\partial e_1^2}(\bar{b}_{11} p) + \frac{1}{2}\frac{\partial^2}{\partial e_2^2}(\bar{b}_{22} p). \tag{5.158}$$

along with the following first and second derivative moments

$$\bar{a}_1 = \gamma T - \gamma e_1, \quad \bar{a}_2 = \gamma T - \gamma e_2, \quad \bar{b}_{11} = 2\gamma T e_1, \quad \bar{b}_{22} = 2\gamma T e_2. \tag{5.159}$$

The mean first-passage time of reacting oscillator energy is governed by the following Pontryagin equation

$$\frac{1}{2}\bar{b}_{11}(e_{10})\frac{\partial^2 \tau}{\partial e_{10}^2} + \frac{1}{2}\bar{b}_{22}(e_{20})\frac{\partial^2 \tau}{\partial e_{20}^2} + \bar{a}_1(e_{10})\frac{\partial \tau}{\partial e_{10}} + \bar{a}_2(e_{20})\frac{\partial \tau}{\partial e_{20}} = -1. \tag{5.160}$$

where $\tau(e_{10}, e_{20})$ is the mean first-passage time of the energy process $E_1(t)$ with threshold E_C. Since the first oscillator is reacting oscillator while the second one is exciting oscillator, the solution domain is an enclosed region where the initial energy $0 \le e_{10} < E_C$ and $0 < e_{20}$ (as shown in Figure 5.30). The boundary conditions for Eq. (5.160) are

$$\tau(e_{10} = 0, e_{20}) = \text{finite}, \quad \tau(e_{10} = E_C, e_{20}) = 0. \tag{5.161}$$

$$\tau(e_{10} < E_C, e_{20} = 0) = \text{finite}, \quad \tau(e_{10} < E_C, e_{20} \to \infty) = \text{finite}. \tag{5.162}$$

Note that the initial energy e_{20} on exciting oscillator is not bounded by any threshold, which means it theoretically has an infinite boundary. Fortunately, under the boundary

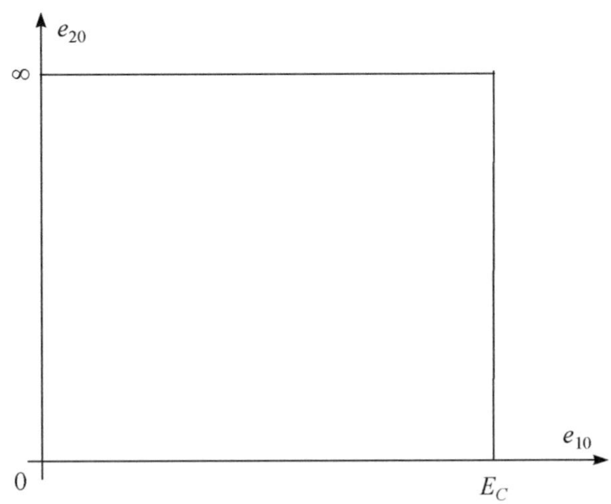

Fig. 5.30 Definition domain of $\tau(e_{10}, e_{20})$ in Eq. (5.160).

conditions (5.161) and (5.162), the following exact analytical solution for the Pontryagin equation (5.160) can be obtained

$$\tau(e_{10}, e_{20}) = \frac{1}{\gamma} \int_{\frac{e_{10}}{T}}^{\frac{E_C}{T}} \frac{e^t - 1}{t} \mathrm{d}t. \tag{5.163}$$

In the internal resonant case, the resonant region is $\omega_1 : \omega_2 = 1 : (2 + \sigma)$, where σ is a small quantity. The amplitudes A_1 and A_2, the phase angle difference $\Psi(t) = 2\Phi_1 - \Phi_2$ are also slowly-varying processes. The governing differential equation for Ψ is obtained from Eq. (5.151) as follows

$$\dot{\Psi} = -\sigma\omega_1 + F_3^A + G_{31}^A W_1(t) + G_{32}^A W_2(t), \tag{5.164}$$

where $F_3^A = 2F_1^\Phi - F_2^\Phi$, $G_{31}^A = 2G_{11}^\Phi$, $G_{32}^A = -G_{22}^\Phi$. Substituting the frequency ratio into system (5.148) and combing it with damping force and the nonlinear force yields

$$\begin{aligned} F_1(\boldsymbol{X}, \dot{\boldsymbol{X}}) &= \gamma \dot{X}_1 - 2c(\omega_2^2 + 2\sigma\omega_1\omega_2) X_1 X_2, \\ F_2(\boldsymbol{X}, \dot{\boldsymbol{X}}) &= \gamma \dot{X}_2 - c(\omega_2^2 + 2\sigma\omega_1\omega_2) X_1^2 + 2\sigma\omega_1\omega_2 X_2, \end{aligned} \tag{5.165}$$

where the terms containing c^2 and σ^2 have been ignored. As $\gamma, c \to 0$, the vector process $[A_1, A_2, \Psi]^\mathrm{T}$ of amplitude $\mathbf{A}$ and phase angle difference Ψ converges weakly to a 3-dimensional Markov diffusion process. Applying the stochastic averaging method of quasi-integrable and resonant Hamiltonian systems under wideband random excitation in Sect. 8.1.2, the following FPK equation for the transition PDF $p = p(\boldsymbol{a}, \psi, t|\boldsymbol{a}_0, \psi_0, t_0)$ (Deng and Zhu 2008a) is obtained.

$$\begin{aligned} \frac{\partial p}{\partial t} &= -\sum_{i=1}^{2} \frac{\partial}{\partial a_i}(\overline{a}_i p) - \frac{\partial}{\partial \psi}(\overline{a}_3 p) + \frac{1}{2}\sum_{i,j=1}^{2} \frac{\partial^2}{\partial a_i \partial a_j}(\overline{b}_{ij} p) + \frac{1}{2}\frac{\partial^2}{\partial \psi^2}(\overline{b}_{33} p) \\ &+ \frac{1}{2}\sum_{i-1}^{2} \frac{\partial^2}{\partial a_i \partial \psi}(\overline{b}_{i3} p) + \frac{1}{2}\sum_{j=1}^{2} \frac{\partial^2}{\partial \psi \partial a_j}(\overline{b}_{3j} p), \quad i, j = 1, 2. \end{aligned} \tag{5.166}$$

The first and second derivative moments are

$$\begin{aligned} \overline{a}_i(\boldsymbol{a}, \psi) &= \left\langle F_i^A + \gamma T \sum_{n,k=1}^{2} \frac{\partial G_{ik}^A}{\partial a_n} G_{nk}^A + \gamma T \sum_{k=1}^{2} \frac{\partial G_{ik}^A}{\partial \psi} G_{3k}^A + \gamma T \sum_{k=1}^{2} \frac{\partial G_{ik}^A}{\partial \phi_2} G_{2k}^\Phi \right\rangle_t, \\ \overline{b}_{ij}(\boldsymbol{a}, \psi) &= \left\langle 2\gamma T \sum_{k=1}^{2} G_{ik}^A G_{jk}^A \right\rangle_t, \quad i, j = 1, 2, 3. \end{aligned} \tag{5.167}$$

According to the relations $\phi_1 = (\phi_2 + \psi)/2$, ϕ_1 in G_{ik}^A, G_{ik}^Φ is replaced by ϕ_2 and ψ. The time averaging is replaced by averaging with respect to ϕ_2, i.e.,

$$\langle[\cdot]\rangle_t = \frac{1}{2\pi}\int_0^{2\pi}[\cdot]\mathrm{d}\phi_2. \tag{5.168}$$

Substituting Eqs. (5.152) and (5.164) into Eq. (5.167) yields

$$\begin{aligned}
\overline{a}_1 &= -\frac{\gamma a_1}{2} + \frac{\gamma T}{2a_1\omega_1^2} - 2c(1+\sigma)a_1 a_2\omega_1 \sin\psi,\\
\overline{a}_2 &= -\frac{\gamma a_2}{2} + \frac{\gamma T}{8a_2\omega_1^2} + \frac{1}{2}c(1+\sigma)a_1^2\omega_1 \sin\psi,\\
\overline{a}_3 &= -\sigma\omega_1 + \frac{c(1+\sigma)\omega_1(a_1^2 - 8a_2^2)\cos\psi}{2a_2},\\
\overline{b}_{11} &= \frac{\gamma T}{\omega_1^2},\quad \overline{b}_{22} = \frac{\gamma T}{4\omega_1^2},\quad \overline{b}_{33} = \frac{4\gamma T}{a_1^2\omega_1^2} + \frac{\gamma T}{4a_2^2\omega_1^2},\quad \overline{b}_{12} = \overline{b}_{13} = \overline{b}_{23} = 0.
\end{aligned} \tag{5.169}$$

According to Itô differential rule and the deterministic relations (5.157), the averaged Itô equations and associated averaged FPK equation governing both process $\mathbf{E} = [E_1, E_2]^{\mathrm{T}}$ and phase angle difference Ψ can be derived from Itô stochastic differential equation associated with Eqs. (5.151) and (5.164). The first and second derivative moments of the averaged FPK equation are

$$\begin{aligned}
\overline{\overline{a}}_1 &= \gamma T - \gamma E_1 - 2c(1+\sigma)E_1\sqrt{2E_2}\sin\psi,\\
\overline{\overline{a}}_2 &= \gamma T - \gamma E_2 + 2c(1+\sigma)E_1\sqrt{2E_2}\sin\psi,\\
\overline{\overline{a}}_3 &= -\sigma\omega_1 + \frac{\sqrt{2}c(1+\sigma)(E_1 - 2E_2)\cos\psi}{\sqrt{E_2}},\\
\overline{\overline{b}}_{11} &= 2\gamma T E_1,\quad \overline{\overline{b}}_{22} = 2\gamma T E_2,\quad \overline{\overline{b}}_{33} = \gamma T\left(\frac{2}{E_1} + \frac{1}{2E_2}\right),\quad \overline{\overline{b}}_{12} = \overline{\overline{b}}_{13} = \overline{\overline{b}}_{23} = 0.
\end{aligned} \tag{5.170}$$

Define $\tau(e_{10}, e_{20}, \psi_0)$ as the mean first-passage time of reacting oscillator energy $E_1(t)$ crossing threshold E_C under the condition that the initial energy $0 \le e_{10} < E_C$, $0 \le e_{20}$. The mean first-passage time satisfies the following Pontryagin equation

$$\frac{1}{2}\overline{\overline{b}}_{11}\frac{\partial^2\tau}{\partial e_{10}^2} + \frac{1}{2}\overline{\overline{b}}_{22}\frac{\partial^2\tau}{\partial e_{20}^2} + \frac{1}{2}\overline{\overline{b}}_{33}\frac{\partial^2\tau}{\partial\psi_0^2} + \overline{\overline{a}}_1\frac{\partial\tau}{\partial e_{10}} + \overline{\overline{a}}_2\frac{\partial\tau}{\partial e_{20}} + \overline{\overline{a}}_3\frac{\partial\tau}{\partial\psi_0} = -1. \tag{5.171}$$

where the derivative moments $\overline{\overline{a}}_i, \overline{\overline{b}}_i$ $(i, j = 1, 2, 3)$ are the $\overline{\overline{a}}_i, \overline{\overline{b}}_i$ in Eq. (5.170) with their variables E_1, E_2, ψ replaced by e_{10}, e_{20}, ψ_0. The boundary conditions are

$$\tau(e_{10} = 0, e_{20}, \psi_0) = \text{ finite}, \quad \tau(e_{10} = E_C, e_{20}, \psi_0) = 0. \tag{5.172}$$

$$\tau(e_{10} < E_C, e_{20} = 0, \psi_0) = \text{ finite}, \quad \tau(e_{10} < E_C, e_{20} \to \infty, \psi_0) = \text{ finite}. \tag{5.173}$$

$$\tau(e_{10} < E_C, e_{20}, \psi_0 = 2\pi) = \tau(e_{10} < E_C, e_{20}, \psi_0 = 0). \tag{5.174}$$

The exact analytical solution for the 3-dimensional parabolic partial differential equation (5.171) has not been found. The numerical methods such as finite difference method can be employed to solve Eq. (5.171). For computational convenience, using $\overline{e}_{20} = 1 - \exp(-e_{20})$, the definition domain of e_{20} is transformed from an unbounded domain $[0, \infty)$ to a bounded domain [0, 1] (as shown in Fig. 5.31). Consequently, boundary conditions (5.172)–(5.174) are transformed into

$$\begin{aligned}
&\tau(e_{10} = 0, \overline{e}_{20} < 1, \psi_0) = \text{finite}, \quad \tau(e_{10} = E_C, \overline{e}_{20} < 1, \psi_0) = 0, \\
&\tau(e_{10} < E_C, \overline{e}_{20} = 0, \psi_0) = \text{finite}, \quad \tau(e_{10} < E_C, \overline{e}_{20} = 1, \psi_0) = \text{finite}, \\
&\tau(e_{10} < E_C, \overline{e}_{20} < 1, \psi_0 = 2\pi) = \tau(e_{10} < E_C, \overline{e}_{20} < 1, \psi_0 = 0).
\end{aligned} \tag{5.175}$$

The Pontryagin equation (5.171) is transformed into

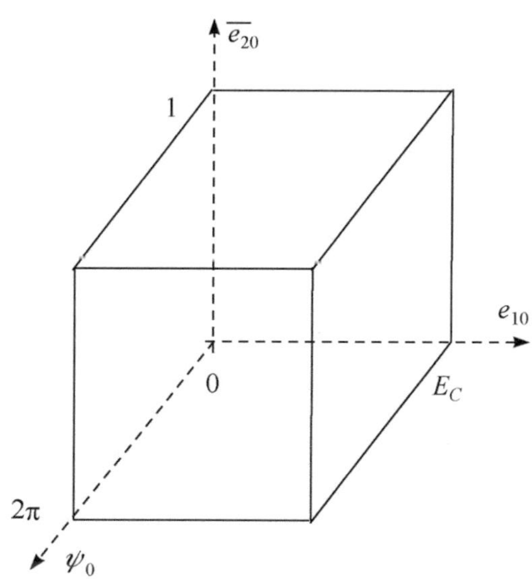

Fig. 5.31 Definition domain of $\tau(e_{10}, \overline{e}_{20}, \psi_0)$ in Eq. (5.176).

$$\frac{1}{2}\overline{\overline{b}}_{11}\frac{\partial^2\tau}{\partial e_{10}^2}+\frac{1}{2}\tilde{b}_{22}\frac{\partial^2\tau}{\partial \overline{e}_{20}^2}+\frac{1}{2}\overline{\overline{b}}_{33}\frac{\partial^2\tau}{\partial \psi_0^2}+\overline{\overline{a}}_1\frac{\partial\tau}{\partial e_{10}}+\tilde{a}_2\frac{\partial\tau}{\partial \overline{e}_{20}}+\overline{\overline{a}}_3\frac{\partial\tau}{\partial \psi_0}=-1, \tag{5.176}$$

where the coefficients $\tilde{a}_2(e_{10}, \overline{e}_{20}, \psi_0)$ and $\tilde{b}_{22}(e_{10}, \overline{e}_{20}, \psi_0)$ can be derived from the first and second derivative moments $\overline{\overline{a}}_2$ and $\overline{\overline{b}}_{22}$ in Eq. (5.170) as follows

$$\begin{aligned}\tilde{a}_2 &= \overline{\overline{a}}_2\frac{\mathrm{d}}{\mathrm{d}e_2}[1-\exp(-e_2)]+\frac{1}{2}\overline{\overline{b}}_{22}\frac{\mathrm{d}^2}{\mathrm{d}e_2^2}[1-\exp(-e_2)],\\ \tilde{b}_{22} &= \{\frac{\mathrm{d}}{\mathrm{d}e_2}[1-\exp(-e_2)]\}^2\overline{\overline{b}}_{22}.\end{aligned} \tag{5.177}$$

Figure 5.32 shows the results from numerical solutions of Eq. (5.176) under boundary conditions (5.175), using contour values to represent the mean first-passage time τ as functions of e_{10} and $\overline{e}_{20}$ with ψ_0 as parameter. As the initial energy e_{10} of reacting oscillator decreases, the influence of the initial energy $\overline{e}_{20}$ (which is a monotonically increasing function with an increasing initial energy e_{20}) of exciting oscillator on the passage time depends on the initial phase difference ψ_0 and energy e_{10}. For small values of e_{10}, an increase in the initial energy $\overline{e}_{20}$ of exciting oscillator results in a decrease in the mean first-passage time τ. Figure 5.32b–d show that when the energy e_{10} of reacting oscillator approaches the threshold $E_C = 1$, the influence of initial energy $\overline{e}_{20}$ is not monotonous. In the case of $\psi_0 = \pi/2$ (see Figure 5.32c), it can be analyzed from Eq. (5.146) that $\mathrm{d}e_1/\mathrm{d}t$ is negative, indicating that energy flows out of reacting oscillator. The energy $\overline{e}_{20}$ of exciting oscillator inhibits the passage behavior of reacting oscillator, leading to an increase of mean first-passage time. Figure 5.32 shows that mean first-passage time τ is symmetric with respect to $\psi_0 = \pi/2$ and $\psi_0 = 3\pi/2$. Given e_{10} and $\overline{e}_{20}$, τ reaches its maximum and minimum values at $\psi_0 = \pi/2$ and $\psi_0 = 3\pi/2$, respectively. Correspondingly, it is known from Eq. (5.146) that energy in the reacting oscillator reaches the maximum output and maximum input, respectively. To verify the theoretical results in Figure 5.32, Monte Carlo simulation was performed on system (5.148). Figure 5.33 presents simulation results from 2000 samples. It is seen that simulation results closely match the theoretical results, confirming the validity of the stochastic averaging method.

In the case of resonance, the coupling coefficient c plays a crucial role in the energy exchange between the reacting oscillator and the exciting oscillator. Figure 5.34 shows the mean first-passage time τ as function of c for two cases of different passage thresholds E_C. It can be observed that τ increases as E_C increasing, and decreases as c increasing. This is because increasing passage threshold implies that the reacting oscillator needs more time to reach the threshold. Decreasing the coupling strength may reduce the energy supplied to the reacting oscillator from the exciting oscillator, leading to longer passage time. It is important to note that the theoretical results in this section are based on the assumption of weak coupling, so as the coupling coefficient c increases, the accuracy of the theoretical results gradually decreases.

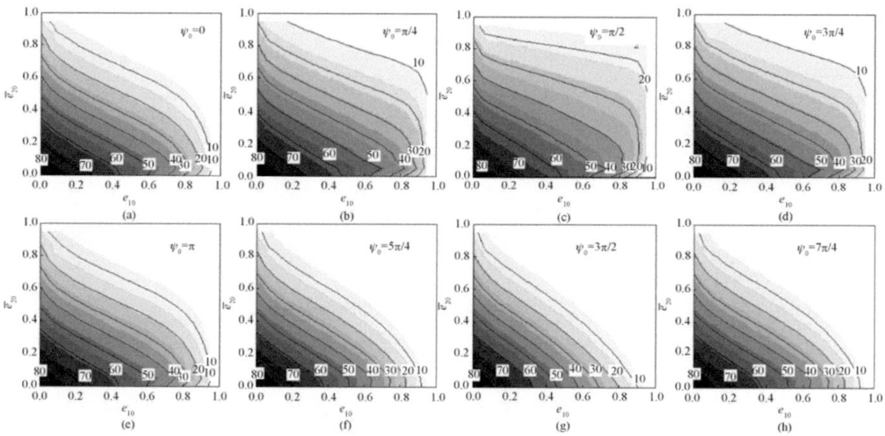

Fig. 5.32 Mean first-passage time τ obtained from solving equation (5.176) for given initial energy $e_{10}, \bar{e}_{20}$ and initial phase difference ψ_0, $c = 0.1$, $\sigma = 0$, $\gamma = 0.01$, $\gamma T = 0.01$, $E_C = 1$, $\omega_1 = 1$, $\omega_2 = 2\omega_1$ (Deng and Zhu 2008a).

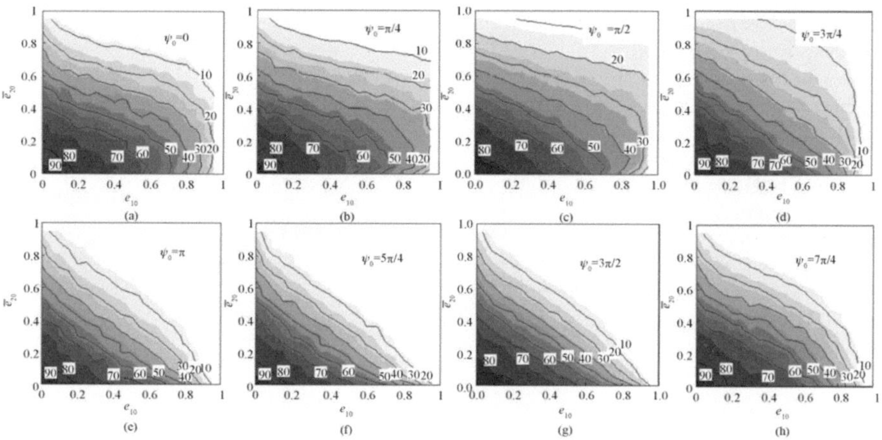

Fig. 5.33 Monte Carlo simulation results for comparison with the theoretical results in Fig. 5.32 (Deng and Zhu 2008a).

Figure 5.35 directly shows the mean first-passage time as the linear frequency ratio $\omega_1 : \omega_2$ changing over a wide range (ω_2 changes from $10^{-2}\omega_1$ to $10^2\omega_1$). It can be seen that the mean first-passage time reaches its minimum value in the case of the frequency relationship of Fermi resonance $\omega_2 = 2\omega_1$. In a small region near the resonant frequency, theoretical predictions can be obtained by solving Eq. (5.176). Other frequency ranges correspond to non-resonant cases, and theoretical predictions for the mean first-passage time can be obtained using Eq. (5.163). The above two theoretical predictions have been verified by using Monte Carlo simulation results.

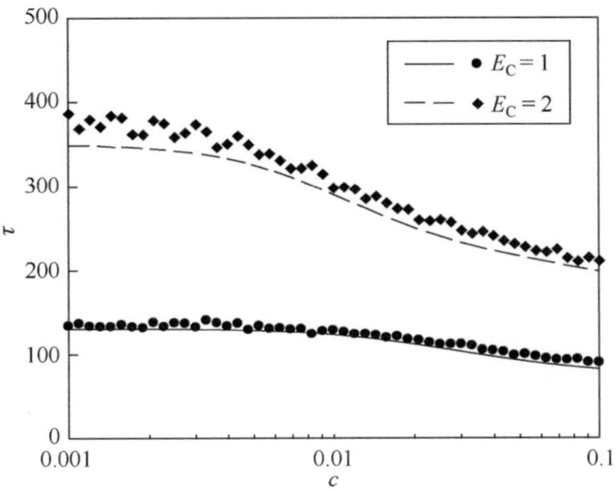

Fig. 5.34 Mean first-passage time changes with coupling coefficient c. Lines represent results from the stochastic averaging method, symbols •, ◆ represent simulation results. $e_{10} = 0, \overline{e}_{20} = 0, \psi_0 = \pi/2$. Other parameters are the same as those in Fig. 5.32 (Deng and Zhu 2008a).

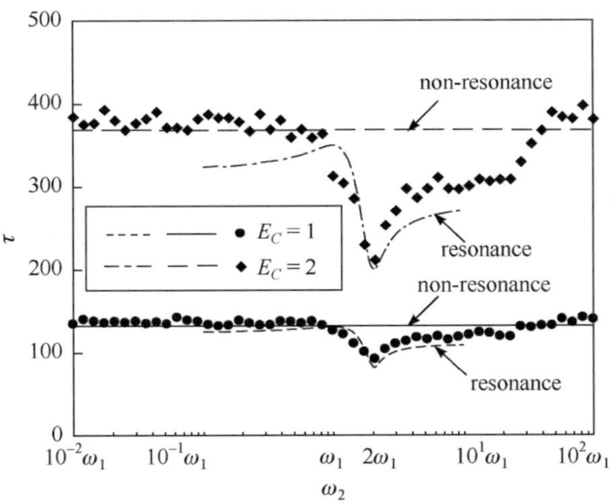

Fig. 5.35 Mean first-passage time τ changes with frequency ratio. Lines represent results from the stochastic averaging method. Results for the resonant region come from solving equation (5.176), and results for the non-resonant region come from Eq. (5.163). Symbols •, ◆ represent simulation results. $e_{10} = 0$, $\overline{e}_{20} = 0$, $\psi_0 = \pi/2$. Other parameters are the same as those in Fig. 5.32 (Deng and Zhu 2008a).

5.3.3 Reaction Rate of Fermi Resonance Under Stochastic Excitation

As described in Fig. 5.19, the classical Kramers reaction rate theory uses a model of a stochastically excited reacting particle moving in a double-well potential energy landscape. In this section, the influence of exciting oscillator on the reaction rate of reacting oscillator in the Fermi resonance is studied. To do this, a two-dimensional potential energy needs to be established. Ebeling constructed the following two-dimensional potential energy (Ebeling et al. 2003)

$$U(x_1, x_2) = U_0(x_1) + \alpha[x_2 - cU_0(x_1)]^2. \tag{5.178}$$

where $U_0(x_1) = ax_1^2 + \cdots$ is another adjustable one-dimensional potential energy. Figure 5.36 illustrates the two-dimensional potential energy (5.178), where it can be observed that the potential energy is of double-wells and symmetric in the x_1 direction. Substituting the potential energy (5.178) into system (5.141) results in a two-dimensional system describing the coupling between reacting oscillator and exciting oscillator. With the potential energy given by (5.178), c represents the coupling intensity, and the linear frequencies of the two oscillators are $\omega_1 = \sqrt{2a}$ and $\omega_2 = \sqrt{2\alpha}$. When $\alpha = 4a$, it corresponds to a frequency ratio of $\omega_1 : \omega_2 = 1 : 2$. When the reacting particle is at bottom of the left well shown in Fig. 5.36, this frequency ratio will induce Fermi resonance. As discussed in Sect. 5.2, in the case of weak damping, the resulting Kramers reaction rate expression (5.81) is derived based on the parabolic potential energy approximation of $U_0(x_1) \approx ax_1^2$ and the high barrier assumption of $\Delta U / k_B T \gg 1$ (Deng and Zhu 2007a). To illustrate the application of the theoretical results obtained in Sect. 5.3.2 to reaction rate theory, the parabolic potential energy approximation and the high barrier assumption will be still used.

By replacing ω_1, c, E_C in Eqs. (5.169), (5.170) and (5.172) with $\sqrt{2a}$, ca, ΔU, the mean first-passage time $\tau(e_{10}, \overline{e}_{20}, \psi_0)$ of the reacting oscillator in the resonance region with potential (5.178) can be obtained by solving Pontryagin equation (5.176) with boundary conditions (5.172)–(5.175). In the resonant case, $\tau(0, 0, 0)$ denotes the mean first-passage time under the initial conditions $e_{10} = 0, \overline{e}_{20} = 0$ and $\psi_0 = 0$. The mean first-passage time in the non-resonant case is obtained from Eq. (5.163), and denoted by $\tau(0, 0)$ under the initial conditions $e_{10} = 0$ and $\overline{e}_{20} = 0$. The reaction rate k is obtained from the reciprocal of the mean first-passage time, i.e., $k = 1/\tau$ (Reimann et al. 1999).

Figure 5.37 shows the reaction rate k as function of α. It can be seen that the reaction rate reaches its peak value near $\alpha = 4a$ (corresponding to $\omega_1 : \omega_2 = 1 : 2$). Figure 5.37 also demonstrates that increasing the barrier height ΔU has an effect of decreasing the reaction rate.

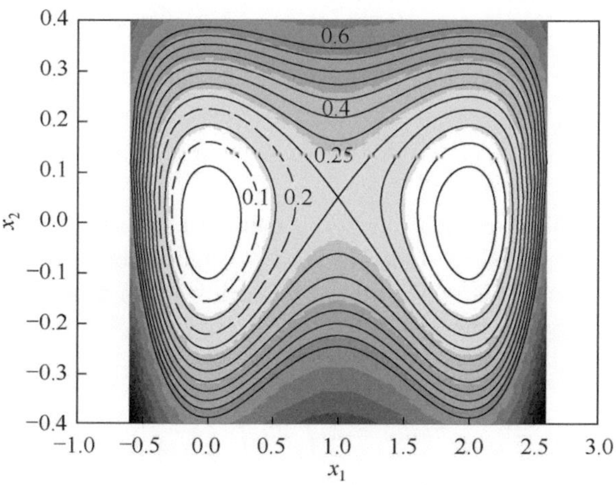

Fig. 5.36 Contour of potential energy in Eq. (5.178), $a = 1, b = 1$, $\alpha = 4a, c = 0.2$ (Deng and Zhu 2008a).

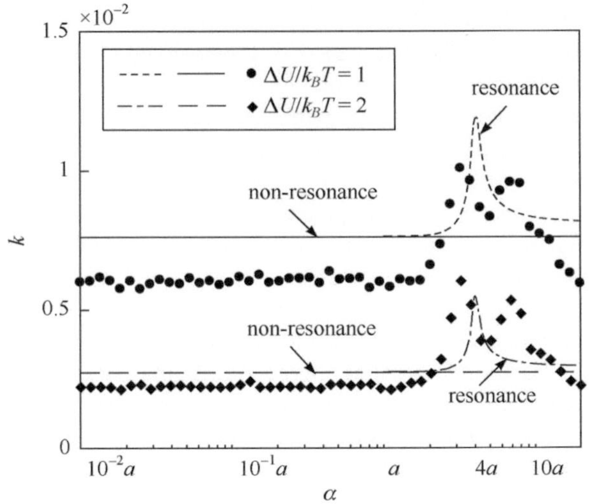

Fig. 5.37 Reaction rate k as function of α. Lines represent results from the stochastic averaging method. Results for the resonant region come from solving equation (5.176), and results for the non-resonant region come from Eq. (5.163). Symbols •, ◆ represent simulation results. Parameters are the same as those in Fig. 5.36 (Deng and Zhu 2008a).

The Fermi resonance studied before are based on the assumption of weak coupling, and the theoretical results are only applicable to the cases with small coupling coefficients c. In the case of strong coupling, i.e., large c, the reacting oscillator and the exciting oscillator move together like a single system. In this case, the entire two-degree-of-freedom system can be considered as a quasi-non-integrable Hamiltonian system. The system Hamiltonian function or total energy is given by

$$H = v_1^2/2 + v_2^2/2 + U(x_1, x_2). \tag{5.179}$$

where $U(x_1, x_2)$ is the potential energy in Eq. (5.178). It is evident that the total energy expression (5.179) is inseparable. Thus, the system associated with Hamiltonian H in Eq. (5.179) cannot be treated as a resonant system. Its motion equation is

$$\begin{aligned}
\ddot{X}_1 + \gamma \dot{X}_1 + \frac{\partial U(X_1, X_2)}{\partial X_1} &= \sqrt{2\gamma k_B T} W_{g1}(t), \\
\ddot{X}_2 + \gamma \dot{X}_2 + \frac{\partial U(X_1, X_2)}{\partial X_2} &= \sqrt{2\gamma k_B T} W_{g2}(t).
\end{aligned} \tag{5.180}$$

The stochastic averaging method of quasi-non-integrable Hamiltonian systems was applied to study the first-passage problem of system (5.180) (Deng and Zhu 2008a). This results in the following mean time for the system when total energy first passes threshold E_C

$$\tau(h_0) = 2\int_{h_0}^{E_C} du \int_0^u \frac{1}{\sigma^2(v)} \exp\left[-2\int_v^u \frac{m(w)}{\sigma^2(w)} dw\right] dv, \tag{5.181}$$

where the initial energy $h_0 < E_C$, drift coefficient $m(H)$ and diffusion coefficient $\sigma^2(H)$ are given by

$$m(H) = 2\gamma k_B T - 2\gamma H + 2\gamma G(H), \quad \sigma^2(H) = 4\gamma k_B TH - 4\gamma k_B TG(H),$$
$$G(H) = \iint\limits_{\Sigma} U(x_1, x_2)\mathrm{d}x_1\mathrm{d}x_2 \Big/ \iint\limits_{\Sigma} \mathrm{d}x_1\mathrm{d}x_2, \quad \Sigma = \{(x_1, x_2)|U(x_1, x_2) \le H\}. \tag{5.182}$$

Substituting the potential energy (5.178) into Eqs. (5.182) and (5.181) yields the mean first-passage time $\tau(h_0)$ and then the reaction rate $k = 1/\tau(0)$.

Figures 5.38 and 5.39 show the theoretical results and Monte Carlo simulation results under condition of strong coupling. It can be observed that they match well, and the influences of several parameters E_C, α, c on the reaction rate are displayed. All figures demonstrate that raising the threshold E_C has a decreasing effect on the reaction rate. Figure 5.38 indicates that the theoretical results are only applicable within a small damping range $\gamma < 1$. Figure 5.39 shows that the linearized frequency ratio of the two oscillators does not affect the reaction rate, even under resonance condition $\alpha = 4a$. Figure 5.39 also demonstrates that the coupling strength between the two oscillators does not affect the reaction rate. As a non-integrable system, there is no resonance and the coupling intensity only influence the energy ratio of the two oscillators. It does not affect the reaction rate of the system total energy.

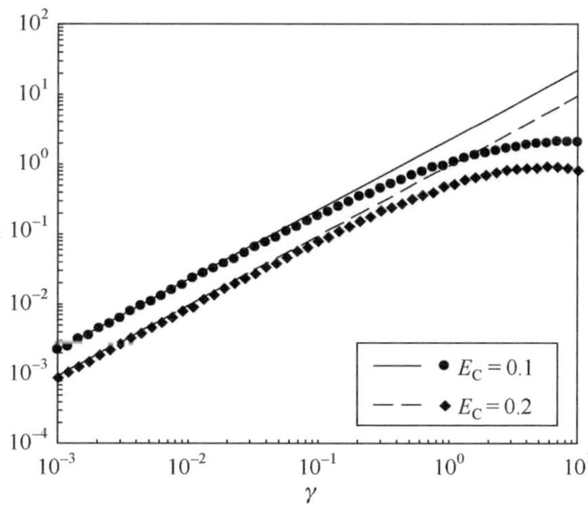

Fig. 5.38 Reaction rate k dominated by total energy of system (5.148) with strong coupling and weak damping. Lines represent results from the stochastic averaging method, symbols •, ◆ represent simulation results. Parameters $c = 1$, $a = 1$, $b = 1$, $\alpha = 4a$, $\Delta U / k_B T = 2$ (Deng and Zhu 2008a)

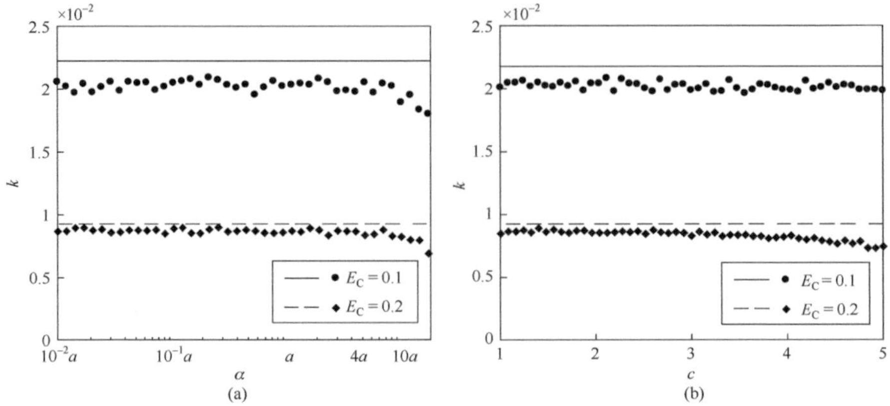

Fig. 5.39 Reaction rate k dominated by total energy of system (5.148) with strong coupling and weak damping. **a** Reaction rate changes with parameter α, **b** reaction rate changes with coupling parameter c. Lines represent results from the stochastic averaging method, and symbols •, ◆ represent simulation results. $\gamma = 0.01$, other parameters are the same as those in Fig. 5.38 (Deng and Zhu 2008a).

5.4 Thermal Motion of DNA Molecule

The static structure of deoxyribonucleic acid (DNA) molecules is insufficient to explain their biological properties. The opening and closing of DNA double strands are crucial for life behaviors like replication, transcription, and protein binding. The motion of local opening and closing DNA double strands is named by DNA denaturation or melting. This continuous motion is an incessant process, vividly referred as DNA breathing. Currently, most theoretical methods for studying DNA dynamics are based on two models, the Poland-Scherage free energy model (Poland and Scheraga 1966) and the Peyrard-Bishop-Dauxois (PBD) model (Dauxois 1991; Peyrard and Bishop 1989). The PBD model can be used to estimate the DNA melting temperatures, thermal stability, and the recognition of transcription start sites (Campa and Giansanti 1998; Kalosakas et al. 2004; Theodorakopoulos et al. 2004; van Erp et al. 2005). In this section, the PBD model is treated as a multi-degree-of-freedom strongly nonlinear dynamical system under thermal perturbations, and the stochastic averaging methods are employed to study the stationary motion and denaturation process of DNA molecule (Deng and Zhu 2008b; Deng 2007).

5.4.1 PBD Model of DNA Molecule

Figure 5.40 sketches the PBD model. The two backbones of model represent the two sugar-phosphate backbones of a DNA molecule, with numerous side chains on the backbones representing the bases. In a base-pair, one base on a backbone interacts

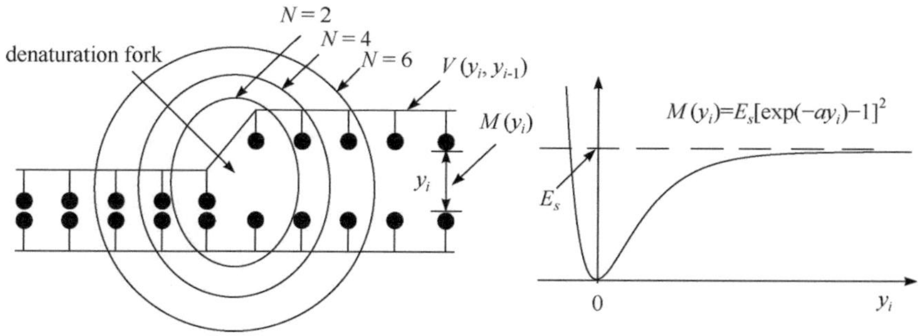

Fig. 5.40 Schematic diagram of DNA denaturation using PBD model. The two bases (denoted by dark circles) in a base-pair interact through the Morse potential in Eq. (5.183). Adjacent base-pairs interact through the PBD stacking potential in Eq. (5.184) (Deng and Zhu, 2008b).

with another base on the other backbone through hydrogen bonds. The thermal denaturation of DNA molecules, i.e., the separation of the two nucleotide strands under thermal effect, is also known as DNA melting. Under electron microscope, regions where the opened strands exhibit a loose structure, known as denaturation bubbles. The size and number of denaturation bubbles increase as the temperature rises, and at the denaturation temperature, the entire double strand of DNA unwind (as shown in Fig. 5.41).

In the PBD model of DNA denaturation dynamics, the interaction between bases is described by the following Morse potential energy

$$M(y_i) = E_S(e^{-ay_i} - 1)^2 \tag{5.183}$$

where y_i represents the relative distance from equilibrium between bases in the i th base-pair (Dauxois, 1991; Peyrard and Bishop, 1989), E_S is the separation energy of base-pair, a is a parameter describing the scale of potential space (Peyrard, 2004). The number of hydrogen bond is two for base-pair A-T (adenine-thymine) and three for base-pair C-G (cytosine-guanine), resulting in different link strength and different values of E_S and a in the Morse potential energy. Therefore, considering the influence of base-pair sequence information on the motion of PBD model would bring high complexity. In this section, only DNA molecule with uniform base pair sequence and its corresponding PBD model are considered.

On the backbone, only the bending effect between neighboring base-pairs is considered (as shown in Fig. 5.40), and the resulting potential energy $V(y_i, y_{i-1})$ is used to describe the stacking interaction between adjacent base-pairs

$$V(y_i, y_{i-1}) = \frac{1}{2}K\{1 + \rho\exp[-\alpha(y_i + y_{i-1})]\}(y_i - y_{i-1})^2. \tag{5.184}$$

It can be seen from potential energy (5.184) that it modified the harmonic potential between neighboring base-pairs through an exponential term which introduces

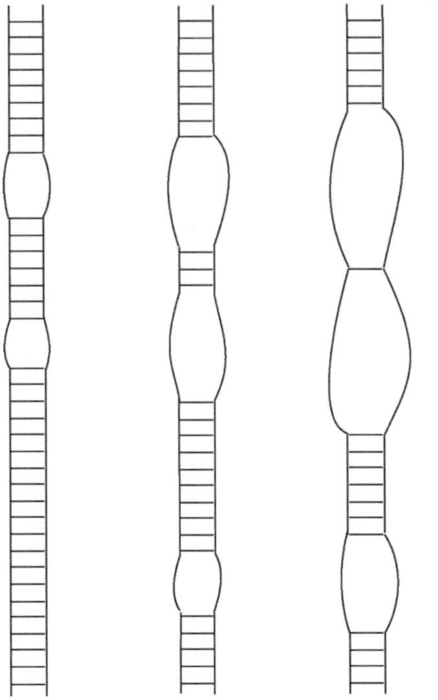

Fig. 5.41 Denaturation process of DNA molecules. The denaturation bubbles grow as temperature increases (Deng and Zhu, 2008b).

nonlinearity into the system. The potential energy (5.184) has been proven to be a successful expression which can not only provide a qualitative description of DNA denaturation but also enable quantitative calculations. In addition, the stacking energy is highly dependent on base-pair sequence information (Saenger 1984), but this section only considers a uniform sequence.

To simulate the thermal denaturation of DNA molecules, it is assumed that the local opening of base-pairs is caused by the separation of the backbones, and the helical effect is ignored to simplify the theoretical analysis. Because of the uniform base-pair sequence, all base-pairs have the same mass, allowing us to derive the following motion equation for PBD model

$$\frac{\mathrm{d}^2 y_i}{\mathrm{d}t^2} = -\frac{\partial U(\mathbf{y})}{\partial y_i}, \quad U(\mathbf{y}) = \sum_{i=1}^{N} M(y_i) + \sum_{i=2}^{N} V(y_i, y_{i-1}), \quad i = 1, 2, \ldots, N. \tag{5.185}$$

where the mass of very base is normalized to 1, $\mathbf{y}$=$[y_1, y_2, \ldots, y_N]^{\mathrm{T}}$, N is the number of base-pairs. The Hamiltonian function or the total energy of the motion system (5.185) is

$$H = \frac{1}{2} \sum_{j=1}^{N} \left(\frac{\mathrm{d}y_j}{\mathrm{d}t} \right)^2 + U(\mathbf{y}). \tag{5.186}$$

5.4.2 *Stationary Motion of DNA Molecules*

As mentioned previously, the continuous local unwinding and rewinding of the double-strands of DNA molecules, known as DNA breathing, are essentially the stationary motion of DNA molecules under thermal perturbation. In research, to introduce a thermal bath environment to the PBD model, random perturbation forces $\sqrt{2D}W_{gi}(t)$, $(i = 1, 2, \ldots, N)$ and friction forces $\gamma \mathrm{d}y_i / \mathrm{d}t$ are typically added to each degree-of-freedom, where $W_{gi}(t)$ represents independent unit Gaussian white noise, and γ is the linear damping coefficient. The fluctuation-dissipation theorem or Einstein's relation $D = \gamma k_B T$ (where T is the thermal bath absolute temperature) is applied. After adding random forces $\sqrt{2D}W_{gi}(t)$ and friction forces $\gamma \mathrm{d}y_i / \mathrm{d}t$ to system (5.185), the system becomes the following stochastically excited dynamical system

$$\frac{\mathrm{d}^2 y_i}{\mathrm{d}t^2} + \gamma \frac{\mathrm{d}y_i}{\mathrm{d}t} + \frac{\partial U(\mathbf{y})}{\partial y_i} = \sqrt{2\gamma k_B T} W_{gi}(t), \quad i = 1, 2, \cdots, N. \tag{5.187}$$

Figures 5.42 and 5.43 show the motion snapshots and grayscale images of the denaturation process simulated by the PBD model with 50 base-pairs. In Fig. 5.42a, the excitation intensity $\gamma k_B T = 0.01$ remains constant, the motion of base-pairs reaches a stationary state, with some base-pairs opening and then closing, demonstrating a dynamical stationary motion, known as DNA breathing. In Fig. 5.42b, with the damping coefficient γ kept constant, the temperature T gradually increases, causing $\gamma k_B T$ linearly increase from 0 to 0.02. It can be observed that as time t progresses, more base-pairs open, and bubbles become larger. Figure 5.43a, b show the distance of base-pair opening as a function of temperature. In addition to the distance of base-pair opening, the PBD mean energy E (the model total energy divided by the total number of base-pairs) can also be used to quantify the strength of DNA breathing. Figure 5.43c illustrates the evolution of PBD mean energy. It is seen that at room temperature (corresponding to Fig. 5.43a), the PBD mean energy fluctuates around a certain constant, whereas with increasing temperature (corresponding to Fig. 5.43b), denaturation bubbles increase in number and size, and the PBD mean energy gradually rises.

Letting $Q_i = y_i$, $P_i = \mathrm{d}y_i / \mathrm{d}t$, system (5.187) can be transformed into the following Itô stochastic differential equation

$$\mathrm{d}Q_i = P_i \mathrm{d}t,$$

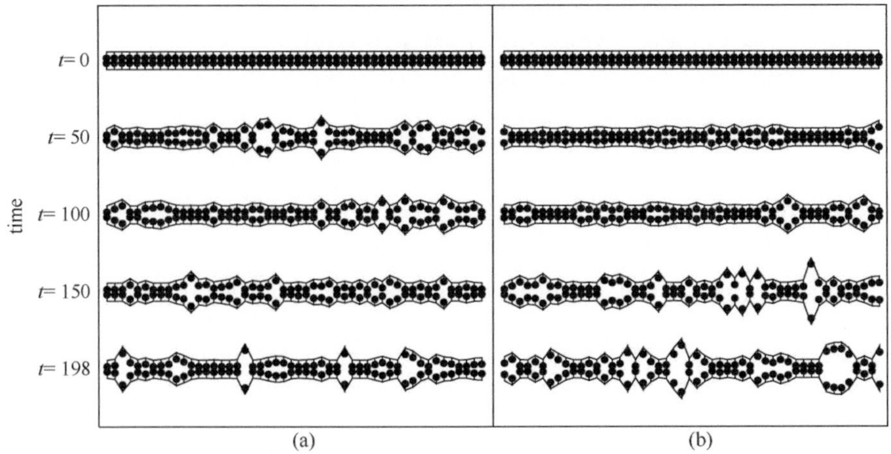

Fig. 5.42 Snapshots of thermal denaturation obtained from Monte Carlo simulation of a PBD model with 50 base-pairs. **a** $\gamma k_B T = 0.01$, **b** $\gamma k_B T$ increases from 0 to 0.02 (Deng and Zhu 2008b).

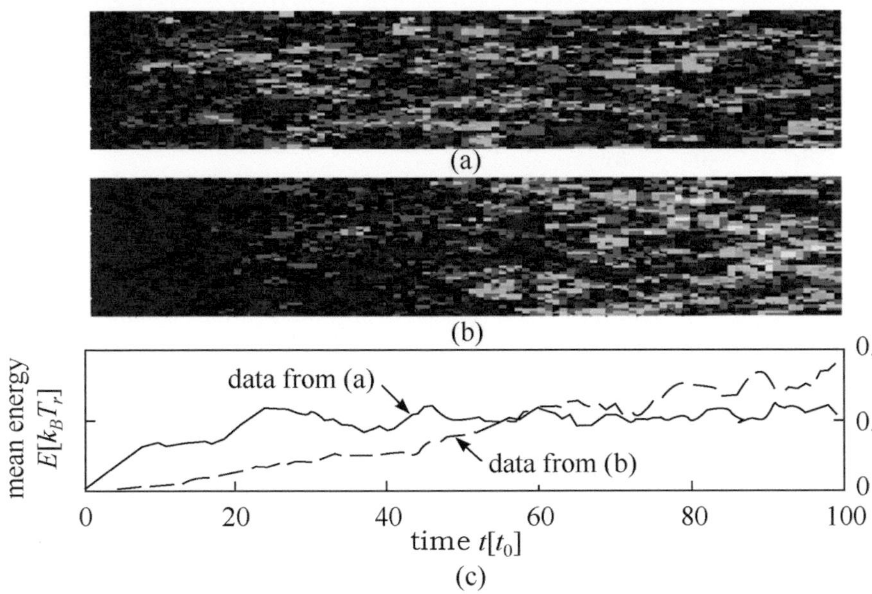

Fig. 5.43 Observations of thermal denaturation for a PBD model with 50 base-pairs. Data are obtained from Monte Carlo simulation. Lighter color represents the separated base-pairs, $\gamma = 1$. **a** $\gamma k_B T = 0.02$, **b** $\gamma k_B T$ increases from 0 to 0.04, **c** Time chart of mean energy of base-pairs (Deng and Zhu 2008b).

$$\mathrm{d}P_i = -(\frac{\partial U(\boldsymbol{Q})}{\partial Q_i} + \gamma P_i)\mathrm{d}t + \sqrt{2\gamma k_B T}\mathrm{d}B_i(t),$$
$$i = 1, 2, \ldots, N, \tag{5.188}$$

where $B_i(t)$ are independent unit Wiener processes. The Hamiltonian function H or total system energy is

$$H = \sum_{j=1}^{N} \frac{1}{2} P_j^2 + U(\boldsymbol{Q}). \tag{5.189}$$

Assuming γ to be small, Eq. (5.188) describes a quasi-Hamiltonian system.

Applying the Itô differential rule and Eq. (5.189), the following Itô stochastic differential equation for H can be derived from Eq. (5.188)

$$\mathrm{d}H = (N\gamma k_B T - \gamma \sum_{j=1}^{N} P_i^2)\mathrm{d}t + \sqrt{2\gamma k_B T} \sum_{j=1}^{N} P_j \mathrm{d}B_j(t), \tag{5.190}$$

where P_1 should be replaced by $\boldsymbol{Q}, H, P_2, \cdots, P_N$ according to

$$P_1 = \pm \sqrt{2\left[H - U(\boldsymbol{Q})\right] - \sum_{j=2}^{N} P_j^2}. \tag{5.191}$$

Using the stochastic averaging method of quasi-non-integrable Hamiltonian systems described in Sect. 5.1, the averaged Itô equation for $H(t)$ can be obtained

$$\mathrm{d}H = m(H)\mathrm{d}t + \sigma(H)\mathrm{d}B(t), \tag{5.192}$$

where the drift coefficient $m(H)$ and diffusion coefficient $\sigma^2(H)$ are given by

$$m(H) = \frac{1}{T(II)} \int_{\Omega} (N\gamma k_B T - \gamma \sum_{j=1}^{N} p_j^2)\mathrm{d}q_1 \cdots \mathrm{d}q_N \mathrm{d}p_2 \cdots \mathrm{d}p_N,$$
$$\sigma^2(H) = \frac{1}{T(H)} \int_{\Omega} (2\gamma k_B T \sum_{j=1}^{N} p_j^2)\mathrm{d}q_1 \cdots \mathrm{d}q_N \mathrm{d}p_2 \cdots \mathrm{d}p_N,$$
$$T(H) = \int_{\Omega} (\frac{1}{p_1})\mathrm{d}q_1 \cdots \mathrm{d}q_N \mathrm{d}p_2 \cdots \mathrm{d}p_N. \tag{5.193}$$

In Eq. (5.193), the integral domain Ω for the $(2N-1)$-fold integrals is determined by

$$\Omega = \left\{ (q_1, q_2, \ldots, q_N, p_2, \ldots, p_N) | \frac{1}{2} \sum_{j=2}^{N} p_i^2 + U(\boldsymbol{q}) \leq H \right\}. \tag{5.194}$$

Completing the integration in Eq. (5.193) with respect to P_i by using the generalized elliptic coordinate transformation (5.32) in vol. 1, the coefficients $m(H)$ and $\sigma^2(H)$ are obtained as follows

$$\begin{aligned}
m(H) &= N\gamma k_B T - 2\gamma A(H) \big/ B(H), \quad \sigma^2(H) = 4\gamma k_B T A(H) \big/ B(H),\\
T(H) &= \frac{\pi^{N/2} 2^{3N/2-2}}{\Gamma(N/2)} B(H),\\
A(H) &= \int_\Sigma [H - U(\boldsymbol{q})]^{N/2} \mathrm{d}q_1 \cdots \mathrm{d}q_N, \quad B(H) = \int_\Sigma [H - U(\boldsymbol{q})]^{N/2-1} \mathrm{d}q_1 \cdots \mathrm{d}q_N,
\end{aligned} \tag{5.195}$$

where $\Gamma(\cdot)$ is the Gamma function, and the integral domain Σ of the N-fold integral is determined by

$$\Sigma = \{(q_1, q_2, \ldots, q_N) | U(\boldsymbol{q}) \leq H\}. \tag{5.196}$$

Using the generalized elliptic coordinate transformation (5.48) in vol. 1, the N–fold domain integral in Eq. (5.195) can be changed into a N-fold integral of the form of Eqs. (5.51)–(5.53) in vol. 1.

The averaged FPK equation associated with the Itô equation (5.192) is given by

$$\frac{\partial p}{\partial t} = -\frac{\partial}{\partial h}\left[m(h)p\right] + \frac{1}{2}\frac{\partial^2}{\partial h^2}\left[\sigma^2(h)p\right], \tag{5.197}$$

where p can be either the transition PDF $p(h, t|h_0)$ with the initial condition

$$p(h, 0|h_0) = \delta(h - h_0). \tag{5.198}$$

or the PDF $p(h, t)$ of H with initial condition

$$p(h, 0) = p(h_0). \tag{5.199}$$

FPK equation (5.197) has the following boundary condition

$$\begin{aligned}
&p \text{ finite}, \quad \text{when } h = 0,\\
&\partial p \big/ \partial h \to 0, \quad \text{when } h \to \infty.
\end{aligned} \tag{5.200}$$

Generally, FPK equation (5.197) need to be solved numerically. Letting $\partial p / \partial t = 0$, it has the following exact stationary solution (5.14) of vol. 1, i.e.,

$$p(h) = \frac{C}{\sigma^2(h)} \exp\left[2 \int_z^h \frac{m(x)}{\sigma^2(x)} \mathrm{d}x \right], \tag{5.201}$$

where z is a nonzero positive constant, C is a normalization constant.

The solution (5.201) is the stationary PDF of the Hamiltonian function for system (5.188). The joint stationary PDF $p(\boldsymbol{q}, \boldsymbol{p})$ can then be obtained by using Eq. (5.18) in vol. 1 as follows

$$p(\boldsymbol{q}, \boldsymbol{p}) = \left[p(h) / T(h) \right] \Big|_{h=H(\boldsymbol{q}, \boldsymbol{p})}. \tag{5.202}$$

Other stationary statistics, such as the stationary PDF $p(e)$ of mean energy $E = H / N$, the stationary PDF $p(y_i)$ of distance between the two bases in base-pair, and the mean-square $E[Y_i^2]$ can be obtained from $p(h)$ or $p(\boldsymbol{q}, \boldsymbol{p})$ as follow

$$p(e) = N p(h)|_{h=Ne}, \quad p(y_i) = \int_{-\infty}^{\infty} p(\boldsymbol{q}, \boldsymbol{p}) \mathrm{d}q_1 \cdots \mathrm{d}q_{i-1} \mathrm{d}q_{i+1} \cdots \mathrm{d}q_N \mathrm{d}p_1 \cdots \mathrm{d}p_N,$$

$$E[Y_i^2] = \int_{-\infty}^{\infty} q_i^2 p(\boldsymbol{q}, \boldsymbol{p}) \mathrm{d}q_1 \cdots \mathrm{d}q_N \mathrm{d}p_1 \cdots \mathrm{d}p_N. \tag{5.203}$$

To verify the theoretical results in Eq. (5.203), Monte Carlo simulation of the original system (5.187) were conducted. The system response was obtained by solving the system motion equations using the fourth-order Runge-Kutta method, with a time step of 0.02. After 20,000 steps, the system response was considered stationary, and the statistics were obtained bases on the ergodicity of the response.

Figure 5.44 shows the stationary PDF $p(e)$ of mean energy. Figure 5.45 displays the PDF $p(y_i)$ of distance between two bases in base-pairs. Figure 5.46 illustrates the mean square distance $E[Y_i^2]$. The theoretical results are represented by solid lines and dashed lines, and the simulation results are denoted by symbols ● ◆. The agreement between the two results is excellent. Therefore, it can be concluded that Eqs. (5.201)–(5.203) are good analytical expressions for stationary statistics of PBD model under thermal perturbations.

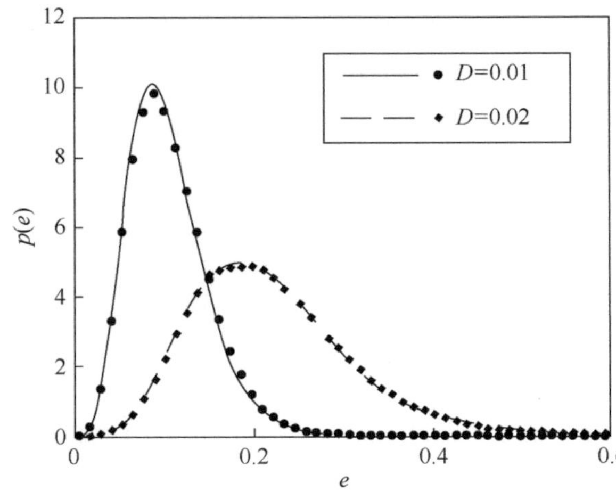

Fig. 5.44 Stationary PDF $p(e)$ of mean energy E in PBD model, $N = 6$, $\varepsilon = 0.1$, $\gamma = 0.6$ (Deng and Zhu 2008b).

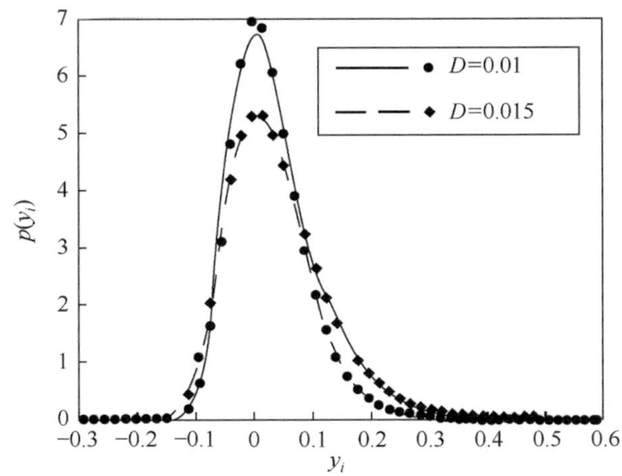

Fig. 5.45 Stationary PDF $p(y_i)$ of base-pair separation distance y_i, $N = 3$. Other parameters are the same as those in Fig. 5.44 (Deng and Zhu 2008b).

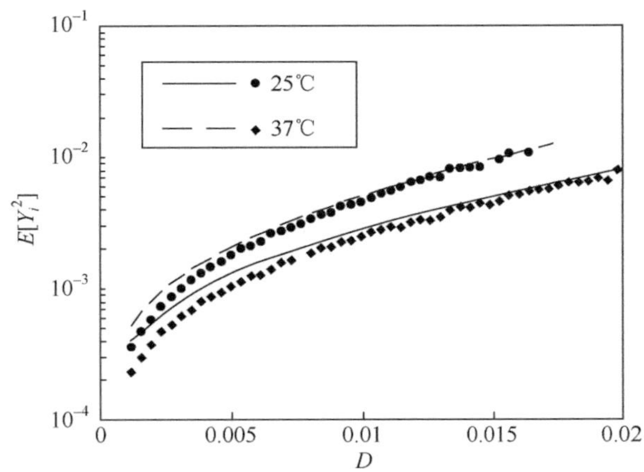

Fig. 5.46 Mean-square $E[Y_i^2]$ of base-pair separation distance y_i changes with excitation intensity, $N = 3$ for two cases of $\gamma = 0.6$ at 37 °C and $\gamma = 1$ at 25 °C. Other parameters are the same as those in Fig. 5.44 (Deng and Zhu 2008b).

5.5 Conformational Transformation of Biomacromolecule

The conformation of biomacromolecule is the foundation of their biological functions. Many life processes involve conformational transformation of biomacromolecule. To study the life activities of biomacromolecule, it is essential to investigate the dynamics of their conformational transformation (Frauenfelder and Wolynes 1985; McCammon and Harvey 1987). Traditionally, the study of biomacromolecular conformational transformation has been based on deterministic models (Mezić 2006a, b). However, in reality, a biomacromolecule has thousands DOFs and exhibits highly nonlinearity. It is known that biomolecules are constantly subjected to collisions with surrounding molecules (corresponding to thermal noise), and thermal perturbations. Since the temperature plays a crucial role in the conformational transformation of biomacromolecules, it is more meaningful to study the process under thermal perturbations. Research indicates that the conformational transformation of biomacromolecule is a stochastic process (Ebeling et al. 2003). In this section, based on previous research, the stochastic averaging methods of quasi-Hamiltonian systems is applied to investigate the conformational transformation dynamics of biomacromolecules under thermal perturbations.

5.5.1 *Model and Motion of Conformational Transformation*

Generally, the number of degrees-of-freedom for the motion of biomacromolecule is huge, and molecular structure is highly nonlinear. This complexity often leads to chaotic motion, and it is only possible to understand molecular dynamic behavior and perform theoretical analysis by establishing relatively simple models. Mezić studied a model for the conformational transformation of biomacromolecules, as shown in Fig. 5.47 (Mezić 2006a, b). The model consists of upper and lower supports with attached rigid pendulums. There is an attractive or repulsive force between the small balls on the upper and lower pendulums. The lower support and its pendulums are stationary, while the upper support can undergo elastic torsion, and the pendulums on it can swing. The conformational transformation of biomacromolecules can be understood as the swing of the pendulums on the upper support from one potential well located in pendulum angle range of $(-\pi, 0)$ to another potential well located in the range of $(0, \pi)$.

The attractive and repulsive forces between the pendulums on the upper support are produced by Morse potential energy. The interaction between adjacent pendulums is governed by the elastic torsion potential energy. The motion equations for the pendulums on the upper support are

$$\begin{aligned} mh^2\ddot{\theta}_i &= D_\mathrm{b}\{\exp[-a_\mathrm{d}(h(1-\cos\theta_i)-x_0)]-1\} \\ &\quad \times \exp[-a_\mathrm{d}(h(1-\cos\theta_i)-x_0)]\sin\theta_i + k(\theta_{i+1}-2\theta_i+\theta_{i-1}), \\ i &= 1, 2, \cdots, N. \end{aligned} \tag{5.204}$$

Fig. 5.47 Model of conformational transformation of biomacromolecule

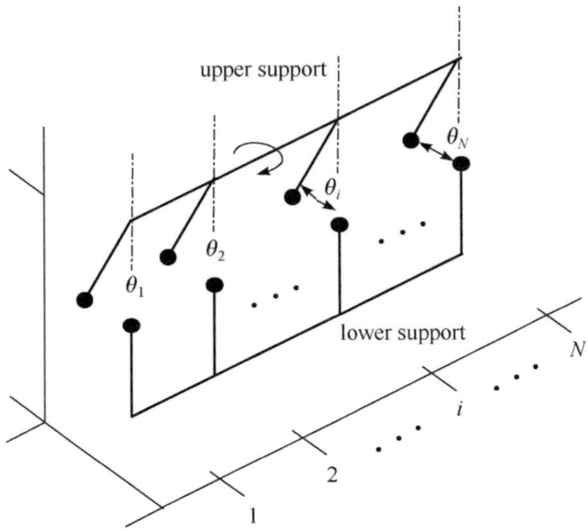

In Eq. (5.204), the first and second terms on the right-hand side represent the restoring forces produced by the Morse potential energy and the torsional potential energy; $i = 1, 2, \cdots, N$ is serial number of pendulum on the upper support; θ_i is the angle of the i th pendulum deviates from the vertical direction, and $\theta_{N+1} = \theta_1$ makes the support to be a closed one; h is the pendulum length; x_0 is the equilibrium distance of hydrogen bonds between pendulums; a_d is attenuation coefficient of the Morse potential; D_b is magnitude of the Morse potential; k is the torsion stiffness of the upper support. Mezić conducted numerical simulation for system (5.204) (Mezić 2006a, b).

To understand the model in Fig. 5.47 and how Eq. (5.204) is used to describe conformational transformation, let $N = 1$ in system (5.204). Figure 5.48 presents a series of equipotential curves of the model. The two local minima, corresponding to potential well positions $\theta_1 = \cos^{-1}(1 - x_0/h)$ and $\theta_1 = 2\pi - \cos^{-1}(1 - x_0/h)$, are symmetric with respect to the unstable position $\theta_1 = \pi$. The motion between two potential wells has to passes $\theta_1 = \pi$. When $N > 1$, if every pendulum remains in one potential well, it corresponds to a stable conformation of biomacromolecule (Mezić 2006a, b).

An important characteristic of this model is its ability to extend local perturbations to a global cooperative motion. Figure 5.49 displays a series of snapshots obtained from Monte Carlo simulation of the free vibrations of the model with 30 pendulums. Figure 5.49a represents the initial moment when only one pendulum is off its equilibrium position, while the other pendulums remain stationary at their equilibrium positions, indicating a single perturbation. As time goes, the energy on a single pendulum gradually spreads to more pendulums, initiating the motion of several pendulums (see Fig. 5.49b–d). Figure 5.49e shows that when a sufficient number of pendulums participate in the motion, they spontaneously organize into cooperative

Fig. 5.48 Phase plane $(\theta_1, \dot{\theta}_1)$ of the conformation transformation model (5.204), $N = 1$

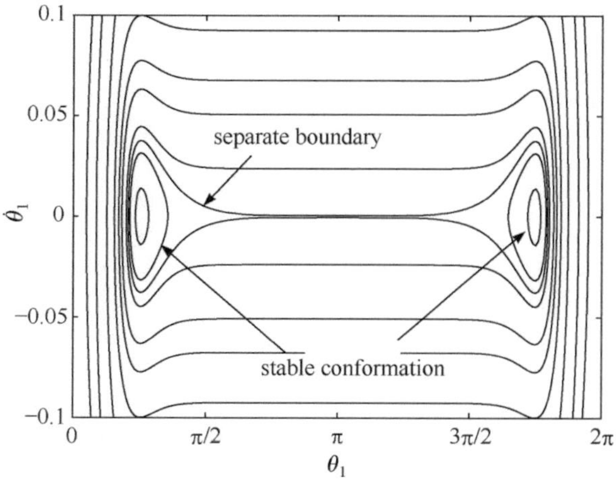

motion, as shown in Fig. 5.49f. Finally, the model flips as a whole to the right side of the support, as shown in Fig. 5.49g, h, completing a conformational transformation.

The time history diagram of three pendulum angles is depicted in Fig. 5.50, illustrating this cooperative motion. At the initial moment, apart from $\theta_1 = 0$, all pendulums rcmain stationary at their equilibrium positions. As time goes, each pendulum sequentially begin moving until all pendulums exhibit cooperative motion. When the angles exceed π, it means overall flip or conformational transformation.

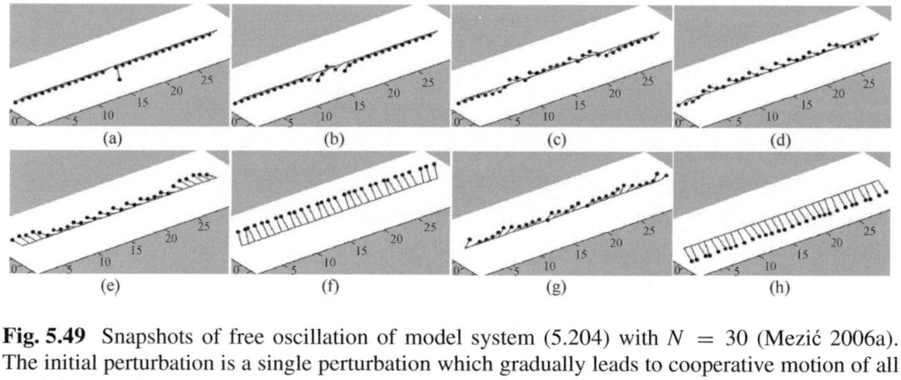

Fig. 5.49 Snapshots of free oscillation of model system (5.204) with $N = 30$ (Mezić 2006a). The initial perturbation is a single perturbation which gradually leads to cooperative motion of all pendulums and finally achieves conformational transformation

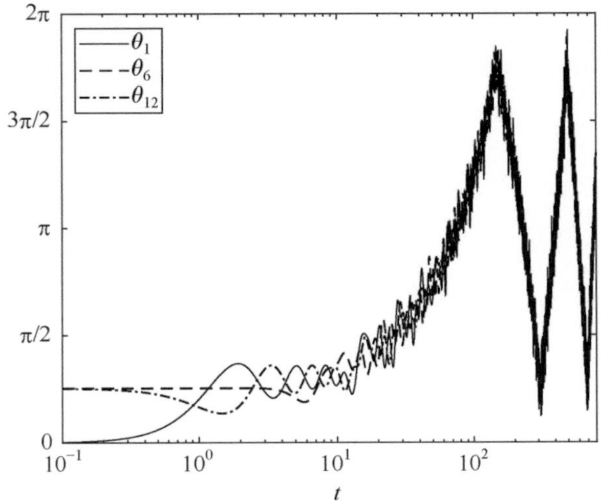

Fig. 5.50 Evolutions of angles $\theta_1, \theta_6, \theta_{12}$ of pendulums 1, 6, and 12 in the model system (5.204) with $N = 12$

5.5.2 *Stochastic Dynamics of Conformational Transformation*

The random forces acting on biomacromolecule mainly arise from collisions with environmental molecules and thermal fluctuations. In this section, the stochastic dynamics of conformational transformation under random forces is studied. For easy analysis, the mass of the pendulum is set to be $m \equiv 1$. Introduce the following new parameters L and S

$$L = h\Big/\sqrt{D_{\mathrm{b}}}, \quad S = k\big/h^2. \tag{5.205}$$

Suppose that each pendulum on the support is excited by a random excitation. Then, the system (5.204) for deterministic motion is transformed into the following stochastically excited system

$$\ddot{\Theta}_i + \gamma\dot{\Theta}_i + \frac{\partial U(\mathbf{\Theta})}{\partial \Theta_i} = W_{gi}(t), \quad i = 1, 2, \ldots, N. \tag{5.206}$$

where $\mathbf{\Theta} = [\Theta_1, \Theta_2, \ldots, \Theta_N]^{\mathrm{T}}$; γ is linear damping coefficients. The potential energy $U(\mathbf{\Theta})$ in Eq. (5.206) includes the Morse potential $M(\Theta_i)$ and the torsional potential $V(\mathbf{\Theta})$, i.e.,

$$U(\mathbf{\Theta}) = V(\mathbf{\Theta}) + \sum_{j=1}^{N} M(\Theta_j),$$

$$M(\Theta_i) = \frac{1}{2L^2 a_d h}\{\exp[-a_d(h(1-\cos\Theta_i) - x_0)] - 1\}^2, \quad i = 1, 2, \cdots, N,$$

$$V(\boldsymbol{\Theta}) = \sum_{j=1}^{N} [\frac{1}{2} S(\Theta_{j-1} - \Theta_j)^2]. \tag{5.207}$$

The $W_{gi}(t)$ in Eq. (5.206) are independent Gaussian white noises with a same intensity $2D$. In all theoretical analyses and Monte Carlo simulations in this section, the following parameters $L = 1$, $h = 4$, $a_d = 0.5$ and $x_0 = 0.3$ are used.

Letting $Q_i = \Theta_i$, $P_i = \dot{\Theta}_i$, system (5.206) can be transformed into a stochastically excited and dissipated Hamiltonian system and then into the following Itô stochastic differential equation (Deng and Zhu 2007c; Deng 2007):

$$\mathrm{d}Q_i = P_i \mathrm{d}t, \quad \mathrm{d}P_i = -(\frac{\partial U(\boldsymbol{Q})}{\partial Q_i} + \gamma P_i)\mathrm{d}t + \sqrt{2D}\mathrm{d}B_i(t),$$
$$i = 1, 2, \cdots, N, \tag{5.208}$$

where $B_i(t)$ is an independent unit Wiener process. The Hamiltonian function H associated with Eq. (5.208) is

$$H = \sum_{i=1}^{N} \frac{P_i^2}{2} + U(\mathbf{Q}). \tag{5.209}$$

The Hamiltonian system associated with Eq. (5.208) is non-integrable. The stochastic averaging method of quasi-non-integrable Hamiltonian system described in Sect. 5.1 can be applied. Using the Itô differential rule (Itô, 1951a) and Eq. (5.209), the following Itô stochastic differential equation for $H(t)$ can be derived

$$\mathrm{d}H = (ND - \gamma \sum_{j=1}^{N} P_j^2)\mathrm{d}t + \sqrt{2D} \sum_{j=1}^{N} P_i \mathrm{d}B_i(t). \tag{5.210}$$

P_1 in Eqs. (5.208) and (5.210) is replaced by

$$P_1 = \pm \sqrt{2[H - U(\boldsymbol{Q})] - \sum_{j=2}^{N} P_i^2}. \tag{5.211}$$

Now the system is composed of Eqs. (5.210) and (5.208) excluding the equation of $\mathrm{d}P_1$. As γ and $D \to 0$, $H(t)$ converges weakly to a Markov diffusion process. The stochastic averaging method of quasi-non-integrable Hamiltonian systems described in Sect. 5.1 can be used to obtain the following averaged Itô equation

$$\mathrm{d}H = m(H)\mathrm{d}t + \sigma(H)\mathrm{d}B(t). \tag{5.212}$$

Using Eqs. (5.10)–(5.13) of vol. 1, the drift coefficient $m(H)$ and the diffusion coefficient $\sigma^2(H)$ are obtained as follows

$$
\begin{aligned}
m(H) &= \frac{1}{T(H)} \int_{\Omega} (ND - \gamma \sum_{j=1}^{N} p_j^2) \mathrm{d}q_1 \cdots \mathrm{d}q_N \mathrm{d}p_2 \cdots \mathrm{d}p_N, \\
\sigma^2(H) &= \frac{1}{T(H)} \int_{\Omega} (2D \sum_{j=1}^{N} p_i^2) \mathrm{d}q_1 \cdots \mathrm{d}q_N \mathrm{d}p_2 \cdots \mathrm{d}p_N, \\
T(H) &= \int_{\Omega} (\frac{1}{p_1}) \mathrm{d}q_1 \cdots \mathrm{d}q_N \mathrm{d}p_2 \cdots \mathrm{d}p_N.
\end{aligned}
\tag{5.213}
$$

The integration domain Ω is

$$
\Omega = \left\{ (q_1, q_2, \ldots, q_N, p_2, \ldots, p_N) | \frac{1}{2} \sum_{j=2}^{N} p_i^2 + U(\boldsymbol{q}) \leq H \right\}. \tag{5.214}
$$

After substituting the expression of p_1 in Eq. (5.211) into Eq. (5.213), using the generalized elliptic coordinate transformation (5.32) in vol. 1, and completing the $(N-1)$-fold integrals with respect to $p_2, \cdots, p_N$ yield the following two coefficients

$$
\begin{aligned}
& m(H) = ND - 2\gamma A(H) \big/ B(H), \quad \sigma^2(H) = 4DA(H) \big/ B(H), \\
& T(H) = \pi^{N/2} 2^{3N/2-2} B(H) / \Gamma(N/2), \\
& A(H) = \int_{\Sigma} [H - U(\boldsymbol{q})]^{N/2} \mathrm{d}q_1 \cdots \mathrm{d}q_N, \\
& B(H) = \int_{\Sigma} [H - U(\boldsymbol{q})]^{N/2-1} \mathrm{d}q_1 \cdots \mathrm{d}q_N.
\end{aligned}
\tag{5.215}
$$

where $\Gamma(\cdot)$ is the Gamma function; Σ is the integration domain

$$
\Sigma = \{(q_1, q_2, \ldots, q_N) | U(\boldsymbol{q}) \leq H\}. \tag{5.216}
$$

Another generalized elliptic coordinate transformation (5.48) in vol. 1 can be used to change the N–dimensional domain integral with respect to q_i into a N-fold integral with respect to angles φ_i. Then the integrations in Eq. (5.215) can be completed.

A conformational transformation is equivalent to the first-passage behavior of energy. For study the later, the probability distribution function $W(t|h_0)$ of waiting time t is introduced. It is defined as the probability that the energy process $H(t)$ with initial energy $h_0 < H_C$ does not exceed H_C before time t, i.e.,

$$W(t|h_0) = \text{Prob}\{\ H(t') < H_C,\ t' \in (0,t] \,\big|\, 0 \le h_0 < H_C\}\,. \tag{5.217}$$

To calculate the probability distribution function $W(t|h_0)$, the conditional transition PDF $p = p(h,t|h_0)$ is introduced. It is the transition PDF of the random process $H(t)$, which starts from initial state h_0 ($0 \le h_0 < H_C$) and remains in the range $[0, H_C)$ at time t. It satisfies the following backward Kolmogorov equation

$$\frac{\partial p}{\partial t} = m(h_0)\frac{\partial p}{\partial h_0} + \frac{1}{2}\sigma^2(h_0)\frac{\partial^2 p}{\partial h_0^2}. \tag{5.218}$$

where $m(h_0)$ and $\sigma^2(h_0)$ are obtained from Eq. (5.215) with h replaced by h_0. The initial condition is

$$p(h, t = 0|h_0) = \delta(h - h_0). \tag{5.219}$$

and boundary conditions are

$$p(h,t|h_0 = 0) = \text{finite}, \quad p(h,t|h_0 = h_C) = 0. \tag{5.220}$$

Equation (5.220) indicates that once the energy $H(t)$ reaches the threshold H_C, it is absorbed by the boundary. The probability distribution function $W(t|h_0)$ of waiting time is the integral of the conditional PDF $p(h,t|h_0)$, i.e.,

$$W(t|h_0) = \int_0^{H_C} p(h,t|h_0)\mathrm{d}h. \tag{5.221}$$

Integrating both sides of Eq. (5.218) over field $(0, H_C)$ and using Eq. (5.221) yields the backward Kolmogorov equation

$$\frac{\partial W}{\partial t} = m(h_0)\frac{\partial W}{\partial h_0} + \frac{1}{2}\sigma^2(h_0)\frac{\partial^2 W}{\partial h_0^2}. \tag{5.222}$$

Applying similar integration to the boundary conditions (5.220) yields the following boundary conditions for Eq. (5.222)

$$W(t|h_0 = 0) = \text{finite}, \quad W(t|h_0 = H_C) = 0. \tag{5.223}$$

The initial condition for Eq. (5.222) is

$$W(t = 0|h_0 < H_C) = 1. \tag{5.224}$$

Combined with the initial condition (5.224) and boundary conditions (5.223), Eq. (5.222) can be solved numerically using the Crank-Nicolson finite difference method. Once the probability distribution function $W(t|h_0)$ of waiting time is obtained, the PDF $\rho(t|h_0)$ of the first-passage time

$$\rho(t|h_0) = -\partial W(t|h_0) \big/ \partial t. \tag{5.225}$$

and the mean first-passage time

$$\tau(h_0) = \int_0^\infty t\rho(t|h_0)\mathrm{d}t = \int_0^\infty W(t|h_0)\mathrm{d}t. \tag{5.226}$$

$\tau(h_0)$ is governed by the following Pontryagin equation (Pontryagin et al. 1933)

$$\frac{1}{2}\sigma^2(h_0)\frac{\partial^2\tau}{\partial h_0^2} + m(h_0)\frac{\partial\tau}{\partial h_0} = -1. \tag{5.227}$$

with boundary condition

$$\tau(h_0 = 0) = \text{finite}, \quad \tau(H_C = 0) = 0. \tag{5.228}$$

By solving Eq. (5.227) with boundary conditions in Eq. (5.228), the analytical expression for the mean first-passage time $\tau(h_0)$ can be obtained as follows

$$\tau(h_0) = 2\int_{h_0}^{H_C} \mathrm{d}u \int_0^u \frac{1}{\sigma^2(v)} \exp[-2\int_v^u \frac{m(w)}{\sigma^2(w)}\mathrm{d}w]\mathrm{d}v. \tag{5.229}$$

In Monte Carlo simulation and theoretical calculation, h_0 is set to be zero, corresponding to the initial state of stable conformation. The probability distribution $W(t|h_0)$ of waiting time, PDF of first-passage time $\rho(t|h_0)$, and mean first-passage time $\tau(h_0)$ are abbreviated to $W(t)$, $\rho(t)$ and τ, respectively. Figures 5.51 and 5.52 display some analytical results from Eqs. (5.222), (5.225) and (5.229) respectively and simulation results from system (5.206), showing good agreement between theoretical results and simulation results. Figure 5.52 indicates that the theoretical prediction for the first-passage time τ is slightly lower than the simulation results.

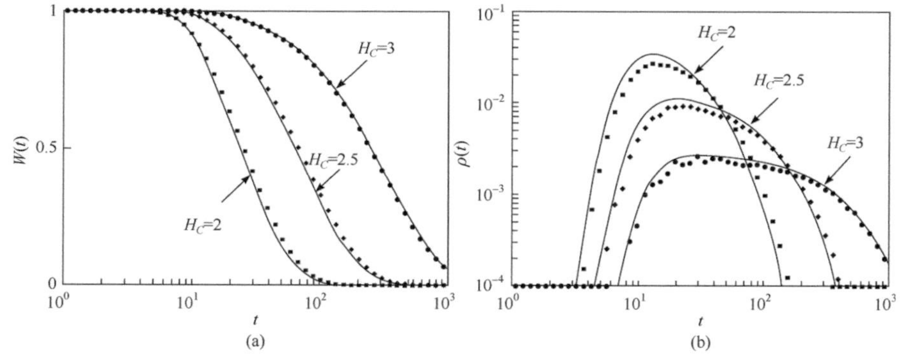

Fig. 5.51 The statistics of the first-passage for system (5.206) with different passage thresholds H_C. **a** Waiting time probability distribution, and **b** PDF of first-passage time. Line ——, results from the stochastic averaging method, and symbols ●, ◆, ■, results from simulation. $S = 1$, $D = 0.032$, $\gamma = 0.2$, $N = 8$ (Deng and Zhu 2007c).

Fig. 5.52 Mean first-passage time of system (5.206) with different passage thresholds H_C. Line ——, results from the stochastic averaging method, and symbols ●, ◆, ■, results from simulation. $S = 1$, $\gamma = 0.2$, $N = 8$ (Deng and Zhu 2007c).

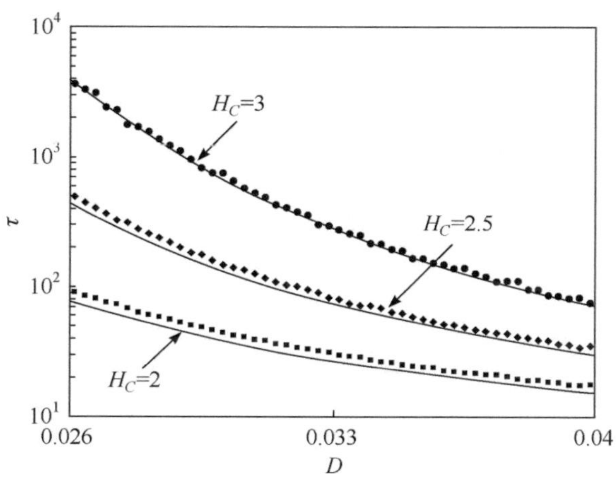

5.5.3 *Denaturation of DNA Molecule*

DNA denaturation is a kind of conformational transformation, which transform the biomacromolecule structure from a closed double-stranded structure to a loosely open structure. As shown in Fig. 5.41, during the denaturation process, a fork-like structure composed of several closed and open base-pairs can be observed at both ends of denaturation bubbles (see Fig. 5.40). The base-pairs near fork structure have a significant impact on the opening of base-pair, while the base-pairs far from the fork structure have a weaker influence. Therefore, to study the mean opening time of base-pair, it is reasonable to establish a local and finite degrees-of-freedom dynamical system, which is composed of several base-pairs near fork structure.

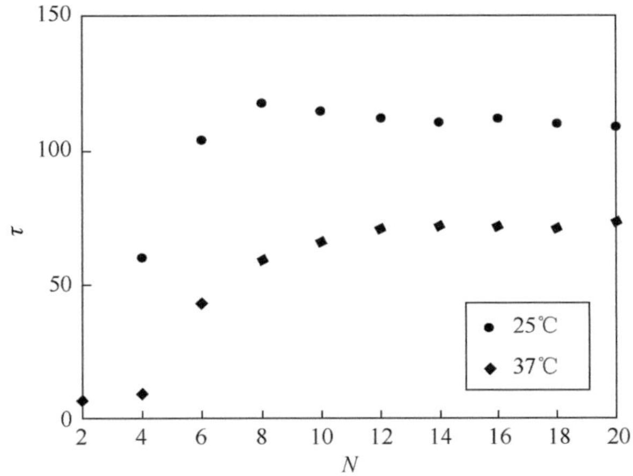

Fig. 5.53 Relationship between the opening time τ of one base-pair and the number N of involving base-pairs on the denaturation fork (Deng and Zhu 2008b)

Firstly, it is necessary to find out how many base-pair near fork structure (see Fig. 5.40) are needed for an acceptable accuracy. Simulation results in Fig. 5.53 show that as the number N of base-pair near the fork structure increases, the mean opening time of base-pair gradually approaches a constant, and $N = 10$ will be accurate enough. Due to the limitations of computational resources and the purpose of this section is just to explain the theoretical method, $N = 6$ will be taken (see Fig. 5.40). Another question is how far apart of base-pair can be identified as open state. To keep the agreement with other literature (Kalosakas et al. 2004), an opening threshold $y_{th} = 1$ Å is used. When the distance of the i th base-pair reaches y_{th}, the base-pair is identified as open state.

Taking $N = 6$ as the number of base-pair, a 6-degree-of-freedom quasi-non-integrable Hamiltonian system (5.188) is obtained. The initial state of the system is three closed base-pairs and three open base-pairs (see Fig. 5.54a), and the final state of the system is two closed base-pairs and four open base-pairs (see Fig. 5.54b). In theoretical research, the problem of the opening time of base-pairs is transformed into the first-passage problem of energy in the fork structure. In other words, the opening time of a base-pair is replaced by the time for energy to exceed the threshold H_C (corresponding to the energy in Fig. 5.54b) with the initial state H_0 (corresponding to the energy in Fig. 5.54a) for the first time.

Similar to the derivation from Eqs. (5.221)–(5.229), the probability distribution function $W(t)$ of waiting time, the PDF $\rho(T)$ of the first-passage time, and the mean opening time τ of the base-pair can be derived. The denaturation rate k can be taken as the reciprocal of the mean first-passage time, i.e., $k = 1/\tau$.

Figures 5.55 and 5.56 show the results for the waiting time distribution $W(t)$ and the first-passage time density $\rho(T)$ where the analytical results are shown by

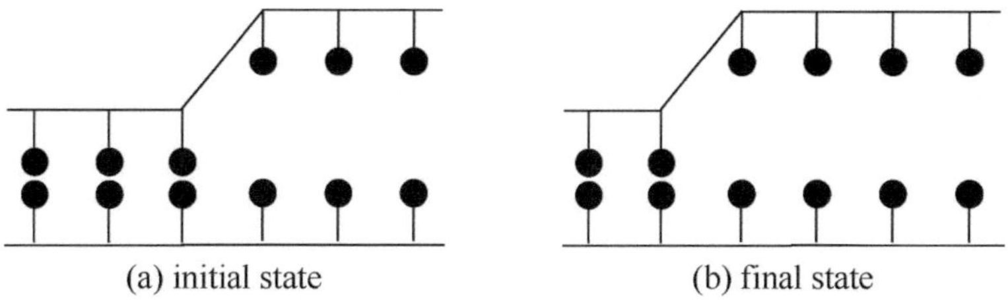

Fig. 5.54 Denaturation fork (Deng and Zhu 2008b)

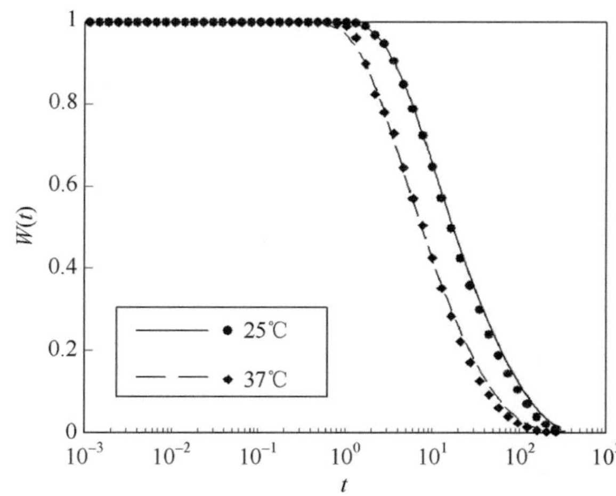

Fig. 5.55 Probability distributions $W(t)$ of waiting time of denaturation fork system, $N = 6$, $\gamma = 0.01$ (Deng and Zhu 2008b)

—— and -----, while the simulation results are denoted by symbol ● ◆. It is seen that the theoretical results match well with simulation results. The waiting time distribution $W(t)$ in Fig. 5.55 is similar to the profile of survival probability function and experimental results (Altan-Bonnet et al. 2003). Figure 5.57 shows that when the damping coefficient γ is between 10^{-3} and 10^{-1}, the calculated mean opening time of base-pair is between 10-400 ps (picosecond). This value is about 2–3 orders lower than that obtained by nuclear magnetic resonance technology and 5–6 orders lower than the experimental measurement of 20–100 μs (microsecond) (Altan-Bonnet et al. 2003). From the trend in Fig. 5.57, it can be estimated that if the friction coefficient is of the order of 10^{-6}, the mean opening time of a base-pair will be likely on the same order of magnitude as the experimental measurement. In practice, to match experimental data, it is necessary to adjust parameters in theoretical analysis. (Altan-Bonnet et al. 2003; Campa and Giansanti 1998).

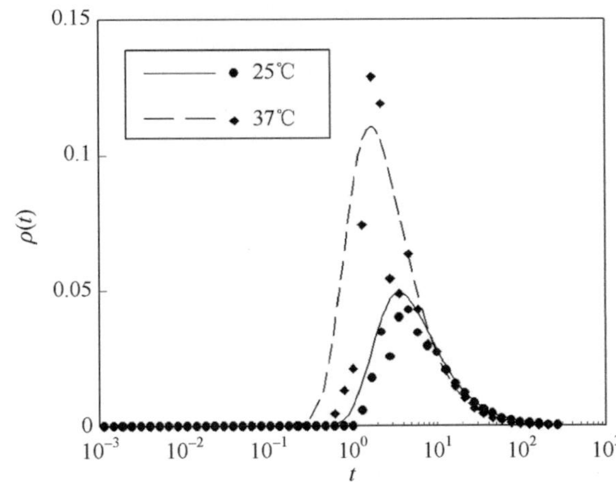

Fig. 5.56 PDF $\rho(T)$ of first-passage time of denaturation fork system. Parameters are the same as those in Fig. 5.55 (Deng and Zhu, 2008b)

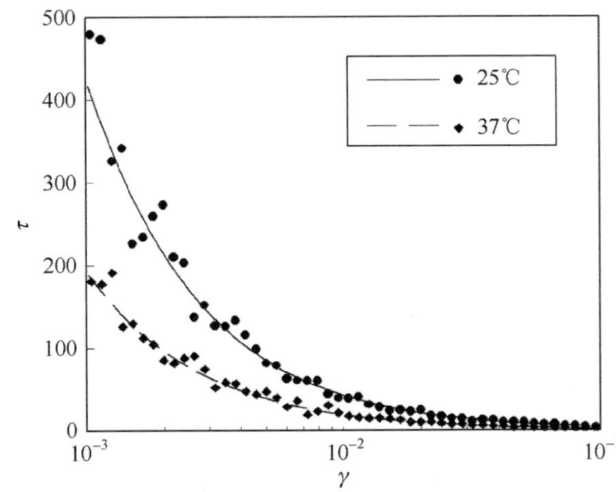

Fig. 5.57 Mean opening time τ of base-pair with versus damping coefficient γ (Deng and Zhu 2008b)

References

Alt W. 1980. Biased random walk models for chemotaxis and related diffusion approximations. Journal of Mathematical Biology, 9: 147–177.

Altan-Bonnet G, Libchaber A, Krichevsky O. 2003. Bubble dynamics in double-stranded DNA. Physical Review Letters, 90: 138101.

Bogolyubov N N, Mitroplskii Yu A. 1974. Asymptotic methods in nonlinear vibration theory. Nauka, Moscow. (in Russian)

Cai G Q, Zhu W Q. 2016. Elements of Stochastic Dynamics. Singapore: World Scientific Publisher.

Campa A, Giansanti A. 1998. Experimental tests of the Peyrard-Bishop model applied to the melting of very short DNA chains. Physical Review E, 58: 3585.

Czirok A, Vicsek T. 2000. Collective behavior of interacting self-propelled particles. Physica A, 281(1-4): 17–29.

Dauxois T. 1991. Dynamics of breather modes in a nonlinear helicoidal model of DNA. Physics Letters A, 159(8–9): 390–395.

Deng M L. 2005. Stochastic Dynamics of Active Brownian Particles. Ph.D. Dissertation. Hangzhou: Zhejiang University. (in Chinese)

Deng M L. 2007. Stochastic Dynamics of Biological Macromolecule. Post-doctoral Work Report. Hangzhou: Zhejiang University. (in Chinese)

Deng M L, Zhu W Q. 2004. Stationary motion of active Brownian particles. Physical Review E, 69(4): 046105.

Deng M L, Zhu W Q. 2007a. Energy diffusion controlled reaction rate in dissipative Hamiltonian system. Chinese Physics, 16: 1510–1515.

Deng M L, Zhu W Q. 2007b. Energy diffusion controlled reaction rate of reacting particle driven by broad-band noise. European Physical Journal B, 59(3): 391–397.

Deng M L, Zhu W Q. 2007c. On the stochastic dynamics of molecular conformation. Journal of Zhejiang University Science A, 8(9): 1401–1407.

Deng M L, Zhu W Q. 2008a. Fermi resonance and its effect on the mean transition time and rate. Physical Review E, 77(6): 061114.

Deng M L, Zhu W Q. 2008b. Stochastic dynamics and denaturation of thermalized DNA. Physical Review E, 77(2): 021918.

Deng M L, Zhu W Q. 2009a. Some applications of stochastic averaging method of quasi-Hamiltonian systems. Science in China Series G: Physics, Mechanics & Astronomy, 39(6): 821–829. (in Chinese)

Deng M L, Zhu W Q. 2009b. Some applications of stochastic averaging method for quasi Hamiltonian systems in physics. Science in China Series G: Physics, Mechanics & Astronomy, 52(8): 1213–1222.

Deng M L, Zhu W Q. 2010. Energy transition rate at peptide bond using stochastic averaging method. Proceeding of IUTAM Symposium on Nonlinear Stochastic Dynamics and Control, May 10–14, Hangzhou, China.

Deng M L, Zhu W Q. 2012. Stochastic energy transition of peptide-bond under action of hydrolytic enzyme. Probabilistic Engineering Mechanics, 27(1): 8–13.

Dickinson R, Tranquillo R T. 1993. A stochastic model for adhesion-mediated cell random motility and haptotaxis. Journal of Mathematical Biology, 31: 563–600.

Ebeling W, Kargovsky A, Netrebko A, Romanovsky Yu M. 2004. Fermi resonance – new applications of an old effect. Fluctuation and Noise Letters, 4: L183–L193.

Ebeling W, Röpke G. 2004. Statistical mechanics of confined systems with rotational excitations. Physica D, 187(1–4): 268–280.

Ebeling W, Schimansky-Geier L, Romanovsky Yu M. 2003. Stochastic Dynamics of Reacting Biomolecules. Singapore: World Scientific.

Ebeling W, Schweitzer F. 2001. Swarms of particle agents with harmonic interactions. Theory in Bioscience, 120: 207–224.

Ebeling W, Schweitzer F, Tilch B. 1999. Active Brownian particles with energy depots modeling animal mobility. Biosystems, 49: 17–29.

Erdmann U, Ebeling W, Anishchenko V S. 2002. Excitation of rotational modes in two-dimensional systems of driven Brownian particles. Physical Review E, 65: 061106.

Erdmann U, Ebeling W, Schimansky-Geier L, Schweitzer F. 2000. Brownian particles far from equilibrium. European Physical Journal B, 15(1): 105–112.

Fermi E. 1931. Über den Ramaneffekt des Kohlendioxyds. Zeitschrift für Physik A, 71: 250–259. (in German)

Fermi E, Rasetti F. 1931. Über den Ramaneffekt des Steinsalzes. Zeitschrift für Physik A, 71: 689–695. (in German)

Franke K, Gruler H. 1990. Galvanotaxis of human granulocytes: electric field jump studies. European Biophysics Journal, 18: 335–346.

Frauenfelder H, Wolynes P G. 1985. Rate theories and the puzzles of hemoprotein kinetics. Science, 229: 337–345.

Hänggi P. Marchesoni F. Grigolini P. 1984. Bistable flow driven by colored Gaussian noise: a critical study. Zeitschrift für Physik B, 56: 333–339.

Hänggi P, Talkner P. Borkovec M. 1990. Reaction-rate theory: fifty years after Kramers. Reviews of Modern Physics, 62: 251–341.

Itô K. 1951a. On stochastic differential equations. Memoirs of the American Mathematical Society, 4: 289–302.

Jung P, Hänggi P. 1988. Bistability and colored noise in nonequilibrium systems: theory versus precise numerics. Physical Review Letters, 61(1): 11–14.

Kalosakas G, Rasmussen K Ø, Bishop A R, Choi C H, Usheva A. 2004. Sequence-specific thermal fluctuations identify start sites for DNA transcription. Europhysics Letters, 68(1): 127–133.

Kramers H A. 1940. Brownian motion in field of force and the diffusion model of chemical reactions. Physica, 4: 284–304.

Langer J S. 1969. Statistical theory of the decay of metastable states. Annals of Physics (N.Y.), 54: 258.

Luciani J F, Verga A D. 1988. Bistability driven by correlated noise: functional integral treatment. Journal of Statistical Physics, 50: 567–597.

Marchesoni F, Menichella-Saetta E, Pochini M, Santucci S. 1988a. Analog simulation of underdamped stochastic systems driven by colored noise: spectral densities. Physical Review A, 37(8): 3058–3066.

Marchesoni F, Menichella-Saetta E, Pochini M, Santucci S. 1988b. Escape rates underdamped bistable potentials driven by colored noise. Physics Letters A, 130: 467–470.

Mezić I. 2006a. On the dynamics of molecular conformation. Proceedings of the National Academy of Sciences of USA, 103: 7542–7547.

Mezić I. 2006b. Biomolecules as Nonlinear Oscillators: Life-enabling Dynamics. The 2nd International Conference On Dynamics, Vibration and Control, August 23-26, Beijing, China.

McCammon J A, Harvey S C. 1987. Dynamics of Proteins and Nucleic Acids. Cambridge University Press, Cambridge.

Mikhailov A, Zanette D H. 1999. Noise-induced breakdown of coherent collective motion in swarms. Physical Review E, 60: 4571.

Netrebko A, Netrebko N, Romanovsky Yu M, Khurgin Yu, Shidlovskaya E. 1994. Complex modulation regimes and vibration stochastization in cluster dynamics models of macromolecules. Izv. Vuzov: Prikladnaya Nelineinaya Dinamika, 2: 26-43. (in Russian)

Peyrard M. 2004. Nonlinear dynamics and statistical physics of DNA. Nonlinearity, 17: R1-R40.

Peyrard M, Bishop A R. 1989. Statistical mechanics of a nonlinear model for DNA denaturation. Physical Review Letters, 62(23): 2755–2758.

Pippard A B. 1983. The Physics of Vibration. The simple vibrator in quantum mechanics. Cambridge University Press, Cambridge, London, New York, Melbourne, Sidney.

Poland D, Scheraga H A. 1966. Phase transitions in one dimension and the helix-coil transition in polyamino acids. Journal of Chemical Physics, 45: 1456–1463.

Pontryagin L S, Andronov A A, Vitt A A. 1933. On the statistical treatment of dynamical systems. Journal of Experimental and Theoretical Physics, 3: 165–180.

Reimann P, Schmid G J, Hänggi P. 1999. Universal equivalence of mean first-passage time and Kramers rate. Physical Review E, 60(1): R1–R4.

Romanczuk P, Bär M, Ebeling W, Lindner B, Schimansky-Geier L. 2012. Active Brownian particles. The European Physical Journal Special Topics, 202: 1–162.

Saenger W. 1984. Principles of Nucleic Acid Structure. Berlin: Springer.

Schienbein M, Gruler H. 1993. Langevin equation, Fokker-Planck equation and cell migration. Bulletin of Mathematical Biology, 55: 585-608.

Shidlovskaya E, Schimansky-Geier L, Romanovsky Yu M. 2000. Nonlinear vibrations in 2-dimensional protein cluster model with linear bonds. Zeitschrift für Physikalische Chemie, 214: 65–82.

Theodorakopoulos N, Peyrard M, Mackay R S. 2004. Nonlinear structures and thermodynamic instabilities in a one-dimensional lattice system. Physical Review Letters, 93: 258101.

van Erp T S, Cuesta-Lopez S, Hagmann J G, Peyrard M. 2005. Can one predict DNA transcription start sites by studying bubbles? Physical Review Letters, 95: 218104.
Volkenstein M V. 1947. Structure of molecules. Izd. Akad. Nauk., Moscow (in Russian).
Zhu W Q, Deng M L. 2005. Stationary swarming motion of active Brownian particles in parabolic external potential. Physica A, 354(2–4): 127–142.

Chapter 6
Several Applications of the Stochastic Averaging Methods in Technical Sciences

Many engineering structures are subjected to various random disturbances, such as bridges and high-rise buildings subjected to random wind loading, power systems affected by the new energy resources and load fluctuations, and ships subjected to wave impacts, etc. In this chapter, the applications of the stochastic averaging methods in three examples of technical sciences and in stability and optimal control of nonlinear stochastic systems are described.

6.1 Vortex-Induced Random Vibration

Vortex-induced vibration is vibration of slender structures such as transmission lines, cables, chimneys, and marine risers etc., when they are subjected to wind or water flow. The flow perpendicular to the structural axis becomes vortices in the structural backside, causing alternating forces on the structure. The resulting forced vibration is known as vortex-induced vibration (Simiu and Scanlan 1996). Under certain conditions, vortex-induced vibration can evolve into vortex-induced resonance, which may cause structural damage. Essentially, vortex-induced vibration is a complex nonlinear vibration caused by fluid–structure interaction. The wake oscillator models have been successful mechanical models for studying vortex-induced vibration for a long time (Facchinetti et al. 2004; Gabbai and Benaroya 2005). These models consist of two oscillators, the structural oscillator describing the structural vibration and the excitation oscillator representing the lift forces produced by the fluid flow. Apart from qualitatively describing of frequency lock-in effect in vortex-induced resonance, these models can also make quantitative prediction of vibration response. Since wind field in the atmosphere are essentially random field, it is reasonable to change the classical wake oscillator models into stochastically excited wake oscillator models and apply the stochastic averaging methods to study vortex-induced random vibration (Deng et al. 2021).

W. Zhu et al., *Stochastic Averaging*, https://doi.org/10.1007/978-981-96-3833-8_6

6.1.1 Hartlen-Currie Wake Oscillator Model

Figure 6.1 is a schematic diagram of vortex-induced vibration of a cylinder. A slender cylindrical rigid body is elastically mounted on a fixed foundation. When the incoming wind blows over the cylinder, alternating vortices detach from the cylinder backside, and the forces from these vortices induce cylinder vibrations. To limit the study to crosswind vibrations, the cylinder is constrained to displace only in the crosswind direction. A classical wake oscillator model, known as the Hartlen-Currie model, has been widely used to describe such crosswind vibration system. The equations governing the model are (Hartlen and Currie 1970).

$$\begin{aligned} &M(\ddot{y}+2\zeta\omega_n\dot{y}+\omega_n^2 y)=\frac{1}{2}\rho_a V^2 LDC_L,\\ &\ddot{C}_L+(\frac{4}{3}\frac{G}{\omega_s}\dot{C}_L^2-\omega_s G C_{L0}^2)\dot{C}_L+\omega_s^2 C_L=\omega_s F\frac{\dot{y}}{D}, \end{aligned} \tag{6.1}$$

where y represents the crosswind displacement of the cylinder, and C_L denotes the dimensionless lift coefficient. The first equation in Eq. (6.1) governs the structural vibration, known as the structural oscillator, while the second equation governs the dynamics of the lift coefficient, known as the excitation oscillator. ω_n is the natural frequency of the structural oscillator; ζ is the damping ratio; D is the cylinder diameter; L is the cylinder length; M is the cylinder mass; $\rho_s = 4M\big/(\pi D^2 L)$ is cylinder density; ρ_a is air density; V is the constant wind speed; F is a dimensionless coupling parameter that reflects the influence of structural oscillator on the excitation oscillator; $\omega_s = 2\pi VS_t\big/D$ is the wind-excited frequency where S_t is the dimensionless Strouhal number; C_{L0} represents the amplitude of the periodically varying lift coefficient when the cylinder is fixed. The parameters of the excitation oscillator in Eq. (6.1) are specially designed so that the amplitude of C_L at $y=\dot{y}=0$ and $t\to\infty$ exactly is C_{L0} (Skop and Griffin 1973; Gabbai and Benaroya 2005). It is important to note that wake oscillator model is semi-empirical, and some parameters in Eq. (6.1) lack

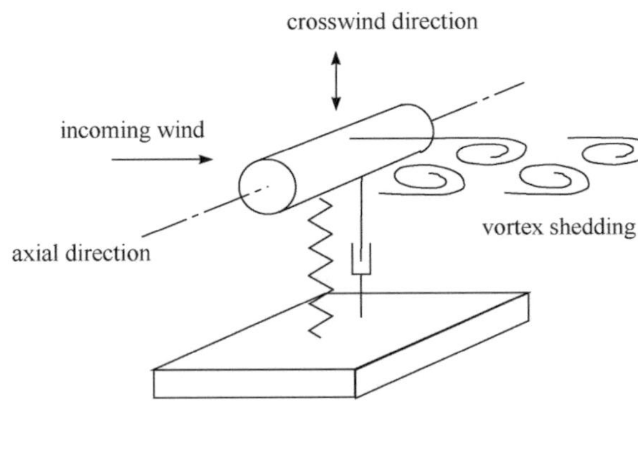

Fig. 6.1 Schematic diagram of vortex-induced vibration of an elastically-mounted cylinder

clear physical interpretations, and their values must be verified by using experimental data.

Let $x_1 = y/D$, $x_2 = C_L$ and introduce some new parameters $\mu = \rho_a LD^2/(8\pi^2 MS_t^2)$, $\alpha = GC_{L0}^2$, $\gamma = 4G/3$, $b = F\omega_s/\omega_n$. Equation (6.1) is transformed into the following equations

$$\begin{aligned} &\ddot{x}_1 + 2\zeta\omega_n\dot{x}_1 + \omega_n^2 x_1 = \mu\omega_s^2 x_2, \\ &\ddot{x}_2 + (\frac{\gamma}{\omega_s}\dot{x}_2^2 - \alpha\omega_s)\dot{x}_2 + \omega_s^2 x_2 = b\omega_n\dot{x}_1. \end{aligned} \tag{6.2}$$

In the case of breeze or weak wind excitation, the two oscillators in system (6.2) are weakly coupled, i.e., the coupling parameters μ and b are both small. In fact, due to $\mu = \rho_a/(2\pi^3\rho_s S_t^2)$, μ can be found to be a small quantity since densities $\rho_a \ll \rho_s$. Experimental results have shown that S_t and C_{L0} remain relatively constant within a wide range of Reynolds numbers. In theoretical studies, model parameters are selected to align with actual parameters and previously reported experimental data (Deng et al. 2021). The parameter values of system (6.2) are $\zeta = 0.0043$, $\mu = 0.0086$, $\omega_n = 75.4rad/s$, $b = 0.26$, $\alpha = 0.045$, $\gamma = 2/3$, $D = 0.06m$..

6.1.2 Hartlen-Currie Model with Fluctuating Wind Excitation—Resonance Case

In the classical Hartlen-Currie model (6.2), the wind speed V is assumed to be constant, while real wind speed is a stochastic process. In wind engineering, real wind speed is the superposition of mean wind speed and fluctuating wind speed. By replacing the wind speed V in the Hartlen-Currie model (6.2) with

$$V = \overline{V} + \sqrt{2\eta}\xi(t). \tag{6.3}$$

where $\overline{V}$ is the mean wind speed; $\sqrt{2\eta}\xi(t)$ is the fluctuating wind speed, the Hartlen-Currie system (6.2) becomes

$$\begin{aligned} &\ddot{X}_1 + 2\zeta\omega_n\dot{X}_1 + \omega_n^2 X_1 = \mu\omega_s^2 X_2 + \frac{4\overline{V}\rho_a}{\pi D^2\rho_s} X_2\sqrt{2\eta}\xi(t), \\ &\ddot{X}_2 + (\frac{\gamma}{\omega_s}\dot{X}_2^2 - \alpha\omega_s)\dot{X}_2 + \omega_s^2 X_2 = b\omega_n\dot{X}_1. \end{aligned} \tag{6.4}$$

Compared with the mean wind speed $\overline{V}$, the fluctuating wind speed $\sqrt{2\eta}\xi(t)$ is small. Serval power spectral density functions have been proposed to describe fluctuating wind speed. In this section, the following Davenport power spectral density $S_D(f)$ (Davenport 1961) will be used

$$S_D(f) = \frac{4k\overline{V}_{10}^2\overline{f}^2}{f(1+\overline{f}^2)^{4/3}}, \quad \overline{f} = \frac{1200f}{\overline{V}_{10}},$$
$$S(\omega) = \frac{1}{2\pi}S_D(\frac{\omega}{2\pi}). \tag{6.5}$$

Here, k represents ground roughness, and $\overline{V}_{10}$ represents the mean wind speed at a height of 10m above the ground. Figure 6.2 shows the Davenport spectrum $S_D(f)$ of the fluctuating wind speed, and its equivalent power spectral density $S(\omega)$ of radian frequency. In this section, the natural frequency ω_n of the structural oscillator is assumed to be $\omega_n = 75.4rad/s$. It can be observed from Fig. 6.2 that the wind spectrum $S(\omega)$ near ω_n changes slowly with ω. Therefore, the stochastic excitation from fluctuating wind can be regarded as a wideband stochastic process. Figure 6.3a shows a wind speed sample generated based on Eqs. (6.3) and (6.5). Figures 6.1b and 6.3c show samples of structural displacement response with and without fluctuation wind excitation, respectively. Figure 6.3b indicates that the displacement response of the structure remains constant in the absence of fluctuation wind. Figure 6.3c indicates that in the presence of fluctuating wind, the structural displacement response appears as an amplitude-modulated signal, where the rapidly changing displacement signal is modulated by a slowly varying amplitude signal, indicating a typical narrowband stochastic process (see Fig. 2.3, vol. 1).

For the Hartlen-Currie model (6.4) with fluctuating wind excitation, the stochastic averaging method of quasi-integrable Hamiltonian systems under wideband noise excitation, as described in Sect. 8.1, can be applied. Let generalized displacement $Q_1 = X_1, \ Q_2 = X_2$ and generalized momentum $P_1 = \dot{X}_1, \ P_2 = \dot{X}_2$. Equation (6.4) can be transformed into the following stochastically excited and dissipated Hamiltonian system

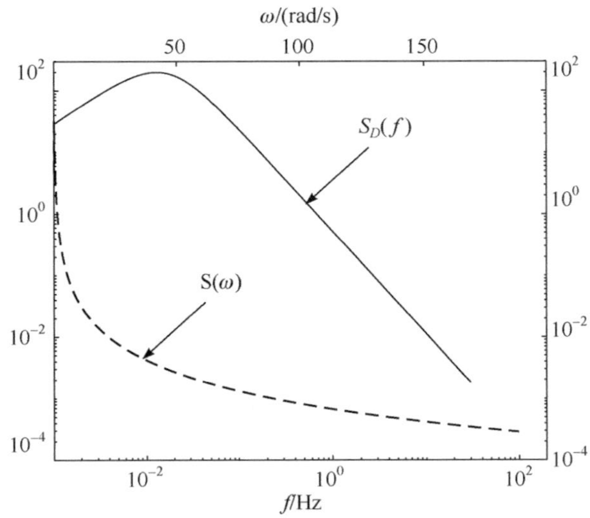

Fig. 6.2 Davenport power spectral density of fluctuation wind speed

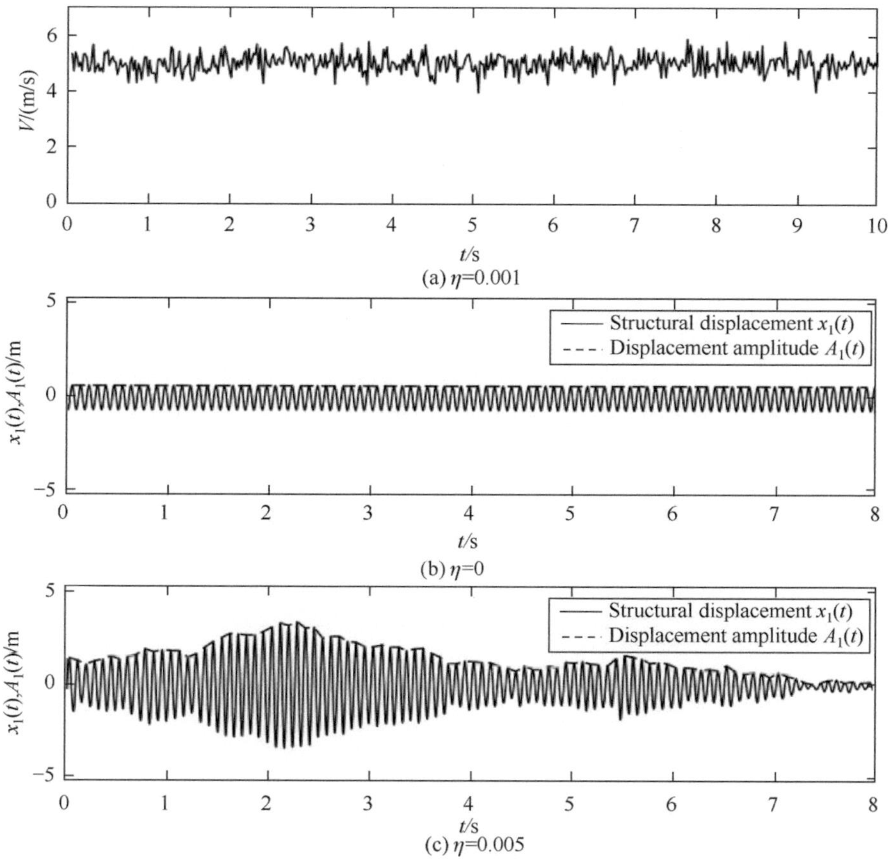

Fig. 6.3 **a** Sample of fluctuation wind speed; **b** Sample of structural displacement response without fluctuation wind; **c** Sample of structural displacement response with fluctuation wind (Deng et al. 2021)

$$
\begin{aligned}
\dot{Q}_1 &= P_1, \quad \dot{P}_1 = -\frac{\partial H}{\partial Q_1} - c_1 + f\xi(t), \\
\dot{Q}_2 &= P_2, \quad \dot{P}_2 = -\frac{\partial H}{\partial Q_2} - c_2,
\end{aligned} \tag{6.6}
$$

where $\mathbf{Q} = [Q_1, Q_2]^{\mathrm{T}}$, $\mathbf{P} = [P_1, P_2]^{\mathrm{T}}$; $H = H(\mathbf{Q}, \mathbf{P})$ is the Hamiltonian function; $c_1(\mathbf{Q}, \mathbf{P}) = 2\zeta\omega_n P_1 - \mu\omega_s^2 Q_2$ and $c_2(\mathbf{Q}, \mathbf{P}) = -\alpha\omega_s P_2 + \gamma P_2^3 / \omega_s - b\omega_n P_1$ are weak damping forces and coupling forces; $f = 4\overline{V}\rho_a Q_2\sqrt{2\eta} / \pi D^2 \rho_s$ is the amplitude of the fluctuating wind speed; $f\xi(t)$ is a random parametric excitation. Since μ, b are small, the Hamiltonian system associated with system (6.6) can be regarded as integrable, and the Hamiltonian function is separable, i.e.,

$$
\begin{aligned}
&H = H_1 + H_2, \quad H_1 = \frac{1}{2}P_1^2 + U_1(Q_1), \quad H_2 = \frac{1}{2}P_1^2 + U_2(Q_2), \\
&U_1(Q_1) = \frac{1}{2}\omega_n^2 Q_1^2, \quad U_2(Q_2) = \frac{1}{2}\omega_s^2 Q_2^2,
\end{aligned}
\tag{6.7}
$$

where H_1 is the energy of the structural oscillator, and H_2 is the first integral of the excitation oscillator. Under conditions of weak coupling, weak damping, and weak excitation, the following transformation can be applied.

$$
\begin{aligned}
&Q_1(t) = \frac{\sqrt{2H_1(t)}}{\omega_n}\cos\Phi_1(t), \quad P_1(t) = -\sqrt{2H_1(t)}\sin\Phi_1(t), \quad \Phi_1(t) = \omega_n t + \Theta_1(t), \\
&Q_2(t) = \frac{\sqrt{2H_2(t)}}{\omega_s}\cos\Phi_2(t), \quad P_2(t) = -\sqrt{2H_2(t)}\sin\Phi_2(t), \quad \Phi_2(t) = \omega_s t + \Theta_2(t).
\end{aligned}
\tag{6.8}
$$

where the Hamiltonian function vector $\mathbf{H} = [H_1(t), H_2(t)]^{\mathrm{T}}$ and the phase angle vector $\boldsymbol{\Phi} = [\Phi_1(t), \Phi_2(t)]^{\mathrm{T}}$ are both random processes. Using transformation (6.8), system (6.6) can be converted into the following equivalent equations

$$
\begin{aligned}
&\dot{H}_1 = F_1^{(H)} + G_1^{(H)}\xi(t), \quad \dot{\Phi}_1 = \omega_n + F_1^{(\Phi)} + G_1^{(\Phi)}\xi(t), \\
&\dot{H}_2 = F_2^{(H)} + G_2^{(H)}\xi(t), \quad \dot{\Phi}_2 = \omega_s + F_2^{(\Phi)} + G_2^{(\Phi)}\xi(t),
\end{aligned}
\tag{6.9}
$$

where

$$
\begin{aligned}
&F_1^{(H)} = c_1\sqrt{2H_1}\sin\Phi_1, \quad F_2^{(H)} = c_2\sqrt{2H_2}\sin\Phi_2, \quad G_1^{(H)} = -\sqrt{2H_1}f\sin\Phi_1, \\
&F_1^{(\Phi)} = \frac{c_1}{\sqrt{2H_1}}\cos\Phi_1, \quad F_2^{(\Phi)} = \frac{c_2}{\sqrt{2H_2}}\cos\Phi_2, \quad G_1^{(\Phi)} = \frac{-1}{\sqrt{2H_1}}f\cos\Phi_1, \\
&G_2^{(H)} = 0, \quad G_2^{(\Phi)} = 0.
\end{aligned}
\tag{6.10}
$$

System (6.9) will be studied for the resonance case in this section and for the non-resonance case in the next section (Deng et al. 2021).

In the case of $\omega_s : \omega_n \approx 1 : 1$, i.e., the wind-excited frequency ω_s is close to the structural natural frequency ω_n, internal resonance may occur between the two oscillators in the stochastic Hartlen-Currie model (6.4). Besides the Hamiltonian process $H_1(t), H_2(t)$ being the slowly-varying processes, the phase difference $\Psi(t) = \Phi_1(t) - \Phi_2(t) = \tan^{-1}(P_1/\omega_n Q_1) - \tan^{-1}(P_2/\omega_s Q_2)$ is also a slowly-varying process. It is governed by the following equation

$$
\begin{aligned}
&\dot{\Psi} = F_3^{(H)} + G_3^{(H)}\xi(t), \\
&F_3^{(H)} = \omega_n - \omega_s + F_1^{(\Phi)} - F_2^{(\Phi)}, \quad G_3^{(H)} = G_1^{(\Phi)} - G_2^{(\Phi)}.
\end{aligned}
\tag{6.11}
$$

Let $\omega_s : \omega_n = (1 + \sigma) : 1$, where σ is a small parameter. The c_1 and c_2 in system (6.6) become

$$\begin{aligned} c_1(\mathbf{Q},\mathbf{P}) &= 2\zeta\omega_n P_1 - \mu(1+2\sigma)\omega_n^2 Q_2, \\ c_2(\mathbf{Q},\mathbf{P}) &= -\alpha\omega_n(1+\sigma)P_2 + \frac{\gamma}{\omega_n}(1-\sigma)P_2^3 - b\omega_n P_1 + 2\sigma\omega_n^2 Q_2, \end{aligned} \tag{6.12}$$

where the terms of orders higher than σ are neglected.

Applying the stochastic averaging method of quasi-integrable and resonant Hamiltonian systems under wideband stochastic excitation described in Sect. 8.1.2, the following averaged Itô stochastic differential equations governing $\overline{\mathbf{H}} = [H_1, H_2, \Psi]^{\mathrm{T}}$ can be obtained.

$$\begin{aligned} \mathrm{d}H_1 &= a_1(\overline{\mathbf{H}})\mathrm{d}t + \sigma_1(\overline{\mathbf{H}})\mathrm{d}B(t), \\ \mathrm{d}H_2 &= a_2(\overline{\mathbf{H}})\mathrm{d}t + \sigma_2(\overline{\mathbf{H}})\mathrm{d}B(t), \\ \mathrm{d}\Psi &= a_3(\overline{\mathbf{H}})\mathrm{d}t + \sigma_3(\overline{\mathbf{H}})\mathrm{d}B(t), \end{aligned} \tag{6.13}$$

where $B(t)$ is a unit Wiener process, and the drift and diffusion coefficients are given by

$$\begin{aligned} a_i(\overline{\boldsymbol{H}}) &= \left\langle F_i^{(H)} + \int_{-\infty}^{0} \Big[\sum_{l=1}^{3}\Big(\frac{\partial G_i^{(H)}}{\partial H_l}\Big|_t G_l^{(H)}\Big|_{t+\tau}\Big) + \frac{\partial G_i^{(H)}}{\partial \Phi_2}\Big|_t G_2^{(\Phi)}\Big|_{t+\tau}\Big] R(\tau)\mathrm{d}\tau \right\rangle_t, \\ b_{ij}(\overline{\boldsymbol{H}}) &= \sigma_i\sigma_j = \left\langle \int_{-\infty}^{\infty} G_i^{(H)}\Big|_t G_j^{(H)}\Big|_{t+\tau} R(\tau)\mathrm{d}\tau \right\rangle_t, \\ i,j &= 1,2,3. \end{aligned} \tag{6.14}$$

In Eq. (6.14), Φ_1 in coefficients $F_i^{(H)}, G_i^{(H)}, G_i^{(\Phi)}$ are replaced by $\Psi + \Phi_2$, and time averaging $\langle[\cdot]\rangle_t$ can be replaced by averaging with respect to Φ_2, i.e., $\langle[\cdot]\rangle_t = \int_0^{2\pi}[\cdot]d\Phi_2/2\pi$. The $R(\tau)$ in Eq. (6.14) is the autocorrelation function of the fluctuating wind. Corresponding to Itô Eq. (6.13), the reduced averaged FPK equation is

$$\begin{aligned} &-\sum_{i=1}^{2}\frac{\partial}{\partial h_i}\big[a_i(h_1,h_2,\psi)p\big] \quad \frac{\partial}{\partial \psi}\big[a_3(h_1,h_2,\psi)p\big] + \frac{1}{2}\sum_{i,j=1}^{2}\frac{\partial^2}{\partial h_i \partial h_j}\big[b_{ij}(h_1,h_2,\psi)p\big] \\ &+\frac{\partial^2}{\partial h_1\partial\psi}\big[b_{13}(h_1,h_2,\psi)p\big] + \frac{\partial^2}{\partial h_2\partial\psi}\big[b_{23}(h_1,h_2,\psi)p\big] = 0. \end{aligned} \tag{6.15}$$

where the first and second derivative moments are

$$\begin{aligned} &a_i(h_1,h_2,\psi) = a_i(\overline{H})\big|_{\overline{H}=[h_1,h_2,\psi]^T}, \quad b_{ij}(h_1,h_2,\psi) = b_{ij}(\overline{H})\big|_{\overline{H}=[h_1,h_2,\psi]^T}, \\ &i,j = 1,2,3. \end{aligned} \tag{6.16}$$

The boundary conditions of Eq. (6.15) are

$$\begin{aligned} &p(h_1 = 0, h_2, \psi) = \text{finite}, \quad p(h_1, h_2 = 0, \psi) = \text{finite}, \\ &p(h_1 \to \infty, h_2, \psi) = 0, \quad p(h_1, h_2 \to \infty, \psi) = 0, \\ &p(h_1, h_2, \psi + 2\pi) = p(h_1, h_2, \psi). \end{aligned} \tag{6.17}$$

By substituting Eqs. (6.10) and (6.11) into Eq. (6.14), expanding $F_i^{(H)}, G_i^{(H)}, G_i^{(\Phi)}$ as Fourier series of Φ_1, Φ_2, replacing Φ_1 with $\Psi + \Phi_2$, then performing averaging over Φ_2, the first and second derivative moments can be obtained as follows

$$\begin{aligned} a_1 &= -2\zeta\omega_n h_1 - \omega_n \mu(1+2\sigma)\sqrt{h_1 h_2}\sin\psi + \frac{16\eta \overline{V}^2 \rho_a^2 h_2}{\pi D^4 \rho_s^2 \omega_n^2}[S(0) + S(2\omega_n)], \\ a_2 &= \alpha(1+\sigma)\omega_n h_2 - \frac{3\gamma}{2\omega_n}(1-\sigma)h_2^2 + \omega_n b\sqrt{h_1 h_2}\cos\psi, \\ a_3 &= -\sigma\omega_n - \frac{b\omega_n \sin\psi}{2}\sqrt{h_1/h_2} - \frac{\mu(1+2\sigma)\omega_n\cos\psi}{2}\sqrt{h_2/h_1} - \frac{8\eta\overline{V}^2\rho_a^2 h_2 \sin 2\psi}{\pi D^4 \rho_s^2 \omega_n^2 h_1}S(0), \\ b_{11} &= \frac{32\eta\overline{V}^2\rho_a^2 h_1 h_2}{\pi D^4 \rho_s^2 \omega_n^2}[2\sin^2\psi S(0) + S(2\omega_n)], \\ b_{13} &= b_{31} = \frac{16\eta\overline{V}^2\rho_a^2 h_2 \sin 2\psi}{\pi D^4 \rho_s^2 \omega_n^2}S(0), \quad b_{33} = \frac{8\eta\overline{V}^2\rho_a^2 h_2}{\pi D^4 \rho_s^2 \omega_n^2 h_1}[2\cos^2\psi S(0) + S(2\omega_n)], \\ b_{12} &= b_{21} = b_{23} = b_{32} = b_{22} = 0, \end{aligned} \tag{6.18}$$

where $S(\omega)$ is the power spectral density of fluctuating wind speed $\xi(t)$ from Eq. (6.5). Equation (6.15) can be solved numerically to obtain the stationary PDF $p(\overline{\mathbf{h}})$. Following Eq. (5.127) the joint stationary PDF for displacement and velocity can be obtained as follows

$$p(q_1, q_2, p_1, p_2) = Cp(\overline{\mathbf{h}})\Big|_{h_1 = \frac{1}{2}p_1^2 + \frac{1}{2}\omega_n^2 q_1^2,\ h_2 = \frac{1}{2}p_2^2 + \frac{1}{2}\omega_s^2 q_2^2,\ \psi = \tan^{-1}\left(\frac{p_1}{\omega q_1}\right) - \tan^{-1}\left(\frac{p_2}{\omega q_2}\right)}. \tag{6.19}$$

where C is a normalization constant. The other stationary marginal PDFs and statistics can then be obtained from $p(\overline{\mathbf{h}})$ or $p(q_1, q_2, p_1, p_2)$.

Figures 6.4, 6.5, 6.6 and 6.7 provide the approximate analytical solutions for several PDFs and statistics, and results from Monte Carlo simulations of the original system (6.6). It is seen that the analytical solutions match the simulation results closely.

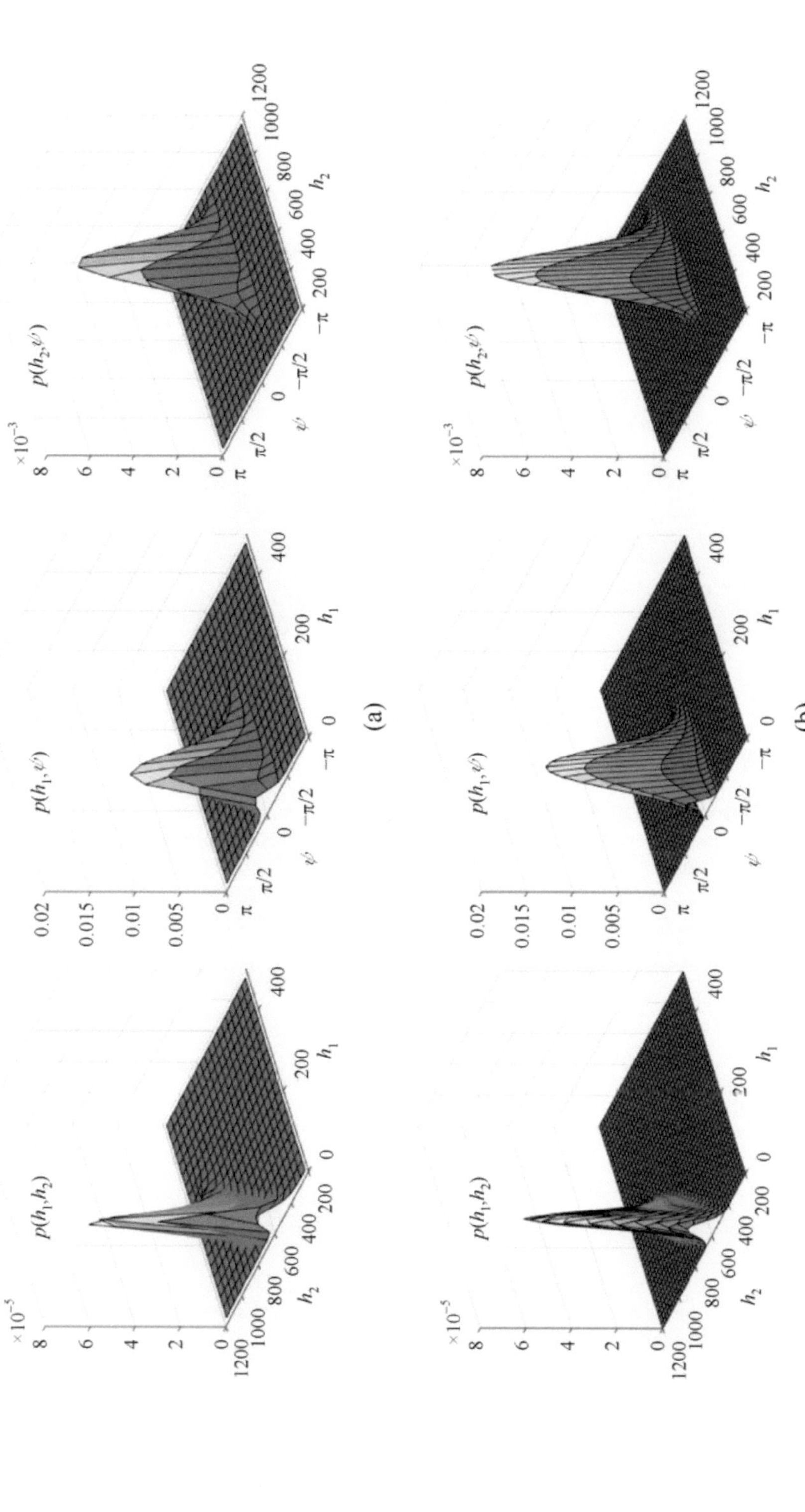

Fig. 6.4 Stationary joint PDFs,, $p(h_1. \psi)$, $p(h_2, \psi)$ of system (6.6) in resonant case, **a** results from Monte Carlo simulation; **b** results from stochastic averaging method (Deng et al. 2021)

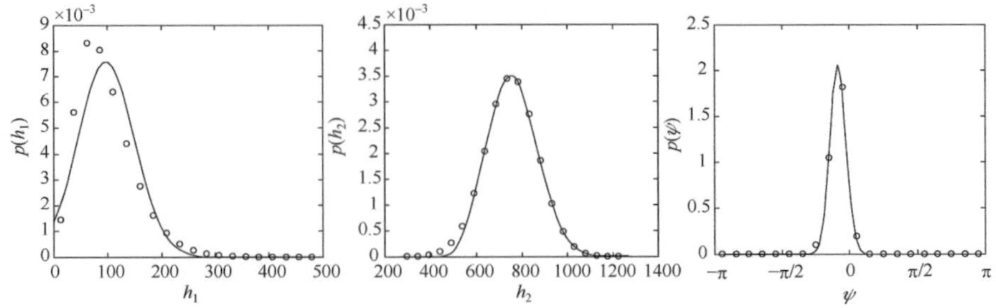

Fig. 6.5 Stationary marginal PDF $p(h_1)$, $p(h_2)$, $p(\psi)$ of system (6.6) in resonant case (Deng et al. 2021), ○, result from Monte Carlo simulation; ——, result from stochastic averaging method

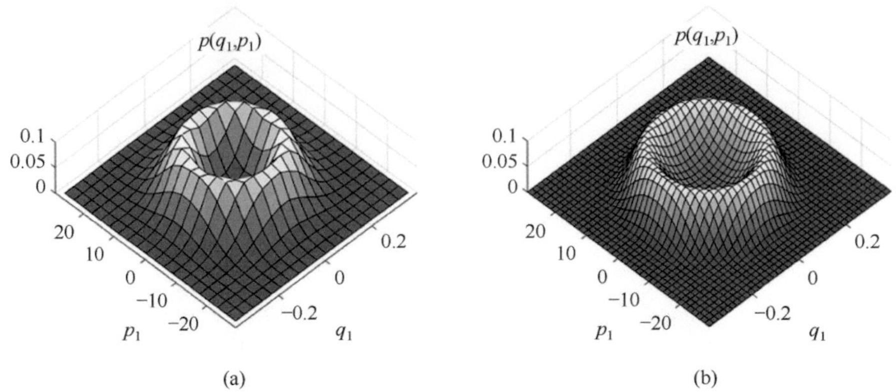

Fig. 6.6 Stationary joint PDF $p(q_1, p_1)$ of structural oscillator displacement and velocity of system (6.6) in resonant case, **a** result from Monte Carlo simulation; **b** result from stochastic averaging method (Deng et al. 2021)

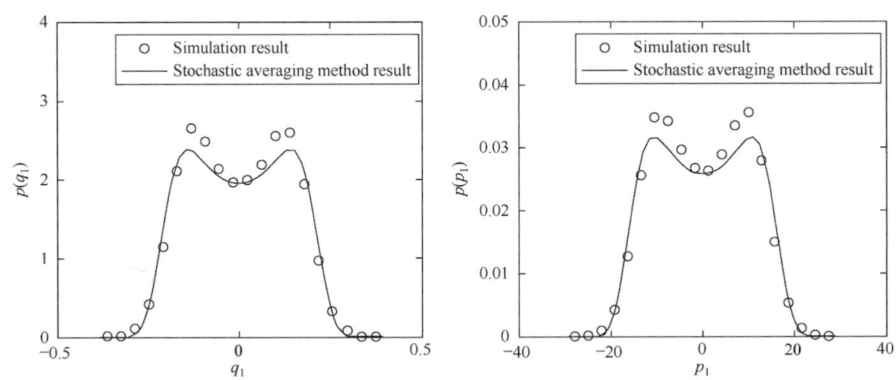

Fig. 6.7 Stationary marginal PDFs $p(q_1)$, $p(p_1)$ of structural oscillator displacement and velocity of system (6.6) in resonant case (Deng et al. 2021)

6.1.3 Hartlen-Currie Model Under Fluctuating Wind Excitation—Non-resonance Case

When the wind-excited frequency ω_s is not near structural frequency ω_n, system (6.6) is the case of non-resonance. Experiments show that in the non-resonance case, the structure response amplitude remains at a lower level. Figure 6.8 displays the PDF of the phase difference $\Psi = \Phi_1 - \Phi_2$, indicating that in the case of $\omega_s / \omega_n = 1$ (i.e., the resonance case), there is a probability peak in the interval $[-\pi / 2, 0)$, suggesting that in the resonance case, the excitation oscillator and the structural oscillator are mostly synchronized. When $\omega_s / \omega_n = 1.5$ (i.e., the non-resonance case), the phase difference Ψ is almost uniformly distributed over the interval $[-\pi, \pi]$, indicating that in the non-resonance case, the excitation oscillator and structural oscillator are out of tune.

In the non-resonance case, the stochastic averaging method of quasi-integrable and non-resonance Hamiltonian systems under wideband stochastic excitation described in Sect. 8.1.2 can also be applied to predict the response of the stochastic Hartlen-Currie model (6.6). In this case, the vector process $\mathbf{H} = [H_1, H_2]^{\mathrm{T}}$ in Eq. (6.9) converges to a 2-dimensional vector Markov diffusion process. By using the stochastic averaging method, the following averaged Itô stochastic differential equation can be obtained

$$\mathrm{d}H_i = a_i(\mathbf{H})\mathrm{d}t + \sigma_i(\mathbf{H})\mathrm{d}B(t), \quad i = 1, 2. \tag{6.20}$$

where the drift and diffusion coefficients are

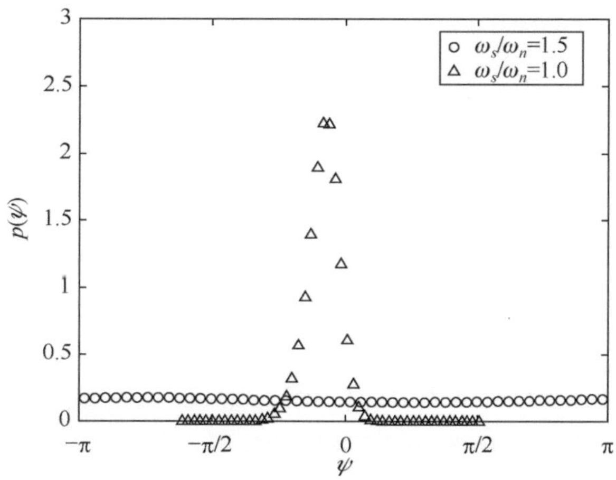

Fig. 6.8 Stationary PDF of the phase difference ψ between two oscillators in system (6.6) (Deng et al. 2021)

$$\begin{aligned}
& a_i(\boldsymbol{H}) = \left\langle F_i^{(H)} + \int_{-\infty}^{0} \sum_{l=1}^{2} \left(\left.\frac{\partial G_i^{(H)}}{\partial H_l}\right|_t \left. G_l^{(H)}\right|_{t+\tau} + \left.\frac{\partial G_i^{(H)}}{\partial \Phi_l}\right|_t \left. G_l^{(\Phi)}\right|_{t+\tau} \right) R(\tau)\mathrm{d}\tau \right\rangle_t, \\
& b_{ij}(\boldsymbol{H}) = \sigma_i \sigma_j = \left\langle \int_{-\infty}^{\infty} \left. G_i^{(H)}\right|_t \left. G_j^{(H)}\right|_{t+\tau} R(\tau)\mathrm{d}\tau \right\rangle_t, \\
& i, j = 1, 2.
\end{aligned} \tag{6.21}$$

where $\langle[\cdot]\rangle_t$ denotes the time averaging, which can be replaced by the spatial averaging with respect to Φ_1, Φ_2, i.e., $\langle[\cdot]\rangle_t = \int_0^{2\pi}\int_0^{2\pi}[\cdot]\mathrm{d}\Phi_1\mathrm{d}\Phi_2 / 4\pi^2$. The $R(\tau)$ in Eq. (6.21) is autocorrelation function of the fluctuating wind excitation.

The reduced averaged FPK equation associated with averaged Itô stochastic differential equation (6.20) and its boundary conditions are

$$\begin{aligned}
& -\sum_{i=1}^{2} \frac{\partial}{\partial h_i}\left[a_i(\boldsymbol{h})p(\boldsymbol{h})\right] + \frac{1}{2}\sum_{i,j=1}^{2} \frac{\partial^2}{\partial h_i \partial h_j}\left[b_{ij}(\boldsymbol{h})p(\boldsymbol{h})\right] = 0, \\
& p(h_1 = 0, h_2) = \text{finite}, \quad p(h_1, h_2 = 0) = \text{finite}, \\
& p(h_1 \to \infty, h_2) = 0, \quad p(h_1, h_2 \to \infty) = 0,
\end{aligned} \tag{6.22}$$

where $\boldsymbol{h} = [h_1, h_2]^{\mathrm{T}}$. The first and second derivative moments are

$$\begin{aligned}
& a_i(\boldsymbol{h}) = a_i(\boldsymbol{H})|_{\boldsymbol{H}=\boldsymbol{h}}, \quad b_{ij}(\boldsymbol{h}) = b_{ij}(\boldsymbol{H})\big|_{\boldsymbol{H}=\boldsymbol{h}}, \\
& i, j = 1, 2.
\end{aligned} \tag{6.23}$$

Further derivation yields

$$\begin{aligned}
a_1 &= -2\zeta\omega_n h_1 + \frac{16\eta \overline{V}^2 \rho_a^2 h_2}{\pi D^4 \rho_s^2 \omega_s^2}[S(\omega_n - \omega_s) + S(\omega_n + \omega_s)], \\
a_2 &= \alpha\omega_s h_2 - \frac{3\gamma}{2\omega_s} h_2^2, \\
b_{11} &= \frac{32\eta \overline{V}^2 \rho_a^2 h_1 h_2}{\pi D^4 \rho_s^2 \omega_s^2}[S(\omega_n - \omega_s) + S(\omega_n + \omega_s)], \\
b_{12} &= b_{21} = b_{22} = 0.
\end{aligned} \tag{6.24}$$

Applying the moment equation method with Gaussian truncation, the first and second moments of the stationary H_1, H_2 processes can be obtained as follows

$$\begin{aligned}
\mu_1 &= E[H_1] = \frac{16\eta\overline{V}^2\rho_a^2\alpha[S(\omega_n-\omega_s)+S(\omega_n+\omega_s)]}{3\pi D^4\rho_s^2\gamma\zeta\omega_n},\\
\mu_2 &= E[H_2] = \frac{2\alpha\omega_s^2}{3\gamma},\quad E[H_1^2]=2\mu_1^2,\\
E[H_2^2] &= \mu_2^2,\quad E[H_1H_2]=\mu_1\mu_2.
\end{aligned} \tag{6.25}$$

Substituting $H_2 = \mu_2$ into Eq. (6.24), the following exact solution of Eq. (6.22) is obtained.

$$p(h_1,h_2) = \frac{1}{\mu_1}\exp(\frac{-h_1}{\mu_1})\delta(h_2-\mu_2). \tag{6.26}$$

Using Eq. (5.106), the stationary PDF and statistical quantities of the structural response can be derived from Eq. (6.26) as follows

$$\begin{aligned}
p(q_1,p_1) &= \tfrac{\omega_n}{2\pi\mu_1}\exp[\tfrac{-1}{2\mu_1}(p_1^2+\omega_n^2q_1^2)],\\
p(q_1) &= \frac{\omega_n}{\sqrt{2\pi\mu_1}}\exp(\frac{-\omega_n^2q_1^2}{2\mu_1}),\quad p(p_1)=\frac{1}{\sqrt{2\pi\mu_1}}\exp(\frac{-p_1^2}{2\mu_1}),\\
E[Q_1^2] &= \frac{\mu_1}{\omega_n^2},\quad E[P_1^2]=\mu_1.
\end{aligned} \tag{6.27}$$

Figures 6.9 and 6.10 display the stationary PDF of structural oscillator of system (6.6) in the non-resonant case. It is seen that the analytical solution (6.27) agrees well with Monte Carlo simulation results. In contrast to the volcano-shape (Fig. 6.6) and double peaks (Fig. 6.7) in the resonant case, the response PDF of structural oscillator in the non-resonant case exhibits a single peak. Figure 6.11 show $E[Q_1^2]$ and $E[P_1^2]$ from analytical solution (6.19) of the resonant case, analytical solution (6.27) of the non-resonant case, and Monte Carlo simulation. It is seen that, over a wide range of frequency and a wide range of wind speed since $\omega_s = 2\pi\overline{V}S_t/D$, the stochastic averaging method provides satisfactory response prediction.

6.1.4 Nonlinear Structural Oscillator

In many wake oscillator models, it is assumed that the structural oscillator is linear. Replacing the linear structural oscillator with nonlinear one leads to nonlinear wake oscillator model. The stochastic averaging method of quasi-integrable Hamiltonian systems under wideband random excitation described in Sect. 8.1 remains applicable to this new model. Consider a structural oscillator with the following nonlinear potential and restoring force

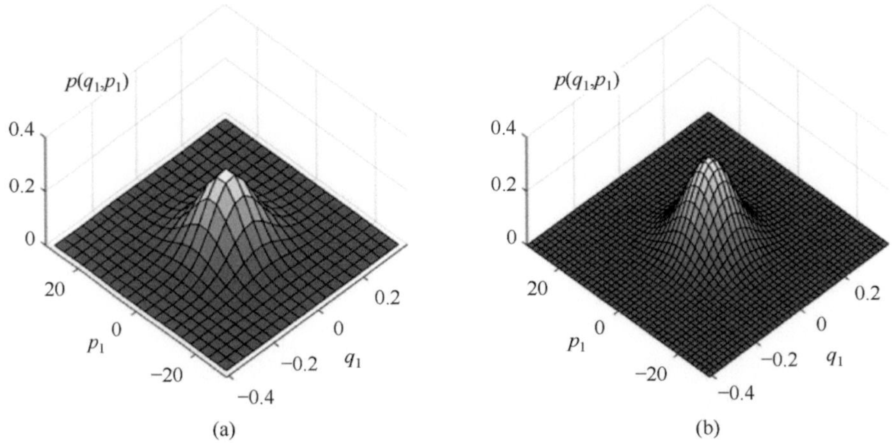

Fig. 6.9 Stationary joint PDF $p(q_1, p_1)$ of structural oscillator displacement and velocity of system (6.6) in non-resonant case, **a** results from Monte Carlo simulation; **b** results from stochastic averaging method (Deng et al. 2021)

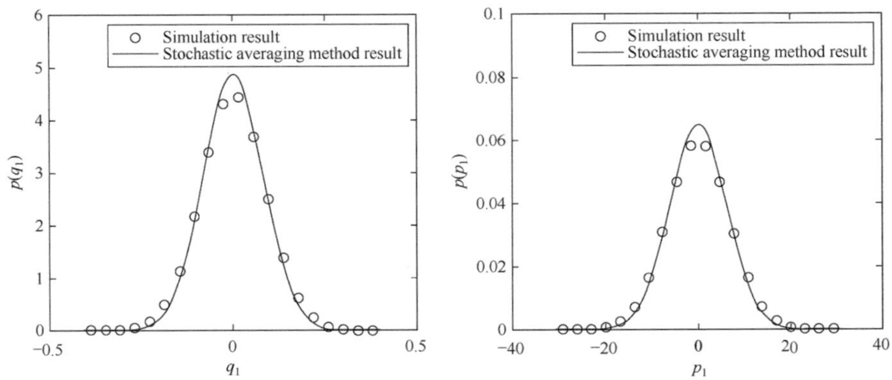

Fig. 6.10 Stationary marginal PDF of structural oscillator displacement and velocity of system (6.6) in non-resonant case (Deng et al. 2021)

$$U_1(x) = \frac{1}{2}\omega_n^2 x^2 + \frac{1}{4}kx^4, \quad g_1(x) = \frac{\mathrm{d}U_1(x)}{\mathrm{d}t} = \omega_n^2 x + kx^3, \tag{6.28}$$

where the nonlinear stiffness coefficient k is assumed to be positive. The Hartlen-Currie motion Eq. (6.1) is modified as

$$\begin{aligned} &\frac{\ddot{y}}{D} + 2\zeta\omega_n\frac{\dot{y}}{D} + \omega_n^2\frac{y}{D} + k\left(\frac{y}{D}\right)^3 = \frac{1}{2}\rho_a V^2 \frac{L}{M} C_L, \\ &\ddot{C}_L + \left(\frac{4}{3}\frac{G}{\omega_s}\dot{C}_L^2 - \omega_s G C_{L0}^2\right)\dot{C}_L + \omega_s^2 C_L = \omega_s F\frac{\dot{y}}{D}. \end{aligned} \tag{6.29}$$

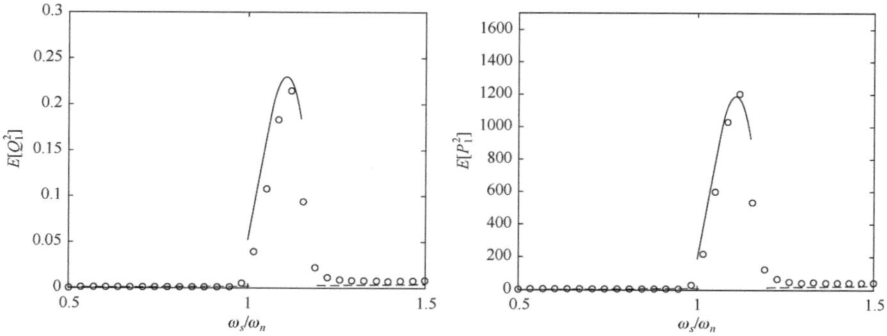

Fig. 6.11 Mean-square response of structural oscillator of system (6.6) with varying frequency ratio ω_s / ω_n (Deng et al. 2021), ○, result from Monte Carlo simulation; ——, result from stochastic averaging method in resonant case; ----, result from stochastic averaging method in non-resonant case

Figure 6.12 shows the effect of the nonlinear stiffness coefficient k on the motion of the free and undamped structural oscillator.

Let $X_1 = y/D,\ X_2 = C_L$ and $V = \overline{V} + \sqrt{2\eta}W(t)$. The nonlinear Hartlen-Currie model subjected to a fluctuating wind excitation is governed by

$$\begin{aligned} &\ddot{X}_1 + 2\zeta\omega_n\dot{X}_1 + \omega_n^2 X_1 + kX_1^3 = \mu\omega_s^2 X_2 + \frac{4\overline{V}\rho_a}{\pi D^2\rho_s}X_2\sqrt{2\eta}\xi(t), \\ &\ddot{X}_2 + (\frac{\gamma}{\omega_s}\dot{X}_2^2 - \alpha\omega_s)\dot{X}_2 + \omega_s^2 X_2 = b\omega_n\dot{X}_1. \end{aligned} \tag{6.30}$$

Let $Q_1 = X_1,\ Q_2 = X_2$ and $P_1 = \dot{X}_1,\ P_2 = \dot{X}_2$, The motion Eq. (6.30) can be written in the form of a quasi-Hamiltonian system. The Hamiltonian function H_2 and

Fig. 6.12 Trajectories of the conservative structural oscillator in phase space $(y, \dot{y})$, total energy $h_1 = 5$ (Deng et al. 2021)

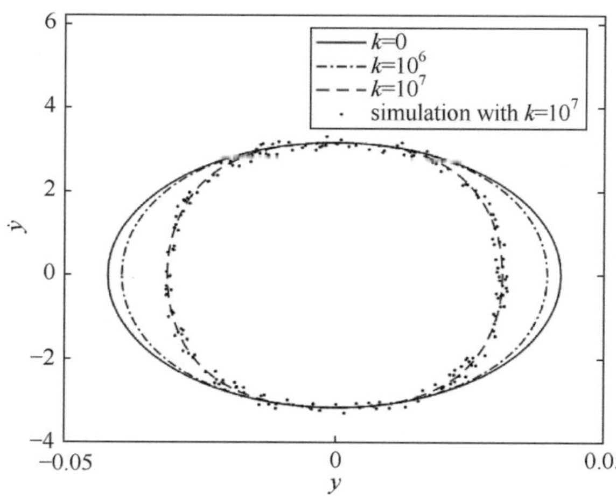

potential energy $U_2(Q_2)$ of the excited oscillator are the same as those in Eq. (6.7), while the Hamiltonian function H_1 and potential energy $U_1(Q_1)$ of the structural oscillator is changed to

$$H_1 = \frac{1}{2}P_1^2 + U_1(Q_1), \quad U_1(Q_1) = \frac{1}{2}\omega_n^2 Q_1^2 + \frac{1}{4}kQ_1^4. \tag{6.31}$$

The Hamiltonian system associated with Eq. (6.30) is integrable. Since the structural oscillator frequency $\omega_1(H_1)$ varies with H_1, the internal resonance condition $\omega_s : \omega_1(H_1) = 1 : 1$ are generally not met. Therefore, the stochastic averaging method of quasi-integrable and non-resonant Hamiltonian systems under wideband noise excitation is applied to system (6.30). Introduce the transformation

$$\begin{aligned}
&Q_1 = U_1^{-1}(H_1)\cos\Phi_1(t), \quad P_1 = -U_1^{-1}(H_1)\nu_1 \sin\Phi_1(t), \quad \Phi_1(t) = \Gamma_1(t) + \Theta_1(t),\\
&Q_2 = \frac{\sqrt{2H_2(t)}}{\omega_s}\cos\Phi_2(t), \quad P_2 = -\sqrt{2H_2(t)}\sin\Phi_2(t), \quad \Phi_2(t) = \omega_s t + \Theta_2(t),
\end{aligned} \tag{6.32}$$

where $U_1^{-1}(H_1) = A_1$ is the amplitude of the structural oscillator,

$$\nu_1 = \nu_1(H_1, \Phi_1) = \frac{\mathrm{d}\Phi_1}{\mathrm{d}t} = \sqrt{\frac{2[H_1 - U_1[U_1^{-1}(H_1)\cos\Phi_1]}{(U_1^{-1}(H_1))^2 \sin^2\Phi_1}} \tag{6.33}$$

is the instantaneous frequency of the structural oscillator. Note that ν_1 is an even function of Φ_1 and can be expanded into the following Fourier series

$$\begin{aligned}
&\nu_1(H_1, \Phi_1) = \frac{1}{2}\kappa_0(H_1) + \sum_{i=1}^{\infty}\kappa_{2i}(H_1)\cos(2i\Phi_1),\\
&\kappa_0 = \nu(2 - \frac{\lambda^2}{8}), \quad \kappa_2 = \nu(\frac{\lambda}{2} + \frac{3\lambda^3}{64}), \quad \kappa_4 = \nu(\frac{-\lambda^2}{16}), \quad \kappa_6 = \nu(\frac{\lambda^3}{64}), \quad \cdots\\
&\lambda = \frac{kA_1^2}{4\nu^2}, \quad \nu = (\omega_n^2 + \frac{3kA_1^2}{4})^{1/2}, \quad A_1 = U^{-1}(H_1).
\end{aligned} \tag{6.34}$$

where $\omega_1(H_1) = \kappa_0(H_1)/2$ is the mean frequency of the structural oscillator. The following approximate relation will be used during the stochastic averaging operation,

$$\Phi_1(t) = \omega_1(H_1)t + \Theta_1(t). \tag{6.35}$$

Under the influence of fluctuating wind $\xi(t)$, the motion equations for H_1, H_2 are of the form of Eq. (6.9) with some new coefficients

$$F_1^{(H)} = c_1 U_1^{-1}(H_1) v_1 \sin\Phi_1, \quad F_1^{(\Phi)} = \frac{c_1 v_1}{g_1[U_1^{-1}(H_1)]}\cos\Phi_1,$$
$$G_1^{(H)} = -U_1^{-1}(H_1) v_1 f \sin\Phi_1, \quad G_1^{(\Phi)} = \frac{-v_1}{g_1[U_1^{-1}(H_1)]} f\cos\Phi_1. \tag{6.36}$$

Other coefficients remain the same as those in Eq. (6.10). The averaged Itô stochastic differential equations governing the diffusion process $[H_1, H_2]^{\mathrm{T}}$ are of the form of Eq. (6.20), and the reduced averaged FPK equation is of the form of Eq. (6.22) with the following modified first and second derivative moments

$$a_1 = -\frac{1}{2}\xi\omega_n[U_1^{-1}(h_1)]^2[2\omega_1^2 + \sum_{i=0}^{\infty}(\kappa_{2i+2}^2 - \kappa_{2i}\kappa_{2i+2})]$$
$$+\frac{2\eta\overline{V}^2\rho_a^2 h_2}{\pi D^4\rho_s^2 g_1(U_1^{-1}(h_1))\omega_s^2}\sum_{i=0}^{\infty}\beta_i\{S[(2i+1)\omega_1-\omega_s]+S[(2i+1)\omega_1+\omega_s]\},$$
$$a_2 = \alpha\omega_s h_2 - \frac{3\gamma}{2\omega_s}h_2^2, \tag{6.37}$$
$$b_{11} = \frac{4\eta\overline{V}^2\rho_a^2[U_1^{-1}(h_1)]^2 h_2}{\pi D^4\rho_s^2\omega_s^2}\sum_{i=0}^{\infty}(\kappa_{2i}-\kappa_{2i+2})^2\{S[(2i+1)\omega_1-\omega_s]+S[(2i+1)\omega_1+\omega_s]\},$$
$$b_{12} = b_{21} = b_{22} = 0,$$
$$\beta_i = (2i+1)U_1^{-1}(h_1)(\kappa_{2i}^2 - \kappa_{2i+2}^2) + \frac{1}{2}g_1(U_1^{-1}(h_1))\frac{\mathrm{d}\{[U_1^{-1}(h_1)]^2(\kappa_{2i}-\kappa_{2i+2})^2\}}{\mathrm{d}h_1}.$$

When $k \to 0$, the first and second derivative moments in Eq. (6.37) tend to those in Eq. (6.24). Solving the reduced averaged FPK Eq. (6.122) with the first and second derivative moments given by Eq. (6.37) yields the following stationary PDF of the Hamiltonian function (energy) of structural oscillator

$$p(h_1) = C\exp\{-\int_0^{h_1}[(\frac{\mathrm{d}b_{11}}{\mathrm{d}h_1} - 2a_1)\Big/ b_{11}]_{h_2=\frac{2\alpha\omega_s^2}{3\gamma}}\mathrm{d}h_1\}. \tag{6.38}$$

where C is the normalization constant. According to Eq. (5.106) in vol. 1, the approximate stationary PDF and statistics of the structural oscillator in Eq. (6.30) are

$$p(q_1, p_1) = C' p(h_1)\big|_{h_1=\frac{1}{2}p_1^2+\frac{1}{2}\omega_n^2 q_1^2+\frac{1}{4}kq_1^4},$$
$$p(q_1) = \int_{-\infty}^{\infty} p(q_1, p_1)\mathrm{d}p_1, \quad p(p_1) = \int_{-\infty}^{\infty} p(q_1, p_1)\mathrm{d}q_1, \tag{6.39}$$
$$E[Q_1^2] = \int_{-\infty}^{\infty} q_1^2 p(q_1)\mathrm{d}q_1, \quad E[P_1^2] = \int_{-\infty}^{\infty} p_1^2 p(p_1)\mathrm{d}p_1.$$

Figure 6.13 shows the PDFs $p(q_1), p(p_1)$ of structural oscillator with strong nonlinear parameter $k = 10^7$. Figure 6.14 indicates that the mean-square responses $E[Q_1^2]$ and $E[P_1^2]$ of structural oscillator increase with increasing mean wind speed. Figure 6.15 illustrates the change of mean-square responses as the nonlinear parameter k increases. In all these figures, theoretical results agree well with simulation results, indicating the applicability of the stochastic averaging method in studying vortex-induced vibration of nonlinear structure. Figure 6.15 also demonstrates that, with other parameters keeps unchanged, increasing the structural nonlinearity significantly reduces the mean-square displacement $E[Q_1^2]$ while almost no influence on the mean-square velocity $E[P_1^2]$.

The stochastic averaging method of quasi-integrable Hamiltonian systems under wideband noise excitation can also be applied to the other wake oscillator models

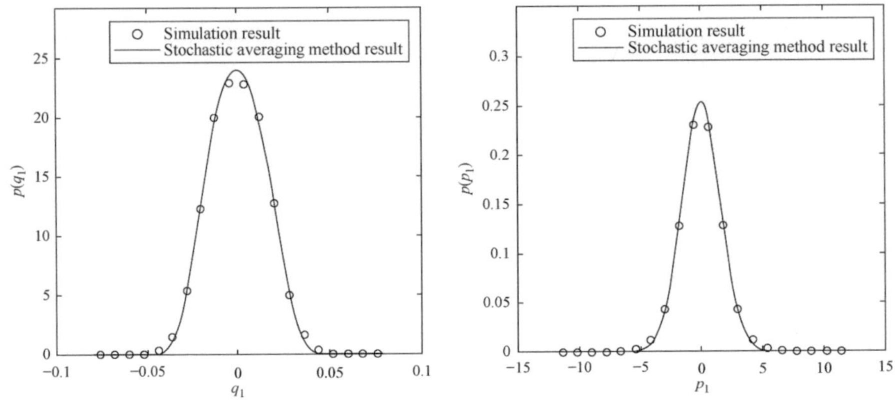

Fig. 6.13 Stationary PDF of structural oscillator of the nonlinear stochastic system (6.30) (Deng et al. 2021)

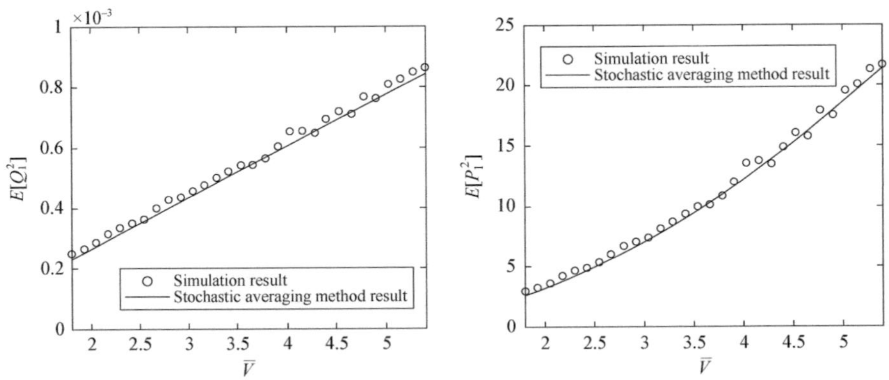

Fig. 6.14 Mean-square response of structural oscillator of the nonlinear stochastic system (6.30) with varying mean wind speed $\overline{V}$ (Deng et al. 2021)

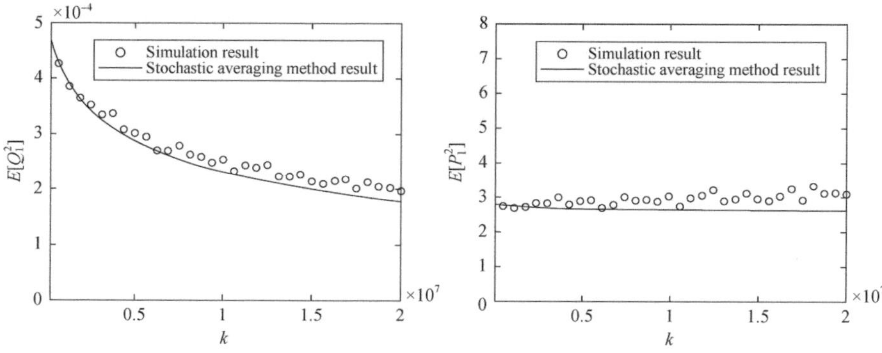

Fig. 6.15 Mean-square response of structural oscillator of the nonlinear stochastic system (6.30) with varying nonlinear parameter k (Deng et al. 2021)

subjected to real wind excitations, such as the Skop-Griffin model (Skop and Griffin 1973) and the Krenk-Nielsen model (Krenk and Nielsen 1999), etc.. The procedure of application is similar to that in handling the Hartlen-Currie model in this section.

6.2 Multi-machine Power Systems with Stochastic Excitations

As new energy sources such as wind and solar power along with electric vehicles are integrated into the power grid, the impact of stochastic excitations on power systems is gaining more attention. Power systems are giant and complex dynamical systems, and the mathematical models of multi-machine power systems subjected to random factors involve highly nonlinear and random perturbations (Ju 2010). Due to the high dimensionality and strong nonlinearity of the systems, the prevailing researches are mainly using numerical methods. However, by applying the stochastic averaging method to the single and multi-machine power systems subjected to stochastic excitations, analytical solutions are sometimes attainable, and they often exhibit good agreement with numerical simulations (Chen and Zhu 2010; Li et al. 2015; Ju 2018).

6.2.1 Model of Single/Multi-machine Power Systems Subjected to Stochastic Excitations

The model of a single-machine infinite bus system is a basis of multi-machine power systems. The single-machine infinite bus system refers to a situation where the synchronous machine's capacity in the power grid is much greater than that of the electric machine studied. In this situation, the external power grid is viewed as a large voltage source with nearly constant voltage amplitude and frequency. This section primarily explores the transient processes of electric machines under stochastic excitation. The generator is modeled using second-order swing equation. Other assumptions include: (i) constant internal voltage E'; (ii) neglecting transient saliency effect; (iii) constant mechanical power P_m.

In the deterministic case, the rotor motion equation (swing equation) for a generator is given by

$$M\frac{\mathrm{d}^2\delta}{\mathrm{d}t^2}+D\frac{\mathrm{d}\delta}{\mathrm{d}t}=P_m-P_e, \tag{6.40}$$

where δ represents the rotor angle, P_m is the mechanical power, P_e is the electromagnetic power, M is the inertia time constant, D is the damping coefficient. The expression for electromagnetic power P_e is given by

$$P_e=P_{\max}\sin(\delta-\nu), \tag{6.41}$$

where $P_{\max}$ is the maximum electrical power, and ν is the impedance angle, typically a positive value. If the electrical resistance is neglected, then

$$P_{\max}=\frac{E'U}{X_\Sigma}. \tag{6.42}$$

where E' is the inner electric potential; U is the infinite bus voltage; X_Σ is the system total reactance.

In stationary state,

$$P_m=P_e(0)=P_{\max}\sin(\delta_0-\nu), \tag{6.43}$$

where $P_e(0)$ and δ_0 represent the stationary values of P_e and δ, respectively. All the variables and parameters in Eqs. (6.40)–(6.43) are per-unit values. Time t was scaled to be dimensionless by multiplying by the rated angular velocity ω_N,

$$\omega_N=2\pi f_N=314.16\big/\mathrm{s}. \tag{6.44}$$

where $f_N=50\mathrm{Hz}$ is the rated frequency.

Stochastic excitations can be the power fluctuation generated by the new energy sources or loads like electric vehicles. These fluctuations result in an imbalance between mechanical power and electromagnetic power. P_L represents the total stochastic excitation which can be either positive or negative since stochastic excitations usually fluctuate around a mean value within a relatively short time (Haesen 2009). Thus, the fluctuation can usually be assumed to be Gaussian process. P_L is added to the right side of the rotor motion Eq. (6.40), i.e.,

$$M\frac{\mathrm{d}^2\delta}{\mathrm{d}t^2}+D\frac{\mathrm{d}\delta}{\mathrm{d}t}=P_m-P_{\max}\sin(\delta-\nu)-P_L. \tag{6.45}$$

The stochastic excitation term is given by

$$P_L=\sigma W_g(t), \tag{6.46}$$

where $W_g(t)$ is a unit Gaussian white noise, and σ^2 specifies the excitation intensity of $P_L(t)$.

Substituting Eqs. (6.44) and (6.46) into Eq. (6.45) yields

$$M\frac{\mathrm{d}^2\delta}{\mathrm{d}t^2}+D\frac{\mathrm{d}\delta}{\mathrm{d}t}=P_{\max}\sin(\delta_0-\nu)-P_{\max}\sin(\delta-\nu)-\sigma W(t). \tag{6.47}$$

Equation (6.47) is the model of a single-machine infinite bus system subjected to stochastic excitation.

Multi-machine power system subjected to stochastic excitation is complex multi-DOF nonlinear stochastic system. To simplify it, the following assumptions are introduced (Li et al. 2015):

Assumption 1: In a multi-machine power system, there must be an infinite bus system as reference machine, and its bus voltage is used as reference voltage. Or, if there isn't an infinite bus system or a reference machine, the motion equations become singular. Thus, a finite reference machine must be designated, and the equation for each generator is subtracted the reference machine equation. But this process will be cumbersome.

Assumption 2: (i) E' is constant; (ii) Transient saliency effects are neglected; (iii) mechanical power P_m is constant.

If the loads are located near the generator or equivalent after partitioning, a multi-machine power system subjected to stochastic excitations can be written as follows

$$\frac{\mathrm{d}\delta_i}{\mathrm{d}t}=\omega_N\omega_i,\quad M_i\frac{\mathrm{d}\omega_i}{\mathrm{d}t}=P_{mi}-P_{ei}-D_i\omega_i+\sigma_i W_{gi}(t),\quad i=1,2,\cdots,n. \tag{6.48}$$

where D, ω_i are per-unit values, M_i, t are the nominal value, ω_i is a value of the real value's per-unit value abstracting its stationary value's per-unit value, δ_i is the real value's per-unit value. $W_{g1}(t), W_{g2}(t), \cdots, W_{gn}(t)$ are mutual independent unit Gaussian white noises, and

$$P_{ei}=\sum_{j=1,j\neq i}^{n}\frac{E_iE_j}{x_{ij}}\sin(\delta_i-\delta_j). \tag{6.49}$$

6.2.2 Stochastic Averaging

It can be seen that the conservative system associated with Eq. (6.48) is a non-integrable Hamiltonian system. The Hamiltonian function is the energy function commonly used in power systems which reads

$$\begin{aligned} H(\boldsymbol{\delta}, \boldsymbol{\omega}) &= \frac{1}{2}\sum_{i=1}^{n} M_i \omega_N \omega_i^2 - \sum_{i=1}^{n} P_{mi}(\delta_i - \delta_{i0}) \\ &- [\sum_{i=1}^{n}\sum_{j=i+1}^{n} \frac{E_i E_j}{x_{ij}} \cos(\delta_i - \delta_j) - \sum_{i=1}^{n}\sum_{j=i+1}^{n} \frac{E_i E_j}{x_{ij}} \cos(\delta_{i0} - \delta_{j0})]. \end{aligned} \tag{6.50}$$

where $\boldsymbol{\delta} = [\delta_1, \delta_2, \cdots, \delta_n]^{\mathrm{T}}$, $\boldsymbol{\omega} = [\omega_1, \omega_2, \cdots, \omega_n]^{\mathrm{T}}$; $\frac{1}{2}\sum_{i=1}^{n} M_i \omega_N \omega_i^2$ is the kinetic energy of all rotors in the system; $\sum_{i=1}^{n} P_{mi}(\delta_i - \delta_{i0})$ is the potential energy of all rotors in the system, and $[\sum_{i=1}^{n}\sum_{j=i+1}^{n} \frac{E_i E_j}{x_{ij}} \cos(\delta_i - \delta_j) - \sum_{i=1}^{n}\sum_{j=i+1}^{n} \frac{E_i E_j}{x_{ij}} \cos(\delta_{i0} - \delta_{j0})]$ is the magnetic field energy stored in all branches of the system. It needs to be clarified that the above energies are not the actual energies of the system, i.e., their units are not Joules (J), and the real energy needs to be multiplied by the fixed coefficients $1/T_N$, where T_N is the reference value of the system torque.

The partial derivatives of H with respect to state variables are

$$\begin{aligned} &\frac{\partial H}{\partial \omega_i} = M_i \omega_N \omega_i, \quad \frac{\partial H}{\partial \delta_i} = -P_{mi} + \sum_{j=1, j\neq i}^{n} \frac{E_i E_j}{x_{ij}} \sin(\delta_i - \delta_j) = -P_{mi} + P_{ei}, \\ &\frac{\partial^2 H}{\partial \omega_i^2} = M_i \omega_N, \quad i = 1, 2, \cdots, n. \end{aligned} \tag{6.51}$$

Physically, as long as the difference of energy input to the Hamiltonian system by stochastic excitation and energy dissipated by damping over one vibration cycle is much smaller than the system total energy, it can be considered as a quasi-Hamiltonian system. In actual power systems, stochastic excitations are not too large, and system damping is relatively small, so multi-machine power systems with stochastic excitations are indeed quasi-Hamiltonian systems.

Using Eqs. (6.51), (6.48) can be rewritten as

$$\frac{\mathrm{d}\delta_i}{\mathrm{d}t} = \frac{1}{M_i}\frac{\partial H}{\partial \omega_i}, \quad \frac{\mathrm{d}\omega_i}{\mathrm{d}t} = -\frac{\partial H}{\partial \delta_i} - \frac{D_i}{M_i^2 \omega_N}\frac{\partial H}{\partial \omega_i} + \frac{\sigma_i}{M_i} W_{gi}(t), \quad i = 1, 2, \cdots, n, \tag{6.52}$$

and then can be further written as an Itô stochastic differential equation

$$\begin{aligned}
\mathrm{d}\delta_i &= \frac{1}{M_i}\frac{\partial H}{\partial \omega_i}\mathrm{d}t, \\
\mathrm{d}\omega_i &= \left(-\frac{1}{M_i}\frac{\partial H}{\partial \delta_i} - \frac{D_i}{M_i^2\omega_N}\frac{\partial H}{\partial \omega_i}\right)\mathrm{d}t + \frac{\sigma_i}{M_i}\mathrm{d}B_i(t), \quad i = 1, 2, \cdots, n.
\end{aligned} \tag{6.53}$$

This is the Itô stochastic differential equation for multi-machine power systems as a quasi-Hamiltonian systems.

By using Itô differential rule and the relation between H and δ_i, ω_i in Eq. (6.50), the Itô stochastic differential equation for $H(t)$ can be obtained from Eq. (6.51) and (6.53) as follows

$$\begin{aligned}
\mathrm{d}H &= \sum_{i=1}^{n}[(-D_i\omega_N\omega_i^2 + \frac{\sigma_i^2\omega_N}{2M_i})\mathrm{d}t + \sigma_i\omega_N\omega_i\mathrm{d}B_i(t)] \\
&= \sum_{i=1}^{n}(-D_i\omega_N\omega_i^2 + \frac{\sigma_i^2\omega_N}{2M_i})\mathrm{d}t + \sum_{i=1}^{n}\sigma_i\omega_N\omega_i\mathrm{d}B_i(t).
\end{aligned} \tag{6.54}$$

It can be seen that if both the system damping D_i and the intensity σ_i^2 of random excitation are both small, according to Khasminskii's theorem (Khasminskii 1968), the Hamiltonian process $H(t)$ in Eq. (6.54) converges weakly to a one-dimensional Markov diffusion process as $D_i,\ \ \sigma_i^2 \to 0$. According to the stochastic averaging method in Sect. 5.1, the averaged Itô stochastic differential equation for this process is

$$\mathrm{d}H = m(H)\mathrm{d}t + \sigma(H)\mathrm{d}B(t). \tag{6.55}$$

where

$$\begin{aligned}
m(H) &= \frac{1}{T(H)}\int_{\Omega}[\sum_{i=1}^{n}(-D_i\omega_N\omega_i^2 + \frac{\sigma_i^2\omega_N}{2M_i})\Big/\frac{\partial H}{\partial \omega_1}]\mathrm{d}\delta_1\cdots\mathrm{d}\delta_n\mathrm{d}\omega_2\cdots\mathrm{d}\omega_n, \\
\sigma^2(H) &= \frac{1}{T(H)}\int_{\Omega}[\sum_{i=1}^{n}(\sigma_i\omega_N\omega_i)^2\Big/\frac{\partial H}{\partial \omega_1}]\mathrm{d}\delta_1\cdots\mathrm{d}\delta_n\mathrm{d}\omega_2\cdots\mathrm{d}\omega_n, \\
T(H) &= \int_{\Omega}(1\Big/\frac{\partial H}{\partial \omega_1})\mathrm{d}\delta_1\cdots\mathrm{d}\delta_n\mathrm{d}\omega_2\cdots\mathrm{d}\omega_n, \\
\Omega &= \{(\delta_1,\cdots,\delta_n,\omega_2,\cdots,\omega_n)|H(\delta_1,\cdots,\delta_n,0,\omega_2,\cdots,\omega_n) \le H\}.
\end{aligned} \tag{6.56}$$

In Eq. (6.56), the integration domain Ω and the integration result can be obtained directly for simple systems. For complex systems, they can be obtained from numerical calculations.

6.2.3 Reliability of Multi-machine Power Systems

If the energy of a multi-machine power system is used as the reliability assessment criterion (Li et al. 2015), then the following conditional reliability function $R(t|h_0)$ can be used. Denote Σ as the safety region of energy. $R(t|h_0)$ is defined as the probability of energy process $H(t)$ starting from safety region $H(0) = h_0 \in \Sigma$ remains within the region Σ in the time interval $(0, t]$, i.e.,

$$R(t|h_0) = \text{Prob}\{H(s) \in \Sigma,\ s \in (0, t]|H(0) = h_0 \in \Sigma\}. \tag{6.57}$$

Based on the averaged Itô stochastic differential equation (6.55) for system energy $H(t)$, the backward Kolmogorov equation for the conditional reliability function is obtained as (Zhu et al. 2002c)

$$\frac{\partial R(t|h_0)}{\partial t} = m(h_0)\frac{\partial R(t|h_0)}{\partial h_0} + \frac{1}{2}\sigma^2(h_0)\frac{\partial^2 R(t|h_0)}{\partial h_0^2}, \tag{6.58}$$

where the first and second derivative moments $m(h_0)$ and $\sigma^2(h_0)$ are obtained from Eq. (6.56) with H replaced by h_0.

For one-dimensional system (6.54), the safety region Σ is defined as $[0, h_{cr})$, where h_{cr} is the critical value. For a specific system, h_{cr} is difficult to determine. In the current researches, the critical potential energy h_{pcr} is often used as a substitution for h_{cr}, which may lead to more conservative results. Although more methods for determining the critical energy h_{cr} and critical potential energy h_{pcr} have been proposed (Kundur et al. 1994; Ni et al. 2002), here the critical potential energy h_{pcr} is used to replace the critical energy h_{cr}.

The initial and boundary conditions for Eq. (6.58) are

$$R(0|h_0) = 1, \quad 0 < h_0 < h_{pcr}. \tag{6.59}$$

$$\begin{aligned} &R(t|h_0) = 0, \quad h_0 = h_{pcr}, \\ &\tfrac{\partial R}{\partial t} = m(0)\tfrac{\partial R}{\partial h_0}, \quad h_0 = 0. \end{aligned} \tag{6.60}$$

Equation (6.58) is a parabolic partial differential equation, and its analytical solution is difficult to obtain, and numerical methods are usually needed.

Example 6.1 Consider a 4-machine 2-area system (Ju et al. 2013). The original system load is considered as a constant impedance load, and a second-order rotor motion equation is used for a generator. After network simplification of maintaining the potential node of generators and transferring the random power fluctuations to each internal node of generators (Kundur et al. 1994), the following Itô stochastic differential equations for a 4-machine power system with stochastic excitations are obtained (Li et al. 2015).

$$
\begin{aligned}
&\mathrm{d}\delta_1 = \omega_N\omega_1\mathrm{d}t,\\
&\mathrm{d}\delta_2 = \omega_N\omega_2\mathrm{d}t,\\
&\mathrm{d}\delta_3 = \omega_N\omega_3\mathrm{d}t,\\
&\mathrm{d}\delta_4 = \omega_N\omega_4\mathrm{d}t,\\
&\mathrm{d}\omega_1 = \frac{1}{M_1}[P_{m1} - D_1\omega_1 - E_1E_2B_{12}\sin(\delta_1-\delta_2) - E_1E_3B_{13}\sin(\delta_1-\delta_3)\\
&-E_1E_4B_{14}\sin(\delta_1-\delta_4)]\mathrm{d}t + \frac{\sigma_1}{M_1}\mathrm{d}B_1(t),\\
&\mathrm{d}\omega_2 = \frac{1}{M_2}[P_{m2} - D_2\omega_2 - E_2E_1B_{21}\sin(\delta_2-\delta_1) - E_2E_3B_{23}\sin(\delta_2-\delta_3)\\
&-E_2E_4B_{24}\sin(\delta_2-\delta_4)]\mathrm{d}t + \frac{\sigma_2}{M_2}\mathrm{d}B_2(t),\\
&\mathrm{d}\omega_3 = \frac{1}{M_3}[P_{m3} - D_3\omega_3 - E_3E_1B_{31}\sin(\delta_3-\delta_1) - E_3E_2B_{32}\sin(\delta_3-\delta_2)\\
&-E_3E_4B_{34}\sin(\delta_3-\delta_4)]\mathrm{d}t + \frac{\sigma_3}{M_3}\mathrm{d}B_3(t),\\
&\mathrm{d}\omega_4 = \frac{1}{M_4}[P_{m4} - D_4\omega_4 - E_4E_1B_{41}\sin(\delta_4-\delta_1) - E_4E_1B_{42}\sin(\delta_4-\delta_2)\\
&-E_4E_3B_{43}\sin(\delta_4-\delta_3)]\mathrm{d}t + \frac{\sigma_4}{M_4}\mathrm{d}B_4(t).
\end{aligned}
\tag{6.61}
$$

In numerical calculations, parameters are $M_1 = 13s$, $M_2 = 13s$, $M_3 = 12.35s$, $M_4 = 12.35s$, $E_1 = 1.03s$, $E_2 = 1.01s$, $E_3 = 1.03s$, $E_4 = 1.01s$, $P_{m1} = 0.2$, $P_{m2} = 0.25$, $P_{m3} = -0.22$, $P_{m4} = -0.23$, $B_{12} = B_{21} = 2.843$, $B_{13} = B_{31} = 0.19$, $B_{14} = B_{41} = 0.187$, $B_{23} = B_{32} = 0.187$, $B_{24} = B_{42} = 0.184$, $B_{34} = B_{43} = 2.843$.

Letting the excitations in Eq. (6.61) to be zero, assuming appropriate initial states, and using the Runge–Kutta method, the stationary response of system can be calculated. They are $\delta_1 - \delta_4 = 1.784 \times 10^{-2}$, $\delta_2 - \delta_4 = 2.503 \times 10^{-2}$, $\delta_3 - \delta_4 = -3.086 \times 10^{-3}$.

According to Eq. (6.50), the system energy function is

$$
H = \frac{1}{2}M_1\omega_N\omega_1^2 + \frac{1}{2}M_2\omega_N\omega_2^2 + \frac{1}{2}M_3\omega_N\omega_3^2 + \frac{1}{2}M_4\omega_N\omega_4^2 + H_p, \tag{6.62}
$$

where H_p is the potential energy

$$
\begin{aligned}
H_p = &-P_{m1}(\delta_1-\delta_{1s}) - P_{m2}(\delta_2-\delta_{2s}) - P_{m3}(\delta_3-\delta_{3s}) - P_{m4}(\delta_4-\delta_{4s})\\
&-E_1E_2B_{12}\cos(\delta_1-\delta_2) - E_1E_3B_{13}\cos(\delta_1-\delta_3) - E_1E_4B_{14}\cos(\delta_1-\delta_4)\\
&-E_2E_3B_{23}\cos(\delta_2-\delta_3) - E_2E_4B_{24}\cos(\delta_2-\delta_4) - E_3E_4B_{34}\cos(\delta_3-\delta_4)\\
&+E_1E_2B_{12}\cos(\delta_{1s}-\delta_{2s}) + E_1E_3B_{13}\cos(\delta_{1s}-\delta_{3s}) + E_1E_4B_{14}\cos(\delta_{1s}-\delta_{4s})\\
&+E_2E_3B_{23}\cos(\delta_{2s}-\delta_{3s}) + E_2E_4B_{24}\cos(\delta_{2s}-\delta_{4s}) + E_3E_4B_{34}\cos(\delta_{3s}-\delta_{4s}).
\end{aligned}
\tag{6.63}
$$

According to Eq. (6.54), the Itô stochastic differential equation for system energy is obtained as follows

$$\mathrm{d}H = (-D_1\omega_N\omega_1^2 - D_2\omega_N\omega_2^2 - D_3\omega_N\omega_3^2 - D_4\omega_N\omega_4^2 + \frac{\sigma_1^2\omega_N}{2M_1} + \frac{\sigma_2^2\omega_N}{2M_2} + \frac{\sigma_3^2\omega_N}{2M_3}$$
$$+\frac{\sigma_4^2\omega_N}{2M_4})\mathrm{d}t + \sigma_1\omega_N\omega_1\mathrm{d}B_1(t) + \sigma_2\omega_N\omega_2\mathrm{d}B_2(t) + \sigma_3\omega_N\omega_3\mathrm{d}B_3(t) + \sigma_4\omega_N\omega_4\mathrm{d}B_4(t). \tag{6.64}$$

Because of $P_{m4} = -P_{m1} - P_{m2} - P_{m3}$, taking the 4th generator as reference machine, the potential energy H_p can be written as

$$\begin{aligned}H_p &= -P_{m1}\delta_{1'} - P_{m2}\delta_{2'} - P_{m3}\delta_{3'} + P_{m1}\delta_{1s'} + P_{m2}\delta_{2s'} + P_{m3}\delta_{3s'}\\ &-E_1E_2B_{12}\cos(\delta_{1'} - \delta_{2'}) - E_1E_3B_{13}\cos(\delta_{1'} - \delta_{3'}) - E_2E_3B_{23}\cos(\delta_{2'} - \delta_{3'})\\ &+E_1E_2B_{12}\cos(\delta_{1s'} - \delta_{2s'}) + E_1E_3B_{13}\cos(\delta_{1s'} - \delta_{3s'}) + E_2E_3B_{23}\cos(\delta_{2s'} - \delta_{3s'})\\ &-E_1E_4B_{14}\cos\delta_{1'} - E_2E_4B_{24}\cos\delta_{2'} - E_3E_4B_{34}\cos\delta_{3'}\\ &+E_1E_4B_{14}\cos\delta_{1s'} + E_2E_4B_{24}\cos\delta_{2s'} + E_3E_4B_{34}\cos\delta_{3s'},\end{aligned} \tag{6.65}$$

where $\delta_{1'} = \delta_1 - \delta_4$, $\delta_{2'} = \delta_2 - \delta_4$, $\delta_{3'} = \delta_3 - \delta_4$, $\delta_{1s'} = 1.784 \times 10^{-2}$, $\delta_{2s'} = 2.503\times10^{-2}$, $\delta_{3s'} = -3.086\times10^{-3}$. It can be seen that H_p is a function of $\delta_{1'}$, $\delta_{2'}$, $\delta_{3'}$, and the relationship is shown in Fig. 6.16.

When the system equipotential surfaces just meet, the corresponding value is the critical potential energy. Figure 6.16 shows that when the potential energy $H_p = 14$, one potential energy region is connected to the other, while it is not connected at $H_p = 13$, so the critical potential energy h_{cr} lies between 13 and 14. Here, set h_{pcr} being 6.

By substituting the parameters $D_1, D_2, D_3, D_4, \sigma_1, \sigma_2, \sigma_3, \sigma_4, h_{pcr}$ into Eq. (6.56) and performing numerical integration, $m(H)$ and $\sigma^2(H)$ can be obtained, as shown in

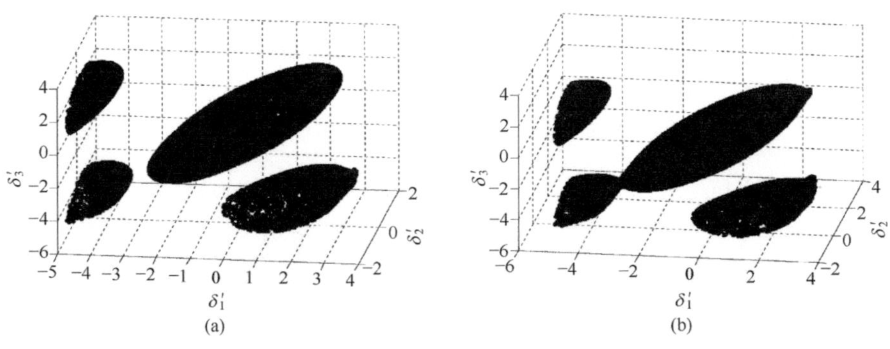

Fig. 6.16 The black region represents the functional relationship between potential energy H_p and $\delta_{1'}$, $\delta_{2'}$, $\delta_{3'}$, **a** state of $H_p = 13$; **b** state of $H_p = 14$ (Li et al. 2015)

Fig. 6.17. Equation (6.58) with initial condition in Eq. (6.59) and boundary conditions in Eq. (6.60) can be solved numerically by using Grank-Nicholson finite difference method. For comparison, Monte Carlo simulation of the original system (6.61) was completed by using the Heun algorithm (Liu et al. 2014). In calculations, time step is 0.001, initial values are $\delta_{10} = 1.784 \times 10^{-2}$, $\delta_{20} = 2.503 \times 10^{-2}$, $\delta_{30} = -3.086 \times 10^{-3}$, $\delta_{40} = 0$, $\omega_{10} = 0$, $\omega_{20} = 0$, $\omega_{30} = 0$, $\omega_{40} = 0$, $H(0) = 0$ and $D_1 = 0.02$, $D_1 = 0.025$, $D_3 = 0.03$, $D_4 = 0.035$, $\sigma_4 = 0.15$, $\sigma_3 = 0.14$, $\sigma_2 = 0.13$, $\sigma_1 = 0.12$. The results obtained using both two methods are shown in Fig. 6.18. It can be seen from Fig. 6.18 that the theoretical solution agrees well with the Monte Carlo simulation results. Further research on the impact of more system parameters on the reliability of system (6.61) can be found in the reference (Li et al. 2015). By using the available computer system, the calculation time for Monte Carlo simulations with 2000 samples is about 20 s, while that for obtaining the theoretical solution is only about 0.1 s.

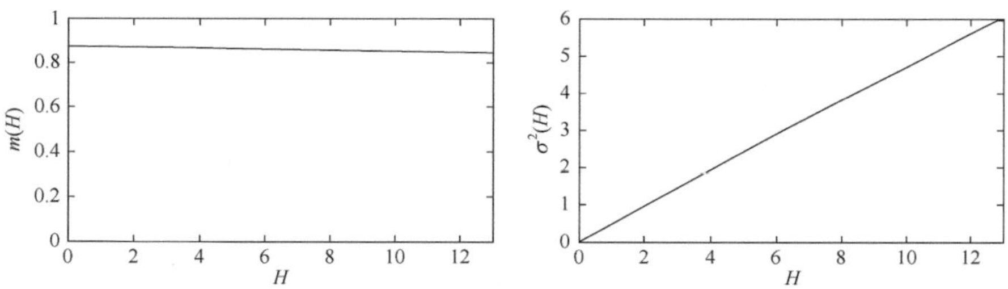

Fig. 6.17 $m(H)$, $\sigma^2(H)$ as functions of H (Li et al. 2015)

Fig. 6.18 Reliability function (Li et al. 2015)

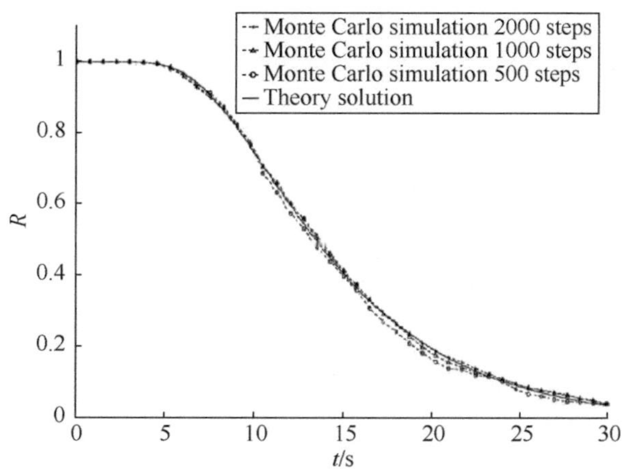

6.3 Ship Rolling Motion

The motion of a ship on irregular waves is essentially a nonlinear stochastic motion, and the rolling motion of a ship is coupled of a translation motion and a pitch motion. Additionally, the excitation due to irregular waves is also highly complex. Therefore, a complete description of the ship motion under irregular wave excitation requires establishing complex multi-DOF nonlinear stochastic dynamical equations. For theoretical analysis, simplifications are necessary. When the ship is subjected only to lateral wave excitation, rolling motion can be decoupled from other motions, allowing the study of rolling motion alone (Roberts 1982). The wave excitation can be modeled as a Gaussian random process (Ochi 1986). Thus, a single-DOF stochastic dynamical system under Gaussian random excitations can be established.

6.3.1 Rolling Motion Equation of Ship Under Irregular Wave Excitation

As shown in Fig. 6.19, for a ship under lateral wave excitation, the ship rolling motion equation can be established as follows (Cai et al. 1994)

$$\ddot{X} + \alpha\dot{X} + \beta\left|\dot{X}\right|\dot{X} + \gamma X - \delta X^3 = X\xi_1(t) + \xi_2(t). \tag{6.66}$$

where X represents the rolling angle, α, β, γ and δ are positive parameters, $\xi_1(t)$ and $\xi_2(t)$ are stationary Gaussian processes with the following correlation functions

$$E[\xi_j(t)\xi_k(t+\tau)] = R_{jk}(\tau), \quad j, k = 1, 2. \tag{6.67}$$

The damping force $\alpha\dot{X} + \beta\left|\dot{X}\right|\dot{X}$ in Eq. (6.66) was introduced by Froude (Froude 1955) and has been proven to be a good form of damping for ship rolling motion

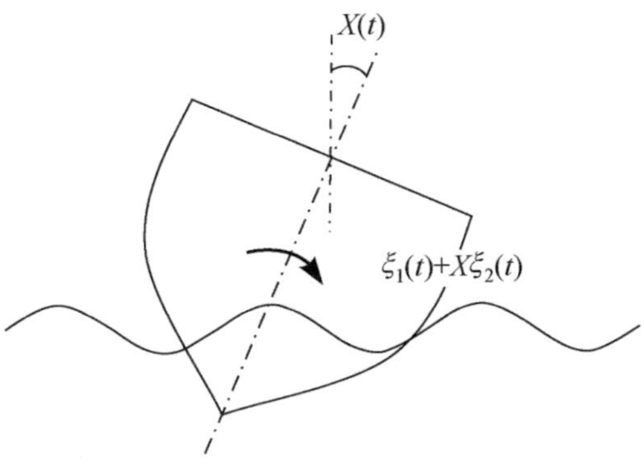

Fig. 6.19 Rolling motion of a ship under lateral wave excitation

in many studies (Dalzell 1973; Roberts 1985). The parametric excitation $X\xi_1(t)$ in Eq. (6.66) was proposed by (Grim 1952) to account for the fact that the restoring moment depends on the wave level relative to rolling motion. The cubic softening nonlinearity $-\delta X^3$ captures the essential feature that, there is a critical value for ship rolling motion, and the ship will capsize when the rolling angle exceeds this critical value (Dalzell 1971, 1973). When the stationary random excitations $\xi_1(t)$ and $\xi_2(t)$ in Eq. (6.1) are wideband processes, the stochastic averaging method can be applied (Roberts 1982; Cai et al. 1994).

Firstly, consider the free undamped rolling motion of ship, which is governed by the equation

$$\ddot{x} + \gamma x - \delta x^3 = 0. \tag{6.68}$$

where the period of rolling motion is given by

$$T = 4T_{1/4} = 4\int_0^a \frac{\mathrm{d}x}{\sqrt{2e - 2U(x)}}, \tag{6.69}$$

where $U(x)$ and e are the potential energy and system energy, which can be expressed as

$$U(x) = \frac{1}{2}\gamma x^2 - \frac{1}{4}\delta x^4. \tag{6.70}$$

$$e = \frac{1}{2}\dot{x}^2 + U(x). \tag{6.71}$$

The upper limit a of integral in Eq. (6.69) is the amplitude determined by equation $U(a) = e$. Equation (6.69) shows that the free oscillation period T varies with energy e. Figure 6.20 shows the potential energy function $U(x)$ and its restoring force function, indicating that the potential energy function forms a potential well with a maximum value

$$U_{\max} = \frac{\gamma^2}{4\delta}. \tag{6.72}$$

When the system energy $e < U_{\max}$, the free rolling of ship is periodic motion, and it remains within the potential well. Once the energy e reaches $U_{\max}$, the rolling angle will increase to the critical value of ship capsizing. Thus, the critical system energy is $e_c = U_{\max}$.

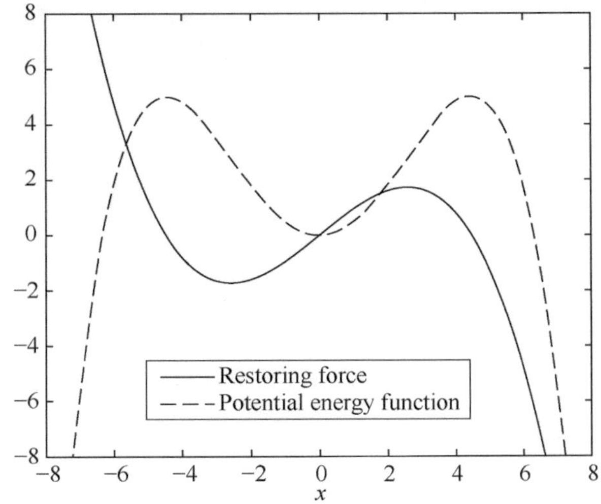

Fig. 6.20 Restoring force and potential energy function in ship rolling motion, $\gamma = 1,\ \delta = 0.05$.

6.3.2 Averaged Itô Stochastic Differential Equation

Now consider the stochastic motion of system (6.66). Introduce transformation

$$\begin{aligned} \sqrt{U(X)} &= \sqrt{E}\cos\Phi, \quad 0 \le \Phi < 2\pi, \\ \dot{X} &= -\sqrt{2E}\sin\Phi. \end{aligned} \tag{6.73}$$

By using Eq. (6.73), the governing equations for $E(t)$ and $\Phi(t)$ can be obtained from (6.66) as follows

$$\dot{E} = f_1(E,\Phi) + g_{11}(E,\Phi)\xi_1(t) + g_{12}(E,\Phi)\xi_2(t). \tag{6.74}$$

$$\dot{\Phi} = f_2(E,\Phi) + g_{21}(E,\Phi)\xi_1(t) + g_{22}(E,\Phi)\xi_2(t), \tag{6.75}$$

where

$$f_1(E,\Phi) = -2E\sin^2\Phi(\alpha + \beta\sqrt{2E}|\sin\Phi|). \tag{6.76}$$

$$f_2(E,\Phi) = -\sin\Phi\cos\Phi(\alpha + \beta\sqrt{2E}|\sin\Phi|) + \frac{\gamma X - \delta X^3}{\sqrt{2E}\cos\Phi}. \tag{6.77}$$

$$g_{11}(E,\Phi) = -\sqrt{2E}X\sin\Phi. \tag{6.78}$$

$$g_{12}(E,\Phi) = -\sqrt{2E}\sin\Phi. \tag{6.79}$$

$$g_{21}(E,\Phi) = -\frac{1}{\sqrt{2E}} X \cos\Phi. \tag{6.80}$$

$$g_{22}(E,\Phi) = -\frac{1}{\sqrt{2E}} \cos\Phi. \tag{6.81}$$

in which X is replaced by E and Φ according to Eq. (6.73).

According to Sect. 4.1, when the correlation times of $\xi_1(t)$ and $\xi_2(t)$ are much shorter than the system relaxation time, $\xi_1(t)$ and $\xi_2(t)$ can be regarded as wide-band processes. By applying the stochastic averaging method, $[E(t), \Phi(t)]^{\mathrm{T}}$ can be regarded as a vector Markov diffusion process, governed by Itô stochastic differential equations

$$\mathrm{d}E = m_E \mathrm{d}t + \sigma_{EE}\mathrm{d}B_1(t) + \sigma_{E\Phi}\mathrm{d}B_2(t). \tag{6.82}$$

$$\mathrm{d}\Phi = m_\Phi \mathrm{d}t + \sigma_{\Phi E}\mathrm{d}B_1(t) + \sigma_{\Phi\Phi}\mathrm{d}B_2(t). \tag{6.83}$$

In these equations, $B_1(t)$ and $B_2(t)$ are independent unit Wiener processes, m_E and m_Φ are drift coefficients, σ_{EE}, $\sigma_{E\Phi}$, $\sigma_{\Phi E}$ and $\sigma_{\Phi\Phi}$ are diffusion coefficients. The drift coefficients are given by

$$m_E(E,\Phi) = f_1(E,\Phi) + \sum_{k,l=1}^{2} \int_{-\infty}^{0} [g_{1k}(t+\tau)\frac{\partial g_{1l}(t)}{\partial E} + g_{2k}(t+\tau)\frac{\partial g_{1l}(t)}{\partial \Phi}]R_{kl}(\tau)\mathrm{d}\tau. \tag{6.84}$$

$$m_\Phi(E,\Phi) = f_2(E,\Phi) + \sum_{k,l=1}^{2} \int_{-\infty}^{0} [g_{1k}(t+\tau)\frac{\partial g_{2l}(t)}{\partial E} + g_{2k}(t+\tau)\frac{\partial g_{2l}(t)}{\partial \Phi}]R_{kl}(\tau)\mathrm{d}\tau. \tag{6.85}$$

The diffusion coefficients are determined from the following equations

$$\sigma_{EE}^2 + \sigma_{E\Phi}^2 = \sum_{k,l=1}^{2} \int_{-\infty}^{\infty} g_{1k}(t+\tau)g_{1k}(t)R_{kl}(\tau)\mathrm{d}\tau. \tag{6.86}$$

$$\sigma_{EE}\sigma_{\Phi E} + \sigma_{E\Phi}\sigma_{\Phi\Phi} = \sum_{k,l=1}^{2} \int_{-\infty}^{\infty} g_{1k}(t+\tau)g_{2l}(t)R_{kl}(\tau)\mathrm{d}\tau. \tag{6.87}$$

$$\sigma_{\Phi E}^2 + \sigma_{\Phi\Phi}^2 = \sum_{k,l=1}^{2} \int_{-\infty}^{\infty} g_{2k}(t+\tau)g_{2l}(t)R_{kl}(\tau)\mathrm{d}\tau. \tag{6.88}$$

In general, damping and wave excitation are weak, $\Phi(t)$ is a fast-varying process. While energy $E(t)$ is a slowly-varying process and converges to a one-dimensional

Markov diffusion process. By carrying out time averaging over Eq. (6.82), the following averaged Itô stochastic differential equation for $E(t)$ is obtained

$$\mathrm{d}E = m(E)\mathrm{d}t + \sigma(E)\mathrm{d}B(t). \tag{6.89}$$

where

$$\begin{aligned} m(E) &= \langle m_E(E, \Phi)\rangle_t \\ &= \langle f_1(E, \Phi)\rangle_t + \left\langle \sum_{k,l=1}^{2} \int_{-\infty}^{0} [g_{1k}(t+\tau)\frac{\partial g_{1l}(t)}{\partial E} + g_{2k}(t+\tau)\frac{\partial g_{1l}(t)}{\partial \Phi}]R_{kl}(\tau)\mathrm{d}\tau \right\rangle_t . \end{aligned} \tag{6.90}$$

$$\sigma^2(E) = \langle \sigma_{EE}^2 + \sigma_{E\Phi}^2 \rangle_t = \left\langle \sum_{k,l=1}^{2} \int_{-\infty}^{\infty} g_{1k}(t+\tau)g_{1k}(t)R_{kl}(\tau)\mathrm{d}\tau \right\rangle_t . \tag{6.91}$$

where $\langle[\cdot]\rangle_t$ represents the time averaging operator.

The time averaging of any function $F(E, \Phi)$ in Eqs. (6.90) and (6.91) can be carried out as follows

$$\langle F(E, \Phi)\rangle_t = \frac{1}{T}\int_0^T F(E, \Phi)\mathrm{d}t = \frac{1}{T}\int_0^T F[E(X, \dot{X}), \Phi(X, \dot{X})]\mathrm{d}t. \tag{6.92}$$

In calculating the integral in Eq. (6.92), E is treated as a constant, T is a function of E, and $X, \dot{X}$ is obtained as a function of time t from the free vibration Eq. (6.68) for a given E. The period T as a function of E can be calculated from Eq. (6.69).

The first term on the right side of Eq. (6.90) is

$$\langle f_1(E, \Phi)\rangle_t = \frac{-1}{T_{1/4}}\int_0^{T_{1/4}} 2E\sin^2\Phi(\alpha + \beta\sqrt{2E}\sin\Phi)\mathrm{d}t = \frac{-1}{T_{1/4}}\int_0^{T_{1/4}} \dot{X}^2(\alpha + \beta\dot{X})\mathrm{d}t. \tag{6.93}$$

which represents the energy dissipated by damping per unit time. To calculate the other terms in Eq. (6.90), the following expressions are obtained first

$$g_{11}(t+\tau)\frac{\partial g_{11}(t)}{\partial E} = (X\sin\Phi)_{t+\tau}(X\sin\Phi)_t + (X\sin\Phi)_{t+\tau}(2E\sin\Phi\frac{\partial X}{\partial E})_t. \tag{6.94}$$

$$g_{11}(t+\tau)\frac{\partial g_{12}(t)}{\partial E} = (X\sin\Phi)_{t+\tau}(\sin\Phi)_t. \tag{6.95}$$

$$g_{12}(t+\tau)\frac{\partial g_{11}(t)}{\partial E} = (\sin\Phi)_{t+\tau}(X\sin\Phi)_t + (\sin\Phi)_{t+\tau}(2E\sin\Phi\frac{\partial X}{\partial E})_t. \tag{6.96}$$

$$g_{12}(t+\tau)\frac{\partial g_{12}(t)}{\partial E} = (\sin\Phi)_{t+\tau}(\sin\Phi)_t. \tag{6.97}$$

$$g_{21}(t+\tau)\frac{\partial g_{11}(t)}{\partial \Phi} = (X\cos\Phi)_{t+\tau}(X\cos\Phi)_t + (X\cos\Phi)_{t+\tau}(\sin\Phi\frac{\partial X}{\partial \Phi})_t. \quad (6.98)$$

$$g_{21}(t+\tau)\frac{\partial g_{12}(t)}{\partial \Phi} = (X\cos\Phi)_{t+\tau}(\cos\Phi)_t. \quad (6.99)$$

$$g_{22}(t+\tau)\frac{\partial g_{11}(t)}{\partial \Phi} = (\cos\Phi)_{t+\tau}(X\cos\Phi)_t + (\cos\Phi)_{t+\tau}(\sin\Phi\frac{\partial X}{\partial \Phi})_t. \quad (6.100)$$

$$g_{22}(t+\tau)\frac{\partial g_{12}(t)}{\partial \phi} = (\cos\phi)_{t+\tau}(\cos\phi)_t. \quad (6.101)$$

All terms in Eqs. (6.94)–(6.101) are periodic function with period T for the free motion of Eq. (6.68). Based on this, the following Fourier expansions can be made

$$\sin\Phi(t) = \sum_{n=1}^{\infty} a_n \sin[(2n-1)\omega_T t]. \quad (6.102)$$

$$\cos\Phi(t) = \sum_{n=1}^{\infty} b_n \cos[(2n-1)\omega_T t]. \quad (6.103)$$

$$X(t)\sin\Phi(t) = \sum_{n=1}^{\infty} c_n \sin(2n\omega_T t). \quad (6.104)$$

$$X(t)\cos\Phi(t) = \sum_{n=1}^{\infty} d_n \cos(2n\omega_T t). \quad (6.105)$$

$$2E\sin\Phi(t)\frac{\partial X(t)}{\partial E} = \sum_{n=1}^{\infty} e_n \sin(2n\omega_T t). \quad (6.106)$$

$$\sin\Phi(t)\frac{\partial X(t)}{\partial \Phi} = \sum_{n=1}^{\infty} f_n \cos(2n\omega_T t). \quad (6.107)$$

where $\omega_T = 2\pi / T$, and the coefficients a_n to f_n are calculated as follows

$$a_n = \frac{-2}{\sqrt{2E}T_{1/4}}\int_0^{T_{1/4}} \dot{X}\sin[(2n-1)\omega_T t]\mathrm{d}t. \quad (6.108)$$

$$b_n = \frac{1}{\sqrt{E}T_{1/4}}\int_0^{T_{1/4}} X\sqrt{2\gamma-\delta X^2}\cos[(2n-1)\omega_T t]\mathrm{d}t. \quad (6.109)$$

$$c_n = \frac{-2}{\sqrt{2E}T_{1/4}}\int_0^{T_{1/4}} X\dot{X}\sin(2n\omega_T t)\mathrm{d}t. \quad (6.110)$$

$$d_n = \frac{1}{\sqrt{E}T_{1/4}} \int_0^{T_{1/4}} X^2\sqrt{2\gamma - \delta X^2}\cos(2n\omega_T t)\mathrm{d}t. \tag{6.111}$$

$$e_n = \frac{-1}{\sqrt{2E}T_{1/4}} \int_0^{T_{1/4}} \frac{\dot{X}X(2\gamma - \delta X^2)}{\gamma - \delta X^2}\sin(2n\omega_T t)\mathrm{d}t. \tag{6.112}$$

$$f_n = \frac{-1}{\sqrt{E}T_{1/4}} \int_0^{T_{1/4}} \frac{\dot{X}^2\sqrt{2\gamma - \delta X^2}}{\gamma - \delta X^2}\cos(2n\omega_T t)\mathrm{d}t. \tag{6.113}$$

Substituting (6.94) to (6.107) into (6.90) and (6.91) results in

$$m(E) = \frac{-1}{T_{1/4}} \int_0^{T_{1/4}} \dot{X}^2(\alpha + \beta\dot{X})\mathrm{d}t + \frac{1}{2}\pi\sum_{n=1}^{\infty}(c_n^2 + c_n e_n + d_n^2 + d_n f_n)S_{11}(2n\omega_T)$$
$$+\frac{1}{2}\pi\sum_{n=1}^{\infty}(a_n^2 + b_n^2)S_{22}[(2n-1)\omega_T]. \tag{6.114}$$

$$\sigma^2(E) = 2\pi E\sum_{n=1}^{\infty}\{c_n^2 S_{11}(2n\omega_T) + a_n^2 S_{22}[(2n-1)\omega_T]\}. \tag{6.115}$$

where $S_{11}(\omega)$ and $S_{22}(\omega)$ are the power spectral densities of $\xi_1(t)$ and $\xi_2(t)$, respectively. For a given E, first solve Eq. (6.68) numerically to obtain $X(t)$ and $\dot{X}(t)$ within $0 \le t < T_{1/4}$, then calculate a_n to f_n from Eqs. (6.108)–(6.113), and finally use Eq. (6.114) and (6.115) to obtain $m(E)$ and $\sigma^2(E)$. In numerical calculations, only first several terms are needed for a required accuracy. In most cases, retaining 3 to 4 terms is enough.

The corresponding FPK equation can be derived from the averaged Itô stochastic differential equation (6.89). Due to the absorbing boundary $E = e_c$, the FPK equation can only be solved numerically to obtain the transient PDF of energy.

6.3.3 *Ship Capsize Probability*

When the ship rolling angle X reaches the critical value x_c for the first time, which is equivalent to that the energy E reaches the critical value e_c, the ship will capsize. This is a first-passage problem in stochastic dynamics. Denote T_f as the time when capsize occurs under given initial energy, $E(t = 0) = e_0$, $e_0 < e_c$ where $e_c = \gamma^2/4\delta$ is the critical value given in Eq. (6.72). Obviously, T_f is a random variable. Since $E(t)$ is a one-dimensional diffusion process, the mean value of the first-passage time T_f, denoted by $\mu_1 = E[T_f]$, is governed by the following Pontryagin equation (Cai and Zhu 2016)

$$\frac{1}{2}\sigma^2(e_0)\frac{\mathrm{d}^2\mu_1}{\mathrm{d}e_0^2} + m(e_0)\frac{\mathrm{d}\mu_1}{\mathrm{d}e_0} = -1. \tag{6.116}$$

One boundary condition for Pontryagin equation is

$$\mu_1(e_c) = 0. \tag{6.117}$$

It is seen from Eqs. (6.114) and (6.115) that when the initial energy is $e_0 = 0$,

$$\omega_T = \sqrt{\gamma}, \quad a_n = b_n = 1, \quad c_n = d_n = e_n = f_n = 0.$$

$$m(0) = \pi S_{22}(\sqrt{\gamma}), \quad \sigma(0) = 0. \tag{6.118}$$

Thus, another boundary conditions are obtained from Eq. (6.116) as

$$\left.\frac{\mathrm{d}\mu_1}{\mathrm{d}e_0}\right|_{e_0=0} = \frac{-1}{\pi S_{22}(\sqrt{\gamma})}. \tag{6.119}$$

Under boundary conditions (6.117) and (6.119), Eq. (6.116) can be solved numerically to obtain the mean time μ_1 for ship capsizing.

In numerical calculations, the power spectral density function of wave excitation $\xi_1(t)$ and $\xi_2(t)$ is required. Here, the following form of the power spectral density is used (Dalzell 1971, 1973)

$$S_{kk}(\omega) = \frac{P_k}{\omega/\omega_p}\exp[-(\frac{1+\pi/8}{4\omega^4/\omega_p^4} + \frac{\pi\omega^2/\omega_p^2}{16} - \frac{1}{4} - \frac{3\pi}{32})], \quad k = 1, 2. \tag{6.120}$$

where P_k is the spectral peak value, and for convenience, P_k is called excitation intensity; ω_p is the frequency at the spectral peak. The power spectral density functions of parametric excitation $\xi_1(t)$ and external excitation $\xi_2(t)$ are Eq. (6.120) with different parameter values. Figure 6.21 shows curves of the power spectral density (6.120) for different parameter values, indicating that larger values of ω_p correspond to a wider spectrum.

Given parameters $\alpha = 0.1$, $\beta = 0.1$, $\gamma = 1.0$, $\delta = 0.05$, $\omega_p = 1$ and 4 parametric excitation intensity parameters $P_1 = 0$, 0.01, 0.02 and 0.05, while keeping the external excitation intensity parameter $P_2 = 0.05$ constant, Figs. 6.22 and 6.23 present the drift coefficients and squares of the diffusion coefficient in Eqs. (6.144) and (6.115), respectively. It can be observed that with increasing parametric excitation intensity P_i $(i = 1, 2, 3, 4)$, the drift coefficients and diffusion coefficients increase. Figure 6.24 shows the mean first capsize time μ_1 versus the external excitation intensity P_2 for different values of the parametric excitation intensity P_1. Figure 6.25 shows the mean first capsizing time μ_1 varying with parametric excitation intensity P_1 for different values of the external excitation intensity P_2. The other

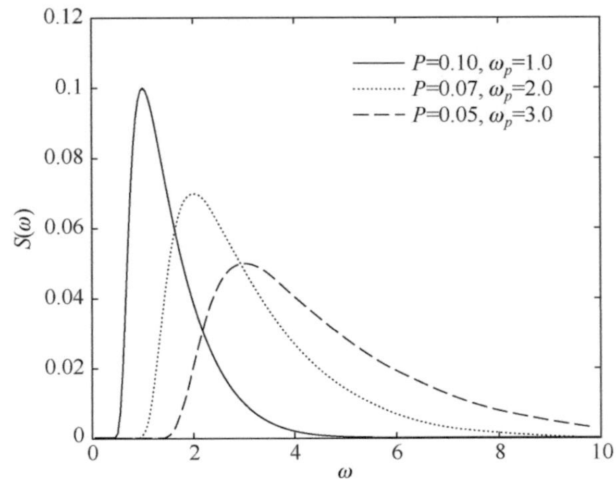

Fig. 6.21 Power spectral densities of random wave

parameters in both figures are the same as those in Fig. 6.22. Both figures indicate that when evaluating the ship capsizing, both external and parametric excitations must be considered. The parameter δ in Eq. (6.66) reflects the intensity of nonlinear term in restoring force. Figure 6.26 shows the mean first capsizing time μ_1 versus the excitation intensity $P_1 = P_2$ for different nonlinear levels, i.e. different δ values. As expected, with an increasing δ value, the mean capsize time is shorter.

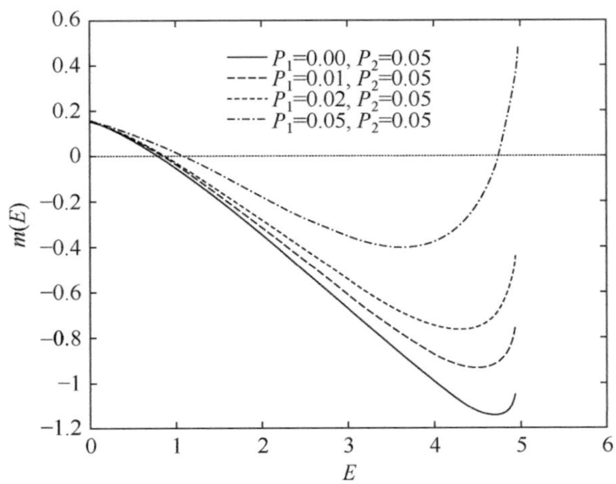

Fig. 6.22 Drift coefficient of energy $E(t)$ after averaging (Cai et al. 1994)

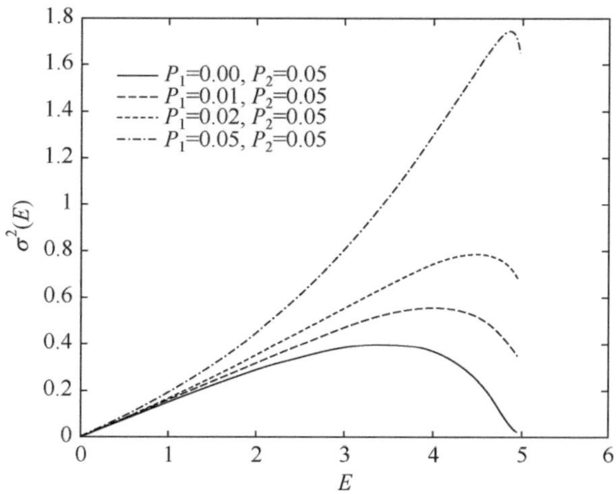

Fig. 6.23 Square of diffusion coefficient of energy $E(t)$ after averaging (Cai et al. 1994)

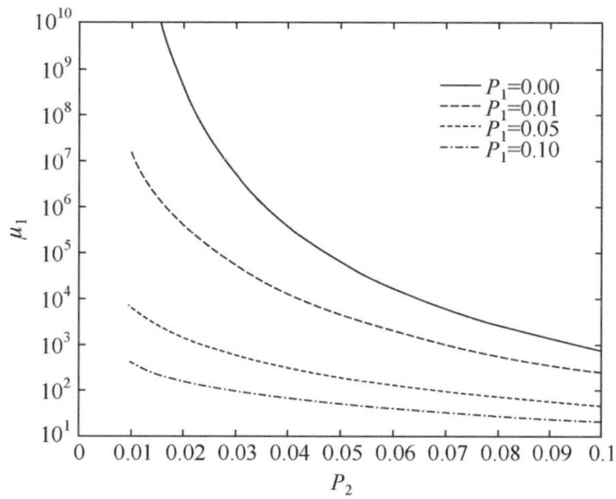

Fig. 6.24 The mean first capsize time μ_1 as function of external excitation intensity P_2 (Cai et al. 1994)

6.4 Asymptotic Lyapunov Stability with Probability 1 of Quasi-Hamiltonian Systems

The theory of stochastic stability explores that, when the initial state deviates from its equilibrium or stationary state, whether a stochastic dynamical system remains bounded or converges to the equilibrium or stationary state as time tends to infinity. Due to various convergence modes, the stochastic stability has several definitions (Kozin 1969), one of which is the asymptotic Lyapunov stability with probability 1 (almost sure stability). The most effective method to investigate this stability is the method of maximum Lyapunov exponent. This section primarily discusses how to utilize the stochastic averaging methods of quasi-Hamiltonian systems to find the

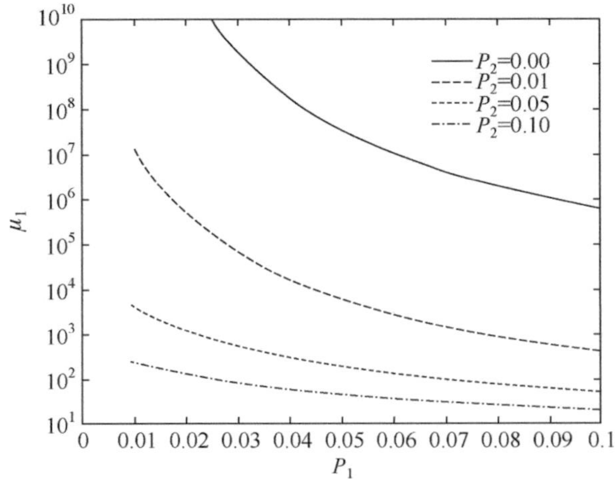

Fig. 6.25 The mean first capsize time μ_1 as function of parametric excitation intensity P_1 (Cai et al. 1994)

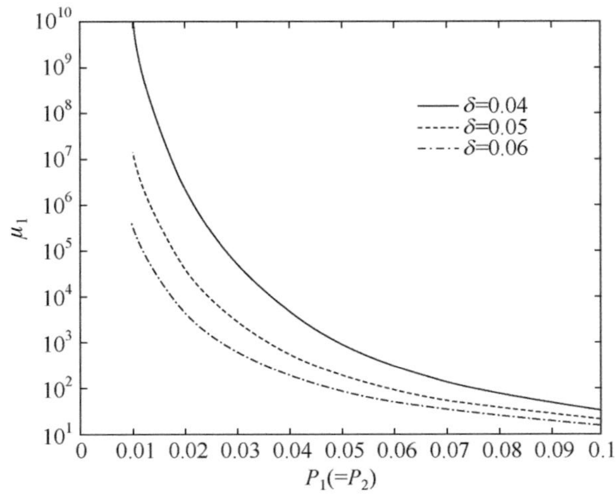

Fig. 6.26 The mean first capsize time μ_1 with 3 different nonlinear intensity parameters δ (Cai et al. 1994)

maximum Lyapunov exponent and analyze the asymptotic Lyapunov stability with probability 1 of quasi-integrable and quasi-non-integrable Hamiltonian systems. For more applications of the stochastic averaging methods in stochastic stability research, see reference (Zhu 2003).

6.4.1 *Asymptotic Lyapunov Stability with Probability 1 of Stochastic Systems*

In stability studies, it is common to use variation equation describing the motion of deviations from equilibriums or stationary states. Therefore, in the stochastic stability studies, the stability of trivial solutions of the stochastic dynamical equations is investigated.

Consider a vector stochastic dynamical equation

$$\dot{\mathbf{X}} = \mathbf{F}(\mathbf{X}, t) + \mathbf{G}(\mathbf{X}, t)\boldsymbol{\xi}(t), \ \ \mathbf{X}(t_0) = \mathbf{x}_0, \tag{6.121}$$

where $\mathbf{X}(t)$ represents a n-dimensional vector stochastic process, $\mathbf{F}$ is a n–dimensional vector function, $\mathbf{G}$ is a $(n \times m)$-dimensional matrix function, and $\boldsymbol{\xi}(t)$ is a m–dimensional vector stochastic process. Let $\mathbf{0}$ be the trivial solution of Eq. (6.121), which requires

$$\mathbf{F}(0, t) = 0, \ \ \mathbf{G}(0, t) = 0. \tag{6.122}$$

In Eq. (6.121), it is assumed that $\mathbf{G}(\mathbf{X}, t)\boldsymbol{\xi}(t)$ are multiplicative (parametric) excitations. The distance of solution $\mathbf{X}(t)$ deviating from the trivial solution is represented by norm $\|\mathbf{X}(t)\|$ which is commonly defined in the following several ways

$$\|\mathbf{X}(t)\| = \sum_{i=1}^{n} |X_i(t)|, \tag{6.123}$$

$$\|\mathbf{X}(t)\| = \sum_{i=1}^{n} [X_i X_i]^{1/2}, \tag{6.124}$$

$$\|\mathbf{X}(t)\| = \sum_{i,j=1}^{n} \left[a_{ij} X_i(t) X_j(t)\right]^{1/2}, \tag{6.125}$$

where a_{ij} are the elements of a positive definite matrix. Clearly, $\|\mathbf{X}(t)\|$ is a stochastic process. Denote $\mathbf{X}(t; \mathbf{x}_0, t_0)$ as the solution of Eq. (6.121) with initial conditions $\mathbf{x}_0$ at t_0. The definitions of Lyapunov stability with probability 1 are given as follows (Lin and Cai 1995):

Lyapunov stability with probability 1. The trivial solution of Eq. (6.121) is said to be Lyapunov stability with probability 1 if for any $\varepsilon_1, \varepsilon_2 > 0$, there exists $\delta(\varepsilon_1, \varepsilon_2, t_0) > 0$ such that

$$\text{Prob}\left\{ \underset{\|\mathbf{x}_0\| < \delta}{U} \left[\sup_{t \ge t_0} \|\mathbf{X}(t; \mathbf{x}_0, t_0)\| \ge \varepsilon_1 \right] \right\} \le \varepsilon_2, \tag{6.126}$$

The definition (6.126) indicates that almost all samples are stable in Lyapunov sense. Since $\varepsilon_1, \varepsilon_2$ can be arbitrarily small, this stability is also known as almost sure stability.

Lyapunov asymptotic stability with probability 1. If Eq. (6.126) holds, and for every $\varepsilon > 0$ there exists $\delta(\varepsilon, t_0) > 0$ as long as $\|\mathbf{x}_0\| \le \delta$, such that

$$\lim_{t_1 \to \infty} \text{Prob}\left\{\sup_{t \geq t_1} \|X(t; \mathbf{x}_0, t_0)\| \geq \varepsilon\right\} = 0. \tag{6.127}$$

then, the trivial solution of Eq. (6.121) is Lyapunov asymptotic stability with probability 1 or almost sure asymptotic stability.

Since Eqs. (6.126) and (6.127) only hold for $\|\mathbf{x}_0\| \leq \delta$, if δ is small the aforementioned stability is a local Lyapunov asymptotic stability with probability 1. If δ is finite value and Eqs. (6.126) and (6.127) still hold, they are termed as large-scale Lyapunov asymptotic stability with probability 1. When $\mathbf{x}_0$ can be any point in n-dimensional space R^n and Eqs. (6.126) and (6.127) hold, it is known as global Lyapunov asymptotic stability with probability 1. For linear stochastic systems, these three types of stability are equivalent.

In Eqs. (6.126) and (6.127), the focus is on the probability of the norm of solution process of Eq. (6.121) in a time interval $[t_0, \infty)$ reaching its upper bound. It is a random variable since the norm reaches its upbound at different times for different sample functions, Therefore, the stabilities defined by Eqs. (6.126) and (6.127) are essentially about the sample stability. In fact, Eqs. (6.126) and (6.127) indicate that almost all sample functions of the solution process of Eq. (6.121) are Lyapunov stable and asymptotic stable, respectively. Therefore, the Lyapunov stability with Probability 1 is an extension of Lyapunov stability from deterministic systems to stochastic systems, and is the most natural and reasonable definition, which has been widely researched.

In Eq. (6.121), $\xi(t)$ can be a stationary ergodic real noise or Gaussian white noise. In the former case, only the sufficient conditions for the Lyapunov asymptotic stability with probability 1 of linear systems (primarily two-dimensional) has been obtained. These results have been summarized in Lin and Cai's monograph (Lin and Cai 1995). For the Lyapunov stability with probability 1 of Itô stochastic differential equations, several theorems exist (Khasminskii 1980), which enable to determine the stability using Lyapunov functions. However, the main challenge is how to construct Lyapunov functions, and so far, this method has primarily been applied to linear Itô stochastic differential equations. A drawback of the Lyapunov function method is that the obtained results are often sufficient conditions for stability rather than necessary conditions.

6.4.2 *Maximum Lyapunov Exponent*

A recent tendency over the past thirty years is to study the Lyapunov asymptotic stability with probability 1 using maximum Lyapunov exponent, based on the Oseledec multiplicative ergodic theorem (Oseledec 1968). Consider the linearized equation of Eq. (6.121) at the trivial solution

$$\dot{\mathbf{X}} = \mathbf{AX} + \sum_{k=1}^{m} \mathbf{B}_k \mathbf{X} \xi_k(t), \ \ \mathbf{X}(0) = \mathbf{x}_0. \tag{6.128}$$

where

$$\mathbf{A} = \left[\frac{\partial F_i}{\partial x_j}\right]_{\mathbf{x}=0} \ \mathbf{B}_k = \left[\frac{\partial G_{ik}}{\partial x_j}\right]_{\mathbf{x}=0} \tag{6.129}$$

The component form of Eq. (6.128) is

$$\begin{aligned} X_i &= \sum_{j=1}^{n} a_{ij} X_j + \sum_{j=1}^{n} \sum_{k=1}^{m} b_{ijk} X_j \xi_k(t), \quad X_i(0) = X_{0i}, \\ i &= 1, 2, \cdots, n. \end{aligned} \tag{6.130}$$

The Lyapunov exponent is defined as the mean exponential growth rate of the ergodic solution of Eq. (6.128) or Eq. (6.130), i.e.,

$$\lambda = \lim_{t \to \infty} \frac{1}{t} \ln \|\mathbf{X}(t; \mathbf{x}_0)\| \tag{6.131}$$

According to the Oseledec multiplicative ergodic theorem, when $\boldsymbol{\xi}(t)$ is an ergodic process, there exist $1 \leq r \leq n$ deterministic Lyapunov exponents $\lambda_s \quad (s = 1, 2, \cdots, r)$, r linear subspaces (Oseledec subspaces) E_s, and r integers d_s, such that for almost all initial values $\mathbf{x}_0$, $\lambda(\mathbf{x}_0) \in \{\lambda_1, \lambda_2, \cdots, \lambda_r\}$ and

$$\lambda_s = \lim_{t \to \infty} \frac{1}{t} \ln \|\mathbf{X}(t; \mathbf{x}_0)\|, \mathbf{x}_0 \in E_s \backslash \{\mathbf{0}\}, s = 1, 2, \cdots, r. \tag{6.132}$$

where $E_s \backslash \{\mathbf{0}\}$ means the subspaces E_s except zero point. Equation (6.132) denotes that λ_s is the mean exponential growth rate of the solution of Eq. (6.128) or Eq. (6.130) in the Oseledec subspace E_s. Oseledec subspaces and Lyapunov exponents of linear stochastic systems are the stochastic counterparts for eigenspaces and eigenvalues of deterministic linear systems. $\lambda_1, \lambda_2, \cdots, \lambda_r$ make up the Lyapunov spectrum, and they are independent of the choice of the norm definition. The system phase space is the sum of r Oseledec subspaces, i.e., $R^n = \oplus_{s=1}^{r} E_s$. The multiplicity of λ_s and the dimension of E_s are both d_s. The r Lyapunov exponents are arranged in descending order as $\lambda_1 \geq \lambda_2 \geq \cdots \geq \lambda_r$ with λ_1 being the maximum Lyapunov exponent.

$$\lambda_{\Sigma} = \frac{1}{n} \mathrm{tr} E[\mathbf{A} + \mathbf{B}_k \xi_k(t)] \tag{6.133}$$

is called mean Lyapunov exponent. $n\lambda_{\Sigma}$ represents the mean rate of change of the phase space volume. The relationship between the maximum, mean, and minimum

Lyapunov exponents is

$$\lambda_1 \geq \lambda_{\sum} \geq \lambda_r \tag{6.134}$$

If $\mathbf{X}(t)$ is an ergodic Markov process in whole phase space, then almost all solution trajectories of Eq. (6.128) or Eq. (6.130) exhibit the same exponential growth rate, namely the maximum Lyapunov exponent. Therefore, if only the Markov solution of Eq. (6.128) or Eq. (6.130) is interested, then only the maximum Lyapunov exponent is needed. If $\lambda_1 < 0$, the trivial solution of Eq. (6.128) or Eq. (6.130) is Lyapunov asymptotic stability with probability 1. If $\lambda_1 > 0$, the solution is Lyapunov unstability with probability 1. If $\lambda_1 = 0$, the solution of Eqs. (6.128) and (6.130) can be arbitrarily large or arbitrarily small.

Khasminskii has given an expression for the maximum Lyapunov exponent of linear Itô stochastic differential equation (Khasminskii 1967). Consider the following linear Itô stochastic differential equation with constant coefficients

$$\begin{aligned} &\mathrm{d}X_i = \sum_{j=1}^{n} m_{ij} X_j \mathrm{d}t + \sum_{j=1}^{n}\sum_{k=1}^{m} \sigma_{ijk} X_j \mathrm{d}B_k(t), \quad X_i(0) = x_{0i}, \\ &i = 1, 2, \cdots, n. \end{aligned} \tag{6.135}$$

The generating operator of $\mathbf{X}$ is

$$L(\mathbf{X}) = \sum_{i=1}^{n} a_i \frac{\partial}{\partial x_i} + \frac{1}{2}\sum_{i,j=1}^{n} b_{ij} \frac{\partial^2}{\partial x_i \partial x_j} \tag{6.136}$$

where

$$a_i = \sum_{j=1}^{n} m_{ij} x_j, \quad b_{ij} = \sum_{r,s}^{n}\sum_{k=1}^{m} \sigma_{irk}\sigma_{jsk} x_r x_s \tag{6.137}$$

Assuming that $\mathbf{B} = [b_{ij}]$ is non-negative and non-degenerate, i.e., for any n-dimensional vector $\boldsymbol{\alpha}$, there exists

$$(\mathbf{B}(\mathbf{x})\boldsymbol{\alpha}, \boldsymbol{\alpha}) \geq c\|\mathbf{x}\|^2 \|\boldsymbol{\alpha}\|^2, \tag{6.138}$$

where $(\cdot, \cdot)$ denotes the inner product of two vectors, c is a finite positive constant. Introduce transformation

$$\rho = \ln\|\mathbf{X}\| = \ln \sum_{i=1}^{n} (X_i X_i)^{1/2}, \tag{6.139}$$

$$U_s = X_s / \|\mathbf{X}\|, \quad s = 1, 2, \cdots, n-1. \tag{6.140}$$

The transformation (6.140) generates a diffusion process $\mathbf{U}(t)$ on the $(n-1)$–dimensional unit spherical surface. The diffusion matrix of $\mathbf{U}(t)$ is non-singular under the condition (6.138). Using Itô differential rule and Eqs. (6.139) and (6.140), the following Itô stochastic differential equations governing ρ and U_s can be derived from Eq. (6.135)

$$\mathrm{d}\rho = Q(U)\mathrm{d}t + \sum_{i,j=1}^{n}\sum_{k=1}^{m}\sigma_{ijk}U_iU_j\mathrm{d}B_k(t), \tag{6.141}$$

$$\mathrm{d}U_s = L(U_s)\mathrm{d}t + \sum_{i,j=1}^{n}\sum_{k=1}^{m}\sigma_{ijk}U_j(\delta_{is} - U_iU_s)\mathrm{d}B_k(t), \tag{6.142}$$

where

$$\begin{aligned} Q(\mathbf{U}) = L(\rho) &= \sum_{i,j=1}^{n} m_{ij}U_iU_j + \frac{1}{2}\sum_{i,j,p,q=1}^{n}\sum_{k=1}^{m}\sigma_{ijk}\sigma_{pqk}U_jU_q(\delta_{ip} - 2U_iU_p) \\ L(U_s) &= \sum_{i,j=1}^{n} m_{ij}U_j(\delta_{is} - U_iU_s) + \frac{1}{2}\sum_{i,j,p,q=1}^{n}\sum_{k=1}^{m}\sigma_{ijk}\sigma_{pqk}U_jU_q(3U_iU_sU_p - U_s\delta_{ip} - U_p\delta_{ps}) \end{aligned} \tag{6.143}$$

in which $L(\rho)$ and $L(U_s)$ are the generating operators of diffusion processes $\rho(t)$ and $U_s(t)$, respectively.

Integrating Eq. (6.141) from 0 to t and dividing it by t yields

$$\frac{1}{t}\ln\|\mathbf{X}(t;\mathbf{x}_0)\| = \frac{1}{t}\ln\|\mathbf{x}(0)\| + \frac{1}{t}\int_0^t Q(\mathbf{U}(\tau))\mathrm{d}\tau + \sum_{i,j=1}^{n}\sum_{k=1}^{m}\sigma_{ijk}\frac{1}{t}\int_0^t U_i(\tau)U_j(\tau)\mathrm{d}B_k(\tau). \tag{6.144}$$

Since $\mathbf{x}(0)$, U_i, U_j are bounded, $B_k(\tau)$ grows as $[2\tau\ln(\ln\tau)]^{1/2}$. When $t \to \infty$, the first and third terms in the right side of Eq. (6.144) approach to zero. Thus,

$$\lambda = \lim_{t\to\infty}\ln\frac{1}{t}\|\mathbf{X}(t;\mathbf{x}_0)\| = \lim_{t\to\infty}\frac{1}{t}\int_0^t Q(\mathbf{U}(\tau))\mathrm{d}\tau. \tag{6.145}$$

Since $\mathbf{U}$ is ergodic on all $(n-1)$-dimensional unit spherical surface,

$$\lambda = E\big[Q(\mathbf{U})\big] = \int Q(\mathbf{U})p(\mathbf{U})\mathrm{d}\mathbf{U} \tag{6.146}$$

Based on the Oseledec multiplicative ergodic theorem (Oseledec 1968), λ in Eqs. (6.145) and Eq. (6.146) is λ_1, i.e., the maximum Lyapunov exponent; where

$p(\mathbf{U})$ is the stationary PDF of $\mathbf{U}(t)$. $\lambda_1 < 0$ is the necessary and sufficient condition of Lyapunov asymptotic stability with probability 1 of the trivial solution of Eq. (6.135). This is the advantage of using the maximum Lyapunov exponent for determining stability compared with the Lyapunov function method.

When the condition (6.138) is not satisfied, $\mathbf{U}$ may become singular, and Eq. (6.145) may not necessarily converge to λ_1 (Lin and Cai 1995). Condition (6.138) can be relaxed (Arnold et al. 1968).

The results mentioned above can be extended to the following nonlinear autonomous Itô stochastic differential equations with linear homogeneous drift and diffusion coefficients (Kozin and Zhang 1991),

$$\mathrm{d}\mathbf{X} = \mathbf{F}(\mathbf{X})\mathrm{d}t + \mathbf{G}(\mathbf{X})\mathrm{d}B(t), \tag{6.147}$$

where $\mathbf{F}$, $\mathbf{G}$ satisfy Eq. (6.122) and the following linear homogeneous conditions

$$k\mathbf{F}(\mathbf{X}) = \mathbf{F}(k\mathbf{X}), k\mathbf{G}(\mathbf{X}) = \mathbf{G}(k\mathbf{X}), \tag{6.148}$$

and non-degeneracy condition

$$(\mathbf{G}(\mathbf{X})\mathbf{G}^T(\mathbf{X})\boldsymbol{\alpha}, \boldsymbol{\alpha}) \geq c\|\mathbf{X}\|^2\|\boldsymbol{\alpha}\|^2. \tag{6.149}$$

By introducing the transformations (6.139) and (6.140), similar derivation yields the equations similar to Eqs. (6.141) and (6.142) with

$$\begin{aligned} Q(\mathbf{U}) &= \mathbf{U}^T\mathbf{F}(\mathbf{U}) + \frac{1}{2}\sum_{i=1}^{n}\left[\mathbf{G}(\mathbf{U})\mathbf{G}^T(\mathbf{U})\right]_{ii}\left(\sum_{j=1}^{n}U_j^2 - 2U_i^2\right) \\ &- \sum_{\substack{i,j=1 \\ i\neq j}}^{n}\left[(\mathbf{G}(\mathbf{U})\mathbf{G}^T(\mathbf{U}))_{ij}U_iU_j\right]. \end{aligned} \tag{6.150}$$

Substituting it into Eq. (6.145) or Eq. (6.146) yields the maximum Lyapunov exponent.

When using Eq. (6.146) to calculate the maximum Lyapunov exponent, it is necessary to have the stationary PDF of $\mathbf{U}(t)$. If the dimension of $\mathbf{U}(t)$ is larger than 1, it is difficult to find the stationary PDF. In such cases, numerical methods described in Sect. 2.5.4 can be used to solve Eq. (6.142), and then the maximum Lyapunov exponent can be calculated according to Eq. (6.145) (Kloeden and Platen 1992; Talay 1999). When the noise intensity is small, perturbation methods can be used to calculate the maximum Lyapunov exponent (Wihstutz 1999).

6.4.3 Lyapunov Asymptotic Stability with Probability 1 for Quasi-non-integrable Hamiltonian Systems

Consider a n-dimensional quasi-non-integrable Hamiltonian system given in Eq. (5.5) of vol. 1. Assume that this system only contains parametric excitations Gaussian white noises. The purpose of this section is to explore how to determine the Lyapunov asymptotic stability with probability 1 of the trivial solutions of the system by calculating the maximum Lyapunov exponent. According to Sect. 6.4.2, Eq. (5.5) in vol. 1 is linearized at the trivial solution, then the maximum Lyapunov exponent of the linearized Itô stochastic differential equation can be obtained. This linearized equation is $2n$–dimensional and its diffusion matrix is degenerate, making the condition (6.138) dissatisfied. Thus, it is difficult to obtain the maximum Lyapunov exponent when $n > 1$.

To overcome this difficulty, introduce a new norm definition. Denote $\mathbf{Z} = [\mathbf{Q}^{\mathrm{T}}\mathbf{P}^{\mathrm{T}}]^{\mathrm{T}}$ and define the norm of $\mathbf{Z}$ as

$$\|\mathbf{Z}\| = H^{1/2}(\mathbf{Q}, \mathbf{P}) \tag{6.151}$$

For mechanical/structural systems, $H(\mathbf{Q}, \mathbf{P})$ represents the total energy of system, which is non-negative. For linear Hamiltonian systems, H is the form of a quadratic homogeneous function of Q_i, P_i, and Eq. (6.151) is similar to Eq. (6.125). For nonlinear Hamiltonian systems, $H(\mathbf{Q}, \mathbf{P})$ is usually of the form of a quadratic homogeneous function of Q_i, P_i, plus higher order terms of Q_i, P_i. In the vicinity of the trivial solution, the main contribution to Eq. (6.151) comes from the quadratic terms of Q_i, P_i in $H(\mathbf{Q}, \mathbf{P})$. Thus, the norm definition in Eq. (6.151) is reasonable in physics.

Replacing norm $\|\mathbf{X}(t, \mathbf{x}_0)\|$ in Eqs. (6.126), (6.127) and (6.131) with $H^{1/2}(\mathbf{Q}, \mathbf{P})$, the new definitions of the Lyapunov stability with probability 1, the Lyapunov asymptotic stability with probability 1, and the Lyapunov exponent for quasi-non-integrable Hamiltonian systems are obtained and then are used in the following investigations.

To calculate the Lyapunov exponent for quasi-non-integrable Hamiltonian systems, linearize the averaged Itô stochastic differential equation (5.7) of vol. 1 at $H = 0$

$$\mathrm{d}H = m'(0)H\mathrm{d}t + \sigma'(0)H\mathrm{d}B(t). \tag{6.152}$$

The linear homogeneous stochastic Itô differential Eq. (6.152) has the solution

$$H(t) = H(0)\exp\left\{\int_0^t \left[m'(0) - (\sigma'(0))^2/2\right]\mathrm{d}s + \int_0^t \sigma'(0)\mathrm{d}B(s)\right\} \tag{6.153}$$

Hence, the Lyapunov exponent of quasi-non-integrable Hamiltonian systems is approximated as

$$
\begin{aligned}
\lambda &= \lim_{t\to\infty} \frac{1}{t} \ln H^{1/2}(\mathbf{Q}, \mathbf{P}) = \lim_{t\to\infty} \frac{1}{2t} \ln H(\mathbf{Q}, \mathbf{P}) \\
&= \lim_{t\to\infty} \frac{1}{2t} \left\{ \ln H(0) + \int_0^t \left[m'(0) - (\sigma'(0))^2/2 \right] \mathrm{d}s + \int_0^t \sigma'(0) \mathrm{d}B(s) \right\} \\
&= \left[m'(0) - (\sigma'(0))^2/2 \right] \Big/ 2.
\end{aligned} \tag{6.154}
$$

The Lyapunov exponent of the aforementioned system can also be derived from the averaged Itô stochastic differential equation for $Y(t) = H^{1/2}(t)$. Using the Itô differential rule, the Itô stochastic differential equation governing $Y(t)$ can be derived from Eq. (5.7) of vol. 1 as follows

$$
\mathrm{d}Y = \overline{m}(Y)\mathrm{d}t + \overline{\sigma}^2(Y)\mathrm{d}B(t) \tag{6.155}
$$

where

$$
\begin{aligned}
&\overline{m}(Y) = Y^{-1}m(Y)/2 - Y^{-3}\sigma^2(Y)/8,\ \overline{\sigma}(Y) = Y^{-2}\sigma^2(Y)/4, \\
&m(Y) = m(H)|_{H=Y^2},\ \sigma^2(Y) = \sigma^2(H)|_{H=Y^2}.
\end{aligned} \tag{6.156}
$$

A derivation similar to Eqs. (6.153) and (6.154) yields

$$
\lambda = \lim_{t\to\infty} \frac{1}{t} \ln Y(t) = \overline{m}'(0) - (\overline{\sigma}'(0))^2 \big/ 2 \tag{6.157}
$$

Both Eqs. (6.154) and (6.157) yield the same Lyapunov exponent.

Either Eq. (6.154) or Eq. (6.157) can be used to approximately determine the local Lyapunov asymptotic stability with probability 1 of quasi-non-integrable Hamiltonian systems (5.5) in vol. 1. It greatly simplify the studying of Lyapunov asymptotic stability with probability 1 of quasi-non-integrable Hamiltonian systems.

Example 6.2 Consider two nonlinearly coupled and nonlinearly damped oscillators under Gaussian white noise parametric excitation. The motion equation is given by (Zhu and Huang 1998).

$$
\begin{aligned}
&\dot{Q}_1 = P_1, \\
&\dot{P}_1 = -\omega_1^2 Q_1 - aQ_2 - b|Q_1 - Q_2|^{\delta}\mathrm{sgn}(Q_1 - Q_2) - \beta_1 P_1 - \alpha_1 Q_1^2 P_1 \\
&+ f_1 Q_1 W_{g1}(t), \\
&\dot{Q}_2 = P_2, \\
&\dot{P}_2 = -\omega_2^2 Q_2 - aQ_1 - b|Q_1 - Q_2|^{\delta}\mathrm{sgn}(Q_1 - Q_2) - \beta_2 P_2 - \alpha_2 Q_2^2 P_2 \\
&+ f_2 Q_2 W_{g2}(t),
\end{aligned} \tag{6.158}
$$

where ω_1, ω_2 are the natural frequencies of the uncoupled oscillators, β_i, α_i are linear and nonlinear damping coefficients, a, b are linear and nonlinear coupling coefficients, δ and f_i are positive parameters; $W_{gk}(t)$ are independent Gaussian white noises with intensity $2D_k$. It is assumed that β_i, α_i, D_k are small values of the same order.

There is no Wong-Zakai correction terms for the system (6.158). The associated Hamiltonian function is

$$H = \left(p_1^2 + p_2^2\right)/2 + U(q_1, q_2), \tag{6.159}$$

where

$$U(q_1, q_2) = \left(\omega_1^2 q_1^2 + \omega_2^2 q_2^2\right)/2 + a q_1 q_2 + b|q_1 - q_2|^{1+\delta}/(1+\delta). \tag{6.160}$$

When $b \neq 0, \delta \neq 0, 1, 3$, $U(q_1, q_2)$ are inseparable, and therefore, Eq. (6.158) is a quasi-non-integrable Hamiltonian system. $q_1 = q_2 = p_1 = p_2 = 0$ is trivial solution. To investigate the Lyapunov asymptotic stability with probability 1 for this trivial solution, the stochastic averaging method of quasi-non-integrable Hamiltonian systems in Sect. 5.1 is first applied to obtain the averaged Itô equation (5.7) of vol. 1 for Eq. (6.158). According to Eqs. (5.8) to (5.13) of vol. 1, the drift and diffusion coefficients are obtained as (Zhu and Huang 1998)

$$\begin{aligned} m(H) = &\frac{1}{T(H)} \int_\Omega \{[-(\beta_1 + \alpha_1 q_1^2)p_1^2 - (\beta_2 + \alpha_2 q_2^2)p_2^2 \\ &+ f_1^2 D_1 q_1^2 + f_2^2 D_2 q_2^2]/p_1\} \mathrm{d}q_1 \mathrm{d}q_2 \mathrm{d}p_2, \\ \sigma^2(H) = &\frac{1}{T(H)} \int_\Omega \left[2\left(f_1^2 D_1 q_1^2 p_1^2 + f_2^2 D_2 q_2^2 p_2^2\right)/p_1\right] \mathrm{d}q_1 \mathrm{d}q_2 \mathrm{d}p_2, \end{aligned} \tag{6.161}$$

where

$$\begin{aligned} T(H) &= \int_\Omega (1/p_1) \mathrm{d}q_1 \mathrm{d}q_2 \mathrm{d}p_2, \\ \Omega &= \{(q_1, q_2, p_2) | H(q_1, q_2, 0, p_2) \le H\}. \end{aligned} \tag{6.162}$$

After integrating Eq. (6.161) with respect to p_2 and then introducing polar coordinates

$$q_1 = (R/\omega_1)\cos\theta, \quad q_2 = (R/\omega_2)\sin\theta \tag{6.163}$$

Equation (6.161) becomes

$$
\begin{aligned}
m(H) = &\frac{2\pi}{T(H)}\int_0^{\pi}\left[-(\beta_1+\beta_2)A(H,\theta) - \left(\frac{\alpha_1}{\omega_1^2}\cos^2\theta\right.\right. \\
&\left.\left.+\frac{\alpha^2}{\omega_2^2}\sin^2\theta\right)B(H,\theta) + \frac{R^4}{2}\left(\frac{f_1^2D_1}{\omega_1^2}\cos^2\theta + \frac{f_2^2D_2}{\omega_2^2}\sin^2\theta\right)\right]\mathrm{d}\theta, \\
\sigma^2(H) = &\frac{4\pi}{T(H)}\int_0^{\pi}\left(\frac{f_1^2D_1}{\omega_1^2}\cos^2\theta + \frac{f_2^2D_2}{\omega_2^{2,}}\sin^2\theta\right)B(H,\theta)\mathrm{d}\theta,
\end{aligned}
\tag{6.164}
$$

where

$$
\begin{aligned}
T(H) &= \frac{2\pi}{\omega_1\omega_2}\int_0^{\pi}R^2\mathrm{d}\theta, \\
A(H,\theta) &= HR^2 - \frac{R^4}{4}\left(1+\frac{a}{\omega_1\omega_2}\sin 2\theta\right) \\
&\quad - \frac{2b}{(1+\delta)(3+\delta)}R^{3+\delta}\left|\frac{\cos\theta}{\omega_1} - \frac{\sin\theta}{\omega_2}\right|^{1+\delta}, \\
B(H,\theta) &= HR^4 - \frac{R^6}{4}\left(1+\frac{a}{\omega_1\omega_2}\sin 2\theta\right) \\
&\quad - \frac{2b}{(1+\delta)(5+\delta)}R^{5+\delta}\left|\frac{\cos\theta}{\omega_1} - \frac{\sin\theta}{\omega_2}\right|^{1+\delta},
\end{aligned}
\tag{6.165}
$$

and where R is the solution of the following equation

$$
H - \frac{R^2}{4}\left(1+\frac{a}{\omega_1\omega_2}\sin 2\theta\right) - \frac{b}{(1+\delta)}R^{1+\delta}\left|\frac{\cos\theta}{\omega_1} - \frac{\sin\theta}{\omega_2}\right|^{1+\delta} = 0. \tag{6.166}
$$

The following two cases of $0 < \delta < 1$ and $\delta > 1$ are considered, separately.

Case 1: $0 < \delta < 1$. In this case, the linearized equation for the averaged Itô equation (6.158) at $H = 0$ is

$$
\mathrm{d}H = \mu_1 H\mathrm{d}t + \mu_2^{1/2}H\mathrm{d}B(t), \tag{6.167}
$$

where

$$
\begin{aligned}
\mu_1 &= -\frac{1}{3}(\beta_1+\beta_2)\eta_1 + \frac{7}{9}\frac{f_1^2D_1+f_2^2D_2}{2a+\omega_1^2+\omega_2^2},\\
\mu_2 &= \frac{2}{3}\frac{f_1^2D_1+f_2^2D_2}{2a+\omega_1^2+\omega_2^2}\eta_2,\\
\eta_1 &= \frac{11}{6} - \frac{1+\delta}{4(4+\delta)} - \frac{2(2-\delta)}{(2+\delta)(3+\delta)} + \frac{\delta^2-1}{24(5+\delta)},\\
\eta_2 &= \frac{13}{12} - \frac{16}{(5+\delta)(2+\delta)} + \frac{4}{3+\delta} - \frac{3+\delta}{2(4+\delta)} + \frac{(3+\delta)(1+\delta)}{24(5+\delta)},
\end{aligned}
\tag{6.168}
$$

According to Eq. (6.154), the Lyapunov exponent of the averaged Itô equation (6.158) is given by

$$
\lambda = \frac{1}{2}\mu_1 - \frac{1}{4}\mu_2 = -\frac{1}{6}(\beta_1+\beta_2)\eta_1 + \frac{1}{18}(7-3\eta_2)\frac{f_1^2D_1+f_2^2D_2}{2a+\omega_1^2+\omega_2^2}. \tag{6.169}
$$

Therefore, the necessary and sufficient condition of the Lyapunov asymptotic stability with probability 1 of the trivial solution of Eq. (6.158) is approximately $\lambda < 0$, i.e.,

$$
\beta_1+\beta_2 > \left(\frac{7}{3\eta_1} - \frac{\eta_2}{\eta_1}\right)\frac{f_1^2D_1+f_2^2D_2}{2a+\omega_1^2+\omega_2^2}. \tag{6.170}
$$

Note that the order less than one term $b|Q_1-Q_2|^{\delta}\mathrm{sgn}(Q_1-Q_2)$ in Eq. (6.158) has contribution to the stability of system (6.158).

Case 2: $\delta > 1$. In this case, the linearized equation for the averaged Itô equation (6.158) at $H=0$ is

$$
\mathrm{d}H = \mu_3 H\mathrm{d}t + \mu_4^{1/2}H\mathrm{d}B(t), \tag{6.171}
$$

where

$$
\begin{aligned}
\mu_3 &= -\frac{1}{2}(\beta_1+\beta_2) + \frac{1}{2}\left(\frac{f_1^2D_1}{\omega_1^2} + \frac{f_2^2D_2}{\omega_2^2}\right)\eta,\\
\mu_4 &= \frac{1}{3}\left(\frac{f_1^2D_1}{\omega_1^2} + \frac{f_2^2D_2}{\omega_2^2}\right)\eta,\\
\eta &= \int_0^{\pi}[1+(a/\omega_1\omega_2)\sin\theta]^{-2}\mathrm{d}\theta \Big/ \int_0^{\pi}[1+(a/\omega_1\omega_2)\sin\theta]^{-1}\mathrm{d}\theta.
\end{aligned}
\tag{6.172}
$$

According to Eq. (6.155), the Lyapunov exponent of the averaged Itô stochastic differential equation (6.158) is given by

$$\lambda = \frac{\mu_3}{2} - \frac{\mu_4}{4} = -\frac{1}{4}(\beta_1 + \beta_2)\eta_1 + \frac{1}{6}\left(\frac{f_1^2 D_1}{\omega_1^2} + \frac{f_2^2 D_2}{\omega_2^2}\right). \tag{6.173}$$

Therefore, the necessary and sufficient condition of the Lyapunov asymptotic stability with probability 1 of the trivial solution of Eq. (6.158) is approximately $\lambda < 0$, i.e.,

$$\beta_1 + \beta_2 > \frac{2}{3}\left(\frac{f_1^2 D_1}{\omega_1^2} + \frac{f_2^2 D_2}{\omega_2^2}\right)\eta. \tag{6.174}$$

Note that the order higher than one term $b|Q_1 - Q_2|^\delta \mathrm{sgn}(Q_1 - Q_2)$ in Eq. (6.158) has no contribution to the stability of system (6.158).

6.4.4 *Lyapunov Asymptotic Stability with Probability 1 of Quasi-integrable Hamiltonian Systems*

Consider a n–DOF quasi-Hamiltonian system of form Eq. (5.5) in vol. 1. It is assumed that the associated Hamiltonian system is integrable and non-resonant. The averaged Itô equations are given by Eq. (5.76) or Eq. (5.90) in vol. 1. If Eq. (5.5) in vol. 1 contains only random parametric excitations, $\mathbf{I} = [I_1, I_2, \cdots, I_n]^{\mathrm{T}} = \mathbf{0}$ and $\mathbf{H} = [H_1, H_2, \cdots, H_n]^{\mathrm{T}} = \mathbf{0}$ are the trivial solutions of Eqs. (5.76) or (5.90) in vol. 1, respectively, i.e.,

$$\lim_{\mathbf{I}\to\mathbf{0}} m_r(\mathbf{I}) = 0, \quad \lim_{\mathbf{I}\to\mathbf{0}} \sigma_{rk}(\mathbf{I}) = 0, \tag{6.175}$$

$$\begin{gathered}\lim_{\mathbf{H}\to\mathbf{0}} \overline{m}_r(\mathbf{H}) = 0, \quad \lim_{\mathbf{H}\to\mathbf{0}} \overline{\sigma}_{rk}(\mathbf{H}) = 0, \\ r = 1, 2, \cdots, n, \quad k = 1, 2, \cdots, m.\end{gathered} \tag{6.176}$$

Linearizing m_r, σ_{rk} and $\overline{m}_r$, $\overline{\sigma}_{rk}$ at $\mathbf{I} = \mathbf{0}$ and $\mathbf{H} = \mathbf{0}$, respectively, yields

$$\mathrm{d}I_r = F_r(I)\mathrm{d}t + \sum_{k=1}^{m} G_{rk}(I)\mathrm{d}B_k(t), \tag{6.177}$$

$$\begin{gathered}\mathrm{d}H_r = \overline{F}_r(H)\mathrm{d}t + \sum_{k=1}^{m} \overline{G}_{rk}(H)\mathrm{d}B_k(t), \\ r = 1, 2, \cdots, n.\end{gathered} \tag{6.178}$$

where F_r, G_{rk} or $\overline{F}_r$, $\overline{G}_{rk}$ satisfy the conditions (6.175) and (6.176), and the following conditions

$$kF_r(\mathbf{I}) = F_r(k\mathbf{I}), \quad kG_{rk}(\mathbf{I}) = G_{rk}(k\mathbf{I}), \tag{6.179}$$

$$k\overline{F}_r(\mathbf{H}) = \overline{F}_r(k\mathbf{H}), \quad k\overline{G}_{rk}(\mathbf{H}) = \overline{G}_{rk}(k\mathbf{H}). \tag{6.180}$$

For Eq. (6.177), defining a norm as

$$\|\mathbf{Z}\| = \overline{I}^{1/2} = \left(\sum_{r=1}^{n} I_r\right)^{1/2}, \tag{6.181}$$

Introducing a transformation

$$\rho = \ln \overline{I}^{1/2} = (\ln \overline{I})/2, \tag{6.182}$$

$$\alpha_r = I_r/\overline{I}, \quad r = 1, 2, \cdots, n. \tag{6.183}$$

and using Itô differential rule, the following Itô stochastic differential equations for ρ, α_r are obtained from Eq. (6.177)

$$\mathrm{d}\rho = Q(\alpha)\mathrm{d}t + \sum_{k=1}^{m} \overline{\sigma}_k(\alpha)\mathrm{d}B_k(t), \tag{6.184}$$

$$\begin{aligned} \mathrm{d}\alpha_r &= m_r(\alpha)\mathrm{d}t + \sum_{k=1}^{m} \sigma_{rk}(\alpha)\mathrm{d}B_k(t), \\ r &= 1, 2, \cdots, n, \end{aligned} \tag{6.185}$$

where $\boldsymbol{\alpha} = [\alpha_1, \alpha_2, \cdots, \alpha_n]^{\mathrm{T}}$ and

$$\begin{aligned} Q(\boldsymbol{\alpha}) &= \frac{1}{2}\sum_{s=1}^{n} F_s(\boldsymbol{\alpha}) - \frac{1}{4}\sum_{s,s'=1}^{n}\sum_{k=1}^{m} G_{sk}(\boldsymbol{\alpha})G_{s'k}(\boldsymbol{\alpha}), \\ m_r(\boldsymbol{\alpha}) &= -\alpha_r \sum_{s=1}^{n} F_s(\boldsymbol{\alpha}) + F_r(\boldsymbol{\alpha}) + \frac{1}{2}\alpha_r \sum_{s,s'=1}^{n}\sum_{k=1}^{m} G_{sk}(\boldsymbol{\alpha})G_{s'k}(\boldsymbol{\alpha}) \\ &\quad -\tfrac{1}{2}\sum_{s=1}^{n}\sum_{k=1}^{m} G_{rk}(\boldsymbol{\alpha})G_{sk}(\boldsymbol{\alpha}), \\ \sigma_{rk}(\boldsymbol{\alpha}) &= G_{rk}(\boldsymbol{\alpha}) - \alpha_r \sum_{s=1}^{n} G_{sk}(\boldsymbol{\alpha}). \end{aligned} \tag{6.186}$$

Note that $\sum_{r=1}^{n} \alpha_r = 1$, and only $n-1$ equations in Eq. (6.185) are independent. Take the first $n-1$ equations, denote $\boldsymbol{\alpha}' = [\alpha_1, \alpha_2, \cdots, \alpha_{n-1}]^{\mathrm{T}}$ and replace α_n with $1 - \sum_{r=1}^{n-1} \alpha_r$.

Similar to Eq. (6.131), the Lyapunov exponent of Eq. (6.177) is defined as the mean exponential growth rate of $\overline{I}^{1/2}$, i.e.,

$$\lambda = \lim_{t\to\infty} \frac{1}{t} \ln \overline{I}^{1/2} = \lim_{t\to\infty} \frac{1}{2t} \ln \overline{I}. \tag{6.187}$$

Integrating Eq. (6.184) from 0 to t and dividing it by t result in

$$\frac{1}{2t} \ln \overline{I}(t) = \frac{1}{2t} \ln \overline{I}(0) + \frac{1}{t} \int_0^t Q(\alpha'(\tau))\mathrm{d}\tau + \frac{1}{t} \int_0^t \sum_{k=1}^{m} (\alpha'(\tau))\mathrm{d}B_k(\tau). \tag{6.188}$$

when $t \to \infty$, the first and third terms on the right-hand side of the above equation tend to zero. Thus,

$$\lambda = \lim_{t\to\infty} \frac{1}{t} \int_0^t Q(\boldsymbol{\alpha}'(\tau))\mathrm{d}\tau. \tag{6.189}$$

$\boldsymbol{\alpha}'$ is a diffusion process with $0 \le \alpha_r \le 1$, $r = 1, 2, \cdots, n-1$. When the diffusion coefficients in Eq. (6.177) satisfy the non-degenerate condition similar to Eq. (6.149), $\boldsymbol{\alpha}'$ is ergodic Markov diffusion process. Based on the Oseledec multiplicative ergodic theorem (Oseledec 1968), $\lambda \to \lambda_1$ (the maximum Lyapunov exponent of the averaged Itô equations (5.76) in vol. 1). Thus,

$$\lambda_1 = \lim_{t\to\infty} \frac{1}{t} \int_0^t Q(\boldsymbol{\alpha}'(\tau))\mathrm{d}\tau = E\big[Q(\boldsymbol{\alpha}')\big] = \int Q(\boldsymbol{\alpha}')p(\boldsymbol{\alpha}')\mathrm{d}\boldsymbol{\alpha}', \tag{6.190}$$

where $p(\boldsymbol{\alpha}')$ is the stationary PDF of $\boldsymbol{\alpha}'$, obtained by solving the reduced FPK equation associated with the first $n-1$ equations of Eq. (6.185).

For Eq. (5.90) in vol.1, define norm as

$$\|\mathbf{Z}\| = \overline{H} = (\sum_{r=1}^{n} H_r)^{1/2} \tag{6.191}$$

which leads to an expression for the maximum Lyapunov exponent similar to Eq. (6.190).

The method for calculating the maximum Lyapunov exponent can also be extended to the quasi-integrable Hamiltonian system (1.4) under stationary wide-band random excitation. According to Sect. 8.1.2, in the case of non-resonance, the averaged Itô equations for displacement amplitudes are given by Eq. (1.55). To calculate the maximum Lyapunov exponent, first transform Eq. (1.55) into the averaged Itô equation (1.66), i.e.,

$$\begin{aligned} \mathrm{d}H_i &= \overline{m}_i(\mathbf{H})\mathrm{d}t + \sum_{k=1}^{m} \overline{\sigma}_{ik}(\mathbf{H})\mathrm{d}B_k(t), \\ i &= 1, 2, \cdots, n. \end{aligned} \tag{6.192}$$

where

$$\begin{aligned} \overline{m}_i(\mathbf{H}) &= \Big[g_i(A_i + B_i)(1 + r_i)m_i(\mathbf{A}) \\ &\quad + \frac{1}{2}\big[g_i(A_i + B_i)(1 + r_i)\big]' \sum_{k=1}^{m} \sigma_{ik}(\mathbf{A})\sigma_{ik}(\mathbf{A}) \Big]\Big|_{A_j = U_j^{-1}(H_j) - B_j}, \\ &\sum_{k=1}^{m} \overline{\sigma}_{ik}(\mathbf{H})\overline{\sigma}_{jk}(\mathbf{H}) = \\ &g_i(A_i + B_i)g_j\big(A_j + B_j\big)(1 + r_i)(1 + r_j) \sum_{k=1}^{m} \sigma_{ik}(\mathbf{A})\,\sigma_{jk}(\mathbf{A})\big|_{A_j = U_j^{-1}(H_j) - B_j}. \end{aligned} \tag{6.193}$$

Equation (6.192) is similar to Eq. (5.90) in vol. 1, and the maximum Lyapunov exponent can be calculated using Eq. (6.190).

For quasi-integrable and resonant Hamiltonian systems, the n frequencies of quasi-integrable Hamiltonian systems satisfy α resonant relations as described in Eq. (5.75) of vol. 1, α combinations Ψ_u of angular variables can be introduced based on Eq. (5.118) in vol. 1. The averaged Itô equations are of the form of Eq. (5.50). If its linearized equations of I_r $(r = 1, 2, \cdots, n)$ in Eq. (5.120) of vol. 1 at $I_r = 0$ satisfy conditions in Eqs. (6.175) and (6.179), then use the norm in Eq. (6.181) and transformation in Eqs. (6.182) and (6.183) to derive the maximum Lyapunov exponent similar to Eq. (6.190). But in this case, both Ψ_u $(u = 1, 2, \cdots, \alpha)$ and α_r $(r = 1, 2, \cdots, n-1)$ in Eq. (6.185) should be treated as $\boldsymbol{\alpha}'$, i.e., $\boldsymbol{\alpha}' = [\alpha_1, \cdots, \alpha_{n-1}, \Psi_1, \cdots, \Psi_\alpha]^T$. If the corresponding Itô stochastic differential equation for $\boldsymbol{\alpha}'$ is ergodic Markov diffusion process, Eq. (6.190) can still be used to calculate the maximum Lyapunov exponent λ_1.

Example 6.3 As an application, consider a n-DOF non-gyroscopic linear system excited by stationary stochastically parametric excitation (Zhu and Huang 1999). The governing motion equations are.

$$\begin{aligned} \dot{Q}_i &= P_i, \\ \dot{P}_i &= -\omega_i^2 Q_i - 2\sum_{j=1}^{n} \beta_{ij}P_j - \omega_i\xi(t)\sum_{j=1}^{n} k_{ij}Q_j, \\ i &= 1, 2, \cdots, n, \end{aligned} \tag{6.194}$$

where ω_i, β_{ij}, k_{ij} are constants, $\xi(t)$ is a stationary wideband process with zero mean and correlation function $R(\tau)$, and both β_{ij} and $R(0)$ are small. The Hamiltonian system associated with system (6.194) is linear. The Hamiltonian function is

$$H = \sum_{i=1}^{n} H_i = \sum_{i=1}^{n} (p_i^2 + \omega_i^2 q_i^2)/2. \tag{6.195}$$

Equation (6.194) is of the form of Eq. (1.4). Applying the stochastic averaging method described in Sect. 8.1.2 to Eq. (6.194) in non-internal resonance case yields the averaged Itô equation of the form of Eq. (1.55), where A_i is the amplitude of the i th oscillator. Then, it can be converted into the averaged Itô equations in terms of H_i of the form of Eq. (1.66). Thus, using Eq. (6.193) (where $g_i(A_i) = \omega_i^2 A_i$, $B_i = 0$) results in the following drift and diffusion coefficients

$$\begin{aligned}
&\overline{m}_r(\mathbf{H}) = \left[-2\beta_{rr} + \frac{1}{2}k_{rr}^2 S(2\omega_r) + \frac{1}{4}\sum_{\substack{s=1\\ s\neq r}}^{n} k_{rs}k_{sr}S_{rs}^- \right] H_r + \frac{1}{4}\omega_r^2 \sum_{\substack{s=1\\ s\neq r}}^{n} \frac{k_{rs}^2}{\omega_s^2} S_{rs}^+ H_s,\\
&\sum_{k=1}^{m} \overline{\sigma}_{rk}\overline{\sigma}_{rk}(\mathbf{H}) = \frac{1}{2}k_{rr}^2 S(2\omega_r) H_r^2 + \frac{1}{2}\omega_r^2 H_r \sum_{\substack{s=1\\ s\neq r}}^{n} \frac{k_{rs}^2}{\omega_s^2} S_{rs}^+ H_s,\\
&\sum_{k=1}^{m} \overline{\sigma}_{rk}\overline{\sigma}_{sk}(\mathbf{H}) = \frac{1}{2}k_{rs}k_{sr}S_{rs}^- H_r H_s, \quad s \neq r,\\
&S_{rs}^{\pm} = S(\omega_r + \omega_s) \pm S(\omega_r - \omega_s),\\
&S(\omega) = 2\int_0^{\infty} R(\tau)\cos w\tau \,\mathrm{d}\tau,\\
&r, s = 1, 2, \cdots, n.
\end{aligned} \tag{6.196}$$

They satisfy the conditions (6.176) and (6.180), and thus, the maximum Lyapunov exponent can be calculated based on Eq. (6.190).

Taking a 2-DOF system as an example, Eq. (6.196) can be simplified to

$$\begin{aligned}
\overline{m}_1(H_1, H_2) &= \left[-2\beta_{11} + \frac{1}{2}k_{11}^2 S(2\omega_1) \pm \frac{1}{4}k^2 S^- \right] H_1 + \frac{k^2}{4}\left(\frac{\omega_1}{\omega_2}\right)^2 S^+ H_2,\\
\overline{m}_2(H_1, H_2) &= \left[-2\beta_{22} + \frac{1}{2}k_{22}^2 S(2\omega_2) \pm \frac{1}{4}k^2 S^- \right] H_2 + \frac{k^2}{4}\left(\frac{\omega_2}{\omega_1}\right)^2 S^+ H_1,
\end{aligned}$$

$$\begin{aligned}\sum_{k=1}^{m}\overline{\sigma}_{1k}\overline{\sigma}_{1k}(H_1,H_2) &= \frac{1}{2}k_{11}^2 S(2\omega_1)H_1^2 + \frac{k^2}{2}\left(\frac{\omega_1}{\omega_2}\right)^2 S^+ H_1 H_2,\\ \sum_{k=1}^{m}\overline{\sigma}_{2k}\overline{\sigma}_{2k}(H_1,H_2) &= \frac{1}{2}k_{22}^2 S(2\omega_2)H_2^2 + \frac{k^2}{2}\left(\frac{\omega_2}{\omega_1}\right)^2 S^+ H_1 H_2,\end{aligned} \tag{6.197}$$

$$\sum_{k=1}^{m}\overline{\sigma}_{1k}\overline{\sigma}_{2k}(H_1,H_2) = \pm\frac{k^2}{2}S^- H_1 H_2,$$
$$k_{12} = \pm k_{21} = k > 0,\ S^\pm = S_{12}^\pm.$$

Introducing a norm

$$\|\mathbf{Z}\| = H^{1/2} = (H_1 + H_2)^{1/2}, \tag{6.198}$$

where $\mathbf{Z} = [Q_1, Q_2, P_1, P_2]^{\mathrm{T}}$. Defining a transformation

$$\rho = \ln H^{1/2} = (\ln H)/2, \tag{6.199}$$

$$\alpha_r = H_r/H,\ r = 1, 2, \tag{6.200}$$

and applying Itô differential rule, the Itô stochastic differential equation for ρ, α_1 can be derived from the Itô equation for H_r in Eq. (6.192), i.e.,

$$\mathrm{d}\rho = Q(\alpha_1)\mathrm{d}t + \overline{\sigma}(\alpha_1)\mathrm{d}B(t), \tag{6.201}$$

$$\mathrm{d}\alpha_1 = m(\alpha_1)\mathrm{d}t + \sigma(\alpha_1)\mathrm{d}B(t), \tag{6.202}$$

where

$$\begin{aligned}Q(\alpha_1) &= \mu_1\alpha_1 + \mu_2(1-\alpha_1) \pm k^2 S^-/8 + \varphi(\alpha_1)/4,\\ m(\alpha_1) &= \left(1/2 - \alpha_1\right)\varphi(\alpha_1) + 2(\mu_1 - \mu_2)\alpha_1(1-\alpha_1),\\ \sigma^2(\alpha_1) &= \alpha_1(1-\alpha_1)\varphi(\alpha_1),\\ \varphi(\alpha_1) &= a\alpha_1^2 + b\alpha_1 + c,\end{aligned} \tag{6.203}$$

$$a=\frac{1}{2}k^2\left[\left(\frac{\omega_1}{\omega_2}\right)^2+\left(\frac{\omega_2}{\omega_1}\right)^2\right]S^+-\frac{1}{2}k_{11}^2S(2\omega_1)-\frac{1}{2}k_{22}^2S(2\omega_2)\pm k^2S^-,$$
$$b=\frac{1}{2}k_{11}^2S(2\omega_1)+\frac{1}{2}k_{22}^2S(2\omega_2)\mp k^2S^+-k^2\left(\frac{\omega_1}{\omega_2}\right)^2S^+,$$
$$c=\frac{1}{2}k^2\left(\frac{\omega_1}{\omega_2}\right)^2S^+,$$
$$\mu_1=-\beta_{11}+(k_{11}^2/8)S(2\omega_1),\quad \mu_2=-\beta_{22}+(k_{22}^2/8)S(2\omega_2).$$

It is noted that $\alpha_2=1-\alpha_1$, and no equation for α_2 is needed.

Equation (6.202) shows that process $\alpha_1(t)$ itself is a diffusion process with two boundaries 0 and 1. According to (Lin and Cai, 1985), both boundaries are singular of the first kind since $\sigma=0$ at two boundaries. The behavior of process $\alpha_1(t)$ depends on the characteristics of the two boundaries. According to Eq. (6.203), the diffusion exponents at the two boundaries are both 1, and $m(0)>0$, $m(1)<0$. Therefore, both boundaries are entrance, and the stationary PDF of α_1 exists and can be obtained by solving the reduced FPK equation associated with Eq. (6.202) as

$$p(\alpha_1)=CF(\alpha_1)/\varphi(\alpha_1), \tag{6.204}$$

where the normalization constant is

$$C=4(\mu_1-\mu_2)\big/[F(1)-F(0)], \tag{6.205}$$

$$F(\alpha_1)=\begin{cases}\exp\left[\dfrac{8(\mu_1-\mu_2)}{\sqrt{\Delta}}\tanh^{-1}\dfrac{b+2a\alpha_1}{\sqrt{\Delta}}\right], & \Delta>0,\\ \exp\left[\dfrac{8(\mu_1-\mu_2)}{\sqrt{-\Delta}}\tan^{-1}\dfrac{b+2a\alpha_1}{\sqrt{-\Delta}}\right], & \Delta<0,\\ \exp\left[-\dfrac{8(\mu_1-\mu_2)}{b+2a\alpha_1}\right], & \Delta=0,\end{cases} \tag{6.206}$$

$$\begin{aligned}\Delta&=b^2-4ac\\&=\left[k_{11}^2S(2\omega_1)+k_{22}^2S(2\omega_2)\right]^2\big/4+k^2\left[k_{11}^2S(2\omega_1)+k_{22}^2S(2\omega_2)\right]\\&\times[S(\omega_1\mp\omega_2)S(\omega_1\pm\omega_2)]-4k^4S^2(\omega_1\mp\omega_2)S(\omega_1\pm\omega_2).\end{aligned} \tag{6.207}$$

Substituting $Q(\alpha_1)$ in Eq. (6.203) and $p(\alpha_1)$ in Eq. (6.204) into Eq. (6.190) results in the following maximum Lyapunov exponent of the 2-DOF system (6.194)

$$\lambda_1 = \int_0^1 Q(\alpha_1)p(\alpha_1)\mathrm{d}\alpha_1 = \frac{1}{2}(\mu_1 + \mu_2) \pm \frac{1}{8}k^2 S^- \\ + \frac{1}{2}(\mu_1 - \mu_2)\frac{F(1) + F(0)}{F(1) - F(0)}, \tag{6.208}$$

This result is the same as that obtained by using the standard stochastic averaging method (Ariaratnam and Xie 1992).

Example 6.4 The method described above is applied to random linear gyroscopic systems, such as a gyroscopic pendulum (Huang and Zhu 2000). Gyroscopic pendulums are essential component in navigation systems, providing a vertical reference axis or an artificial platform. Therefore, the direction stability of a gyroscopic pendulum is crucial.

As a parametric excitation, carrier's vertical vibrations affect gyroscopic pendulum. It's necessary to investigate the influence of the excitations on the direction stability of a gyroscopic pendulum. Several studies have examined the stability of gyroscopic pendulums under harmonic excitations. (Asokanthan and Ariaratnam 2000; Sri Namchchivaya 1987) used the classical stochastic averaging method to study the moment stability of gyroscopic pendulums under stationary wideband random excitations and the Lyapunov asymptotic stability under Gaussian white noise excitations. Here, we investigate the Lyapunov asymptotic stability with probability 1 of gyroscopic pendulums under stationary wideband random excitations is studied.

The Lagrangian motion equation for gyroscopic pendulum under vertical excitations is given by (Arnold 1961)

$$\mathbf{M}\ddot{\mathbf{Q}} + \mathbf{G}\dot{\mathbf{Q}} + \mathbf{D}\dot{\mathbf{Q}} + [\mathbf{K} + \mathbf{K}_1\xi(t)]\mathbf{Q} = 0, \tag{6.209}$$

where $\mathbf{Q} = [\theta_2, \theta_1]^{\mathrm{T}}$ represents a generalized displacement vector of deflection angles of rotor axis, and

$$\mathbf{M} = \begin{bmatrix} A_0 & 0 \\ 0 & B_0 \end{bmatrix}, \quad \mathbf{G} = \begin{bmatrix} 0 & -cn \\ cn & 0 \end{bmatrix}, \quad \mathbf{D} = \begin{bmatrix} d_1 & 0 \\ 0 & d_2 \end{bmatrix}, \\ \mathbf{K} = \begin{bmatrix} k & 0 \\ 0 & k \end{bmatrix}, \quad \mathbf{K}_1 = \begin{bmatrix} k_1 & 0 \\ 0 & k_1 \end{bmatrix}. \tag{6.210}$$

Note that $\mathbf{G}$ is the coefficient matrix of gyroscopic force, and $\xi(t)$ is a stationary wideband process with power spectral density $S(\omega)$. Assume that d_i, k_1^2, $S(\omega)$ are small quantities. Applying a Legendre transformation (3.11) of vol. 1 to Eq. (6.209) yields the corresponding Hamiltonian equation

$$\dot{\mathbf{Z}} = \mathbf{JSZ} - [\mathbf{B}_1 + \mathbf{B}_2\xi(t)]\mathbf{Z}, \tag{6.211}$$

where $\mathbf{Z} = [Q_1, Q_2, P_1, P_2]^{\mathrm{T}}$; Q_1, Q_2 are generalized displacements; P_1, P_2 are generalized momenta; $\mathbf{J}$ is a 4×4 symplectic matrix (see Eq. (3.17) in vol. 1),

$$\mathbf{S} = \begin{bmatrix} (1/A_0)(cn/2)^2 + k & 0 & 0 & -cn/2A_0 \\ 0 & (1/B_0)(cn/2)^2 + k & cn/2B_0 & 0 \\ 0 & cn/2B_0 & 1/B_0 & 0 \\ -cn/2A_0 & 0 & 0 & 1/A_0 \end{bmatrix},$$
$$\mathbf{B}_1 = \begin{bmatrix} \mathbf{0} & \mathbf{0} \\ \mathbf{DM}^{-1}\mathbf{G}^T/2 & \mathbf{DM}^{-1} \end{bmatrix}, \quad \mathbf{B}_2 = \begin{bmatrix} \mathbf{0} & \mathbf{0} \\ \mathbf{K}_1 & \mathbf{0} \end{bmatrix}, \tag{6.212}$$

and where $\mathbf{0}$ is a 2×2 zero matrix. Assuming that $\mathbf{M}, \mathbf{K}$ is positive definite, the trivial solution of unperturbed system is stable. From the characteristic equation $|\mathbf{JS} \mp j\omega_r \mathbf{I}| = 0$, the natural frequencies are obtained as

$$\omega_r = \left\{ \left(k_{11} + k_{22} + g^2\right) \pm \left[g^4 + 2g^2(k_{11} + k_{22}) + (k_{11} - k_{22})^2\right]^{1/2} \right\}^{1/2} \Big/ \sqrt{2} \tag{6.213}$$

where

$$k_{11} = k/B_0, \quad k_{22} = k/A_0, \quad g^2 = (cn)^2/A_0 B_0. \tag{6.214}$$

Using the canonical transformation in Sect. 3.2.4, in which the transformation matrix is

$$\mathbf{T} = \begin{bmatrix} a_1 & -a_2\gamma_2 & -a_1 & -a_2\gamma_2 \\ a_1\gamma_1 & a_2 & a_1\gamma_1 & -a_2 \\ a_1\beta_1 & -a_2\alpha_2 & a_1\beta_1 & a_2\alpha_2 \\ a_1\alpha_1 & a_2\beta_2 & -a_1\alpha_1 & a_2\beta_2 \end{bmatrix} \tag{6.215}$$

where

$$\begin{aligned} &a_r = 1/2(\beta_r - \alpha_r\gamma_r), \quad \gamma_1 = \omega_1 cn/\left(B_0\omega_1^2 - k\right), \\ &\gamma_2 = \omega_2 cn/\left(A_0\omega_2^2 - k\right), \quad \alpha_1 = (cn/2) - B_0\omega_1\gamma_1, \\ &\alpha_2 = (cn/2) - A_0\omega_2\gamma_2, \quad \beta_1 = A_0\omega_1 - (cn/2)\gamma_1, \\ &\beta_2 = B_0\omega_2 - (cn/2)\gamma_2, \quad r = 1, 2, \cdots, \end{aligned} \tag{6.216}$$

Equation (6.211) is transformed into

$$\begin{aligned} \dot{\overline{\mathbf{Q}}} &= \Omega_1\overline{\mathbf{P}} - \left(\mathbf{A}_1^{11}\overline{\mathbf{Q}} + \mathbf{A}_1^{12}\overline{\mathbf{P}}\right) - \left(\mathbf{A}_2^{11}\overline{\mathbf{Q}} + \mathbf{A}_2^{12}\overline{\mathbf{P}}\right)\xi(t), \\ \dot{\overline{\mathbf{P}}} &= -\Omega_1\overline{\mathbf{Q}} - \left(\mathbf{A}_1^{21}\overline{\mathbf{Q}} + \mathbf{A}_1^{22}\overline{\mathbf{P}}\right) - \left(\mathbf{A}_2^{21}\overline{\mathbf{Q}} + \mathbf{A}_2^{22}\overline{\mathbf{P}}\right)\xi(t). \end{aligned} \tag{6.217}$$

The following relations are used in the above derivation,

$$\mathbf{T}^{-1} = -\mathbf{J}\mathbf{T}^{\mathrm{T}}\mathbf{J}, \tag{6.218}$$

$$\mathbf{T}^{-1}\mathbf{JST} = \mathbf{J}\boldsymbol{\Omega}, \tag{6.219}$$

$$\mathbf{A}_i = \mathbf{J}^T\mathbf{T}^T\mathbf{J}\mathbf{B}_i\mathbf{T} = \begin{bmatrix} A_i^{11} & A_i^{12} \\ A_i^{21} & A_i^{22} \end{bmatrix}, \tag{6.220}$$

$$\boldsymbol{\Omega} = \mathrm{diag}[\omega_1, \omega_2, \omega_1, \omega_2] = \begin{bmatrix} \boldsymbol{\Omega}_1 & 0 \\ 0 & \boldsymbol{\Omega}_1 \end{bmatrix}. \tag{6.221}$$

The Hamiltonian function associated with the quasi-Hamiltonian system (6.217) is

$$H(\overline{\mathbf{Q}}, \overline{\mathbf{P}}) = \sum_{r=1}^{2} H_r(\overline{Q}_r, \overline{P}_r) = \sum_{r=1}^{2} \omega_r\left(\overline{Q}_r^2 + \overline{P}_r^2\right)\Big/2.. \tag{6.222}$$

Introducing transformations

$$\overline{Q}_r = \sqrt{2H_r/\omega_r}\cos\theta, \quad \overline{P}_r = \sqrt{2H_r/\omega_r}\sin\theta, \tag{6.223}$$

$$H_r = \omega_r\left(\overline{Q}_r^2 + \overline{P}_r^2\right)/2, \quad \theta = tg^{-1}\overline{P}_r/\overline{Q}_r, \tag{6.224}$$

the differential equations for H_r, θ can be derived from Eq. (6.217). In the non-resonant case, applying the stochastic averaging method to the resulting equations yields the averaged Itô stochastic differential equations for H_r

$$\mathrm{d}H_r = \overline{m}_r(H_1, H_2)\mathrm{d}t + \sum_{k=1}^{2} \overline{\sigma}_{rk}(H_1, H_2)\mathrm{d}B_k(t), \tag{6.225}$$

where

$$\overline{m}_1(H_1, H_2) = \Delta_1^2\left\{-\frac{2\omega_1}{\Delta_1}\left(\frac{d_1}{\gamma_1} + \gamma_1 d_2\right)\right.$$
$$\left. + k_1^2\left[\delta_1 + 2\left(\frac{1}{\gamma_1} - \gamma_1\right)^2 S(2\omega_1)\right]\right\}H_1 - k_1^2\Delta_1^2\left(\frac{\omega_1}{\omega_2}\right)\delta_2 H_2,$$

$$\overline{m}_2(H_1, H_2) = -k_1^2\Delta_1^2\left(\frac{\omega_2}{\omega_1}\right)\delta_2 H_1 + \Delta_1^2\left\{-\frac{2\omega_1}{\Delta_1}\left(\frac{d_2}{\gamma_2} + \gamma_2 d_1\right)\right.$$
$$\left. + k_1^2\left[\delta_1 + 2\left(\tfrac{1}{\gamma_2} - \gamma_2\right)^2 S(2\omega_2)\right]\right\} H_2,$$

$$\sum_{k=1}^{2}\overline{\sigma}_{1k}\overline{\sigma}_{1k}(H_1, H_2) = 2k_1^2\Delta_1^2(1/\gamma_1 - \gamma_1)^2 S(2\omega_1)H_1^2 - 2k_1^2\Delta_1^2(\omega_1/\omega_2)\delta_2 H_1 H_2,$$

$$\sum_{k=1}^{2}\overline{\sigma}_{2k}\overline{\sigma}_{2k}(H_1, H_2) = 2k_1^2\Delta_1^2(1/\gamma_2 - \gamma_2)^2 S(2\omega_2)H_2^2 - 2k_1^2\Delta_1^2(\omega_2/\omega_1)\delta_2 H_1 H_2,$$

$$\sum_{k=1}^{2}\overline{\sigma}_{1k}\overline{\sigma}_{2k}(H_1, H_2) = 2k_1^2\Delta_1^2\delta_1 H_1 H_2,$$

(6.226)

$$\delta_1 = \left[(\gamma_1 + \gamma_2)^2/\gamma_1\gamma_2\right]S(\omega_1 - \omega_2) - \left[(-\gamma_1 + \gamma_2)^2/\gamma_1\gamma_2\right]S(\omega_1 + \omega_2),$$
$$\delta_2 = \left[(\gamma_1 + \gamma_2)^2/\gamma_1\gamma_2\right]S(\omega_1 - \omega_2) + \left[(-\gamma_1 + \gamma_2)^2/\gamma_1\gamma_2\right]S(\omega_1 + \omega_2),$$
$$\Delta_1 = cn/2A_0B_0\left(\omega_1^2 - \omega_2^2\right).$$

Introducing the norm in Eq. (6.198) and the transformations in Eqs. (6.199) and (6.200) result in the equations of the form of Eqs. (6.201) and (6.202), where

$$Q(\alpha_1) = (\mu_1 + \mu_2)/2 + (\mu_1 - \mu_2)(\alpha_1 - 1/2) + \varphi(\alpha_1)/4,$$
$$m(\alpha_1) = (1/2 - \alpha_1)\varphi(\alpha_1) + 2(\mu_1 - \mu_2)\alpha_1(1 - \alpha_1),$$
$$\sigma^2(\alpha_1) = \alpha_1(1 - \alpha_1)\varphi(\alpha_1),$$
$$\varphi(\alpha_1) = a\alpha_1^2 + b\alpha_1 + c,$$

$$a = -2k_1^2\Delta_1^2\delta_2\left(\omega_1/\omega_2 + \omega_2/\omega_1\right) + 4k_1^2\Delta_1^2\delta_1$$
$$-2k_1^2\Delta_1^2\left[\left(1/\gamma_1 - \gamma_1\right)^2 S(2\omega_1) + \left(1/\gamma_2 - \gamma_2\right)^2 S(2\omega_2)\right],$$
$$b = 4k_1^2\Delta_1^2\delta_2(\omega_1/\omega_2) - 4k_1^2\Delta_1^2\delta_1$$
$$+2k_1^2\Delta_1^2\left[\left(1/\gamma_1 - \gamma_{11}\right)^2 S(2\omega_1) + \left(1/\gamma_2 - \gamma_2\right)^2 S(2\omega_2)\right],$$
$$c = -2k_1^2\Delta_1^2\delta_2\left(\omega_1/\omega_2\right), \tag{6.227}$$
$$\mu_1 = \Delta_1^2\left\{-\frac{\omega_1}{\Delta_1}\left(\frac{d_1}{\gamma_1} + \gamma_1 d_2\right) + k_1^2\left[\delta_1 + \frac{1}{2}\left(\frac{1}{\gamma_1} - \gamma_1\right)^2 S(2\omega_1)\right]\right\},$$
$$\mu_1 = \Delta_1^2\left\{-\frac{\omega_2}{\Delta_1}\left(\frac{d_2}{\gamma_2} + \gamma_2 d_1\right) + k_1^2\left[\delta_1 + \frac{1}{2}\left(\frac{1}{\gamma_2} - \gamma_2\right)^2 S(2\omega_2)\right]\right\}.$$

Similar to α_1 in Eq. (6.202), it can be proved that α_1 in Eq. (6.227) is also ergodic in $0 < \alpha_1 < 1$. The stationary PDF of the form of Eq. (6.204) with different $\varphi(\alpha_1)$ can be obtained. Finally, the maximum Lyapunov exponent is

$$\lambda_1 = (\mu_1 + \mu_2)/2 + (\mu_1 - \mu_2)[F(1) + F(0)]/2[F(1) - F(0)], \tag{6.228}$$

where $F(\alpha_1)$ is given in Eq. (6.208), a, b are determined by Eq. (6.227), and

$$\begin{aligned} \Delta &= 4k_1^4\Delta_1^4 \\ &\times\left[(\frac{1}{\gamma_1} - \gamma_1)^2 S(2\omega_1) + (\frac{1}{\gamma_2} - \gamma_2)^2 S(2\omega_2) + \frac{4(\gamma_2 - \gamma_1)^2}{\gamma_1\gamma_2} S(\omega_1 + \omega_2)\right] \\ &\times\left[(\frac{1}{\gamma_1} - \gamma_1)^2 S(2\omega_1) + (\frac{1}{\gamma_2} - \gamma_2)^2 S(2\omega_2) - \frac{4(\gamma_2 + \gamma_1)^2}{\gamma_1\gamma_2} S(\omega_1 - \omega_2)\right]. \end{aligned} \tag{6.229}$$

When $cn = 0$ (see Eq. (6.210), it means no gyroscopic term), Eq. (6.228) becomes

$$\lambda_1 = \begin{cases} \mu_1 \; \mu_1 > \mu_2 \\ \mu_2 \; \mu_2 > \mu_1 \end{cases}. \tag{6.230}$$

And when $A_0 = B_0 = A$, $\gamma_1 = -\gamma_2 = 1$, $\Delta = 0$,

$$\begin{aligned} \mu_1 &= -(d_1 + d_2)\Delta_1\omega_1 + 2\Delta_1^2 k_1^2 S(\omega_1 + \omega_2) \\ \mu_2 &= -(d_1 + d_2)\Delta_1\omega_2 + 2\Delta_1^2 k_1^2 S(\omega_1 + \omega_2) \\ \varphi(\alpha_1) &= \left(8k_1^2\Delta_1^2/\omega_1\omega_2\right) S(\omega_1 + \omega_2)[(\omega_1 + \omega_2)\alpha_1 - \omega_1]^2 \end{aligned} \tag{6.231}$$

It is known from Eqs. (6.227) and (6.231) that, besides $\alpha_1 = 0, 1$, $\alpha_1^* = \omega_1/(\omega_1 + \omega_2)$ is also a singular point of the first kind for α_1, and $m(\alpha_1^*) < 0$, the diffusion exponent is 2. It is known from (Lin and Cai 1995) that α_1^* is an entrance boundary in the range of $[0, \alpha_1^*]$, indicating a stationary PDF exists. Solving the corresponding reduced FPK equation yields

$$p(\alpha_1) = CF_1(\alpha_1)/\varphi(\alpha_1), 0 < \alpha_1 < \alpha_1^*, \tag{6.232}$$

where

$$\begin{aligned} F_1(\alpha_1) &= \exp\left\{\frac{(\mu_1 - \mu_2)\omega_1\omega_2}{2k_1^2\Delta_1^2(\omega_1 + \omega_2)[\omega_1 - (\omega_1 + \omega_2)\alpha_1]S(\omega_1 + \omega_2)}\right\}, \\ C &= -4(\mu_1 - \mu_2)/F_1(0). \end{aligned} \tag{6.233}$$

and the maximum Lyapunov exponent is

$$\lambda_1 = (\mu_1 + \mu_2)/2 + (\mu_1 - \mu_2)\left[F_1(\alpha_1^*) + F_1(0)\right]/2\left[F_1(\alpha_1^*) - F_1(0)\right] = \mu_2. \tag{6.234}$$

The necessary and sufficient condition for the Lyapunov asymptotic stability with probability 1 of the trivial solution of system (6.209) is $\mu_2 < 0$, i.e.,

$$d_1 + d_2 > (2\Delta_1 k_1^2/\omega_2)S(\omega_1 + \omega_2) = \left[\omega_1 k_1^2/k(\omega_1 + \omega_2)\right]S(\omega_1 + \omega_2). \tag{6.235}$$

Based on this,0 the stability domains in parameter space can be determined (Huang and Zhu 2000). In (Sri Namchchivaya 1987), the condition for mean-square asymptotic stability of the trivial solution of system (6.209) is

$$d_1 + d_2 > \left(k_1^2/k\right)S(\omega_1 + \omega_2) \tag{6.236}$$

Equation (6.236) is stricter than Eq. (6.235), since mean-square asymptotic stability is stronger than the Lyapunov asymptotic stability with probability 1.

Example 6.5 Stability of nonlinear stochastic systems (Zhu and Huang 1999).

The main advantage of the method proposed in this section for calculating the maximum Lyapunov exponent is that it can be used to study the asymptotic stability with probability 1 of nonlinear stochastic dynamical systems. In the following, this will be demonstrated using a simple example (Zhu and Huang 1999). Consider the system

$$\begin{aligned}
&\dot{Q}_1 = P_1,\\
&\dot{P}_1 = -\omega_1^2 Q_1 - \beta_{11}P_1 - \beta_{22}P_2 + f_{11}Q_1 W_{g1}(t) + f_{12}Q_2^2 W_{g2}(t),\\
&\dot{Q}_2 = P_2,\\
&\dot{P}_2 = -\alpha Q_2^3 - \beta_{21}P_1 - \beta_{22}P_2 + f_{21}Q_1 W_{g1}(t) + f_{22}Q_2^2 W_{g2}(t),
\end{aligned} \tag{6.237}$$

where constants $\omega_1, f_{ik}, \alpha, \beta_{ij} > 0$, $W_{gk}(t)$ are independent Gaussian white noises with intensities $2D_k$, and both β_{ij} and D_k are small quantities. The Hamiltonian associated with quasi-Hamiltonian system (6.237) is

$$H = H_1 + H_2, \tag{6.238}$$

where

$$H_1 = \left(p_1^2 + \omega_1^2 q_1^2\right)/2, H_2 = \left(p_2^2 + \alpha Q_2^4/2\right)/2. \tag{6.239}$$

Applying the stochastic averaging method of quasi-integrable and non-resonant Hamiltonian systems described in Sect. 5.2, the averaged Itô equations for H_i are of the form of Eq. (5.90) in vol. 1 with

$$\begin{aligned}
&\overline{m}_1(H_1,H_2) = -\beta_{11}H_1 + f_{11}^2D_1H_1/\omega_1^2 + \eta_1 f_{12}^2D_2H_2,\\
&\overline{m}_2(H_1,H_2) = -\eta_2\beta_{22}H_2 + f_{21}^2D_1H_1/\omega_1^2 + \eta_1 f_{22}^2D_2H_2,\\
&\sum_{k=1}^{m}\overline{\sigma}_{1k}\overline{\sigma}_{1k}(H_1,H_2) = f_{11}^2D_1H_1^2/\omega_1^2,\\
&\sum_{k=1}^{m}\overline{\sigma}_{2k}\overline{\sigma}_{2k}(H_1,H_2) = 2\eta_3 f_{22}^2D_2H_2^2,\\
&\sum_{k=1}^{m}\overline{\sigma}_{1k}\overline{\sigma}_{2k}(H_1,H_2) = 0,
\end{aligned} \tag{6.240}$$

where

$$\eta_1 = \frac{2\sqrt{2}\Gamma(3/4)}{3\sqrt{\pi}\Gamma(5/4)\alpha},\quad \eta_2 = \frac{4}{3},\quad \eta_3 = \frac{16}{21\alpha}. \tag{6.241}$$

where $\Gamma(\cdot)$ is the gamma function. Equation (6.240) satisfy the conditions (6.176) and (6.180). Introducing the norm in Eq. (6.198) and transformations (6.199) and (6.200), the Itô stochastic differential equations for ρ, α_1 are of the same form as Eqs. (6.201) and (6.202) with

$$\begin{aligned}
&Q(\alpha_1) = (\mu_1 + f_{21}^2D_1/\omega_1^2)\alpha_1 + (\mu_2 + f_{12}^2D_2\eta_1)(1-\alpha_1) + A\alpha_1(1-\alpha_1)/4,\\
&m(\alpha_1) = [2(\mu_1-\mu_2) + A(1-2\alpha_1)/2]\alpha_1(1-\alpha_1)\\
&\quad + f_{12}^2D_2\eta_1(1-\alpha_1)^2 - f_{21}^2D_1\alpha_1^2/\omega_1^2,\\
&\sigma^2(\alpha_1) = A\alpha_1^2(1-\alpha_1)^2,
\end{aligned} \tag{6.242}$$

$$\begin{aligned}
&\mu_1 = -\beta_{11}/2 + f_{11}^2D_1/4\omega_1^2,\\
&\mu_2 = -\beta_{22}\eta_2/2 + (\eta_1-\eta_3)f_{22}^2D_2/2,\\
&A = f_{11}^2D_1/\omega_1^2 + 2\eta_3 f_{22}^2D_2.
\end{aligned} \tag{6.243}$$

$\sigma^2(\alpha_1)$ isn't zero at $0 < \alpha_1 < 1$, $\sigma^2(0) - \sigma^2(1) = 0$ and $m(0) \neq 0$, $m(1) \neq 0$. According to Table 1 for the first singular boundary in (Lin and Cai 1995), both $\alpha_1 = 0$ and $\alpha_1 = 1$ are the first class singular boundary and non-trivial PDF exists at $0 < \alpha_1 < 1$. Solving the reduced FPK equation associated with Itô equation of $\alpha_1(t)$ yields the following stationary PDF

$$p(\alpha_1) = \frac{C}{\alpha_1(1-\alpha_1)}\left(\frac{\alpha_1}{1-\alpha_1}\right)^{4(\mu_1-\mu_2)/A}\exp\left\{-\frac{2}{A}\left[\frac{\eta_1 f_{12}^2D_2}{\alpha_1} + \frac{f_{21}^2D_1}{\omega_1^2(1-\alpha_1)}\right]\right\}. \tag{6.244}$$

Then the maximum Lyapunov exponent is

$$\lambda_1 = \int_0^1 Q(\alpha_1)p(\alpha_1)d\alpha_1. \tag{6.245}$$

The necessary and sufficient condition for the Lyapunov asymptotic stability with probability 1 of the trivial solution of Eq. (6.237) is $\lambda_1 < 0.\lambda_1 < 0$.

6.5 Nonlinear Stochastic Optimal Control of Quasi-Hamiltonian Systems

The most used method for stochastic optimal control is Bellman dynamical programming (Bellman 1957; Fleming and Rishel 1975). According to the method, the optimal control of stochastic dynamical systems can be carried out by building and solving the stochastic dynamical programming (Hamilton–Jacobi-Bellman, HJB) equation. For controlled linear systems under Gaussian white noise excitation with quadratic cost function, both the uncontrolled and controlled processes are Gaussian processes, and the corresponding stochastic dynamical programming equations can be solved exactly. However, for controlled nonlinear systems, both the uncontrolled and controlled processes are non-Gaussian processes, solving the corresponding dynamical programming equations becomes challenging, especially for high-dimensional controlled nonlinear systems. In this section, the combination of the stochastic averaging methods and the stochastic dynamical programming is used for the stochastic optimal control (Zhu 2003, 2006). It will be seen that this combination procedure will greatly simplify the optimal control problem and prediction of dynamical responses of both uncontrolled and controlled quasi-Hamiltonian systems.

6.5.1 *Controlled Quasi-Hamiltonian Systems*

Consider a controlled quasi-Hamiltonian system governed by the following motion equations

$$\begin{aligned} \dot{Q}_i &= \frac{\partial H'}{\partial P_i}, \\ \dot{P}_i &= -\frac{\partial H'}{\partial Q_i} - \varepsilon \sum_{j=1}^{n} c'_{ij}(\mathbf{Q}, \mathbf{P}) \frac{\partial H'}{\partial P_j} + u_i(\mathbf{Q}, \mathbf{P}) + \varepsilon^{1/2} \sum_{k=1}^{m} f'_{ik}(\mathbf{Q}, \mathbf{P}) \xi_k(t), \\ i &= 1, 2, \cdots, n, \end{aligned} \tag{6.246}$$

where $H' = H'(\mathbf{Q}, \mathbf{P})$ represents the Hamiltonian function of free undamped system, $\varepsilon c_{ij}{}'$ are quasi-linear damping coefficients, $\xi_k(t)$ are stochastic processes, $\varepsilon^{1/2} f'_{ik}$ are amplitudes of excitations. When all $\xi_k(t)$ are Gaussian white noises with correlation functions $2\pi K_{kl}\delta(\tau)$, Eq. (6.246) leads to the following Itô stochastic differential equations

$$
\begin{aligned}
dQ_i &= \frac{\partial H''}{\partial P_i}dt, \\
dP_i &= -[\frac{\partial H''}{\partial Q_i} + \varepsilon \sum_{j=1}^{n} m'_{ij}(\mathbf{Q},\mathbf{P})\frac{\partial H''}{\partial P_j} - \varepsilon u_i(\mathbf{Q},\mathbf{P})]dt + \varepsilon^{1/2}\sum_{k=1}^{m}\sigma'_{ik}(\mathbf{Q},\mathbf{P})dB_k(t), \\
i &= 1, 2, \cdots, n.
\end{aligned}
\tag{6.247}
$$

where $H'' = H''(\mathbf{Q},\mathbf{P})$ and $\varepsilon m'_{ij}$ are the Hamiltonian function and quasi-linear damping coefficients after added Wong-Zakai correction terms; $\boldsymbol{\sigma}'\boldsymbol{\sigma}'^T = 2\pi \mathbf{f}'\mathbf{K}\mathbf{f}'^T$.. In Eqs. (6.246) and (6.247), u_i $(i = 1, 2, \cdots, n)$ is the generalized feedback control forces. If there are β control actuators in the system, and each actuator generates a control force $U_\alpha = U_\alpha(\mathbf{Q},\mathbf{P})$, then $u_i = \sum_{\alpha=1}^{\beta} g_{i\alpha}U_\alpha$ where $g_{i\alpha}$ are coefficients of control forces. If $g_{i\alpha}$ does not depend on $\mathbf{Q}$,$\mathbf{P}$ for an i, u_i is called external control force, while if $g_{i\alpha}$ is function of $\mathbf{Q}, \mathbf{P}$, u_i is called parametric control force. For example, in the lateral vibration control of a cable, the control forcc in thc dircction perpendicular to the cable axis is an external control force, while the control force along the cable axis is a parametric control force.

The goal of studying stochastic optimal control is to find an optimal feedback control law which enhances a certain system performance, such as minimizing response, ensuring response following a given PDF, maximizing stability, or maximizing reliability. Here, the objective is only to minimize the system response and predict the response of the optimal control system. The basic idea of stochastic optimal control strategy proposed by Zhu et al. (1999, 2001) is to divide the feedback control force u_i into a conservative control force $u_i^{(1)}$ and a dissipative control force $u_i^{(2)}$. $u_i^{(1)}$ is used to change the Hamiltonian system structure and the distribution of energy and response within the system. $u_i^{(2)}$ is used to consume system energy, thereby reduce the system response and improve the system stability. $u_i^{(1)}$ is determincd using optimal control theory of the corresponding conservative Hamiltonian system, while $u_i^{(2)}$ is determined using stochastic dynamical programming based on the stochastic averaged equation of the quasi-Hamiltonian system.

Usually, the objective of the conservative control force $u_i^{(1)}$ is to modify the kinetic and/or potential energy to improve the system stability (Murray 1997). However, in the present study, the uncontrolled Hamiltonian system is stable, and the conservative control force $u_i^{(1)}$ is used to modify the system integrability and resonance characteristics in order to re-distribute the system energy and response. This can be formulated as the following optimal control problem. Let H'' be the Hamiltonian function of the uncontrollable system, H be the desired Hamiltonian function. The optimal control input $\mathbf{u}^{(1)} = [u_1^{(1)}, u_2^{(1)}, \cdots, u_n^{(1)}]^{\mathrm{T}}$ is selected to minimize the following performance index

$$\min_{\mathbf{u}^{(1)}} J\left(H'', H, \mathbf{u}^{(1)}\right). \tag{6.248}$$

where $J(H'', H, \mathbf{u}^{(1)})$ is a functional of $H'', H, \mathbf{u}^{(1)}$, indicating the desired performance. It is noted that so far there is no general method to find $\mathbf{u}^{(1)}$, and it may be determined by intuition for a specific problem as shown in the following example.

Once $\mathbf{u}^{(1)}$ is determined, it may change the Hamiltonian function $H''(\mathbf{Q}, \mathbf{P})$ to $H(\mathbf{Q}, \mathbf{P})$.. Then Eq. (6.247) becomes

$$\begin{aligned} dQ_i &= \frac{\partial H}{\partial P_i} dt, \\ dP_i &= -[\frac{\partial H}{\partial Q_i} + \varepsilon \sum_{j=1}^{n} m'_{ij}(\mathbf{Q}, \mathbf{P}) \frac{\partial H}{\partial P_j} - u_i^{(2)}(\mathbf{Q}, \mathbf{P})]dt + \varepsilon^{1/2} \sum_{k=1}^{m} \sigma'_{ik}(\mathbf{Q}, \mathbf{P}) dB_k(t), \\ i &= 1, 2, \cdots, n. \end{aligned} \tag{6.249}$$

Although it is possible to apply the stochastic dynamical programming directly to Eq. (6.249), however, the dynamical programming equation is $2n$-dimensional and has no classical solution but viscosity solution, since the diffusion matrix of Eq. (6.249) is degenerate. One method to avoid this situation is first to apply the stochastic averaging method of quasi-Hamiltonian system to Eq. (6.249) and then to apply stochastic dynamical programming to the averaged Itô equation. This method not only reduces the dimensionality of the dynamical programming equation but also makes it having classical solutions since the diffusion matrix of averaged Itô equations is non-degenerate.

In this section, the application of stochastic averaging method is illustrated only for the response control of quasi-non-integrable Hamiltonian systems. For the other cases, such as other control objectives and other types of quasi-Hamiltonian systems, applications of stochastic averaging method in stochastic optimal control problems refer to (Zhu 2003, 2006; Zhu and Ying 2013).

6.5.2 *Optimal Control of Quasi-non-integrable Hamiltonian Systems*

Assume that the Hamiltonian system associated with Eq. (6.249) is non-integrable. Applying the stochastic averaging method described in Sect 5.1, the following partially averaged Itô stochastic differential equation is obtained

$$dH = [m(H) + \varepsilon\left\langle \sum_{i=1}^{n} u_i^{(2)} \frac{\partial H}{\partial P_i} \right\rangle_t]dt + \sigma(H)dB(t), \tag{6.250}$$

where $m(H)$ and $\sigma(H)$ are determined according to Eqs. (5.8) and (5.9) in vol. 1, and

$$\langle[\cdot]\rangle_t = \frac{1}{T(H)} \int_{\Omega} \left([\cdot] \Big/ \frac{\partial H}{\partial p_1}\right) dq_1 \cdots dq_n dp_2 \cdots dp_n. \tag{6.251}$$

In Eq. (6.250), the term including control forces has not been averaged since the control forces have not been undetermined. So, Eq. (6.250) is called partially averaged Itô equation. Note that $u_i^{(2)}$ are functions of $\mathbf{Q}$, $\mathbf{P}$, and $\langle u_i^{(2)} \partial H / \partial P_i \rangle_t$ is function of $\mathbf{H}$ after averaging.

Consider the control of Eq. (6.250) on a finite time interval $[0, t_f]$. The goal of the optimal control is to minimize the cost functional

$$J(\mathbf{u}^{(2)}) = E\left[\int_0^{t_f} f\left(H(s), \left\langle \mathbf{u}^{(2)}(s) \right\rangle_t\right) ds + g\left(H(t_f)\right)\right]. \tag{6.252}$$

where f and g are given functions and called running cost and final cost, respectively. Introduce the value function

$$V(h, t) = \min_{\mathbf{u}^{(2)}} E\left[\int_t^{t_f} f\left(H(s), \left\langle \mathbf{u}^{(2)}(s) \right\rangle_t\right) ds + g\left(H(t_f)\right)\right]. \tag{6.253}$$

For the control problem in Eqs. (6.250) and (6.252), the following dynamical programming equation can be established (Zhu 2003)

$$\frac{\partial V}{\partial t} = -\min_{\mathbf{u}^{(2)}} \left\{ \frac{1}{2}\overline{\sigma}^2(h) \frac{\partial^2 V}{\partial h^2} + \left[\overline{m}(h) + \left\langle \sum_{i=1}^{n} u_i^{(2)} \frac{\partial h}{\partial p_i} \right\rangle_t \right] \frac{\partial V}{\partial h} + f\left(h, \left\langle \mathbf{u}^{(2)} \right\rangle_t\right) \right\} \tag{6.254}$$

with the final condition

$$V(h, t_f) = g(h(t_f)). \tag{6.255}$$

The necessary condition for minimizing the right side of Eq. (6.254) is

$$\frac{\partial}{\partial u_i^{(2)}} \left[\left\langle \sum_{i=1}^{n} u_i^{(2)} \frac{\partial h}{\partial p_i} \right\rangle_t \frac{\partial V}{\partial h} + f\left(h, \left\langle \mathbf{u}^{(2)} \right\rangle_t\right) \right] = 0, \quad i = 1, 2, \cdots, n. \tag{6.256}$$

Equation (6.266) can be used to determine the optimal control law. For instance, if

$$f\left(h,\left\langle \mathbf{u}^{(2)}\right\rangle_t\right) = f_1(h) + \left\langle \mathbf{u}^{(2)\mathrm{T}}\mathbf{R}\mathbf{u}^{(2)}\right\rangle_t, \tag{6.257}$$

where **R** is a positive definite symmetric constant matrix. The optimal control law becomes

$$u_i^{(2)*} = -\frac{1}{2}\sum_{j=1}^{n}\left(\mathbf{R}^{-1}\right)_{ij}\frac{\partial V}{\partial h}\frac{\partial h}{\partial p_j} = -\frac{1}{2}\sum_{j=1}^{n}\left(\mathbf{R}^{-1}\right)_{ij}\frac{\partial V}{\partial h}\dot{q}_j. \tag{6.258}$$

When $\partial V/\partial h > 0$, $u_i^{(2)*}$ becomes dissipative control force. Note that $\partial V/\partial h$ is generally a function of h, $u_i^{(2)*}$ is generally a quasi-linear function of generalized velocity. Equation (6.258) is generally nonlinear negative feedback control.

The sufficient condition that Eq. (6.268) surely is the optimal control law is

$$\left.\frac{\partial^2}{\partial u_i^{(2)2}} f\left(h,\left\langle \mathbf{u}^{(2)}\right\rangle_t\right)\right|_{u_i^{(2)}=u_i^{(2)*}} \geq 0. \tag{6.259}$$

where f is given by Eq. (6.257). If Eq. (6.259) holds, then Eq. (6.258) is indeed the optimal control law.

Substituting $u_i^{(2)*}$ in Eq. (6.258) into Eq. (6.254) to replace $u_i^{(2)}$, and completing the averaging of the terms including control forces according to Eq. (6.261), the final dynamical programming equation is obtained. For example, when **R** is a diagonal matrix, substituting $u_i^{(2)*}$ in Eq. (6.258) into Eq. (6.254) to replace $u_i^{(2)}$ and completing the averaging of the terms including $u_i^{(2)*}$ yield the following final dynamical programming equation

$$\frac{\partial V}{\partial t} = -\left\{\frac{1}{2}\overline{\sigma}^2(h)\frac{\partial^2 V}{\partial h^2} + \overline{m}(h)\frac{\partial V}{\partial h} - \sum_{i=1}^{n}\frac{1}{4R_i}\left\langle\left(\frac{\partial h}{\partial p_i}\right)^2\right\rangle_t\left(\frac{\partial V}{\partial h}\right)^2 + f_1(h)\right\}. \tag{6.260}$$

It is a one-dimensional, nonhomogeneous and nonlinear parabolic partial differential equation with the final condition Eq. (6.255). Solving this equation yields the optimal control forces $u_i^{(2)*}$.

In the above derivation, the magnitude of control force is not restricted, resulting in optimal unbounded control. If the control force is bounded, that is

$$|u_i| \leq b_i,\ b_i > 0,\ i = 1, 2, \cdots, n. \tag{6.261}$$

The corresponding optimal control is called optimal bounded control. In this case, a cost function that doesn't dependent on $\mathbf{u}^{(2)}$ can be considered as follows

$$J_1 = E\left[\int_0^{t_f} f_1(H(s))\mathrm{d}s + g(H(t_f))\right]. \tag{6.262}$$

where f_1 and g are given running cost function and final cost function, respectively. The value function is defined as

$$V_1(h, t) = \min_{\mathbf{u}^{(2)}} E\left[\int_t^{t_f} f_1(H(s))\mathrm{d}s + g\big(H(t_f)\big)\right]. \tag{6.263}$$

The corresponding dynamical programming equation is

$$\frac{\partial V_1}{\partial t} = -\min_{\mathbf{u}^{(2)}}\left\{\frac{1}{2}\overline{\sigma}^2(h)\frac{\partial^2 V_1}{\partial h^2} + \left[\overline{m}(h) + \left\langle\sum_{i=1}^{n} u_i^{(2)}\frac{\partial h}{\partial p_i}\right\rangle_t\right]\frac{\partial V_1}{\partial h} + f_1(h)\right\} \tag{6.264}$$

with the final condition

$$V_1(h, t_f) = g(h(t_f)). \tag{6.265}$$

Minimizing the $u_i^{(2)}$ in the right side of Eq. (6.264) under condition in Eq. (6.261) leads to

$$u_i^{(2)*} = -b_i \mathrm{sgn}\left(\frac{\partial h}{\partial p_i}\frac{\partial V_1}{\partial h}\right). \tag{6.266}$$

As Eq. (6.263) shows, the sign of $\partial V_1 / \partial H$ depends on $f_1'(H)$. If $f_1'(H) > 0$, then

$$u_i^{(2)*} = -b_i \mathrm{sgn}\left(\frac{\partial H}{\partial P_i}\right) = -b_i \mathrm{sgn}(\dot{Q}_i). \tag{6.267}$$

This is the dry friction control or the Bang-Bang control. The control force remains unchanged in magnitude, but its direction is opposite to the direction of the generalized velocity, and it switches direction at $\dot{Q} = 0$. Since $u_i^{(2)*}$ is independent of V_1, there is no need to solve the dynamical programming equation.

For the unbounded ergodic control for system (6.250) over a semi-infinite time interval, let the cost function be

$$J_2(\mathbf{u}^{(2)}) = \lim_{T\to\infty}\frac{1}{T}\int_0^{T} f\big(H(s), \langle\mathbf{u}^{(2)}(s)\rangle_t\big)\mathrm{d}s. \tag{6.268}$$

Its dynamical programming equation is as follows

$$\min_{\mathbf{u}^{(2)}}\left\{\frac{1}{2}\overline{\sigma}^2(h)\frac{\mathrm{d}^2V_2}{dh^2}+\left[\overline{m}(h)+\left\langle\sum_{i=1}^{n}u_i^{(2)}\frac{\partial h}{\partial p_i}\right\rangle_t\right]\frac{\mathrm{d}V_2}{\mathrm{d}h}+f\left(h,\left\langle\mathbf{u}^{(2)}\right\rangle_t\right)\right\}=\gamma. \tag{6.269}$$

where γ denotes the optimal average cost given by

$$\gamma=\lim_{T\to\infty}\frac{1}{T}\int_0^T f\left(h(s),\left\langle\mathbf{u}^{(2)*}(s)\right\rangle_t\right)\mathrm{d}s. \tag{6.270}$$

If $f(h, \langle\mathbf{u}^{(2)}\rangle_t)$ has a form of Eq. (6.257), then an optimal control law $u_i^{(2)*}$ similar to Eq. (6.258) is obtained. When $\mathbf{R}$ is a diagonal matrix, substituting it into Eq. (6.269) to replace $u_i^{(2)}$, resulting in the following final dynamical programming equation

$$\frac{1}{2}\overline{\sigma}^2(h)\frac{\mathrm{d}^2V_2}{dh^2}+\overline{m}(h)\frac{\mathrm{d}V_2}{\mathrm{d}h}-\sum_{i=1}^{n}\frac{1}{4R_i}\left\langle\left(\frac{\partial h}{\partial p_i}\right)^2\right\rangle_t\left(\frac{\mathrm{d}V_2}{\mathrm{d}h}\right)^2+f_1(h)=\gamma. \tag{6.271}$$

This is a one-dimensional, nonhomogeneous and nonlinear ordinary differential equation. Substituting its solution into Eq. (6.258) yields the optimal control forces $u_i^{(2)*}$.

For the bounded ergodic control of Eq. (6.250) over a semi-infinite time interval, the cost function can be

$$J_3=\lim_{T\to\infty}\frac{1}{T}\int_0^T f_1(H(s))ds. \tag{6.272}$$

where f_1 is a given function. The corresponding dynamical programming equation is as follows

$$\min_{u^{(2)}}\left\{\frac{1}{2}\overline{\sigma}^2(h)\frac{d^2V_3}{dh^2}+\left[\overline{m}(h)+\left\langle\sum_{i=1}^{n}u_i^{(2)}\frac{\partial h}{\partial p_i}\right\rangle_t\right]\frac{dV_3}{dh}+f_1(h)\right\}=\gamma_1, \tag{6.273}$$

where

$$\gamma_1=\lim_{T\to\infty}\frac{1}{T}\int_0^T f_1(H(s))ds. \tag{6.274}$$

When control force is under the constraint in Eq. (6.261), minimizing γ_1 leads to the optimal control force in Eq. (6.267).

Finally, substituting the optimal control forces $u_i^{(2)*}$ in Eq. (6.258) or Eq. (6.267) into the averaged Itô equation (6.260) to replace $u_i^{(2)}$, completing the averaging $\langle\sum_{i=1}^{n}u_i^{(2)*}\partial H/\partial P_i\rangle_t$ result in the averaged Itô equation for the optimally controlled system

$$dH = \overline{m}(H)dt + \sigma(H)dB(t), \tag{6.275}$$

where

$$\overline{m}(H) = m(H) + \left\langle \sum_{i=1}^{n} u_i^{(2)*} \frac{\partial H}{\partial P_i} \right\rangle_t . \tag{6.276}$$

Solving the averaged FPK equation associated with Itô equation (6.275) yields PDF for the response of optimally controlled system.

To assess the effectiveness of the optimal control strategy, two criteria are introduced. One is the control effectiveness defined as

$$k_w = \frac{\sigma_w^u - \sigma_w^c}{\sigma_w^u}, \tag{6.277}$$

where σ_w^u, σ_w^c represents the standard deviation of some response quantities, the superscripts u, c denote the uncontrolled and controlled systems, respectively. k_w represents the reduction percentage of system response standard deviation due to optimal control. Another criterion is the control efficiency defined as

$$\mu_w = k_w \big/ \sigma_{\mathbf{u}^{(2)*}}. \tag{6.278}$$

where $\sigma_{\mathbf{u}^{(2)*}}$ is the standard deviation of optimal control force. It represents the control effectiveness per unit standard deviation of the control force. Obviously, the larger k_w and μ_w are, the better the control strategy is.

Example 6.6 Consider a controlled Duffing oscillator excited by Gaussian white noise (Zhu et al. 2001a). The system is given by

$$\dot{Q} = P, \quad \dot{P} = -a_0 Q - b_0 Q^3 - cP + u + eW_g(t), \tag{6.279}$$

where a_0, b_0, c, e are constants, $W_g(t)$ is a Gaussian white noise with an intensity $2\pi K$. The control force u in Eq. (6.279) is separated into two parts

$$u = u^{(1)} + u^{(2)}, \tag{6.280}$$

$$u^{(1)} = -(a_1 Q + b_1 Q^3). \tag{6.281}$$

It is clear that $u^{(1)}$ is used to adjust the linear and nonlinear stiffness of the oscillator. Substituting Eqs. (6.280) and (6.281) into Eq. (6.279) results in

$$\dot{Q} = P, \quad \dot{P} = -aQ - bQ^3 - cP + u^{(2)} + eW_g(t), \tag{6.282}$$

where $a = a_0 + a_1$, $b = b_0 + b_1$. Assuming $a, b > 0$, the Hamiltonian associated with system (6.282) is

$$H = p^2/2 + aq^2/2 + bq^4/4. \tag{6.283}$$

By applying the stochastic averaging method of quasi-non-integrable Hamiltonian systems described in Sect. 5.1, the following partially averaged Itô stochastic differential equation can be derived

$$dH = \left[\overline{m}(H) + \left\langle u^{(2)} \frac{\partial H}{\partial P} \right\rangle_t \right] dt + \overline{\sigma}(H) dB(t). \tag{6.284}$$

Using Eqs. (5.8) and (5.9) in vol. 1 yields

$$\overline{m}(H) = \pi e^2 K - cG(H), \quad \overline{\sigma}^2(H) = 2\pi e^2 KG(H), \tag{6.285}$$

where

$$\begin{aligned}
&\langle [\cdot] \rangle_t = \frac{1}{T(H)} \int_\Omega \left([\cdot] \Big/ \frac{\partial H}{\partial p} \right) dq, \\
&G(H) = [4/T(H)] \int_0^A (2H - aq^2 - bq^4/2)^{1/2} dq, \\
&T(H) = \int_\Omega \left(1 \Big/ \frac{\partial H}{\partial p} \right) dq = 4 \int_0^A (2H - aq^2 - bq^4/2)^{-1/2} dq, \\
&\Omega = \left\{ q \middle| \left(aq^2/2 + bq^4/4 \right) \le H \right\}, \\
&A = \left\{ \left[(a^2 + 4bH)^{1/2} - a \right] \Big/ b \right\}^{1/2}.
\end{aligned} \tag{6.286}$$

For the unbounded ergodic control over a semi-infinite time interval, function f in Eq. (6.268) is defined as

$$f(H, \langle u^{(2)} \rangle_t) = f_1(H) + R \langle u^{(2)^2} \rangle_t, \tag{6.287}$$

$$f_1(H) = s_0 + s_1 H + s_2 H^2 + s_3 H^3. \tag{6.288}$$

According to Eq. (6.279), the dynamical programming equation is

$$\min_{u^{(2)}} \left\{ \frac{1}{2} \sigma^2(h) \frac{d^2 V_2}{dh^2} + \left[m(h) + \left\langle u^{(2)} \frac{\partial h}{\partial p} \right\rangle_t \right] \frac{dV_2}{dh} + f\left(h, \left\langle u^{(2)} \right\rangle\right) \right\} = \gamma, \tag{6.289}$$

where

$$\gamma = \lim_{T\to\infty} \frac{1}{T} \int_0^T f\left(H, \left\langle u^{(2)*} \right\rangle_t\right) dt. \tag{6.290}$$

By minimizing the left side of Eq. (6.289) with respect to $u^{(2)}$, the optimal control law is obtained as

$$u^{(2)*} = -\frac{1}{2R}\frac{dV_2}{dh}\frac{\partial h}{\partial p} = -\frac{1}{2R}\frac{dV_2}{dh}p. \tag{6.291}$$

Substituting Eq. (6.301) into Eq. (6.299) to replace $u^{(2)}$ and completing averaging result in the final dynamical programming equation

$$\frac{1}{2}\sigma^2(h)\frac{d^2V_2}{dh^2} + m(h)\frac{dV_2}{dh} - \frac{G(h)}{4R}\left(\frac{dV_2}{dh}\right)^2 + f_1(h) = \gamma. \tag{6.292}$$

This is a nonhomogeneous and nonlinear ordinary differential equation. Since Eq. (6.299) only depends on dV_2/dh, it is only needed to solve Eq. (6.292) to get dV_2/dh. When $h \to 0$, $G(h), \overline{\sigma}^2(h) \to 0$, $\overline{m}(h) \to e^2D$, Eq. (6.292) reduces to

$$\left.\frac{dV_2}{dh}\right|_{h=0} = \frac{\gamma - s_0}{\pi e^2 K}. \tag{6.293}$$

This is the boundary conditions for $\mathrm{d}V_2/\mathrm{d}h$ at $h = 0$. Substituting Eq. (6.293) into Eq. (6.291) yields the optimal control force $u^{(2)*}$.

Substituting the optimal control force $u^{(2)*}$ into Eq. (6.284), and noting

$$\left\langle u^{(2)*}\frac{\partial h}{\partial p}\right\rangle_t = -\frac{G(h)}{2R}\frac{dV_2}{dh}. \tag{6.294}$$

lead to the averaged Itô equation for the optimal control system

$$dH = \overline{m}(H)dt + \sigma(H)dB(t), \tag{6.295}$$

where

$$\overline{m}(H) = m(H) - \left[G(H)/2R\right]dV_2/dH. \tag{6.296}$$

Solving the reduced FPK equation associated with Itô Eq. (6.295) yields the stationary PDF of the Hamiltonian process of the optimally controlled system

$$p^c(h) = C^c \exp\left\{-\int_0^h [(-2m(x) + d\sigma^2(x)/dx + (G(x)/R)dV_2/dx)/\sigma^2(x)]dx\right\}. \tag{6.297}$$

Letting $u^{(2)} = 0$ in Eq. (6.284) and similar derivation yield the stationary PDF of the uncontrolled system

$$p^u(h) = C^u \exp\left\{-\int_0^h [(-2m(x) + d\sigma^2(x)/dx)/\sigma^2(x)]dx\right\}. \tag{6.298}$$

According to Eq. (5.18) in vol. 1, the joint stationary PDFs of the generalized displacements and momenta of the controlled and uncontrolled systems are given by

$$p^c(q,p) = p^c(h)/T^c(h)\big|_{h=h(q,p)}, \tag{6.299}$$

$$p^u(q,p) = p^u(h)/T^u(h)\big|_{h=h(q,p)}. \tag{6.300}$$

where $T^c(h)$ and $T^u(h)$ are the periods of controlled and uncontrolled systems. From Eqs. (6.299) and (6.300), the stationary variances $(\sigma_Q^2)_c$, $(\sigma_P^2)_c$, $(\sigma_Q^2)_u$, $(\sigma_P^2)_u$ of displacements and momenta of the optimally controlled and uncontrolled systems can be obtained. While the variance $\sigma^2_{u^{(2)*}}$ of the optimal control force can then be obtained from Eqs. (6.291) and (6.299). Finally, Eqs. (6.277) and (6.278) can be used to calculate the control effectiveness k_Q, k_P and control efficiency μ_Q, μ_P for evaluation of the control strategy. For example, the control effectiveness of displacement is

$$k_Q = \frac{E[Q^2]_u - E[Q^2]_c}{E[Q^2]_u}, \tag{6.301}$$

and the control efficiency of displacement is

$$\mu_Q = k_Q/E[u^2], \tag{6.302}$$

When $\xi(t)$ is stationary wideband process, the same procedure applies, with the only difference is that the expressions for $m(H)$, $\sigma(H)$ in Eq. (6.284) can be obtained from Eq. (1.36) in Example 1.1 by setting $\omega_0^2 = a$, $\alpha = b$, $-\beta_1 = c$, $\beta_2 = 0$, $S_1(\omega) = 0$, $S_2(\omega) = e^2 S(\omega)$..

Let the parameters of system (6.282) be $a = 1$, $b = 0.2$, $c = 0.1$, $e = 0.4$. Two types of random excitations are used. One is a Gaussian white noise with intensity $2nK = 2n$, and another is a stationary wideband noise with power spectral density

$$S(\omega) = \frac{\omega_g^4 + 4\zeta_g^2\omega_g^2\omega^2}{(\omega_g^2 - \omega^2)^2 + 4\zeta_g^2\omega_g^2\omega^2} S_0, \tag{6.303}$$

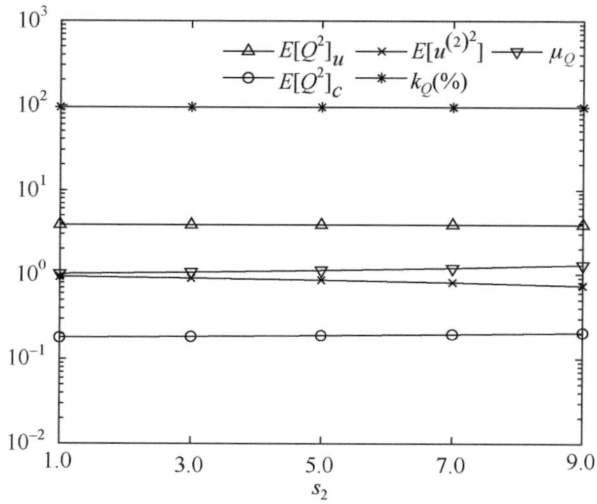

Fig. 6.27 Results obtained by using the present method for system (6.279) under Gaussian white noise excitation with control parameters s_2, $s_1 = s_3 = 0$, $R = 1$, $dV_2/dh|_{h=0} = 1.4$ (Zhu et al. 2001a)

where $S_0 = 1$, $\zeta_g = 0.32$, and $\omega_g = 20.3$. The stochastic optimal control method described in this section is applied. Figures 6.27, 6.28, 6.29, 6.30 and 6.31 show the mean square values $E[Q^2]_c$ and $E[Q^2]_u$ of displacement Q for the optimally controlled and uncontrolled systems, the mean square values of the control force $E[u^{(2)^2}]$, and the control criteria k_Q and μ_Q for the Gaussian white noise and stationary wideband noise excitations. From these figures, several observations can be made. (i) The comparison of Figs. 6.27, 6.29, and 6.30 shows that the value of k_Q, μ_Q is not sensitive. (ii) It can be seen from comparison of Figs. 6.27 and 6.31, as well as comparison of Figs. 6.28 and 6.32 that the optimal control method described in this section is effective in both cases of the two noises. (iii) Comparison of Figs. 6.27 and 6.28 shows that increasing $dV_2/dH|_{H=0}$ leads to an increase of k_Q and a significant decrease of μ_Q. (iv) Fig. 6.28 shows that increasing R/s_2 results in a decrease of k_Q and a significant increase of μ_Q.

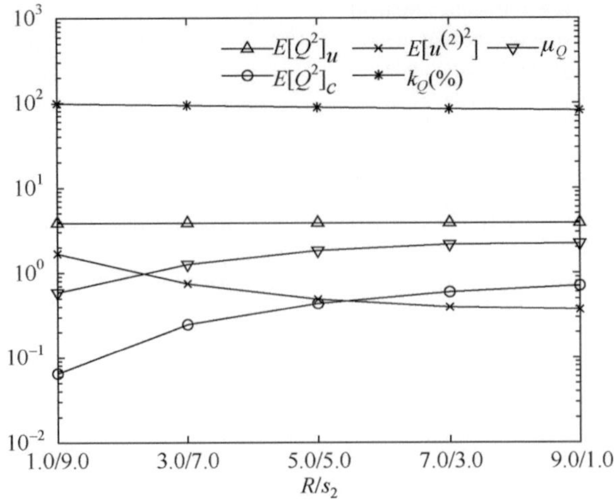

Fig. 6.28 Results obtained by using the present method for system (6.279) under Gaussian white noise excitation with control parameters R, s_2, $s_1 = s_3 = 0$, $dV_2/dh|_{h=0} = 3.3$ (Zhu et al. 2001a)

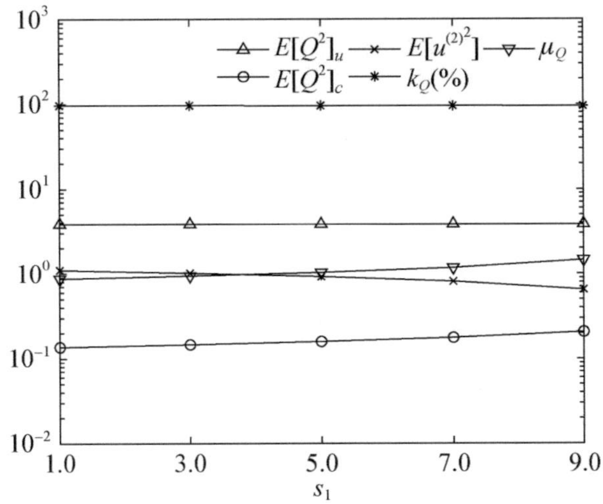

Fig. 6.29 Results obtained by using the present method for system (6.279) under Gaussian white noise excitation with control parameters s_1, $s_2 = s_3 = 0$, $R = 1$, $dV_2/dh|_{h=0} = 1.8$ (Zhu et al. 2001a)

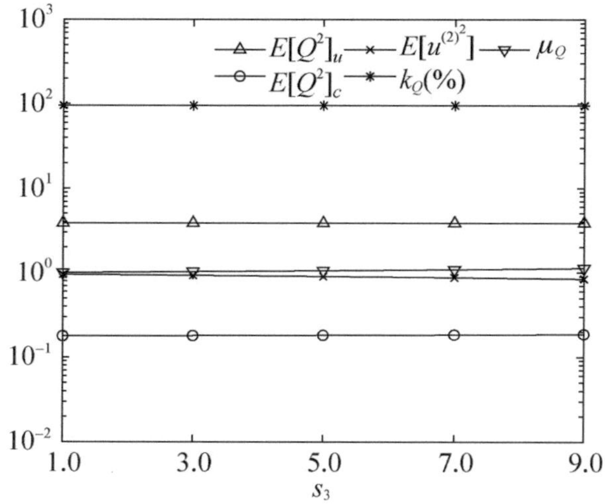

Fig. 6.30 Results obtained by using the present method for system (6.279) under Gaussian white noise excitation with control parameters s_3, $s_1 = s_2 = 0$, $R = 1$, $dV_2/dh\big|_{h=0} = 1.4$ (Zhu et al. 2001a)

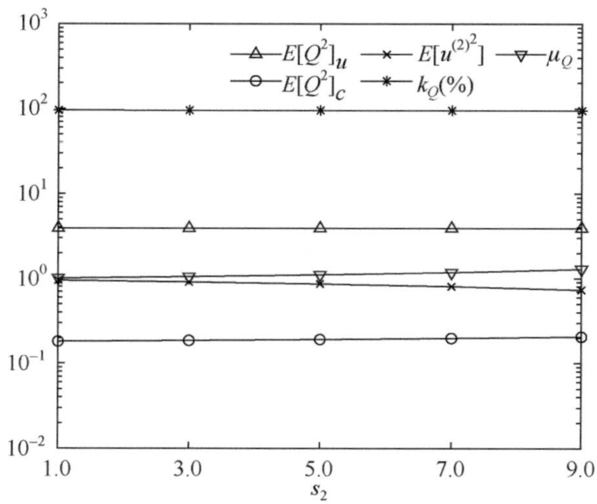

Fig. 6.31 Results obtained by using the present method for system (6.279) under stationary wideband noise excitation with control parameters s_2, $s_1 = s_3 = 0$, $R = 1$, $dV_2/dh\big|_{h=0} = 1.4$ (Zhu et al. 2001a)

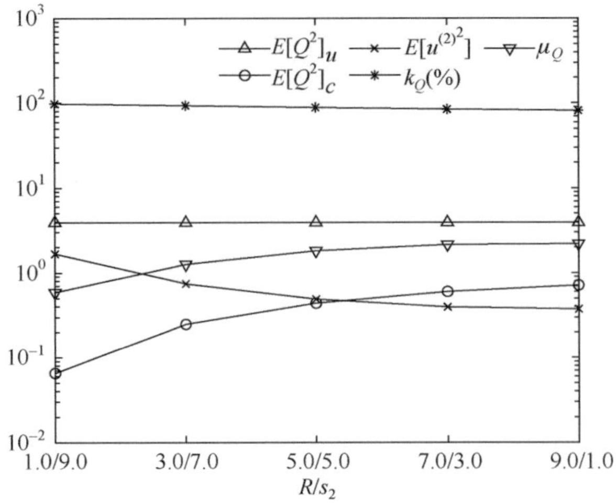

Fig. 6.32 Results obtained by using the present method for system (6.279) under stationary wideband noise excitation with control parameters R, s_2, $s_1 = s_3 = 0$, $dV_2 / dh|_{h=0} = 3.3$ (Zhu et al. 2001a)

References

Ariaratnam S T, Xie W C. 1992. Lyapunov exponent and stochastic stability of coupled linear systems under real noise excitation. ASME Journal of Applied Mechanics, 59: 664-673.

Arnold R N. 1961. Maunder L. Gyrodynamics and Its Engineering Applications. New York: Academic Press.

Arnold L, Kliemann W, Oeljeklaus E. 1968. Lyapunov exponents of linear stochastic systems. Lecture Notes in Mathematics, Vol, 1186, New York: Springer-Verlag.

Asokanthan S F, Ariaratnam S T. 2000. Almost-sure stability of a gyro pendulum subjected to white-noise random support motion. Journal of Sound and Vibration, 235: 801-812.

Bellman R. 1957. Dynamical programming. Princeton: Princeton University Press.

Cai G Q, Yu J S, Lin Y K. 1994. Ship rolling in random sea. DE-Vol. 77, Stochastic Dynamics and Reliability of Nonlinear Ocean Systems, ASME.

Cai G Q, Zhu W Q. 2016. Elements of Stochastic Dynamics. Singapore: Word Scientific Publishing Co. Pte. Ltd.

Chen L C, Zhu W Q. 2010. Feedback minimization of reliability of a simple power system under random perturbations. Journal of Dynamics and Control, 8(1): 19-23. (in Chinese)

Dalzell J F. 1973. A note on the distribution of maxima of ship rolling. Journal of ship research, 17(4): 217-226.

Dalzell J F. 1971. A study of the distribution of maxima of non-linear ship rolling in a seaway. Report DL-71–1562, Stevens Institute of Technology, Hoboken, N.J.

Davenport A G. 1961, The spectrum of horizontal gustiness near the ground in high winds. Journal of the Royal Meteorological Society, 87: 194-211.

Deng, M.L., Mu, G.J., and Zhu, W.Q., 2021. Random response of wake-oscillator excited by fluctuating wind. Journal of Applied Mechanics, 88(10): 101002.

Facchinetti M L, de Langre E, Biolley F. 2004. Coupling of structure and wake-oscillators in vortex-induced vibrations. Journal of Fluids and Structures, 19: 123-140.

Fleming W H, Rishel R W. 1975. Deterministic and Stochastic Optimal Control. New York: Springer-Verlag.

Froude W. 1955. The Papers of W. Froude. Institution of Naval Architects.

Gabbai R D, Benaroya H. 2005. An overview of modeling and experiments of vortex-induced vibration of circular cylinders. Journal of Sound and Vibration, 282(3–5): 575–616.

Grim O. 1952. Roll oscillation, stability and safety in sea states. Ship Technology Research, 1: 10-21. (in Germany)
Haesen E, Bastiaensen C, Driesen J, Belmans R. 2009. A probabilistic formulation of load margins in power systems with stochastic generation. IEEE Transactions on Power Systems, 24(2): 951-958.
Hartlen R T, Currie I G. 1970. Lift oscillator model for vortex-induced vibrations. American Society of Civil Engineers, Journal of Engineering Mechanics, 96: 577-591.
Huang Z L, Zhu W Q. 2000. Lyapunov exponent and almost sure asymptotic stability of quasi-linear gyroscopic systems. International Journal of Non-Linear Mechanics, 35: 645-655.
Ju P. 2010. Modeling Theory and Method of Power System. Beijing: Science Press. (in Chinese)
Ju P, Li H Y, Xue Y S, Liu Y F, Wu F. 206. Stochastic models for study of electromechanical transient process in power systems. Journal of Hohai University (Natural Sciences), 41(6): 536-541. (in Chinese)
Ju P. 2018. Stochastic Dynamics of Power System. Singapore: Springer Nature Singapore Pte. Ltd.
Khasminskii R Z. 1967. Necessary and sufficient conditions for the asymptotic stability of linear stochastic systems. Theory of Probability and Application, 11: 144-147.
Khasminskii R Z. 1968. On the averaging principle for Itô stochastic differential equations. Kibernetika, 3(4): 260-279. (in Russian)
Khasminskii R Z. 1980. Stochastic Stability of Differential Equations. Alphen aan den Rijn: Sijthoff & Noordhoff.
Khasminskii R Z. 1996. On robustness of some concepts in stability of stochastic differential equations. Fields Institute Communications, 9: 131-137.
Kloeden P, Platen E. 1992. Numerical Solution of Stochastic Differential Equations. Berlin: Springer-Verlag.
Kozin F. 1969. A survey of stability of stochastic systems. Automatica, 5: 95-112.
Kozin F, Zhang Z Y. 1991. On almost sure sample stability of nonlinear Itô differential equations. Probabilistic Engineering Mechanics, 6: 92-95.
Krenk S, Nielsen R K. 1999. Energy balanced double oscillator model for vortex-induced vibrations. Journal of Engineering Mechanics, 125(3): 263-271.
Kundur P, Balu N J, Lauby M G. 1994. Power system stability and control. New York: McGraw-hill.
Li H Y, Ju P, Chen X Q, Sun W Z, Wu F. 2015. Stochastic averaging method for quasi Hamiltonian system of multi-machine power systems. Scientia Sinica (Technologica), 45(7): 766-772. (in Chinese)
Lin Y K, Cai G Q. 1995. Probabilistic structural dynamics: advanced theory and applications. New York: McGraw-Hill.
Liu Y F, Ju P, Xue Y S, Wu F, Zhang J Y. 2014. Calculation analysis on power system characteristics under random excitation. Automation of Electric Power Systems, 38(9): 137-142. (in Chinese)
Murray R M. 1997. Nonlinear control of mechanical system: a Lagrangian perspective. A Rev. Control, 21: 31-42.
Ni Y X, Chen S S, Zhang B L. 2002. Theory and Analysis of Dynamic Power Systems. Beijing: Tsinghua University Publishing House Co. Ltd. (in Chinese)
Ochi M K. 1986. Non-Gaussian random process in ocean engineering. Probabilistic Engineering Mechanics, 1(1): 28-38.
Oseledec V I. 1968. A multiplicative ergodic theorem, Lyapunov characteristic numbers for dynamical systems. Transactions of Moscow Mathematical Society, 19: 197-231.
Roberts J B. 1982. A stochastic theory for nonlinear ship rolling in irregular seas. Journal of Ship Research, 26(4): 229-245.
Roberts J B. 1985. Estimation of nonlinear ship roll damping from free-decay data. Journal of ship research, 29(2): 127-138.
Simiu E, Scanlan R H. 1996. Wind effects on structures: Fundamentals and applications to design, 3rd Ed., Wiley, New York.
Skop R A, Griffin O M. 1973. A model for the vortex-excited resonant response of bluff cylinders. Journal of Sound and Vibration, 27(2): 225-233.

Sri Namchchivaya N. 1987. Stochastic stability of a gyropendulum under random vertical support excitation. Journal of Sound and Vibration, 119: 363-373.
Talay D. 1999. The Lyapunov exponent of Euler scheme for stochastic differential equations. Stochastic Dynamics, Crauel H, Gundlach M. (Eds.), New York: Springer-Verlag, 241–258.
Wihstutz W. 1999. Perturbation methods for Lyapunov exponents. Stochastic Dynamics, Crauel H, Gundlach M. (Eds.), New York: Springer-Verlag, 210–239.
Zhu W Q. 2003. Nonlinear Stochastic Dynamics and Control – Hamilton's Theoretical Framework. Beijing: Science Press. (in Chinese)
Zhu W Q. 2006. Nonlinear stochastic dynamics and control in Hamiltonian formulation. ASME Appl. Mech. Rev., 59(4): 230-248.
Zhu W Q, Cai G Q. 2017. Introduction to the stochastic dynamics. Beijing: Science Press. (in Chinese)
Zhu W Q, Huang Z L. 1998. Stochastic stability of quasi–non–integrable -Hamiltonian systems. Journal of Sound and Vibration, 218: 769-789.
Zhu W Q, Huang Z L. 1999. Lyapunov exponent and stochastic stability of quasi-Hamiltonian systems. ASME Journal of Applied Mechanics, 66: 211-217.
Zhu W Q, Ying Z G. 1999. Optimal nonlinear feedback control of quasi-Hamiltonian systems. Sciences in China, Series A, 42: 1213-1219.
Zhu W Q, Ying Z G. 206. Advances in research on nonlinear stochastic optimal control of quasi-Hamiltonian systems. Advances in Mechanics, 43(1): 39-55. (in Chinese)
Zhu W Q, Ying Z G, Soong T T. 2001a. An optimal nonlinear feedback control strategy for randomly excited structural systems. Nonlinear Dynamics, 24: 31-51.
Zhu W Q, Huang Z L, Deng M L. 2002c. Feedback minimization of first-passage failure of quasi non-integrable Hamiltonian systems. International Journal of Non-Linear Mechanics, 37(6): 1057-1071.

Index

W. Zhu et al., *Stochastic Averaging*, https://doi.org/10.1007/978-981-96-3833-8

W

MIX
Papier aus verantwortungsvollen Quellen
Paper from responsible sources
FSC® C105338

If you have any concerns about our products, you can contact us on
ProductSafety@springernature.com

In case Publisher is established outside the EU, the EU authorized representative is:
Springer Nature Customer Service Center GmbH
Europaplatz 3, 69115 Heidelberg, Germany

Printed by Libri Plureos GmbH
in Hamburg, Germany